ELEMENTS OF ORDINARY DIFFERENTIAL EQUATIONS

ELEMENTS
OF ORDINARY
DIFFERENTIAL
EQUATIONS

Louis L. Pennisi

University of Illinois

Holt, Rinehart and Winston, Inc.

New York Chicago San Francisco
Atlanta Dallas Montreal Toronto
London Sydney

To Jane

PREFACE

The purpose of this book is to present a systematic approach, with sufficient rigor, to the basic topics encountered in an elementary course in differential equations. The prerequisite is a first-year course in elementary calculus.

The material in the text is presented in such a manner as to render service not only to mathematics students, but also to students in the fields of engineering, physics, and related branches of physical science. The theory and applications have been developed in a detailed fashion.

A glance at the Table of Contents will best reveal the topics covered in each chapter. Chapter 1 defines and gives a classification of differential equations. A solution of a differential equation is defined and two existence theorems (whose proofs are found in Appendix 2) are stated. Chapter 2 is essentially concerned with some basic methods for solving three types of first-order equations, namely, exact equations, separable equations, and linear equations. Chapter 3 is devoted to some important applications of first-order equations. A novel feature of this chapter is Section 3.7, A Problem in the Calculus of Variations. Chapter 4 gives a systematic development of some of the basic results concerning second-order linear differential equations. The introduction of Euler's formula $e^{i\theta} = \cos\theta + i\sin\theta$ simplifies the presentation. Chapter 5 is devoted primarily to the study of the differential equation of the motion of a mass on a spring. Chapter 6 considers higher-order linear differential equations. Essentially, this chapter consists of generalizations of the results of Chapter 4. Chapter 7 deals primarily with series solutions of linear differential equations about a regular point and about a regular singular point. The gamma function has been introduced, making it possible to express the Bessel functions in a convenient compact form. Chapter 8 is devoted to a straightforward development of methods for solving systems of linear differential equations. One of the features of this chapter is the utilization of some basic concepts of matrix algebra to obtain solutions of systems of linear equations.

Appendix 1 is devoted to some basic definitions and concepts of point sets.

These concepts, in particular those of a region and a domain, are utilized in Appendix 2. In Appendix 2, some existence theorems are established. The proofs of these theorems depend upon some basic definitions and results from advanced calculus. These results have been stated without proofs; however, instructive examples are supplied to illustrate their meanings. Some of the existence theorems stated in the text without proofs are obtained as corollaries of some of the existence theorems proved in Appendix 2.

This book may be used with considerable flexibility. For example, in a three-hour, one-semester course or in a three-hour, two-quarter course, after Chapters 1, 2, and 4 have been covered (or partly covered), the instructor may take the remaining chapters (or parts of them) in any order that he desires to conform with the requirements of his course.

The numbering system is according to chapter, section, and specified item. For example, the third section of Chapter 4 is designated Section 4.3; the fourth theorem of Section 4.3 is designated Theorem 4.3.4, and so on.

To provide an opportunity for the student to assimilate and familiarize himself with the subject matter, a considerable number of examples have been incorporated into the text, and a large variety of exercises have been placed at the end of various sections. The exercises constitute an important part of the course. Hints and solutions to many of the exercises and answers to all of the numerical ones will be found in the back of the book.

Since the book has been written in detailed fashion, it is apparent that to avoid having the text become too voluminous, some topics had to be omitted. However, it has been my experience that the material contained in this book is fully adequate for the majority of beginning courses in differential equations.

One of my main objectives in writing the text was to stimulate the reader to a further interest in mathematics and other related fields. If this is accomplished, this volume will substantially have served its purpose.

I wish to express my appreciation to Professor Duane Walker of Lewis College for using the text in mimeographed form and for making valuable suggestions towards the improvement of the book. I would also like to express my utmost appreciation to my former graduate assistant, Mr. Richard Amtman, and to one of my most promising undergraduate students, Mr. James G. Gatz, both of whom read the entire manuscript, checked the answers, made corrections, and contributed some interesting problems on applications. Thanks are also due to graduate assistants Mr. Marius Prapuolenis and Mr. Alex Thannikkary for their invaluable assistance in the preparation of the manuscript.

I am very grateful to my wife, Jane Pennisi, for her excellent and accurate job of typing the manuscript and for her infinite patience. This book is dedicated to her.

Finally, I would like to express my utmost appreciation to the editorial staff of Holt, Rinehart and Winston, Inc. for their continual cooperation during the preparation of the manuscript.

Louis L. Pennisi

Chicago, Illinois
January 1972

CONTENTS

Chapter 4.   SECOND-ORDER LINEAR DIFFERENTIAL EQUATIONS

Chapter 5.   APPLICATIONS OF SECOND-ORDER LINEAR DIFFERENTIAL EQUATIONS

Chapter 6.   HIGHER-ORDER LINEAR DIFFERENTIAL EQUATIONS

ELEMENTS OF ORDINARY DIFFERENTIAL EQUATIONS

chapter 1   FORMATION OF DIFFERENTIAL EQUATIONS AND THEIR SOLUTIONS

1.1. Definitions and Classification of Differential Equations

Definition 1.1.1. A *differential equation* is an equation that involves derivatives or differentials of one or more dependent variables with respect to one or more independent variables.

Differential equations may be divided into two main classes, ordinary and partial differential equations.

Definition 1.1.2. An *ordinary differential equation* is an equation that involves ordinary derivatives of one or more dependent variables with respect to a single independent variable.

Definition 1.1.3. A *partial differential equation* is an equation that involves partial derivatives of one or more dependent variables with respect to two or more independent variables.

Example 1.1.1. Some examples of ordinary and partial differential equations follow.

$$\frac{dy}{dx} = f(x). \tag{1.1.1}$$

$$M(x, y)dx + N(x, y)dy = 0. \tag{1.1.2}$$

$$\frac{dx}{dt} + \frac{dy}{dt} = f(t). \tag{1.1.3}$$

$$y = x\frac{dy}{dx} + \sqrt{1 + \left(\frac{dy}{dx}\right)^2}. \tag{1.1.4}$$

$$\left[1 + \left(\frac{dy}{dx}\right)^2\right]^{3/2} = \frac{d^2y}{dx^2}. \tag{1.1.5}$$

$$y \frac{d^3 y}{dx^3} + k \frac{dy}{dx} = \sin x. \tag{1.1.6}$$

$$m \frac{d^4 y}{dx^4} + ky^2 = f(x). \tag{1.1.7}$$

$$\left(\frac{d^2 y}{dx^2}\right)^3 + k\left(\frac{dy}{dx}\right)^4 = e^x. \tag{1.1.8}$$

$$\frac{d^3 y}{dx^3} + 4\left(\frac{d^2 y}{dx^2}\right)\left(\frac{dy}{dx}\right) + 6y = 0. \tag{1.1.9}$$

$$x \frac{d^2 y}{dx^2} + (1 - x)\frac{dy}{dx} + 2ny = 0. \tag{1.1.10}$$

$$(1 - x^2)\frac{d^2 y}{dx^2} - 2x \frac{dy}{dx} + n(n + 1)y = 0. \tag{1.1.11}$$

$$\frac{\partial^2 u}{\partial x^2} + \frac{\partial^2 u}{\partial y^2} + \frac{\partial^2 u}{\partial z^2} = 0. \tag{1.1.12}$$

$$\frac{\partial^2 u}{\partial r^2} + \frac{2}{r}\frac{\partial u}{\partial r} + \frac{1}{r^2 \sin^2 \theta}\frac{\partial^2 u}{\partial \phi^2} + \frac{1}{r^2}\frac{\partial^2 u}{\partial \theta^2} + \frac{\cot \theta}{r^2}\frac{\partial u}{\partial \theta} = 0. \tag{1.1.13}$$

$$\frac{\partial u}{\partial t} = k\left(\frac{\partial^2 u}{\partial x^2} + \frac{\partial^2 u}{\partial y^2}\right). \tag{1.1.14}$$

$$u \frac{\partial^2 u}{\partial t^2} = k\left(\frac{\partial^4 u}{\partial x^2 \, \partial y^2}\right)^3. \tag{1.1.15}$$

$$\frac{\partial^2 u}{\partial t \, \partial x} + \frac{\partial \phi}{\partial y}\frac{\partial^2 \phi}{\partial x \, \partial y} = -\frac{1}{\rho}\frac{\partial p}{\partial x} + \frac{\partial v}{\partial x}. \tag{1.1.16}$$

Equations (1.1.1) to (1.1.11) are ordinary differential equations while Equations (1.1.12) to (1.1.16) are partial differential equations.

Remark 1.1.1.   We shall not include in the class of differential equations those equations that are differential identities, such as

$$\frac{d}{dx}(\cos ax) = -a \sin ax \quad \text{or} \quad \frac{d}{dx}\left(\frac{u}{v}\right) = \frac{v \dfrac{du}{dx} - u \dfrac{dv}{dx}}{v^2}.$$

Remark 1.1.2.   The symbol

$$\left(\frac{d^m y}{dx^m}\right)^k,$$

where m and k are positive integers, is known as the ordinary derivative of y with respect to x of *order m* and *degree k*. The dependent variable y is a function of the independent variable x. Also, the symbol

$$\left(\frac{\partial^m u}{\partial x^r \, \partial y^s}\right)^k,$$

where r, s are nonnegative integers, m, k, and $r + s$ are positive integers, and $r + s = m$, is known as a partial derivative of u with respect to x and y of *order m* and *degree k*. Here, the dependent variable u is a function of two independent variables x and y. In Equation (1.1.16), the dependent variables u, ϕ, p, and v are functions of the independent variables t, x, and y.

Further classification of differential equations will now be given.

Definition 1.1.4. The *order* of a differential equation is the highest of all the orders of the derivatives occurring in the equation.

In Example 1.1.1, Equations (1.1.1) to (1.1.4) are of order one. Equations (1.1.5), (1.1.8), (1.1.10) to (1.1.14) and (1.1.16) are of order two. Equations (1.1.6) and (1.1.9) are of order three. Equations (1.1.7) and (1.1.15) are of order four.

Remark 1.1.3. Let us recall that a polynomial in the variable X is an expression of the form

$$a_0 X^n + a_1 X^{n-1} + \cdots + a_{n-1}X + a_n,$$

where the coefficients a_i, $i = 0, 1, \ldots, n$, are complex numbers and n is a nonnegative integer. This expression is also called a rational integral function in the variable X. In a similar fashion, one defines a polynomial in two or more variables. For example,

$$3X^2Y^2 - 5X + 7XY + Y^3$$

is a polynomial in X and Y. A rational function is one that can be expressed as a ratio of two polynomials. Thus,

$$\frac{3X^4 - 5X^3 + 6X - 2}{X^3 - 4X + 1}$$

is a rational function in X. The expression

$$3Y^3 + 5Y^2 + \frac{5Y + 6X - 4}{X^2}$$

is a rational integral function in Y and a rational function in X. The degree of the polynomial

$$5X^4Y^3 + 3X^3Y - 2XY^2 + 5$$

is $4 + 3$ or 7; the degree in X is 4 and the degree in Y is 3. The degree of the polynomial

$$X^2 Y^5 - 2X^5 Y + 3X^4 Y^2 - 3Y^6$$

is also 7; the degree in X is 5 and the degree in Y is 6. The polynomial $7X + 8Y$ is linear in X and Y. The polynomial $7XY$ is not linear in X and Y; however, it is linear in X when Y is held fixed, and is linear in Y when X is held fixed.

Remark 1.1.4. By letting derivatives of various orders play the role of indeterminates (like the X and Y preceding), we can discuss polynomials of the form

$$4\left(\frac{d^2 y}{dx^2}\right)^2 - 3\left(\frac{d^2 y}{dx^2}\right)\left(\frac{dy}{dx}\right)^3 - \frac{1}{2}\left(\frac{dy}{dx}\right),$$

corresponding to

$$4X^2 - 3XY^3 - \tfrac{1}{2}Y.$$

Such an understanding allows us to further classify differential equations.

Definition 1.1.5. The *degree* of a differential equation is the degree of the highest-ordered derivative involved in the equation.

Definition 1.1.6. A differential equation is said to be *linear* if the derivatives and the dependent variables occur to the first degree and there are no products of these, while the coefficients in the separate terms are either constants or functions of the independent variables.

A differential equation that is not linear is called a *nonlinear differential equation.*

In Example 1.1.1, the degree of Equation (1.1.4) is two, since it can be written as

$$(x^2 - 1)\left(\frac{dy}{dx}\right)^2 - 2xy\frac{dy}{dx} + y^2 - 1 = 0. \tag{1.1.4$'$}$$

Also, Equation (1.1.5) is of degree two, since it can be written as

$$\left(\frac{d^2 y}{dx^2}\right)^2 - \left(\frac{dy}{dx}\right)^6 - 3\left(\frac{dy}{dx}\right)^4 - 3\left(\frac{dy}{dx}\right)^2 - 1 = 0. \tag{1.1.5$'$}$$

Note that in Equations (1.1.4$'$) and (1.1.5$'$) the highest-ordered derivative is of degree two. Equations (1.1.8) and (1.1.15) are of degree three. The remaining equations are of degree one.

In Example 1.1.1, Equation (1.1.2) may or may not be linear; it depends upon the functions M and N.[1] From Equations (1.1.4′) and (1.1.5′) it is clear that Equations (1.1.4) and (1.1.5) are nonlinear. Note that in Equation (1.1.4′) the presence of any of the three terms, $(dy/dx)^2$, $y(dy/dx)$, or y^2, disqualifies the differential equation from being linear. Also, in Equation (1.1.5′) the presence of any of the four terms, $(d^2y/dx^2)^2$, $(dy/dx)^6$, $(dy/dx)^4$, or $(dy/dx)^2$, renders the differential equation nonlinear.

Equation (1.1.6) is nonlinear owing to the term $y(d^3y/dx^3)$. Equation (1.1.7) is nonlinear owing to the term y^2, that is, the dependent variable is of degree two. Equation (1.1.8) is nonlinear owing to the presence of the term $(d^2y/dx^2)^3$ or $(dy/dx)^4$. Equation (1.1.9) is nonlinear because of the term $(d^2y/dx^2)(dy/dx)$. Equation (1.1.15) is nonlinear because of the term $u(\partial^2u/\partial t^2)$ or $(\partial^4u/\partial x^2\,\partial y^2)^3$. Equation (1.1.16) is nonlinear owing to the term $(\partial\phi/\partial y)(\partial^2\phi/\partial x\,\partial y)$.

The remaining equations in Example 1.1.1 are linear. For instance, in Equation (1.1.11), (d^2y/dx^2), (dy/dx), and y are each of the first degree. Also, the coefficient of (d^2y/dx^2) is $(1 - x^2)$, which is a function of the independent variable x; the coefficient of (dy/dx) is $-2x$, which is a function of x; and the coefficient of y is $n(n + 1)$, which is a constant. Thus, Equation (1.1.11) fulfills all the requirements given in Definition 1.1.6.

Remark 1.1.5. Note that every linear equation is of degree one. However, the converse is false. See, for instance, Equations (1.1.6), (1.1.7), and (1.1.9).

Remark 1.1.6. Since it may not always be possible to reduce a given differential equation to a polynomial, or rational expression, not every differential equation has a degree, nor can every differential equation be classified as either linear or nonlinear. For example,

$$\frac{dy}{dx} + \ln\left(\frac{dy}{dx} + y\right) = 0$$

is a case in point. Also, the differential equations

$$\sin(y''') + \cot y' = \ln\cosh(yy'')$$

and

$$\frac{e^{x+1}}{e-1} = y + \frac{y'}{e} + \frac{y''}{e^2} + \cdots,$$

cannot be made to have rational and integral derivatives.

Remark 1.1.7. When one changes a differential equation by writing it as a polynomial (if this is possible), one may also change the "set of solutions."

---

[1] For $M(x, y)dx + N(x, y)dy = 0$, either x or y could be the independent variable.

Exercises 1.1

For each of the following differential equations, determine (a) whether the equation is ordinary or partial; (b) whether the equation is linear or nonlinear; (c) the order and degree of the equation.

1. $x\,dx + y\,dy = 0.$

2. $\dfrac{dy}{dx} = P(x) + Q(x)y + R(x)y^2.$

3. $\left(\dfrac{d^4y}{dx^4}\right)^2 + \left(\dfrac{d^2y}{dx^2}\right)^4 + y = e^x.$

4. $x^2\,\dfrac{d^3y}{dx^3} + (y-1)\dfrac{dy}{dx} + 4y = \cos x.$

5. $\dfrac{\partial v}{\partial t} + v\dfrac{\partial v}{\partial x} = \dfrac{q}{m}\,E.$

6. $\dfrac{d^2x}{dt^2} + \dfrac{dy}{dt} - 10 = 0.$

7. $m\dfrac{d^2X}{dt^2} + c\dfrac{dX}{dt} + kX = F(t).$

8. $\left[1 + \left(\dfrac{dy}{dx}\right)^2\right]\dfrac{d^3y}{dx^3} - 3\dfrac{dy}{dx}\left(\dfrac{d^2y}{dx^2}\right)^2 = 0.$

9. $(x+1)\dfrac{\partial^2 u}{\partial x^2} + (y+1)^2\dfrac{\partial^2 u}{\partial y^2} - u = 1.$

10. $\left(\dfrac{\partial u}{\partial t}\right)^2 = a(t)\dfrac{\partial^2 u}{\partial x^2}.$

11. $\sqrt[3]{1 + \left(\dfrac{dy}{dx}\right)^4} = \left(\dfrac{d^2y}{dx^2}\right)^{3/2}.$

12. $y^3\dfrac{d^2y}{dx^2} + 3x^3\left(\dfrac{dy}{dx}\right)^4 = 5y^2x.$

13. $y\dfrac{dx}{dt} + \dfrac{dy}{dt} - 10 = 0.$

14. $\dfrac{d^2}{dx^2}\left[E(x)\dfrac{d^2Y(x)}{dx^2}\right] = W(x).$

15. $\dfrac{\partial}{\partial x}\left[k(T)\dfrac{\partial T}{\partial x}\right] = c\dfrac{\partial T}{\partial t}.$

16. $a_0(x)\dfrac{d^ny}{dx^n} + a_1(x)\dfrac{d^{n-1}y}{dx^{n-1}} + \cdots + a_{n-1}(x)\dfrac{dy}{dx} + a_n(x)y = f(x).$

17. $\left(\dfrac{\partial^2 F}{\partial x^2} + \dfrac{\partial^2 F}{\partial y^2}\right)^2 - \dfrac{4}{1 - k^2}\left[\dfrac{\partial^2 F}{\partial x^2}\dfrac{\partial^2 F}{\partial y^2} - \dfrac{\partial^2 F}{\partial x \partial y}\right]^2 = 0.$

18. $a\dfrac{\partial^2 u}{\partial t^2} = \dfrac{b}{\sqrt{1 + \left(\dfrac{\partial u}{\partial x}\right)^2}}\dfrac{\partial^2 u}{\partial x^2}.$

1.2. Elimination of Constants and Formation of Ordinary Differential Equations

Consider a family of plane curves represented by the equation

$$f(x, y, c) = 0. \tag{1.2.1}$$

Note that by varying the parameter, or arbitrary constant c, the corresponding graphs of Equation (1.2.1) make up a one-parameter family of plane curves.

Example 1.2.1. Consider the equation

$$x^2 + y^2 = c^2. \tag{1.2.2}$$

Equation (1.2.2) defines a one-parameter family of concentric circles whose centers are at the origin. Figure 1.2.1 illustrates a few curves of this family. We may characterize this family of circles by a first-order nonlinear differential equation obtained by eliminating the parameter c from (1.2.2). Differentiating (1.2.2) with respect to x, we obtain

$$y\frac{dy}{dx} + x = 0. \tag{1.2.3}$$

From (1.2.3) we have

$$y\,dy + x\,dx = 0,$$

which upon integration gives Equation (1.2.2). Thus, the relation given in

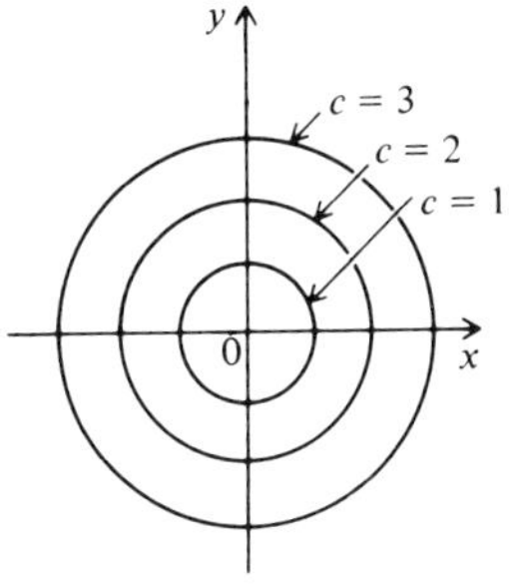

Figure 1.2.1.

(1.2.2) satisfies identically the differential equation (1.2.3). Loosely speaking, the relation in (1.2.2) is said to represent an implicit solution of the differential equation (1.2.3). We shall be more precise on this point (implicit solution) in Section 1.3. Observe that the order of the differential equation (1.2.3) is equal to the number of parameters defining the family of curves in (1.2.2).

Consider a family of plane curves represented by the equation

$$f(x, y, c_1, c_2, \ldots, c_n) = 0. \tag{1.2.4}$$

Note that by varying the parameters, or arbitrary constants c_i, $i = 1, 2, \ldots, n$, we see that Equation (1.2.4) represents an n-parameter family of plane curves.

Example 1.2.2.   Consider the equation

$$(x - h)^2 + (y - k)^2 = r^2. \tag{1.2.5}$$

This equation defines a three-parameter family of circles with centers at (h, k) and radii equal to r $(r > 0)$. Here, h, k, and r are the parameters. Figure 1.2.2 illustrates two curves of this family, namely,

$$S_1 : (x - 1)^2 + (y - 1)^2 = (2)^2, \qquad S_2 : (x + 2)^2 + (y + 1)^2 = (1)^2.$$

The elimination of the three parameters from (1.2.5) leads to an ordinary nonlinear differential equation of order three. To see this, we proceed as follows. First, to eliminate the parameter r, we differentiate (1.2.5) with respect to x, and we obtain

$$x - h + (y - k)y' = 0.^2 \tag{1.2.6}$$

Secondly, to eliminate h, we differentiate (1.2.6) with respect to x, and we get

$$1 + (y - k)y'' + (y')^2 = 0. \tag{1.2.7}$$

Finally, to eliminate k, we first write (1.2.7) as

$$k - y = \frac{1 + (y')^2}{y''} \qquad (y'' \neq 0). \tag{1.2.8}$$

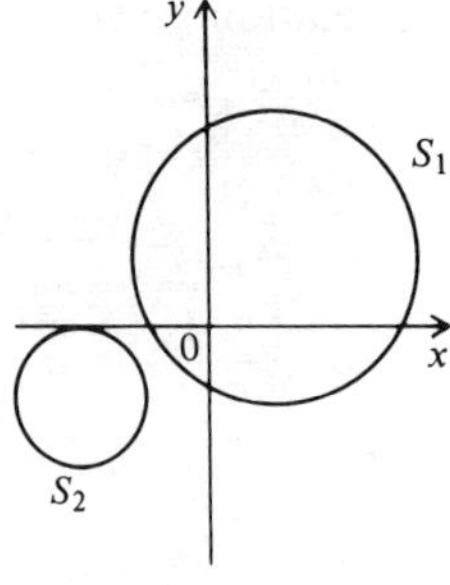

Figure 1.2.2.

---

^{2} We shall, at times, denote $y(x)$ by y, $\dfrac{dy}{dx}$ by y', $\dfrac{d^2y}{dx^2}$ by y'', $\ldots$, and $\dfrac{d^ny}{dx^n}$ by $y^{(n)}$.

Now, differentiate (1.2.8) with respect to x and simplify the result to obtain the desired differential equation

$$[1 + (y')^2]y''' - 3y'(y'')^2 = 0. \tag{1.2.9}$$

Before we proceed to show that integrations of (1.2.9) lead back to (1.2.5), note that (1.2.5) may be written equivalently as

$$x^2 + y^2 + c_1 x + c_2 y + c_3 = 0, \tag{1.2.10}$$

where $c_1 = -2h$, $c_2 = -2k$, and $c_3 = h^2 + k^2 - r^2$.   Now, the parameters are c_1, c_2, and c_3.   Also, in passing from Equation (1.2.7) to (1.2.8), we assumed that $y'' \neq 0$.   What happens if $y'' = 0$?   Observe that the equation

$$y = Ax + B, \tag{1.2.11}$$

where A and B are parameters, represents the set of all straight lines in a plane (except the lines $x = K$ whose slopes are not defined).   We may characterize this two-parameter family of straight lines by the differential equation

$$y'' = 0, \tag{1.2.12}$$

obtained by differentiating (1.2.11) twice with respect to x.

If we integrate (1.2.12) twice with respect to x, we obtain (1.2.11), that is,

$$A = \int y''(x)dx = \int \frac{d}{dx}[y'(x)]dx = \int d[y'(x)] = y'(x),$$

where A is an arbitrary constant of integration; and now integrating this equation, we find

$$Ax + B = \int y'(x)dx = \int \frac{d}{dx}[y(x)]dx = \int d[y(x)] = y(x),$$

where B is another arbitrary constant of integration.

Note that if $y'' = 0$, then $y''' = 0$.   Substituting $y'' = 0$ and $y''' = 0$ in Equation (1.2.9), we see that the left-hand member of (1.2.9) reduces to zero. Thus the two-parameter family of straight lines given in (1.2.11) satisfies the differential equation (1.2.9).   Now, let us assume that $y'' \neq 0$.   We may write (1.2.9) as

$$\frac{y'''}{y''} - \frac{3y'y''}{1 + (y')^2} = 0. \tag{1.2.13}$$

Integrating (1.2.13), we obtain

$$\ln a = \int \frac{y'''(x)dx}{y''(x)} - 3 \int \frac{y'(x)y''(x)dx}{1 + [y'(x)]^2} = \int \frac{d[y''(x)]}{y''(x)} - \frac{3}{2} \int \frac{d\{1 + [y'(x)]^2\}}{1 + [y'(x)]^2}$$

$$= \ln y''(x) - \frac{3}{2} \ln \{1 + [y'(x)]^2\}, \tag{1.2.14}$$

where ln a is an arbitrary constant of integration.[3] Using the properties of logarithms, we may write (1.2.14) as

$$\frac{y''(x)}{\{1 + [y'(x)]^2\}^{3/2}} = a. \tag{1.2.15}$$

From tables on integrals, we know that

$$\int \frac{du}{[B + Au^2]^{3/2}} = \frac{u}{B\sqrt{B + Au^2}} + C.$$

If we let $u = y'(x)$, then $du = y''(x)dx$. Thus, integrating (1.2.15), we obtain

$$\frac{y'(x)}{\sqrt{1 + [y'(x)]^2}} = ax + b, \tag{1.2.16}$$

where b is another arbitrary constant of integration. Squaring both sides of Equation (1.2.16) and solving for $y'(x)$, we find

$$y'(x) = \frac{ax + b}{\sqrt{1 - (ax + b)^2}}. \tag{1.2.17}$$

If we let $v = 1 - (ax + b)^2$, then $dv = -2a(ax + b)dx$. Thus, upon integrating (1.2.17), we get

$$y(x) + c = -\frac{1}{a}\sqrt{1 - (ax + b)^2},$$

or

$$ay(x) + d = -\sqrt{1 - (ax + b)^2}, \tag{1.2.18}$$

where $d = ac$ is an arbitrary constant. Equation (1.2.18) may now be written as

$$a^2x^2 + a^2y^2 + 2abx + 2ady + d^2 + b^2 - 1 = 0, \qquad a \neq 0,$$

or

$$x^2 + y^2 + Ax + By + C = 0,$$

which is the same as Equation (1.2.5).

---

[3] For convenience we denote the arbitrary constant of integration by ln a rather than a. If we use a, the right-hand member of Equation (1.2.15) becomes e^a. We replace e^a by the arbitrary constant c. Finally, we relable c with a. We shall often choose the most convenient constant for the constant of integration. Also, it is to be understood that whenever we write ln x, we mean ln $|x|$. Note that ln $x = $ ln $|x|$ only when $x > 0$. When $x < 0$, ln x is not defined, but ln $|x|$ is defined. Only when the argument of the logarithm is positive for all allowable real values of the variables can the absolute value symbol $|\ |$ be omitted. With this understanding, we have chosen to write, say ln $f(x)$ instead of ln $|f(x)|$ for convenience rather than clarity.

Again, loosely speaking, the relation given in (1.2.5) or (1.2.10) is said to represent an implicit solution of the differential equation (1.2.9). Observe that the order of the differential equation (1.2.9) is equal to the number of parameters defining the family of curves in (1.2.5) or (1.2.10), namely, three. Also, note very carefully that the differential equation (1.2.9), besides having a solution defined by the relation in (1.2.5), also has, as we have seen, another solution defined by the function in (1.2.11). The solution given in (1.2.11) has only two parameters. Moreover, this solution cannot be obtained from the solution given in (1.2.10) by assigning values to the arbitrary constants c_1, c_2, and c_3. The solution in (1.2.11) is said to represent a singular solution of the differential equation (1.2.9). We shall be more precise on this point (singular solution) in Section 1.3. Once again, we repeat, even though a three-parameter family of curves gives rise to a third-order differential equation, it does not necessarily follow that a third-ordered differential equation will have for its solution only a three-parameter family of curves. (It may have a separate two-parameter family of curves.) The statement above also applies to differential equations of orders other than three.

Remark 1.2.1. In defining Equation (1.2.4), it was tacitly assumed that all the parameters c_i, $i = 1, 2, \ldots, n$, were *necessary* or *essential*, that is, the number of parameters cannot be equal to $n - r$, $r > 0$. The following two examples illustrate that the order of a differential equation may be lowered if the parameters are not truly essential.

Example 1.2.3. Consider the equation

$$f[x, y, \phi(c_1, c_2)] = 0. \tag{1.2.19}$$

It appears in (1.2.19) that there are two necessary parameters, namely, c_1 and c_2. Since $\phi(c_1, c_2)$ is but an arbitrary constant, if we put $c = \phi(c_1, c_2)$, we see that the family of curves given by (1.2.19) depends truly on one essential parameter. Thus, for instance, in the equation

$$y = (c_1 + c_2)x, \tag{1.2.20}$$

we may replace the arbitrary constant $c_1 + c_2$ by c. Hence, (1.2.20) becomes

$$y = cx, \tag{1.2.21}$$

which is a one-parameter family of nonvertical straight lines with slope c and passing through the origin. If we differentiate (1.2.21) with respect to x, we obtain $dy/dx = c$. If we substitute this value of c in (1.2.21), we obtain a first-order linear differential equation

$$x\frac{dy}{dx} - y = 0, \tag{1.2.22}$$

which characterizes the family of straight lines given in (1.2.21).

From (1.2.22), we have

$$\frac{dy}{y} = \frac{dx}{x} \qquad (x \neq 0, \quad y \neq 0). \tag{1.2.23}$$

Note that $x = 0$ and $y = 0$ satisfy trivially Equation (1.2.22). Integrating Equation (1.2.23), we obtain

$$\ln y = \ln x + \ln c,$$

or

$$y = cx,$$

which is Equation (1.2.21). Thus, for all x, $y = cx$ satisfies identically the differential equation (1.2.22). That is, for all x, $y = cx$ is termed an explicit solution of the differential equation (1.2.22). We shall be more precise on this point (explicit solution) in Section 1.3. Note that the order of the differential equation (1.2.22) is equal to the number of essential parameters defining the family of curves given in Equation (1.2.20), which is one rather than two. [*See* (1.2.21).]

Example 1.2.4. Consider the equation

$$y = c_1 e^{c_2 + 2x}. \tag{1.2.24}$$

Again it appears in (1.2.24) that we have two necessary parameters, namely, c_1 and c_2. However, we may write (1.2.24) as

$$y = ce^{2x}, \tag{1.2.25}$$

where

$$c = c_1 e^{c_2}$$

is an arbitrary constant. Thus, Equation (1.2.24) represents a one-parameter family of exponential curves all intersecting the y axis at $y = c$. From (1.2.25), we obtain

$$\frac{dy}{dx} = 2ce^{2x}. \tag{1.2.26}$$

If we substitute the value of c, found in (1.2.26), into (1.2.25), we get a first-order linear differential equation

$$\frac{dy}{dx} - 2y = 0, \tag{1.2.27}$$

that characterizes the family of exponential curves given in (1.2.25).

From (1.2.27) we have

$$\frac{dy}{y} - 2\,dx = 0 \qquad (y \neq 0). \tag{1.2.28}$$

Note that if $y = 0$, then $y' = 0$ and Equation (1.2.27) is satisfied. Thus, integrating (1.2.28) we obtain

$$\ln y = 2x + \ln c,$$

or

$$y = ce^{2x},$$

which is Equation (1.2.25). Thus, for all x, $y = ce^{2x}$ satisfies identically the differential equation (1.2.27). Again we say that for all x, $y = ce^{2x}$ is an explicit solution of the differential equation (1.2.27). Note that the order of the differential equation (1.2.27) is equal to the number of essential constants defining the family of curves in Equation (1.2.24), which is one rather than two. [*See* (1.2.25).]

The differential equation (1.2.27), with 2 replaced by a number k, will play an important role in Chapter 3 where we deal with applications of first-order differential equations. It will be utilized to solve problems concerning growth, decay, and compound interest.

Remark 1.2.2. From the above two examples, we see that the order of a differential equation is equal to the number of essential parameters (or constants) defining the family of curves which give rise to the differential equation.

We shall now give an example where the order of the differential equation is not equal to the number of independent parameters defining the family of curves which gives rise to the differential equation.

Example 1.2.5. Consider the equation

$$y^2 - 2axy - bx^2 = 0. \tag{1.2.29}$$

Clearly, we have two independent parameters in (1.2.29), namely, a and b. We may write (1.2.29) as

$$\left(\frac{y}{x}\right)^2 - 2a\frac{y}{x} - b = 0 \qquad (x \neq 0). \tag{1.2.30}$$

Equation (1.2.30) is a quadratic equation in the variable y/x. Thus, using the quadratic formula, we obtain

$$y = k_1 x, \qquad y = k_2 x, \tag{1.2.31}$$

where

$$k_1 = a + \sqrt{a^2 + b}, \qquad k_2 = a - \sqrt{a^2 + b}. \tag{1.2.32}$$

Equation (1.2.30) represents, in view of (1.2.31) and (1.2.32), two one-parameter families of straight lines going through the origin. Each of these lines satisfies the first-order linear differential equation

$$x\frac{dy}{dx} - y = 0. \tag{1.2.33}$$

Remark 1.2.3.   If Equation (1.2.4) is such that it factors, each factor containing y, the order of the resulting differential equation may be less than n.

Remark 1.2.4.   Before we describe a general technique for obtaining an ordinary equation of order n from the family of plane curves given by Equation (1.2.4), we shall recall a basic result from college algebra.   Let a linear system of n equations in $n - 1$ unknowns $x_1, x_2, \ldots, x_{n-1}$ be given by

$$\sum_{j=1}^{n-1} a_{ij} x_j = B_i, \qquad i = 1, 2, \ldots, n, \tag{1.2.34}$$

where the a_{ij} and B_i are either constants or quantities that are independent of the $n - 1$ unknowns.

If n linear equations in $n - 1$ unknowns have a solution, then

$$\begin{vmatrix} a_{11} & a_{12} & \cdots & a_{1\,n-1} & B_1 \\ a_{21} & a_{22} & \cdots & a_{2\,n-1} & B_2 \\ \vdots & \vdots & & \vdots & \vdots \\ a_{n1} & a_{n2} & \cdots & a_{n\,n-1} & B_n \end{vmatrix} = 0. \tag{1.2.35}$$

Note carefully that the above result does not assert that if the determinant given in the left-hand member of Equation (1.2.35) is equal to zero, then the system given in (1.2.34) necessarily has a solution.   It may or may not have a solution.   For example, consider the system

$$\begin{cases} x + 2y = 1, \\ x + 2y = 2, \\ x + 2y = 3. \end{cases}$$

Here

$$\begin{vmatrix} 1 & 2 & 1 \\ 1 & 2 & 2 \\ 1 & 2 & 3 \end{vmatrix} = 0,$$

and yet this system does not have a solution.

Example 1.2.6.   Let us obtain the differential equation of the following family of curves by eliminating the essential constants c_1, c_2, and c_3.

$$y = c_1 e^x + c_2 e^{2x} + c_3 e^{-x}. \tag{1.2.36}$$

SOLUTION.   Note that (1.2.36) is a special case of Equation (1.2.4), namely, when $y = h(x, c_1, c_2, c_3)$, where $h(x, c_1, c_2, c_3)$ is given by the right-hand member of (1.2.36).   Differentiating (1.2.36) three times in succession with respect to x, we obtain

$$\begin{cases} y' = c_1 e^x + 2c_2 e^{2x} - c_3 e^{-x}, \\ y'' = c_1 e^x + 4c_2 e^{2x} + c_3 e^{-x}, \\ y''' = c_1 e^x + 8c_2 e^{2x} - c_3 e^{-x}. \end{cases} \tag{1.2.37}$$

We may consider the system given in (1.2.36) and (1.2.37) as one having four equations in the unknowns c_1, c_2, and c_3. Thus, if this system has a solution, then we must have

$$\begin{vmatrix} e^x & e^{2x} & e^{-x} & y \\ e^x & 2e^{2x} & -e^{-x} & y' \\ e^x & 4e^{2x} & e^{-x} & y'' \\ e^x & 8e^{2x} & -e^{-x} & y''' \end{vmatrix} = 0.$$

Factoring out e^x, e^{2x}, and e^{-x}, respectively, from the first, second, and third columns of the above determinant, and noticing that $e^x \cdot e^{2x} \cdot e^{-x} = e^{2x} > 0$ for all x, we obtain

$$\begin{vmatrix} 1 & 1 & 1 & y \\ 1 & 2 & -1 & y' \\ 1 & 4 & 1 & y'' \\ 1 & 8 & -1 & y''' \end{vmatrix} = 0.$$

Utilizing the rules for expanding a determinant, we obtain

$$y''' - 2y'' - y' + 2y = 0. \tag{1.2.38}$$

Substituting y''', y'', y', and y of (1.2.37) and (1.2.36) into (1.2.38), we find that the family of curves given in (1.2.36) indeed satisfies (1.2.38). Thus, (1.2.38) is the desired differential equation.

Remark 1.2.5.   The above method may be used to eliminate the essential constants $c_1, c_2, \ldots, c_n$ from the family of curves given by

$$y = c_1 e^{m_1 x} + c_2 e^{m_2 x} + \cdots + c_n e^{m_n x}, \tag{1.2.39}$$

where $m_1, m_2, \ldots, m_n$ are distinct constants. It turns out that the elimination of the c_i, $i = 1, 2, \ldots, n$, will always give rise to an nth order linear differential equation

$$a_0 y^{(n)} + a_1 y^{(n-1)} + \cdots + a_{n-1} y' + a_n y = 0, \tag{1.2.40}$$

where the a_i, $i = 1, 2, \ldots, n$, are constants. This very important differential equation will be studied in Chapter 6. For the very essential special case when $n = 2$, we shall give a detailed discussion in Chapter 4.

  Also, one may utilize the method given in Example 1.2.6 to eliminate the essential constants c_i, $i = 1, 2, \ldots, n$, from the family of curves given by

$$y = (c_1 + c_2 x + c_3 x^2 + \cdots + c_n x^{n-1}) e^{mx}. \tag{1.2.41}$$

  We now proceed to describe a general technique for obtaining an ordinary differential equation of order n from the family of plane curves given by Equation (1.2.4). We assume in what follows that the appropriate partial derivatives of f exist and that $\partial f/\partial y$ is not identically zero. Also, let us assign to the parameters c_i, $i = 1, 2, \ldots, n$, definite but arbitrary chosen values.   If we differentiate Equation (1.2.4) n times in succession with respect to x, we obtain

$$
\begin{cases}
\dfrac{\partial f}{\partial x} + \dfrac{\partial f}{\partial y}\, y' = 0, \\[2ex]
\dfrac{\partial^2 f}{\partial x^2} + 2\,\dfrac{\partial^2 f}{\partial x\,\partial y}\, y' + \dfrac{\partial^2 f}{\partial y^2}\,(y')^2 + \dfrac{\partial f}{\partial y}\, y'' = 0, \\[2ex]
\dfrac{\partial^3 f}{\partial x^3} + 3\,\dfrac{\partial^3 f}{\partial x^2\,\partial y}\, y' + 3\,\dfrac{\partial^3 f}{\partial x\,\partial y^2}\,(y')^2 + 3\,\dfrac{\partial^2 f}{\partial x\,\partial y}\, y'' \\[2ex]
\qquad + \dfrac{\partial^3 f}{\partial y^3}\,(y')^3 + 3\,\dfrac{\partial^2 f}{\partial y^2}\, y'y'' + \dfrac{\partial f}{\partial y}\, y''' = 0, \\[2ex]
\hdashline \\[-1ex]
\dfrac{\partial^n f}{\partial x^n} + \cdots + \dfrac{\partial f}{\partial y}\, y^{(n)} = 0.
\end{cases}
\qquad (1.2.42)
$$

Note that the first equation in (1.2.42) is obtained by differentiating (1.2.4) with respect to x.   The second equation in (1.2.42) is obtained by differentiating the first equation in (1.2.42) with respect to x.   The third equation in (1.2.42) is obtained by differentiating the second equation in (1.2.42) with respect to x; and so on.   From Equations (1.2.4) and (1.2.42), we now have $(n + 1)$ relations between x, y, y', y'', $\ldots$, $y^{(n)}$ and the n constants c_1, c_2, $\ldots$, c_n.   The elimination of these n constants leads in general to a relation involving the variables $(x, y, y', y'', \ldots, y^{(n)})$, namely, to an ordinary differential equation of order n:

$$
F(x, y, y', y'', \ldots, y^{(n)}) = 0. \qquad (1.2.43)
$$

Exercises 1.2

In each of the following exercises, obtain the differential equation of the family of curves by eliminating the essential parameters. The letters α, β, and k are not to be eliminated. Sketch a few curves for some of these families.   As usual, r and θ represent polar coordinates.

1. $y = c_1 e^x + c_2 e^{-x}.$
2. $y = c_1 e^x + c_2 e^{-2x} + c_3 e^{-x}.$
3. $y = (c_1 + c_2 x)e^{-2x}.$
4. $y = (c_1 + c_2 x + c_3 x^2)e^{2x}.$
5. $y = c_1 + c_2 e^x + c_3 e^{2x} + c_4 e^{-x}.$
6. $y = c_1 \sin \beta x + c_2 \cos \beta x.$
7. $y = e^{\alpha x}(c_1 \sin \beta x + c_2 \cos \beta x).$
8. $y = e^{\alpha x}[(c_1 + c_2 x) \sin \beta x + (c_3 + c_4 x) \cos \beta x].$
9. $r\theta = c.$
10. $r^2\theta = c^2.$
11. $r = e^{c\theta}.$
12. $r = c \sin 2\theta.$
13. $r^2 = c^2 \cos 2\theta.$
14. $(r - c)^2 = 4c\theta.$
15. $x^2 + y^2 = cx.$
16. $y = c \operatorname{sech} e^x.$

17. $y^2(2c - x) = x^3$.

18. $x^2 y = 4c^2(2c - y)$.

19. $r \cos \theta = \ln (c \sec \theta)$.

20. $(x - h)^2 + (y - h)^2 = h^2$.

21. $(x - h)^2 + (y - c)^2 = h^2$.

22. $r - b - a \cos \theta$.

23. $y^2 = x^2 \dfrac{c + x}{c - x}$.

24. $y(2x + y) = ce^x$.

25. $e^{2x} = y^2(x^2 + c)$.

26. $\dfrac{x^2}{\alpha^2 + c} + \dfrac{y^2}{\beta^2 + c} = 1$.

27. $y^2 + cxy - 12c^2 x^2 = 0$.

28. $y^2(1 - x^2 + ce^{-x^2}) = 1$.

29. $x^3 + y^3 - 3cxy = 0$.

30. $e^x = (x + 1)(c + y^2)$.

31. $y = xe^{cx}$.

32. $y(c - \tan x) = c \tan x + 1$.

33. $y = \dfrac{ce^{\cos x}}{1 + ce^{\cos x}}$.

34. $ay^2 = bx^2 + cx + d$.

35. $\dfrac{(x - h)^2}{a^2} + \dfrac{(y - h)^2}{b^2} = 1$.

36. $r \cos \theta = c(\sec \theta + \tan \theta)^k$.

37. $y = \dfrac{c}{2}\left[\left(\dfrac{x}{c}\right)^{1-k} - \left(\dfrac{x}{c}\right)^{1+k}\right]$.

1.3. Solutions of Ordinary Differential Equations[4]

In the sequel, unless stated to the contrary, the functions considered are real-valued and are defined on a real interval. Also, unless stated to the contrary, the solutions we seek for the differential equations are real-valued functions.

The examples and observations of Section 1.2 motivate, in part, the following definitions.

Definition 1.3.1. Let f be a function defined for every x in an interval I and possessing derivatives up to order n for every x in I. The function f is called an *explicit solution*, or simply a *solution*, of the differential equation

$$F[x, y, y', y'', \ldots, y^{(n)}] = 0, \tag{1.3.1}$$

if

(1) $$F[x, f(x), f'(x), \ldots, f^{(n)}(x)]$$

is defined for every x in I, and if

(2) $$F[x, f(x), f'(x), \ldots, f^{(n)}(x)] = 0$$

for every x in I. That is, if in (1.3.1) we replace y by $f(x)$, y' by $f'(x)$, $\ldots$, $y^{(n)}$ by $f^{(n)}(x)$, Equation (1.3.1) reduces to an indentity on I.

---

[4] For some basic definitions and concepts of point sets, the student is urged to review the material in Appendix 1.

Definition 1.3.2.   A relation $g(x, y) = 0$ is said to represent an *implicit solution* of the differential equation (1.3.1) on an interval I, if (1) $g(x, y) = 0$ defines at least one function f of the variable x on an interval I, that is, $g[x, f(x)] = 0$ for every x in I, and if (2) f is an explicit solution of the differential equation (1.3.1) for every x in I.

Definition 1.3.3.   A *general solution* of the differential equation (1.3.1) is an n-parameter family of curves (or functions)

$$f(x, y, c_1, c_2, \ldots, c_n) = 0, \tag{1.3.2}$$

such that each curve (or function) of the family is a solution of the differential equation (1.3.1).

Definition 1.3.4.   A *particular solution* of the differential equation (1.3.1) is one that is obtained from a general solution of (1.3.1) by assigning definite values to the n-parameters.

Definition 1.3.5.   A *singular solution* of the differential equation (1.3.1) is a solution of (1.3.1) and is one that cannot be obtained from any general solution of (1.3.1) no matter what definite values we assign to the n-parameters.

Let us return to Example 1.2.1 where we considered the one-parameter family of circles given by the relation

$$x^2 + y^2 = c^2. \tag{1.3.3}$$

If we solve for y in terms of x, we obtain

$$y = \pm\sqrt{c^2 - x^2}. \tag{1.3.4}$$

Thus, y is defined on the interval $-c \leq x \leq c$.   Note that for each x on this interval (except, $x = \pm c$), y has two values.   Hence, each circle of the family given in (1.3.3) is not the graph of a function, that is, the set of ordered pairs (x, y) where y is given in (1.3.4) and x belongs to the above interval is not a function.   [*See* Definition 1 of Appendix 1.]   Differentiating (1.3.3) with respect to x, we have

$$x + y \frac{dy}{dx} = 0. \tag{1.3.5}$$

Let $g(x, y) = x^2 + y^2 - c^2$.   According to Definition 1.3.2, in order for the relation $g(x, y) = 0$ to represent an implicit solution of the differential equation (1.3.5) on the interval I: $-c < x < c$, the following two conditions must be satisfied: (1) $g(x, y) = 0$ must define at least one function f of the variable x on I, that is, $g[x, f(x)] = 0$ for every x in I, and (2) f must be an explicit solution of (1.3.5) for every x in I.   The functions f_1 and f_2 given by [*see* Example 4 of Appendix 1]

$$f_1(x) = \sqrt{c^2 - x^2}, \tag{1.3.6}$$

and

$$f_2(x) = -\sqrt{c^2 - x^2}, \tag{1.3.7}$$

each satisfy the above two conditions for all x in I. To see that condition (1) is satisfied, we have for all x in I

$$g[x, f_1(x)] = x^2 + [f_1(x)]^2 - c^2 = x^2 + c^2 - x^2 - c^2 = 0,$$

and

$$g[x, f_2(x)] = x^2 + [f_2(x)]^2 - c^2 = x^2 + c^2 - x^2 - c^2 = 0.$$

Concerning condition (2), we have for all x in I

$$x + f_1(x)f_1'(x) = x + \sqrt{c^2 - x^2}\left(\frac{-x}{\sqrt{c^2 - x^2}}\right) = x - x = 0,$$

and

$$x + f_2(x)f_2'(x) = x - \sqrt{c^2 - x^2}\left(\frac{x}{\sqrt{c^2 - x^2}}\right) = x - x = 0.$$

Thus, $g(x, y) = x^2 + y^2 - c^2 = 0$ represents an implicit solution of (1.3.5).

There are other functions that satisfy the relation

$$g(x, y) = x^2 + y^2 - c^2 = 0. \tag{1.3.8}$$

For instance, the function f_3 defined by

$$f_3(x) = \left\{\begin{array}{ll} \sqrt{c^2 - x^2}, & \text{if } x \in (-c, 0) \\ -\sqrt{c^2 - x^2}, & \text{if } x \in [0, c) \end{array}\right\}^5. \tag{1.3.9}$$

(*See* Figure 1.3.1.)   Observe that the function f_3 defined in (1.3.9) does not satisfy the differential equation (1.3.5).   Since f is not continuous at $x = 0$, $f_3'(0)$ does not exist.

Example 1.3.1.   Consider the differential equation

$$\frac{d^2y}{dx^2} - y = -x. \tag{1.3.10}$$

Let us verify that the function f defined by

$$f(x) = c_1e^x + c_2 e^{-x} + x, \tag{1.3.11}$$

is a general solution of (1.3.10) for all x.

---

[5] Recall that $[a, b)$ stands for the half-open interval from the right, that is, $[a, b) = \{x \mid a \leq x < b\}$. Similarly, the half-open interval from the left is $(a, b] = \{x \mid a < x \leq b\}$. Also $(a, \infty) = \{x \mid x > a\}$; $[a, \infty) = \{x \mid x \geq a\}$; $(-\infty, a) = \{x \mid x < a\}$; $(-\infty, a] = \{x \mid x \leq a\}$; $(-\infty, \infty) = \{x \mid x \in R\}$, where R is the entire real line.   "∞" and "$-\infty$" are convenient symbols, and are not considered as real numbers.   Whenever we write the closed interval $[a, b] = \{x \mid a \leq x \leq b\}$, it is to be understood that a and b are finite and $a < b$.

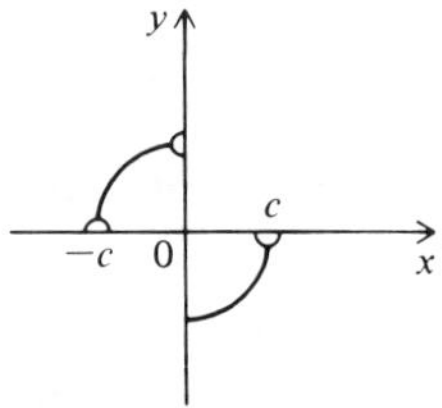

Figure 1.3.1.

The graph of f_3

SOLUTION.   [*Note*: Since $f(x)$ has two essential parameters, namely, c_1 and c_2, in view of Definition 1.3.3, $f(x)$ is a candidate for a general solution of (1.3.10).]   Differentiating $f(x)$ twice with respect to x, we obtain

$$f''(x) = c_1 e^x + c_2 e^{-x}.$$

Substituting $f(x)$ for y and $f''(x)$ for y'' in (1.3.10), we see that the left-hand member of (1.3.10) reduces to $-x$ for all x.   Thus, the function f defined in (1.3.11) is a general solution of (1.3.10) for all x.

Example 1.3.2.   Let us obtain the differential equation for the family of conics given by the equation

$$ax^2 + 2bxy + cy^2 + 2dx + 2ey + k = 0. \tag{1.3.12}$$

SOLUTION.   It appears in Equation (1.3.12) that we have six essential parameters, namely, a, b, c, d, e, and k.   However, if we divide each term in (1.3.12) say, by $c \neq 0$, we obtain

$$y^2 + 2c_1 xy + c_2 x^2 + 2c_3 x + 2c_4 y + c_5 = 0, \tag{1.3.13}$$

which has five essential constants: c_1, c_2, c_3, c_4, and c_5.   Thus, upon eliminating these constants, we expect to obtain a differential equation of order five.   This will be the case if we assume that the conics given in (1.3.13) do not have asymptotes parallel to the y axis.   (Why?)   Let us write (1.3.13) as a quadratic equation in y:

$$y^2 + 2(c_1 x + c_4)y + c_2 x^2 + 2c_3 x + c_5 = 0. \tag{1.3.14}$$

Using the quadratic formula to solve (1.3.14), we see that the functions f and g defined by

$$f(x) = -c_1 x - c_4 + \sqrt{Ax^2 + 2Bx + C}, \tag{1.3.15}$$

and

$$g(x) = -c_1 x - c_4 - \sqrt{Ax^2 + 2Bx + C}, \tag{1.3.16}$$

where

$$A = c_1{}^2 - c_2, \qquad B = c_1 c_4 - c_3, \qquad C = c_4{}^2 - c_5, \qquad \text{and } AC - B^2 \neq 0,$$

each satisfies Equation (1.3.14). Let us consider simply the function f defined in (1.3.15). Differentiating (1.3.15) twice with respect to x, we obtain upon simplification

$$f''(x) = \frac{AC - B^2}{(Ax^2 + 2Bx + C)^{3/2}}. \tag{1.3.17}$$

To facilitate the elimination of the remaining arbitrary constants A, B, and C in (1.3.17), let us write (1.3.17) as

$$[f''(x)]^{-2/3} = (AC - B^2)^{-2/3}(Ax^2 + 2Bx + C). \tag{1.3.18}$$

Now if we differentiate (1.3.18) three times with respect to x, we obtain

$$\frac{d^3}{dx^3}\{[f''(x)]^{-2/3}\} = 0. \tag{1.3.19}$$

Carrying out the differentiation in (1.3.19), we find after simplification that $f(x)$ satisfies the following equation:

$$9\left(\frac{d^2f(x)}{dx^2}\right)^2 \frac{d^5f(x)}{dx^5} - 45 \frac{d^2f(x)}{dx^2}\frac{d^3f(x)}{dx^3}\frac{d^4f(x)}{dx^4} + 40\left(\frac{d^3f(x)}{dx^3}\right)^3 = 0. \tag{1.3.20}$$

Thus, the desired differential equation for the family of conics is given by

$$9(y'')^2 y^{(5)} - 45\, y''y'''y^{(4)} + 40(y''')^3 = 0. \tag{1.3.21}$$

Exercises 1.3

1. Show that the relation $x^4 + 2x^2y^2 - 16 = 0$ defines an implicit solution of the differential equation

$$xy\frac{dy}{dx} + x^2 + y^2 = 0$$

   for all values of x such that $0 < |x| < 2$.

2. Show that the relation $y^2 - x^2y - 2x = 0$ defines an implicit solution of the differential equation

$$x(x^2 - 2y)\frac{dy}{dx} + y(x^2 + y) = 0$$

   for all values of x except for those on the interval $-2 \leq x \leq 0$.

3. Show that the relation $yx^2 - x^3 - 2 = 0$ defines an implicit solution of the differential equation

$$x^2\frac{d^2y}{dx^2} + 2x\frac{dy}{dx} - 2y = 0$$

   for all values of x except for $x = 0$.

4. Verify that the function f defined by

$$f(x) = \sqrt[3]{3kx^2 - x^3}$$

is an explicit solution of the differential equation

$$y^5 \frac{d^2y}{dx^2} + 2k^2x^2 = 0$$

for all values of x except for $x = 0$ and $x = 3k$.

5. Verify that the function f defined by

$$f(x) = e^x + 1 - 2\ln(e^x + 1) - \frac{1}{e^x + 1}$$

is an explicit solution of the differential equation

$$(1 + e^{-x})^2 \frac{dy}{dx} - e^x = 0$$

for all values of x.

6. Follow a method similar to that given in Example 1.3.2 to show that the function f defined by

$$f(x) = \frac{ax + b}{cx + d}$$

satisfies the differential equation

$$2y'y''' - 3(y'')^2 = 0$$

for all values of x except for $x = -d/c$ if $a + d \neq 0$. If $a + d = 0$, show that we then obtain

$$(y - x)y'' - 2y'(1 + y') = 0.$$

7. Verify that the function f defined by

$$f(x) = \frac{\ln x - x}{(1 - x)^2}$$

is a solution of the differential equation

$$x(1 - x)\frac{d^2y}{dx^2} + (1 - 4x)\frac{dy}{dx} - 2y = 0$$

for all values of x on the interval I: $0 < x < 1$.

8. *Leibnitz's Rule.*   Let D be a domain in the xt-plane containing the rectangle $a \leq x \leq b$, $t_1 \leq t \leq t_2$.   Let $g(x, t)$ be continuous in D and have a continuous partial derivative $\partial g/\partial t$ in D.   Let $\alpha(t)$ and $\beta(t)$ be defined and have continuous derivatives for $t_1 < t < t_2$.   Then for $t_1 < t < t_2$,

(1) $$\frac{d}{dt}\int_{\alpha(t)}^{\beta(t)} g(x, t)\,dx = g[\beta(t), t]\beta'(t) - g[\alpha(t), t]\alpha'(t) + \int_{\alpha(t)}^{\beta(t)} \frac{\partial g(x, t)}{\partial t}\,dx.$$

[*See* [24],[6] pp. 220–221, for a proof of this result.]

---

[6] Numbers in brackets refer to the list of references in the Bibliography.

Examples:

$$\frac{d}{dt}\int_{t^2}^{t^3} e^{x^2 t^2}\,dx = e^{t^8}(3t^2) - e^{t^6}(2t) + \int_{t^2}^{t^3} e^{x^2 t^2}(2x^2 t)\,dx;$$

$$\frac{d}{dt}\int_{a}^{t^3} e^{x^2 t^2}\,dx = e^{t^8}(3t^2) + \int_{a}^{t^3} e^{x^2 t^2}(2x^2 t)\,dx;$$

$$\frac{d}{dt}\int_{a}^{b} e^{x^2 t^2}\,dx = \int_{a}^{b} e^{x^2 t^2}(2x^2 t)\,dx; \qquad \frac{d}{dt}\int_{a}^{t} F(x)\,dx = F(t);$$

$$\frac{d}{dt}\int_{0}^{t} te^x\,dx = te^t + \int_{0}^{t} e^x\,dx = te^t + e^t - 1;$$

$$\frac{d}{dt}\int_{0}^{t} F(x)\sinh(x-t)\,dx = F(t)\sinh(t-t) + \int_{0}^{t} F(x)\cosh(x-t)(-1)\,dx$$

$$= -\int_{0}^{t} F(x)\cosh(x-t)\,dx.$$

(a) Verify that the function f defined by

$$f(t) = 2\cos\beta t + \frac{1}{\beta}\sin\beta t + \frac{1}{\beta}\int_{0}^{t} F(x)\sin\beta(t-x)\,dx \qquad (\beta \neq 0)$$

is a solution of the differential equation

$$\frac{d^2 Y(t)}{dt^2} + \beta^2 Y(t) = F(t).$$

(b) Verify that the function f defined by

$$f(t) = \frac{t}{a\beta}\int_{0}^{a}(a-x)F(x)\,dx - \frac{1}{\beta}\int_{0}^{t}(t-x)F(x)\,dx \qquad (\beta \neq 0)$$

is a solution of the differential equation

$$\frac{d^2 Y(t)}{dt^2} + \frac{1}{\beta}F(t) = 0.$$

(c) Verify that the function f defined by

$$f(t) = c_1 \sin\beta t + c_2 \cos\beta t + c_3 \cosh\beta t + c_4 \sinh\beta t$$

$$+ \frac{1}{2\beta^3}\int_{0}^{t} F(x)[\sinh\beta(t-x) - \sin\beta(t-x)]\,dx \qquad (\beta \neq 0)$$

is a general solution of the differential equation

$$\frac{d^4 Y(t)}{dt^4} - \beta^4 Y(t) = F(t).$$

(d) Verify that the function f defined by

$$f(t) = 1 - 2t - 4t^2 + \int_0^t [3 + 6(t - x) - 4(t - x)^2] f(x)\,dx,$$

is a solution of the differential equation

$$\frac{d^3 Y(t)}{dt^3} - 3 \frac{d^2 Y(t)}{dt^2} - 6 \frac{dY(t)}{dt} + 8 Y(t) = 0.$$

9. (a) Verify that the function f defined by

$$(1) \qquad\qquad f(x) = x^r \qquad (x > 0)$$

is a solution of the *Euler homogeneous differential equation*:

$$(2) \qquad\qquad ax^2 y'' + bxy' + cy = 0,$$

where a, b, and c are constants, provided that r is a root of the quadratic equation

$$(3) \qquad\qquad ar^2 + (b - a)r + c = 0.$$

(b) Verify that the functions f_1 and f_2 defined by $f_1(x) = x^2$ and $f_2(x) = x^3$ are solutions of the Euler equation

$$(4) \qquad\qquad x^2 y'' - 4xy' + 6y = 0.$$

(c) In part (b), verify that the function f defined by

$$f(x) = c_1 f_1(x) + c_2 f_2(x) = c_1 x^2 + c_2 x^3,$$

where c_1 and c_2 are arbitrary constants, is also a solution of Equation (4).

(d) Suppose that r_1 and r_2 are two distinct real roots of Equation (3). Verify that the function f defined by

$$(5) \qquad\qquad f(x) = c_1 x^{r_1} + c_2 x^{r_2},$$

where c_1 and c_2 are arbitrary constants, is a general solution of Equation (2).

In Chapter 4, Section 10, we shall consider the Euler equation in greater detail. Our discussion will include the cases when the roots in Equation (3) are real and equal, and also when they are complex. Euler equations arise in connection with solving partial differential equations when one utilizes the method of "separation of variables." Partial differential equations play a very important role in mathematical physics and other branches of applied mathematics and engineering.

(e) Find a general solution of each of the following Euler equations on the interval I: $x > 0$.

   (i) $x^2 y'' - xy' - 3y = 0.$
   (ii) $2x^2 y'' + 3xy' - y = 0.$
   (iii) $3x^2 y'' - 4xy' - 20y = 0.$

1.4. Existence Theorems

In solving differential equations, we shall be interested not only in their general solutions, but also in those solutions that satisfy certain other conditions. We shall concern ourselves primarily with those conditions that are known as initial conditions and boundary conditions of an ordinary differential equation.

Definition 1.4.1.   An *initial condition* is a condition on the solution of the differential equation at *one point*; a *boundary condition* is a condition on the solution of the differential equation at *two (or more) points.*

Definition 1.4.2.   A differential equation with its initial conditions will be called an *initial-value problem*; one that involves its boundary conditions will be called a *boundary-value problem.*

Example 1.4.1.   Let us find a function f that satisfies the initial-value problem

$$\frac{dy}{dx} - ky = 0, \tag{1.4.1}$$

and at $x = 0$, the function f has the value 2.

SOLUTION.   A general solution of (1.4.1) is given by [*see* Example 1.2.4 with 2 replaced by k]

$$f(x) = ce^{kx}, \tag{1.4.2}$$

where c is an arbitrary constant.   We shall now determine c so that at $x = 0$, the function f has the value 2.   Thus,

$$2 = f(0) = ce^0 = c.$$

Hence,

$$f(x) = 2e^{kx} \tag{1.4.3}$$

is a particular solution that satisfies the initial-value problem for all values of x.

Remark 1.4.1.   It is customary to write the initial-value problem of Example 1.4.1 in the following abbreviated form

$$\begin{cases} \dfrac{dy}{dx} - ky = 0, \\[2mm] y(0) = 2. \end{cases} \tag{1.4.4}$$

The symbolic notation $y(0) = 2$ means that the solution of the differential equation in (1.4.4) has the value 2 at $x = 0$.   Also, we shall write the solution given in (1.4.3) as

$$y(x) = 2e^{kx},$$

or simply

$$y = 2e^{kx}. \tag{1.4.5}$$

We shall often write an initial-value problem and a boundary-value problem in the abbreviated form.

Example 1.4.2.   Let us solve the boundary-value problem

$$\begin{cases} [1 + (y')^2]y''' - 3y'(y'')^2 = 0, \\ y(3) = 3 + \sqrt{3}, \\ y(4) = 5, \\ y(5) = 3 + \sqrt{3}. \end{cases} \tag{1.4.6}$$

SOLUTION.   A general solution of the above differential equation is given by [*see* Example 1.2.2]

$$x^2 + y^2 + c_1 x + c_2 y + c_3 = 0, \tag{1.4.7}$$

where c_1, c_2, and c_3 are arbitrary constants.   To determine c_1, c_2, and c_3 in (1.4.7), we use the boundary conditions, namely, $y = 3 + \sqrt{3}$ at $x = 3$, $y = 5$ at $x = 4$, and $y = 3 + \sqrt{3}$ at $x = 5$.   Thus, we obtain

$$\begin{cases} 3c_1 + (3 + \sqrt{3})c_2 + c_3 = -21 - 6\sqrt{3}, \\ 4c_1 + \phantom{(3 + \sqrt{3})}5c_2 + c_3 = -41, \\ 5c_1 + (3 + \sqrt{3})c_2 + c_3 = -37 - 6\sqrt{3}. \end{cases}$$

On solving these equations for c_1, c_2, and c_3, we find that

$$c_1 = -8, \qquad c_2 = -6, \qquad \text{and} \qquad c_3 = 21.$$

When these values of c_1, c_2, and c_3 are substituted in Equation (1.4.7), we obtain

$$y^2 - 6y + x^2 - 8x + 21 = 0. \tag{1.4.8}$$

Equation (1.4.8) may be considered as a quadratic equation in y.   Thus, in view of our boundary conditions, the function f defined by

$$f(x) = 3 + \sqrt{4 - (x - 4)^2},$$

or simply

$$y = 3 + \sqrt{4 - (x - 4)^2}, \tag{1.4.9}$$

is a particular solution that satisfies the boundary-value problem for all values of x in the interval $2 < x < 6$. [*Note*: The solution given in (1.4.9) consists of the set of all points (x, y) whose coordinates satisfy

$$(x - 4)^2 + (y - 3)^2 = 4 \quad \text{and} \quad 2 < x < 6, \quad 3 < y < 5.]$$

The following example shows that a boundary-value problem need not have a solution.

Example 1.4.3. The boundary-value problem

$$\begin{cases} \dfrac{d^2 y}{dx^2} + 4y = 0, \\[2mm] y(0) = 1, \\[2mm] y(\pi/2) = 2, \end{cases} \tag{1.4.10}$$

has no solution.

PROOF. The differential equation obtained in solving Exercise 1.2.6 shows that a general solution of the above differential equation is given by

$$y = c_1 \sin 2x + c_2 \cos 2x, \tag{1.4.11}$$

where c_1 and c_2 are arbitrary constants. Utilizing the boundary conditions, namely, $y = 1$ at $x = 0$ and $y = 2$ at $x = \pi/2$, we see that

$$1 = c_1 \sin 0 + c_2 \cos 0 = c_2$$

and

$$2 = c_1 \sin \pi + c_2 \cos \pi = -c_2 .$$

Thus $c_2 = 1$ and $c_2 = -2$. Since the arbitrary constant c_2 cannot be uniquely determined, it follows that the boundary-value problem has no solution.

The following example shows that an initial-value problem need not have a *unique* solution, that is, *only one* solution.

Example 1.4.4. The initial-value problem

$$\begin{cases} \dfrac{dy}{dx} = y^{2/3}, \\[2mm] y(0) = 0, \end{cases} \tag{1.4.12}$$

has more than one solution.

VERIFICATION. One may easily verify that the *null function*

$$y(x) = 0 \quad \text{for all } x,$$

and

$$y(x) = \frac{x^3}{27}$$

are both solutions of the initial-value problem for all values of x.

Remark 1.4.2. It is of the utmost importance to have some basic theorems that will enable one to determine whether or not a given differential equation with initial conditions has a unique solution. Fortunately there are existence theorems that will help us (not always, however) to answer this question. For the present, we shall state two existence theorems, one on first-order differential equations and the other on second-order differential equations.[7] The proofs of these theorems will be given in Appendix 2. [*See* Theorems 6 and 8 and the material immediately following the proofs of these theorems.]

Let us denote the partial derivatives of $f(x, y, z)$ with respect to x, y, and z by $f_x(x, y, z)$, $f_y(x, y, z)$, and $f_z(x, y, z)$, respectively.

Theorem 1.4.1. *Let the functions f and f_y be continuous in a domain D of the xy-plane containing the point (x_0, y_0).* *Then an interval*

$$I_0 : \quad |x - x_0| \leqq h \qquad (h > 0) \tag{1.4.13}$$

exists on which there is a unique solution $y = y(x)$ satisfying the differential equation

$$\frac{dy}{dx} = f(x, y) \tag{1.4.14}$$

and the initial condition

$$y(x_0) = y_0. \tag{1.4.15}$$

Theorem 1.4.2. *Let the functions f, f_y, and f_z be continuous in a three-dimensional domain D containing the point (x_0, y_0, z_0).* *Then an interval*

$$I_0 : \quad |x - x_0| \leqq h \qquad (h > 0) \tag{1.4.16}$$

exists on which there is a unique solution $y = y(x)$ satisfying the differential equation

$$y'' = f(x, y, y') \tag{1.4.17}$$

and the initial conditions

$$y(x_0) = y_0 \quad \text{and} \quad y'(x_0) = z_0. \tag{1.4.18}$$

Remark 1.4.3. We may write the interval I_0 in its equivalent form as

$$I_0 : \quad x_0 - h \leqq x \leqq x_0 + h \qquad (h > 0). \tag{1.4.19}$$

Note that the above theorems do not tell us the size of the interval I_0. They

---

[7] The existence theorem for a second-order differential equation will involve a three-dimensional region R. The definition of such a region may be formulated in a similar fashion as that given in Definition 35 of Appendix 1. The interiors of a cube, a rectangular parallelopiped, and a sphere are examples of open regions or domains in a three-dimensional space.

merely assert that there exists an interval I_0. Furthermore, the theorems do not indicate a method for finding the unique solution. The theorems provide what are known as *sufficient conditions*[8] for the existence of a unique solution. That is, if the conditions stated in the hypothesis of the theorems are satisfied, then we are assured that there exists a unique solution of the initial-value problem on some interval I_0. However, the conditions stated in the hypothesis of the theorems are not *necessary*; that is, if these conditions are not all satisfied, there may still exist a unique solution. The reader should be fully aware of what information the above existence theorems really offer. For these or other existence theorems may save one from doing many hours of needless work in vain. Also note the geometric significance of the initial conditions in (1.4.18). They assert that the curve $y(x)$ passes through the point (x_0, y_0), and at $x = x_0$ the slope of the curve is z_0.

Example 1.4.5. Let us show that the initial-value problem

$$\frac{dy}{dx} = ky, \tag{1.4.20}$$

$$y(1) = e^k, \tag{1.4.21}$$

has a unique solution in the interval I_0 : $|x - 1| \leq h$.[9]

SOLUTION. Here $f(x, y) = ky$ and $f_y(x, y) = k$. Clearly the functions f and f_y both satisfy the hypothesis of Theorem 1.4.1 throughout the xy-plane. In particular, we may apply the theorem to any domain D containing the point $(x_0, y_0) = (1, e^k)$. Thus, an interval

$$I_0: \quad |x - x_0| = |x - 1| \leq h,$$

exists on which there is a unique solution $y(x)$ satisfying the differential equation (1.4.20) and the initial condition (1.4.21). A general solution of (1.4.20) is given by [*see* Example 1.2.4 with 2 replaced by k]

$$y = ce^{kx}$$

where c is an arbitrary constant. Since $y = e^k$ at $x = 1$, we find $c = 1$. Thus

$$y = e^{kx} \tag{1.4.22}$$

is the unique solution of the initial-value problem on the interval

$$I_0: \quad |x - 1| \leq h.$$

In this example, the unique solution actually exists over the interval $-\infty < x < \infty$. [*See* Figure 1.4.1.]

---

[8] See the footnote on page 622 for the meaning of a "necessary and sufficient condition."
[9] Henceforth, it is to be understood that when we write I_0 : $|x - x_0| \leq h$, we assume $h > 0$, *except when otherwise indicated.*

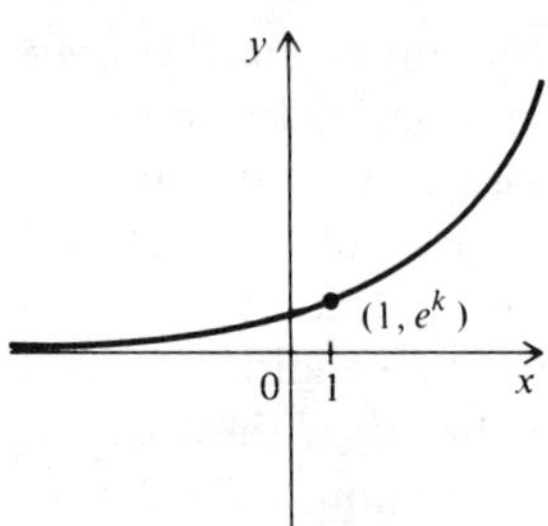

Figure 1.4.1.

$y = e^{kx}$

Example 1.4.6.   Let us show that $y = e^{2x} - e^{-x}$ is the unique solution of the initial-value problem

$$\begin{cases} y'' - y' - 2y = 0, & (1.4.23) \\ y(0) = 0, & (1.4.24) \\ y'(0) = 3, & (1.4.25) \end{cases}$$

in the interval $I_0: \quad |x| \leqq h.$

SOLUTION.   From (1.4.23), we see that

$$y'' = f(x, y, y'),$$

where

$$f(x, y, y') = 2y + y'.$$

Thus,

$$f(x, y, z) = 2y + z,$$

and

$$f_y(x, y, z) = 2, \qquad f_z(x, y, z) = 1.$$

Clearly the functions f, f_y, and f_z satisfy the hypothesis of Theorem 1.4.2 for all values of x, y, and z.   In particular, we may apply the theorem to any domain D containing the point $(x_0, y_0, z_0) = (0, 0, 3)$.   Thus, an interval

$$I_0: \quad |x - x_0| = |x - 0| = |x| \leqq h,$$

exists on which there is a unique solution $y(x)$ satisfying the differential equation (1.4.23) and the initial conditions (1.4.24) and (1.4.25).   One may easily verify that

$$y = e^{2x} - e^{-x} \tag{1.4.26}$$

satisfies (1.4.23), (1.4.24), and (1.4.25).   This solution is then the unique solution of the initial-value problem in the interval $I_0: \quad |x| \leqq h.$

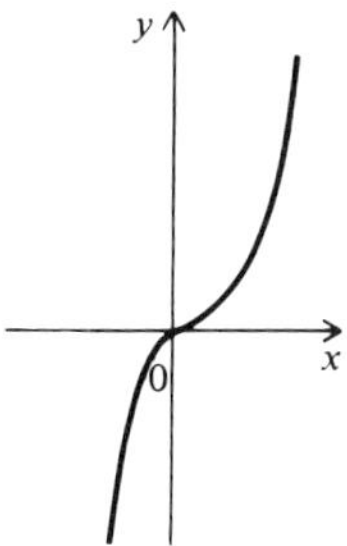

Figure 1.4.2.

$y = e^{2x} - e^{-x}$

Again, in this example, the unique solution actually exists over the interval: $-\infty < x < \infty$. [*See* Figure 1.4.2.]

Example 1.4.7. Let us show that the initial-value problem

$$\begin{cases} y' = 4 + y^2, & (1.4.27) \\ y(0) = 0, & (1.4.28) \end{cases}$$

has a unique solution in the interval $I_0 : \quad |x| \leqq h$.

SOLUTION. From (1.4.27), we see that $f(x, y) = 4 + y^2$ and $f_y(x, y) = 2y$. The functions f and f_y satisfy the hypothesis of Theorem 1.4.1 throughout the xy-plane. For a domain D containing the point $(x_0, y_0) = (0, 0)$, we are assured by the theorem of an interval

$$I_0 : \quad |x - x_0| = |x - 0| = |x| \leqq h,$$

on which there is a unique solution $y(x)$ satisfying (1.4.27) and (1.4.28). Separating the variables in (1.4.27) and then integrating, we obtain

$$c + x = \int \frac{dy}{4 + y^2} = \frac{1}{2} \arctan \frac{y}{2}, \qquad (1.4.29)$$

where c is an arbitrary constant of integration. From (1.4.28), $y = 0$ at $x = 0$; consequently, $c = 0$. Thus, (1.4.29) becomes

$$x = \frac{1}{2} \arctan \frac{y}{2},$$

or

$$y = 2 \tan 2x. \qquad (1.4.30)$$

One may easily verify that the function y in (1.4.30) satisfies (1.4.27) and (1.4.28). Thus, this function is then the unique solution of the initial-value problem in the interval $I_0 : \quad |x| \leqq h$.

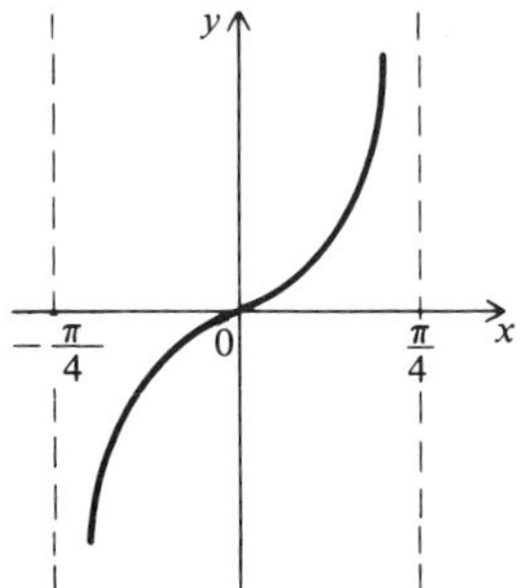

Figure 1.4.3.

$y = 2 \tan 2x$

In this example, the unique solution actually exists over the interval $-\pi/4 < x < \pi/4$. [*See* Figure 1.4.3.] Note, however, that the unique solution cannot be extended beyond this interval. Any larger interval would contain the points $x = \pi/4$ or $x = -\pi/4$, and the function y given by (1.4.30) becomes infinite at these points.

Example 1.4.8. Let us show that the initial-value problem

$$y' = 3y^{2/3}(2x + 1), \tag{1.4.31}$$

$$y(0) = 1, \tag{1.4.32}$$

has a unique solution in the interval $I_0: |x| \leq h$.

SOLUTION.   From (1.4.31), we see that $f(x, y) = 3y^{2/3}(2x + 1)$ and $f_y(x, y) = 2(2x + 1)/y^{1/3}$. The function f is continuous for all values of x and y; while the function f_y is continuous for all values of x and for all values of $y \neq 0$, that is, f_y is not continuous along the entire x axis. From (1.4.32),

$$(x_0, y_0) = (0, 1).$$

For the domain D in Theorem 1.4.1, let us take any domain in the xy-plane that contains the point $(0, 1)$, and any part of the x axis that is contained in its exterior. Clearly, the functions f and f_y satisfy the hypothesis of Theorem 1.4.1 in D. Thus, an interval $I_0: |x| \leq h$ exists, on which there is a unique solution $y(x)$ satisfying (1.4.31) and (1.4.32). Since $y \neq 0$ in D, from (1.4.31), we have

$$\tfrac{1}{3}y^{-2/3}\, dy = (2x + 1)dx,$$

and an integration gives

$$y^{1/3} = x^2 + x + c, \tag{1.4.33}$$

where c is an arbitrary constant of integration. Since $y = 1$ at $x = 0$, we find that $c = 1$. We may now write (1.4.33) as

$$y = (x^2 + x + 1)^3. \tag{1.4.34}$$

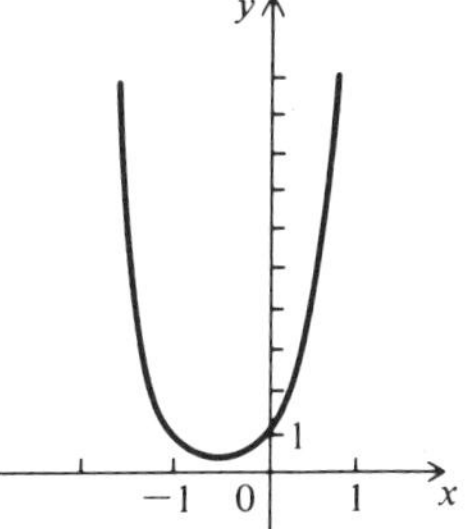

Figure 1.4.4.

$y = (x^2 + x + 1)^3$

One can easily verify that the function y in (1.4.34) satisfies (1.4.31) and (1.4.32). Thus, this function is then the unique solution of the initial-value problem in the interval $I_0 : |x| \leq h$. In this example, the unique solution actually exists over the interval $-\infty < x < \infty$. [*See* Figure 1.4.4.]

Exercises 1.4

1. Show that $y = e^{x+1} - 3(x+1)$ is the unique solution of the initial-value problem
$$\begin{cases} y' = 3x + y, \\ y(-1) = 1, \end{cases}$$
   for some interval I_0: $|x + 1| \leq h$.

2. Show that $y = \dfrac{(x + 2 - \cos x)\cos x}{1 + \sin x}$ is the unique solution of the initial-value problem
$$\begin{cases} y' + y \sec x = \cos x, \\ y(0) = 1, \end{cases}$$
   for some interval I_0: $|x| \leq h$.

3. Show that $y = (1 + x)e^{-2x}$ is the unique solution of the initial-value problem
$$\begin{cases} y'' + 4y' + 4y = 0, \\ y(0) = 1, \\ y'(0) = -1, \end{cases}$$
   for some interval I_0: $|x| \leq h$.

4. Show that $y = 1 + 2x - 3x^2 - e^{-3x}$ is the unique solution of the initial-value problem
$$\begin{cases} y'' + 3y' = -18x, \\ y(0) = 0, \\ y'(0) = 5, \end{cases}$$
   for some interval I_0: $|x| \leq h$.

5. Show that $y = 1 - 1/x + \ln x$ is the unique solution of the initial-value problem

$$\begin{cases} x^2 y'' + 2xy' = 1, \\ y(1) = 0, \\ y'(1) = 2, \end{cases}$$

for some interval I_0: $|x - 1| \leq h$.

6. Show that $y = \sqrt[3]{x}$ is the unique solution of the initial-value problem

$$\begin{cases} yy'' + 2(y')^2 = 0, \\ y(1) = 1, \\ y'(1) = 1/3, \end{cases}$$

for some interval I_0: $|x - 1| \leq h$.

7. By integration, find the unique solution of the initial-value problem

$$\begin{cases} y' = y \cot x, \\ y\left(\dfrac{\pi}{4}\right) = \dfrac{\sqrt{2}}{2}, \end{cases}$$

for some interval I_0: $|x - \pi/4| \leq h$.

8. By integration, find the unique solution of the initial-value problem

$$\begin{cases} (e^{2x} + 1)y' - 2y = 0, \\ y(0) = 1, \end{cases}$$

for some interval I_0: $|x| \leq h$.

9. By integrations, find the unique solution of the initial-value problem

$$\begin{cases} (x^2 + 1)y'' - 2xy' = 0, \\ y(2) = 11/3, \\ y'(2) = 5, \end{cases}$$

for some interval I_0: $|x - 2| \leq h$.

10. By integrations, find the unique solution of the initial-value problem

$$\begin{cases} (e^x + e^{-x})y'' - 2e^x y' = 0, \\ y(0) = 3/2, \\ y'(0) = 2, \end{cases}$$

for some interval I_0: $|x| \leq h$.

11. By integrations, find the unique solution of the initial-value problem

$$\begin{cases} (x^2 + 1)^2 y'' - x(y')^3 = 0, \\ y(0) = 0, \\ y'(0) = 1, \end{cases}$$

for some interval I_0: $|x| \leq h$.

12. (a) Verify that the function u defined by

$$u(x, y, z) = \sin mx \cos ny \exp\left(-\sqrt{m^2 + n^2}\, z\right)$$

is a solution of the *Laplace partial differential equation*:

$$u_{xx}(x, y, z) + u_{yy}(x, y, z) + u_{zz}(x, y, z) = 0.$$

This equation is utilized in the study of steady-state temperature distribution in a three-dimensional region.

(b) If m and n are positive integers, verify the following boundary conditions:

$$u(0, y, z) = u(\pi, y, z) = u_y(x, 0, z) = u_y(x, \pi, z) = 0.$$

(c) The Laplace equation for the *xyz*-coordinate system is often written as $\nabla^2 u = 0$, where ∇^2 represents the *Laplacian operator*:

$$\frac{\partial^2}{\partial x^2} + \frac{\partial^2}{\partial y^2} + \frac{\partial^2}{\partial z^2}.$$

The symbol $\nabla^2 u$ is read as "del-squared of u" or "del-squared u."

If $u = \sqrt{x^2 + y^2 + z^2}$, then

$$\nabla^2 u = \left(\frac{\partial^2}{\partial x^2} + \frac{\partial^2}{\partial y^2} + \frac{\partial^2}{\partial z^2}\right)\sqrt{x^2 + y^2 + z^2}$$

$$= \frac{\partial^2(\sqrt{x^2 + y^2 + z^2})}{\partial x^2} + \frac{\partial^2(\sqrt{x^2 + y^2 + z^2})}{\partial y^2} + \frac{\partial^2(\sqrt{x^2 + y^2 + z^2})}{\partial z^2}.$$

Performing the indicated partial differentiations, show that the above expression simplifies to

$$\nabla^2 u = \frac{2}{u}.$$

(d) If $u = \frac{1}{2}\ln(x^2 + y^2 + z^2)$, show that $\nabla^2 u = \dfrac{1}{x^2 + y^2 + z^2}$.

13. (a) Verify that the function u defined by

$$u(x, y) = \sum_{n=1}^{m} c_n \cos nx \sinh ny$$

is a solution of Laplace partial differential equation

$$u_{xx}(x, y) + u_{yy}(x, y) = 0$$

or

$$\nabla^2 u = 0,$$

where

$$\nabla^2 = \frac{\partial^2}{\partial x^2} + \frac{\partial^2}{\partial y^2},$$

is the Laplacian operator for the *xy*-coordinate system.

(b) Also verify the following boundary conditions:

$$u(x, 0) = u_x(0, y) = u_x(\pi, y) = 0.$$

(c) A function u is said to be *harmonic* in a domain D contained in the xy-plane if for all points x, y in D, all second partial derivatives exist, are continuous and $\nabla^2 u = 0$.

Thus harmonic functions satisfy the Laplace equation. The Laplace equation is also known as the *potential equation* and plays an important role in mathematical physics and other branches of applied mathematics.

Show that the functions u and v defined by

$$u(x, y) = e^x x \cos y - e^x y \sin y \quad \text{and} \quad v(x, y) = e^{(x^2 - y^2)} \cos 2xy$$

are harmonic in any domain D contained in the xy-plane.

(d) In applications, sometimes it is more convenient to use polar coordinates (r, θ) rather than rectangular coordinates (x, y). The relation between the two coordinate systems, when the polar axis is taken along the positive x-axis with the pole coinciding with the origin, are

$$x = r \cos \theta, \qquad y = r \sin \theta.$$

Using the chain rule for partial derivatives, show that the *polar form of Laplace equation* is given by

$$r^2 \frac{\partial^2 u}{\partial r^2} + r \frac{\partial u}{\partial r} + \frac{\partial^2 u}{\partial \theta^2} = 0.$$

(e) Verify that the function u defined by

$$u(r, \theta) = \sum_{n=1}^{m} c_n r^n \sin n\theta$$

is a solution of Laplace equation in polar form.

(f) Also verify the following boundary conditions:

$$u(r, 0) = u(r, \pi) = 0.$$

14. Verify that the function u defined by

$$u(x, t) = \sum_{n=0}^{m} e^{-kn^2 \pi^2 t} \cos n\pi x$$

is a solution of the *one-dimensional heat equation*:

$$u_t(x, t) = k u_{xx}(x, t) \qquad (k \neq 0).$$

15. (a) Verify that the function u defined by

$$u(x, t) = \sum_{n=1}^{m} c_n e^{\{-[n-(1/2)]^2 \pi^2 t\}} \sin[(n - \tfrac{1}{2})\pi x]$$

is a solution of the *heat equation*:

$$u_t(x, t) = u_{xx}(x, t).$$

(b) Verify the following boundary conditions:

$$u(0, t) = u_x(1, t) = 0.$$

(c) Determine the constants c_n in part (a) when

$$u(x, 0) = \sin \frac{3\pi x}{2} - 3 \sin \frac{5\pi x}{2} + 6 \sin \frac{7\pi x}{2}.$$

16. (a) Verify that the function z defined by

$$z(x, y, t) = \sum_{n=1}^{r} \sum_{m=1}^{s} c_{mn} \sin \frac{n\pi x}{a} \sin \frac{m\pi y}{b} \cos \left(\pi c \sqrt{\frac{n^2}{a^2} + \frac{m^2}{b^2}}\, t \right)$$

is a solution of the *vibrating membrane equation*:

$$z_{tt}(x, y, t) = c^2 [z_{xx}(x, y, t) + z_{yy}(x, y, t)] \qquad (c \neq 0).$$

(b) Verify the following boundary conditions:

$$z(0, y, t) = z(a, y, t) = z(x, 0, t) = z(x, b, t) = 0.$$

17. In solving partial differential equations by integrations, the general solution depends on arbitrary functions of one or more variables rather than on arbitrary constants. For example, to solve

$$u_{xy}(x, y) = 0,$$

we proceed as follows. Integrating with respect to x, and keeping y fixed, we obtain

$$u_y(x, y) = g(y)$$

where g is an arbitrary function. Integrating now with respect to y, and holding x constant, we get

$$(1) \qquad u(x, y) = \int_{y_0}^{y} g(\zeta)d\zeta + f(x),$$

where f is another arbitrary function. In many problems, we may take $y_0 = 0$ in Equation (1).

By integrations, find the function u that satisfies each of the following boundary-value problems.

(a)
$$\begin{cases} u_{xy}(x, y) = 0, \\ u(0, y) = y^2, \\ u(x, 0) = x^2. \end{cases}$$

(b)
$$\begin{cases} u_{xy}(x, y) = 3x^2, \\ u(0, y) = y, \\ u(x, 0) = \sin x. \end{cases}$$

(c)
$$\begin{cases} u_{xy}(x, y) = \cos x, \\ u(0, y) = \cos y, \\ u(x, 0) = x \sin x. \end{cases}$$

(d)
$$\begin{cases} u_{xy}(x, y) = 3x^2, \\ u(0, y) = y, \\ u_x(x, 0) = \sin x. \end{cases}$$

(e)
$$\begin{cases} u_{xy}(x, y) = 2x + 2y, \\ u_y(0, y) = 2 \sin 2y, \\ u(x, 0) = \cos x/2. \end{cases}$$

(f)
$$\begin{cases} u_{yy}(x, y) = 0, \\ u(x, 0) = x^2, \\ u(x, 1) = x. \end{cases}$$

$$(g) \quad \begin{cases} u_{xx}(x, y, z) = 6xyz, \\ u(0, y, z) = y^2 + z^2, \\ u(1, y, z) = 2yz. \end{cases} \qquad (h) \quad \begin{cases} u_{xyz}(x, y, z) = 8xyz, \\ u(x, y, 0) = y \sin x, \\ u_z(x, 0, z) = x \cos z, \\ u_{yz}(0, y, z) = 9y^2 z^2. \end{cases}$$

18. (a) Verify that the function[10] y defined by

$$y(x, t) = F(x + at) + G(x - at)$$

is a solution of the *one-dimensional wave equation*:

$$y_{tt}(x, t) = a^2 y_{xx}(x, t) \qquad (a \neq 0).$$

(b) Does the function y defined by

$$y(x, t) = e^{\sin (x + at)} + \ln (x + at) + \cos^2 (x - at) + \sinh (x - at)$$

satisfy the wave equation?

19. (a) Let

$$\xi = x + at, \qquad \zeta = x - at,$$

where ξ and ζ are new independent variables. Utilize the chain rule for partial derivatives to show that the wave equation

$$y_{tt}(x, t) = a^2 y_{xx}(x, t)$$

reduces to the form

$$(1) \qquad \frac{\partial^2 y}{\partial \xi\, \partial \zeta} = 0.$$

(b) By integrations, show that the general solution of Equation (1) is given by

$$(2) \qquad y(x, t) = F(x + at) + G(x - at),$$

where F and G are functions with continuous first and second partial derivatives. This method of solving the wave equation is due to D'Alembert.

(c) Utilize Equation (2) to show that the solution of the initial-value problem (*Vibrating String*)

$$(3) \qquad \begin{cases} y_{tt}(x, t) = a^2 y_{xx}(x, t), \\ y(x, 0) = f(x), \\ y_t(x, 0) = 0, \end{cases}$$

is given by

$$(4) \qquad y(x, t) = \tfrac{1}{2}\,[f(x + at) + f(x - at)].$$

[The solution given in (4) represents two waves traveling in the opposite direction with velocity a.]

---

[10] We are implicitly assuming that the function y has continuous first and second partial derivatives in its domain of definition. This assumption also applies to Exercise 19.

(d) Solve the initial-value problem given in (3) for each function f defined by

   (i) $f(x) = x^2$.   (iv) $f(x) = x^2 \cosh x$.

   (ii) $f(x) = \sin x$.   (v) $f(x) = \sin^3 x$.

   (iii) $f(x) = x \sin x$.   (vi) $f(x) = x \cos^3 x$.

Suggested Readings

Berg and McGregor [5]   Leighton [28]
Goursat [17]   Ross [40]
Ince [23]

chapter 2   EQUATIONS OF THE FIRST-ORDER

2.1. Introduction

In this chapter, we shall concern ourselves with some basic methods and techniques of solving three types of first-order differential equations. These are: exact equations, separable equations, and linear equations. Also, as we shall see, there are certain classes of first-order differential equations that do not belong to any one of these three types; however, under a suitable transformation they may be changed into one of these types.

2.2. Exact Equations and Forms

A differential equation of order one and degree one may be written in the form

$$\frac{dy}{dx} = f(x, y). \tag{2.2.1}$$

If

$$f(x, y) = -\frac{M(x, y)}{N(x, y)} \qquad [N(x, y) \neq 0],$$

we may then write (2.2.1) equivalently as

$$M(x, y)dx + N(x, y)dy = 0. \tag{2.2.2}$$

Definition 2.2.1. The differential equation

$$M(x, y)dx + N(x, y)dy = 0 \tag{2.2.2}$$

is called *exact* if there exists a function u such that

$$du(x, y) = M(x, y)dx + N(x, y)dy. \tag{2.2.3}$$

Notice that for exactness, the left-hand member of Equation (2.2.2) is the total differential of the function u.   Thus, the general solution of (2.2.2) is a one-parameter family of curves given by

$$u(x, y) = C. \tag{2.2.4}$$

For, upon taking the differential of (2.2.4), we obtain

$$du = 0,$$

and in view of (2.2.3), we then have

$$M\,dx + N\,dy = 0$$

as desired.   The one-parameter family of curves given by (2.2.4) is also known as the *level curves* of $u(x, y)$.   These level curves may then be depicted graphically.

The following theorem will give us a very useful criterion for determining whether or not the differential equation (2.2.2) is exact; and, if it is exact, how to determine the function u.

Theorem 2.2.1.   *Let the functions M and N be continuous and possess continuous first partial derivatives in the interior of a rectangle R: $|x - x_0| \leq a$, $|y - y_0| \leq b$.   Then a necessary and sufficient condition for the differential equation*

$$M(x, y)dx + N(x, y)dy = 0 \tag{2.2.5}$$

to be exact is that

$$\frac{\partial M}{\partial y} = \frac{\partial N}{\partial x} \tag{2.2.6}$$

in R.   When condition (2.2.6) is satisfied, the general solution of Equation (2.2.5) is given by

$$u(x, y) = C \tag{2.2.7}$$

where

$$u(x, y) = \int_{x_0}^{x} M(s, y)ds + \int_{y_0}^{y} N(x_0, t)dt, \tag{2.2.8}$$

C is an arbitrary constant, (x_0, y_0) is a fixed point in R, and (x, y) is an arbitrary point in R.

PROOF.   *Necessity.*   This condition asserts that if Equation (2.2.5) is exact, then (2.2.6) is true.   To establish this result, we proceed as follows.   In view of Definition 2.2.1, there exists a function u such that

$$du = M\,dx + N\,dy. \tag{2.2.9}$$

From calculus, the total differential of the function u is given by

$$du = \frac{\partial u}{\partial x}\,dx + \frac{\partial u}{\partial y}\,dy. \tag{2.2.10}$$

From (2.2.9) and (2.2.10), it follows by the usual argument that

$$\frac{\partial u}{\partial x} = M \qquad \text{and} \qquad \frac{\partial u}{\partial y} = N. \tag{2.2.11}$$

Since by hypothesis, M and N have continuous first partial derivatives, it follows from (2.2.11) that

$$\frac{\partial^2 u}{\partial y\,\partial x} \qquad \text{and} \qquad \frac{\partial^2 u}{\partial x\,\partial y}$$

are continuous. Again, from calculus, we then have

$$\frac{\partial^2 u}{\partial y\,\partial x} = \frac{\partial^2 u}{\partial x\,\partial y}. \tag{2.2.12}$$

In view of (2.2.11) and (2.2.12), we see that

$$\frac{\partial M}{\partial y} = \frac{\partial N}{\partial x},$$

and the necessity condition is established.

Sufficiency. This condition asserts that if (2.2.6) holds, then Equation (2.2.5) is exact. Thus, we must exhibit a function u such that

$$\frac{\partial u(x, y)}{\partial x} = M(x, y), \tag{2.2.13}$$

and

$$\frac{\partial u(x, y)}{\partial y} = N(x, y). \tag{2.2.14}$$

Then we will have

$$du = \frac{\partial u}{\partial x}\,dx + \frac{\partial u}{\partial y}\,dy = M\,dx + N\,dy, \tag{2.2.15}$$

and the differential equation (2.2.5) will be exact in view of Definition 2.2.1.

To this end, if a function u exists, then it must satisfy (2.2.13). Hence, integrating (2.2.13) with respect to x, we obtain

$$u(x, y) = \int_{x_0}^{x} M(s, y)\,ds + \lambda(y), \tag{2.2.16}$$

where x_0 is a convenient constant and λ is an arbitrary function of y only. [*Note:* If we differentiate (2.2.16) with respect to x we obtain (2.2.13).] We shall determine $\lambda(y)$ so that the function u in (2.2.16) will also satisfy (2.2.14). Differentiating (2.2.16) with respect to y, we obtain

$$\frac{\partial u(x,\, y)}{\partial y} = \frac{\partial}{\partial y} \int_{x_0}^{x} M(s,\, y)\, ds + \lambda'(y),$$

and utilizing (2.2.14), we may write the above expression as

$$N(x,\, y) = \frac{\partial}{\partial y} \int_{x_0}^{x} M(s,\, y)\, ds + \lambda'(y). \tag{2.2.17}$$

Since by hypothesis the functions M and $\partial M / \partial y$ are continuous in R, in view of Exercise 1.3.8 we have

$$N(x,\, y) = \int_{x_0}^{x} \frac{\partial M(s,\, y)}{\partial y}\, ds + \lambda'(y). \tag{2.2.18}$$

We are given that

$$\frac{\partial M(x,\, y)}{\partial y} = \frac{\partial N(x,\, y)}{\partial x}.$$

Thus, (2.2.18) becomes

$$N(x,\, y) = \int_{x_0}^{x} \frac{\partial N(s,\, y)}{\partial s}\, ds + \lambda'(y)$$

$$= N(s,\, y)\, \Big|_{x_0}^{x} + \lambda'(y) = N(x,\, y) - N(x_0,\, y) + \lambda'(y),$$

from which we immediately obtain

$$\lambda'(y) = N(x_0,\, y). \tag{2.2.19}$$

Integrating (2.2.19) with respect to y yields

$$\lambda(y) = \int_{y_0}^{y} N(x_0,\, t)\, dt, \tag{2.2.20}$$

where y_0 is a convenient constant. [The constant of integration is suppressed in (2.2.20), since it is absorbed ultimately with the constant C.] [*See* (2.2.22).] Substituting $\lambda(y)$ of (2.2.20) in (2.2.16), we obtain

$$u(x,\, y) = \int_{x_0}^{x} M(s,\, y)\, ds + \int_{y_0}^{y} N(x_0,\, t)\, dt. \tag{2.2.21}$$

The function u given by (2.2.21) clearly satisfies Equations (2.2.13) and (2.2.14), for

$$\frac{\partial u(x,\, y)}{\partial x} = \frac{\partial}{\partial x} \int_{x_0}^{x} M(s,\, y)\, ds + \frac{\partial}{\partial x} \int_{y_0}^{y} N(x_0,\, t)\, dt$$

$$= M(x,\, y) + 0 = M(x,\, y).$$

Also

$$\frac{\partial u(x, y)}{\partial y} = \frac{\partial}{\partial y} \int_{x_0}^{x} M(s, y)\,ds + \frac{\partial}{\partial y} \int_{y_0}^{y} N(x_0, t)\,dt$$

$$= \int_{x_0}^{x} \frac{\partial M(s, y)}{\partial y}\,ds + N(x_0, y)$$

$$= \int_{x_0}^{x} \frac{\partial N(s, y)}{\partial s}\,ds + N(x_0, y)$$

$$= N(s, y)\Big|_{x_0}^{x} + N(x_0, y)$$

$$= N(x, y) - N(x_0, y) + N(x_0, y)$$

$$= N(x, y),$$

and the sufficiency condition is thus established.

In view of (2.2.21) and (2.2.4), the general solution of the differential equation (2.2.5) is then given by

$$u(x, y) = C,$$

where   (2.2.22)

$$u(x, y) = \int_{x_0}^{x} M(s, y)\,ds + \int_{y_0}^{y} N(x_0, t)\,dt,$$

C is an arbitrary constant, (x_0, y_0) is a fixed point in R, and (x, y) is an arbitrary point in R. The proof of the theorem is now established.

Remark 2.2.1.   We could have started the sufficiency proof by asserting that if a function u exists, then it must satisfy Equation (2.2.14).   Then, integrating (2.2.14) with respect to y and proceeding as above, the general solution of Equation (2.2.5) would be given by

$$u(x, y) = C,$$

where   (2.2.23)

$$u(x, y) = \int_{y_0}^{y} N(x, t)\,dt + \int_{x_0}^{x} M(s, y_0)\,ds.$$

The verification of (2.2.23) will be left as an exercise for the reader.   [*See* Exercise 2.2.19.]

Example 2.2.1.   Let us solve the equation

$$\left(4x^3 y^3 + \frac{1}{x}\right) dx + \left(3x^4 y^2 - \frac{1}{y}\right) dy = 0. \tag{2.2.24}$$

SOLUTION.   Here

$$M(x, y) = 4x^3y^3 + \frac{1}{x} \quad \text{and} \quad N(x, y) = 3x^4y^2 - \frac{1}{y}, \tag{2.2.25}$$

$$\frac{\partial M}{\partial y} = 12x^3y^2 \quad \text{and} \quad \frac{\partial N}{\partial x} = 12x^3y^2. \tag{2.2.26}$$

In view of Equations (2.2.26), Equation (2.2.24) is exact.   The functions M_y and N_x are continuous for all values of x and y.   The functions M and N are also continuous for all values of x and y except that M is not continuous along the line $x = 0$ (the y axis) and N is not continuous along the line $y = 0$ (the x-axis).   For the rectangle R of Theorem 2.2.1, we may take any rectangle in which the lines $x = 0$ and $y = 0$ are exterior.   Utilizing (2.2.7) and (2.2.8) with $M(s, y) = 4s^3y^3 + 1/s$ and $N(x_0, t) = 3x_0{}^4t^2 - 1/t$, we have

$$\int_{x_0}^{x} \left(4s^3y^3 + \frac{1}{s}\right)ds + \int_{y_0}^{y} \left(3x_0{}^4t^2 - \frac{1}{t}\right)dt = C,$$

where (x_0, y_0) is a fixed point in R and (x, y) is an arbitrary point in R. Carrying out the above integration, we find

$$x^4y^3 + \ln x - x_0{}^4y^3 - \ln x_0 + x_0{}^4y^3 - \ln y - x_0{}^4y_0{}^3 + \ln y_0 = C.$$

Thus,

$$x^4y^3 + \ln x - \ln y = C + \ln x_0 - \ln y_0 + x_0{}^4y_0{}^3.$$

The above expression can be written as

$$x^4y^3 + \ln \frac{x}{y} = c, \tag{2.2.27}$$

where $c = C + \ln (x_0/y_0) + x_0{}^4y_0{}^3$ is an arbitrary constant.   The general solution of (2.2.24) is given by (2.2.27).

Remark 2.2.2.   From the contents of the sufficiency proof of Theorem 2.2.1, we have seen that the differential equation

$$M(x, y)dx + N(x, y)dy = 0 \tag{2.2.5}$$

is *exact* if there exists a function u such that

$$\frac{\partial u(x, y)}{\partial x} = M(x, y) \quad \text{and} \quad \frac{\partial u(x, y)}{\partial y} = N(x, y). \tag{2.2.28}$$

We shall also use (2.2.28) *in conjunction with* (2.2.4) *to find the general solution of* (2.2.5).

Remark 2.2.3.   Suppose that the differential equation

$$M(x, y)dx + N(x, y)dy = 0 \tag{2.2.5}$$

is exact, that is,

$$\frac{\partial M}{\partial y} = \frac{\partial N}{\partial x}.$$ [2.2.6]

Then, in view of Theorem 2.2.1, the general solution of (2.2.5) is given by

$$\int_{x_0}^{x} M(s, y)ds + \int_{y_0}^{y} N(x_0, t)dt = C.$$ (2.2.29)

Observe that the second integral in (2.2.29), evaluated at its lower limit y_0, simply depends upon the fixed point (x_0, y_0), and is therefore a constant which can be absorbed in C.   We then have

$$\int_{x_0}^{x} M(s, y)ds + \int N(x_0, y)dy = C,$$ (2.2.30)

where C is again arbitrary, and the second integral is now indefinite.   Notice that each integral depends upon x_0.   If (2.2.30) is used to solve (2.2.5) and x_0 is left arbitrary, then those x_0-terms that cannot be absorbed in C must necessarily cancel, since only one arbitrary constant is allowed.   However, it is usually easier to assign to x_0 that (permissible) value for which the second integral in (2.2.30) assumes its simplest form.   In most cases that value would be $x_0 = 0$, whence

$$\int_{0}^{x} M(s, y)ds + \int N(0, y)dy = C.$$ (2.2.31)

In other cases, the value might be π, e, $\ln 2$, and so on.   In all cases, there must correspond to x_0, values of y such that (x_0, y) belongs to the interior of the rectangle R described in Theorem 2.2.1.

Recalling (2.2.23), we can see that a symmetrical argument holds, with the roles of M and N interchanged.   In this case, we have

$$\int M(x, y_0)dx + \int_{y_0}^{y} N(x, t)dt = C,$$ (2.2.32)

and when $y_0 = 0$,

$$\int M(x, 0)dx + \int_{0}^{y} N(x, t)dt = C.$$ (2.2.33)

Here again, to y_0 there must correspond values of x such that (x, y_0) belongs to the interior of R.

If one attempts to solve (2.2.5) using (2.2.30), but finds the second integral untenable for any permissible x_0, he is likely to do no better with (2.2.32), since the same integration is involved.   The idea to remember is that (2.2.30) may simplify N, and (2.2.32) may simplify M.   It is largely a matter of convenience (sometimes necessity) and one must use his judgment.

The following two examples illustrate this remark.

Example 2.2.2.   Let us solve the equation

$$\left[\frac{\ln(\ln y)}{x} + \frac{2}{3}xy^3\right]dx + \left[\frac{\ln x}{y\ln y} + x^2y^2\right]dy = 0.$$

SOLUTION.   Here

$$M(x, y) = \frac{\ln(\ln y)}{x} + \frac{2}{3}xy^3 \qquad \text{and} \qquad N(x, y) = \frac{\ln x}{y\ln y} + x^2y^2.$$

Thus,

$$\frac{\partial M}{\partial y} = \frac{1}{xy\ln y} + 2xy^2 \qquad \text{and} \qquad \frac{\partial N}{\partial x} = \frac{1}{xy\ln y} + 2xy^2.$$

Since

$$\frac{\partial M}{\partial y} = \frac{\partial N}{\partial x},$$

the given differential equation is exact.   Now, (x_0, y_0) can be any point interior to the first quadrant such that $x_0 > 0$ and $y_0 > 1$. $x_0 = 1$ is certainly permissible, and one would make this choice to avoid integrating the $1/y\ln y$ term.   Thus, (2.2.30) becomes

$$\int_1^x M(s, y)ds + \int N(1, y)dy = C.$$

With $N(1, y) = y^2$, we have

$$\int_1^x \left[\frac{\ln(\ln y)}{s} + \frac{2}{3}y^3s\right]ds + \int y^2 dy = C.$$

Hence, the general solution is

$$\ln x \ln(\ln y) + \frac{1}{3}x^2y^3 = C.$$

Example 2.2.3.   Let us solve the equation

$$\left(\frac{\ln y - 1}{\sqrt{x^2 + 1}} + x^2\right)dx + \frac{1}{y}\ln(x + \sqrt{x^2 + 1})dy = 0.$$

SOLUTION.   Here

$$M(x, y) = \frac{\ln y - 1}{\sqrt{x^2 + 1}} + x^2 \qquad \text{and} \qquad N(x, y) = \frac{1}{y}\ln(x + \sqrt{x^2 + 1}).$$

Thus,

$$\frac{\partial M}{\partial y} = \frac{1}{y\sqrt{x^2 + 1}} \qquad \text{and} \qquad \frac{\partial N}{\partial x} = \frac{1}{y\sqrt{x^2 + 1}},$$

and the given differential equation is exact.   Here (x_0, y_0) may be taken as any point for which $y_0 > 0$ and any x_0.   To avoid integrating the term $1/\sqrt{x^2 + 1}$, we choose $y_0 = e$.   Therefore, $M(x, e) = x^2$.   Thus, (2.2.32) becomes

$$\int M(x, e)dx + \int_e^y N(x, t)dt = C.$$

Consequently, we have

$$\int x^2 \, dx + \int_e^y \frac{1}{t} \ln (x + \sqrt{x^2 + 1})dt = C,$$

and the general solution is

$$\frac{x^3}{3} + \ln (x + \sqrt{x^2 + 1})(\ln y - 1) = C.$$

Example 2.2.4.   Let us solve the equation

$$(3x^2y^2 - \cos 2y)dx + (2x^3y + 2x \sin 2y - \cos 2y)dy = 0. \quad (2.2.34)$$

Here

$$M(x, y) = 3x^2y^2 - \cos 2y \qquad \text{and} \qquad N(x, y) = 2x^3y + 2x \sin 2y - \cos 2y.$$

$$(2.2.35)$$

$$\frac{\partial M}{\partial y} = 6x^2y + 2 \sin 2y \qquad \text{and} \qquad \frac{\partial N}{\partial x} = 6x^2y + 2 \sin 2y. \quad (2.2.36)$$

Thus,

$$\frac{\partial M}{\partial y} = \frac{\partial N}{\partial x},$$

and (2.2.34) is exact.   The functions M, N, M_y, and N_x are continuous for all values of x and y.

SOLUTION 1.   From the first equations in (2.2.28) and (2.2.35), we have

$$\frac{\partial u(x, y)}{\partial x} = M(x, y) = 3x^2y^2 - \cos 2y.$$

Integrating this expression with respect to x, we obtain

$$u(x, y) = x^3y^2 - x \cos 2y + \lambda(y), \qquad\qquad (2.2.37)$$

where λ is a function of y which we shall presently determine.   Differentiating (2.2.37) with respect to y gives

$$\frac{\partial u(x, y)}{\partial y} = 2x^3y + 2x \sin 2y + \lambda'(y),$$

and in view of the second equation in (2.2.28), we then have

$$N(x, y) = 2x^3y + 2x \sin 2y + \lambda'(y).$$

From the second equation in (2.2.35), we may write the above expression as

$$2x^3y + 2x \sin 2y - \cos 2y = 2x^3y + 2x \sin 2y + \lambda'(y),$$

from which we immediately obtain

$$\lambda'(y) = -\cos 2y.$$

Thus,

$$\lambda(y) = -\int \cos 2y \, dy = -\tfrac{1}{2} \sin 2y.$$

Substituting $\lambda(y)$ in (2.2.37), we have

$$u(x, y) = x^3y^2 - x \cos 2y - \tfrac{1}{2} \sin 2y. \qquad (2.2.38)$$

From (2.2.4) and (2.2.38), the general solution of (2.2.34) is

$$x^3y^2 - x \cos 2y - \tfrac{1}{2} \sin 2y = C. \qquad (2.2.39)$$

SOLUTION 2.   Again, from (2.2.35),

$$N(x, y) = 2x^3y + 2x \sin 2y - \cos 2y.$$

Also, we have by taking $x_0 = 0$,

$$N(x_0, y) = N(0, y) = -\cos 2y.$$

Thus the general solution of (2.2.34) is also given by (2.2.31):

$$\int_0^x (3s^2y^2 - \cos 2y)\,ds - \int \cos 2y \, dy = C,$$

or

$$x^3y^2 - x \cos 2y - \tfrac{1}{2} \sin 2y = C. \qquad [2.2.39]$$

SOLUTION 3.   *Method of Grouping.*   We may write (2.2.34) as

$$(3x^2y^2 \, dx + 2x^3y \, dy) - (\cos 2y \, dx - 2x \sin 2y \, dy) - \cos 2y \, dy = 0.$$

[*Note:*   Each term in the parenthesis is an exact differential; clearly the last term is also an exact differential.]   Thus we may write the above expression as

$$d(x^3y^2) - d(x \cos 2y) - \tfrac{1}{2}d(\sin 2y) = 0,$$

from which we obtain upon integration

$$x^3y^2 - x \cos 2y - \tfrac{1}{2} \sin 2y = C. \qquad [2.2.39]$$

Remark 2.2.4.   Theorem 2.2.1 is also valid if we replace "in the interior of a rectangle $R: |x - x_0| \leqq a, \quad |y - y_0| \leqq b$" by "in a simply connected region R." [*See* [45], pp. 434–440.] It turns out that if the region is not simply connected [*see* Definition 36 of Appendix 1], the line integral[1]

$$\int_{\Gamma_1} M(x, y)dx + N(x, y)dy \tag{2.2.40}$$

is not necessarily equal to the line integral

$$\int_{\Gamma_2} M(x, y)dx + N(x, y)dy, \tag{2.2.41}$$

where Γ_1 and Γ_2 are two simple curves [*see* Definition 30 of Appendix 1] having the same initial point (x_0, y_0) and terminal point (x, y) as shown in Figure 2.2.1.

The line integrals in (2.2.40) and (2.2.41) may also be expressed as ordinary integrals, namely,

$$\int_{y_0}^{y} N(x_0, t)dt + \int_{x_0}^{x} M(s, y)ds, \tag{2.2.40$'$}$$

and

$$\int_{x_0}^{x} M(s, y_0)ds + \int_{y_0}^{y} N(x, t)dt. \tag{2.2.41$'$}$$

Thus, in view of (2.2.40$'$) and (2.2.41$'$), the value of $u(x, y)$ given in either (2.2.8) or (2.2.23) may depend upon which path of integration we take (either Γ_1 or Γ_2), so that $u(x, y)$ may be multiple-valued, and consequently it may not be a function as asserted in Theorem 2.2.1. We shall illustrate this point in Example 2.2.6. However, when in Theorem 2.2.1, R is a simply connected region, it turns out that we always have

$$\int_{\Gamma_1} M\, dx + N\, dy = \int_{\Gamma_2} M\, dx + N\, dy,$$

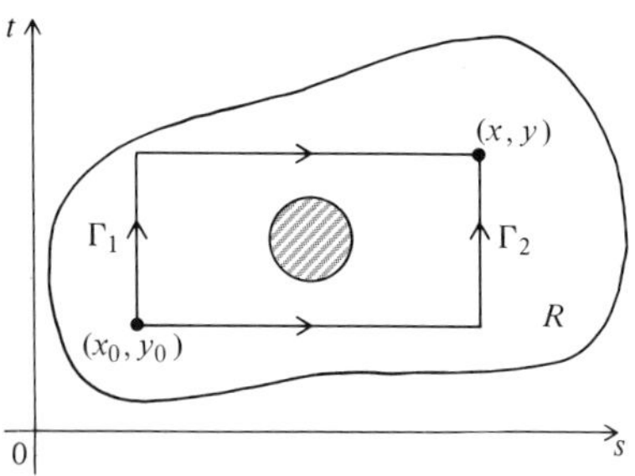

Figure 2.2.1.

Multiply Connected Region R

---

[1] We are assuming that the reader has been exposed to some basic concepts of line integrals in calculus.

where Γ_1 and Γ_2 are any two simple curves in R having the same initial point and terminal point. Thus, the value of u given by

$$u(x, y) = \int_\Gamma M(x, y)dx + N(x, y)dy, \tag{2.2.42}$$

where Γ is any simple curve in R, does not depend upon the path of integration. Consequently, u is single-valued for all points (x, y) in R and is indeed a function.

Definition 2.2.2. The expression

$$M(x, y)dx + N(x, y)dy \tag{2.2.43}$$

is called a *first-order differential form in the variables x and y.*

Definition 2.2.3. The differential form (2.2.43) is said to be *exact* if there exists a function u such that

$$du = M\, dx + N\, dy. \tag{2.2.44}$$

One may state a theorem for differential forms that is identical to Theorem 2.2.1, and its proof is essentially the same.

Theorem 2.2.2. *Let the functions M and N be continuous and possess continuous first partial derivatives in the interior of a rectangle R: $|x - x_0| \leqq a$, $|y - y_0| \leqq b$. Then a necessary and sufficient condition for the differential form*

$$M(x, y)dx + N(x, y)dy \tag{2.2.45}$$

to be exact is that

$$\frac{\partial M}{\partial y} = \frac{\partial N}{\partial x} \tag{2.2.46}$$

for all points (x, y) in R. When condition (2.2.46) is satisfied, a function u such that

$$du = M\, dx + N\, dy \tag{2.2.47}$$

is given by

$$u(x, y) = \int_{x_0}^{x} M(s, y)ds + \int_{y_0}^{y} N(x_0, t)dt, \tag{2.2.48}$$

where (x_0, y_0) is a fixed point in R, and (x, y) is any arbitrary point in R.

Remark 2.2.3 is essentially the same for differential forms. We shall now evaluate line integrals of differential forms that are exact.

Theorem 2.2.3. *Let R be a region in the xy-plane. Let Γ be a simple curve in*

R with initial point (x_1, y_1) and terminal point (x_2, y_2). Let the functions M and N be continuous in R, and suppose there is a function u such that

$$du(x, y) = M(x, y)dx + N(x, y)dy, \qquad (2.2.49)$$

for all points (x, y) in R. Then

$$\int_\Gamma M\, dx + N\, dy = u(x_2, y_2) - u(x_1, y_1). \qquad (2.2.50)$$

PROOF.   Let us express Γ by its parametric equations [*see* Definitions 27 and 30 of Appendix 1]

$$\Gamma: \quad x = x(t), \qquad y = y(t), \qquad t_1 \leqq t \leqq t_2. \qquad (2.2.51)$$

As t varies from t_1 to t_2, the point (x, y) moves along Γ from its initial point $(x_1, y_1) = [x(t_1), y(t_1)]$ to its terminal point $(x_2, y_2) = [x(t_2), y(t_2)]$. Utilizing Equations (2.2.49) and (2.2.51), we then have

$$\int_\Gamma M\, dx + N\, dy = \int_{t_1}^{t_2} \left(M\frac{dx}{dt} + N\frac{dy}{dt} \right) dt = \int_{t_1}^{t_2} \frac{d\{u[x(t), y(t)]\}}{dt}\, dt$$

$$= u[x(t), y(t)] \Big|_{t_1}^{t_2} = u[x(t_2), y(t_2)] - u[x(t_1), y(t_1)]$$

$$= u(x_2, y_2) - u(x_1, y_1),$$

and the theorem is established.

Example 2.2.5.   Let us evaluate the line integral

$$\int_\Gamma y \cos x\, dx + \sin x\, dy, \qquad (2.2.52)$$

where Γ is a simple curve with initial point $(1, 0)$ and terminal point $(1/2, \sqrt{3}/2)$.

SOLUTION.   Here

$$M(x, y) = y \cos x, \qquad N(x, y) = \sin x,$$

$$\frac{\partial M}{\partial y} = \cos x, \qquad \frac{\partial N}{\partial x} = \cos x. \qquad (2.2.53)$$

The functions M, N, M_y, and N_x are continuous for all values of x and y. Let R be any rectangle in the xy-plane containing in its interior the simple curve Γ. In view of (2.2.53) and Theorem 2.2.2, we see that the differential form

$$M\, dx + N\, dy = y \cos x\, dx + \sin x\, dy$$

is exact for all values of x and y in R. Thus, a function u exists such that

$$du = M\, dx + N\, dy = y \cos x\, dx + \sin x\, dy. \qquad (2.2.54)$$

Moreover, utilizing Formula (2.2.48) with $(x_0, y_0) = (1, 0)$, we obtain

$$u(x, y) = \int_1^x y \cos s \, ds + \int_0^y \sin 1 \, dt = y \sin x. \tag{2.2.55}$$

Clearly, the function u given by (2.2.55) satisfies (2.2.54) for all values of x and y in R. Utilizing Formula (2.2.50), the value of the line integral (2.2.52) is

$$\int_\Gamma y \cos x \, dx + \sin x \, dy = u\left(\frac{1}{2}, \frac{\sqrt{3}}{2}\right) - u(1, 0) = \frac{\sqrt{3}}{2} \sin \frac{1}{2} - 0 \sin 1 = 0.415.$$

The following popular example is very instructive.

Example 2.2.6.   Let us evaluate the line integral

$$\int_\Gamma -\frac{y}{x^2 + y^2} \, dx + \frac{x}{x^2 + y^2} \, dy, \tag{2.2.56}$$

along various simple curves Γ, all having the same initial point $(1, 0)$ and terminal point $(-1, 0)$.

SOLUTION.   Here

$$M(x, y) = -\frac{y}{x^2 + y^2}, \qquad N(x, y) = \frac{x}{x^2 + y^2}, \tag{2.2.57}$$

$$\frac{\partial M}{\partial y} = \frac{y^2 - x^2}{(x^2 + y^2)^2}, \qquad \frac{\partial N}{\partial x} = \frac{y^2 - x^2}{(x^2 + y^2)^2}. \tag{2.2.58}$$

The functions M, N, M_y, and N_x are continuous for all values of x and y except for $(x, y) = (0, 0)$. Let R be the region given by the interior of the rectangle $|x| \leq a$, $|y| \leq b$ with the point $(0, 0)$ removed. R is then a multiply connected region. [*See* Definition 36 of Appendix 1 and Figure 2.2.2.]

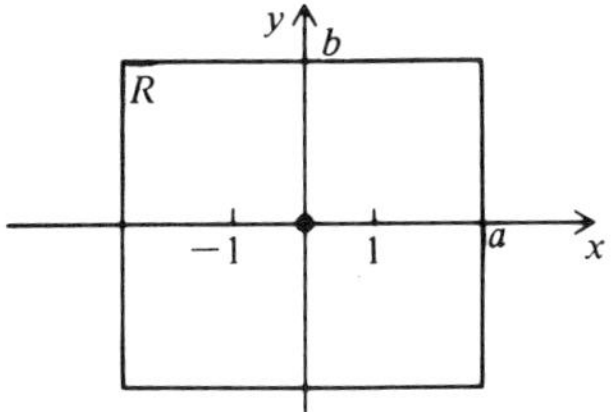

Figure 2.2.2.

Multiply Connected Region R

Consider the differential form

$$M\,dx + N\,dy = -\frac{y}{x^2 + y^2}\,dx + \frac{x}{x^2 + y^2}\,dy. \qquad (2.2.59)$$

In view of (2.2.58), the above differential form is exact for all points (x, y) in R. Thus in view of Theorem 2.2.2 (where in this theorem, R is now the above rectangle with the origin removed), there exists a function such that its total differential is equal to the differential form given in (2.2.59). In fact, if r and θ are polar coordinates,

$$x = r\cos\theta, \qquad y = r\sin\theta,$$

$$dx = \cos\theta\,dr - r\sin\theta\,d\theta, \qquad dy = \sin\theta\,dr + r\cos\theta\,d\theta,$$

then substituting these values in (2.2.59), we obtain upon simplification

$$\frac{-y\,dx + x\,dy}{x^2 + y^2} = d\theta. \qquad (2.2.60)$$

We shall now attempt to find the value of the line integral (2.2.56) by utilizing Formula (2.2.50) with $\theta = u(x, y)$. Since Theorem 2.2.3 requires that θ be a single-valued and differentiable function of x and y for all points (x, y) in R, how shall we standardize the definition of θ to achieve this result? We shall see in a moment that we cannot achieve this result for all points (x, y) in R. Suppose we standardize the definition of θ by requiring that $0 \leq \theta < 2\pi$. Then, clearly, θ is defined and is single-valued for all points (x, y) in R. However, θ is not continuous for all points (x, y) in R, in particular, when (x, y) belongs to the interval I: $\;\;0 < x < a, \;\; y = 0$. For, if $(x_0, 0)$ is any point in I, then as $(x, y) \to (x_0, 0)$ from above the x axis, $\theta \to 0$, while as $(x, y) \to (x_0, 0)$ from below the x axis, $\theta \to 2\pi$. Thus θ is not continuous and consequently it is not a differentiable function for any point (x, y) in I. That is, for this particular standardization of θ, we cannot achieve simultaneously the fact that θ is single-valued and is a differentiable function of x and y for all points (x, y) in R. We also could have standardized the definition of θ by requiring that $-\pi < \theta \leq \pi$ for all points (x, y) in R. For this case, once again, θ is defined and is single-valued for all points (x, y) in R, however, θ is now not continuous for any point (x, y) belonging to the interval I: $\;\;-a < x < 0, \;\; y = 0$.

Actually, there are infinitely many ways of standardizing the definition of θ so that θ is defined and is single-valued for all points (x, y) in R. For example, $\alpha < \theta \leq \alpha + 2\pi$, where α is any real number; however, θ is not continuous for any point (x, y) on the ray $\theta = \alpha$. Thus, no matter how we try to standardize the definition of θ, we cannot make θ simultaneously single-valued and a differentiable function of x and y for all points (x, y) in R. There will always exist at least one point of discontinuity on any simple

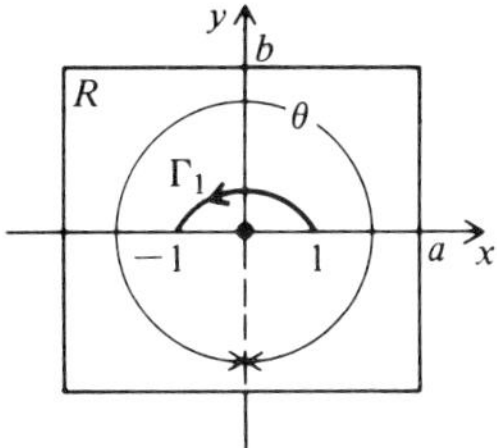

Figure 2.2.3.

closed curve surrounding the origin. (This difficulty arises because our region R is not simply connected.)

Let us return to the line integral (2.2.56). In R, let Γ_1 be a simple curve contained in the upper-half part of the region R with initial point at $(1, 0)$ and terminal point at $(-1, 0)$. [*See* Figure 2.2.3.] Let us standardize the definition of θ by requiring that $-\pi/2 < \theta \leq 3/2\pi$. Thus, in R, the only discontinuities of θ are located in the interval $x = 0$, $-b < y < 0$, which is also to be excluded from the region R. [*Note:* For this particular standardization of θ, if we introduce a boundary line along the interval $x = 0$, $-b \leq y < 0$, the region R has now been rendered into a simply connected region.] [*See* Remark 9 of Appendix 1.] We may now utilize Formula (2.2.50) to evaluate the line integral (2.2.56). In view of (2.2.60), we find its value to be

$$\int_0^\pi d\theta = \pi. \tag{2.2.61}$$

In R, let Γ_2 be a simple curve contained in the lower-half part of the region R with initial point at $(1, 0)$ and terminal point at $(-1, 0)$. [*See* Figure 2.2.4.] Let us standardize the definition of θ by requiring that $(-\tfrac{3}{2})\pi < \theta \leq \pi/2$. Thus, in R, the only discontinuities of θ are located in the interval $x = 0$, $0 < y < b$, which is also to be excluded from the region R. [As above, we may now render R simply connected by drawing a boundary

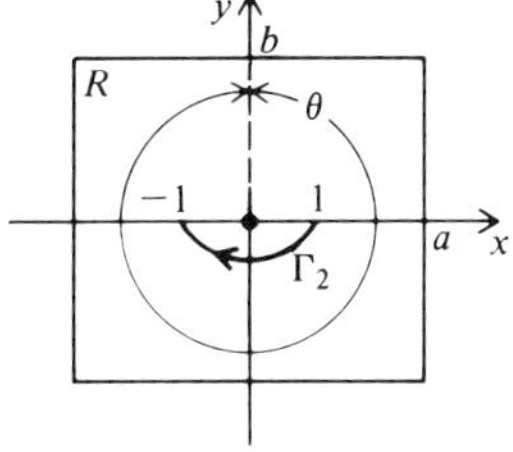

Figure 2.2.4.

line along the interval $x = 0$, $0 < y \leq b$.]   The value of the line integral is

$$\int_0^{-\pi} d\theta = -\pi. \tag{2.2.62}$$

Remark 2.2.5.   Clearly, the values of the above line integrals for the paths Γ_1 and Γ_2 are independent of the method used to standardize the definition of θ.   These two standardizations of the definition of θ were chosen so that the discontinuities of θ would lie along the y axis rather than along a ray $\theta = \alpha$.   However, as we have seen, the value of the line integral (2.2.56) does depend upon the path.   The above example also shows that if the differential form $M\,dx + N\,dy$ is such that $\partial M/\partial y = \partial N/\partial x$ for all points (x, y) in a region R, then it is not necessarily true that $M\,dx + N\,dy$ is the differential of a function u which is differentiable for all points (x, y) in R. As remarked earlier, however, this result is always true if the region R is simply connected.

Exercises 2.2

In Exercises 1 through 15, test each equation for exactness, and solve those that are exact.

1. $(x - 2y)dx - 2x\,dy = 0$.
2. $(x + y)dx + (y - x)dy = 0$.
3. $(x^2 + xy)dx + xy\,dy = 0$.
4. $(x - y\cos x)dx - \sin x\,dy = 0$.
5. $\sec^2 x \tan y\,dx + \sec^2 y \tan x\,dy = 0$.
6. $e^{-x}\,dx + ye^{-y}dy = 0$.
7. $(y - x\sqrt{x^2 + y^2})dx + (x - y\sqrt{x^2 + y^2})dy = 0$.

8. $\left(x + \dfrac{1}{y}\right)dx - \left(\dfrac{x}{y^2} - y\right)dy = 0$.

9. $[3x^2 + 2x + y\cos(xy)]dx + [\sin y + x\cos(xy)]dy = 0$.
10. $\cosh 2x \cosh 2y\,dx + \sinh 2x \sinh 2y\,dy = 0$.
11. $(y + xe^{-y^2})dx + (x + x^2ye^{-y^2})dy = 0$.
12. $(\tan x - 2x^2y^3)dx + (\tan y + 2x^3y^2)dy = 0$.
13. $(2xye^{x^2y} + y^2e^{xy^2} + 2x)dx + (x^2e^{x^2y} + 2xye^{xy^2} + 2y)dy = 0$.
14. $(e^{y^2} - \csc y \csc^2 x)dx + (2xye^{y^2} - \csc y \cot y \cot x)dy = 0$.

15. $\left(y^2 - \dfrac{y}{x(x + y)} + 2x\right)dx + \left(\dfrac{1}{x + y} + 2yx + 2y\right)dy = 0$.

In exercises 16 through 18, solve each initial-value problem.

16. $\begin{cases} \dfrac{dy}{dx} = \dfrac{2x - 3y + 5}{3x - 4y + 2}, \\ y(1) = 1. \end{cases}$

17. $\begin{cases} \dfrac{dy}{dx} = -\dfrac{ye^{xy} + 4y^3}{xe^{xy} + 12xy^2 - 2y}, \\ y(0) = 2. \end{cases}$

18.
$$\begin{cases} \dfrac{dy}{dx} = \dfrac{3x^2 \ln x + x^2 - y}{x}, \\ y(1) = 5. \end{cases}$$

19.   Derive Equation (2.2.23).

In Exercises 20 through 23, follow the method given in Example 2.2.5 to evaluate each line integral.

20. $\displaystyle\int_\Gamma (2xy^2 + y)dx + (2x^2y + x)dy$, where Γ is a simple curve with initial point $(1, 1)$ and terminal point $(4, 3)$.

21. $\displaystyle\int_\Gamma (2xy + y^2 e^{xy^2})dx + (x^2 + 2xye^{xy^2})dy$, where Γ is a simple curve with initial point $(0, 0)$ and terminal point $(4, \frac{1}{2})$.

22. $\displaystyle\int_\Gamma [y \cos (xy) + 2xy \cos y]dx + [x \cos (xy) + x^2 \cos y - x^2y \sin y]dy$, where Γ is a simple curve with initial point $(0, 0)$ and terminal point $(\frac{1}{2}, 1)$.

23. $\displaystyle\int_\Gamma \dfrac{(y^3 + 2xy^2 + x^2y)dx + (x^3 + 2x^2y + xy^2)dy}{(x + y)^2}$, where Γ is a simple curve not passing through $(0, 0)$, with initial point $(1, 0)$ and terminal point $(3, 1)$.

24.   Let Γ denote the perimeter of the rectangle $|x| \leq 4,\ \ |y| \leq 2$. Utilize polar coordinates to show that
$$\int_\Gamma \frac{x^2y}{(x^2 + y^2)^2}\, dx - \frac{x^3}{(x^2 + y^2)^2}\, dy = -\pi.$$

25.   Let Γ denote the perimeter of the rectangle $1 \leq x \leq 4,\ \ 0 \leq y \leq 2$. Show that
$$\int_\Gamma \frac{x^2y}{(x^2 + y^2)^2}\, dx - \frac{x^3}{(x^2 + y^2)^2}\, dy = 0.$$

26.   Find all values of the line integrals
$$\int_\Gamma \frac{x^2y}{(x^2 + y^2)^2}\, dx - \frac{x^3}{(x^2 + y^2)^2}\, dy,$$
where Γ is a simple curve with initial point $(0, 1)$ and terminal point $(3, 3)$ and not passing through the origin.

27.   In Theorem 2.2.3, suppose that Γ is a simple closed curve contained with its interior in R. Show that
$$\int_\Gamma M\, dx + N\, dy = 0.$$

In Exercises 28 through 30, evaluate each line integral. Γ is any simple closed curve.

28. $\int_\Gamma (1 + y^2)dx + (1 + 2xy + 3y^2)dy.$

29. $\int_\Gamma (3y^3 + y^2 + 2xy^2 + 4xy)dx + (2x^2 + 2x^2y + 9xy^2 + 2xy)dy.$

30. $\int_\Gamma [(1 + x^2y) \cos x + (2xy - x - y^2) \sin x]dx + [2y \cos x + x^2 \sin x]dy.$

2.3. Separation of Variables

Let us again consider the differential equation of order one and degree one of the form

$$M(x, y)dx + N(x, y)dy = 0.$$

Definition 2.3.1.   The differential equation

$$M(x, y)dx + N(x, y)dy = 0 \tag{2.3.1}$$

is said to have its *variables separated* if M is a function of x only and N is a function of y only; that is, if Equation (2.3.1) is of the form

$$X(x)dx + Y(y)dy = 0.^2 \tag{2.3.2}$$

Clearly, the differential equation (2.3.2) is exact, since

$$\frac{\partial M}{\partial y} = \frac{\partial X(x)}{\partial y} = 0,$$

and

$$\frac{\partial N}{\partial x} = \frac{\partial Y(y)}{\partial x} = 0.$$

In view of Theorem 2.2.1, *the general solution of Equation (2.3.2) is given by either*

$$\int_{x_0}^x X(s)ds + \int_{y_0}^y Y(t)dt = C, \tag{2.3.3}$$

or

$$\int X(x)dx + \int Y(y)dy = C. \tag{2.3.4}$$

---

² A differential equation of this form is also called a *separable equation*.

Let us consider the differential equation of the form

$$X(x)Y_1(y)dx + X_1(x)Y(y)dy = 0, \qquad (2.3.5)$$

where X and X_1 are functions of x only while Y and Y_1 are functions of y only. Equation (2.3.5) may be reduced to the form given in (2.3.2) by dividing (2.3.5) by $X_1(x)Y_1(y)$. Thus, we obtain

$$\frac{X(x)}{X_1(x)}dx + \frac{Y(y)}{Y_1(y)}dy = 0, \qquad (2.3.6)$$

in which the variables are now separated. *The general solution of Equation (2.3.6) is given by either*

$$\int_{x_0}^{x}\frac{X(s)}{X_1(s)}ds + \int_{y_0}^{y}\frac{Y(t)}{Y_1(t)}dt = C, \qquad (2.3.7)$$

or

$$\int\frac{X(x)}{X_1(x)}dx + \int\frac{Y(y)}{Y_1(y)}dy = C. \qquad (2.3.8)$$

Remark 2.3.1. In dividing (2.3.5) by $X_1(x)Y_1(y)$ a number of solutions of (2.3.5) may be lost; namely, the solutions arising from the equations

$$X_1(x) = 0 \quad \text{and} \quad Y_1(y) = 0. \qquad (2.3.9)$$

Clearly, the solutions arising from (2.3.9) are constants. If $y = a$ is a solution of $Y_1(y) = 0$, then $y = a$ satisfies (2.3.5), while in general it will not be included in the general solution of (2.3.6). Such solutions may be singular, or particular solutions of the general solution of (2.3.6).

Example 2.3.1. Solve the equation

$$\sqrt{a^2 - y^2}\, dx + y\, dy = 0. \qquad (2.3.10)$$

SOLUTION. Note that $y = \pm a$ are solutions of (2.3.10). Dividing (2.3.10) by $\sqrt{a^2 - y^2}$, we have

$$dx + \frac{y\, dy}{\sqrt{a^2 - y^2}} = 0.$$

Integrating the above expression, we obtain

$$x - \sqrt{a^2 - y^2} = C. \qquad (2.3.11)$$

Note that $y = \pm a$ are not included in the general solution given by (2.3.11). That is, we cannot assign definite values to the parameter C to obtain $y = a$ and $y = -a$. Thus $y = \pm a$ are singular solutions of (2.3.10). [*See* Definition 1.3.5 for singular solutions.]

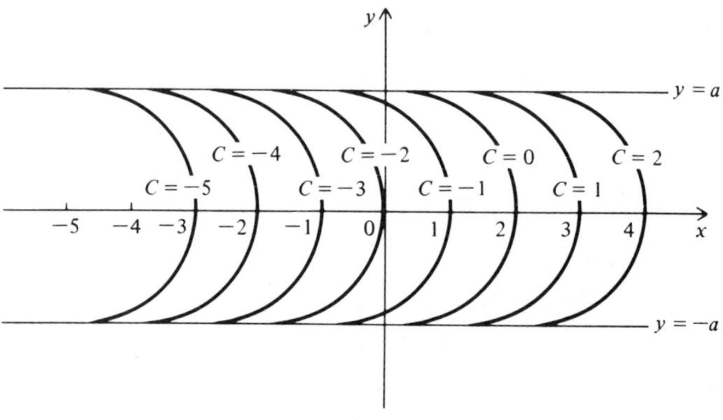

Figure 2.3.1.

The one-parameter family of curves given by (2.3.11) are right-hand semi-circles of radius a, centered along the x axis. The singular solutions $y = a$ and $y = -a$ are tangent to these semicircles as shown in Figure 2.3.1.

Example 2.3.2. Solve the equation

$$(y - y^2)dx - dy = 0. \qquad (2.3.12)$$

SOLUTION. Note that $y = 0$ and $y = 1$ are solutions of (2.3.12). Dividing (2.3.12) by $(y - y^2)$, we have

$$dx - \frac{dy}{y - y^2} = 0. \qquad (2.3.13)$$

Partial fraction decomposition gives

$$\frac{1}{y(1 - y)} = \frac{A}{y} + \frac{B}{1 - y},$$

from which we readily obtain that $A = 1$ and $B = 1$. Thus, (2.3.13) may be written as

$$dx - \frac{dy}{y} - \frac{dy}{1 - y} = 0.$$

Integrating the above expression, we get

$$x - \ln y + \ln (1 - y) = \ln C. \qquad (2.3.14)$$

Using properties of logarithms, we may write (2.3.14) as

$$\ln \left(\frac{1 - y}{y}\right) = \ln C - x,$$

or

$$\frac{1 - y}{y} = Ce^{-x}. \qquad (2.3.15)$$

Solving for y in (2.3.15), the general solution of (2.3.12) is given by

$$y = \frac{1}{1 + Ce^{-x}}. \tag{2.3.16}$$

Since C is an arbitrary constant, we may also write (2.3.16), replacing C by $1/B$ as

$$y = \frac{B}{B + e^{-x}}. \tag{2.3.17}$$

We readily see now that $y = 1$ and $y = 0$ are particular solutions. For, if we set $C = 0$ in (2.3.16), we obtain $y = 1$; while if we set $B = 0$ in (2.3.17) we obtain $y = 0$, namely, the x axis.

[*Note:* We could also have obtained (2.3.17) by writing (2.3.14) as

$$x - \ln y + \ln (1 - y) + \ln B = 0,$$

and then solving for y. The reader should carry out these steps.]

Figure 2.3.2 illustrates a few level curves of the one-parameter family of curves given by (2.3.16). Also, the particular solutions have been drawn.

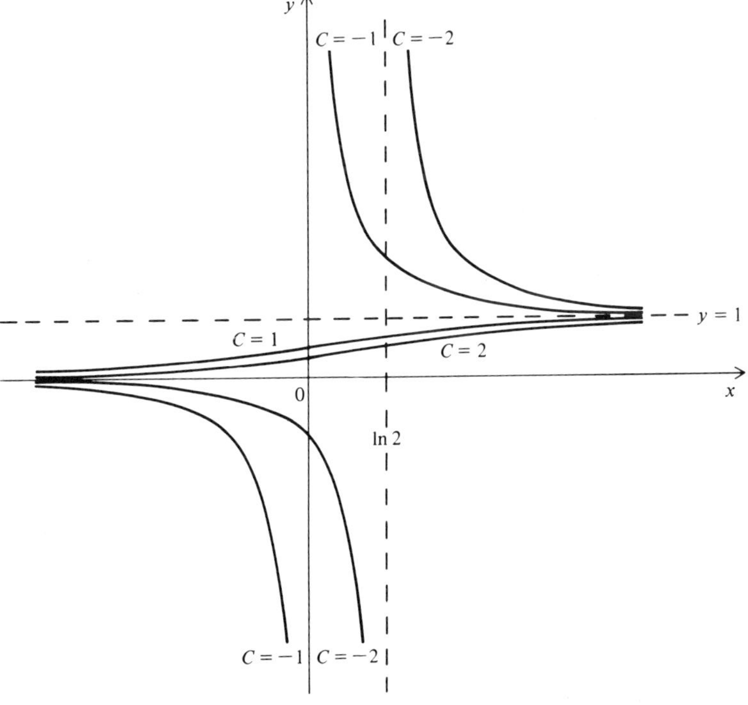

Figure 2.3.2.

Exercises 2.3

In each of the following exercises find the general solution.   Also determine the singular and particular solutions.

1. $\sqrt{y}\,dx - \sqrt{x}\,dy = 0.$
2. $y^2\,dx + (x - 1)dy = 0.$
3. $dx + (x^3 - x)dy = 0.$
4. $(1 - y^2)dx + (1 - x^2)dy = 0.$
5. $(1 - y^2)dx + x^3\,dy = 0.$
6. $\sqrt{1 - y^2}\,dx + \sqrt{1 - x^2}\,dy = 0.$
7. $(1 + y^2)dx + (1 + x^2)dy = 0.$
8. $(x + 3)y\,dx + (6x - x^2)dy = 0.$
9. $xe^x\,dx - 2(x + 1)^2 y\,dy = 0.$
10. $y\ln x\,dx - (x + 1)^2\,dy = 0.$
11. $e^x\,dx - (1 + e^{-x})^2\,dy = 0.$
12. $(\cot x + \tan x)y\,dx - (\ln\tan x)dy = 0.$

2.4. Homogeneous Equations

Definition 2.4.1.   Let D be a domain in the xy-plane.   Suppose that the points (tx, ty), $t > 0$, are in D whenever the points (x, y) are in D.   A function f defined in D is said to be *homogeneous of degree k* if

$$f(tx, ty) = t^k f(x, y), \tag{2.4.1}$$

for all (x, y) in D and all $t > 0$.

This definition can be extended to functions of more than two variables.

Example 2.4.1.   The function f defined by

(a)
$$f(x, y) = x^2 - 2xy + y^2 \tag{2.4.2}$$

is homogeneous of degree 2 in any domain D contained in the xy-plane, since

$$f(tx, ty) = (tx)^2 - 2(tx)(ty) + (ty)^2 = t^2(x^2 - 2xy + y^2) = t^2 f(x, y).$$

(b)   The function f defined by

$$f(x, y, z) = \sqrt[5]{x^3 + xy^2 + y^3 - xyz - z^3} \tag{2.4.3}$$

is homogeneous of degree 3/5 in any domain D contained in the xyz-space, since

$$f(tx, ty, tz) = \sqrt[5]{(tx)^3 + (tx)(ty)^2 + (ty)^3 - (tx)(ty)(tz) - (tz)^3}$$
$$= \sqrt[5]{t^3(x^3 + xy^2 + y^3 - xyz - z^3)}$$
$$= t^{3/5}\sqrt[5]{x^3 + xy^2 + y^3 - xyz - z^3} = t^{3/5}f(x, y, z).$$

(c)   The function f defined by

$$f(x, y) = \frac{x^2 - xy + y^2}{x^2 + xy + y^2} + \sin\left(\frac{x}{y}\right) \tag{2.4.4}$$

is homogeneous of degree 0 in a domain D in the xy-plane, $xy \neq 0$, since

$$f(tx, ty) = \frac{(tx)^2 - (tx)(ty) + (ty)^2}{(tx)^2 + (tx)(ty) + (ty)^2} + \sin\left(\frac{tx}{ty}\right)$$

$$= \frac{t^2}{t^2} \cdot \frac{x^2 - xy + y^2}{x^2 + xy + y^2} + \sin\left(\frac{x}{y}\right) = t^0 f(x, y).$$

(d)   The function f defined by

$$f(x, y) = e^{y^2/x^2} + \cos\left(\frac{x^3 + y^3}{xy^2}\right) + \ln\left(\frac{x + y}{x - y}\right) + \frac{x + y}{\sqrt{x^2 - y^2}} \qquad (2.4.5)$$

is homogeneous of degree 0 in a domain D: $x > |y|$, $y \neq 0$, in the xy-plane, since

$$f(tx, ty) = e^{t^2 y^2/t^2 x^2} + \cos\left(\frac{t^3 x^3 + t^3 y^3}{t^3 xy^2}\right) + \ln\left(\frac{tx + ty}{tx - ty}\right) + \frac{tx + ty}{\sqrt{t^2 x^2 - t^2 y^2}}$$

$$= e^{y^2/x^2} + \cos\left(\frac{x^3 + y^3}{xy^2}\right) + \ln\left(\frac{x + y}{x - y}\right) + \frac{x + y}{\sqrt{x^2 - y^2}} = t^0 f(x, y).$$

We may write the above functions which are homogeneous of degree zero in the following form.

(c')
$$f(x, y) = \frac{1 - \left(\frac{y}{x}\right) + \left(\frac{y}{x}\right)^2}{1 + \left(\frac{y}{x}\right) + \left(\frac{y}{x}\right)^2} + \sin\left(\frac{y}{x}\right)^{-1} \equiv F\left(\frac{y}{x}\right). \qquad (2.4.4')$$

(d') $$f(x, y) = e^{(y/x)^2} + \cos\left[\frac{1 + \left(\frac{y}{x}\right)^3}{\left(\frac{y}{x}\right)^2}\right] + \ln\left[\frac{1 + \left(\frac{y}{x}\right)}{1 - \left(\frac{y}{x}\right)}\right] + \frac{1 + \left(\frac{y}{x}\right)}{\sqrt{1 - \left(\frac{y}{x}\right)^2}}$$

$$\equiv F\left(\frac{y}{x}\right). \qquad (2.4.5')$$

Remark 2.4.1.   In general, if the function $f = f(x, y)$ is homogeneous of degree zero, then f can be expressed as a function of y/x only.   To see this, we let

$$\frac{y}{x} = v \quad \text{or} \quad y = vx. \qquad (2.4.6)$$

Then with x playing the role of t in Definition 2.4.1, we have

$$f(x, y) = f(x, vx) = x^0 f(1, v) = x^0 f\left(1, \frac{y}{x}\right) \equiv F\left(\frac{y}{x}\right). \qquad (2.4.7)$$

Observe that the converse of the above statement is also true, that is, a function of y/x only is necessarily a homogeneous function of degree zero.

Remark 2.4.2. Observe that the functions given by (2.4.4) and (2.4.5) can also be expressed as functions of x/y only. Also, Remark 2.4.1 can be formulated in terms of functions of x/y only.

Definition 2.4.2. The differential equation,

$$M(x, y)dx + N(x, y)dy = 0, \tag{2.4.8}$$

is called *homogeneous* if the functions M and N are homogeneous of the same degree.

Example 2.4.2. The following differential equations are homogeneous.

(a)
$$(x^2 + y^2) \tan\left(\frac{y}{x}\right)dx + xy \cos (e^{x/y})dy = 0, \tag{2.4.9}$$

since the functions $M = (x^2 + y^2) \tan\left(\dfrac{y}{x}\right)$ and $N = xy \cos (e^{x/y})$ are homogeneous of the same degree, namely, two.

(b)
$$\sqrt{x^2 + y^2} \ln\left(\frac{x^2 + y^2}{xy}\right)dx + \frac{x^{3/2}}{\sqrt{x + y}} dy = 0, \tag{2.4.10}$$

since the functions $M = \sqrt{x^2 + y^2} \ln\left(\dfrac{x^2 + y^2}{xy}\right)$ and $N = \dfrac{x^{3/2}}{\sqrt{x + y}}$ are homogeneous of the same degree, namely, one.

Let us write the homogeneous differential equation (2.4.8) in the form

$$\frac{dy}{dx} + \frac{M(x, y)}{N(x, y)} = 0 \qquad [N(x, y) \neq 0]. \tag{2.4.11}$$

Since the functions M and N are homogeneous of the same degree, it readily follows that the function f, defined by

$$f(x, y) = \frac{M(x, y)}{N(x, y)}, \tag{2.4.12}$$

is homogeneous of degree zero. In view of Remark 2.4.1, f can be expressed as a function of y/x only, say, $F(y/x)$. Thus we may write (2.4.11) in the form

$$\frac{dy}{dx} + F\left(\frac{y}{x}\right) = 0. \tag{2.4.13}$$

We may now separate the variables in Equation (2.4.13) by utilizing (2.4.6), namely,

$$\frac{y}{x} = v \quad \text{or} \quad y = vx, \tag{2.4.14}$$

where the new variables are x and v. From (2.4.14), we obtain

$$\frac{dy}{dx} = v + x\frac{dv}{dx}. \tag{2.4.15}$$

Substituting (2.4.14) and (2.4.15) in (2.4.13), we get

$$x\frac{dv}{dx} + v + F(v) = 0,$$

which can be written as

$$\frac{dx}{x} + \frac{dv}{v + F(v)} = 0. \tag{2.4.16}$$

Clearly, Equation (2.4.16) has its variables x and v separated. Integrating (2.4.16), we obtain

$$x = C\exp\left[-\int\frac{dv}{v + F(v)}\right], \tag{2.4.17}$$

where C is an arbitrary constant of integration. If we replace v by y/x in the answer given in (2.4.17), we obtain the *general solution of Equation* (2 4.8) in the form

$$x = C\phi(y/x). \tag{2.4.18}$$

Remark 2.4.3. The one-parameter family of curves given by (2.4.18) are similar to each other with respect to the origin, the ratio of magnification being a single variable. To see this, observe that we can derive (2.4.18) from

$$x = \phi(y/x) \tag{2.4.19}$$

by simply replacing x by x/C and y by y/C.

Remark 2.4.4. We may also write (2.4.11) as [*see* Remark 2.4.2]

$$\frac{dy}{dx} + G\left(\frac{x}{y}\right) = 0. \tag{2.4.20}$$

We may separate the variables in (2.4.20) by making the substitution

$$\frac{x}{y} = v \quad \text{or} \quad x = vy, \tag{2.4.21}$$

where the new variables are now v and y.   We leave it for the reader [*see* Exercise 2.4.2] to show that the *general solution of Equation* (2.4.8) can also be written as

$$y = C\Phi\left(\frac{x}{y}\right). \tag{2.4.22}$$

Example 2.4.3.   Find the general solution of the equation

$$(\sqrt{x^2 + y^2} - y)dx + x\,dy = 0. \tag{2.4.23}$$

SOLUTION.   Equation (2.4.23) is homogeneous, since the functions $M = \sqrt{x^2 + y^2} - y$ and $N = x$ are homogeneous of the same degree, namely, one. Let

$$y = vx. \tag{2.4.24}$$

Then

$$dy = v\,dx + x\,dv,$$

and we may write (2.4.23) upon simplification as

$$\frac{dx}{x} + \frac{dv}{\sqrt{1 + v^2}} = 0.$$

Integrating the above equation, we obtain

$$\ln x + \ln\left(v + \sqrt{1 + v^2}\right) = \ln C,$$

or

$$x\left(v + \sqrt{1 + v^2}\right) = C. \tag{2.4.25}$$

Substituting y/x for v in (2.4.25), we find that the general solution of (2.4.23) is

$$y + \sqrt{x^2 + y^2} = C. \tag{2.4.26}$$

Figure 2.4.1 shows a few level curves of the one-parameter family given by (2.4.26).

Example 2.4.4.   Find the general solution of the equation

$$y \cos\frac{x}{y}\,dx + \left(y \sin\frac{x}{y} - x \cos\frac{x}{y}\right)dy = 0. \tag{2.4.27}$$

SOLUTION.   Equation (2.4.27) is homogeneous, since the functions $M = y \cos\dfrac{x}{y}$ and $N = y \sin\dfrac{x}{y} - x \cos\dfrac{x}{y}$ are homogeneous of the same degree, namely, one.

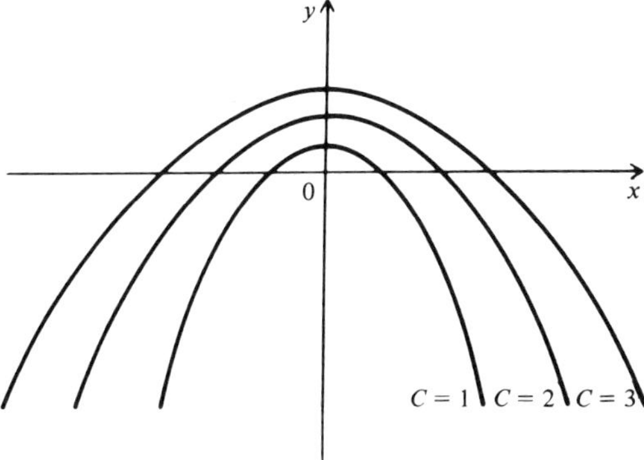

Figure 2.4.1.

Let

$$\frac{x}{y} = v \quad \text{or} \quad x = vy. \tag{2.4.28}$$

Then

$$dx = v\,dy + y\,dv. \tag{2.4.29}$$

Substituting x and dx of (2.4.28) and (2.4.29) in (2.4.27) and simplifying, we obtain

$$\frac{\cos v\,dv}{\sin v} + \frac{dy}{y} = 0. \tag{2.4.30}$$

Integrating (2.4.30), we get

$$\ln \sin v + \ln y = \ln C,$$

or

$$y = C \csc v. \tag{2.4.31}$$

Substituting x/y for v in (2.4.31), the general solution of (2.4.27) is

$$y = C \csc \frac{x}{y}. \tag{2.4.32}$$

Exercises 2.4

1. Express the functions given in (2.4.4) and (2.4.5) as functions of x/y only.
2. Derive Equation (2.4.22).

In Exercises 3 through 7, obtain the general solution.

3. $(x - 2y)dx + y\,dy = 0.$ 4. $(y + xe^{y/x})dx - x\,dy = 0.$

5. $[\sqrt{x^2 - y^2} + y]dx - x\,dy = 0.$ 6. $\dfrac{dy}{dx} = \dfrac{y}{x}\ln\dfrac{y}{x} + \dfrac{y}{x}.$

7. $(y^2 - x^2)dx + xy\,dy = 0.$

In Exercises 8 through 10, solve the initial-value problem.

8. $\begin{cases}\dfrac{dy}{dx} = \dfrac{y^2 + xy}{x^2}, \\[2mm] y(1) = 1.\end{cases}$ 9. $\begin{cases}\dfrac{dy}{dx} = \dfrac{y - x\cos^2\dfrac{y}{x}}{x}, \\[2mm] y(1) = \dfrac{\pi}{4}.\end{cases}$

10. $\begin{cases}\dfrac{dy}{dx} = \dfrac{4y^3 + 3xy^2}{x^3}, \\[2mm] y(3) = 1.\end{cases}$

11.   Suppose that

(a) $$\frac{dy}{dx} = f\left(\frac{ax + by + c}{Ax + By + C}\right),$$

where a, b, c, A, B, and C are constants.   In order that Equation (a) be homogeneous it is sufficient that $c = C = 0$.   If we apply a translation of axes

(b) $$x = X + h, \qquad y = Y + k,$$

where X and Y are the new variables and h and k are any two constants, Equation (a) may be written as

(c) $$\frac{dY}{dX} = f\left(\frac{aX + bY + ah + bk + c}{AX + BY + Ah + Bk + C}\right).$$

Equation (c) will be homogeneous if

(d) $$\begin{cases}ah + bk + c = 0, \\ Ah + Bk + C = 0.\end{cases}$$

Case 1. $\Delta \equiv \begin{vmatrix} a & b \\ A & B \end{vmatrix} \neq 0.$ Then

(e) $$h = \frac{\begin{vmatrix} -c & b \\ -C & B \end{vmatrix}}{\Delta} \quad \text{and} \quad k = \frac{\begin{vmatrix} a & -c \\ A & -C \end{vmatrix}}{\Delta}.$$

Substituting these values of h and k into (c) renders Equation (c) homogeneous.

Case 2. $\Delta = 0$. Then $aB - bA = 0$. Suppose that $b \neq 0$. Show that

(f) $$Ax + By = K(ax + by),$$

where K is a constant. Let

(g) $$ax + by = u.$$

Show that Equation (a) may be written as

(h) $$\frac{1}{b}\frac{du}{dx} = \frac{a}{b} + f\left(\frac{u + c}{Ku + C}\right).$$

in which the variables are now separated.

In Exercises 12 through 14, use the results of Exercise 11 to find the general solution.

12. $\dfrac{dy}{dx} = -\dfrac{x - 2y + 1}{2x - 4y + 3}.$

13. $\dfrac{dy}{dx} = \dfrac{2x + 3y + 4}{4x + 6y + 5}.$

14. $\dfrac{dy}{dx} = \dfrac{6x - 3y + 2}{2x + y - 1}.$

15. Solve the equation

$$\frac{dy}{dx} - \cos^2(x - y) = 0.$$

16. Solve the equation

$$\frac{dy}{dx} + (x - y + 1)^2 = 0.$$

17. Suppose that

$$\frac{dy}{dx} = f(x, y),$$

where the function f is homogeneous of degree zero. Show that the polar coordinate transformation

$$x = r \cos \theta \quad \text{and} \quad y = r \sin \theta$$

transforms the above differential equation into one that has its variables separated.

18. Use the method given in Exercise 17 to solve the following equations.

(a) $(x^2 + y^2)dx - xy\, dy = 0.$

(b) $\dfrac{dy}{dx} = \dfrac{y}{x}\ln\dfrac{y}{x} + \dfrac{y}{x}.$

19. (a) Suppose that the function $f = f(x, y, z, \ldots)$ is homogeneous of degree k and that f possesses continuous first partial derivatives. Show that

$$xf_x + yf_y + zf_z + \cdots = kf.$$

This expression is known as *Euler's relation*.

(b) Conversely, show that if the Euler's relation is satisfied, then the function is homogeneous of degree k. (Thus, Euler's relation is a necessary and sufficient condition for the homogeneity of the function.)

20.   Suppose that the function $f = f(x, y, z, \ldots)$ is homogeneous of degree one and that f possesses continuous second partial derivatives.   Show that

$$x^2 f_{xx} + y^2 f_{yy} + z^2 f_{zz} + \cdots + 2xy f_{xy} + 2xz f_{xz} + \cdots = 0.$$

21.   Suppose that the function $f = f(x, y, z, \ldots)$ is homogeneous of degree k and that f possesses continuous partial derivatives of order r.   Show that any rth derivative of f is a homogeneous function of degree $k - r$.

2.5. Integrating Factors

Definition 2.5.1.   A nonzero function μ is said to be an *integrating factor* of the differential equation

$$M(x, y)dx + N(x, y)dy = 0 \tag{2.5.1}$$

if the equation

$$\mu(x, y)M(x, y)dx + \mu(x, y)N(x, y)dy = 0 \tag{2.5.2}$$

is exact.

Thus an integrating factor transforms a nonexact differential equation into one that is exact.   If the functions μM and μN satisfy the hypothesis of Theorem 2.2.1 in the interior of a rectangle R: $|x - x_0| \leq a,$ $|y - y_0| \leq b,$ then a necessary and sufficient condition for Equation (2.5.2) to be exact is that

$$\frac{\partial(\mu M)}{\partial y} = \frac{\partial(\mu N)}{\partial x} \tag{2.5.3}$$

in R.

Example 2.5.1.   The equation

$$y\, dx + (x + x^2 y^2)dy = 0 \tag{2.5.4}$$

is not exact, since

$$1 = \frac{\partial M}{\partial y} \neq \frac{\partial N}{\partial x} = 1 + 2xy^2 \qquad (x \neq 0, \quad y \neq 0).$$

If we multiply (2.5.4) by [*see* Example 2.5.3]

$$\mu(x, y) = \frac{1}{x^2 y^2} \qquad (x \neq 0, \quad y \neq 0),$$

we obtain

$$\frac{dx}{x^2 y} + \left(\frac{1}{xy^2} + 1\right) dy = 0, \qquad (2.5.5)$$

which is now exact, since

$$\frac{\partial}{\partial y}(\mu M) = \frac{\partial}{\partial y}\left(\frac{1}{x^2 y}\right) = -\frac{1}{x^2 y^2} \quad \text{and} \quad \frac{\partial(\mu N)}{\partial x} = \frac{\partial}{\partial x}\left(\frac{1}{xy^2} + 1\right) = -\frac{1}{x^2 y^2},$$

and thus

$$\frac{\partial(\mu M)}{\partial y} = \frac{\partial(\mu N)}{\partial x} \qquad (x \neq 0, \quad y \neq 0).$$

We may now utilize (2.2.7) and (2.2.8) to solve (2.5.5). Thus, the general solution of (2.5.4) is

$$\int_{x_0}^{x} \frac{ds}{s^2 y} + \int_{y_0}^{y}\left(\frac{1}{x_0 t^2} + 1\right) dt = c,$$

or

$$xy^2 = Cxy + 1. \qquad (2.5.6)$$

Remark 2.5.1. *If there exists an integrating factor $\mu(x, y)$ of Equation (2.5.1), then (2.5.1) has an infinite number of integrating factors given by*

$$\mu(x, y)f(u), \qquad (2.5.7)$$

where f is an arbitrary continuous function of u. To see this, note that (2.5.2) is now exact. Then in view of Definition 2.2.1, there exists a function u such that

$$\mu M \, dx + \mu N \, dy = du, \qquad (2.5.8)$$

and thus $u(x, y) = C$ is a solution of (2.5.2). Multiplying (2.5.8) by f, where f is an arbitrary continuous function u, we obtain

$$\mu(x, y)f(u)(M \, dx + N \, dy) = f(u)du. \qquad (2.5.9)$$

Clearly, the right-hand member of (2.5.9) is an exact differential, and thus the left-hand member is also an exact differential. Consequently, it follows that $\mu(x, y)f(u)$ is an integrating factor of (2.5.1).

Sometimes we may determine by inspection integrating factors of Equation (2.5.1). This inspection process is facilitated somewhat by utilizing some formulas from differential calculus. We shall give here a few of these formulas.

$$d(xy) = x \, dy + y \, dx. \qquad (2.5.10)$$

$$d\left(\frac{x}{y}\right) = \frac{y\,dx - x\,dy}{y^2}. \tag{2.5.11}$$

$$d\left(\arctan\frac{x}{y}\right) = \frac{y\,dx - x\,dy}{x^2 + y^2}. \tag{2.5.12}$$

$$d\left(\ln\frac{x}{y}\right) = \frac{y\,dx - x\,dy}{xy}. \tag{2.5.13}$$

The reader should also write the above formulas interchanging the role of x and y.

Example 2.5.2.   Solve the equation

$$y\,dx - x\,dy = 0. \tag{2.5.14}$$

SOLUTION.   In view of Formulas (2.5.11), (2.5.12), and (2.5.13), Equation (2.5.14) has the following integrating factors.

$$\mu_1(x, y) = \frac{1}{y^2}; \qquad \mu_2(x, y) = \frac{1}{x^2 + y^2}; \qquad \mu_3(x, y) = \frac{1}{xy}; \qquad (x \neq 0, \quad y \neq 0).$$
$$\tag{2.5.15}$$

If we multiply (2.5.14) by μ_1, μ_2, and μ_3, respectively, we obtain

$$\frac{y\,dx - x\,dy}{y^2} = 0; \qquad \frac{y\,dx - x\,dy}{x^2 + y^2} = 0; \qquad \frac{y\,dx - x\,dy}{xy} = 0.$$

The solutions of the above equations are given, respectively, by

$$\frac{x}{y} = C_1; \qquad \arctan\frac{x}{y} = \arctan C_2; \qquad \ln\frac{x}{y} = \ln C_3.$$

Thus, the solution of (2.5.14) is given by

$$x = Cy. \tag{2.5.16}$$

In view of Remark 2.5.1, Equation (2.5.14) has an infinite number of integrating factors.   Once we have found one integrating factor, the others may be found also.   In the above example, suppose that $\mu_1(x, y) = 1/y^2$ is known.   By properly specifying the function f, we may find the other two integrating factors given in (2.5.15).   From (2.5.16), $u(x, y) = x/y = C$ is the general solution of

$$\mu_1(x, y)M(x, y)dx + \mu_1(x, y)N(x, y)dy = \frac{y\,dx - x\,dy}{y^2} = 0.$$

Thus, if we take $f(u) = \dfrac{1}{u^2 + 1}$, where $u = \dfrac{x}{y}$, then

$$\mu_1(x, y)f(u) = \frac{1}{y^2} \cdot \frac{1}{u^2 + 1} = \frac{1}{y^2} \cdot \frac{1}{\left(\dfrac{x}{y}\right)^2 + 1} = \frac{1}{x^2 + y^2} = \mu_2(x, y).$$

If we now take $f(u) = 1/u$, then

$$\mu_1(x, y)f(u) = \frac{1}{y^2} \cdot \frac{1}{u} = \frac{1}{y^2} \cdot \frac{1}{\dfrac{x}{y}} = \frac{1}{xy} = \mu_3(x, y).$$

If we take $f(u) = u^2$, then

$$\mu_1(x, y)f(u) = \frac{1}{y^2} \cdot u^2 = \frac{1}{y^2} \cdot \left(\frac{x}{y}\right)^2 = \frac{x^2}{y^4} = \mu_4(x, y)$$

is another integrating factor of Equation (2.5.14).

As an exercise, the reader should give five more integrating factors of Equation (2.5.14).

Clearly, the above method of finding the other integrating factors of Equation (2.5.14) depends upon knowing $u(x, y) = x/y$, from which the general solution of (2.5.14) is determined, namely, $x = Cy$. Thus, the fact that Equation (2.5.1) has an infinite number of integrating factors is of no special assistance in solving the equation. However, if we are able to find two integrating factors of Equation (2.5.1) either by inspection or by some other method, then we are able to find the general solution of Equation (2.5.1) without using integrations. [*See* Exercise 2.5.19(b).] We shall have occasions to utilize this method [*see*, for instance, Exercise 2.8.29(b) and (c)].

Example 2.5.3.   Let us determine an integrating factor of Equation (2.5.4).

SOLUTION.   We may write Equation (2.5.4) as

$$x \, dy + y \, dx + (xy)^2 \, dy = 0,$$

which also can be written, in view of (2.5.10), as

$$d(xy) + (xy)^2 \, dy = 0.$$

The presence of the differential of xy in the above equation suggests that if we multiply this term by any function of the product xy, it will not disturb its integrability. Thus, if we multiply the above equation by $(xy)^{-2}$, we obtain

$$(xy)^{-2} \, d(xy) + dy = 0. \tag{2.5.17}$$

The left-hand member in (2.5.17) is the differential of

$$-(xy)^{-1} + y,$$

and therefore Equation (2.5.17) is exact. Thus, an integrating factor of Equation (2.5.4) is given by $\mu(x, y) = (xy)^{-2}$.

Although one often guesses integrating factors, there are a few situations where a particular procedure may be followed. Such is the case when the function μ is a function of x alone, or a function of y alone (or even a function of xy or x/y). The following pattern, while not always applicable, deserves to be mentioned.

We have seen that $\mu(x, y)$ is an integrating factor of Equation (2.5.1) if and only if Equation (2.5.3) is satisfied. Performing the indicated partial differentiation in (2.5.3), we obtain

$$\mu \frac{\partial M}{\partial y} + M \frac{\partial \mu}{\partial y} = \mu \frac{\partial N}{\partial x} + N \frac{\partial \mu}{\partial x},$$

or

$$\mu\left(\frac{\partial M}{\partial y} - \frac{\partial N}{\partial x}\right) = N \frac{\partial \mu}{\partial x} - M \frac{\partial \mu}{\partial y}. \tag{2.5.18}$$

Thus μ is an integrating factor of Equation (2.5.1) if and only if μ satisfies the partial differential equation (2.5.18). In this course, we are not in a position to solve Equation (2.5.18). However, by specializing the functions μ, M, N, and their derivatives, we may obtain some interesting results as a consequence of Equation (2.5.18).

First of all, suppose that μ is a function of x only. Then

$$\frac{\partial \mu}{\partial y} = 0 \quad \text{and} \quad \frac{\partial \mu}{\partial x} = \frac{d\mu}{dx}. \tag{2.5.19}$$

In view of (2.5.19), we may write (2.5.18) as

$$\frac{1}{N}\left(\frac{\partial M}{\partial y} - \frac{\partial N}{\partial x}\right) dx = \frac{d\mu}{\mu} \qquad [N(x, y) \neq 0]. \tag{2.5.20}$$

Now, if we further assume that the left-hand member in (2.5.20) is also a function of x only, say,

$$\frac{1}{N}\left(\frac{\partial M}{\partial y} - \frac{\partial N}{\partial x}\right) = f(x), \tag{2.5.21}$$

then Equation (2.5.20) becomes

$$\frac{d\mu}{\mu} = f(x)dx. \tag{2.5.22}$$

Thus, an integrating factor of Equation (2.5.1) is obtained by integrating (2.5.22). We find

$$\mu(x) = \exp\left[\int f(x)dx.\right] \tag{2.5.23}$$

By a similar argument as given above, if we assume that μ is a function of y only and if

$$\frac{1}{M}\left(\frac{\partial M}{\partial y} - \frac{\partial N}{\partial x}\right) \qquad [M(x, y) \neq 0]$$

is also a function of y only, say $g(y)$, that is,

$$\frac{1}{M}\left(\frac{\partial M}{\partial y} - \frac{\partial N}{\partial x}\right) = g(y), \tag{2.5.24}$$

then an integrating factor of Equation (2.5.1) is given by

$$\mu(y) = \exp\left[-\int g(y)dy\right]. \tag{2.5.25}$$

The verification of (2.5.25) will be left as an exercise for the reader. [*See* Exercise 2.5.1.]

Example 2.5.4. Solve the equation

$$(x^2 + x - y^2)dx - xy\,dy = 0. \tag{2.5.26}$$

SOLUTION.   Here

$$M(x, y) = x^2 + x - y^2 \quad \text{and} \quad N(x, y) = -xy.$$

Thus,

$$\frac{\partial M}{\partial y} = -2y \quad \text{and} \quad \frac{\partial N}{\partial x} = -y.$$

Utilizing (2.5.21), we obtain

$$f(x) = \frac{1}{N}\left(\frac{\partial M}{\partial y} - \frac{\partial N}{\partial x}\right) = -\frac{1}{xy}(-2y + y) = \frac{1}{x}.$$

In view of (2.5.23), an integrating factor of (2.5.26) is given by

$$\mu(x) = \exp\left[\int f(x)dx\right] = \exp\left[\int \frac{dx}{x}\right] = e^{\ln x} = x.$$

If we multiply Equation (2.5.26) by x, we obtain the exact equation

$$(x^3 + x^2 - xy^2)dx - x^2y\,dy = 0. \tag{2.5.27}$$

Utilizing (2.2.31) to solve (2.5.27), the general solution of Equation (2.5.26) is

$$\int_0^x (s^3 + s^2 - y^2 s)ds = C,$$

or

$$3x^4 + 4x^3 - 6x^2y^2 = C.$$

Exercises 2.5

1. Verify Equation (2.5.25).

In Exercises 2 through 10, find the general solution.

2. $(y + x^4 \ln x)dx - x\,dy = 0.$
3. $2xy\,dx + (2y^2 - 3x^2)dy = 0.$

4. $\left[y + \dfrac{1}{x^2 y^2}\cos(\ln x)\right]dx + x\,dy = 0.$

5. $(y^4 + 2y)dx + (xy^3 + 2y^4 - 4x)dy = 0.$
6. $[y + (x^2 + y^2)e^x \sin x]dx - x\,dy = 0.$
7. $(2y^3 + 3xe^{x^3})dx + 3xy^2 dy = 0.$
8. $(y + x^3 ye^{2x})dx - (x + xy)dy = 0.$

9. $\left(1 + \dfrac{x^3}{y}\sin^2 x\right)dx + \dfrac{1}{y}\left(x + \dfrac{1}{\cos^2 2y}\right)dy = 0.$

10. $[e^y + xe^y + \tan(e^x)]dx + xe^y\,dy = 0.$
11. Suppose that the differential equation

    (1) $$M(x, y)dx + N(x, y)dy = 0$$

    is homogeneous of degree k.
    (a) If $Mx + Ny$ is not identically zero, show that $1/(Mx + Ny)$ is an integrating factor of Equation (1).
    (b) If $Mx + Ny$ is identically zero, show that the general solution of Equation (1) is given by $y = Cx$.
12. Use the result of Exercise 11 to obtain the general solution of the following equations.
    (a) $(x^3 + y^3)dx - xy^2\,dy = 0.$
    (b) $(y + xe^{y/x})dx - x\,dy = 0.$

    (c) $\left(x\cosh\dfrac{y}{x} + 2y\sinh\dfrac{y}{x}\right)dx - 2x\sinh\dfrac{y}{x}\,dy = 0.$

13. Suppose that the differential equation is of the form

    (1) $$M\,dx + N\,dy \equiv yf(xy)dx + xg(xy)dy = 0.$$

    (a) If $Mx - Ny$ is not identically zero, show that $1/(Mx - Ny)$ is an integrating factor of Equation (1).
    (b) If $Mx - Ny$ is identically zero, show that the general solution of Equation (1) is given by $xy = C.$
14. Use the result of Exercise 13 to obtain the general solution of the following equations.
    (a) $y(x^3 y^3 + 3)dx + x(3 - 3x^3 y^3)dy = 0.$
    (b) $y(x^2 y^2 + xy + 1)dx + x(x^2 y^2 - xy - 1)dy = 0.$
15. Show that the differential equation

    (1) $$x^a y^\beta (Ay\,dx + Bx\,dy) + x^\gamma y^\delta (ay\,dx + bx\,dy) = 0,$$

α, β, γ, δ, A, B, a, and b are constants and $Ab - Ba \neq 0$, has an integrating factor of the form

$$x^r y^s,$$

where

$$r = \frac{kA - Ka}{Ab - Ba}, \qquad s = \frac{kB - Kb}{Ab - Ba},$$

and

$$k = a(\delta + 1) - b(\gamma + 1), \qquad K = A(\beta + 1) - B(\alpha + 1).$$

Discuss the solution of Equation (1) when $Ab - Ba = 0$.

16. Use the result of Exercise 15 to find the general solution of the following equations.
    (a) $x^2(y\,dx + 2x\,dy) + y^3(3y\,dx + 4x\,dy) = 0.$
    (b) $x^2 y^2(2y\,dx + 3x\,dy) - (y\,dx + 6x\,dy) = 0.$

17. Show that $\dfrac{1}{x^2} f\left(\dfrac{y}{x}\right)$ is an integrating factor of $x\,dy - y\,dx = 0$.

18. Suppose that $f(x, y) = C$ is the general solution of the equation

    (1) $$M(x, y)dx + N(x, y)dy = 0.$$

    Show that the common value of the two ratios given by

    $$\frac{\dfrac{\partial f}{\partial x}}{M} = \frac{\dfrac{\partial f}{\partial y}}{N}$$

    is an integrating factor of Equation (1).   Verify this result for Example 2.5.1.
19. (Review Remark 2.5.1.)
    (a) Show that if $\mu_1(x, y)$ and $\mu_2(x, y)$ are integrating factors of the equation

    (1) $$M(x, y)dx + N(x, y)dy = 0,$$

    then $\mu_2(x, y)/\mu_1(x, y)$ is a function of u only.
    (b) Also show that if μ_2/μ_1 is not equal to a constant, the general solution of Equation (1) is given by

    $$\frac{\mu_2(x, y)}{\mu_1(x, y)} = C.$$

    where C is an arbitrary constant. Verify this result for Example 2.5.2 by using the integrating factors given in (2.5.15).

2.6. Linear Differential Equations of First Order

Definition 2.6.1.   A differential equation of the first order is said to be *linear* if it can be written in the form

$$\frac{dy}{dx} + P(x)y = Q(x). \tag{2.6.1}$$

We shall assume that the functions P and Q are continuous over the intervals on which the solutions of Equation (2.6.1) are sought.   [*Note:* This definition is consistent with Definition 1.1.6.]

Theorem 2.6.1.   *The general solution of a linear differential equation of the first order*

$$\frac{dy}{dx} + P(x)y = Q(x), \qquad\qquad [2.6.1]$$

is given by

$$y = \exp\left[-\int P(x)dx\right]\left[\int \exp\left(\int P(x)dx\right)Q(x)dx + C\right], \qquad (2.6.2)$$

where C is an arbitrary constant.

PROOF.   Suppose first that $Q(x) \equiv 0$ in (2.6.1).   Then Equation (2.6.1) may be written as

$$\frac{dy}{dx} + P(x)y = 0. \qquad\qquad (2.6.3)$$

Equation (2.6.3) is one in which its variables are readily separated.   Thus we have

$$\frac{dy}{y} = -P(x)dx. \qquad\qquad (2.6.4)$$

Integrating (2.6.4), the general solution of Equation (2.6.3) is given by

$$y = c\,\exp\left[-\int P(x)dx\right]. \qquad\qquad (2.6.5)$$

   To obtain the general solution of Equation (2.6.1), we shall attempt to satisfy (2.6.1) by taking y to be of the form given in (2.6.5) with the parameter c replaced by an unknown function of x.   That is, we shall make the following change in variable.

$$y = Yu, \qquad\qquad (2.6.6)$$

where u is a function of x to be determined and Y is any integral given in (2.6.5).   Then

$$Y = \exp\left[-\int P(x)dx\right], \qquad\qquad (2.6.7)$$

and Y satisfies also the equation

$$\frac{dY}{dx} + P(x)Y = 0. \qquad\qquad (2.6.8)$$

From (2.6.6) we obtain by differentiation

$$\frac{dy}{dx} = \frac{dY}{dx}u + Y\frac{du}{dx}.$$ (2.6.9)

Substituting (2.6.6) and (2.6.9) in Equation (2.6.1), we obtain

$$\left[\frac{dY}{dx} + P(x)Y\right]u + Y\frac{du}{dx} = Q(x).$$ (2.6.10)

In view of (2.6.8), Equation (2.6.10) simplifies to

$$Y\frac{du}{dx} = Q(x),$$

or

$$\frac{du}{dx} = \frac{Q(x)}{Y}.$$ (2.6.11)

Integrating (2.6.11) and then using (2.6.7), we find

$$u = \int \frac{Q(x)}{Y}\,dx + C$$

$$= \int \exp\left(\int P(x)dx\right)Q(x)dx + C.$$ (2.6.12)

Substituting (2.6.7) and (2.6.12) in (2.6.6), the general solution of Equation (2.6.1) is given by

$$y = \exp\left[-\int P(x)dx\right]\left[\int \exp\left(\int P(x)dx\right)Q(x)dx + C\right],$$

and the theorem is thus established.

Remark 2.6.1. The method employed above to solve Equation (2.6.1) when $Q(x) \not\equiv 0$ is sometimes referred to as the method of *variation of parameters.* This method is also often used to solve linear equations of order greater than one.

Remark 2.6.2. One may also arrive at (2.6.2) by multiplying (2.6.1) by the integrating factor[3]

$$\mu(x) = \exp\left[\int P(x)dx\right].$$ (2.6.13)

---

[3] This integrating factor is obtained as follows. Rewriting (2.6.1) in the form $[P(x)y - Q(x)]dx + dy = 0$, we have that $M = P(x)y - Q(x)$ and $N = 1$. Hence $\partial M/\partial y = P(x)$ and $\partial N/\partial x = 0$. Now use (2.5.21) and (2.5.23).

We may then write Equation (2.6.1) as

$$\exp\left[\int P(x)dx\right]dy + y\exp\left[\int P(x)dx\right]P(x)dx = \exp\left[\int P(x)dx\right]Q(x)dx,$$

which can also be written as

$$d\left(y\exp\left[\int P(x)dx\right]\right) = \exp\left[\int P(x)dx\right]Q(x)dx. \tag{2.6.14}$$

Clearly, Equation (2.6.14) is exact.   Integrating (2.6.14) we obtain

$$y\exp\left[\int P(x)dx\right] = \int\exp\left[\int P(x)dx\right]Q(x)dx + C, \tag{2.6.15}$$

which is the same as (2.6.2).

[*Note:* The left-hand member in (2.6.15) is the product of the dependent variable and the integrating factor.]

Example 2.6.1.   Find the general solution of the equation

$$\cos^2 x\,\frac{dy}{dx} + y = \tan x. \tag{2.6.16}$$

SOLUTION.   Let us write (2.6.16) in the form given by Equation (2.6.1).   We have

$$\frac{dy}{dx} + y\sec^2 x = \sec^2 x \tan x.$$

Thus,

$$P(x) = \sec^2 x \quad \text{and} \quad Q(x) = \sec^2 x \tan x,$$

and an integrating factor of (2.6.16) is given by

$$\exp\left[\int P(x)dx\right] = \exp\left[\int \sec^2 x\,dx\right] = e^{\tan x}.$$

Utilizing (2.6.2), the general solution of Equation (2.6.16) is

$$y = e^{-\tan x}\int e^{\tan x}\sec^2 x \tan x\,dx + Ce^{-\tan x}. \tag{2.6.17}$$

To evaluate the above integral, we integrate by parts.   Let

$$u = \tan x \quad \text{and} \quad dv = e^{\tan x}\sec^2 x\,dx,$$

then

$$du = \sec^2 x\,dx \quad \text{and} \quad v = e^{\tan x}.$$

Thus,

$$\int e^{\tan x}\sec^2 x \tan x \, dx = \tan x e^{\tan x} - \int e^{\tan x}\sec^2 x \, dx$$

$$= \tan x e^{\tan x} - e^{\tan x}.$$

Substituting this expression in (2.6.17), we obtain

$$y = \tan x - 1 + Ce^{-\tan x}.$$

Example 2.6.2.   Obtain the general solution of the equation

$$y \, dx + (2x + xy - 4)dy = 0. \tag{2.6.18}$$

SOLUTION.   Since the product $y \, dy$ occurs, Equation (2.6.18) is not linear in y. [*See* Definition 1.1.6.]   However, Equation (2.6.18) is linear in x, for we may write (2.6.18) as

$$\frac{dx}{dy} + \left(\frac{2}{y} + 1\right)x = \frac{4}{y}. \tag{2.6.19}$$

An integrating factor of Equation (2.6.19) is given by

$$\exp\left[\int P(y)dy\right] = \exp\left[\int \left(\frac{2}{y} + 1\right)dy\right] = e^{2\ln y + y} = e^{\ln y^2}\cdot e^y = y^2 e^y.$$

If we multiply (2.6.19) by $y^2 e^y$, we obtain the exact equation

$$y^2 e^y \, dx + ye^y(2 + y)x \, dy = 4ye^y \, dy,$$

or

$$d(xy^2 e^y) = 4ye^y \, dy,$$

which upon integration gives

$$xy^2 e^y = 4\int ye^y \, dy + C = 4e^y(y - 1) + C.$$

Thus, we may write the general solution of Equation (2.6.18) as

$$xy^2 = 4(y - 1) + Ce^{-y}.$$

Exercises 2.6

In Exercises 1 through 20, find the general solution of each differential equation.

1. $\dfrac{dy}{dx} + y = x.$

2. $\dfrac{dy}{dx} + y = \sin x.$

3. $\dfrac{dy}{dx} + 4y = \dfrac{1}{3 + 2e^{4x}}.$

4. $\dfrac{dy}{dx} = \dfrac{1}{y^3 - 2xy}.$

5. $\dfrac{dx}{dy} - x \ln y = y^y$.

6. $(x^2 + 1)\dfrac{dy}{dx} + 2xy = x^2 e^x$.

7. $dx + 2x\,dy = e^{-2y}\sec^2 y\,dy$.

8. $\dfrac{dx}{dy} + \dfrac{6y}{y^2 + 1}\,x = \dfrac{y^2}{(y^2 + 1)^4}$.

9. $\dfrac{dy}{dx} - \dfrac{2}{x}\,y = \arctan 2x$.

10. $x \ln x\,dy + (y - 2\ln x)dx = 0$.

11. $\dfrac{dx}{dy} + x\cos y = \sin 2y$.

12. $\dfrac{dy}{dx} = \dfrac{\sin x - (x - y)\cos x}{\sin x}$.

13. $\dfrac{dy}{dx} + \dfrac{n}{x}\,y = \dfrac{\tan ax\,\sec^n ax}{x^n}$ $(n \neq 0)$.

14. $\dfrac{dr}{d\theta} + 3r\cot\theta = -5\sin 2\theta$.

15. $x\cos x\,dy + [y(x\sin x + \cos x) - 1]dx = 0$.

16. $x(x^2 - 1)\dfrac{dy}{dx} + y = \dfrac{x^2}{\sqrt{x^2 - 1}}$.

17. $\dfrac{dy}{dx} + y\dfrac{du}{dx} = u(x)\dfrac{du}{dx}$.

18. $\dfrac{dy}{dx} + y\sec^2 x = \tan x\,\sec^2 x$.

19. $x \ln x\dfrac{dy}{dx} + y = \ln(\ln x)$.

20. $\dfrac{dr}{d\theta} + 2r\cos 2\theta = \sin 4\theta$.

21. The *Bernoulli equation* is given by

(1)
$$\frac{dy}{dx} + P(x)y = Q(x)y^n,$$

where n is any real number and the functions P and Q are continuous over the intervals on which the solutions of Equation (1) are sought. Suppose that n is neither zero nor unity. We may write Equation (1) as

(2)
$$y^{-n}\frac{dy}{dx} + P(x)y^{-n+1} = Q(x).$$

By means of the substitution,

(3)
$$z = y^{-n+1},$$

show that Equation (2) may be written as

(4)
$$\frac{dz}{dx} + (1 - n)P(x)z = (1 - n)Q(x),$$

which is a linear equation of the first order.   The general solution of Equation (1) is then given by (3), where in (3), z is the general solution of the linear Equation (4).   Thus in view of Theorem 2.6.1, *the general solution of Equation (1) is given by*

$$(5) \qquad y^{-n+1} = \exp\left[-\int (1-n)P(x)dx\right]$$

$$\times \left[\int \exp\left(\int (1-n)P(x)dx\right)(1-n)Q(x)dx + C\right].$$

If $n = 0$, then Equation (1) reduces to Equation (2.6.1).

If $n = 1$, then Equation (1) reduces to a type given in Equation (2.6.3), namely, in (2.6.3) we replace $P(x)$ by $P(x) - Q(x)$.

22.   Use the result of Exercise 21 to find the general solution of each of the following equations.

(a) $\dfrac{dy}{dx} - xy = x^3 y^3.$

(b) $\dfrac{dy}{dx} + \dfrac{xy}{1-x^2} = xy^{1/2}.$

(c) $\dfrac{dx}{dy} + 2x \csc 2y = x^2.$

(d) $\dfrac{dy}{dx} + \dfrac{x}{2(x^2+1)} y = x^3 y^5.$

23.   We may write the general solution of Equation (2.6.1) which is given by (2.6.2) as

$$(1) \qquad\qquad y = Cf(x) + g(x),$$

where

$$f(x) = \exp\left[-\int P(x)dx\right], \qquad g(x) = \exp\left[-\int P(x)dx\right]\left[\int \exp\left(\int P(x)dx\right)Q(x)dx\right].$$

We may consider (1) as a *linear function of the constant of integration C.*

(a) Conversely, show that the differential equation obtained by eliminating C between (1) and the derived function

$$y' = Cf'(x) + g'(x)$$

is a linear equation of order one.

(b) If y_1 and y_2 are particular solutions of Equation (2.6.1), show that the general solution of (2.6.1) is given by

$$\frac{y - y_1}{y - y_2} = K,$$

where K is an arbitrary constant.   [*Note:* No integration is needed to obtain the general solution of Equation (2.6.1).]

(c) If y_1 is a particular solution of Equation (2.6.1), show that the substitution

$$y = y_1 + Y$$

reduces Equation (2.6.1) to

$$\frac{dY}{dx} + P(x)\,Y = 0.$$

The general solution of Equation (2.6.1) is then given by

$$y = y_1 + C \exp\left[- \int P(x)dx \right].$$

24.   The equation

$$\frac{dy}{dx} = \frac{\sin x - (x - y)\cos x}{\sin x}$$

has particular solutions

$$y_1 = x \quad \text{and} \quad y_2 = x + \sin x.$$

(a) Use Exercise 23(b) to obtain the general solution of the equation.

(b) Use Exercise 23(c) to obtain the general solution of the equation.

2.7. Riccati's Equation

A differential equation of the first order and degree one of the form

$$\frac{dy}{dx} = P(x) + Q(x)y + R(x)y^2 \tag{2.7.1}$$

is called a *Riccati equation.*   Owing to the term y^2, Equation (2.7.1) is not linear in y.   We shall assume that the functions P, Q, and R are continuous over the intervals on which the solutions of Equation (2.7.1) are sought.

It is known that, as a rule, the general solution of Equation (2.7.1) cannot be obtained by a sequence of integrations.   However, as we shall presently see, if we know a particular solution of Equation (2.7.1), then Equation (2.7.1) can be reduced to a linear equation of the type given in Section 2.6.   Thus, (2.7.1) is solvable by a sequence of integrations, whenever a particular solution is known.

Theorem 2.7.1.   *Let $y_1(x)$ be a particular solution of the Riccati equation*

$$\frac{dy}{dx} = P(x) + Q(x)y + R(x)y^2. \tag{2.7.1}$$

Then the general solution of Equation (2.7.1) is given by

$$y = y_1(x) + \frac{1}{z}, \tag{2.7.2}$$

where

$$z = \exp\left[- \int (Q + 2Ry_1)dx \right]\left[C - \int \exp\left(\int (Q + 2Ry_1)dx \right) R\,dx \right],$$

$$\tag{2.7.3}$$

and C is an arbitrary constant.

PROOF.   Let

$$y = y_1(x) + u, \tag{2.7.4}$$

where u is a new variable and $y_1(x)$ is a particular solution of Equation (2.7.1), that is,

$$\frac{dy_1}{dx} = P + Qy_1 + Ry_1{}^2. \tag{2.7.5}$$

Differentiating (2.7.4), we obtain

$$\frac{dy}{dx} = \frac{dy_1}{dx} + \frac{du}{dx}. \tag{2.7.6}$$

Substituting (2.7.4) and (2.7.6) in (2.7.1), we get

$$\frac{dy_1}{dx} + \frac{du}{dx} = P + Qy_1 + Ry_1{}^2 + Qu + 2Ry_1u + Ru^2,$$

which in view of (2.7.5) reduces to

$$\frac{du}{dx} = (Q + 2Ry_1)u + Ru^2. \tag{2.7.7}$$

Clearly, (2.7.7) is a Bernoulli equation.   [*See* Exercise 2.6.21.]   If we let

$$u = z^{-1}, \tag{2.7.8}$$

Equation (2.7.7) may be written as

$$\frac{dz}{dx} + (Q + 2Ry_1)z = -R. \tag{2.7.9}$$

Thus, we have reduced Equation (2.7.1) to a linear equation in the variable z. Using the result of Theorem 2.6.1 with the functions P replaced by $Q + 2Ry_1$ and Q replaced by $-R$, the general solution of Equation (2.7.9) is given by

$$z = \exp\left[-\int (Q + 2Ry_1)dx\right]\left[C - \int \exp\left(\int (Q + 2Ry_1)dx\right)R\,dx\right]. \tag{2.7.10}$$

The theorem now follows from (2.7.4), (2.7.8), and (2.7.10).

From Exercise 2.6.23, we know that the general solution of Equation (2.7.9) can be written as a linear function of the constant of integration C, namely,

$$z = Cf(x) + g(x).^4 \tag{2.7.11}$$

---

[4] Note that the functions f and g given in (2.7.11) are not necessarily the same as those given in Exercise 2.6.23.

In view of (2.7.2) and (2.7.11), *the general solution of Riccati's equation may also be written as a rational function of the first degree in the constant of integration C, namely,*

$$y = \frac{CF(x) + G(x)}{Cf(x) + g(x)}, \qquad (2.7.12)$$

where

$$F(x) = y_1(x)f(x) \quad and \quad G(x) = y_1(x)g(x) + 1. \qquad (2.7.13)$$

Conversely, every differential equation of the first order whose general solution can be expressed in the form given in (2.7.12), satisfies a Riccati equation. For, solving for C in (2.7.12), we have

$$C = \frac{G - gy}{fy - F}. \qquad (2.7.14)$$

Differentiating (2.7.14), we obtain

$$(fy - F)(G' - gy' - g'y) - (G - gy)(f'y + fy' - F') = 0. \qquad (2.7.15)$$

Simplifying (2.7.15), we see that the resulting equation is of the type given by Equation (2.7.1). The reader should carry out this simplification.

Definition 2.7.1. The *cross-ratio* (or *anharmonic ratio*) of four distinct functions p, q, r, and s, each defined on a common interval I: $a \leqq x \leqq b$, is given by

$$\frac{(p - q)(r - s)}{(p - s)(r - q)}. \qquad (2.7.16)$$

Property 1. *The cross-ratio of four distinct particular solutions of a Riccati equation is independent of x.*

PROOF. Let y_1, y_2, y_3, and y_4 be four distinct particular solutions corresponding, respectively, to the values C_1, C_2, C_3, and C_4 of the constant C in (2.7.12). Then,

$$y_1 - y_2 = \frac{C_1 F + G}{C_1 f + g} - \frac{C_2 F + G}{C_2 f + g}$$

$$= \frac{C_1 C_2 Ff + C_2 Gf + C_1 Fg + Gg - C_1 C_2 Ff - C_1 Gf - C_2 Fg - Gg}{(C_1 f + g)(C_2 f + g)}.$$

Thus,

$$y_1 - y_2 = \frac{(C_1 - C_2)(Fg - Gf)}{(C_1 f + g)(C_2 f + g)}.$$

Similarly, we get

$$y_1 - y_4 = \frac{(C_1 - C_4)(Fg - Gf)}{(C_1 f + g)(C_4 f + g)},$$

$$y_3 - y_4 = \frac{(C_3 - C_4)(Fg - Gf)}{(C_3 f + g)(C_4 f + g)},$$

$$y_3 - y_2 = \frac{(C_3 - C_2)(Fg - Gf)}{(C_3 f + g)(C_2 f + g)}.$$

Forming the cross-ratio, we obtain upon simplification

$$\frac{(y_1 - y_2)(y_3 - y_4)}{(y_1 - y_4)(y_3 - y_2)} = \frac{(C_1 - C_2)(C_3 - C_4)}{(C_1 - C_4)(C_3 - C_2)} \equiv c, \qquad (2.7.17)$$

where c is independent of x.

Property 2. *Suppose that y_2, y_3, and y_4 are three distinct particular solutions of a Riccati equation.* *Then the general solution of Equation (2.7.1) is given by*

$$\frac{(y - y_2)(y_3 - y_4)}{(y - y_4)(y_3 - y_2)} = c. \qquad (2.7.18)$$

PROOF.   This result follows immediately from (2.7.17) with y_1 replaced by y. [*Note:* For this case, the general solution is obtained without integrations.]

Property 3. *Suppose that y_1 and y_2 are two distinct particular solutions of a Riccati equation.* *Then the general solution of Equation (2.7.1) is given by*

$$\ln\left(\frac{y - y_1}{y - y_2}\right) = \int (y_1 - y_2)R \, dx + C, \qquad (2.7.19)$$

where C is an arbitrary constant.

PROOF.   Since y_1 and y_2 are solutions of Equation (2.7.1), we have

$$y_1' = P + Qy_1 + Ry_1{}^2,$$

and

$$y_2' = P + Qy_2 + Ry_2{}^2.$$

Then

$$y' - y_1' = P + Qy + Ry^2 - P - Qy_1 - Ry_1{}^2 = (y - y_1)[Q + (y + y_1)R],$$

or

$$\frac{y' - y_1'}{y - y_1} = Q + (y + y_1)R. \qquad (2.7.20)$$

Similarly, we obtain

$$\frac{y' - y_2'}{y - y_2} = Q + (y + y_2)R. \tag{2.7.21}$$

Subtracting (2.7.21) from (2.7.20), we get

$$\frac{y' - y_1'}{y - y_1} - \frac{y' - y_2'}{y - y_2} = (y_1 - y_2)R,$$

which upon integration gives

$$\ln(y - y_1) - \ln(y - y_2) = \int (y_1 - y_2)R\, dx + C,$$

and thus (2.7.19) is established.

Example 2.7.1.   Find the general solution of the Riccati equation

$$\frac{dy}{dx} = 1 + y^2. \tag{2.7.22}$$

SOLUTION.   Comparing (2.7.22) with (2.7.1), we see that

$$P(x) = 1, \qquad Q(x) = 0, \qquad \text{and} \qquad R(x) = 1. \tag{2.7.23}$$

Clearly, a particular solution of Equation (2.7.22) is given by

$$y_1 = \tan x. \tag{2.7.24}$$

Utilizing (2.7.23) and (2.7.24), we may write (2.7.3) as

$$z = \exp\left[-\int 2\tan x\, dx\right]\left[C - \int \exp\left(\int 2\tan x\, dx\right)dx\right],$$

or

$$z = \cos^2 x\left[C - \int \sec^2 x\, dx\right].$$

Thus,

$$z = \cos^2 x[C - \tan x]. \tag{2.7.25}$$

Substituting y_1 of (2.7.24) and z of (2.7.25) in Equation (2.7.2), we obtain the general solution of (2.7.22), namely,

$$y = \tan x + \frac{1}{\cos^2 x[C - \tan x]}$$

$$= \tan x + \frac{\sec^2 x}{C - \tan x}$$

$$= \tan x + \frac{\tan^2 x + 1}{C - \tan x},$$

which can be written as

$$y(C - \tan x) = C \tan x + 1. \tag{2.7.26}$$

We can also write the general solution of (2.7.22) as

$$y(1 - B \tan x) = \tan x + B. \tag{2.7.27}$$

In particular, when $B = 0$ in (2.7.27), we obtain the particular solution given by (2.7.24).

Exercises 2.7

In Exercises 1 through 9, find the general solution of each Riccati equation.

1. $\dfrac{dy}{dx} = (1 + xy)P(x) + y^2.$

$$\left[\textit{Note: } y_1 = -\frac{1}{x} \text{ is a particular solution.} \right]$$

2. $\dfrac{dy}{dx} = \dfrac{1 + xy}{x^3} + y^2.$

3. $\dfrac{dy}{dx} = \sin x + y \cos x + y^2.$

   [*Note:* $y_1 = -\cos x$ is a particular solution.]

4. $\dfrac{dy}{dx} = -\dfrac{2}{x^2} + \dfrac{2y}{x} - y^2.$

$$\left[\textit{Note: } y_1 = \frac{2}{x} \text{ is a particular solution.} \right]$$

5. $\dfrac{dy}{dx} = x^2 + \dfrac{y}{x} - y^2.$

   [*Note:* $y_1 = x$ is a particular solution.]

6. $\dfrac{dy}{dx} = \dfrac{1}{x} - 2y + (x - 1)y^2.$

$$\left[\textit{Note: } y_1 = \frac{1}{x} \text{ is a particular solution.} \right]$$

7. $\dfrac{dy}{dx} = -\dfrac{3}{2x^3} + \dfrac{1}{2} - x^2 y + \dfrac{x}{2}(x^3 - 1)y^2.$

$$\left[\textit{Note: } y_1 = \frac{1}{x^2} \text{ is a particular solution.} \right]$$

8. $\dfrac{dy}{dx} = (\sin x - 1) \sin^2 x + \cos x + y(1 - 2 \sin x) \sin x + y^2 \sin x.$

[*Note:* $y_1 = \sin x$ is a particular solution.]

9. $\dfrac{dy}{dx} = x (\ln x)^2 - 2x \ln x + \dfrac{1}{x} + 2x(1 - \ln x)y + xy^2.$

[*Note:* $y_1 = \ln x$ is a particular solution.]

10. Verify that $y_1 = -\cot x$ and $y_2 = (1 - \cot x)/(1 + \cot x)$ are particular solutions of the equation

$$\frac{dy}{dx} = 1 + y^2.$$

Use Formula (2.7.19) to obtain the general solution of the equation.

11. Verify that

$$y_2 = x, \qquad y_3 = \frac{x^3}{x^2 - 2}, \qquad \text{and} \qquad y_4 = \frac{x^3 + x}{x^2 - 1}$$

are three particular solutions of the equation

$$\frac{dy}{dx} = - x^2 + \left(2x + \frac{1}{x}\right)y - y^2.$$

Use Formula (2.7.18) to find the general solution of the equation.

12. An equation of the type

(1) $$\frac{d^2y}{dx^2} + Q(x)\frac{dy}{dx} + P(x)y = 0$$

is called a *homogeneous linear differential equation of the second order.* We shall assume that the functions Q and P are continuous over the intervals on which the solutions of Equation (1) are sought.

(a) Show that the substitution

(2) $$y = \exp\left[\int u(x)dx\right]$$

reduces Equation (1) to a Riccati equation, namely,

(3) $$\frac{du}{dx} = -P(x) - Q(x)u - u^2.$$

[*Note:* From (2) we obtain by differentiation $y' = yu$, and thus $y'' = y(u' + u^2)$.]

(b) Conversely, show that every Riccati equation can be transformed into a homogeneous linear equation of the second order.

(c) Suppose that $y_1(x)$ and $y_2(x)$ are solutions of Equation (1). Show that

(4) $$y = c_1y_1(x) + c_2 y_2 (x),$$

where c_1 and c_2 are arbitrary constants, is also a solution of Equation (1). If the functions y_1 and y_2 are linearly independent [*see* Definition 4.3.6], that is, if

$$(5) \qquad c_1 y_1(x) + c_2 y_2(x) \equiv 0$$

for all x in some interval I implies that $c_1 = c_2 = 0$, then the function y given in (4) is known as the general solution of Equation (1) in I.

13. Use the result of Exercise 12 to find the general solution of the equation

$$\frac{d^2y}{dx^2} + \frac{2}{x}\frac{dy}{dx} - \frac{2}{x^2}y = 0 \qquad x > 0.$$

[*Note:* $y_1 = x$ is a particular solution.]

2.8. Miscellaneous Exercises

In this section we shall give additional exercises which can be solved by utilizing the methods given in this chapter. These additional exercises are intended to develop further the skill of recognizing what type of differential equation is given and what method (or methods) should be used to solve it.

Exercises 2.8

In Exercises 1 through 15, solve each differential equation.

1. $(x^2y + y^3 + x \ln x)dx + \left(\dfrac{x^3}{3} + 3xy^2 - ye^y\right)dy = 0.$

2. $(6y^2 + 3xy^3)dx + (9xy + 4x^2y^2)dy = 0.$
3. $(1 + y^2)dx - xy(1 + x^2)dy = 0.$
4. $(xy + 1)dx + x^3\,dy = 0.$
5. $(xe^{-y/x} - y)dx + x\,dy = 0.$
6. $[\cos(x - y) + \cos(x + y)]dx - dy = 0.$
7. $(2x \tan y \sec x + y^2 \sec y)dx + (2y \tan x \sec y + x^2 \sec x)dy = 0.$
8. $\sqrt{x^2 + y^2 + x^2y^2 + 1}\,dx - dy = 0.$
9. $(2x^2 + y^2)dx + (2xy + 3y^2)dy = 0.$
10. $\cos\theta\,dr + (r - \sin 2\theta)d\theta = 0.$
11. $2xy^2(y^2 - 1)dx - (x^2 + 1)(y^2 + 1)dy = 0.$
12. $(2y^4 + y \tan x)dx - dy = 0.$
13. $(e^{2x} + ye^x)dx - (e^x + 1)dy = 0.$
14. $[y - x(\ln x)^2]dx + x \ln x\,dy = 0.$
15. $[2x^3y + 2(x^4 - 1)(x^2 + x + 1)y^3]dx + (x^4 - 1)dy = 0.$

In Exercises 16 through 28, solve each initial-value problem.

16. $\begin{cases} \dfrac{dy}{dx} = -\dfrac{3x^2y + xy^2 + 2xy}{x^3 + x^2y + x^2}, \\[2mm] y(1) = 2. \end{cases}$

17. $\begin{cases} \dfrac{dy}{dx} = 2y + \cos(e^{-x}), \\[2mm] y(0) = -1. \end{cases}$

18. $\begin{cases} \dfrac{dy}{dx} = \dfrac{y^2 \ln x - 2xy}{x^2}, \\[2ex] y(1) = \dfrac{9}{2}. \end{cases}$

19. $\begin{cases} \dfrac{dy}{dx} = \sqrt{\dfrac{1+x}{1-x}}, \\[2ex] y(0) = -1. \end{cases}$

20. $\begin{cases} \dfrac{dy}{dx} = -\dfrac{10x + 8y}{7x + 5y}, \\[2ex] y(0) = -2. \end{cases}$

21. $\begin{cases} \dfrac{dy}{dx} = 2y + \ln (e^{2x} + 1), \\[2ex] y(0) = \ln (\tfrac{1}{2}). \end{cases}$

22. $\begin{cases} \dfrac{dy}{dx} = \dfrac{y - k\sqrt{x^2 + y^2}}{x}, \\[2ex] y(a) = 0, \quad a > 0. \end{cases}$

23. $\begin{cases} \dfrac{dr}{d\theta} = r(k \sec \theta + \tan \theta), \\[2ex] r = a \quad \text{when } \theta = 0, \quad a \neq 0. \end{cases}$

24. $\begin{cases} \dfrac{dy}{dx} = \dfrac{y + y^2 x^5 \sin x^3}{x}, \\[2ex] y(\sqrt[3]{\pi}) = \sqrt[3]{\pi}. \end{cases}$

25. $\begin{cases} \dfrac{dy}{dx} = \dfrac{1}{x^2} + \dfrac{y}{x} + y^2, \\[2ex] y(1) = 2. \end{cases}$

26. $\begin{cases} \dfrac{dy}{dx} = \dfrac{3y - 2x + 3}{x + y - 4}, \\[2ex] y(4) = 1. \end{cases}$

27. $\begin{cases} \dfrac{dy}{dx} = x - x^6 + \left(2x^4 + \dfrac{1}{x}\right) y - x^2 y^2. \\[2ex] y(1) = 4. \end{cases}$

28. $\begin{cases} \dfrac{dy}{dx} = x \sin^2 (\alpha \ln x), \\[2ex] y(b) = 0, \\[2ex] \text{where } \alpha = \dfrac{n\pi}{\ln b}, \quad b > 1, \quad \text{and} \quad n = 0, 1, 2, \ldots. \end{cases}$

29. Let $P(x)$ and $P(y)$ be two polynomials of the fourth degree in x and y, respectively, and having the same coefficients, that is,

(1) $\qquad \begin{cases} P(x) = ax^4 + bx^3 + cx^2 + dx + e, \\ P(y) = ay^4 + by^3 + cy^2 + dy + e, \end{cases}$

where a, b, c, d, and e are constants. A differential equation of the type

(2) $$\frac{dx}{\sqrt{P(x)}} + \frac{dy}{\sqrt{P(y)}} = 0,$$

where $P(x)$ and $P(y)$ are given in (1) is known as *Euler's equation.*

Case 1. Let $a = b = 0$ in (1).   Then we have

(3)
$$\begin{cases} P(x) = cx^2 + dx + e, \\ P(y) = cy^2 + dy + e. \end{cases}$$

Suppose that these polynomials of the second degree are not perfect squares.
(a) Show that transformations of the type

(4)
$$\begin{cases} x = \xi + h, & y = \zeta + h, \\ \xi = ku, & \zeta = kv, \end{cases}$$

where ξ, ζ, u, v are new variables and h, k are constants, transform (3) into the form

(5)
$$\begin{cases} P(u) = K(u^2 - 1), \\ P(v) = K(v^2 - 1), \end{cases}$$

where K is a constant. Thus, for Case 1, we may write Equation (2) as

(6)
$$\frac{dx}{\sqrt{1 - x^2}} + \frac{dy}{\sqrt{1 - y^2}} = 0.$$

(b) Clearly, 1 is an integrating factor of Equation (6). Show that the general solution of Equation (6) is given by

$$\text{arc sin } x + \text{arc sin } y = C,$$

or

(7)
$$x\sqrt{1 - y^2} + y\sqrt{1 - x^2} = B,$$

where B is an arbitrary constant.
(c) Verify the following differential identity.

$$d(x\sqrt{1 - y^2} + y\sqrt{1 - x^2}) = (\sqrt{(1 - x^2)(1 - y^2)} - xy)$$
$$\times \left(\frac{dx}{\sqrt{1 - x^2}} + \frac{dy}{\sqrt{1 - y^2}} \right).$$

Thus,

(8)
$$\sqrt{(1 - x^2)(1 - y^2)} - xy$$

is also an integrating factor of Equation (6), and the general solution of Equation (6) is also given by [*see* Exercise 2.5.19]

(9)
$$\sqrt{(1 - x^2)(1 - y^2)} - xy = C,$$

where C is an arbitrary constant.
(d) Verify the following identity.

$$[x\sqrt{1 - y^2} + y\sqrt{1 - x^2}]^2 + [\sqrt{(1 - x^2)(1 - y^2)} - xy]^2 = 1.$$

Utilize this identity to show that the solutions given in (7) and (9) are equivalent.

(e) By rationalizing the expression in (9), show that the general solution of Equation (6) is given by

(10)
$$x^2 + y^2 + 2Cxy + C^2 - 1 = 0.$$

(f) Show that Equation (10) represents a one-parameter family of conics that are tangent to the four lines $x = \pm 1$ and $y = \pm 1$.

Case 2. It can be shown that the general solution of the differential Equation (2), where $P(x)$ and $P(y)$ are given by (1), is given by

(11)
$$\left(\frac{\sqrt{P(x)} - \sqrt{P(y)}}{x - y}\right)^2 - a(x + y)^2 - b(x + y) - c = C.$$

(g) Use this result to solve the system

$$\begin{cases} P(x) = K(x^2 - 1), \\ P(y) = K(y^2 - 1). \end{cases}$$

30. The general linear first-order partial differential equation with constant coefficients is given by

(1)
$$au_x(x, y) + bu_y(x, y) + cu(x, y) = f(x, y),$$

where a, b, c are constants and f is a continuous function of x and y. Consider the transformation

(2)
$$\begin{cases} x = \alpha\xi + \beta\eta, \\ y = \gamma\xi + \delta\eta, \end{cases} \quad \begin{vmatrix} \alpha & \beta \\ \gamma & \delta \end{vmatrix} \neq 0,$$

where ξ, η are new independent variables and α, β, γ, δ are constants to be determined. Then

(3)
$$u(x, y) = u(\alpha\xi + \beta\eta, \quad \gamma\xi + \delta\eta) = U(\xi, \eta).$$

Show that

(4)
$$\frac{\partial U}{\partial \xi} = \alpha u_x + \gamma u_y.$$

If we take $\alpha = a$ and $\gamma = b$, then in view of the above equations, we may write (1) as

(5)
$$U_\xi(\xi, \eta) + cU(\xi, \eta) = F(\xi, \eta).$$

By treating η as a parameter, Equation (5) may be considered as an ordinary linear differential equation of the first order. Thus, show that the general solution of Equation (5) is given by

(6)
$$U(\xi, \eta) = e^{-c\xi}[G(\xi, \eta) + h(\eta)],$$

where

(7)
$$G(\xi, \eta) = \int e^{c\xi} F(\xi, \eta) d\xi,$$

and h is an arbitrary function of η rather than an arbitrary constant.

Suppose that the solution of Equation (1) satisfies the auxiliary condition

$$(8) \qquad u(x, 0) = u_0(x),$$

where the function u_0 is differentiable at x. Also, *suppose that $b \neq 0$ in (1)*. In view of the auxiliary condition, we shall choose β and δ in (2) so that the line $\xi = 0$ corresponds to the line $y = 0$. Thus $\delta = 0$, and since β is arbitrary, we may conveniently choose it to be 1. We may now write Equation (2) as

$$(9) \qquad \begin{cases} x = a\xi + \eta, \\ y = b\xi, \end{cases}$$

or

$$(10) \qquad \begin{cases} \xi = \dfrac{1}{b}\,y, \\[2mm] \eta = x - \dfrac{a}{b}\,y. \end{cases}$$

Notice now that in (7)

$$(11) \qquad F(\xi, \eta) \equiv f(a\xi + \eta, b\xi) = f(x, y).$$

Utilizing the above equations, show that *the general solution of Equation (1) is given by*

$$(12) \qquad u(x, y) = e^{-(c/b)y}\left[G\!\left(\frac{1}{b}\,y,\, x - \frac{a}{b}\,y\right) + h\!\left(x - \frac{a}{b}\,y\right)\right].$$

To find the particular solution of (1) which satisfies the auxiliary condition (8), we first substitute (12) into (8), and we get

$$(13) \qquad G(0, x) + h(x) = u_0(x).$$

From (13), we see that

$$(14) \qquad h\!\left(x - \frac{a}{b}\,y\right) = u_0\!\left(x - \frac{a}{b}\,y\right) - G\!\left(0, x - \frac{a}{b}\,y\right).$$

Thus, the unique particular solution of Equation (1) satisfying the auxiliary condition given in (8) is given by

$$(15) \quad u(x, y) = e^{-(c/b)y}\left[G\!\left(\frac{1}{b}\,y,\, x - \frac{a}{b}\,y\right) - G\!\left(0, x - \frac{a}{b}\,y\right) + u_0\!\left(x - \frac{a}{b}\,y\right)\right].$$

As an example, let us solve

$$(16) \qquad \begin{cases} u_x - u_y + u = x + y, \\ u(x, 0) = x \sin x. \end{cases}$$

$$(17)$$

SOLUTION. From (1) and (16), we see that $a = 1$, $b = -1$, $c = 1$, and $f(x, y) = x + y$. From (9) or (10), we have

$$\begin{cases} x = a\xi + \eta = \xi + \eta, \\ y = b\xi = -\xi, \end{cases} \quad \text{or} \quad \begin{cases} \xi = \dfrac{1}{b}y = -y, \\[2mm] \eta = x - \dfrac{a}{b}y = x + y. \end{cases}$$

From (11), $F(\xi, \eta) \equiv f(\xi + \eta, -\xi) = f(x, y)$. Thus, $F(\xi, \eta) = (\xi + \eta) - \xi = \eta$. In view of (7), we then have

$$G(\xi, \eta) = \int e^{c\xi}F(\xi, \eta)d\xi = \int e^{\xi}\eta \, d\xi = \eta \int e^{\xi} \, d\xi = e^{\xi}\eta.$$

Hence,

$$G\left(\frac{1}{b}y, \, x - \frac{a}{b}y\right) = G(-y, \, x + y) = e^{-y}(x + y); \quad \text{and} \quad G\left(0, \, x - \frac{a}{b}y\right)$$

$$= G(0, x + y) = e^{0}(x + y) = x + y.$$

From (8) and (17), we have $u_0(x) = u(x, 0) = x \sin x$. Thus,

$$u_0\left(x - \frac{a}{b}y\right) = u_0(x + y) = (x + y) \sin(x + y).$$

From (15), the particular solution of (16) satisfying the auxiliary condition (17) is then given by

$$u(x, y) = e^{y}[(x + y)e^{-y} + (x + y)\sin(x + y) - (x + y)]$$
$$= (x + y)e^{y}[e^{-y} + \sin(x + y) - 1].$$

Find the solution $u(x, y)$ of each equation subject to its auxiliary condition.

(a) $\begin{cases} u_x - u_y + u = 2, \\ u(x, 0) = \cos x. \end{cases}$

(b) $\begin{cases} 2u_x - u_y + 4u = x + 2y, \\ u(x, 0) = e^{x}. \end{cases}$

(c) $\begin{cases} au_x + bu_y + cu = 0, \\ u(x, 0) = u_0(x), \\ \text{where } a, b, c \text{ are constants,} \\ b \neq 0, \text{ and } u_0 \text{ is a dif-} \\ \text{ferentiable function of } x. \end{cases}$

(d) $\begin{cases} 3u_x - 4u_y + 2u = 0, \\ u(x, 0) = x^2 \cos x. \end{cases}$

(f) $\begin{cases} u_y - 4u = 0, \\ u(x, 0) = x^2. \end{cases}$

(e) $\begin{cases} 3u_x - 4u_y = 0, \\ u(x, 0) = e^{\sin x}. \end{cases}$

(g) $\begin{cases} u_x - u_y + u = e^{x+y} \\ u(x, 0) = \sinh x. \end{cases}$

Suppose that $b = 0$ in (1). Then we have

(18)
$$au_x(x, y) + cu(x, y) = f(x, y).$$

Suppose that we wish to find the solution of (18) subject to the auxiliary condition

$$(19) \qquad u(x, 0) = u_0(x)$$

where the function u_0 is differentiable at x. This problem will now have solutions not for arbitrary functions u_0, but only for functions u_0 that are of a special form. Even for these special functions u_0, the problem will not have a unique solution but rather infinitely many solutions.

As an example, let us determine all functions u_0 for which the problem

$$(20) \qquad \qquad u_x + 3u = 0,$$

$$(21) \qquad \qquad u(x, 0) = u_0(x),$$

has a solution, and for such functions let us determine all solutions of the problem.

From (20), we have $u_x/u = -3$. An integration of this equation with respect to x gives

$$(22) \qquad \ln u(x, y) = -3x + \ln g(y),$$

where $\ln g$ is an arbitrary function. We may write (22) as

$$(23) \qquad u(x, y) = g(y)e^{-3x}.$$

From (21) and (23), we have

$$u_0(x) = u(x, 0) = g(0)e^{-3x},$$

or

$$(24) \qquad u_0(x) = Ce^{-3x},$$

where C is any constant such that $g(0) = C$. Thus, all functions u_0 for which the problem has a solution are given by (24); and for such functions, all the solutions of the problem are given by (23), where g is any function that satisfies the condition $g(0) = C$.

In problems (h) through (k), determine all functions u_0 for which each problem has a solution, and for such functions, determine all solutions of each problem.

(h) $\begin{cases} au_x(x, y) + cu(x, y) = 0, \\ u(x, 0) = u_0(x), \\ \text{where } a \text{ and } c \text{ are constants.} \end{cases}$

(i) $\begin{cases} u_x + u = 2, \\ u(x, 0) = u_0(x). \end{cases}$

(j) $\begin{cases} u_x + u = x + y, \\ u(x, 0) = u_0(x). \end{cases}$

(k) $\begin{cases} 2u_x - 4u = ye^{3x}, \\ u(x, 0) = u_0(x). \end{cases}$

Suggested Readings

Berg and McGregor [5]
Coddington [12]
Goursat [17]

Ince [23]
Taylor [45]

chapter 3 APPLICATIONS OF FIRST-ORDER DIFFERENTIAL EQUATIONS

3.1. Growth, Decay, and Interest Problems

Let us consider the first-order differential equation

$$\frac{dx}{dt} = kx. \tag{3.1.1}$$

Equation (3.1.1) asserts that the time rate of change of the variable x is proportional to x. The constant of proportionality is k. In this chapter we shall concern ourselves primarily with the time rate of change in the quantity of a substance so that $x > 0$. When k is positive, then $dx/dt > 0$, and thus x is increasing. We may refer to this case as one in which the quantity x is growing, and consider the problem as one of growth. When k is negative, then $dx/dt < 0$, and thus x is decreasing. We may refer to this case as one in which the quantity x is decaying, and consider this problem as one of decay or decomposition.

Remark 3.1.1. In practice, many growth, decomposition, and other types of problems are depicted graphically by discontinuous functions of the variable t, and thus the derivative dx/dt, say in (3.1.1), has in general no meaning. To circumvent this difficulty somewhat, we approximate the graph of a discontinuous function by a smooth curve or function, that is, by a function that is continuous and has continuous derivative (dx/dt) in any given interval of time. [*See* Figure 3.1.1. The discontinuous curve is represented by the solid line segments, while the smooth curve is represented by a dotted curve.]

Also, in practice, we are not usually given the differential equation, but rather a certain collection of experimental data. When these data are plotted, the graph of the function is usually represented by a discontinuous function. For example, in the growth of some culture of bacteria, the increase in the number of bacteria does not necessarily occur continuously, but rather after certain intervals of time. For instance, in Figure 3.1.1, $x(t)$ could represent the growth in the number of bacteria present at any time t. The jump discontinuities of the function represent a sudden growth in the number of

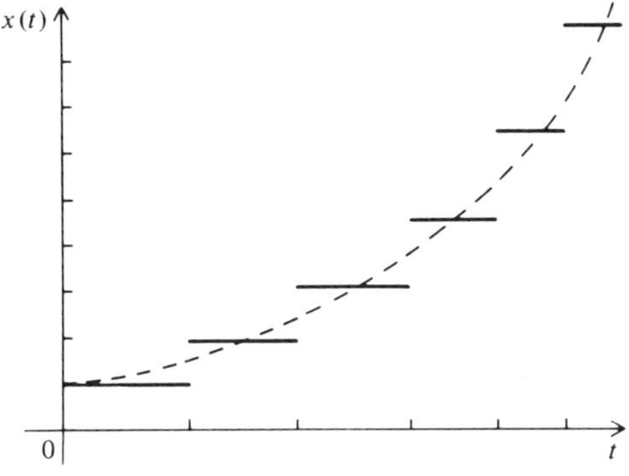

Figure 3.1.1.

bacteria, and the horizontal lines represent a dormant stage. A similar situation could occur in the decomposition of some radioactive substances. After we have graphed the discontinuous function depicting the experimental data, our task is to find (if possible) a smooth curve that approximates it. Thus, we are led to seek a differential equation whose solution will give rise to the smooth curve.

Equation (3.1.1) is one having separable variables. Thus,

$$\frac{dx}{x} = k \, dt,$$

from which we get upon integration

$$\ln x = kt + \ln C,$$

or

$$x = Ce^{kt}. \tag{3.1.2}$$

The general solution of Equation (3.1.1) is given by Equation (3.1.2).

Remark 3.1.2. Note that Equation (3.1.2) involves two constants, namely, the constant of integration C and the proportionality constant k. Thus, two conditions are required to determine these constants. For convenience we shall write Equation (3.1.1) with the two conditions required to determine the constants C and k in the form of a boundary-value problem.

Also, we shall take the constant of proportionality k to be greater than zero. Thus, growth and decomposition problems will be given, respectively, by

$$\frac{dx}{dt} = kx$$

and

$$\frac{dx}{dt} = -kx. \tag{3.1.3}$$

Note that if $k > 0$, then $-k < 0$.

Example 3.1.1. The number of bacteria in a yeast culture grows at a rate which is proportional to the number present. If the population of a colony of yeast bacteria triples in 1 hour, find the number of bacteria which will be present at the end of 5 hours.

SOLUTION 1. This is a growth problem. Let x_0 be equal to the number of bacteria initially (that is, at $t = 0$), and let x be equal to the number of bacteria at any time t. Thus, we have

$$\frac{dx}{dt} = kx, \tag{3.1.4}$$

$$x(0) = x_0, \qquad x(1) = 3x_0. \tag{3.1.5}$$

The solution of (3.1.4) is given by

$$x = Ce^{kt}. \tag{3.1.6}$$

To determine C, we utilize the first condition in (3.1.5). Thus,

$$x_0 = x(0) = Ce^{k \cdot 0} = C,$$

and we may now write (3.1.6) as

$$x = x_0 e^{kt}. \tag{3.1.7}$$

From (3.1.7) and the second condition in (3.1.5), we determine k. Thus,

$$3x_0 = x(1) = x_0 e^{k \cdot 1},$$

from which we obtain

$$e^k = 3 \quad \text{or} \quad k = \ln 3. \tag{3.1.8}$$

Note that we may write (3.1.7) as

$$x = x_0(e^k)^t. \tag{3.1.9}$$

Instead of substituting the value of k given in the second member of (3.1.8), we prefer to substitute the value of e^k given in the first member of (3.1.8). We obtain

$$x = x_0 \, 3^t. \tag{3.1.10}$$

If we had substituted $k = \ln 3$ in (3.1.9), we would have obtained

$$x = x_0 \, e^{t \ln 3} = x_0 \, e^{\ln 3^t} = x_0 \, 3^t,$$

which is the same as Equation (3.1.10). The solution of (3.1.4) subject to the conditions in (3.1.5) is given by Equation (3.1.10).

To find the number of bacteria present at the end of 5 hours, we substitute $t = 5$ in Equation (3.1.10). We find

$$x = x_0 3^5 = 243 x_0. \tag{3.1.11}$$

Equation (3.1.11) asserts that the number of bacteria present at the end of the fifth hour equals 243 times the initial amount.

SOLUTION 2. Integrating (3.1.4) between the limits $t = 0,\quad x = x_0$ and $t = 1,\quad x = 3x_0$ (which gives us the net change in the population), we have

$$\int_{x_0}^{3x_0} \frac{dx}{x} = k \int_0^1 dt.$$

Thus,

$$\ln 3x_0 - \ln x_0 = k \quad \text{or} \quad k = \ln 3. \tag{3.1.12}$$

Again integrating (3.1.4) between the limits $t = 0,\quad x = x_0$ and $t = 5,\quad x = x$, we obtain

$$\ln x - \ln x_0 = 5k \quad \text{or} \quad \ln x = \ln x_0 + 5k. \tag{3.1.13}$$

Substituting the value of k found in (3.1.12) into (3.1.13), we get

$$\ln x = \ln x_0 + 5 \ln 3 = \ln x_0 + \ln 3^5 = \ln (x_0 3^5),$$

or

$$x = x_0 3^5,$$

which is the same answer as in (3.1.11).

[*Note*: This method of solving the problem bypassed the evaluation of the constant of integration C.]

Example 3.1.2. A quantity of radioactive substance originally weighing x_0 grams decomposes at a rate which is proportional to the amount present.

(a) If half of the original quantity is present after p years, find the amount x of the substance remaining after t years.
   In part (a), suppose it is known that $p = 2000$ years.
(b) Find the time required to decompose three-fourths of the original amount.
(c) Find the amount of substance left after 3000 years.

SOLUTION 1. *Concerning* (a). This problem is one of decomposition. Thus, we have

$$\frac{dx}{dt} = -kx, \tag{3.1.14}$$

$$x(0) = x_0, \qquad x(p) = \tfrac{1}{2}x_0. \tag{3.1.15}$$

The general solution of (3.1.14) is given by

$$x = Ce^{-kt}. \tag{3.1.16}$$

Utilizing the first condition in (3.1.15), we determine C.   Thus,

$$x_0 = x(0) = Ce^{-k \cdot 0} = C,$$

and we may now write Equation (3.1.16) as

$$x = x_0 e^{-kt}. \tag{3.1.17}$$

From (3.1.17) and the second condition in (3.1.15), we have

$$\tfrac{1}{2}x_0 = x(p) = x_0 e^{-kp},$$

or

$$e^{-kp} = 2^{-1}. \tag{3.1.18}$$

Let us write Equation (3.1.17) as

$$x = x_0(e^{-kp})^{t/p}. \tag{3.1.19}$$

If we substitute the value of e^{-kp} found in (3.1.18) into (3.1.19), we obtain

$$x = x_0 2^{-(t/p)}, \tag{3.1.20}$$

which is the answer to part (a).

Concerning (b).   With $p = 2000$ in Equation (3.1.20), we have

$$x = x_0 2^{-t/2000}. \tag{3.1.21}$$

Since three-fourths of the original amount has been decomposed, the amount x that is now present is

$$x_0 - \tfrac{3}{4}x_0 = \tfrac{1}{4}x_0 .$$

Substituting $x = \tfrac{1}{4}x_0$ in Equation (3.1.21), we get

$$\tfrac{1}{4} x_0 = x_0 2^{-t/2000} \quad \text{or} \quad 2^{-2} = 2^{-t/2000},$$

and upon equating the exponents we obtain

$$\frac{t}{2000} = 2 \quad \text{or} \quad t = 4000 \text{ years.} \tag{3.1.22}$$

Thus, it requires 4000 years to decompose three-fourths of the original amount.

Concerning (c).   Substituting $t = 3000$ in Equation (3.1.21), we obtain

$$x = x_0 2^{-3000/2000} = x_0 2^{-3/2} \approx 0.353\, x_0 . \tag{3.1.23}$$

Thus, after 3000 years the amount of radioactive substance left is approximately 0.353 times the original amount.   The symbol "$\approx$" is read "is approximately."

SOLUTION 2. *Concerning* (a).   Integrating (3.1.14) between the limits $t = 0$, $x = x_0$ and $t = p$, $x = \frac{1}{2}x_0$, we have

$$\int_{x_0}^{(1/2)x_0} \frac{dx}{x} = -k \int_0^p dt.$$

Thus,

$$\ln \frac{1}{2} x_0 - \ln x_0 = -kp \quad \text{or} \quad k = \frac{1}{p} \ln 2. \tag{3.1.24}$$

Again integrating (3.1.14) between the limits $t = 0$, $x = x_0$ and $t = t$, $x = x$, we obtain

$$\ln x - \ln x_0 = -kt \quad \text{or} \quad \ln x = \ln x_0 - kt. \tag{3.1.25}$$

Substituting the value of k found in (3.1.24) into (3.1.25), we obtain

$$\ln x = \ln x_0 - \frac{t}{p} \ln 2 = \ln x_0 + \ln 2^{-(t/p)} = \ln (x_0 2^{-(t/p)}),$$

or

$$x = x_0 2^{-(t/p)},$$

which is the same answer as in (3.1.20).

Concerning (b).   Integrating (3.1.14) between the limits $t = 0$, $x = x_0$ and $t = t$, $x = \frac{1}{4}x_0$, we obtain

$$\ln x_0 - \ln \frac{x_0}{4} = kt \quad \text{or} \quad t = \frac{2}{k} \ln 2. \tag{3.1.26}$$

Substituting the value of k found in (3.1.24) with $p = 2000$ into (3.1.26), we get

$$t = \frac{4000 \ln 2}{\ln 2} = 4000 \text{ years},$$

which is the same answer as in (3.1.22).

Concerning (c).   Integrating (3.1.14) between the limits $t = 0$, $x = x_0$ and $t = 3000$, $x = x$, we find

$$\ln x - \ln x_0 = -3000\, k \quad \text{or} \quad \ln x = \ln x_0 - 3000\, k. \tag{3.1.27}$$

Substituting the value of k found in (3.1.24) with $p = 2000$ into (3.1.27), we obtain

$$\ln x = \ln x_0 - \frac{3000}{2000} \ln 2 = \ln x_0 + \ln 2^{-(3/2)} = \ln \left(x_0 \, 2^{-(3/2)} \right),$$

or

$$x = x_0 \, 2^{-(3/2)},$$

which is the same answer as in (3.1.23).

Remark 3.1.3.   The *half-life* of a radioactive substance is defined as the time required to decompose one-half (or 50 percent) of the substance.   Example 3.1.2 illustrates the usefulness of this concept.

Interest Problems.   We shall now consider a problem involving compound interest.   Suppose that A_0 dollars is invested at a rate k per year.   The total amount of principal and interest at the end of one year if compounded n times per year is given by

$$A_0 \left(1 + \frac{k}{n} \right)^n. \tag{3.1.28}$$

At the end of t years it will be

$$A_0 \left[\left(1 + \frac{k}{n} \right)^n \right]^t. \tag{3.1.29}$$

Keeping k and t fixed, let us see what happens to this amount as n tends to infinity.   From calculus, we have that

$$\lim_{n \to \infty} \left(1 + \frac{k}{n} \right)^{n/k} = e.$$

Consequently,

$$A_0 \lim_{n \to \infty} \left[\left(1 + \frac{k}{n} \right)^n \right]^t = A_0 \lim_{n \to \infty} \left[\left(1 + \frac{k}{n} \right)^{n/k} \right]^{kt}$$

$$= A_0 \left[\lim_{n \to \infty} \left(1 + \frac{k}{n} \right)^{n/k} \right]^{kt} = A_0 \, e^{kt}. \tag{3.1.30}$$

Thus, we see that as the number n times of compounding in a year tends to infinity, the amount at the end of t years approaches a definite limit given by (3.1.30).   Let us denote the value of this limit by A, that is,

$$A = A_0 \lim_{n \to \infty} \left[\left(1 + \frac{k}{n} \right)^n \right]^t = A_0 \, e^{kt}. \tag{3.1.31}$$

From the above analysis, we are led to make the following definition.

Definition 3.1.1.   Money is said to increase at a rate k per year *compounded continuously* (or *instantaneously*) if an amount A_0 at a given time increases to the amount A, where

$$A = A_0 e^{kt}, \tag{3.1.32}$$

after t years from the given time.

Eliminating A_0 from Equation (3.1.32), we obtain the differential equation

$$\frac{dA}{dt} = kA, \tag{3.1.33}$$

which is the same as Equation (3.1.1).

Remark 3.1.4.   When in Definition 3.1.1, "at a rate k per year" is expressed "at a rate of k percent per annum," we replace k by $k/100$ in (3.1.32) and (3.1.33).

Example 3.1.3.   How long will it take 100 dollars to triple itself
(a) if it is compounded continuously at a rate of 5 percent per annum?
(b) if it is compounded quarterly at a rate of 5 percent per annum?

SOLUTION 1.   *Concerning* (a).   We may utilize either Equation (3.1.32) or (3.1.33) to solve this problem.   Substituting $A_0 = 100$, $A = 300$ (since the amount A increases to the amount $3A_0$) and $k = 0.05$ in Equation (3.1.32), we have

$$3A_0 = A_0 e^{0.05t}.$$

Solving for t, we obtain

$$t = 20 \ln 3 \approx 20(1.09861) \approx 21.97 \text{ years.} \tag{3.1.34}$$

Concerning (b).   Let us denote the expression given in (3.1.29) by A_n, that is,

$$A_n = A_0\left(1 + \frac{k}{n}\right)^{nt} \qquad (n > 0). \tag{3.1.35}$$

For interest compounded quarterly, $n = 4$.   Thus with $A_4 = 300$, $A_0 = 100$, and $k = 0.05$, Equation (3.1.35) becomes

$$3 = (1.0125)^{4t}$$

which, when solved for t, gives

$$t = \frac{\log 3}{4 \log 1.0125} = \frac{0.47712}{0.02156} \approx 22.13 \text{ years.}^{[1]} \tag{3.1.36}$$

---

[1] Note that $\ln x$ denotes the "natural logarithm" of x, that is, the logarithm of x to the base e, where $e = 2.71828\ldots$, while $\log x$ denotes the "common logarithm" of x, that is, the logarithm of x to the base 10. Also, $\ln x = 2.30259 \log x$ and $\log x = 0.43429 \ln x$, approximately.

From (3.1.36) and (3.1.34) we see that the time required to achieve the result in part (b) differs from that in part (a) by approximately 0.16 year, or 58 days.

SOLUTION 2.   *Concerning* (a).   Integrating (3.1.33) between the limits $t = 0$, $A = 100$ and $t = t$, $A = 300$, we have

$$\int_{100}^{300} \frac{dA}{A} = 0.05 \int_{0}^{t} dt.$$

Thus,

$$\ln 300 - \ln 100 = 0.05t \quad \text{or} \quad t = 20 \ln 3 \approx 21.97 \text{ years},$$

which is the same answer as in (3.1.34).

Exercises 3.1

In Exercises 1 through 8, assume that the rate of growth is proportional to the number of bacteria (or population) present, and that the substances decompose at a rate proportional to the amount present.

1. If the number of bacteria in a culture is 10^6 at the end of 4 hours and $9 \cdot 10^6$ at the end of 6 hours, how many bacteria were there initially?
2. If the number of bacteria in a culture triples in 5 hours and at the end of 20 hours the number is $2 \cdot 10^6$, how many bacteria were there initially?
3. If the population of a city doubles in 40 years and the present population is 400,000, when will the population reach 700,000?
4. If the number of bacteria in a culture is initially n_0 and the number at the end of times t_1 and t_2 $(t_1 < t_2)$ are n_1 and n_2, respectively, show that

$$\left(\frac{n_2}{n_0}\right)^{t_1} = \left(\frac{n_1}{n_0}\right)^{t_2}.$$

5. If one-half of a radioactive substance is decomposed in 2000 years, what percentage is lost in 200 years?
6. If 3 percent of a radioactive substance is decomposed in 100 years and initially there are 20 grams, what amount is present at the end of 1000 years?
7. If 4 percent of a radioactive substance is decomposed in 200 years, what percentage will be present at the end of 800 years?
8. If the amount present of a radioactive substance at times t_1 and t_2 $(t_1 < t_2)$ are x_1 and x_2, respectively, show that the half-life of the substance is

$$\frac{(t_2 - t_1) \ln 2}{\ln \dfrac{x_1}{x_2}}.$$

9. A sponge dries in open air at a rate that is proportional to its moisture content. A sponge hung in open air loses half of its moisture content in 30 minutes. How long will it take to lose 95 percent of its moisture content? (Assume that the weather conditions remain uniform during this process.)

10. At what rate of interest per year compounded continuously will 100 dollars triple itself in 20 years?

11. (a) A man has A dollars in a fund paying interest at a rate of k percent per annum compounded continuously. If he withdraws money continuously at a rate of B dollars ($B > kA/100$) per year, show that the time required to deplete the fund is

$$ t = \frac{1}{\dfrac{k}{100}} \ln\left(\frac{B}{B - \dfrac{k}{100}A} \right). $$

(b) In part (a), find the time required to deplete the fund when $A = 100{,}000$, $B = 6000$, and $k = 5$ percent.

12. If a man saves 10 dollars a day in a bank that pays interest at a rate of $5\frac{1}{2}$ percent per annum compounded continuously, how long will it take him to save 100,000 dollars?

13. Let x_0 and x be equal, respectively, to the number of bacteria in a culture at time $t = 0$ and at time t. Also, let the rate of increase due to reproduction and the rate of decrease due to mortality be given, respectively, by

$$ a + bt \text{ percent} \quad \text{and} \quad \alpha + \beta t \text{ percent,} $$

where a, b, α, and β are positive constants.

(a) Show that

$$ x = x_0 \, e^{(1/100)[(a-\alpha)t + (1/2)(b-\beta)t^2]}. $$

(b) Show that if $a > \alpha$ and $b < \beta$, then x will first increase but ultimately will tend to zero.

(c) Show that if $a > \alpha$, $b < \beta$, then the maximum value of x is given by

$$ x_{\max} = x_0 e^{(a-\alpha)^2/200(\beta-b)}. $$

3.2. Temperature and Mixture Problems

It is known from experimental results that under certain conditions one may obtain a good approximation to the temperature of an object by utilizing *Newton's Law of Cooling:*

The temperature of a body changes at a rate which is proportional to the difference in temperature between the body and its surrounding medium.[2]

---

[2] The law holds for differences of temperature up to 20°C or 36°F.

Let u denote the temperature of a body at time t, the time being measured from the instant the body has been placed in a given surrounding medium whose temperature is maintained at T, $T < u$.[3]   Then according to Newton's law of cooling, we have

$$\frac{du}{dt} = \alpha(u - T), \tag{3.2.1}$$

where α is a constant of proportionality.   Since $du/dt < 0$ when $(u - T) > 0$, we take $\alpha = -k$, $k > 0$, and we write Equation (3.2.1) as

$$\frac{du}{dt} = -k(u - T). \tag{3.2.2}$$

Equation (3.2.2) is one whose variables are easily separated.   Thus,

$$\frac{du}{u - T} = -k\, dt$$

and integration gives[4]

$$u = T + Ce^{-kt}. \tag{3.2.3}$$

The general solution of Equation (3.2.2) is given by Equation (3.2.3).   Note that (3.2.3) involves two constants, namely, the constant of integration C and the proportionality constant k.   Thus two conditions are required to determine these constants.

Example 3.2.1.   A thermometer reading 80°F in a house is taken outside where the temperature of the air is 45°F.   After 3 minutes, the thermometer reading is 60°F.

(a) Find the thermometer reading 6 minutes after it was brought outside.
(b) Find the time required to drop the reading of the thermometer from 80°F
    to 46°F.

SOLUTION 1.   We have

$$\begin{cases} \dfrac{du}{dt} = -k(u - 45), & \tag{3.2.4} \\[2mm] u(0) = 80, \qquad u(3) = 60. & \tag{3.2.5} \end{cases}$$

The solution of (3.2.4) is given by

$$u = 45 + Ce^{-kt}. \tag{3.2.6}$$

---

[3] In temperature problems, we shall asssume, unless stated to the contrary, that the surrounding medium is always kept at a constant temperature.
[4] Equation (3.2.3) also holds when $T > u$, $k > 0$.   In this case, the body absorbs heat from the medium.

From (3.2.6) and the first condition in (3.2.5), we have

$$80 = u(0) = 45 + Ce^{-k \cdot 0} = 45 + C,$$

from which $C = 35$.   Thus, we may write (3.2.6) as

$$u = 45 + 35e^{-kt}. \tag{3.2.7}$$

From (3.2.7) and the second condition in (3.2.5), we have

$$60 = u(3) = 45 + 35e^{-3k},$$

from which we obtain

$$e^{-3k} = \tfrac{3}{7} \quad \text{or} \quad k = -\tfrac{1}{3} \ln \tfrac{3}{7} = \tfrac{1}{3} \ln \tfrac{7}{3} \approx 0.2824.$$

Thus we may write (3.2.7) as

$$u = 45 + 35(\tfrac{3}{7})^{t/3}, \tag{3.2.8}$$

or approximately

$$u = 45 + 35e^{-0.2824t}. \tag{3.2.9}$$

If one has a table of e^{-t}, he may prefer to utilize Equation (3.2.9) over Equation (3.2.8) to determine u when t is given, or to determine t when u is given,

Concerning (a).   Substituting $t = 6$ into (3.2.8), we obtain

$$u = 45 + 35 \left(\frac{9}{49} \right) \approx 51.4°\text{F}. \tag{3.2.10}$$

Thus, after 6 minutes the thermometer reading is approximately 51.4°F.

Concerning (b).   Substituting $u = 46$ into (3.2.9), we obtain

$$e^{-0.2824t} = \frac{1}{35},$$

or

$$t = -\frac{1}{0.2824} \ln \frac{1}{35} = \frac{1}{0.2824} \ln 35 = \frac{1}{0.2824} (3.55535) \approx 12.6 \text{ minutes}. \tag{3.2.11}$$

Thus, it takes approximately 12.6 minutes to drop the reading of the thermometer from 80°F to 46°F.

SOLUTION 2.   *Concerning* (a).   Integrating (3.2.4) between the limits $t = 0$, $u = 80$ and $t = 3$, $u = 60$, we have

$$\int_{80}^{60} \frac{du}{u - 45} = -k \int_{0}^{3} dt.$$

Thus,

$$\ln 15 - \ln 35 = -3k \quad \text{or} \quad k = \tfrac{1}{3} \ln \tfrac{7}{3}. \tag{3.2.12}$$

Again integrating (3.2.4) between the limits of $t = 0, \quad u = 80$ and $t = 6,$ $u = u,$ we obtain

$$\ln (u - 45) - \ln 35 = -6k. \tag{3.2.13}$$

Substituting the value of k found in (3.2.12) into (3.2.13), we obtain

$$\ln (u - 45) = \ln 35 - 2 \ln \tfrac{7}{3} = \ln 35 + \ln \left(\tfrac{3}{7}\right)^2 = \ln \left[35\left(\tfrac{3}{7}\right)^2\right],$$

or

$$u = 45 + 35\left(\tfrac{3}{7}\right)^2 \approx 51.4°F,$$

which is the same answer as in (3.2.10).

Concerning (b).   Integrating (3.2.4) between the limits $t = 0, \quad u = 80$ and $t = t, \quad u = 46,$ we obtain

$$\ln 1 - \ln 35 = -kt \quad \text{or} \quad t = \frac{1}{k} \ln 35. \tag{3.2.14}$$

Substituting the value of k found in (3.2.12) into (3.2.14), we get

$$t = 3 \frac{\ln 35}{\ln \tfrac{7}{3}} \approx \frac{3(3.55535)}{0.84730} \approx 12.6 \text{ minutes},$$

which is the same answer as in (3.2.11).

Mixture Problems.   Suppose that a substance X flows at a certain rate into a container having a certain mixture, and the resulting mixture is now kept uniform (in concentration) by a stirring device.[5]   Suppose that simultaneously the uniform mixture is allowed to run out of the container at a rate that may or may not be the same as the inflow rate of the substance X. [*See* Figure 3.2.1.]   Our problem is to determine the amount of substance X in the container at any given time.

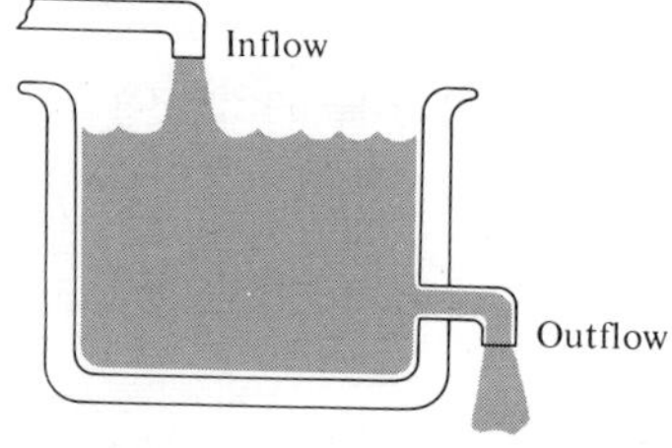

Figure 3.2.1.

---

[5] In mixture problems we shall assume, unless stated to the contrary, that the mixture in the container is always kept at uniform concentration by some stirring device.

Let x denote the amount of substance X in the container at time t.   Then dx/dt is the time rate of change in this amount.   If we denote by α and β, respectively, the rate of inflow and outflow of the substance X, we then have

$$\frac{dx}{dt} = (\textit{the rate of inflow of } X) - (\textit{the rate of outflow of } X)$$

or

$$\frac{dx}{dt} = \alpha - \beta. \tag{3.2.15}$$

Equation (3.2.15) will enable us to determine the amount x of X at any given time.

Example 3.2.2.   Suppose that a container whose capacity is 200 gallons is filled with a 40 percent solution of a certain dye.   A 30 percent solution of the dye is allowed to flow into the container at a rate of 15 gallons per minute and the resulting uniform mixture flows out of the container at the same rate.

(a) Find the amount of dye in the solution at time t.
(b) Find the amount of dye in the solution at the end of 40 minutes.
(c) Find the equilibrium state approached by the mixture.

SOLUTION.   *Concerning* (a).   Let x denote the amount of dye in gallons in the container at time t.   The rate α of inflow of dye is

$$\alpha = (15 \text{ gal/min})\left(\frac{30}{100}\right) = \frac{9}{2} \text{ gal/min}. \tag{3.2.16}$$

Since the rate of inflow is the same as the rate of outflow, there are always 200 gallons in the container.   Also, since there are x gallons of dye at any time t, the composition of the solution is x gallons of dye in 200 gallons of solution.   Thus the rate β of the outflow of dye is

$$\beta = \left(\frac{x \text{ gal}}{200 \text{ gal}}\right)(15 \text{ gal/min}) = \frac{3}{40} x \text{ gal/min}. \tag{3.2.17}$$

Also when $t = 0$, there are $(200 \text{ gal})\left(\frac{40}{100}\right) = 80$ gallons of dye in the container.

Substituting α and β given in (3.2.16) and (3.2.17) into (3.2.15), we have the following initial-value problem

$$\frac{dx}{dt} = \frac{9}{2} - \frac{3}{40} x, \tag{3.2.18}$$

$$x(0) = 80. \tag{3.2.19}$$

We may write (3.2.18) as

$$\frac{dx}{x - 60} = -\frac{3}{40} \, dt,$$

and integration gives

$$\ln (x - 60) = -\frac{3}{40} t + \ln C,$$

or

$$x = 60 + Ce^{-(3/40)t}. \tag{3.2.20}$$

From (3.2.20) and (3.2.19), we have

$$80 = x(0) = 60 + Ce^{-(3/40)\cdot 0} = 60 + C.$$

Thus $C = 20$, and we may write Equation (3.2.20) as

$$x = 60 + 20e^{-(3/40)t}, \tag{3.2.21}$$

which is the answer to part (a).

Concerning (b).   Substitute $t = 40$ in (3.2.21) and we obtain

$$x = 60 + 20e^{-3} \approx 60 + 20(0.04979) \approx 60.996 \text{ gal.}$$

Concerning (c).   When no further change in the solution takes place, we say that the solution has reached its *equilibrium state*.   At the equilibrium state the quantity x present does not depend on the time, so that $dx/dt = 0$. Clearly, the equilibrium state occurs when we left $t \to \infty$ in Equation (3.2.21). We obtain $x = 60$ gallons.   Thus, at equilibrium we have a 30 percent solution of dye in the container.   [*Note:* $x = 60$ gallons also comes from (3.2.18) when we set $x' = 0$.]

Example 3.2.3.   A container contains 100 gallons of brine made by dissolving 70 lb of salt in water.   Brine containing 2 lb of dissolved salt per gallon flows into the container at a rate of 3 gallons per minute and the resulting uniform mixture flows out at a rate of 5 gallons per minute.

(a) Find the amount of salt in the container at time t.
(b) Find the amount of salt in the container at the end of 32 minutes.

SOLUTION.   *Concerning* (a).   Let x denote the amount of salt in the container in pounds at time t.   The rate α of inflow of salt is

$$\alpha = (2 \text{ lb/gal})(3 \text{ gal/min}) = 6 \text{ lb/min.} \tag{3.2.22}$$

In this example, the rate of inflow is different from the rate of outflow.   Since the rate of inflow is 3 gallons per minute, while the rate of outflow is 5 gallons per minute, it is evident that there is a net loss of 2 gallons per minute of brine in the container. Initially the container contains 100 gallons of brine. Clearly, the amount of brine in the container at the end of t minutes is

$$100 - 2t \text{ gal.}$$

Consequently, the concentration at the end of t minutes is

$$\frac{x}{100 - 2t} \text{ lb/gal.}$$

Thus, the rate β of outflow of salt is

$$\beta = \left(\frac{x}{100 - 2t} \text{ lb/gal}\right)(5 \text{ gal/min}) = \frac{5x}{100 - 2t} \text{ lb/min.} \qquad (3.2.23)$$

At $t = 0$, there are 70 lb of salt in the container. Substituting α and β given in (3.2.22) and (3.2.23) into (3.2.15), we have the following initial-value problem

$$\begin{cases} \dfrac{dx}{dt} + \dfrac{5x}{100 - 2t} = 6, & (3.2.24) \\[2mm] x(0) = 70. & (3.2.25) \end{cases}$$

Equation (3.2.24) is a linear differential equation of the type discussed in Section 2.6.1. An integrating factor of Equation (3.2.24) is given by

$$\exp\left[\int \frac{5\,dt}{100 - 2t}\right] = \exp\left[-\frac{5}{2}\ln(100 - 2t)\right] = \exp\left[\ln(100 - 2t)^{-(5/2)}\right]$$

$$= \frac{1}{(100 - 2t)^{5/2}}.$$

Thus, the general solution of Equation (3.2.24) is given by [*see* (2.6.15)]

$$\frac{x}{(100 - 2t)^{5/2}} = \int \frac{6\,dt}{(100 - 2t)^{5/2}} + C = \frac{2}{(100 - 2t)^{3/2}} + C,$$

or

$$x = 2(100 - 2t) + C(100 - 2t)^{5/2}. \qquad (3.2.26)$$

From (3.2.25) and (3.2.26), we have

$$70 = x(0) = 200 + 10^5 C.$$

Thus $C = -13/10{,}000$, and we may write (3.2.26) as

$$x = 2(100 - 2t) - \frac{13}{10{,}000}(100 - 2t)^{5/2}, \qquad (3.2.27)$$

which is the answer to part (a).

Concerning (b).   Substituting $t = 32$ in (3.2.27), we get

$$x = 72 - \frac{13}{10,000} (6)^5 \approx 61.9 \text{ lb.}$$

Thus, at the end of 32 minutes, there are approximately 61.9 pounds of salt in the container.

Exercises 3.2

1. An alloy at temperature 100°C is brought into a medium whose temperature is kept at 80°C.   After 20 minutes, the temperature of the alloy is 90°C.   Find the time required to drop the temperature of the alloy from 100°C to 85°C.
2. An alloy at temperature 100°C is brought into a medium whose temperature is kept at T°C.   After 5 minutes, the temperature of the alloy is 90°C.   Five minutes later its temperature is 86°C.   What is the temperature of the medium?
3. A liquid initially at temperature 35°F requires 10 minutes to raise its temperature to 50°F in a medium kept at 70°F.   (a) Find the temperature of the liquid at the end of 30 minutes.   (b) Find the time required to bring the temperature of the liquid to 65°F.
4. A liquid at temperature 50°F requires 5 minutes to lower its temperature to 40°F in a medium kept at 25°F.   In the same medium kept at 30°F, how long will it take to lower the temperature of the liquid from 60°F to 48°F?
5. At 1:00 p.m., an alloy at temperature 10°C is placed into a freezer kept at -10°C.   At 1:15 p.m., the temperature of the alloy is -2°C.   Later, the alloy is brought into a room whose temperature is kept at 10°C.   At 1:50 p.m., the temperature of the alloy is 8°C.   At what time was the alloy placed in the room?
6. A law which may be used for greater temperature ranges than Newton's Law of Cooling is known as *Stefan's Law of Cooling* which states:

The temperature of a body changes at a rate which is proportional to the difference of the fourth powers of the temperature of the body and that of the surrounding medium.

   Using the notation given for the temperature problem of Newton's Law of Cooling, we have

(1)
$$\frac{du}{dt} = -k(u^4 - T^4).$$

(a) Show that the general solution of (1) is given by

(2)
$$\ln\left(\frac{u + T}{u - T}\right) + 2 \text{ arc tan}\left(\frac{u}{T}\right) = 4T^3 kt + C,$$

where C is a constant of integration.

    (b) An alloy at temperature 200°C is brought into a medium whose temperature is kept at 100°C. After 20 minutes, the temperature of the alloy is 150°C. Find the time required to drop the temperature of the alloy from 200°C to 120°C.

7. A container is filled to its capacity with 200 gallons of water. Pure water flows into the container at a rate of 2 gallons per minute and simultaneously brine containing 1/2 lb of salt per gallon flows into the container at a rate of 2 gallons per minute. The resulting uniform mixture flows out of the container at a rate of 4 gallons per minute. (a) Find the amount of salt in the container at the end of 30 minutes. (b) Find the equilibrium state approached by the mixture.

8. A container whose capacity is 100 gallons is filled with a 60 percent solution of a certain dye. Pure water flows into the container at a rate of 3 gallons per minute. The resulting uniform mixture flows into a second container initially containing 100 gallons of pure water at a rate of 3 gallons per minute. The resulting uniform mixture in the second container flows out also at a rate of 3 gallons per minute. Find the amount of dye in the second container at the end of 50 minutes.

9. A tank contains 200 gallons of pure water. Brine containing 1/3 lb of dissolved salt per gallon flows from a pipe into the tank at a rate of 3 gallons per minute, and simultaneously from another pipe, brine containing 3/5 lb of dissolved salt per gallon flows into the tank at a rate of 5 gallons per minute. The resulting uniform mixture flows out of the tank at a rate of 6 gallons per minute. Find the amount of salt and the concentration in the tank at the end of $3\frac{1}{3}$ hours.

10. The air in a recently used office room 60 ft $\times$ 60 ft $\times$ 30 ft tested 0.20 percent by volume of carbon dioxide. How many cubic feet of fresh air must be admitted per minute in order that 20 minutes later the office room will contain 0.08 percent by volume of carbon dioxide? (Assume that fresh air contains 0.04 percent by volume of carbon dioxide. Also assume that there is uniform mixture of fresh air and stale air, and that the rate of stale air leaving the room by leakage through doors and windows is the same as the rate at which fresh air enters the room.)

11. Two identical tanks have capacity G gallons each. Initially, the first is filled with brine containing p pounds of salt, while the second is filled with water. Solution is pumped from each tank to the other at the same rate of r gallons per minute. Find the amount of salt $x(t)$ (in the first tank) and $y(t)$ (in the second tank) at time t.

3.3. The Law of Mass Action and Chemical Reactions

The law of mass action states that *the rate (or velocity) of a chemical reaction is proportional to the molar concentration of the reacting substances.* This law is due to Guldberg and Waage. In the application of this law to chemical reactions, one needs additional assumptions, for example, that during a chemical reaction the volume and temperature are kept constant.

Recall from elementary chemistry that the number of grams equal to the molecular weight of a substance is called a *mole*. The (*molar*) *concentration* of a substance is equal to the number of moles per unit volume, usually moles per liter.[6]

Example 3.3.1. Let us determine how many moles are represented by 10.5 grams (g) of SO_2 (sulfur dioxide).

SOLUTION. Atomic weight of $S = 32.06$ and of $O = 16.00$. The molecular weight of $SO_2 = 32.06 + 2 \times 16 = 64.06$. Thus, 1 mole of $SO_2 = 64.06$ g of SO_2.

$$\text{Moles of } SO_2 = \frac{\text{grams of } SO_2}{\text{molecular weight of } SO_2}$$

$$= \frac{10.5 \text{ g}}{64.06 \text{ g/mole}} = 0.164 \text{ moles } SO_2.$$

Now if we dissolve 10.5 g of SO_2 in one liter (l) of a solution, the concentration of the solution [also known as the *molarity* (M) of the solution] is 0.164 molar or 0.164 M. Also

$$\text{molarity of a solution} = \frac{\text{number of moles of solute}}{\text{number of liters of solution}}. \qquad (3.3.1)$$

For example, if we dissolve 10.5 g of SO_2 in 250 ml or 0.250 l of a solution, then

$$\text{molarity of the solution} = \frac{0.164 \text{ mole}}{0.250 \text{ l}} = 0.656 \text{ mole/l} = 0.656 \text{ M}.$$

Equilibrium constant. Let us consider a perfectly general reversible reaction

$$A + B \; \rightleftarrows \; F + G, \qquad (3.3.2)$$

where A and B denote the *reactants*, F and G denote the *products*. Also the symbol "$\rightarrow$" denotes the *forward reaction*, while the symbol "$\leftarrow$" denotes the *reverse reaction*. Let C_A, C_B, C_F, and C_G denote, respectively, the (molar) concentrations of A, B, F, and G. In view of the Law of Mass Action, we have

$$\text{rate of forward reaction} = kC_A C_B$$

and $\qquad\qquad\qquad\qquad\qquad\qquad\qquad\qquad\qquad\qquad\qquad\qquad (3.3.3)$

$$\text{rate of reverse reaction} = k'C_F C_G,$$

where k and k' denote, respectively, the proportionality constants of the forward and reverse reactions.

---

[6] 1 liter (l) $= 1000$ milliliters (ml) $= 1000$ cubic centimeters (cc). 1 ml $= 1$ cc $= \dfrac{1}{1000}$ l.

By definition, a state of *chemical equilibrium* is attained when the rates of forward and reverse reactions are equal.   Thus, at a state of chemical equilibrium, we must have in view of Equation (3.3.3) that

$$kC_A C_B = k'C_F C_G, \tag{3.3.4}$$

where C_A, C_B, C_F, and C_G are now, respectively, the (molar) concentrations of A, B, F, and G at equilibrium.   Equation (3.3.4) may be written as

$$\frac{C_F C_G}{C_A C_B} = \frac{k}{k'} = K_C, \tag{3.3.5}$$

where K_C denotes the *equilibrium constant* of the reaction (3.3.2).

Example 3.3.2.   Suppose that 2 moles of C_2H_5OH (ethyl alcohol) and 2 moles of CH_3COOH (acetic acid) are mixed at 25°C.   When equilibrium is reached there are 4/3 moles of $CH_3COOC_2H_5$ (ethyl acetate, commonly known as ester) and 4/3 moles of H_2O (water).   Let us determine the equilibrium constant K_C.

SOLUTION.   We have the chemical reaction

$$C_2H_5OH + CH_3COOH \quad \rightleftharpoons \quad CH_3COOC_2H_5 + H_2O.$$

Let V denote the volume of the solution in liters.   At equilibrium, we have

$$C_{\text{alcohol}} = C_{\text{acid}} = \frac{(2 - \frac{4}{3})}{V} = \frac{\frac{2}{3}}{V} \text{ mole/liter},$$

$$C_{\text{ester}} = C_{\text{water}} = \frac{\frac{4}{3}}{V} \text{ mole/liter}.$$

Utilizing Formula (3.3.5), we have

$$K_C = \frac{C_{\text{ester}} C_{\text{water}}}{C_{\text{alcohol}} C_{\text{acid}}} = \frac{\left(\frac{\frac{4}{3}}{V}\right)\left(\frac{\frac{4}{3}}{V}\right)}{\left(\frac{\frac{2}{3}}{V}\right)\left(\frac{\frac{2}{3}}{V}\right)} = 4.00. \tag{3.3.6}$$

In the foregoing discussion, the process considered involved only a single molecule of each of the reactants and products.   Consider now the following reaction

$$A + 3B \quad \rightleftharpoons \quad 2F. \tag{3.3.7}$$

Equation (3.3.7) may be written equivalently as

$$A + B + B + B \quad \rightleftharpoons \quad F + F. \tag{3.3.8}$$

Applying the Law of Mass Action to Equation (3.3.8), we have

$$\text{rate of forward reaction} = kC_A C_B C_B C_B = kC_A(C_B)^3,$$

and   (3.3.9)

$$\text{rate of reverse reaction} = k'C_F C_F = k'(C_F)^2.$$

At the state of chemical equilibrium, we then have in view of (3.3.9) that

$$kC_A(C_B)^3 = k'(C_F)^2,$$

or

$$\frac{(C_F)^2}{C_A(C_B)^3} = \frac{k}{k'} = K_C.$$ (3.3.10)

The above method for determining the equilibrium constant K_C extends to a general reaction.   Let us consider the reaction

$$aA + bB + \cdots \; \rightleftharpoons \; mM + nN + \cdots,$$ (3.3.11)

where $a, b, \ldots, m, n \ldots$ are positive integers signifying the number of molecules taking part in the reaction.   Then the expression of the equilibrium constant K_C, in terms of its concentration, is given by

$$\frac{C_M{}^m C_N{}^n \cdots}{C_A{}^a C_B{}^b \cdots} = \frac{k}{k'} = K_C.$$ (3.3.12)

Classification of chemical reactions.   One method of classifying chemical reactions is according to their *molecularity*.   By molecularity we mean the number of atoms or molecules taking part in each act leading to a chemical reaction.   A *unimolecular* reaction is one in which only one molecule reacts at a time.   For example, the dissociation of gaseous bromine (Br_2):

$$Br_2 \longrightarrow Br + Br.$$

A *bimolecular* reaction is one in which two molecules react.   For example, the dissociation of gaseous hydrogen iodide (HI):

$$2HI \longrightarrow H_2 + I_2.$$

A *termolecular* reaction is one which involves the reaction of three molecules. For example,

$$2NO + O_2 \longrightarrow 2NO_2.$$

Here 2 molecules of gaseous nitric oxide (NO) react with 1 molecule of gaseous oxygen (O_2) to form 2 molecules of gaseous nitric dioxide (NO_2).

First-order reactions.   A reaction is said to be of the *first order* if the rate (or velocity) of the reaction is proportional to the concentration of the reacting substance.   In view of the Law of Mass Action, we then have

$$\frac{dx}{dt} = \alpha x, \tag{3.3.13}$$

where x denotes the concentration of the reactant at time t and α is the proportionality constant.   Since the concentration of the reacting substance decreases with time, we take $\alpha = -k, \quad k > 0$, and write Equation (3.3.13) as

$$\frac{dx}{dt} = -kx. \tag{3.3.14}$$

The constant k is called the *velocity constant* or *rate constant*.   The general solution of Equation (3.3.14) is given by

$$x = Ce^{-kt}, \tag{3.3.15}$$

where C is an arbitrary constant of integration.   Suppose that at $t = 0$, the concentration of the reactant is x_0.   Then

$$x_0 = Ce^{-k \cdot 0} = C.$$

Thus we may write Equation (3.3.15) as

$$x = x_0 e^{-kt}. \tag{3.3.16}$$

Solving for the velocity constant k in (3.3.16), we obtain

$$k = \frac{1}{t} \ln\left(\frac{x_0}{x}\right). \tag{3.3.17}$$

[*Note:* The unit of k is $(\text{time})^{-1}$.]

Remark 3.3.1.   In general, whenever one shows that a given chemical reaction is of the first order, one may assume the reaction to be unimolecular.   Also, a reaction is said to be of the first order if whenever a set of values, determined by experimental data, are substituted into the right-hand member of Equation (3.3.17),[7] the value of k remains essentially (or sufficiently) constant.   We shall illustrate this point with the following example.

Example 3.3.3.   Let us verify that the experimental data given below obtained by Van't Hoft, by observing the decomposition of 5.11 grams of dibromsuccinic acid in hot water, gives rise to a reaction of the first order.

---

[7] Or one may use the equations in Exercises 3.3.1(a) and 3.3.2.

t	0	10	20	30	40	50	60
x	5.11	3.77	2.74	2.02	1.48	1.08	0.80

Here the time t is expressed in minutes.

SOLUTION.   Since we assume that the volume remains constant during this experiment, it is not necessary to change the grams of dibromsuccinic acid, at various times, to moles per liter.   The expression x_0/x, in Equation (3.3.17), will be the same whether x_0 and x are expressed in grams or moles per liter.   Thus, we have

$$
\left.
\begin{aligned}
k &= \frac{1}{10} \ln\left(\frac{5.11}{3.77}\right) = 0.03041. \\[2ex]
k &= \frac{1}{20} \ln\left(\frac{5.11}{2.74}\right) = 0.03116. \\[2ex]
k &= \frac{1}{30} \ln\left(\frac{5.11}{2.02}\right) = 0.03094. \\[2ex]
k &= \frac{1}{40} \ln\left(\frac{5.11}{1.48}\right) = 0.03098. \\[2ex]
k &= \frac{1}{50} \ln\left(\frac{5.11}{1.08}\right) = 0.03108. \\[2ex]
k &= \frac{1}{60} \ln\left(\frac{5.11}{0.80}\right) = 0.03093.
\end{aligned}
\right\}
\qquad (3.3.18)
$$

Since k is essentially (or sufficiently) constant (within the limit of experimental error), the above reaction may be considered of the first order.

Second-order reactions.   The expression that z is proportional to the product of x and $\cdot y$ means that $z = \alpha xy$, where α is a constant of proportionality.   A reaction is said to be of the *second order* if the rate of the reaction is proportional to the product of the concentrations of two reacting substances.   If

$$
A + B \quad \longrightarrow \quad AB, \qquad (3.3.19)
$$

then

$$
\frac{dC_A}{dt} = \frac{dC_B}{dt} = -kC_A C_B, \qquad k > 0. \qquad (3.3.20)
$$

If α and β denote, respectively, the number of moles per liter of A and B initially present, and if x denotes the number of moles per liter of A or B that have reacted in time t, then $\alpha - x$ and $\beta - x$ are the concentrations of A

and B remaining at time t.   Since $dx/dt > 0$, we have in view of Equation (3.3.20) that

$$\frac{dx}{dt} = k(\alpha - x)(\beta - x). \tag{3.3.21}$$

Case 1.   *When $\alpha \neq \beta$ in Equation* (3.3.21).   We may write (3.3.21) as

$$k\,dt = \frac{dx}{(\alpha - x)(\beta - x)}.$$

Utilizing partial fraction decomposition on the right-hand member of the above equation, we obtain

$$k\,dt = \frac{1}{\beta - \alpha}\frac{dx}{\alpha - x} + \frac{1}{\alpha - \beta}\frac{dx}{\beta - x}. \tag{3.3.22}$$

Integrating (3.3.22) and then simplifying the result, we get

$$kt + C = \frac{1}{\alpha - \beta}\ln\!\left(\frac{\alpha - x}{\beta - x}\right), \tag{3.3.23}$$

where C is an arbitrary constant of integration.   Since $x = 0$ when $t = 0$, we readily find from (3.3.23) that

$$C = \frac{1}{\alpha - \beta}\ln\!\left(\frac{\alpha}{\beta}\right).$$

Substituting this value of C in (3.3.23) and then solving for the velocity constant k, we obtain upon simplification

$$k = \frac{1}{t(\alpha - \beta)}\ln\!\left[\frac{\beta(\alpha - x)}{\alpha(\beta - x)}\right]. \tag{3.3.24}$$

[*Note:* If t is given in seconds, the velocity constant k has units liters/mole-seconds.]

Remark 3.3.2.   In general, whenever one shows that a given chemical reaction is of the second order, one may assume the reaction to be bimolecular. Also, a reaction is said to be of the second order if whenever a set of values, determined by experimental data, are substituted into the right-hand member of Equation (3.3.24), the value of k remains essentially (or sufficiently) constant.

Case 2.   *When $\alpha = \beta$ in Equation* (3.3.21).   This case will be left as an exercise for the reader.   [*See* Exercise 3.3.4.]   Also, Remark 3.3.2 applies to Case 2.

Example 3.3.4. In the reaction of ethyl acetate with sodium hydroxide,

$$CH_3COOC_2H_5 + NaOH \longrightarrow C_2H_5OH + CH_3COONa,$$

| ethyl acetate | sodium hydroxide | ethyl alcohol | sodium acetate |

Reicher compiled the following experimental data.

t	0	393	699	1010	1265
$\alpha - x$	0.5638	0.4866	0.4467	0.4113	0.3879
$\beta - x$	0.3114	0.2342	0.1943	0.1589	0.1354

Here α and β denote, respectively, the initial concentration in moles per liter of sodium hydroxide and ethyl acetate, and x is the concentration in moles per liter of ethyl alcohol at time t. Let us verify that the reaction is of the second order. Here t is expressed in minutes.

SOLUTION. Since at $t = 0$, we have $x = 0$, we see from the above table that $\alpha = 0.5638$ and $\beta = 0.3114$. Thus,

$$\frac{1}{\alpha - \beta} = 3.9619.$$

Utilizing Equation (3.3.24), we have, in view of the above experimental data,

$$k = \frac{3.9619}{393} \ln\left[\frac{(0.3114)(0.4866)}{(0.5638)(0.2342)}\right] = 0.00139.$$

$$k = \frac{3.9619}{699} \ln\left[\frac{(0.3114)(0.4467)}{(0.5638)(0.1943)}\right] = 0.00135.$$

$$k = \frac{3.9619}{1010} \ln\left[\frac{(0.3114)(0.4113)}{(0.5638)(0.1589)}\right] = 0.00140.$$

$$k = \frac{3.9619}{1265} \ln\left[\frac{(0.3114)(0.3879)}{(0.5638)(0.1354)}\right] = 0.00144.$$

$$(3.3.25)$$

Since k is sufficiently constant, the above reaction may be considered to be of the second order.

Exercises 3.3

1. (a) Consider the unimolecular or first-order reaction given by

$$\begin{cases} \dfrac{d(\alpha - x)}{dt} = -k(\alpha - x), \\[2mm] x(0) = 0, \end{cases}$$

where α denotes the initial concentration of a given reactant and x denotes the amount (per unit volume) reacted by time t.   [*Note:* $\alpha - x$ denotes the concentration of the reactant at time t.]   Show that the velocity constant is given by

$$k = \frac{1}{t} \ln\left(\frac{\alpha}{\alpha - x}\right).$$

(b) *Half-life.*   In part (a), let $t_{1/2}$ denote the time necessary for half (or 50 percent) of the reactant to react.   Show that

$$t_{1/2} = \frac{1}{k} \ln 2.$$

2. Consider the unimolecular reaction or first-order reaction given by

$$\begin{cases} \dfrac{dx}{dt} = -kx, \\[2mm] x(t_1) = x_1, \qquad x(t_2) = x_2, \end{cases}$$

where x_1 and x_2 denote, respectively, the concentration of the given reactant at times t_1 and t_2, $t_1 < t_2$.   Show that the velocity constant k is given by

$$k = \frac{1}{t_2 - t_1} \ln\left(\frac{x_1}{x_2}\right).$$

3. Refer back to Example 3.3.3.   Take the average of k given in (3.3.18).   With this value of k, find the amount of dibromsuccinic acid when $t = 38$ minutes.

4. (a) Show that if $\alpha = \beta$ in Equation (3.3.21), then the velocity constant k is given by

$$k = \frac{x}{t\alpha(\alpha - x)}.$$

(b) *Half-life.*   In part (a), let $t_{1/2}$ denote the time necessary for half (or 50 percent) of the reactant to react.   Show that

$$t_{1/2} = \frac{1}{k\alpha}.$$

5. A certain bimolecular reaction in which $\alpha = \beta = 2$ moles per liter is 20 percent complete in 20 minutes.   Find the time required for the reaction to be 50 percent complete.

6. Refer back to Example 3.3.4.   Take the average of k given in (3.3.25).   With this value of k, find the amount of grams of ethyl alcohol when $t = 855$ minutes. The molecular weight of ethyl alcohol is 46.068 g/mole.

7. Consider the termolecular reaction or the third-order reaction

$$A + B + D \longrightarrow M + N.$$

Then, in view of the Law of Mass Action, we have

$$\text{(1)} \qquad \frac{dC_A}{dt} = \frac{dC_B}{dt} = \frac{dC_D}{dt} = -kC_A C_B C_D, \qquad k > 0.$$

If α, β, and γ denote, respectively, the number of moles per liter of A, B, and D initially present, and if x denotes the number of moles per liter of A, B, and D which have reacted at time t, then $\alpha - x$, $\beta - x$, and $\gamma - x$ are the concentrations of A, B, and D at time t.   Thus,

$$(2) \qquad \begin{cases} \dfrac{dx}{dt} = k(\alpha - x)(\beta - x)(\gamma - x), \\[2ex] x(0) = 0. \end{cases}$$

(3)

Case 1.   *When α, β, and γ are equal in* (2).   Show that

$$(4) \qquad kt = \frac{x(2\alpha - x)}{2\alpha^2(\alpha - x)^2} \,.$$

Case 2.   *When only two of α, β, and γ are equal in* (2).   In particular, say, when $\gamma = \beta$, show that

$$(5) \qquad kt = \frac{1}{\alpha - \beta}\left\{ \frac{1}{\beta - x} - \frac{1}{\alpha - \beta}\ln\left[\frac{\beta(\alpha - x)}{\alpha(\beta - x)}\right] - \frac{1}{\beta} \right\}.$$

Equation (2), with $\gamma = \beta$, was studied by Goldschmidt in the chemical reaction between HCl (hydrochloric acid) and C_2H_5OH (ethyl alcohol).

Case 3.   *When α, β, and γ are not equal in* (2).   Show that

$$(6) \qquad \left(1 - \frac{x}{\alpha}\right)^{\gamma - \beta} \left(1 - \frac{x}{\beta}\right)^{\alpha - \gamma} \left(1 - \frac{x}{\gamma}\right)^{\beta - \alpha} = e^{(\gamma - \beta)(\alpha - \gamma)(\beta - \alpha)kt}.$$

8. In the decomposition of dibromsuccinic acid in aqueous solution, Van't Hoft compiled the following experimental data.

t	0	20	30	40	50	60
$\alpha - x$	5.11	2.74	2.02	1.48	1.08	0.80

Here α denotes the number of cc of dibromsuccinic acid initially in the aqueous solution, and x is the number of cc of dibromsuccinic acid that has decomposed in t minutes.

(a) Utilize the result of Exercise 1(a) to verify that the above experimental data gives rise to a first-order differential equation.

(b) Show that the average of the velocity constant k found in part (a) is equal to 0.03101.   With this value of k, find the number of cc of dibromsuccinic acid remaining in the solution after 35 minutes.

9. Under suitable conditions, the velocity of the reaction between $HBrO_3$ (bromic acid) and HBr (hydrobromic acid) is given by

$$\begin{cases} \dfrac{dx}{dt} = k(n\alpha + x)(\alpha - x) \qquad (n \neq -1), \\[2ex] x(0) = 0. \end{cases}$$

Show that

$$kt = \frac{1}{\alpha(n+1)} \ln\left[\frac{n\alpha + x}{n(\alpha - x)}\right].$$

10. Upon mixing a reactant A with a reactant B in the ratio of 4 to 3, a product C is formed. If 48 grams of A react with 24 grams of B to form 36 grams of C in 3 minutes, (a) find the amount C present at time t. (b) How many minutes will it take to form 42 grams of C?

3.4. Electrical Circuits

The following table gives the quantities, symbols, and units that are commonly used in circuit analysis or electrical network theory.

Quantity	Symbol	Unit
Resistance	R	ohm
Inductance	L	henry
Capacitance	C	farad
Voltage	E	volt
Current	I	ampere
Charge	Q	coulomb

The elements (or components) of an electrical network are usually represented by the following conventional symbols:

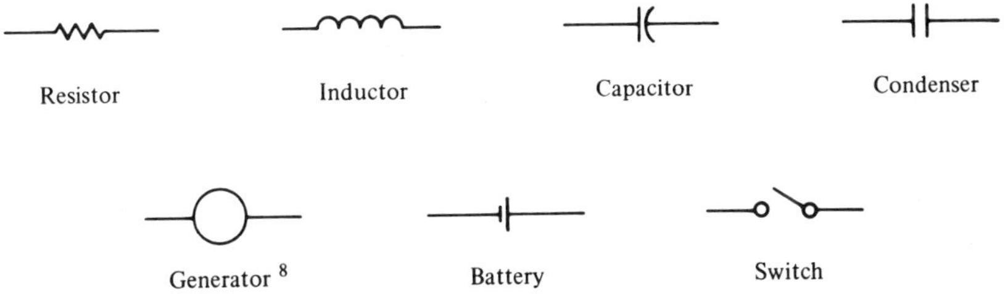

Remark 3.4.1. A resistor is an element that transforms electrical energy into heat. An inductor is an element that opposes a change in current; it stores electromagnetic energy. A capacitor is an element that stores electrical potential energy. The agency causing the electric current to flow in an electric circuit is called electromotive force (abbreviated emf). Batteries and electric generators are the usual sources of emf which set up currents in electric circuits. The mks unit of emf is a volt (1 joule per coulomb).

---

[8] A voltage of the form $E_0 \cos \omega t$ (or $E_0 \sin \omega t$) reverses its sign with time. In this case, no polarity $(+, -)$ is indicated on the generator.

The R-L-C series circuit.   We shall develop the differential equation for the circuit (or network) shown in Figure 3.4.1, which is known as an *R-L-C series circuit.*   The following relation is known to exist between $Q(t)$, the charge on the capacitor, and $I(t)$, the current in the circuit,

$$Q'(t) = I(t), \qquad (3.4.1)$$

that is, the current $I(t)$ is the rate of change of the charge $Q(t)$.   Also, the following basic facts are known from fundamental principles of physics.

The voltage drops, due to (1) *the resistor,* (2) *the inductor, and* (3) *the capacitor are equal, respectively, to*

$$(1') \; RI(t) \qquad (2') \; LI'(t), \qquad \text{and} \qquad (3') \frac{1}{C} Q(t), \qquad (3.4.2)$$

where R, L, and C are positive constants.

In the ensuing analysis we shall need Kirchhoff's *Voltage Law* which states:

In a closed circuit the sum of all the voltage drops is equal to the sum of all voltage increases.

Thus, if at $t = 0$, the switch S is closed and simultaneously a voltage source $E(t)$ is applied to the circuit given in Figure 3.4.1, we then have at time t, in view of (3.4.2) and the above law, that

$$RI(t) + LI'(t) + \frac{1}{C} Q(t) = E(t). \qquad (3.4.3)$$

Utilizing (3.4.1), we may write Equation (3.4.3) as

$$LQ''(t) + RQ'(t) + \frac{1}{C} Q(t) = E(t), \qquad (3.4.4)$$

which is a linear second-order differential equation.

For the present, we shall restrict ourselves to circuit problems for which Equation (3.4.4) reduces to a linear first-order differential equation.   Thus,

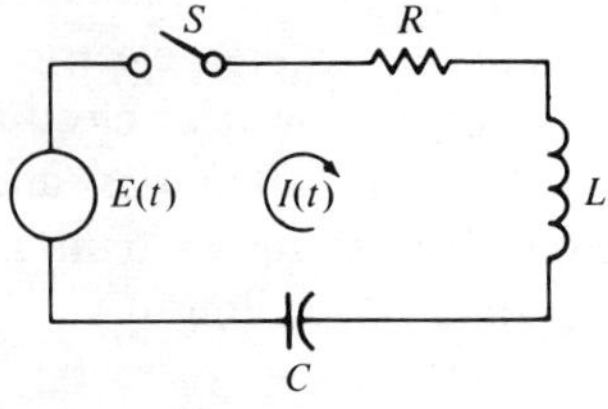

Figure 3.4.1.

we shall be primarily interested in the following two special cases of Equation (3.4.4). [*See* Section 5.7 for the general solution of Equation (3.4.4).]

Case 1. *The R-L series circuit.* Since there is no capacitor in this circuit, the term $Q(t)/C$ in (3.4.2) does not appear. Thus, Equation (3.4.4) may be written as

$$LQ''(t) + RQ'(t) = E(t),$$

and in view of (3.4.1), this equation becomes

$$LI'(t) + RI(t) = E(t), \qquad (3.4.5)$$

which is a linear first-order equation.

Case 2. *The R-C series circuit.* Since there is no inductance in this circuit, the term $LI'(t)\,[=LQ''(t)]$ in (3.4.2) does not appear. Thus (3.4.4) reduces to

$$RQ'(t) + \frac{1}{C}\,Q(t) = E(t), \qquad (3.4.6)$$

which is also a linear first-order equation for the instantaneous charge.

Once $Q(t)$ is determined from Equation (3.4.6), one may obtain $I(t)$ by utilizing (3.4.1). However, if we first differentiate Equation (3.4.6) and then use (3.4.1), we obtain (assuming that E is a differentiable function of t)

$$RI'(t) + \frac{1}{C}\,I(t) = E'(t), \qquad (3.4.7)$$

from which $I(t)$ may now be determined directly in an *R-C* series circuit.

Remark 3.4.2. *Kirchhoff's Voltage Law* is often stated in the following equivalent form:

The algebraic sum of the electromotive forces around a closed circuit in a specific direction is zero.

This law is a statement concerning the conservation of energy in the circuit. Perhaps it would be somewhat instructive to again derive Equation (3.4.4) by utilizing the above law. The quantities, symbols, and units will be the same as those given in the previous discussion.

The algebraic sign of the electromotive force (emf) or current indicates its direction. Let us arbitrarily choose the clockwise direction as the positive direction of the emf and the current. In addition to the electromotive forces (emfs) that are applied externally to the circuit by means of, say, generators and batteries, there are also emfs due to the current in the circuit elements. The following basic facts are known.

The electromotive forces, due to (a) *the resistor,* (b) *the inductor,* (c) *the capacitor, and* (d) *the generator are equal, respectively, to*

$$\text{(a')} -RI(t), \quad \text{(b')} -LI'(t), \quad \text{(c')} -\frac{1}{C}Q(t), \quad \text{and} \quad \text{(d')} E(t). \qquad (3.4.2')$$

In view of the above law, we then have for the circuit given in Figure 3.4.1 (when the switch S is closed)

$$-RI(t) - LI'(t) - \frac{1}{C}Q(t) + E(t) = 0. \qquad (3.4.3')$$

Utilizing (3.4.1), we may write Equation (3.4.3') as

$$LQ''(t) + RQ'(t) + \frac{1}{C}Q(t) - E(t) = 0, \qquad (3.4.4')$$

which is the same as Equation (3.4.4).

Example 3.4.1.   Let us solve for the current $I(t)$ in the initial-value problem

$$\begin{cases} LI'(t) + RI(t) = E_0 \sin \omega t, & (3.4.8) \\ I(0) = I_0, & (3.4.9) \end{cases}$$

where E_0 (volts) and ω (radians/sec) are constants.

SOLUTION.   The above problem is an R-L series circuit in which the initial current is I_0 and the emf, $E = E_0 \sin \omega t$. Note that the emf, $E = E_0 \sin \omega t$, has the following properties.   (1) It is a simple harmonic function of t (that is, the function E satisfies the differential equation $d^2E/dt^2 + \omega^2 E = 0$, $\omega > 0$). (2) Its amplitude (the maximum displacement of the graph of the function E from the t axis) is E_0. (3) Its period [the time interval between two successive maxima (or minima)] is $2\pi/\omega$. (4) Its frequency (the reciprocal of the period or the number of cycles per second) is $\omega/2\pi$.   [*See* Figure 3.4.2 for the case $\omega = 2$. Unless stated to the contrary, the units of ω are rad/sec.]

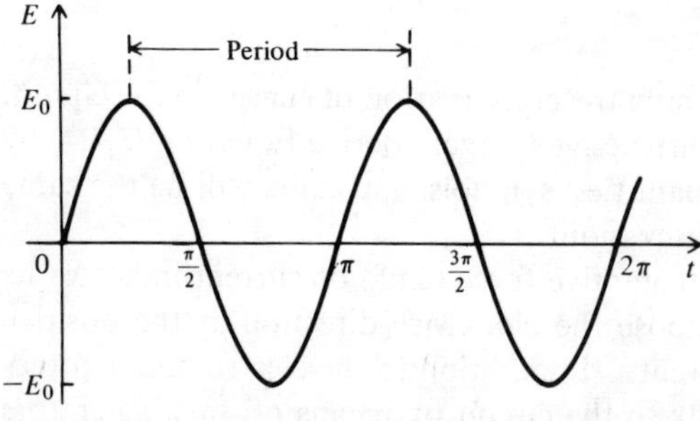

Figure 3.4.2.
$E = E_0 \sin 2t$

Let us write Equation (3.4.8) as

$$I'(t) + \frac{R}{L} I(t) = \frac{E_0}{L} \sin \omega t. \tag{3.4.10}$$

An integrating factor of Equation (3.4.10) is given by [*see* Equation (2.6.13)]

$$\exp\left[\frac{R}{L} \int dt\right] = e^{(R/L)t}.$$

Thus, the general solution of (3.4.10) is given by [*see* Equation (2.6.15) or (2.6.2)]

$$I(t) = e^{-(R/L)t} \int e^{(R/L)t} \frac{E_0}{L} \sin \omega t \, dt + Ce^{-(R/L)t}.$$

Integrating the above integral by parts and simplifying, we find

$$I(t) = E_0 \frac{R \sin \omega t - \omega L \cos \omega t}{R^2 + \omega^2 L^2} + Ce^{-(R/L)t}. \tag{3.4.11}$$

Since from (3.4.9), $I(0) = I_0$, we see that

$$C = I_0 + \frac{E_0 \omega L}{R^2 + \omega^2 L^2},$$

and Equation (3.4.11) becomes

$$I(t) = E_0 \frac{R \sin \omega t - \omega L \cos \omega t}{R^2 + \omega^2 L^2} + \left(I_0 + \frac{E_0 \omega L}{R^2 + \omega^2 L^2}\right) e^{-(R/L)t}. \tag{3.4.12}$$

We shall write Equation (3.4.12) in a more convenient form. First, write Equation (3.4.12) as

$$I(t) = \frac{E_0}{\sqrt{R^2 + \omega^2 L^2}} \left(\frac{R}{\sqrt{R^2 + \omega^2 L^2}} \sin \omega t - \frac{\omega L}{\sqrt{R^2 + \omega^2 L^2}} \cos \omega t\right)$$

$$+ \left(I_0 + \frac{E_0 \omega L}{R^2 + \omega^2 L^2}\right) e^{-(R/L)t}. \tag{3.4.13}$$

Let ϕ, $0 \leqq \phi < \dfrac{\pi}{2}$, be such that

$$\tan \phi = \frac{\omega L}{R}. \tag{3.4.14}$$

From (3.4.14), we readily find $\sin \phi$ and $\cos \phi$ by constructing a right triangle as shown in Figure 3.4.3. Thus,

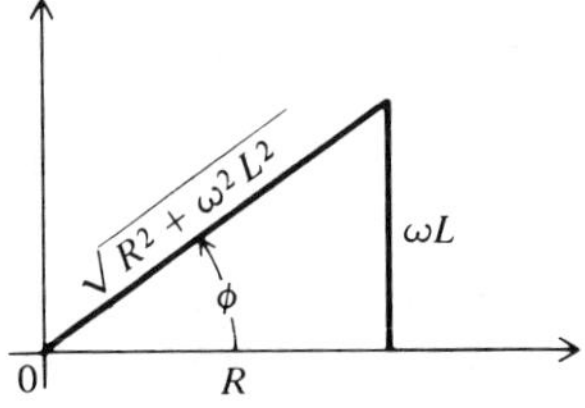

Figure 3.4.3.

$$\sin \phi = \frac{\omega L}{\sqrt{R^2 + \omega^2 L^2}},$$

and   (3.4.15)

$$\cos \phi = \frac{R}{\sqrt{R^2 + \omega^2 L^2}}.$$

Since $\sin (A - B) = \sin A \cos B - \sin B \cos A$, utilizing (3.4.15) we may then write Equation (3.4.13) as

$$I(t) = \frac{E_0}{\sqrt{R^2 + \omega^2 L^2}} \sin (\omega t - \phi) + \left(I_0 + \frac{E_0 \omega L}{R^2 + \omega^2 L^2}\right) e^{-(R/L)t}. \qquad (3.4.16)$$

Equation (3.4.16) is the desired form of the solution for the given initial-value problem.

Remark 3.4.3.   It is apparent that the second term in the right-hand member of Equation (3.4.16) becomes exceedingly small as t becomes large. It is called the *transient* term. The remaining part of the solution is called the *steady-state* term. The steady-state current, namely,

$$I_s(t) = \frac{E_0}{\sqrt{R^2 + \omega^2 L^2}} \sin (\omega t - \phi), \qquad (3.4.17)$$

is periodic with the same period as the emf: $E_0 \sin \omega t$. The constant ϕ in (3.4.17) represents the *phase lag* (or *phase angle*) of the current behind the applied emf. The amplitude of I_s is $E_0/\sqrt{R^2 + \omega^2 L^2}$ which is less than E_0/R, the value it would have been if there were no inductance.

The emf will have maxima, namely, its amplitude E_0, when $\sin \omega t = 1$, that is, when

$$\omega t = \frac{\pi}{2} + 2n\pi, \quad \text{or when} \quad t = \frac{1}{\omega}\left(\frac{\pi}{2} + 2n\pi\right), \qquad n = 0, 1, 2, \ldots . \qquad (3.4.18)$$

The steady-state current I_s will have maxima when $\sin(\omega t - \phi) = 1$, that is, when

$$\omega t - \phi = \frac{\pi}{2} + 2n\pi, \quad \text{or when} \quad t = \frac{1}{\omega}\left(\frac{\pi}{2} + 2n\pi\right) + \frac{\phi}{\omega}, \qquad n = 0, 1, 2, \dots. \tag{3.4.19}$$

Thus, the steady-state current lags behind the emf by a time of ϕ/ω. The steady-state current is alternately positive and negative, and oscillates between $E_0/\sqrt{R^2 + \omega^2 L^2}$ and $-E_0/\sqrt{R^2 + \omega^2 L^2}$. It is a simple harmonic function of time.

The *impedance*, Z, of the circuit is defined by the complex quantity

$$Z \equiv R + i\omega L, \qquad i^2 = -1. \tag{3.4.20}$$

The *absolute value* or *magnitude* of a complex number $z = x + iy$, x, y real numbers, written $|z|$, is given by

$$|z| = \sqrt{x^2 + y^2}, \tag{3.4.21}$$

which is a nonnegative real number. Thus, the magnitude of the impedance is given by

$$|Z| = \sqrt{R^2 + \omega^2 L^2}. \tag{3.4.22}$$

Let

$$Z_L \equiv |Z| = \sqrt{R^2 + \omega^2 L^2}. \tag{3.4.23}$$

Note that as $L \to 0$, $Z_L \to R$. We may think of Z_L as the "apparent resistance" of the circuit, due to the combined effect of the inductance and the actual resistance. In view of (3.4.23), we may also write Equation (3.4.16) as

$$I(t) = \frac{E_0}{Z_L} \sin(\omega t - \phi) + \left(I_0 + \frac{E_0 \omega L}{Z_L^2}\right)e^{-(R/L)t}. \tag{3.4.24}$$

In summary, the steady-state current I_s is a sinusoidal wave with

$$\left. \begin{aligned} &\text{period} = \frac{2\pi}{\omega}; \quad \text{phase angle} = \phi = \tan^{-1}\left(\frac{\omega L}{R}\right), \\[2mm] &\text{frequency} = \frac{\omega}{2\pi}; \quad \text{where} \quad 0 \leqq \phi < \frac{\pi}{2}; \\[2mm] &\text{amplitude} = \frac{E_0}{Z_L}; \quad \text{impedance} = Z_L = \sqrt{R^2 + \omega^2 L^2}. \end{aligned} \right\} \tag{3.4.25}$$

Remark 3.4.4. Consider the function F defined by

$$F(t) = A \sin \beta t + B \cos \beta t, \tag{3.4.26}$$

where A, B, and β are constants.   Let us write (3.4.26) as

$$F(t) = \sqrt{A^2 + B^2}\left(\frac{A}{\sqrt{A^2 + B^2}}\sin \beta t + \frac{B}{\sqrt{A^2 + B^2}}\cos \beta t\right). \quad (3.4.27)$$

Define

$$\sin \phi = \frac{A}{\sqrt{A^2 + B^2}} \quad \text{and} \quad \cos \phi = \frac{B}{\sqrt{A^2 + B^2}}. \quad (3.4.28)$$

Thus, (3.4.27) may be written as

$$F(t) = \sqrt{A^2 + B^2}\,(\sin \beta t \sin \phi + \cos \beta t \cos \phi),$$

or

$$F(t) = \sqrt{A^2 + B^2}\,\cos(\beta t - \phi). \quad (3.4.29)$$

Hence, the amplitude of $F(t)$ is $\sqrt{A^2 + B^2}$, and the graph of F oscillates between $\sqrt{A^2 + B^2}$ and $-\sqrt{A^2 + B^2}$.

Suppose that in (3.4.28), we had defined

$$\sin \phi = \frac{B}{\sqrt{A^2 + B^2}} \quad \text{and} \quad \cos \phi = \frac{A}{\sqrt{A^2 + B^2}}. \quad (3.4.30)$$

Then, proceeding as above, we would obtain

$$F(t) = \sqrt{A^2 + B^2}\,(\sin \beta t \cos \phi + \cos \beta t \sin \phi),$$

or

$$F(t) = \sqrt{A^2 + B^2}\,\sin(\beta t + \phi). \quad (3.4.31)$$

Again, the amplitude of $F(t)$ is $\sqrt{A^2 + B^2}$.

Note that in (3.4.29), ϕ is determined by the equations given in (3.4.28), while in (3.4.31), ϕ is determined by the equations given in (3.4.30).   In the special case when A and B are nonnegative, the angles ϕ in (3.4.29) and (3.4.31) are given, respectively, by

$$\phi = \tan^{-1}\left(\frac{A}{B}\right), \quad A \geqq 0, \quad B > 0, \quad 0 \leqq \phi < \frac{\pi}{2},$$

and $\hfill (3.4.32)$

$$\phi = \tan^{-1}\left(\frac{B}{A}\right), \quad A > 0, \quad B \geqq 0, \quad 0 \leqq \phi < \frac{\pi}{2}.$$

Exercises 3.4

1. Find the current in the circuit at time t in Example 3.4.1 when $L = 0.5$ henry, $R = 10$ ohms, $E_0 = 150$ volts, $\omega = 120\pi$, and $I_0 = 0$.   Also sketch the graph of the function I_s representing the steady-state current. [*Note:* Here the emf, $E = 150 \sin 120\,\pi t$, is a 60 cycle sine wave with amplitude 150 volts.]

2. (a) Show that the solution for the charge at time t of the initial-value problem for an R-C series circuit,

$$\begin{cases} RQ'(t) + \dfrac{1}{C}\, Q(t) = E_0 \sin \omega t, \\[2mm] Q(0) = Q_0, \end{cases}$$

where E_0 and ω are constants, is given by

$$Q(t) = \frac{E_0}{\omega \sqrt{R^2 + \left(\dfrac{1}{\omega C}\right)^2}}\, \sin(\omega t - \phi) + \left[Q_0 + \frac{E_0\, R}{\omega\left(R^2 + \left(\dfrac{1}{\omega C}\right)^2\right)} \right] e^{-t/RC},$$

where

$$\phi = \tan^{-1}(\omega RC), \qquad 0 \le \phi < \frac{\pi}{2}.$$

(b) Show that the solution for the current at time t in the above circuit is given by

$$I(t) = \frac{E_0}{Z_C}\, \cos(\omega t - \phi) - \frac{1}{RC}\left(Q_0 + \frac{E_0\, R}{\omega Z_C^{\,2}} \right) e^{-(t/RC)},$$

where

$$Z_C = \sqrt{R^2 + \left(\frac{1}{\omega C}\right)^2}$$

is the *impedance* or "apparent resistance" due to the combined effect of R and C.

(c) Find the current at time t when $R = 10$ ohms, $C = 0.002$ farad, $E_0 = 150$ volts, $\omega = 120\pi$, and $Q_0 = 0$. [*See* Figure 3.4.4.]

3. Show that the solution for the charge at time t of the initial-value problem for an R-C series circuit,

$$\begin{cases} RQ'(t) + \dfrac{1}{C}\, Q(t) = E_0\, e^{-kt}, \\[2mm] Q(0) = 0, \end{cases}$$

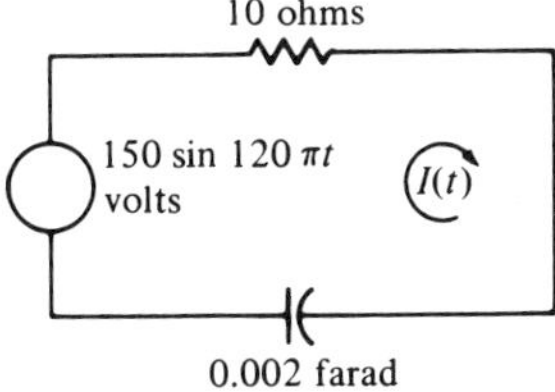

Figure 3.4.4.

where E_0 and k are constants, is given by

(a)
$$Q(t) = \frac{E_0 C}{1 - kRC} (e^{-kt} - e^{-(t/RC)}), \qquad k \neq \frac{1}{RC},$$

(b)
$$Q(t) = \frac{E_0}{R} te^{-(t/RC)}, \qquad k = \frac{1}{RC}.$$

(c) Find the charge and current at time t when $R = 15$ ohms, $C = 0.01$ farad, $E_0 = 225$ volts, $k = 20/3$, and $Q(0) = 0$. Also determine the maximum value of the charge. [*See* Figure 3.4.5.] [*Note:* The emf, $E = E_0 e^{-kt}$, $k > 0$, may be considered as a decaying function, that is, it dies out as t increases.]

4. (a) Show that the solution for the current at time t of the initial-value problem for an R-L series circuit,

$$\begin{cases} LI'(t) + RI(t) = E_0, \\ I(0) = I_0, \end{cases}$$

where E_0 is a constant voltage, is given by

$$I(t) = \frac{E_0}{R} + \left(I_0 - \frac{E_0}{R}\right) e^{-(R/L)t}.$$

$$\left[Note: \text{As } t \to \infty, \quad I(t) \to \frac{E_0}{R} \text{ (constant).} \right]$$

(b) Find the current at the end of 0.1 second when $R = 10$ ohms, $L = 2$ henrys, $E_0 = 100$ volts, and $I_0 = 0$.

5. (a) Show that the solution for the charge at time t of the initial-value problem for an R-C series circuit,

$$\begin{cases} RQ'(t) + \dfrac{1}{C} Q(t) = 0, \\ \\ Q(0) = Q_0, \end{cases}$$

is given by

$$Q(t) = Q_0 e^{-(t/RC)}.$$

[*Note:* As $t \to \infty$, $Q(t) \to 0$.]

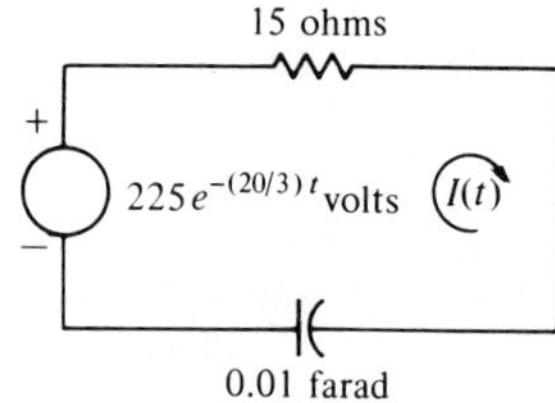

Figure 3.4.5.

(b) Find the charge and current at the end of 0.1 second when $R = 10$ ohms, $C = 0.02$ farad, and $Q_0 = 0.06$ coulomb.

(c) We may consider the above initial-value problem as one in which a condenser of capacitance C is being discharged through a resistance R.   Suppose that a condenser of capacitance 5×10^{-4} farad is discharged through a resistance of 10 ohms.   If the current is -1.5 amps at the end of 0.01 second, what was the initial charge on the condenser?

6. (a) In an R-C series circuit, the resistance R varies with time.   Suppose that $R = a + bt$, where $a \geq 0$ and $b > 0$ are real numbers.   Show that the solution of the charge at time t of the initial-value problem

$$\begin{cases} (a + bt)Q'(t) + \dfrac{1}{C}\, Q(t) = E_0, \\[2mm] Q(0) = Q_0, \end{cases}$$

where E_0 is a constant voltage, is given by

$$Q(t) = E_0 C + a^{1/bC}(Q_0 - E_0 C)(a + bt)^{-1/bC}.$$

(b) Find the charge and current at the end of 0.1 second when $C = 0.08$ farad, $E_0 = 80$ volts, $Q_0 = 4$ coulombs, $a = 1$, and $b = 0.01$.

7. In an R-C series circuit, an uncharged condenser of capacitance C is charged from a source of constant voltage (E_0) through a resistance R.   [*See* Figure 3.4.6.]   Determine the time when the current is equal in magnitude to the charge on the condenser.

8. (a) In Exercise 7, show that the time required for the charge to reach r percent of its theoretical maximum is given by

$$t = RC \ln \left(\dfrac{1}{1 - \dfrac{r}{100}} \right) \text{ sec.}$$

[*Note:* The theoretical maximum of the charge in Exercise 7 is equal to $E_0 C$.]

(b) Suppose that $C = 5 \times 10^{-3}$ farad, $R = 10$ ohms, and $E_0 = 120$ volts.   Find the times required for the charge to reach 50 percent and 99 percent of its theoretical maximum.

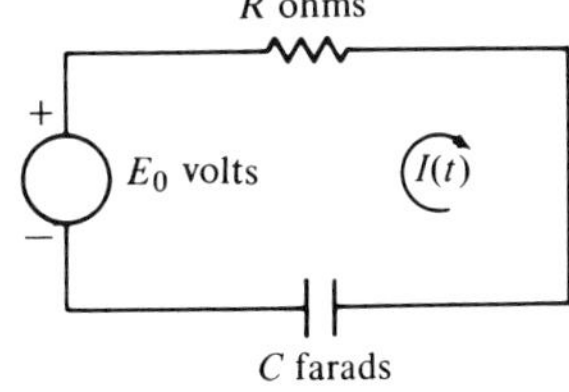

Figure 3.4.6.

9. (a) Show that the solution for the current at time t of the initial-value problem for an R-L series circuit,

$$\begin{cases} LI'(t) + RI(t) = E_0\, e^{-kt} \sin \omega t, \\ I(0) = I_0, \end{cases}$$

where E_0, k, and ω are constants, is given by

$$I(t) = \frac{E_0}{Z_L}\, e^{-kt} \sin (\omega t - \phi) + \left(I_0 + \frac{\omega E_0 L}{Z_L{}^2} \right) e^{-(R/L)t},$$

where

$$Z_L = \sqrt{(R - kL)^2 + \omega^2 L^2}, \qquad \phi = \tan^{-1}\!\left(\frac{\omega L}{R - kL} \right),$$

and

$$0 \leqq \phi < \frac{\pi}{2} \quad \text{if } R > kL.$$

   (b) Find the current at time t when $L = 0.5$ henry, $R = 10$ ohms, $E_0 = 150$ volts, $k = 2$, $\omega = 120\pi$, and $I_0 = 0$.
10. An R-C series circuit contains a resistance of 20 ohms, a condenser of capacitance 0.001 farad, and an emf, $E = 150 \cos 120\,\pi t$ volts. If initially $I = 10$ amps, find the current when $t = 0.1$ second.

3.5. Dynamics: Newton's Laws of Motion

In the study of dynamics, which deals with the laws of motion, Newton's three laws of motion play a very basic role. These laws may be stated as follows.

  I. *A body persists in a state of rest or uniform motion in a straight line, unless it is compelled to change that state by the application of some external force.*
 II. *The rate of change of momentum of a body is proportional to the net force acting on the body and is in the same direction as the force.*
III. *To every action there is an equal and opposite reaction.*

We shall primarily concern ourselves with Newton's Second Law of Motion. The *momentum* of a body is defined as the product of the mass and the velocity. That is,

$$\text{momentum} = \text{mass} \times \text{velocity}. \tag{3.5.1}$$

We may now express Newton's Second Law of Motion as follows.

$$F = k\,\frac{d}{dt}\,(mv) = km\,\frac{dv}{dt} = kma, \tag{3.5.2}$$

where F is the net force acting on the body, m is the mass of the body (assumed to be constant), $a = dv/dt$ is the acceleration of the body, and k is a constant of proportionality whose numerical value is determined by the units in which force, mass, and acceleration are measured.

From this law it follows that if the net force acting on a body is constant, then the body moves with constant acceleration. If the net force varies, then the acceleration also varies in the same proportion. [*Note:* Newton's Frst Law of Motion is a special case of Newton's Second Law of Motion.]

Remark 3.5.1. By specifying properly the units of mass, acceleration, and force, the constant k becomes unity and Equation (3.5.2) may be written as

$$F = ma. \tag{3.5.3}$$

One way to achieve this is to specify the basic unit of mass and the basic unit of acceleration and derive the corresponding basic unit of force. Another way is to specify the basic unit of force and the basic unit of acceleration and derive the corresponding basic unit of mass.

Units of Force
  I. *The meter-kilogram-second or mks absolute system.* By choosing the kilogram (kg) as the basic unit of mass, and the meter per second per second (m/sec²) as the basic unit of acceleration, the corresponding basic unit of force is the *newton* (nt). One *newton* is that force which produces an acceleration of 1 m/sec² in a mass of 1 kg.
 II. *The centimeter-gram-second or cgs absolute system.* By choosing the gram (g) as the basic unit of mass, and the centimeter per second per second (cm/sec²) as the basic unit of acceleration, the corresponding basic unit of force is the *dyne*. One *dyne* is that force which produces an acceleration of 1 cm/sec² in a mass of 1 g.
III. *The foot-pound-second or fps absolute system.* By choosing the pound (lb) as the basic unit of mass, and the foot per second per second (ft/sec²) as the basic unit of acceleration, the corresponding basic unit of force is the *poundal* (pdl). One *poundal* is that force which produces an acceleration of 1 ft/sec² in a mass of 1 lb.
 IV. *The English gravitational system.* By choosing the pound (lb) as the basic unit of force, and the foot per second per second (ft/sec²) as the basic unit of acceleration, the corresponding basic unit of mass is the *slug*. One *slug* is the mass of a body that acquires an acceleration of 1 ft/sec² when acted upon by a force of 1 lb.

Remark 3.5.2. The English gravitational system is also known as the British engineering system. A force of 1 lb gives a 1 lb mass an acceleration of 32.17 ft/sec² (approximately) and gives a 1 slug mass an acceleration of 1 ft/sec². Thus, we see that the slug has a mass slightly greater than 32 times that of the standard pound mass.

The *mass* of a body is the "quantity of matter" it contains.   Thus, the mass is independent of the position of the body on earth.   The *weight* is the force of attraction between the earth and the body.   The weight of a body at a particular place on the earth's surface is proportional to its mass.   The weight of a body depends on where it is located relative to the earth.

If a body of mass m is allowed to fall freely, the net force acting on the body is its weight W, and its acceleration g is due to that weight (or force).   For this case, Equation (3.5.3) becomes

$$W = mg. \tag{3.5.4}$$

Solving for the mass m in (3.5.4), and then substituting it into (3.5.3), we obtain

$$F = \frac{W}{g}\, a. \tag{3.5.5}$$

Remark 3.5.3.   The acceleration constant g due to gravity has the following approximate values.

(a) In the mks absolute system, $g = 9.805$ m/sec^2.
(b) In the cgs absolute system, $g = 980.5$ cm/sec^2.
(c) In the fps absolute system, $g = 32.17$ ft/sec^2.
(d) In the English gravitational system, $g = 32.17$ ft/sec^2.

Remark 3.5.4.   We associate magnitude and direction with a force; therefore, we call force a vector quantity.   In utilizing the above formulas, a force which corresponds to a positive acceleration (in a given frame of reference) is represented by a positive number.   Here the force tends to increase the velocity algebraically.   A force which corresponds to a negative acceleration is represented by a negative number.   Here the force tends to decrease the velocity algebraically.

Remark 3.5.5.   One factor that always affects motion, whatever its direction, is friction.   *Force of friction*, or simply *friction*, is the tangential force on a body which opposes any tendency for its surface to move relative to another surface.   The tangential forces are parallel to the surfaces that are in contact. *Kinetic* or *sliding friction* is the tangential force between two surfaces when one surface is sliding over the other.

Let f_k denote the force required to overcome friction and keep the body moving with uniform velocity across the surface, and let N denote the normal (perpendicular) force pressing the surfaces together.   It has been found experimentally that f_k is proportional to N.   Thus,

$$f_k = \mu_k N, \tag{3.5.6}$$

where μ_k is a constant called the *coefficient of kinetic or sliding friction.* [*Note:* $\mu_k = f_k/N$, being the ratio of two forces, is a pure number.]

Remark 3.5.6.   In the ensuing discussion, we shall need some of the basic definitions concerning rectilinear motion encountered in calculus, namely:

$$\text{distance} \equiv s, \tag{3.5.7}$$

$$\text{velocity} \equiv v = \frac{ds}{dt}, \tag{3.5.8}$$

$$\text{acceleration} \equiv a = \frac{dv}{dt} = \frac{d^2s}{dt^2}. \tag{3.5.9}$$

Also,

$$\frac{dv}{dt} = \frac{dv}{ds}\frac{ds}{dt} = v\frac{dv}{ds}. \tag{3.5.10}$$

For convenience of reference, let us summarize in Table 3.5.1 the consistent basic units discussed above.

Table 3.5.1.

System	Distance $s, x, y, \ldots$	Time t	Mass m	Acceleration a	Force F
mks	Meter (m)	Second (sec)	Kilogram (kg)	m/sec²	Newton (nt)
cgs	Centimeter (cm)	Second (sec)	Gram (g)	cm/sec²	Dyne
fps	Foot (ft)	Second (sec)	Pound (lb)	ft/sec²	Poundal (pdl)
English	Foot (ft)	Second (sec)	Slug	ft/sec²	Pound (lb)

Also, for convenience we give the following partial list of conversion factors.

$$
\left.
\begin{array}{ll}
\textit{Length} & \text{1 kilometer (km)} = 1000\ \text{m}; \qquad \text{1 mile (mi)} = 5280\ \text{ft};\\
& \text{1 m} = 100\ \text{cm}; \qquad\qquad\qquad \text{1 ft} = 30.48\ \text{cm};\\
& \text{1 m} = 3.281\ \text{ft}; \qquad\qquad\quad\ \ \text{1 mi} = 1.609\ \text{km}.\\
& \text{1 km} = 0.6214\ \text{mi};\\
\textit{Mass} & \text{1 kg} = 1000\ \text{g} = 0.06852\ \text{slug}; \qquad \text{1 slug} = 14.59\ \text{kg}.\\
\textit{Force} & \text{1 kilogram weight (kg wt)} = 2.205\ \text{lb} = 9.807\ \text{nt};\\
& \text{1 nt} = 10^5\ \text{dynes} = 0.1020\ \text{kg wt} = 0.2248\ \text{lb};\\
& \text{1 lb} = 4.448\ \text{nt} = 0.4536\ \text{kg wt} = 32.17\ \text{pdl}.
\end{array}
\right\} \tag{3.5.11}
$$

Example 3.5.1.   Find the mass m of a body whose weight W is (a) 29.415 nt; (b) 3922 dynes;   (c) 64.34 pdl;   (d) 321.7 lb.

SOLUTION

(a) In the mks units: $m = \dfrac{W}{g} = \dfrac{29.415 \text{ nt}}{9.805 \text{ m/sec}^2} = 3 \text{ kg}.$

(b) In the cgs units: $m = \dfrac{W}{g} = \dfrac{3922 \text{ dynes}}{980.5 \text{ cm/sec}^2} = 4 \text{ g}.$

(c) In the fps units: $m = \dfrac{W}{g} = \dfrac{64.34 \text{ pdl}}{32.17 \text{ ft/sec}^2} = 2 \text{ lb}.$

(d) In the English units: $m = \dfrac{W}{g} = \dfrac{321.7 \text{ lb}}{32.17 \text{ ft/sec}^2} = 10 \text{ slugs}.$

Example 3.5.2.   A body of mass 4 kg is acted upon by a force of (a) 8 nt and (b) 12,000 dynes.   Find the acceleration in each case.

SOLUTION

(a) In the mks units: $F = ma$ or $a = \dfrac{F}{m} = \dfrac{8 \text{ nt}}{4 \text{ kg}} = 2 \text{ m/sec}^2.$

(b) In the cgs units: $F = ma$ or $a = \dfrac{F}{m} = \dfrac{12{,}000 \text{ dynes}}{4 \text{ kg}} = \dfrac{12{,}000 \text{ dynes}}{4000 \text{ g}}$

$$= 3 \text{ cm/sec}^2.$$

Example 3.5.3.   What force is required to give a body weighing 16.085 lb an acceleration of 12 ft/sec^2?

SOLUTION. $F = \dfrac{W}{g}\, a = \dfrac{16.085 \text{ lb}}{32.17 \text{ ft/sec}^2} \times 12 \text{ ft/sec}^2 = 6 \text{ lb}.$

[*Note:* Here the mass of the body is 1/2 slug.   Thus, the above problem may also be stated as follows.   What force is required to give a body of mass 1/2 slug an acceleration of 12 ft/sec^2?]

Example 3.5.4.   A block weighing 150 lb is being pushed in a straight line along a rough surface by a force of 50 lb.   Suppose that the force of friction is 10 lb and there is an air resistance to motion whose magnitude in pounds is twice the velocity (of the block) in ft per second.   If the block is initially at rest, find (a) the velocity and the displacement at the end of t sec; (b) the velocity and the displacement at the end of 2 sec.

SOLUTION.   Let x (ft) be the displacement from the initial position ($x = 0$) at the end of t sec.   Then $v = dx/dt$ is the velocity.   Since the force acting on the block tends to increase the velocity algebraically, it is positive.   The forces due to friction and air resistance tend to decrease the velocity algebraically, and they are negative. Thus, the net force F (lb), or unbalanced force, is

$$F = 50 \text{ lb} - (10 + 2v) \text{ lb} = (40 - 2v) \text{ lb}. \tag{3.5.12}$$

Since $a = dv/dt$, we may write (3.5.5) as

$$F = \frac{W}{g}\frac{dv}{dt}, \tag{3.5.13}$$

where $g = 32.17$ ft/sec^2. In view of (3.5.12) and (3.5.13), we then have

$$40 - 2v = \frac{150}{g}\frac{dv}{dt}, \tag{3.5.14}$$

which is a first-order equation with separable variables. Thus,

$$\frac{dv}{20 - v} = \frac{g}{75}\,dt. \tag{3.5.15}$$

Integrating Equation (3.5.15) between the limits $t = 0$, $v = 0$ and $t = t$, $v = v$, we get

$$\int_0^v \frac{dv}{20 - v} = \frac{g}{75}\int_0^t dt.$$

Performing the above integrations and simplifying the result, we obtain

$$v = 20(1 - e^{-(g/75)t}) \tag{3.5.16}$$

Since $v = dx/dt$, we have in view of (3.5.16)

$$dx = 20(1 - e^{-(g/75)t})dt.$$

Integrating the above equation between the limits $t = 0$, $x = 0$ and $t = t$, $x = x$, and simplifying the result, we get

$$x = 20\left[t + \frac{75}{g}(e^{-(g/75)t} - 1)\right] \tag{3.5.17}$$

The answers to part (a) are given by (3.5.16) and (3.5.17).

If we substitute $t = 2$ in (3.5.16) and (3.5.17), we find

$$v = 20(1 - e^{-(2 \cdot 32.17)/75}) = 20(1 - e^{-0.858}) = 20(1 - 0.4240) = 11.52 \text{ ft/sec,}$$

$$x = 20\left[2 + \frac{75}{32.17}(e^{-0.858} - 1)\right] = 20[2 + 2.33(-0.576)] = 13.14 \text{ ft,}$$

which are the answers to part (b). (Clearly, these answers are approximate.)

Note from (3.5.16) and (3.5.17) that as t increases, both v and x increase. The value of x increases without limit but v tends to a limiting value:

$$\lim_{t \to \infty} v = 20 \text{ ft/sec.} \tag{3.5.18}$$

This limiting value of v agrees with the value of v found by setting $dv/dt = 0$ in Equation (3.5.14). The *limiting value of v* is also known as the *terminal velocity*.

Example 3.5.5. A particle of mass m is thrown vertically upwards with initial velocity v_0. If the air resists the motion with a force per unit mass of k $(k > 0)$ times the velocity squared, show that the particle returns to its starting point with velocity

$$\frac{v_0}{\sqrt{1 + \dfrac{kv_0^{\,2}}{g}}}. \tag{3.5.19}$$

SOLUTION. In Figure 3.5.1, the x axis is taken to be vertical with origin at O so that $v = v_0$ when $x = 0$. We consider "up" as the positive direction. At the end of time t, let the particle be at P: $OP = x \leq H$, where H is the maximum height attained by the particle. Then $v = dx/dt$ and $a = dv/dt = v\,dv/dx$ are, respectively, the velocity and the acceleration of the particle at P. The forces acting on the particle at P are:

(a) force of gravity mg downwards, that is, $-mg$ along the positive x axis, and

(b) the air resistance $-mkv^2$ along the positive x axis.

Thus, the (total) net force acting on the particle at P along the positive x axis is

$$-(mg + mkv^2). \tag{3.5.20}$$

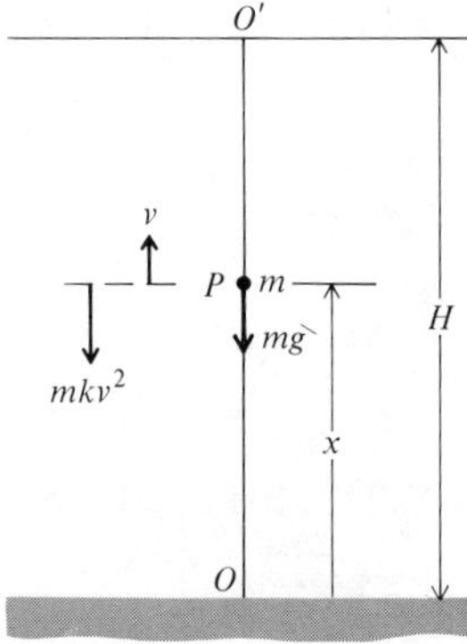

Figure 3.5.1.

Utilizing (3.5.3) and (3.5.20), we find that the equation of motion is

$$-(mg + mkv^2) = mv\,\frac{dv}{dx}, \tag{3.5.21}$$

which is a first-order differential equation with separable variables.   Equation (3.5.21) may be written as

$$\frac{v\,dv}{kv^2 + g} = -dx.$$

Integrating the above equation and simplifying the result, we obtain

$$v^2 = Ce^{-2kx} - \frac{g}{k}, \tag{3.5.22}$$

where C is an arbitrary constant of integration.   Since $v = v_0$ when $x = 0$, we find $C = v_0{}^2 + g/k$, and Equation (3.5.22) becomes

$$v^2 = \left(v_0{}^2 + \frac{g}{k}\right)e^{-2kx} - \frac{g}{k}. \tag{3.5.23}$$

Taking $v = 0$ when $x = H$ in (3.5.23), we find upon simplification

$$H = \frac{1}{2k}\ln\left(1 + \frac{kv_0{}^2}{g}\right). \tag{3.5.24}$$

For the downward motion of the particle see Figure 3.5.2.   Here the y axis is taken vertical with origin at O' so that $v = 0$ when $y = 0$.   We consider "down" as the positive direction.

Reasoning as in the discussion given above for the upward motion of the particle, we find that the net force acting on the particle at P is now

$$mg - mkv^2,$$

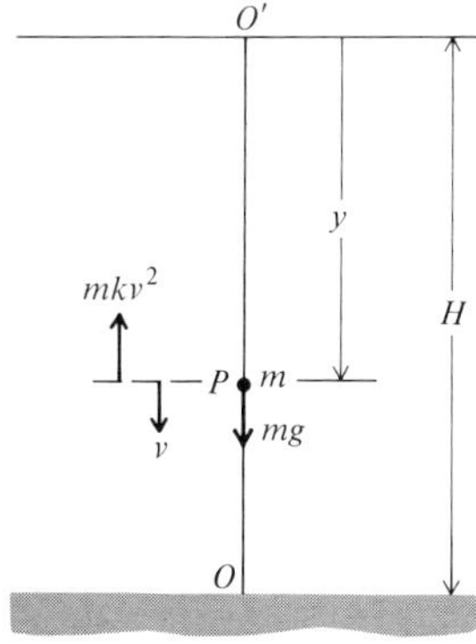

Figure 3.5.2.

and the equation of motion is

$$mg - mkv^2 = mv\,\frac{dv}{dy}.$$ (3.5.25)

Equation (3.5.25) may be written as

$$\frac{v\,dv}{kv^2 - g} = -dy.$$ (3.5.26)

Integrating Equation (3.5.26) and simplifying the result, we get

$$v^2 = C'e^{-2ky} + \frac{g}{k},$$ (3.5.27)

where C' is an arbitrary constant of integration.   Since $v = 0$ when $y = 0$, we find $C' = -g/k$ and Equation (3.5.27) becomes

$$v^2 = \frac{g}{k}(1 - e^{-2ky}).$$ (3.5.28)

If we substitute in (3.5.28), $y = H$, where H is given in (3.5.24), we find

$$v^2 = \frac{g}{k}[1 - e^{-\ln(1 + kv_0^2/g)}] = \frac{g}{k}\left(1 - \frac{1}{1 + \dfrac{kv_0^2}{g}}\right) = \frac{v_0^2}{1 + \dfrac{kv_0^2}{g}},$$

and thus the stated result is established.

Example 3.5.6.   A wooden block weighing 50 lb is released from the top of a plane inclined at 30° with the horizontal. [*See* Figure 3.5.3.]   The coefficient of sliding friction between the block and the plane is 0.4.   Find the velocity and the distance traveled by the block at the end of 4 sec.

SOLUTION.   Let s (ft) be the distance the block has traveled from the initial position $(s = 0)$ at the end of t sec.   Then $v = ds/dt$ and $a = dv/dt$ are, respectively, the velocity and acceleration.   Denote by W the weight of the

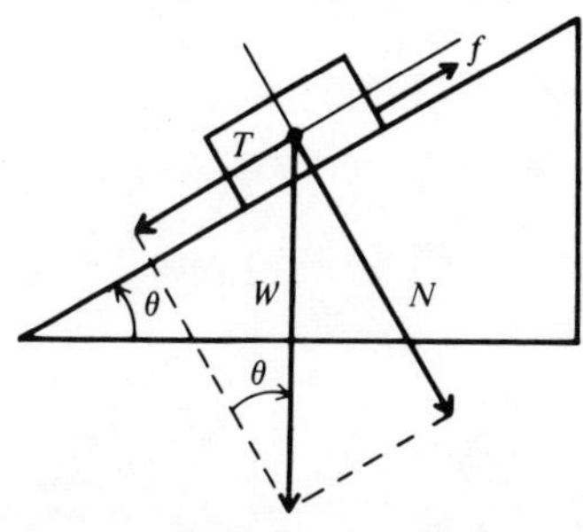

Figure 3.5.3.
$W = 50\text{lb}$ and $\theta = 30°$

50 lb wooden block. Let us resolve W into two components. Let T represent the component of W down the plane, and N the component of W normal to the plane. [*Note:* N is the force pressing the surfaces together.] The force of friction (f) between the block and the plane is parallel to and up the plane.

From Figure 3.5.3, we see that

$$T = W \sin \theta = 50 \sin 30° = 50 \cdot \tfrac{1}{2} = 25 \text{ lb}, \qquad (3.5.29)$$

and

$$N = W \cos \theta = 50 \cos 30° = 50 \cdot \frac{\sqrt{3}}{2} = 25\sqrt{3} \text{ lb}. \qquad (3.5.30)$$

From (3.5.6), the coefficient of friction $\mu = f/N$. Thus

$$f = \mu N. \qquad (3.5.31)$$

Since $\mu = 0.4$, utilizing (3.5.30) and (3.5.31), we find

$$f = 0.4 \cdot 25\sqrt{3} = 10\sqrt{3} \text{ lb}. \qquad (3.5.32)$$

Clearly, in view of (3.5.29) and (3.5.32), the net force (F) (or unbalanced force) acting on the wooden block is

$$F = T - f = (25 - 10\sqrt{3}) \text{ lb}. \qquad (3.5.33)$$

Thus, the equation of motion,

$$F = \frac{W}{g} \frac{dv}{dt},$$

becomes

$$25 - 10\sqrt{3} = \frac{50}{32.17} \frac{dv}{dt},$$

or

$$\frac{dv}{dt} = 4.94. \qquad (3.5.34)$$

Integrating Equation (3.5.34), we get

$$v = 4.94t + C, \qquad (3.5.35)$$

where C is a constant of integration. The initial condition here is that $v = 0$ when $t = 0$. Thus, $C = 0$ and (3.5.35) becomes

$$v = 4.94t. \qquad (3.5.36)$$

If we substitute $t = 4$ in (3.5.36), we find $v = 19.76$ ft/sec. Thus, the block moves with velocity of 19.76 ft/sec at the end of 4 sec.

Since

$$\frac{ds}{dt} = v = 4.94t,$$

we obtain

$$s = 2.47t^2 + C', \tag{3.5.37}$$

where C' is a constant of integration.   The initial condition here is that $s = 0$ when $t = 0$.   Thus, $C' = 0$ and (3.5.37) becomes

$$s = 2.47t^2. \tag{3.5.38}$$

If we substitute $t = 4$ in (3.5.38), we find $s = 39.52$ ft, which is the distance the block has traveled during this time.

Remark 3.5.7.   Suppose now that in the above example the weight of the block, W (lb), is arbitrary, and the plane is inclined at some angle θ with the horizontal.   In view of the first equality in (3.5.29), (3.5.30), (3.5.31), and (3.5.33), we then have

$$F = T - f = W \sin \theta - \mu N = W \sin \theta - \mu W \cos \theta = W(\sin \theta - \mu \cos \theta).$$

$$\tag{3.5.39}$$

In order for the block to slide down the inclined plane when started, we must have $F > 0$ in (3.5.39).   Thus, the following relation exists between the angle θ and the coefficient of sliding friction μ.

$$\sin \theta - \mu \cos \theta > 0,$$

or

$$\tan \theta > \mu. \tag{3.5.40}$$

In the above example, this inequality is clearly satisfied, since $\tan 30° = 0.5773 > 0.4$.

Note that if $T < f$, the block will not slide down the inclined plane.   It will remain at rest at the top of the plane.   In view of (3.5.39), we then have $\tan \theta < \mu$.   If we now begin to vary the angle θ until the block will just slide down the plane with uniform velocity when once started, we then have $T = f$, since the block experiences no acceleration.   Thus, utilizing (3.5.39), we now have

$$\tan \theta = \mu. \tag{3.5.41}$$

Equation (3.5.41) asserts that the tangent of the angle at which a block slides down the plane with uniform velocity is equal to the coefficient of kinetic friction.

Remark 3.5.8. At times [*see*, for example, Exercise 3.5.10], we shall need to obtain the equation of motion of a particle satisfying the differential equation

$$ma = kf(x), \qquad a = \frac{dv}{dt} = \frac{d^2x}{dt^2}, \tag{3.5.42}$$

subject to the initial conditions:

$$\text{when} \quad t = 0, \qquad x = x_0, \quad \text{and} \quad v = v_0. \tag{3.5.43}$$

We may write (3.5.42) as

$$\frac{dv}{dt} = Kf(x), \qquad K = \frac{k}{m}. \tag{3.5.44}$$

Multiplying both sides of (3.5.44) by $v = dx/dt$, and then integrating with respect to t, we have

$$\int v \, dv = K \int f(x)v \, dt + c = K \int f(x) \frac{dx}{dt} \cdot dt + c = K \int f(x) \, dx + c.$$

Thus,

$$v^2 = 2K \int f(x)dx + c_1, \tag{3.5.45}$$

where c_1 is an arbitrary constant of integration. Since $v^2 = (dx/dt)^2$, we see that

$$\int \frac{dx}{\pm \sqrt{2K \int f(x)dx + c_1}} = t + c_2, \tag{3.5.46}$$

where c_2 is another arbitrary constant of integration.

Exercises 3.5

1. A body falls from rest in a medium for which the resistance is proportional to the square of the velocity.
   (a) If the limiting velocity is U, show that the time required for the velocity (v) to become kU, $\;\; 0 < k < 1$, is given by

   $$t = \frac{U}{2g} \ln \left(\frac{1+k}{1-k} \right).$$

   (b) In part (a), if $U = 200$ ft/sec and $g = 32.17$ ft/sec^2, find the time when the velocity of the body is 100 ft/sec.

2. Suppose that the terminal velocity of fall of a human under suitable atmospheric conditions is U. Assume that air resistance is the only factor opposing the downward motion, and that air resistance is proportional to the square of the velocity. Suppose that a man falls from rest from an altitude H and that his parachute must open when he reaches an altitude h.

   (a) Show that the time (t) at which he should open his parachute is given by the equation

   $$\ln\left[\cosh\left(\frac{g}{U}t\right)\right] = \frac{g(H-h)}{U^2}.$$

   (b) In part (a), if $U = 60$ m/sec, $g = 9.805$ m/sec^2, $H = 1600$ m, and $h = 600$ m, find the time when the man should open his parachute.

3. A body of mass m, constrained to move horizontally, is subjected to a periodic force of $A \sin \omega t$ and is retarded in its motion by a frictional drag which is proportional to the velocity. (That is, the frictional drag is given by kv, where v is the velocity and $k > 0$ is the proportionality constant.)

   (a) If the body starts from rest, show that the velocity as a function of time is given by

   $$v = \frac{A}{m\left[\left(\dfrac{k}{m}\right)^2 + \omega^2\right]}\left(\frac{k}{m}\sin \omega t - \omega \cos \omega t + \omega e^{-(k/m)t}\right).$$

   (b) In part (a), find the velocity of the body when $m = 4$ slugs, $t = \dfrac{\pi}{3}$ sec, $A = 120$, $k = 4$, and $\omega = 1$. [In part (b), what units should be given to the constants A, k, and ω so that the unit of v is ft/sec?]

4. A small stone of mass m is thrown vertically upwards with initial velocity v_0. Air resists the motion with a force equal to kmv, where v is the velocity and $k > 0$. Show that the stone returns to its starting point with velocity U given by the equation

   $$g - kU = (g + kv_0)e^{-(k/g)(U + v_0)}.$$

5. A weight (W) is given an initial velocity v_0 down a plane inclined at an angle θ with the horizontal. Let μ denote the coefficient of sliding friction between the weight and the plane. Show that after time t, the velocity (v) and the distance (s) traveled by the weight are given, respectively, by

   $$v = v_0 + g(\sin \theta - \mu \cos \theta)t,$$

   and

   $$s = v_0 t + \frac{g}{2}(\sin \theta - \mu \cos \theta)t^2$$

   $$= \frac{(v_0 + v)}{2}t,$$

   provided that $\tan \theta > \mu$.

6. A man, by exerting a pulling force of A, can climb up a rope a distance of S in time T. If he starts from rest,
   (a) show that the weight of the man, W, is given by the formula

$$W = \frac{gT^2 A}{2S + gT^2}.$$

   (b) In part (a), determine the weight of the man when $A = 220$ lb, $S = 30$ ft, $T = 4$ sec, and $g = 32.17$ ft/sec^2.

7. A body falls with an initial velocity of unity into a medium for which the resistance is proportional to v^n, $n \neq 1$, and v is velocity. Suppose that the density of the medium is the same as the density of the body.[9] Let v_1 and v_2 denote, respectively, the velocity of the body at times t_1 and t_2, $t_1 < t_2$.
   (a) Show that the constant n satisfies the relation

$$\frac{v_1^{1-n} - 1}{v_2^{1-n} - 1} = \frac{t_1}{t_2}.$$

   (b) Determine the constant n if $v_1 = 9/16$ ft/sec, $v_2 = 1/4$ ft/sec, $t_1 = 1$ sec, and $t_2 = 3$ sec.

8. A body falls from rest in a medium for which the resistance is proportional to the velocity. The specific gravity of the medium is $K\,(0 < K < 1)$ times that of the body. If the terminal velocity is U,

[9] Density of a body = mass per unit volume = $\dfrac{\text{mass of the body}}{\text{volume of the body}} = \dfrac{m}{V}$. The density of solids and liquids may be expressed in kg/m^3, g/cm^3, lb/ft^3, or slugs/ft^3.

$$\text{Specific gravity of a body} = \frac{\text{density of the body}}{\text{density of water}}$$

$$= \frac{\text{mass of the body}}{\text{mass of equal volume of water}}$$

$$= \frac{\text{weight of the body}}{\text{weight of equal volume of water}}.$$

The buoyant force is equal in magnitude to the weight of the medium that the body displaces and opposes the force of gravity. In Exercise 7, since the density of the medium is the same as that of the body, it follows that the specific gravity of the medium is the same as that of the body. Thus, the weight of the medium displaced by the body is equal to the weight of the body, and the equation of motion for the above exercise becomes

Net force on the body = weight of the body − buoyant force − resistance,

or

$$m \frac{dv}{dt} = mg - mg - kv^n = -kv^n.$$

If the specific gravity of the medium is $K\,(0 < K < 1)$ times that of the body, then the weight of the medium displaced by the body is equal to K times the weight of the body. The equation of motion of Exercise 7 would then become

$$m \frac{dv}{dt} = mg - Kmg - kv^n.$$

(a) show that after time t, the velocity (v) and distance (s) traveled by the body are given, respectively, by

$$v = U\left[1 - \exp\left(-\frac{(1-K)g}{U}t\right)\right],$$

$$s = Ut - \frac{U^2}{(1-K)g}\left[1 - \exp\left(-\frac{(1-K)g}{U}t\right)\right]$$

$$= U\left[t - \frac{v}{(1-K)g}\right].$$

(b) In part (a), find the velocity and distance traveled by the body when $K = 1/5$, $U = 8$ m/sec, $g = 9.805$ m/sec², and $t = 2$ sec.

(c) If $K > 1$, will the body fall from rest in the medium?

9. *A rectilinear simple harmonic motion* is one in which a particle moves along a straight line under a force that is always directed toward a point O in that line, the magnitude (of the force) being proportional to the particle displacement from O.

   At time t, let a particle of mass m be at P. [*See* Figure 3.5.4.] Then $\overrightarrow{OP} = x$, so that $v = dx/dt$ and $a = dv/dt = v\,dv/dx$ are, respectively, the velocity and acceleration of the particle at P measured in the direction of increasing x. Also, the magnitude of the force (f) that is always directed toward O is equal to $m\omega^2 x$ where $m\omega^2$ is the constant of proportionality. At time t, show that the equation of motion of the particle executing simple harmonic motion is given by

(1) $$x = A \sin(\omega t + \alpha),$$

where A and α are arbitrary constants.

[*Note:* In Equation (1), the amplitude is A, the period is $2\pi/\omega$, the frequency is $\omega/2\pi$, and the phase angle is α. ω is called the angular frequency.]

10. A particle of mass m is repelled from the origin by a force inversely proportional to the cube of its distance from the origin. If the particle is given an initial velocity v_0 (directed towards the origin) at a distance x_0 from the origin, show that the equation of motion of the particle at time t is

$$x = \sqrt{\left(\frac{mx_0^2 v_0^2 + k}{mx_0^2}\right)t^2 + 2x_0 v_0 t + x_0^2},$$

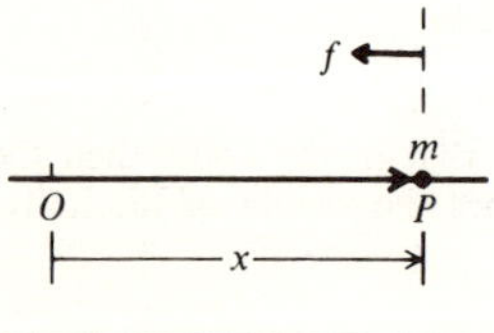

Figure 3.5.4.

where k is the constant of proportionality. In particular, when $v_0 = 0$, we obtain

$$x = \sqrt{\frac{k}{mx_0{}^2}\, t^2 + x_0{}^2}.$$

11. *Velocity of escape from the earth.*   Suppose that a particle is projected in a radial direction outward from the earth and acted upon only by the gravitational attraction of the earth.   *Newton's Law of Gravitational Motion states:*

The acceleration of the particle is inversely proportional to the square of the distance from the particle to (and directed toward) the center of the earth.

Let r be the variable distance from the center of the earth O, to the particle P, at time t (taking the positive direction of r outward), and let R be the radius of the earth.   [*See* Figure 3.5.5.]   Let v denote the velocity of the particle.   Then $a = dv/dt$ is its acceleration which (according to Newton's law above) is negative in our system.   Therefore,

$$(1) \qquad \frac{dv}{dt} = a = -\frac{k}{r^2},$$

where $k > 0$ is the constant of proportionality.   Note that if r is increasing (with time), then $a = dv/dt > 0$.   When $r = R$, $a = -g$, where the number $g > 0$ is the magnitude of the surface acceleration due to gravity.   Thus,

$$(2) \qquad -g = -\frac{k}{R^2} \quad \text{or} \quad k = gR^2,$$

and Equation (1) may be written as

$$(3) \qquad a = -\frac{gR^2}{r^2}.$$

Since $a = \dfrac{dv}{dt} = v\,\dfrac{dv}{dr}$, Equation (3) becomes

$$(4) \qquad v\,\frac{dv}{dr} = -\frac{gR^2}{r^2}.$$

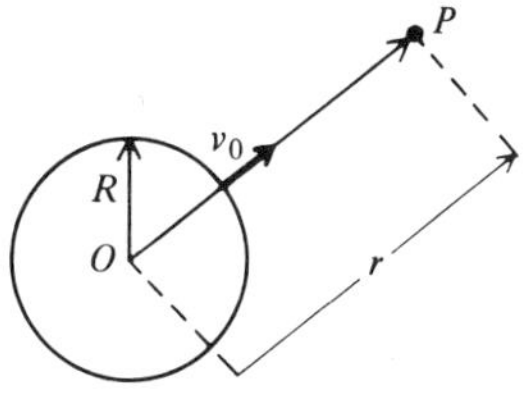

Figure 3.5.5.

Suppose that the particle leaves the earth's surface with velocity v_0, then

(5) $$v = v_0 \quad \text{when} \quad r = R.$$

(a) Show that the solution of Equation (4), subject to the initial conditions stated in (5), is given by

(6) $$v^2 = \frac{2gR^2}{r} + v_0{}^2 - 2gR.$$

(b) Since the velocity of the particle is positive when $r = R$, show that the velocity of the particle will remain positive if

(7) $$v_0{}^2 - 2gR \geq 0.$$

Thus, in order for a particle to escape from the earth's surface, the *minimal escape velocity* required, v_e, is

(8) $$v_e = \sqrt{2gR}.$$

(c) If we take $R = 3960$ miles and $g = 32.17$ ft/sec $= 0.006093$ mi/sec, show that

(9) $$v_e = 6.95 \text{ mi/sec (approximately).}$$

12. *Earth falling into the sun.*   *Newton's Law of Universal Gravitation states:*

Every particle of matter in the universe attracts every other particle with a force which is directly proportional to the product of the masses of the particles and inversely proportional to the square of the distance between them.

In symbols,

$$F = \frac{GMm}{d^2},$$

where M and m are the masses of the particles; d is the distance between them; F is the force of attraction; and G is the constant of proportionality, commonly known as the *gravitational constant.*

A planet has mass M and radius R.   A particle of mass $m \ll M$ (that is, m is much smaller than M) falls from rest at a distance d from the center of the planet.

(a) Show that the time T, required for the particle to strike the surface of the planet is given by

(1) $$T = \sqrt{\frac{d^3}{8GM}} \left[\frac{\pi}{2} + 2\sqrt{\frac{R}{d} - \left(\frac{R}{d}\right)^2} - \sin^{-1}\left(2\frac{R}{d} - 1\right) \right].$$

Notice that if the particle falls from a very great distance $(d \gg R)$, then

(2) $$T \approx \pi \sqrt{\frac{d^3}{8GM}}.$$

Incidentally, the approximation (2) happens to be exactly $\left(\dfrac{1}{4\sqrt{2}}\right)$ times the period of a satellite orbiting at a mean distance of d.

(b) The earth orbits the sun in a near circle at a distance of approximately 9.3×10^7 miles, and at a speed of approximately 18.5 mi/sec. If the earth were suddenly deprived of this speed (not likely), it would fall into the sun. The approximate masses of the sun and earth are, respectively, 2×10^{33} g and 6×10^{27} g, their ratio being over $300,000 : 1$; thus, we can take the earth as a free-falling particle. Use the result of part (a) to show that this free-fall would take approximately $64\frac{1}{2}$ days. (For R and G take the following approximate values:

$$R = 4.32 \times 10^5 \text{ mi and } G = 6.67 \times 10^{-8} \text{ cm}^3/\text{g sec}^2.)$$

13. *Rocket in space.* A small rocket leaves a space station. Its initial mass (including fuel) and the mass of the fuel are, respectively, m_0 and P. The rocket expels mass at a constant rate of $\dot{m} = dm/dt < 0$, and at a speed of u (a positive constant) relative to itself. It continues to fire in a straight line away from the station. In order to return, it must cut off its engine and turn itself around. At this point, it is receding from the station at a constant speed. If the rocket can (with "retro-grade fire") cancel its recession and establish any small closing speed by the time its fuel runs out, it will eventually drift back to the station; otherwise it will never return. (Assume the motion to take place in "force-free" space.)

(a) In view of the above, there is a critical speed of separation v_1, which must not be attained if the rocket is ever to return. Show that

$$(1) \qquad v_1 = \ln\left[1 - \frac{P}{m_0}\right]^{-(u/2)}.$$

(b) Likewise, there is a critical fraction, r, of the initial mass P, of fuel which must be reserved for the return fire. Show that

$$(2) \qquad r > \sqrt{\frac{m_0}{P} - 1}\left[\sqrt{\frac{m_0}{P}} - \sqrt{\frac{m_0}{P} - 1}\right].$$

[*Note:* Equality $(=)$ in (2) would not establish the closing speed mentioned above.]

3.6. Trajectories

Let us consider a one-parameter family of plane curves

$$f(x, y, c) = 0. \tag{3.6.1}$$

A curve which intersects each member of the family of curves given by (3.6.1) at a constant preassigned angle α is called an *isogonal trajectory* or an *α-trajectory* of the family. When $\alpha = 90°$, the curve is called an *orthogonal trajectory* of the family. If two one-parameter families of plane curves are such that each member of one family intersects each member of the other family at a constant preassigned angle α, each family is said to be *a family of isogonal trajectories* with respect to the other. When $\alpha = 90°$, we have *a family of orthogonal trajectories*.

Method for finding the isogonal and orthogonal trajectories; rectangular coordinates. We may utilize the material given in Section 1.2 to eliminate the parameter c from Equation (3.6.1). Suppose that the resulting differential equation is of the form

$$M(x, y)dx + N(x, y)dy = 0,$$

or   (3.6.2)

$$\frac{dy}{dx} = -\frac{M(x, y)}{N(x, y)} \equiv F(x, y).$$

This differential equation is satisfied by a curve of the family given in (3.6.1), going through the point (x, y). The tangent line to the curve at the point (x, y) has slope y'. That is, the angle β that the tangent line makes with the positive x axis satisfies the relation

$$\beta = \arctan F(x, y). \tag{3.6.3}$$

Let the trajectory intersect the above curve at the point (x, y) at an angle α. The angle ϕ, which the tangent line to the trajectory at the point (x, y) makes with the positive x axis, is thus α units greater than or less than β. That is [*see* Figure 3.6.1], $\beta = \phi \pm \alpha$. Thus,

$$\phi = \beta \pm \alpha. \tag{3.6.4}$$

Let the differential equation of the isogonal trajectories of the family of curves given by (3.6.1) be of the form

$$y' = G(x, y). \tag{3.6.5}$$

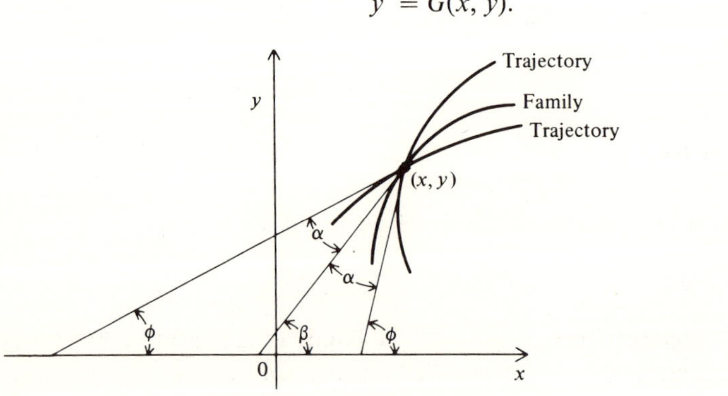

Figure 3.6.1.

Then at the point (x, y), we have

$$\phi = \text{arc tan } G(x, y). \tag{3.6.6}$$

Case 1. *When* $\alpha \neq 90°$. In view of Equations (3.6.2) through (3.6.6), we have

$$y' = G(x, y) = \tan \phi = \tan(\beta \pm \alpha) = \frac{\tan \beta \pm \tan \alpha}{1 \mp \tan \beta \tan \alpha} = \frac{F(x, y) \pm \tan \alpha}{1 \mp F(x, y)\tan \alpha}$$

$$= \frac{-\dfrac{M(x, y)}{N(x, y)} \pm \tan \alpha}{1 \pm \dfrac{M(x, y)}{N(x, y)} \tan \alpha}.$$

If we simplify the above equations, we find

$$[-M(x, y) + \tan \alpha\, N(x, y)]dx - [N(x, y) + \tan \alpha\, M(x, y)]dy = 0,$$
$$\phi = \beta + \alpha, \tag{3.6.7}$$

$$[M(x, y) + \tan \alpha\, N(x, y)]dx + [N(x, y) - \tan \alpha\, M(x, y)]dy = 0,$$
$$\phi = \beta - \alpha. \tag{3.6.8}$$

Equations (3.6.7) and (3.6.8) are the differential equations of the isogonal trajectories of the family of curves given by (3.6.1). *The solution of these differential equations gives the isogonal trajectories of the family of curves given by (3.6.1).*

[*Note:* We have here two one-parameter families of isogonal trajectories.]

Case 2. *When* $\alpha = 90°$. We then have

$$y' = G(x, y) = \tan \phi = \tan (\beta \pm \alpha) = \tan (\beta \pm 90°) = -\cot \beta = -\frac{1}{\tan \beta}$$

$$= -\frac{1}{F(x, y)} = -\frac{1}{-\dfrac{M(x, y)}{N(x, y)}} = \frac{N(x, y)}{M(x, y)}.$$

Thus,

$$N(x, y)dx - M(x, y)dy = 0. \tag{3.6.9}$$

Equation (3.6.9) is the differential equation of the orthogonal trajectories of the family of curves given by (3.6.1). *The solution of this equation gives the orthogonal trajectories of the family of curves given by (3.6.1).*

[*Note:* The slope of the orthogonal trajectory is equal to the negative reciprocal of the slope of the curve of the given family.]

Remark 3.6.1. In electrostatics, hydrodynamics, thermodynamics, and other applied fields, the concept of orthogonal trajectories plays an important role. For example, in applications to electrostatic and gravitational potential, one family of curves, say (3.6.1), are called the lines of force, while the other family of orthogonal trajectories, say $g(x, y, k) = 0$, are called the equipotential lines. Similarly, in hydrodynamics, they are called the stream lines and velocity potential lines. In heat-flow, they are called the heat-flow lines and isothermals.

Example 3.6.1. Find the orthogonal trajectories of the family of ellipses

$$x^2 + 2y^2 = c^2. \tag{3.6.10}$$

SOLUTION. To eliminate the parameter c from the given family, take the differential of Equation (3.6.10). Thus, the differential equation of the family is

$$x \, dx + 2y \, dy = 0. \tag{3.6.11}$$

In Equation (3.6.11), $M(x, y) = x$ and $N(x, y) = 2y$. Equation (3.6.9) thus becomes

$$2y \, dx - x \, dy = 0,$$

or

$$2 \frac{dx}{x} - \frac{dy}{y} = 0. \tag{3.6.12}$$

Integrating Equation (3.6.12) and simplifying the result, we obtain

$$x^2 = ky, \tag{3.6.13}$$

where k is an arbitrary constant (or parameter). Equation (3.6.13) gives the orthogonal trajectories of the family (3.6.10). They are a family of parabolas with vertex at the origin and focus along the y axis. Figure 3.6.2 illustrates a few curves from each family.

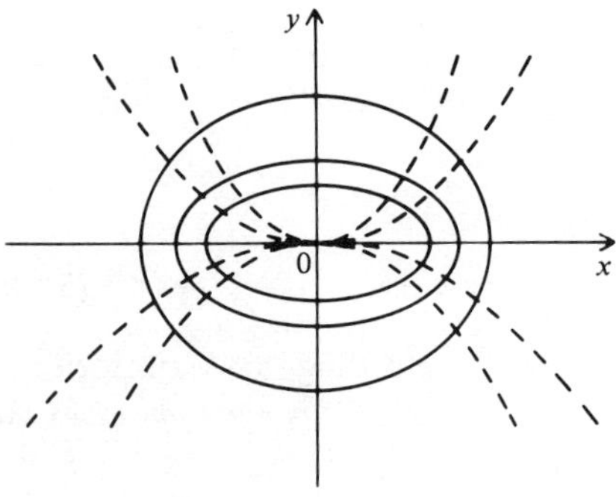

Figure 3.6.2.

Example 3.6.2.   Let us find the trajectories that intersect the one-parameter family of straight lines,

$$y = cx, \tag{3.6.14}$$

isogonally at the angle $\alpha = 60°$.

SOLUTION.   To eliminate the parameter c from the given family, write (3.6.14) as $y/x = c$, and then take the differential.   We thus obtain

$$y\,dx - x\,dy = 0. \tag{3.6.15}$$

In Equation (3.6.15), $M(x, y) = y$ and $N(x, y) = -x$.   Tan $\alpha = \tan 60° = \sqrt{3}$. Substituting these quantities in (3.6.7) and (3.6.8), we have

$$(-y - \sqrt{3}\,x)dx - (-x + \sqrt{3}\,y)dy = 0, \tag{3.6.16}$$

$$(y - \sqrt{3}\,x)dx + (-x - \sqrt{3}\,y)dy = 0. \tag{3.6.17}$$

We shall first solve Equation (3.6.16).   Equation (3.6.16) may be written as

$$(y + \sqrt{3}\,x)dx - (x - \sqrt{3}\,y)dy = 0,$$

which is a first-order homogeneous differential equation.   Using the substitution given in (2.4.14), that is, $y = vx$, the above equation becomes upon simplification

$$\frac{\sqrt{3}\,v\,dv}{v^2 + 1} - \frac{dv}{v^2 + 1} = -\frac{\sqrt{3}\,dx}{x}. \tag{3.6.18}$$

An integration of Equation (3.6.18) gives

$$\frac{\sqrt{3}}{2}\ln(v^2 + 1) - \text{arc tan } v = -\sqrt{3}\ln x - \sqrt{3}\ln k, \tag{3.6.19}$$

where the constant of integration was taken for convenience to be $-\sqrt{3}\ln k$. Equation (3.6.19) may be written as

$$\ln[k^2 x^2 (v^2 + 1)] - \frac{2}{\sqrt{3}}\text{arc tan } v = 0. \tag{3.6.20}$$

Since $v = y/x$, Equation (3.6.20) takes on the following form

$$\ln[k^2(x^2 + y^2)] - \frac{2}{\sqrt{3}}\text{arc tan }\frac{y}{x} = 0. \tag{3.6.21}$$

Similarly, the solution of Equation (3.6.17) is

$$\ln[b^2(x^2 + y^2)] + \frac{2}{\sqrt{3}}\text{arc tan }\frac{y}{x} = 0, \tag{3.6.22}$$

where b^2 is an arbitrary constant.

[*Note:* In polar coordinates, (3.6.21) and (3.6.22) are respectively the exponential spirals

$$r = \frac{1}{k}\, e^{\theta/\sqrt{3}} \quad \text{and} \quad r = \frac{1}{b}\, e^{-(\theta/\sqrt{3})}]$$

Method for finding the isogonal and orthogonal trajectories; polar coordinates.
Let us consider a one-parameter family of plane curves

$$\Phi(r,\, \theta,\, c) = 0, \tag{3.6.23}$$

where $(r,\, \theta)$ represent polar coordinates.   Suppose that upon eliminating the parameter c, the resulting differential equation of the family of curves given by (3.6.23) is of the form

$$P(r,\, \theta)dr + Q(r,\, \theta)d\theta = 0. \tag{3.6.24}$$

We may also write Equation (3.6.24) as

$$r\,\frac{d\theta}{dr} = -\frac{rP}{Q}. \tag{3.6.25}$$

Recall from calculus, that if ψ is the angle between the radius vector OP and the tangent line at P, then

$$\tan \psi = r\,\frac{d\theta}{dr}. \tag{3.6.26}$$

[*See* Figure 3.6.3.]   The angle ψ is measured positive in the counter-clockwise direction from the radius vector to the tangent line at P.   In view of (3.6.26), we may write (3.6.25) as

$$\tan \psi = -\frac{rP}{Q}. \tag{3.6.27}$$

Let Γ be any curve of the family (3.6.23), and let the isogonal trajectory intersect Γ at P at an angle α.   [*See* Figure 3.6.4.]   From Figure 3.6.4, we see that

$$\psi_2 = \psi_1 \pm \alpha. \tag{3.6.28}$$

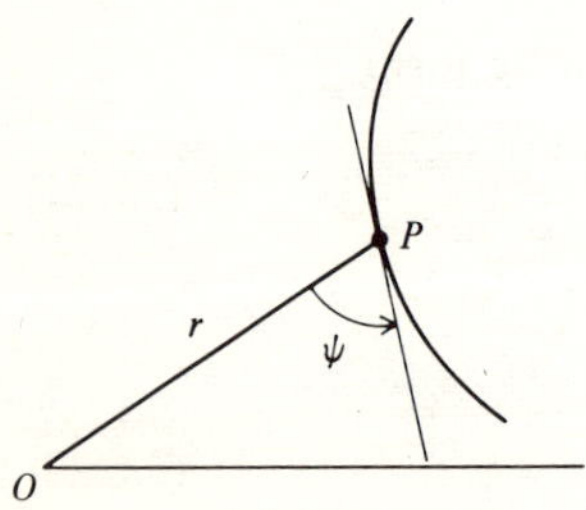

Figure 3.6.3.

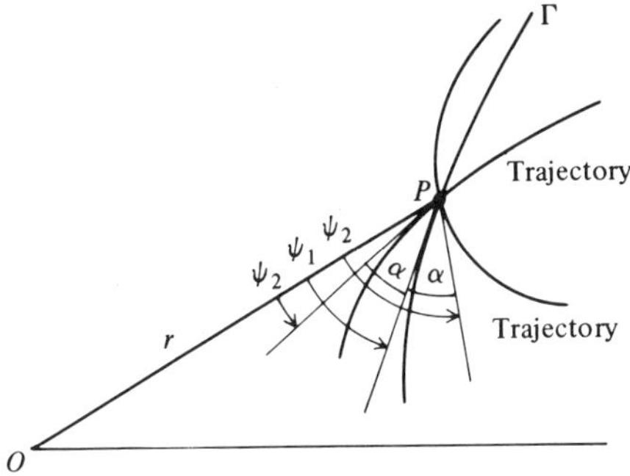

Figure 3.6.4.

Case 1. *When $\alpha \neq 90°$.* Utilizing the above equations, we obtain

$$r\frac{d\theta}{dr} = \tan\psi_2 = \tan(\psi_1 \pm \alpha) = \frac{\tan\psi_1 \pm \tan\alpha}{1 \mp \tan\psi_1 \tan\alpha} = \frac{-\dfrac{rP}{Q} \pm \tan\alpha}{1 \pm \dfrac{rP}{Q}\tan\alpha}.$$

If we simplify the above equations, we find

$$\left(-\frac{rP}{Q} + \tan\alpha\right)dr - r\left(1 + \frac{rP}{Q}\tan\alpha\right)d\theta = 0, \qquad \psi_2 = \psi_1 + \alpha, \qquad (3.6.29)$$

$$\left(\frac{rP}{Q} + \tan\alpha\right)dr + r\left(1 - \frac{rP}{Q}\tan\alpha\right)d\theta = 0, \qquad \psi_2 = \psi_1 - \alpha. \qquad (3.6.30)$$

The solutions of the differential equations (3.6.29) and (3.6.30) give the isogonal trajectories of the family of curves given by (3.6.23).

Case 2. *When $\alpha = 90°$.*

$$r\frac{d\theta}{dr} = \tan\psi_2 = \tan(\psi_1 \pm 90°) = -\cot\psi_1 = -\frac{1}{\tan\psi_1} = -\frac{1}{-\dfrac{rP}{Q}} = \frac{Q}{rP}.$$

Thus,

$$Q\,dr - r^2 P\,d\theta = 0. \qquad (3.6.31)$$

The solution of the differential equation (3.6.31) gives the orthogonal trajectories of the family of curves given by (3.6.23).

Example 3.6.3. Find the orthogonal trajectories of the one-parameter family of cardioids

$$r = c(1 - \sin\theta). \qquad (3.6.32)$$

SOLUTION.   Let us write (3.6.32) as

$$\frac{r}{1 - \sin \theta} = c.$$

Thus, the differential equation of the family of curves given by (3.6.32) is

$$(1 - \sin \theta)dr + r \cos \theta \, d\theta = 0. \tag{3.6.33}$$

In (3.6.33), $P = 1 - \sin \theta$ and $Q = r \cos \theta$.   Substituting these quantities in (3.6.31), we obtain

$$r \cos \theta \, dr - r^2(1 - \sin \theta)d\theta = 0,$$

or

$$\frac{dr}{r} - \sec \theta \, d\theta + \frac{\sin \theta}{\cos \theta} \, d\theta = 0. \tag{3.6.34}$$

An integration of (3.6.34) gives

$$\ln r - \ln (\sec \theta + \tan \theta) - \ln \cos \theta = \ln b,$$

or

$$\ln r = \ln [b(1 + \sin \theta)].$$

Hence, the orthogonal trajectories of the family (3.6.32) are given by

$$r = b(1 + \sin \theta). \tag{3.6.35}$$

Equation (3.6.35) is also a family of cardioids.   Actually, this family is the same as the given family of cardioids.   Thus, the family of cardioids (3.6.32) is *self-orthogonal*.   Figure 3.6.5 illustrates a few curves from each family.

Example 3.6.4.   Let us find the trajectories that intersect the one-parameter family of spirals (the spirals of Archimedes)

$$r = a\theta \tag{3.6.36}$$

isogonally at the angle $\alpha = 45°$.

SOLUTION.   To eliminate the parameter a, write (3.6.36) as $r/\theta = a$ and then take the differential.   We thus obtain

$$\theta \, dr - r \, d\theta = 0. \tag{3.6.37}$$

In (3.6.37), $P = \theta$ and $Q = -r$.   Tan $\alpha = \tan 45° = 1$.   Substituting these quantities in (3.6.29) and (3.6.30), we obtain

$$(\theta + 1)dr - r(1 - \theta)d\theta = 0, \tag{3.6.38}$$

$$(-\theta + 1)dr + r(1 + \theta)d\theta = 0. \tag{3.6.39}$$

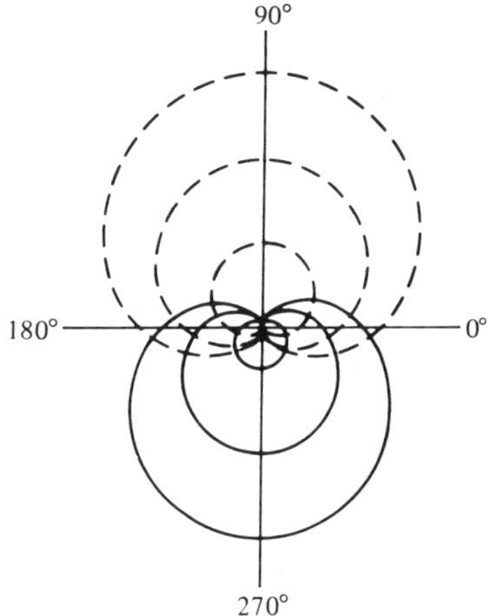

Figure 3.6.5.

We may write Equation (3.6.38) as

$$\frac{dr}{r} = \frac{(1 - \theta)d\theta}{1 + \theta},$$

or

$$\frac{dr}{r} = -d\theta + \frac{2d\theta}{\theta + 1}.$$

(3.6.40)

An integration of (3.6.40) gives

$$\ln r = -\theta + 2\ln (\theta + 1) + \ln c,$$

or

$$r = c(\theta + 1)^2 e^{-\theta},$$

(3.6.41)

where c is an arbitrary constant of integration.

Similarly, the solution of (3.6.39) is

$$r = k(\theta - 1)^2 e^{\theta},$$

(3.6.42)

where k is another arbitrary constant of integration.   Equations (3.6.41) and (3.6.42) give the isogonal trajectories of the family (3.6.36).

Exercises 3.6

In Exercises 1 through 12, find the orthogonal trajectories of the given family of curves.   Draw a few representative curves in each family for Exercises 1, 2, 3, 4, 6, and 7.

1. $x^2 + y^2 = c^2$.
2. $ay^2 = x^3$.
3. $x^2 + y^2 = cx$.
4. $y = ce^{mx}$ with m held fixed.
5. $x^{4/3} + y^{4/3} = c^{4/3}$.
6. $x^2 + 4y^2 = cy$.
7. $y^2 = 2cx + c^2$.   [*Note:* This family is self-orthogonal.]

8. $r = c(\sec \theta + \tan \theta)$.
9. $r = \dfrac{c\theta}{1 + \theta}$.

10. $r = c + \sin n\theta$.
11. $r^n = c^n \cos n\theta$.
12. $r = c(1 + \cos^2 \theta)$.

In Exercises 13 through 16, find the trajectories that intersect the given family isogonally at the angle α.

13. $x^2 + y^2 = a^2$, $\alpha = 60°$.
14. $y^2 = 4ax$, $\alpha = 45°$.

15. $r = 2a \cos \theta$, $\alpha = \dfrac{\pi}{4}$.
16. $r^2\theta = a$, $\alpha = \dfrac{\pi}{6}$.

17. Find the value of the constant b so that the families $cy^2 = x^3$ and $2x^2 + by^2 = k^2$ are orthogonal.

18. Find the value of p such that the curves $x^p + y^p = c^p$ are the orthogonal trajectories of the family $x^{3/2} - y^{3/2} = k^{3/2}$.

19. Show that the family of confocal central conics

$$\frac{x^2}{a^2 + \lambda} + \frac{y^2}{b^2 + \lambda} = 1,$$

with a and b fixed, is self-orthogonal.

20. Find the orthogonal trajectories of the one-parameter family of curves

$$x^2 + y^2 + c^2 = 1 + 2cxy.$$

21. Let us consider the complex-valued function f defined in a domain D of the xy-plane by

(1)
$$f(z) = u(x, y) + iv(x, y),$$

where $z = x + iy$, $i^2 = -1$, x, y are real numbers, u, v are real-valued functions. The functions u and v are called, respectively, the *real* part and the *imaginary* part of f.

Two complex-valued functions are *equal* if and only if their respective real and imaginary parts are equal. For example, if $f_1(z) = x^2 - y^2 + i2xy$ and $f_2(z) = 3 + i4$, then $f_1 = f_2$ means that $x^2 - y^2 = 3$ and $2xy = 4$. The *conjugate* of f denoted by $\bar{f}$ is defined by $\bar{f}(z) = u(x, y) - iv(x, y)$. For example, if $f(z) = x + iy$, then $\bar{f}(z) = x - iy$, or if $f(z) = x^2 - y^2 - i2xy$, then $\bar{f}(z) = x^2 - y^2 + i2xy$. Note that $f(z)\bar{f}(z) = (u + iv)(u - iv) = u^2 + v^2 \geq 0$. The expression $f(z)\bar{f}(z)$ is usually written $|f(z)|^2$. The quantity $|f(z)| = \sqrt{u^2 + v^2} \geq 0$, is called the *absolute value* of the function f.

Suppose that we wish to determine the functions u and v such that

$$u + iv = f(z) = z^2.$$

Since $z = x + iy$, we may write the above expression as

$$u + iv = (x + iy)^2 = x^2 + i2xy + i^2y^2 = x^2 - y^2 + i2xy.$$

Thus,

$$u = x^2 - y^2, \qquad v = 2xy.$$

As another example, let us determine the functions u and v such that

$$u + iv = f(z) = \frac{z}{z + i}, \qquad z \neq -i.$$

Since $z = x + iy$, the above expression becomes

$$u + iv = \frac{x + iy}{x + iy + i} = \frac{x + iy}{x + i(y + 1)}.$$

We now multiply the numerator and denominator by the conjugate of the function g defined by $g(z) = x + i(y + 1)$; we obtain

$$u + iv = \frac{(x + iy)\overline{g(z)}}{g(z)\overline{g(z)}} = \frac{(x + iy)[x - i(y + 1)]}{[x + i(y + 1)][x - i(y + 1)]}$$

$$= \frac{x^2 - i^2y(y + 1) - ix(y + 1) + iyx}{x^2 - i^2(y + 1)^2} = \frac{x^2 + y(y + 1) - ix}{x^2 + (y + 1)^2}$$

$$= \frac{x^2 + y^2 + y}{x^2 + (y + 1)^2} - i\frac{x}{x^2 + (y + 1)^2}.$$

Thus,

$$u = \frac{x^2 + y^2 + y}{x^2 + (y + 1)^2}, \qquad v = -\frac{x}{x^2 + (y + 1)^2} \qquad (x, y) \neq (0, -1).$$

The following result is proved in courses on complex variables. [*See*, for example, [33], p. 85.]

A function f defined by $f(z) = u(x, y) + iv(x, y)$ is said to be analytic in a domain D of the xy-plane, if the four partial derivatives u_x, u_y, v_x, v_y exist, are continuous, and satisfy the Cauchy-Riemann differential equations:

$$(2) \qquad u_x = v_y, \qquad u_y = -v_x$$

at each point of D.

We shall now consider the level curves of the real part u and the imaginary part v of an analytic function f defined in a domain D. That is,

$$(3) \qquad u(x, y) = c, \qquad v(x, y) = k$$

where c and k are arbitrary constants (or parameters).

(a) Suppose that $f(z) = u(x, y) + iv(x, y)$ is analytic in a domain D. Show that the family of level curves $u(x, y) = c$ are the orthogonal trajectories of the family of level curves $v(x, y) = k$ in the domain D.

(b) Suppose that $f(z) = u(x, y) + iv(x, y)$ is analytic in a domain D. Suppose also that the second partial derivatives of u and v exist and are continuous in D. Show that the functions u and v satisfy the *Laplace partial differential equation:*

$$(4) \qquad \frac{\partial^2 u}{\partial x^2} + \frac{\partial^2 u}{\partial y^2} = 0, \qquad \frac{\partial^2 v}{\partial x^2} + \frac{\partial^2 v}{\partial y^2} = 0,$$

at each point of D. [*Note:* In view of Exercise 1.4.13(c), the functions u and v in part (b) are harmonic in D.]

The above results are useful in obtaining the lines of force and the lines of constant potential in electrostatics, or stream lines in hydrodynamics.

In Exercises 22 through 26, the function f defined by $f(z) = u(x, y) + iv(x, y)$ is analytic except at the indicated point. (a) Find the family of level curves of the real part and imaginary part of $f(z)$. (b) Identify by name each family of level curves. (c) Sketch a few level curves from each family. (d) Verify that the functions u and v satisfy the Laplace partial differential equation.

22. $f(z) = z.$ 23. $f(z) = z^2.$ 24. $f(z) = 1/z, \quad z \neq 0.$[10]

25. $f(z) = \dfrac{z+1}{z}, \quad z \neq 0.$ 26. $f(z) = \dfrac{z-1}{z+1}, \quad z \neq -1.$

27. Suppose that $f(z) = u(x, y) + iv(x, y)$ is analytic in a domain D, and that $u(x, y) = x^3 - 3xy^2$. Utilize Equation (2) of Exercise 21 to determine the imaginary part $v(x, y)$ of $f(z)$. (Review the material given in Exercise 1.4.17.) Also, sketch a few of the level curves $u(x, y) = c$ and $v(x, y) = k$ of $f(z)$.

3.7. A Problem in the Calculus of Variations

Let J denote the closed interval $x_1 \leq x \leq x_2$. Let us consider the integral

$$I = \int_{x_1}^{x_2} f[x, y(x), y'(x)]\,dx, \qquad (3.7.1)$$

where the function f has continuous second partial derivatives with respect to any one, or any combination of its arguments x, y, and y' for all $x \in J$, and the function y has a continuous second derivative for all $x \in J$. We shall now describe a problem of considerable importance in the calculus of variations. We consider the problem of determining a function y which renders the integral I an extremum (that is, a minimum or a maximum), and which satisfies the prescribed end conditions $y(x_1) = y_1$ and $y(x_2) = y_2$. Here we are given in advance the function f and the constants x_1, y_1, x_2, and y_2. Also, we know in advance that there exists a function y which renders the integral I

---

[10] $z = x + iy = 0$ if and only if $x = 0$ and $y = 0$. Thus, $z \neq 0$ if and only if $(x, y) \neq (0, 0)$.

an extremum.   As we shall presently see, if the function y renders I an extremum, it must satisfy a certain differential equation from which $y(x)$ may be determined.

Theorem 3.7.1.   *A necessary condition for the integral*

$$I = \int_{x_1}^{x_2} f[x, y(x), y'(x)]dx \qquad [3.7.1]$$

to be an extremum is that

$$\frac{\partial f}{\partial y} - \frac{d}{dx}\left(\frac{\partial f}{\partial y'}\right) = 0. \qquad (3.7.2)$$

Equation (3.7.2) is called the *Euler-Lagrange differential equation.*

PROOF.   We know in advance that there exists a function y, $y(x_1) = y_1$, $y(x_2) = y_2$, which renders the integral I an extremum.   Let

$$Y(x, \alpha) = y(x) + \alpha\eta(x), \qquad (3.7.3)$$

where α is a parameter (and independent of x), and η is any arbitrary function with continuous first derivative for all $x \in J$ and having the property that

$$\eta(x_1) = \eta(x_2) = 0. \qquad (3.7.4)$$

[*See* Figure 3.7.1.]   Define

$$I(\alpha) = \int_{x_1}^{x_2} f[x, Y, Y']dx$$

$$= \int_{x_1}^{x_2} f[x, y + \alpha\eta, y' + \alpha\eta']dx. \qquad (3.7.5)$$

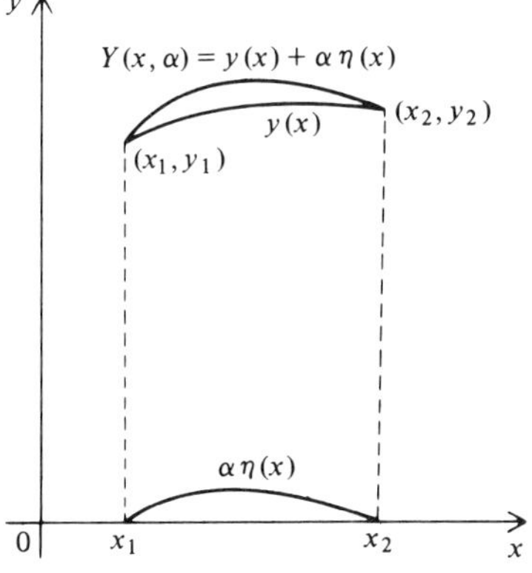

Figure 3.7.1.

By assumption $y(x)$ renders the integral I an extremum. Thus, $I(\alpha)$ is an extremum for $\alpha = 0$. A necessary condition that this be so is that

$$\left.\frac{dI(\alpha)}{d\alpha}\right|_{\alpha=0} = 0. \tag{3.7.6}$$

From (3.7.3) we see that $\partial Y/\partial\alpha = \eta$, $\;Y' = y' + \alpha\eta'$ and thus $\partial Y'/\partial\alpha = \eta'$. Utilizing Exercise 1.3.8, we obtain

$$\frac{dI(\alpha)}{d\alpha} = \frac{d}{d\alpha}\int_{x_1}^{x_2} f(x,\,Y,\,Y')dx = \int_{x_1}^{x_2}\frac{\partial f}{\partial\alpha}\,dx = \int_{x_1}^{x_2}\left(\frac{\partial f}{\partial Y}\frac{\partial Y}{\partial\alpha} + \frac{\partial f}{\partial Y'}\frac{\partial Y'}{\partial\alpha}\right)dx$$

$$= \int_{x_1}^{x_2}\left(\frac{\partial f}{\partial Y}\,\eta + \frac{\partial f}{\partial Y'}\,\eta'\right)dx. \tag{3.7.7}$$

In view of Equation (3.7.3), we have when $\alpha = 0$ that $Y(x,\,0) = y(x)$ and $Y'(x,\,0) = y'(x)$. Thus, when $\alpha = 0$, we have

$$\frac{\partial f}{\partial Y}\,\eta + \frac{\partial f}{\partial Y'}\,\eta' = \frac{\partial f}{\partial y}\,\eta + \frac{\partial f}{\partial y'}\,\eta'.$$

In view of Equations (3.7.6) and (3.7.7), we then obtain

$$\int_{x_1}^{x_2}\left(\frac{\partial f}{\partial y}\,\eta + \frac{\partial f}{\partial y'}\,\eta'\right)dx = \left.\frac{dI(\alpha)}{d\alpha}\right|_{\alpha=0} = 0. \tag{3.7.8}$$

Let us write Equation (3.7.8) as

$$\int_{x_1}^{x_2}\frac{\partial f}{\partial y}\,\eta\,dx + \int_{x_1}^{x_2}\frac{\partial f}{\partial y'}\,\eta'\,dx = 0. \tag{3.7.9}$$

We now integrate by parts the second integral in the left-hand member of Equation (3.7.9). Let

$$u = \frac{\partial f}{\partial y'}, \quad\text{then}\quad du = \frac{d}{dx}\left(\frac{\partial f}{\partial y'}\right)dx;$$

$$dv = \eta'\,dx, \quad\text{then}\quad v = \eta.$$

Thus, we may write Equation (3.7.9) as

$$\int_{x_1}^{x_2}\frac{\partial f}{\partial y}\,\eta\,dx + \left(\eta\,\frac{\partial f}{\partial y'}\right)\Bigg|_{x=x_1}^{x=x_2} - \int_{x_1}^{x_2}\frac{d}{dx}\left(\frac{\partial f}{\partial y'}\right)\eta\,dx = 0. \tag{3.7.10}$$

From (3.7.4), $\eta(x_1) = \eta(x_2) = 0$. Thus, the second term in Equation (3.7.10) is zero, and we have

$$\int_{x_1}^{x_2}\left[\frac{\partial f}{\partial y} - \frac{d}{dx}\left(\frac{\partial f}{\partial y'}\right)\right]\eta\,dx = 0. \tag{3.7.11}$$

Let

$$F(x) = \frac{\partial f}{\partial y} - \frac{d}{dx}\left(\frac{\partial f}{\partial y'}\right). \qquad (3.7.12)$$

By the assumption on the function f, the function F is continuous for all $x \in J$. *We would like to show that $F(x) = 0$ identically for all $x \in J$.* Suppose to the contrary, that $F(x) \neq 0$ for some $x \in J$. Then there exists a point $c \in J$ such that $F(c) \neq 0$. Suppose for definiteness that $F(c) > 0$. Since the function F is continuous at $x = c$, we know that given any $\varepsilon > 0$, there exists a number $\delta > 0$, such that $|F(x) - F(c)| < \varepsilon$ for all points $x \in J$ satisfying $|x - c| < \delta$. This statement can be expressed equivalently as

$$F(c) - \varepsilon < F(x) < F(c) + \varepsilon \quad \text{whenever } c - \delta < x < c + \delta, \quad x \in J.$$

In particular, if we take $\varepsilon = \tfrac{1}{2}F(c)$, we then have

$$\tfrac{1}{2}F(c) < F(x) < \tfrac{3}{2}F(c) \quad \text{whenever } c - \delta < x < c + \delta, \quad x \in J.$$

Since $F(c) > 0$, we see that

$$F(x) > 0 \quad \text{whenever } c - \delta < x < c + \delta, \quad x \in J.$$

Thus, there exists a closed interval J_1: $\alpha \le x \le \beta$, where $\alpha > c - \delta$, $\beta < c + \delta$ and $J_1 \subset J$, such that $F(x) > 0$ for all $x \in J_1$.

The function η [*see* Equation (3.7.3)] was given to be any arbitrary function having a continuous derivative (and hence, it is continuous) for all $x \in J$, and $\eta(x_1) = \eta(x_2) = 0$. Thus, we may construct $\eta(x)$ to be positive for all $x \in J_1$ and zero elsewhere. For example,

$$\eta(x) = \begin{cases} 0, & x_1 \le x \le \alpha, \\ (x - \alpha)^2(x - \beta)^2, & \alpha \le x \le \beta, \\ 0, & \beta \le x \le x_2. \end{cases}$$

(*See* Figure 3.7.2.)   Hence

$$\int_{x_1}^{x_2} F(x)\eta(x)dx = \int_{x_1}^{\alpha} F(x)\eta(x)dx + \int_{\alpha}^{\beta} F(x)\eta(x)dx + \int_{\beta}^{x_2} F(x)\eta(x)dx$$

$$= 0 + \int_{\alpha}^{\beta} F(x)\eta(x)dx + 0.$$

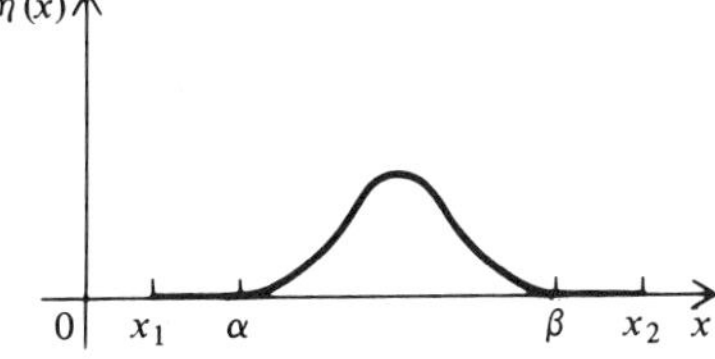

Figure 3.7.2.

Since $F(x)\eta(x) > 0$ for all $x \in J_1$, it follows that

$$\int_{x_1}^{x_2} F(x)\eta(x)dx = \int_{x_1}^{x_2} \left[\frac{\partial f}{\partial y} - \frac{d}{dx}\left(\frac{\partial f}{\partial y'}\right)\right]\eta(x)dx > 0,$$

which contradicts the result given in Equation (3.7.11). A similar contradiction is obtained if we assume that $F(c) < 0$ for at least one point $c \in J$. Thus, $F(x) = 0$ identically for all $x \in J$. Consequently we have that

$$\frac{\partial f}{\partial y} - \frac{d}{dx}\left(\frac{\partial f}{\partial y'}\right) = 0,$$

and the theorem is thus established.

Remark 3.7.1.   A function y which renders the integral I of (3.7.1) an extremum is called an *extremizing function.*

If the function f is explicitly independent of (or is free of) the independent variable x, that is,

$$f = f(y, y'), \tag{3.7.13}$$

then we can arrive at an integral (solution) of Equation (3.7.2). To see this, we first verify

$$\frac{d}{dx}\left(f - y'\frac{\partial f}{\partial y'}\right) = 0. \tag{3.7.14}$$

We have

$$\frac{d}{dx}\left(f - y'\frac{\partial f}{\partial y'}\right) = \frac{\partial f}{\partial y}y' + \frac{\partial f}{\partial y'}y'' - \left[y''\frac{\partial f}{\partial y'} + y'\frac{d}{dx}\left(\frac{\partial f}{\partial y'}\right)\right]$$

$$= y'\left[\frac{\partial f}{\partial y} - \frac{d}{dx}\left(\frac{\partial f}{\partial y'}\right)\right]$$

$$= 0,$$

in view of Equation (3.7.2). Next, we integrate Equation (3.7.14) with respect to x, and we obtain

$$f - y'\frac{\partial f}{\partial y'} = \text{constant}. \tag{3.7.15}$$

Thus, the extremizing function y may be obtained as the solution of a first-order differential equation involving y and y' only.

Remark 3.7.2.   If the function f is explicitly independent of the dependent variable y, then $\partial f/\partial y = 0$, and Equation (3.7.2) reduces to

$$\frac{d}{dx}\left(\frac{\partial f}{\partial y'}\right) = 0. \tag{3.7.16}$$

An integration of Equation (3.7.16) gives

$$\frac{\partial f}{\partial y'} = \text{constant.} \qquad (3.7.17)$$

Thus, the extremizing function y may be obtained as the solution of a first-order differential equation involving y' and x only.

Example 3.7.1.   Let us find the function y which renders the integral

$$I = \int_{x_1}^{x_2} ds = \int_{x_1}^{x_2} \sqrt{1 + y'^2}\, dx$$

a minimum.   [That is, we would like to find the curve $y = y(x)$ of shortest length in the xy-plane, and connecting the points $y(x_1) = y_1$ and $y(x_2) = y_2$.]

SOLUTION.   Here the function f, namely,

$$f = \sqrt{1 + y'^2}$$

is explicitly independent of the dependent variable y.   Utilizing Equation (3.7.17) with constant $= 1/\alpha$ and $\partial f/\partial y' = y'/\sqrt{1 + y'^2}$, we have

$$\frac{y'}{\sqrt{1 + y'^2}} = \frac{1}{\alpha}. \qquad (3.7.18)$$

Simplifying Equation (3.7.18), we get

$$y'^2 = \frac{1}{\alpha^2 - 1}. \qquad (3.7.19)$$

If we solve Equation (3.7.19) subject to the conditions that $y(x_1) = y_1$ and $y(x_2) = y_2$, we find

$$y(x) = \frac{y_2 - y_1}{x_2 - x_1}\, x + \frac{x_2 y_1 - x_1 y_2}{x_2 - x_1}. \qquad (3.7.20)$$

Thus, the shortest plane curve connecting the two given points is a straight line.

Remark 3.7.3.   In view of Equation (3.7.2), we see that if y is an extremizing function for the integral

$$\int_{x_1}^{x_2} Kf[x, y(x), y'(x)]dx,$$

where K is a constant, then y is also an extremizing function for the above integral when the factor K is omitted, and vice versa.   Thus, constant factors of an integrand may be omitted.

Exercises 3.7

1. *Minimum surface of revolution.* Suppose that a smooth curve $y = y(x) > 0$ for all $x \in [x_1, x_2]$.   Let $y = y(x)$ be graphed from A: (x_1, y_1) to B: (x_2, y_2). Rotation of the arc $\overset{\frown}{AB}$ about the x axis generates the surface of a certain solid of revolution.   The area of the surface S generated is given by

$$S = \int_{x_1}^{x_2} 2\pi y \sqrt{1 + y'^2} \, dx.$$

Determine the positive function y which generates the minimum area of the surface S.

2. *Brachistochrome problem.*   This problem is one of determining a curve (or path) along which a particle will fall from one given point to another in the shortest time.   Suppose that a particle of mass m slides down a frictionless wire connecting the points $P_0(0, 0)$ and $P_1(x_1, y_1)$.   Let us take the positive direction of the y axis vertically downward with $P_0(0, 0)$ as origin.   [*See* Figure 3.7.3.]
   Suppose that the particle of mass m starts from rest at $P_0(0, 0)$.   Thus, the initial velocity of the particle is zero, and hence its kinetic energy at $P_0(0, 0)$ is also zero.   The work done by gravity in moving the particle (along the curve) from $(0, 0)$ to any point (x, y) is mgy and this must equal the change in kinetic energy.   Hence,

$$mgy = \tfrac{1}{2}mv^2 - \tfrac{1}{2}m(0)^2 = \tfrac{1}{2}mv^2.$$

Thus, $v = \sqrt{2gy}$.   Since $v = ds/dt$, we have

$$dt = \frac{ds}{v} = \frac{ds}{\sqrt{2gy}} = \frac{\sqrt{1 + y'^2}}{\sqrt{2gy}} \, dx = \frac{1}{\sqrt{2g}} \sqrt{\frac{1 + y'^2}{y}} \, dx.$$

We wish to find a curve $y = y(x)$ which renders the time-integral

$$T = \int_0^T dt = \int_0^{x_1} \frac{1}{\sqrt{2g}} \sqrt{\frac{1 + y'^2}{y}} \, dx$$

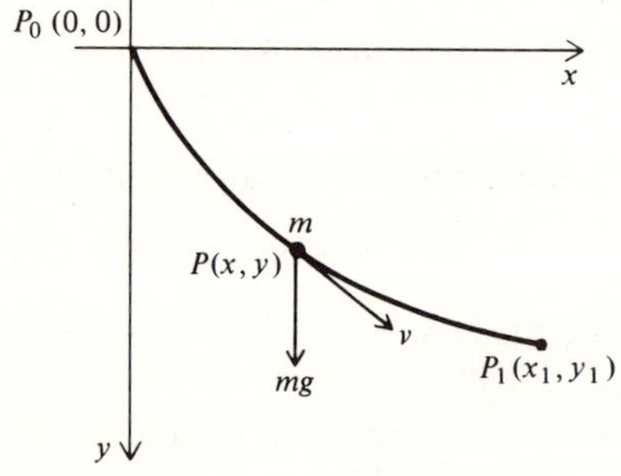

Figure 3.7.3.

a minimum and passes through the points $P_0(0, 0)$ and $P_1(x_1, y_1)$. From Remark 3.7.3, it is sufficient to minimize the integral

$$\int_0^{x_1} \sqrt{\frac{1 + y'^2}{y}} \, dx.$$

Utilize Equation (3.7.15) with constant $= 1/k$, $f = \sqrt{(1 + y'^2)/y}$ and $\partial f/\partial y' = y'/\sqrt{y(1 + y'^2)}$. Show that

(1)
$$dx = \frac{\sqrt{y}}{\sqrt{k^2 - y}} \, dy.$$

Substitute $y = k^2 \sin^2 t$ and $dy = 2k^2 \sin t \cos t \, dt$ in Equation (1), and then integrate between the limits $x = 0$, $t = 0$ and $x = x$, $t = t$. Show that

(2)
$$x = \frac{k^2}{2} (2t - \sin 2t).$$

Show that the solution is given by

(3)
$$x = K^2(2t - \sin 2t),$$
$$y = K^2(1 - \cos 2t),$$

where the constant K is determined by the condition that the curve passes through the point (x_1, y_1).[11]

3. *Fermat's principle.* Suppose that in a plane transparent medium, the velocity of light varies continuously (that is, varies from point to point). Denote respectively by $v(x, y)$ and $\eta(x, y)$ the velocity and index of refraction at an arbitrary point (x, y). By definition, $\eta(x, y) = c/v(x, y)$, where c is constant (velocity of light in a vacuum). Suppose that the disturbance (ray of light) travels along an arc $y = y(x)$. Then,

$$dt = \frac{ds}{v} = \frac{ds}{\dfrac{c}{\eta(x, y)}} = \frac{\eta(x, y)}{c} \, ds = \frac{1}{c} \, \eta(x, y)\sqrt{1 + y'^2} \, dx.$$

Thus, the time required for the disturbance to traverse the arc $y = y(x)$ joining the points P_1: (x_1, y_1) and P_2: (x_2, y_2) is given by

$$T = \int_0^T dt = \int_{x_1}^{x_2} \frac{1}{c} \, \eta(x, y)\sqrt{1 + y'^2} \, dx.$$

It has been verified experimentally that the path of a ray of light in a medium in which the velocity varies continuously is always one on which the time-integral T

---

[11] The equations given in (3) are the parametric equations of a cycloid. Thus, the curve down which a particle starting with initial velocity zero at the point $P_0(0, 0)$ will fall in the shortest time to a second point $P_1(x_1, y_1)$ is necessarily an arc having parametric equations of the form given in (3).

is, for short arcs at least, a minimum.   Thus, the problem of minimizing the time-integral T is that of determining the paths of rays of light in a plane medium whose variable index of refraction is $\eta(x, y)$.   This is known as Fermat's principle.

Suppose that $\eta(x, y) = bx$, where b is a constant.   Determine the curve $y = y(x)$ which passes through the points $(0, 0)$ and (x_2, y_2) and renders the time-integral T a minimum.

In the following exercises, the functions f and y are assumed to satisfy the conditions given in the first paragraph of Section 3.7.   Also, the condition that the curve passes through the points (x_1, y_1) and (x_2, y_2) will enable one to determine the constants k and c appearing in these exercises.   The question as to whether it is always possible to determine the constants k and c and, if so, whether there is a maximum or minimum, will not be considered here.   It is physically evident in many cases that the solution exists.   We shall primarily be interested in those functions y which, from physical considerations, are known to render the given integral an extremum.

4. (a) Show that the function y which renders the integral

$$I = \int_{x_1}^{x_2} g(y)\sqrt{1 + y'^2}\, dx$$

an extremum, satisfies the equation

$$\int \frac{k\, dy}{\sqrt{[g(y)]^2 - k^2}} = x + c.$$

(b) In part (a), if $g(y) = 1/y$, $y > 0$, show that the minimizing function y is a semicircular arc given by

$$y^2 + (x + c)^2 = \frac{1}{k^2}, \qquad y > 0.$$

5. (a) Show that the function y which renders the integral

$$I = \int_{x_1}^{x_2} g(x)\sqrt{1 + y'^2}\, dx$$

an extremum, satisfies the equation

$$y = \int \frac{k\, dx}{\sqrt{[g(x)]^2 - k^2}} + c.$$

(b) In part (a), if $g(x) = e^x$, show that the minimizing function y is given by

$$\cos(y - c) = ke^{-x} \quad \text{or} \quad \cos(y - c) = e^{-(x-k)}.$$

6. Show that the function y which renders the integral

$$I = \int_{x_1}^{x_2} \sqrt{y}\,\sqrt{1 - y'^2}\,dx, \qquad y > 0,$$

a minimum, is a parabolic arc given by

$$(x + c)^2 = -4k(y - k), \qquad y > 0.$$

7. (a) Show that, if the function y renders the integral

$$I = \int_{x_1}^{x_2} [p(x)y'^2 + 2q(x)yy' + r(x)y^2]dx$$

an extremum, then the extremizing function satisfies the second-order linear differential equation

$$py'' + p'y' + (q' - r)y = 0.$$

(b) In part (a), if $q' = r$, show that the extremizing function is given by

$$y = \int \frac{k\,dx}{p(x)} + c.$$

(c) Show that the extremizing function of the integral

$$I = \int_{x_1}^{x_2} [\sqrt{x + 1}\,y'^2 + 2(x^3 + x)yy' + (3x^2 + 1)y^2]dx$$

is given by

$$y = 2k\sqrt{x + 1} + c.$$

8. Show that, if the function y renders the integral

$$I = \int_{x_1}^{x_2} [p^2(x)y'^2 + r^2(x)y^2]dx$$

an extremum, then

$$I = (p^2 yy')\Big|_{x=x_1}^{x=x_2}.$$

[This result is utilized in the calculus of variations to establish other basic results.]

9. *Mirage on a highway.*[12]   Anyone who has driven a car on a hot summer day knows about the "puddles" that emerge on the highway, then vanish as one approaches them.   These are not reflections; they are actually mirages of the distant sky.   The air in contact with the hot road is expanded (that is, less dense) than that say of a few inches above it.   Thus, the refractive index of the air increases rapidly with height to its normal value at perhaps a foot or so above the surface.   Therefore, the closer a light ray approaches to the surface, the more will be its curvature, with practically no bending taking place above the first foot, where the air temperature has attained its normal value.

[12] This problem was proposed and solved by Richard Amtman.

A light ray will bend upward from the highway (in the direction of increasing η; that is $\nabla\eta$) and emerge on a new course [*see* Figure 3.7.4] provided its incident direction θ was not so steep as to strike the highway. [*See* Figure 3.7.5.] Let $\eta(y)$ be the refractive index at a height y above the surface. Let

(1)
$$\begin{cases} \eta(0) = a, \\ \eta(\infty) = b, \end{cases}$$

where by ∞ we mean roughly >1 foot, since η is essentially constant except for the lowest few inches. Note that we could even specify a function $\eta(y)$ satisfying (1), namely,

(2)
$$\eta(y) = (a - b)e^{-\gamma y} + b,$$

where we are still free to adjust γ to conform with some intermediate point. However, we can derive a rather nice general result without even specifying $\eta(y)$ any further than conditions (1). Thus, a depends only upon the temperature of the road, and b depends upon the general air temperature. In fact, $b \approx 1.00029$.

We want to extremize

(3)
$$I = \int_{x_1}^{x_2} \eta(y)\sqrt{1 + y'^2}\, dx,$$

where

(4)
$$f = \eta(y)\sqrt{1 + y'^2}$$

is free of the variable x. Thus, in view of (3.7.15), we have

(5)
$$f - y'\,\frac{\partial f}{\partial y'} = k.$$

Utilize (4) to show that (5) can be expressed as

(6)
$$y'^2 = \frac{[\eta(y)]^2 - k^2}{k^2},$$

where $\eta(0) = a$ and $\eta(\infty) = b$. First of all, we would like to find the maximum initial angle of incidence θ, for which the ray will not strike the surface. The path of such a ray must satisfy $y' = 0$ when $y = 0$. Hence, $y' = 0$ when $\eta(0) = a$. Thus, in view of (6), we have

$$0 = \frac{a^2 - k^2}{a^2} \quad \text{or} \quad a^2 = k^2.$$

Consequently, (6) becomes

(7)
$$y'^2 = \frac{[\eta(y)]^2 - a^2}{a^2}.$$

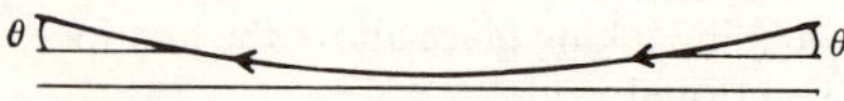

Figure 3.7.4.

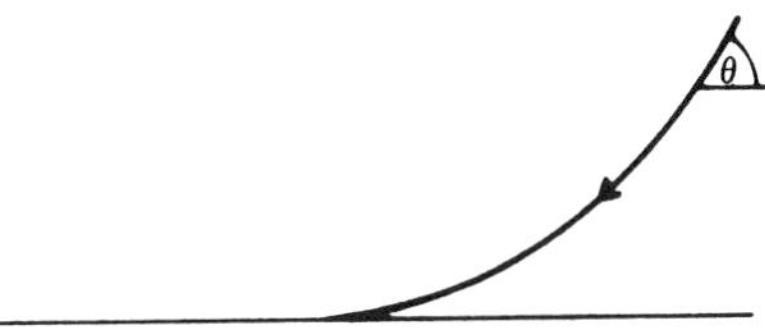

Figure 3.7.5.

For $y > 1$ foot, we have $y'^2 = \tan^2 \theta$ and $\eta = b$.   Hence, (7) becomes

$$(8) \qquad \tan^2 \theta = \frac{b^2 - a^2}{a^2} = \left(\frac{b}{a}\right)^2 - 1.$$

Thus,

$$(9) \qquad \frac{a}{b} = \cos \theta \quad \text{or} \quad \tan \theta = \sqrt{\left(\frac{b}{a}\right)^2 - 1}.$$

Since θ is the maximum angle of incidence that can give rise to a mirage, we can compute the closest distance, d, at which a mirage can be seen by a driver whose eyes are h feet above a flat highway.   We have immediately

$$(10) \qquad d = \frac{h}{\tan \theta} = \frac{h}{\sqrt{\left(\frac{b}{a}\right)^2 - 1}}.$$

Assume that

$$(11) \qquad \eta = 1 + 0.00029\left(\frac{300}{T}\right),$$

where T is the absolute temperature, that is, $T = 300°\text{K} \approx 27°\text{C}$.   Suppose that the road temperature is $320°\text{K} \approx 47°\text{C}$ and the air temperature is $300°\text{K}$.   Show that the closest distance at which a mirage can be seen, if a driver whose eyes are 4 feet above a flat highway, is $d = 632$ ft (approximately).

10. *Geodesic on the surface of a sphere.*   The arc of minimum length connecting two given points on a surface is called a *geodesic* for the surface.   The parameters u and v most convenient for describing position on the sphere are the colatitude v and the longitude u, with

$$(1) \qquad x = a \sin v \cos u, \quad y = a \sin v \sin u, \quad z = a \cos v,$$

where a is the radius of the sphere with center at the origin, $0 \le v \le \pi$ and $0 \le u < 2\pi$.   Utilizing (1), show that $ds^2 = dx^2 + dy^2 + dz^2$ can be expressed as

$$(2) \qquad ds^2 = a^2 \sin^2 v \, du^2 + a^2 \, dv^2.$$

If the two given points on the surface of the sphere are (u_1, v_1) and (u_2, v_2) with $u_2 > u_1$, and we confine our discussion to arcs whose equations are expressible of the form

$$v = v(u), \quad v(u_1) = v_1, \quad v(u_2) = v_2,$$

then in view of (2), the length of the arc is given by

$$(3) \qquad I = a \int_{u_1}^{u_2} \sqrt{\sin^2 v + v'^2}\, du,$$

where $v' = v'(u)$ denotes the derivative (dv/du). Observe that the function f given by

$$(4) \qquad f = \sqrt{\sin^2 v + v'^2}$$

is explicitly independent of the independent variable u. Thus, with u and v playing the respective roles of x and y, we may write Equation (3.7.15) as

$$(5) \qquad f - v'\, \frac{\partial f}{\partial v'} = k,$$

where k is an arbitrary constant. Utilizing (4) and (5), show that

$$(6) \qquad du = \frac{k \csc v\, dv}{\sqrt{\sin^2 v - k^2}} = \frac{k \csc^2 v\, dv}{\sqrt{1 - k^2 \csc^2 v}}$$

$$= \frac{k \csc^2 v\, dv}{\sqrt{1 - k^2(1 + \cot^2 v)}} = \frac{\csc^2 v\, dv}{\sqrt{K^2 - \cot^2 v}},$$

where $K^2 = \dfrac{1}{k^2} - 1$. Integrating (6), we obtain

$$(7) \qquad u = -\arc\sin\left(\frac{\cot v}{K}\right) + c,$$

where c is an arbitrary constant of integration. Show that (7) can be written as

$$(8) \qquad \sin c \sin v \cos u - \cos c \sin v \sin u - \frac{1}{K}\cos v = 0.$$

Multiplying (8) by a and utilizing (1), show that the sphere geodesic lies on the plane

$$(9) \qquad Ax + By + Cz = 0,$$

which passes through the center of the sphere, and where $A = \sin c$, $B = -\cos c$, $C = -1/K = -1/\sqrt{(1/k)^2 - 1}$. Thus, the shortest arc connecting two points on the sphere is the intersection of the sphere with the plane containing the given points and the center of the sphere. That is, the sphere geodesic is the so-called great-circle arc.

11. *Geodesic on the surface of a right circular cylinder.* The parametric equations (in cylindrical coordinates) of a right circular cylinder are given by

$$x = a \cos u, \quad y = a \sin u, \quad z = z \text{ (or } z = v),$$

where a is the radius of the circle in the xy-plane with center at the origin and $0 \leq u < 2\pi$.

(a) Show that $ds^2 = dx^2 + dy^2 + dz^2$ can be expressed as $ds^2 = a^2 du^2 + dv^2$.

(b) Follow the method given in Exercise 10 to show that the parametric equations of the geodesic on a right circular cylinder are given by

$$x = a \cos u, \quad y = a \sin u, \quad z = bu,$$

where b is an arbitrary constant.   Thus, the right circular cylinder geodesic is the so-called helix.

12. *Geodesic on the surface of one nappe of a right circular cone.*   The parametric equations (in spherical coordinates) of a right circular cone are given by

$$x = \rho \sin \alpha \cos u, \quad y = \rho \sin \alpha \sin u, \quad z = \rho \cos \alpha,$$

where $\rho = |OP|$ is the distance from the origin O to a point $P{:}(x, y, z)$ on the surface, $0 \leq u < 2\pi$, and α (the generating angle) is constant.

(a) Show that $ds^2 = dx^2 + dy^2 + dz^2$ can be expressed as

$$ds^2 = d\rho^2 + \rho^2 \sin^2 \alpha \, du^2.$$

(b) Show that the geodesic on one nappe of a right circular cone is given by

$$\rho = C \sec \, [(\sin \alpha)u + B],$$

where C and B are arbitrary constants.

(c) Identify the geodesic.

13. *Geodesic on a surface $g(x, y, z) = 0$.*   Suppose that the parametric equations of the surface are given by

$$(1) \qquad\qquad x = x(u, v), \quad y = y(u, v), \quad z = z(u, v).$$

Utilizing (1), we have from calculus

$$(2) \qquad dx = \frac{\partial x}{\partial u} du + \frac{\partial x}{\partial v} dv, \quad dy = \frac{\partial y}{\partial u} du + \frac{\partial y}{\partial v} dv, \quad dz = \frac{\partial z}{\partial u} du + \frac{\partial z}{\partial v} dv.$$

(a) Use (2) to show that $ds^2 = dx^2 + dy^2 + dz^2$ can be expressed as

$$(3) \qquad ds^2 = P(u, v)du^2 + 2Q(u, v)du \, dv + R(u, v)dv^2,$$

where

$$(4) \qquad P = \left(\frac{\partial x}{\partial u}\right)^2 + \left(\frac{\partial y}{\partial u}\right)^2 + \left(\frac{\partial z}{\partial u}\right)^2, \quad R = \left(\frac{\partial x}{\partial v}\right)^2 + \left(\frac{\partial y}{\partial v}\right)^2 + \left(\frac{\partial z}{\partial v}\right)^2,$$

$$Q = \frac{\partial x}{\partial u}\frac{\partial x}{\partial v} + \frac{\partial y}{\partial u}\frac{\partial y}{\partial v} + \frac{\partial z}{\partial u}\frac{\partial z}{\partial v}.$$

Formula (3) is known as the *first fundamental form of the surface*, and it is utilized in courses in tensor analysis and metric differential geometry. Thus, following the analysis given in Exercise 10, the length of the arc is given by

$$(5) \qquad\qquad I = \int_{u_1}^{u_2} \sqrt{P + 2Qv' + Rv'^2} \, du.$$

(b) With u and v playing the respective roles of x and y, and the function f given by

(6)
$$f = \sqrt{P + 2Qv' + Rv'^2},$$

show that the Euler-Lagrange equation

(7)
$$\frac{\partial f}{\partial v} - \frac{d}{du}\left(\frac{\partial f}{\partial v'}\right) = 0$$

becomes

(8)
$$\frac{\dfrac{\partial P}{\partial v} + 2v'\dfrac{\partial Q}{\partial v} + v'^2\dfrac{\partial R}{\partial v}}{2\sqrt{P + 2Qv' + Rv'^2}} - \frac{d}{du}\left(\frac{Q + Rv'}{\sqrt{P + 2Qv' + Rv'^2}}\right) = 0.$$

It can be shown that a necessary and sufficient condition for the level curves $u = c_1$ to form an orthogonal trajectory with the level curves $v = c_2$ on the given surface $g(x, y, z) = 0$, is that the function Q be identically zero.

(c) If $Q = 0$ and the functions P and R given in (4) are explicitly independent of the independent variable u, utilize Equation (5) of Exercise 10 to show that

(9)
$$u = k \int \frac{\sqrt{R}\,dv}{\sqrt{P^2 - k^2 P}} + c,$$

where k and c are arbitrary constants.

(d) Solve Exercises 10, 11, and 12 by utilizing Formula (9).

Suggested Readings

Betz, Burcham, and Ewing [6]
Bliss [7]
Gladstone and Lewis [16]
Reddick and Miller [37]

Spiegel [43]
Webster [50]
Whittaker [54]

chapter 4   SECOND-ORDER LINEAR DIFFERENTIAL EQUATIONS

4.1. Introduction

In this chapter, we shall develop some basic results concerning second-order linear differential equations. These results extend very nicely to linear differential equations of order greater than two. Linear differential equations, in particular those of order two, play a very basic role in many branches of pure and applied mathematics, engineering, physics and in many other fields of science.

4.2. Definitions. The Fundamental Existence Theorem

Definition 4.2.1. A *linear differential equation of order two* is an equation that can be expressed in the form

$$P(x)\frac{d^2y}{dx^2} + Q(x)\frac{dy}{dx} + R(x)y = F(x). \tag{4.2.1}$$

Remark 4.2.1. In the ensuing discussion, unless stated to the contrary, we shall assume that P, Q, R, and F given in (4.2.1) are continuous real-valued functions defined on a common interval I: $a \leq x \leq b$, and $P(x)$ is not identically zero in I. However, for the sake of emphasis, we shall repeat these conditions in some of the existence theorems involving these functions.

Definition 4.2.2. If $F(x)$ is not identically zero, (4.2.1) is called a *nonhomogeneous* (or *inhomogeneous*) linear differential equation of order two.

Definition 4.2.3. If $F(x)$ is identically zero in (4.2.1), the resulting equation

$$P(x)\frac{d^2y}{dx^2} + Q(x)\frac{dy}{dx} + R(x)y = 0 \tag{4.2.2}$$

is called a *homogeneous* (or *reduced*) linear differential equation of order two.

If $F(x)$ is not identically zero in (4.2.1), then Equation (4.2.2) is called the *related homogeneous differential equation*.

We shall need the following fundamental existence theorem. The proof is omitted. A more general result is established in Appendix 2. [*See* Theorem 10.]

Theorem 4.2.1.   *Let the functions P, Q, R, and F be continuous on the interval I: $a \leq x \leq b$, with $P(x) \neq 0$ for every x in I.   Let x_0 be any fixed point in I, and let y_0 and y_0' be two arbitrary constants.   Then there exists a unique solution $y = y(x)$ satisfying the differential equation*

$$P(x)y'' + Q(x)y' + R(x)y = F(x) \qquad [4.2.1]$$

for all x in I and the initial conditions

$$y(x_0) = y_0 \quad \text{and} \quad y'(x_0) = y_0'. \qquad (4.2.3)$$

Example 4.2.1.   Let us show that the initial-value problem

$$\begin{cases} y'' + (x^2 + 1)y' + xy = \sin x, \\ y(0) = 1, \qquad y'(0) = 3, \end{cases}$$

has a unique solution in the interval I: $-\infty < x < \infty$.

SOLUTION.   Here $P(x) = 1 \neq 0$ for all x in I.   Also, $P(x) = 1$, $Q(x) = x^2 + 1$, $R(x) = x$, and $F(x) = \sin x$ are each continuous for all x in I.   The point $x_0 = 0$ belongs to I.   The two arbitrary constants are $y_0 = 1$ and $y_0' = 3$.   In view of Theorem 4.2.1, our initial-value problem has a unique solution for all x in I: $a \leq x \leq b$, where $a > -\infty$ and $b < \infty$.

As an immediate consequence of Theorem 4.2.1, we have the following result.

Theorem 4.2.2.   *Let the functions P, Q, and R be continuous on the interval I: $a \leq x \leq b$, with $P(x) \neq 0$ for every x in I.   Let x_0 be any fixed point in I. Let $y = y(x)$ be a solution of the differential equation*

$$P(x)y'' + Q(x)y' + R(x)y = 0, \qquad [4.2.2]$$

such that $y(x_0) = 0$ and $y'(x_0) = 0$.   Then $y(x) = 0$ for every x in I.

PROOF.   Clearly, $y(x) = 0$ satisfies the differential equation (4.2.2) and the initial conditions $y(x_0) = 0$ and $y'(x_0) = 0$.   By Theorem 4.2.1, it is the only solution with this property, and the theorem is established.

Example 4.2.2.   The solution of the initial-value problem

$$\begin{cases} 3y'' + x^2y' + 4(x + 1)y = 0, \\ y(3) = 0, \qquad y'(3) = 0, \end{cases}$$

is the trivial solution, namely, $y(x) = 0$ for all x.

4.3. Linear Operators. Linearly Dependent and Linearly Independent Functions

Definition 4.3.1.   An *operator* A is a rule that associates with each function u of one class, another function $v = Au$ of a second class.   (The two classes need not be disjoint.)

Definition 4.3.2. A is a *linear operator* if for any two functions u_1, u_2 and any constants c_1, c_2, for which Au_1, Au_2 are defined, $A(c_1 u_1 + c_2 u_2)$ is also defined, and

$$A(c_1 u_1 + c_2 u_2) = c_1 A u_1 + c_2 A u_2. \tag{4.3.1}$$

Example 4.3.1.   The operator A defined by

$$Au = \frac{\partial^2 u}{\partial x^2} + \frac{\partial^2 u}{\partial y^2} + \frac{\partial^2 u}{\partial z^2} \tag{4.3.2}$$

is a linear operator, since

$$A(c_1 u_1 + c_2 u_2) = \frac{\partial^2}{\partial x^2}(c_1 u_1 + c_2 u_2) + \frac{\partial^2}{\partial y^2}(c_1 u_1 + c_2 u_2) + \frac{\partial^2}{\partial z^2}(c_1 u_1 + c_2 u_2)$$

$$= c_1 \frac{\partial^2 u_1}{\partial x^2} + c_2 \frac{\partial^2 u_2}{\partial x^2} + c_1 \frac{\partial^2 u_1}{\partial y^2} + c_2 \frac{\partial^2 u_2}{\partial y^2} + c_1 \frac{\partial^2 u_1}{\partial z^2} + c_2 \frac{\partial^2 u_2}{\partial z^2}$$

$$= c_1 \left(\frac{\partial^2 u_1}{\partial x^2} + \frac{\partial^2 u_1}{\partial y^2} + \frac{\partial^2 u_1}{\partial z^2} \right) + c_2 \left(\frac{\partial^2 u_2}{\partial x^2} + \frac{\partial^2 u_2}{\partial y^2} + \frac{\partial^2 u_2}{\partial z^2} \right)$$

$$= c_1 A u_1 + c_2 A u_2.$$

Thus, the operator A is linear in view of Definition 4.3.2.   The operator A defined by (4.3.2) is the Laplacian operator for the xyz-coordinate system. [*See* Exercise 1.4.12(c).]

Suppose that the functions u_1 and u_2 are n times differentiable at x.   Then, from calculus, we have

$$[u_1(x) + u_2(x)]^{(n)} \equiv \frac{d^n}{dx^n}[c_1 u_1(x) + c_2 u_2(x)] = c_1 \frac{d^n u_1(x)}{dx^n} + c_2 \frac{d^n u_2(x)}{dx^n}$$

$$= c_1 u_1^{(n)}(x) + c_2 u_2^{(n)}(x). \tag{4.3.3}$$

By definition, when $n = 0$, Formula (4.3.3) becomes

$$[c_1 u_1(x) + c_2 u_2(x)]^{(0)} \equiv c_1 u_1(x) + c_2 u_2(x). \tag{4.3.4}$$

The expressions $[c_1 u_1(x) + c_2 u_2(x)]^{(2)}$ and $[c_1 u_1(x) + c_2 u_2(x)]^{(1)}$ are often written as $[c_1 u_1(x) + c_2 u_2(x)]''$ and $[c_1 u_1(x) + c_2 u_2(x)]'$, respectively.

Theorem 4.3.1. *Let the functions P, Q, and R be continuous on the interval I:*
$a \leq x \leq b$. Let the function y be twice differentiable in I. Then the operator
L defined by

$$L[y] = P(x)y'' + Q(x)y' + R(x)y \qquad (4.3.5)$$

is a linear operator.

PROOF.

$$L[c_1 y_1 + c_2 y_2]$$
$$= P(x)[c_1 y_1 + c_2 y_2]'' + Q(x)[c_1 y_1 + c_2 y_2]' + R(x)[c_1 y_1 + c_2 y_2]$$
$$= P(x)[c_1 y_1'' + c_2 y_2''] + Q(x)[c_1 y_1' + c_2 y_2'] + R(x)[c_1 y_1 + c_2 y_2]$$
$$= c_1[P(x)y_1'' + Q(x)y_1' + R(x)y_1] + c_2[P(x)y_2'' + Q(x)y_2' + R(x)y_2]$$
$$= c_1 L[y_1] + c_2 L[y_2].$$

Thus, in view of Definition 4.3.2, the operator L is linear.

Remark 4.3.1. The operator L of Theorem 4.3.1 is known as a *linear differential operator of order two*. In Section 4.4, we shall develop some basic properties of Equation (4.2.2). Observe that

$$L[y](x) = P(x)y''(x) + Q(x)y'(x) + R(x)y(x)$$

denotes the value of the function $L[y]$ at the point x. Thus, with $y = y(x)$, we shall often write Equation (4.2.2) conveniently as

$$L[y] = P(x)y'' + Q(x)y' + R(x)y = 0, \qquad (4.3.6)$$

or simply $L[y] = 0$. Note that $L[y] = 0$ means that $L[y](x) = 0$ for all x in I. In $L[y] = 0$, the right-hand side actually denotes the function which is equal to zero at all points x in I.

Definition 4.3.3. The *sum $A + B$* of two operators is the operator defined by

$$(A + B)u = Au + Bu \qquad (4.3.7)$$

for all functions u for which both Au and Bu are defined.

Example 4.3.2. Let the operators A and B be defined, respectively, by

$$Au = \frac{\partial^2 u}{\partial x^2} + \frac{\partial^2 u}{\partial y^2} + \frac{\partial^2 u}{\partial z^2} \quad \text{and} \quad Bu = x\frac{\partial u}{\partial x} + y\frac{\partial u}{\partial y} + z\frac{\partial u}{\partial z}.$$

Then

$$(A + B)u = \frac{\partial^2 u}{\partial x^2} + \frac{\partial^2 u}{\partial y^2} + \frac{\partial^2 u}{\partial z^2} + x\frac{\partial u}{\partial x} + y\frac{\partial u}{\partial y} + z\frac{\partial u}{\partial z}.$$

Theorem 4.3.2. *Let A and B be linear operators. Then the sum $A + B$ is also a linear operator.*

PROOF. Utilizing (4.3.7) and (4.3.1), we have

$$(A + B)(c_1 u_1 + c_2 u_2) = A(c_1 u_1 + c_2 u_2) + B(c_1 u_1 + c_2 u_2)$$
$$= c_1 A u_1 + c_2 A u_2 + c_1 B u_1 + c_2 B u_2$$
$$= c_1(A + B)u_1 + c_2(A + B)u_2.$$

The theorem is thus established.

Remark 4.3.2. We now present the definitions of a linear combination, linear dependence, and linear independence of n functions. All the reader needs to do to obtain these definitions for two functions is to set $n = 2$ in the definitions given below. These concepts will play a very basic role in the forthcoming sections.

Definition 4.3.4. The function

$$c_1 u_1 + c_2 u_2 + \cdots + c_n u_n$$

where $u_1, u_2, \ldots, u_n$ are n given functions and $c_1, c_2, \ldots, c_n$ are n constants, is called a *linear combination* of $u_1, u_2, \ldots, u_n$.

Definition 4.3.5. The n functions $u_1, u_2, \ldots, u_n$ are said to be *linearly dependent on the interval* I: $a \leqq x \leqq b$, if there exist constants $c_1, c_2, \ldots, c_n$, not all zero, such that

$$c_1 u_1(x) + c_2 u_2(x) + \cdots + c_n u_n(x) = 0 \qquad (4.3.8)$$

for every x in I.

Definition 4.3.6. The n functions $u_1, u_2, \ldots, u_n$ are said to be *linearly independent* on the interval I: $a \leqq x \leqq b$, if they are not linearly dependent on the interval I.

Thus, the n functions $u_1, u_2, \ldots, u_n$ are linearly independent on the interval I: $a \leqq x \leqq b$, if the only constants $c_1, c_2, \ldots, c_n$ such that

$$c_1 u_1(x) + c_2 u_2(x) + \cdots + c_n u_n(x) = 0, \qquad (4.3.9)$$

for every x in I are the constants $c_1 = 0, \quad c_2 = 0, \ldots, c_n = 0$.

Example 4.3.3. Let us show that the functions u_1, u_2, and u_3 defined by $u_1(x) = \sin x$, $u_2(x) = \cos x$, and $u_3(x) = \sin(x + \beta)$, where β is a constant, are linearly dependent on the interval I: $-\infty < x < \infty$.

SOLUTION. Clearly, the given three functions exist for every x in I. In view of Definition 4.3.5, we must show that there exist constants c_1, c_2, and c_3, not all zero, such that

$$c_1 \sin x + c_2 \cos x + c_3 \sin(x + \beta) = 0 \qquad (4.3.10)$$

for every x in I. Now, from trigonometry, we know that

$$\sin(x + \beta) = \sin x \cos \beta + \cos x \sin \beta.$$

Thus, Equation (4.3.10) may be written as

$$(c_1 + c_3 \cos \beta) \sin x + (c_2 + c_3 \sin \beta) \cos x = 0. \qquad (4.3.11)$$

One possible choice of constants, not all zero, which will satisfy Equation (4.3.11) for every x in I is $c_1 = \cos \beta$, $c_2 = \sin \beta$, and $c_3 = -1$. Thus, the given three functions are linearly dependent on I.

Example 4.3.4. Let us show that the functions u_1 and u_2 defined by $u_1(x) = e^{ax}$ and $u_2(x) = e^{bx}$, where a and b are distinct constants, are linearly independent on the interval I: $-\infty < x < \infty$.

SOLUTION. Clearly, the two given functions exist for every x in I. We shall show that the two given functions are linearly independent on I via a contradiction. (This technique is very useful.) Suppose to the contrary that the functions u_1 and u_2 are linearly dependent on I. Then there exist constants c_1 and c_2, not both zero, such that

$$c_1 e^{ax} + c_2 e^{bx} = 0 \qquad (4.3.12)$$

for every x in I. Since $e^{ax} \neq 0$ for every x in I, multiplying (4.3.12) by e^{-ax}, we obtain

$$c_1 + c_2 e^{(b-a)x} = 0. \qquad (4.3.13)$$

Differentiating (4.3.13), we get

$$c_2(b - a)e^{(b-a)x} = 0. \qquad (4.3.14)$$

Since $b - a \neq 0$ and $e^{(b-a)x} \neq 0$ for every x in I, (4.3.14) implies that $c_2 = 0$. With $c_2 = 0$, (4.3.12) becomes

$$c_1 e^{ax} = 0. \qquad (4.3.15)$$

Equation (4.3.15) implies that $c_1 = 0$. Thus, the functions u_1 and u_2 are not linearly dependent on I; and hence, they are linearly independent on I. [*See* Definition 4.3.6.]

Exercises 4.3

1. Let $f(x) \not\equiv 0$ on an interval I. Show that the functions f and g are linearly dependent on an interval I, if and only if $g(x) = kf(x)$ for every x in I, where k is a constant.
2. Suppose that the n functions $u_1, u_2, \ldots, u_n$ are linearly independent on an interval I. Show that a function v defined on I can be written as a linear combination of $u_1, u_2, \ldots, u_n$ if and only if the functions $v, u_1, u_2, \ldots, u_n$ are linearly dependent on $|I|$

3. The functions $u_1, u_2, \ldots, u_n$ defined below exist on the interval I: $-\infty < x < \infty$. The constants a, b, and c are distinct. Determine whether the given functions are linearly dependent on I.

   (a) $u_1(x) = 1, \quad u_2(x) = 2x$.

   (b) $u_1(x) = 3, \quad u_2(x) = 5x, \quad u_3(x) = 2x^2$.

   (c) $u_1(x) = 2x^3, \quad u_2(x) = 5x^3$.

   (d) $u_1(x) = 1, \quad u_2(x) = x, \quad u_3(x) = x^2, \ldots, u_n(x) = x^{n-1}$.

   (e) $u_1(x) = 1, \quad u_2(x) = e^{ax}$.

   (f) $u_1(x) = x, \quad u_2(x) = xe^{ax}$.

   (g) $u_1(x) = ax, \quad u_2(x) = bx^2, \quad u_3(x) = cx^2 + 3x$.

   (h) $u_1(x) = e^{ax}, \quad u_2(x) = xe^{ax}$.

   (i) $u_1(x) = \sin x, \quad u_2(x) = \cos x$.

   (j) $u_1(x) = 1, \quad u_2(x) = \sin^2 x, \quad u_3(x) = \cos^2 x$.

   (k) $u_1(x) = 2, \quad u_2(x) = \sin^2 x, \quad u_3(x) = \cos^2 x$.

   (l) $u_1(x) = x, \quad u_2(x) = |x|$.

   (m) $u_1(x) = e^{ax} \sin bx, \quad u_2(x) = e^{ax} \cos bx$.

   (n) $u_1(x) = 1, \quad u_2(x) = \cos 2x, \quad u_3(x) = \sin^2 x$.

   (o) $u_1(x) = 2x, \quad u_2(x) = 3x^2 + 6x + 9, \quad u_3(x) = x^2 + 3x + 3$.

   (p) $u_1(x) = x^2, \quad u_2(x) = x|x|$.

   (q) $u_1(x) = e^{ax}, \quad u_2(x) = ae^{ax}, \quad u_3(x) = be^{ax}, \quad u_4(x) = ce^{ax}$.

   (r) $u_1(x) = e^{ax}(1 + x) \cos bx, \quad u_2(x) = e^{ax}(1 + x) \sin bx$.

4. Suppose that the n functions $u_1, u_2, \ldots, u_n$ are linearly independent on an interval I.

   (a) Show that any subcollection of the u_i's are linearly independent on I.

   (b) If $v(x) = 0$ for every x in I (null function), show that the $(n + 1)$ functions $v, u_1, u_2, \ldots, u_n$ are linearly dependent on I.

5. Suppose that the n functions $u_1, u_2, \ldots, u_n$ are linearly dependent on an interval I. Show that at least one of them may be written as a linear combination of the rest.

6. Suppose that the functions f and g are linearly independent on an interval I. Show, by an example, that f and g are not necessarily linearly independent on an interval $J \subset I$.

7. Suppose that the functions f and g are linearly dependent on an interval I. Show that f and g are linearly dependent on an interval $J \subset I$.

8. Let $a_0, a_1, \ldots, a_n$ be continuous functions on the interval I: $a \leq x \leq b$. Let y be a function which is n times differentiable in I. Show that the operator L defined by

$$L[y] = a_0(x)y^{(n)} + a_1(x)y^{(n-1)} + \cdots + a_{n-1}(x)y' + a_n(x)y$$

   is a linear operator, which is known as a *linear differential operator of order n.*

9. Show that the operator A defined by

$$Au = \frac{\partial^2 u}{\partial x^2} + \left(\frac{\partial u}{\partial y}\right)^2$$

   is not a linear operator.

10. Let u be a function of the variables x, y, and z. Show that the operator A defined by

$$Au = u(t, y, z) + u(x, t, z) + u(x, y, t)$$

is a linear operator. [*Note:* The operator A transforms a function u of three variables into a function g of four variables.]

11. Let u be a function of the variables x and y. Show that the operator A defined by

$$Au = u_{xx}(x, 0) + u(x, 0)$$

is a linear operator. [*Note:* The operator A transforms a function u of two variables into a function f of one variable.]

12. Show that the operator A defined by

$$Au = \int_a^b G(x, t)u(x, t)dt$$

is a linear operator.

13. Show that the operator A defined by

$$Au = \int_a^b G(x, t)u^2(x, t)dt$$

is not a linear operator.

14. Let u be a function of the variables x, y, and z. Show that the operator A defined by

$$Au = \int_0^1 \int_0^1 \int_0^1 u(\xi, \eta, \zeta) \sqrt{(x - \xi)^2 + (y - \eta)^2 + (z - \zeta)^2} \, d\xi \, d\eta \, d\zeta$$

is a linear operator. [*Note:* The operator A transforms a function u of three variables into a function h of three variables.]

4.4. Some Basic Results on Homogeneous Linear Differential Equations

In view of Remark 4.3.1, we may write the homogeneous linear differential Equation (4.2.2) as

$$L[y] = P(x)y'' + Q(x)y' + R(x)y = 0. \tag{4.4.1}$$

Theorem 4.4.1. *Let $y_1(x)$ and $y_2(x)$ be two solutions of $L[y] = 0$. Then the linear combination of these two solutions*

$$c_1 y_1(x) + c_2 y_2(x), \tag{4.4.2}$$

where c_1 and c_2 are arbitrary constants, is also a solution of $L[y] = 0$.

PROOF. By hypothesis, $L[y_1] = 0$ and $L[y_2] = 0$. From Theorem 4.3.1, the operator L is linear. Thus,

$$L[c_1 y_1 + c_2 y_2] = c_1 L[y_1] + c_2 L[y_2] = c_1 \cdot 0 + c_2 \cdot 0 = 0.$$

Hence, the linear combination of the two given solutions of $L[y] = 0$ is also a solution of $L[y] = 0$.

Remark 4.4.1.   We readily see from Theorem 4.4.1 with $c_1 = 0$ or $c_2 = 0$, that if $g(x)$ is a solution of $L[y] = 0$, so is any constant multiple of $g(x)$ a solution of $L[y] = 0$.

Remark 4.4.2.   The reader should prove that if $y_1(x)$, $y_2(x)$, $\ldots$, $y_m(x)$ are m solutions of $L[y] = 0$, then the linear combination of these m solutions is also a solution of $L[y] = 0$.

Example 4.4.1.   One may easily verify that e^{2x} and e^{-2x} are solutions of

$$y'' - 4y = 0.$$

Theorem 4.4.1 asserts that $c_1 e^{2x} + c_2 e^{-2x}$, where c_1 and c_2 are arbitrary constants, is also a solution of the above differential equation.

Example 4.4.2.   On any interval I not containing the origin, one may easily verify that x^2 and x^{-3} are solutions of

$$x^2 y'' + 2xy' - 6y = 0. \tag{4.4.3}$$

By Theorem 4.4.1, the linear combination $c_1 x^2 + c_2 x^{-3}$ is also a solution of Equation (4.4.3) on I.

Example 4.4.3.   One may easily verify that $y(x) = x^2 + x$ is a solution of

$$x^3 y'' - x^2 y' + y = x. \tag{4.4.4}$$

One may also verify that it is not necessarily true that any constant multiple of $y(x)$ is a solution of Equation (4.4.4), for instance, $2(x^2 + x)$.   This fact, however, does not violate Remark 4.4.1, since the differential equation (4.4.4) is not homogeneous.

Example 4.4.4.   One may easily verify that x and x^2 are solutions of

$$(x^3 - \tfrac{1}{2}x^4)y'' - x^2 y' + y^2 = 0. \tag{4.4.5}$$

One may also verify that it is not necessarily true that the linear combination $c_1 x + c_2 x^2$ is also a solution of Equation (4.4.5), for instance, $x + x^2$.   This fact, however, does not contradict the result given in Theorem 4.4.1, since the differential equation (4.4.5) is not linear owing to the term y^2.

Remark 4.4.3.   In the proofs of the forthcoming theorems, we shall need some of the basic results on systems of linear equations established in college algebra.   Recall that a system of m linear equations in n unknowns x_1, x_2, $\ldots$, x_n is given by

$$\sum_{j=1}^{n} a_{ij} x_j = B_i, \qquad i = 1, 2, \ldots, m, \tag{4.4.6}$$

where the a_{ij} and B_i are either constants or quantities which are independent of the unknowns. If $B_i = 0$, $i = 1, 2, \ldots, m$, Equation (4.4.6) becomes

$$\sum_{j=1}^{n} a_{ij} x_j = 0, \qquad i = 1, 2, \ldots, m. \tag{4.4.7}$$

We call (4.4.7) a *system of m homogeneous linear equations in n unknowns.* If not all of the B_i are equal to zero in (4.4.6), then the system (4.4.6) is called *nonhomogeneous.* When $m = n$ in (4.4.6), the determinant of coefficients of the unknowns is denoted by det A, that is,

$$\det A = \begin{vmatrix} a_{11} & a_{12} & \cdots & a_{1n} \\ a_{21} & a_{22} & \cdots & a_{2n} \\ \vdots & \vdots & & \vdots \\ a_{n1} & a_{n2} & \cdots & a_{nn} \end{vmatrix}. \tag{4.4.8}$$

Example 4.4.5. When $m = 2$ and $n = 2$, the system (4.4.6) becomes

$$\begin{cases} a_{11}x_1 + a_{12}x_2 = B_1, \\ a_{21}x_1 + a_{22}x_2 = B_2, \end{cases} \tag{4.4.9}$$

and

$$\det A = \begin{vmatrix} a_{11} & a_{12} \\ a_{21} & a_{22} \end{vmatrix} = a_{11}a_{22} - a_{12}a_{21}. \tag{4.4.10}$$

Example 4.4.6. When $m = 2$ and $n = 3$, the system (4.4.7) becomes

$$\begin{cases} a_{11}x_1 + a_{12}x_2 + a_{13}x_3 = 0, \\ a_{21}x_1 + a_{22}x_2 + a_{23}x_3 = 0. \end{cases} \tag{4.4.11}$$

Example 4.4.7. When $m = 2$ and $n = 2$, the system (4.4.7) becomes

$$\begin{cases} a_{11}x_1 + a_{12}x_2 = 0, \\ a_{21}x_1 + a_{22}x_2 = 0. \end{cases} \tag{4.4.12}$$

Theorem I. *Suppose that $m = n$ in Equation* (4.4.6). *If* det $A \neq 0$, *the system* (4.4.6) *has a unique solution.*

Theorem II. *Suppose that $m = n$ in Equation* (4.4.7). *A necessary and sufficient condition for a system of n homogeneous linear equations in n unknowns to have a nontrivial solution is that* det $A = 0$.

Theorem III. *If $m < n$, a system of m homogeneous linear equations in n unknowns always has a nontrivial solution.*

If det $A \neq 0$ in (4.4.10), the system (4.4.9) has a unique solution by Theorem I. The system (4.4.11) has a nontrivial solution by Theorem III. The system (4.4.12) has a nontrivial solution provided that det $A = a_{11}a_{22} - a_{12}a_{21} = 0$ by Theorem II.

The following existence theorem is important.

Theorem 4.4.2.   *Let the functions P, Q, and R be continuous on the interval I:* $a \leq x \leq b$, *and* $P(x) \neq 0$ *for every x in I.*

(a) *The differential equation*

$$L[y] = P(x)y'' + Q(x)y' + R(x)y = 0, \qquad [4.4.1]$$

always possesses two linearly independent solutions on I.

(b) *If* $y_1(x)$ *and* $y_2(x)$ *are two linearly independent solutions of* $L[y] = 0$ *on I, then every solution of* $L[y] = 0$ *can be expressed as a linear combination*

$$c_1 y_1(x) + c_2 y_2(x)$$

of these two linearly independent solutions, for suitably chosen constants c_1 *and* c_2.

PROOF.   *Concerning* (a).   We shall actually exhibit the two linearly independent solutions of $L[y] = 0$ on I by the following useful technique.

(A)
$$\begin{cases}
\text{Denote by } y_1(x) \text{ and } y_2(x) \text{ two solutions of } L[y] = 0 \text{ on } I \text{ satisfying,} \\
\text{respectively, the initial conditions} \\[4pt]
\qquad y_1(x_0) = 1, \qquad y_1'(x_0) = 0, \qquad (4.4.13) \\[4pt]
\text{and} \\[4pt]
\qquad y_2(x_0) = 0, \qquad y_2'(x_0) = 1, \qquad (4.4.14) \\[4pt]
\text{where } x_0 \text{ is a point in } I.
\end{cases}$$

In view of the hypothesis of our theorem, it follows from the existence Theorem 4.2.1 that $y_1(x)$ not only exists but it is also the unique solution of $L[y] = 0$ on I satisfying the initial conditions given in (4.4.13).   A similar statement holds for the solution $y_2(x)$ subject to the initial conditions given in (4.4.14).

We shall next show that the functions y_1 and y_2 as given in (A) are linearly independent on I.   We shall establish this fact via a contradiction.   That is, suppose that y_1 and y_2 are linearly dependent on I.   Then in view of Definition 4.3.5, with $n = 2$, there exist constants c_1 and c_2, not both zero, such that

$$c_1 y_1(x) + c_2 y_2(x) = 0 \qquad (4.4.15)$$

for every x in I.   Differentiating (4.4.15), we get

$$c_1 y_1'(x) + c_2 y_2'(x) = 0 \qquad (4.4.16)$$

for every x in I.   When $x = x_0$ is any fixed point in I, (4.4.15) and (4.4.16) become, respectively,

$$c_1 y_1(x_0) + c_2 y_2(x_0) = 0, \tag{4.4.17}$$

and

$$c_1 y_1'(x_0) + c_2 y_2'(x_0) = 0. \tag{4.4.18}$$

Utilizing (4.4.13) and (4.4.14), we see that (4.4.17) and (4.4.18) become, respectively,

$$c_1 \cdot 1 + c_2 \cdot 0 = 0, \tag{4.4.19}$$

and

$$c_1 \cdot 0 + c_2 \cdot 1 = 0. \tag{4.4.20}$$

Equations (4.4.19) and (4.4.20) imply $c_1 = 0$ and $c_2 = 0$, which is a contradiction.   Thus, the functions y_1 and y_2 are linearly independent on I.

Concerning (b).   We shall now show that if $y_1(x)$ and $y_2(x)$ are two linearly independent solutions of $L[y] = 0$ on I, then any other solution $y_3(x)$ of $L[y] = 0$ on I can be expressed as a linear combination of $y_1(x)$ and $y_2(x)$ for suitably chosen constants.   This fact will be established if we can show that the functions y_1, y_2, and y_3 are linearly dependent on I.   [*See* Exercise 4.3.2.]   To this end, let us consider the system of two homogeneous linear equations in the three unknowns α_1, α_2, and α_3:

$$\begin{cases} y_1(x_0)\alpha_1 + y_2(x_0)\alpha_2 + y_3(x_0)\alpha_3 = 0, \\ y_1'(x_0)\alpha_1 + y_2'(x_0)\alpha_2 + y_3'(x_0)\alpha_3 = 0, \end{cases} \tag{4.4.21}$$

where x_0 is any fixed point in I.   From Theorem III, the system (4.4.21) has a nontrivial solution, say

$$\alpha_1 = c_1, \qquad \alpha_2 = c_2, \quad \text{and} \quad \alpha_3 = c_3, \qquad [(c_1, c_2, c_3) \neq (0, 0, 0)].$$

$$\tag{4.4.22}$$

Let us define a function g by

$$g(x) = c_1 y_1(x) + c_2 y_2(x) + c_3 y_3(x) \tag{4.4.23}$$

for every x in I.   We shall now show that $g(x) = 0$ for every x in I.   This will then prove that the functions y_1, y_2, and y_3 are linearly dependent on I. [*See* Definition 4.3.5 with $n = 3$.]   Since $y_1(x)$, $y_2(x)$, and $y_3(x)$ are solutions of $L[y] = 0$ on I, it follows that $g(x)$ is also a solution of $L[y] = 0$ on I. [*See* Remark 4.4.2.]   Differentiating (4.4.23), we get

$$g'(x) = c_1 y_1'(x) + c_2 y_2'(x) + c_3 y_3'(x) \tag{4.4.24}$$

for every x in I.   When $x = x_0$ (our fixed point in I), (4.4.23) and (4.4.24) become

$$\begin{cases} g(x_0) = c_1 y_1(x_0) + c_2 y_2(x_0) + c_3 y_3(x_0), \\ g'(x_0) = c_1 y_1'(x_0) + c_2 y_2'(x_0) + c_3 y_3'(x_0). \end{cases} \tag{4.4.25}$$

From (4.4.22), we know that c_1, c_2, and c_3 satisfy the system (4.4.21).   Thus, the right-hand members of the system (4.4.25) are each equal to zero.   Hence,

$$g(x_0) = 0 \quad \text{and} \quad g'(x_0) = 0. \tag{4.4.26}$$

Since $g(x)$ is a solution of $L[y] = 0$ on I and also satisfies the initial conditions (4.4.26), it follows from Theorem 4.2.2 that $g(x) = 0$ for every x in I. Thus, part (b) is established and the proof of the theorem is now completed.

Since the general solution of a second-order linear differential equation should possess two essential constants (or parameters), the results given in the above theorems motivate us to make the following definition.

Definition 4.4.1.   Let $y_1(x)$ and $y_2(x)$ be two linearly independent solutions of the differential equation

$$L[y] = P(x)y'' + Q(x)y' + R(x)y = 0 \tag{4.4.27}$$

on the interval I: $a \leqq x \leqq b$.   The *general solution* $y(x)$ of Equation (4.4.27) on I is defined by

$$y(x) = c_1 y_1(x) + c_2 y_2(x), \tag{4.4.28}$$

where c_1 and c_2 are arbitrary constants.

The special solutions defined in (A) in the proof of Theorem 4.4.2 will now be singled out by the following definition.

Definition 4.4.2.   Let x_0 be any fixed point in the interval I: $a \leq x \leq b$. Two linearly independent solutions $y_1(x)$ and $y_2(x)$ of $L[y] = 0$ on I which satisfy the initial conditions

$$y_1(x_0) = 1, \qquad y_1'(x_0) = 0, \tag{4.4.29}$$

and

$$y_2(x_0) = 0, \qquad y_2'(x_0) = 1, \tag{4.4.30}$$

are called the *basic solutions at* x_0 of $L[y] = 0$ on I.

Definition 4.4.3.   Any set of two linearly independent solutions of $L[y] = 0$ on the interval I: $a \leq x \leq b$ is said to form a *fundamental set* or a *fundamental system* of $L[y] = 0$ on I.

Remark 4.4.4.   Suppose that $y(x)$ is the general solution of $L[y] = 0$ on the interval I: $a \leq x \leq b$ and satisfies also the initial conditions

$$y(x_0) = A \quad \text{and} \quad y'(x_0) = B, \tag{4.4.31}$$

where x_0 is any fixed point in I and A and B are constants. Then the general solution $y(x)$ can be readily obtained if we know the basic solutions at x_0 of $L[y] = 0$ on I. To see this, let $y_1(x)$ and $y_2(x)$ be the basic solutions at x_0 of $L[y] = 0$ on I. Then, the solution

$$g(x) = Ay_1(x) + By_2(x) \tag{4.4.32}$$

satisfies the initial conditions given in (4.4.31). For, utilizing (4.4.29) and (4.4.30), we have

$$g(x_0) = Ay_1(x_0) + By_2(x_0) = A \cdot 1 + B \cdot 0 = A,$$
$$g'(x_0) = Ay_1'(x_0) + By_2'(x_0) = A \cdot 0 + B \cdot 1 = B.$$

By Theorem 4.2.1, the solution satisfying $L[y] = 0$ on I and the initial conditions given in (4.4.31) is unique. Thus, $y(x) = g(x)$ on I or

$$y(x) = Ay_1(x) + By_2(x). \tag{4.4.33}$$

The above technique is useful, for example, if one has to solve the same differential equation subject to different initial conditions at x_0. That is,

$$\begin{cases} L[y] = 0, \\ y(x_0) = A_1, \qquad y'(x_0) = B_1. \end{cases}$$

$$\begin{cases} L[y] = 0, \\ y(x_0) = A_2, \qquad y'(x_0) = B_2. \end{cases} \tag{4.4.34}$$

$$\vdots$$

$$\begin{cases} L[y] = 0, \\ y(x_0) = A_n, \qquad y'(x_0) = B_n. \end{cases}$$

Also, when the initial conditions in (4.4.31) are awkward, the above method may save some tedious computations.

Example 4.4.8. In Example 4.4.1, one verified that the functions y_1 and y_2 given by $y_1(x) = e^{2x}$ and $y_2(x) = e^{-2x}$ were solutions of

$$y'' - 4y = 0. \tag{4.4.35}$$

From Example 4.3.4, the functions y_1 and y_2 are linearly independent on the interval I: $-\infty < x < \infty$. Thus, the general solution of Equation (4.4.35) on I is

$$y(x) = c_1 e^{2x} + c_2 e^{-2x},$$

where c_1 and c_2 are arbitrary constants.

Example 4.4.9. Suppose that $y_1(x) = (\sqrt{2}/2)(\cos 2x + \sin 2x)$ and $y_2(x) = (\sqrt{2}/4)(-\cos 2x + \sin 2x)$ are basic solutions at $x_0 = \pi/8$ of the equation

$$L[y] = y'' + 4y = 0$$

on the interval I: $-\infty < x < \infty$.   [*See* Exercise 4.6.13.]   Then, the general solution of $L[y] = 0$ on I satisfying the initial conditions

$$y\left(\frac{\pi}{8}\right) = 1.735 \quad\text{and}\quad y'\left(\frac{\pi}{8}\right) = 0.575$$

is given by [*see* (4.4.33)]

$$y(x) = 1.735\,\frac{\sqrt{2}}{2}\,(\cos 2x + \sin 2x) + 0.575\,\frac{\sqrt{2}}{4}\,(-\cos 2x + \sin 2x)$$

$$= 1.024 \cos 2x + 1.430 \sin 2x.$$

Remark 4.4.5.   The following definition and theorems will give us a very simple criterion for determining whether two solutions $y_1(x)$ and $y_2(x)$ of $L[y] = P(x)y'' + Q(x)y' + R(x)y = 0$ are linearly independent.

Definition 4.4.4.   Let f_1 and f_2 be two differentiable functions in the interval I: $a \leqq x \leqq b$.   The determinant

$$W(f_1, f_2) = \begin{vmatrix} f_1 & f_2 \\ f_1' & f_2' \end{vmatrix} \tag{4.4.36}$$

is called the *Wronskian* of the two functions.   Observe that $W(f_1, f_2)$ is also a function on I.   Its value at any point x in I will be denoted by $W(f_1, f_2; x)$. Thus,

$$W(f_1, f_2; x) = \begin{vmatrix} f_1(x) & f_2(x) \\ f_1'(x) & f_2'(x) \end{vmatrix} = f_1(x)f_2'(x) - f_2(x)f_1'(x). \tag{4.4.37}$$

A notation that is also used for $W(f_1, f_2; x)$ is $W(f_1, f_2)(x)$.

Example 4.4.10.   If $f_1(x) = x$ and $f_2(x) = \cos x$ in I: $-\infty < x < \infty$, then

$$W(f_1, f_2; x) = \begin{vmatrix} x & \cos x \\ 1 & -\sin x \end{vmatrix} = -x \sin x - \cos x.$$

Example 4.4.11.   If $f_1(x) = e^{2x}$ and $f_2(x) = e^{-2x}$ in I: $-\infty < x < \infty$, then

$$W(f_1, f_2; x) = \begin{vmatrix} e^{2x} & e^{-2x} \\ 2e^{2x} & -2e^{-2x} \end{vmatrix} = -2e^0 - 2e^0 = -4.$$

Example 4.4.12.   If $f_1(x) = \cos 2x$ and $f_2(x) = \tfrac{1}{2} \sin 2x$ in I: $-\infty < x < \infty$, then

$$W(f_1, f_2; x) = \begin{vmatrix} \cos 2x & \tfrac{1}{2} \sin 2x \\ -2 \sin 2x & \cos 2x \end{vmatrix} = \cos^2 2x + \sin^2 2x = 1.$$

Example 4.4.13.   If $f_1(x) = x^2$ and $f_2(x) = x^3$ in I: $-\infty < x < \infty$, then

$$W(f_1, f_2; x_0) = \begin{vmatrix} x^2 & x^3 \\ 2x & 3x^2 \end{vmatrix}_{x=x_0} = \begin{vmatrix} x_0^2 & x_0^3 \\ 2x_0 & 3x_0^2 \end{vmatrix} = 3x_0^4 - 2x_0^4 = x_0^4.$$

Theorem 4.4.3. *Let* $y_1 = y_1(x)$ *and* $y_2 = y_2(x)$ *be two solutions of* $L[y] = 0$ *on the interval* I: $a \leq x \leq b$. *A necessary and sufficient condition that* y_1 *and* y_2 *be linearly independent on* I *is that* $W(y_1, y_2; x) \neq 0$ *for every* x *in* I.

PROOF. *Sufficiency.* This condition asserts that if $W(y_1, y_2; x) \neq 0$ for every x in I, then y_1 and y_2 are linearly independent on I. To show that y_1 and y_2 are linearly independent on I, we must show that the only constants c_1 and c_2 such that

$$c_1 y_1(x) + c_2 y_2(x) = 0 \tag{4.4.38}$$

for every x in I are the constants $c_1 = 0$ and $c_2 = 0$. [*See* Definition 4.3.6.] Differentiating (4.4.38), we get

$$c_1 y_1'(x) + c_2 y_2'(x) = 0 \tag{4.4.39}$$

for every x in I. Let $x = x_0$ be any fixed point in I. Equations (4.4.38) and (4.4.39) then become

$$\begin{cases} y_1(x_0)c_1 + y_2(x_0)c_2 = 0, \\ y_1'(x_0)c_1 + y_2'(x_0)c_2 = 0, \end{cases} \tag{4.4.40}$$

which is a system of two homogeneous linear equations in two unknowns, c_1 and c_2. By hypothesis, $W(y_1, y_2; x_0) \neq 0$, that is,

$$\begin{vmatrix} y_1(x_0) & y_2(x_0) \\ y_1'(x_0) & y_2'(x_0) \end{vmatrix} \neq 0. \tag{4.4.41}$$

But the determinant here is that of the coefficients of the system (4.4.40). It follows from Theorem II, with $n = 2$, that the solution of the system (4.4.40) is the trivial solution, namely, $c_1 = 0$ and $c_2 = 0$. Thus, y_1 and y_2 are linearly independent on I.

Necessity. This condition asserts that if y_1 and y_2 are two linearly independent solutions of $L[y] = 0$ on I, then $W(y_1, y_2; x) \neq 0$ for every x in I. We shall establish this result by a contradiction. Suppose that y_1 and y_2 are linearly independent on I and there exists a point x_0 in I such that $W(y_1, y_2; x_0) = 0$; that is,

$$\begin{vmatrix} y_1(x_0) & y_2(x_0) \\ y_1'(x_0) & y_2'(x_0) \end{vmatrix} = 0. \tag{4.4.42}$$

Let us consider the system of two homogeneous linear equations in the unknowns α_1 and α_2:

$$\begin{cases} y_1(x_0)\alpha_1 + y_2(x_0)\alpha_2 = 0, \\ y_1'(x_0)\alpha_1 + y_2'(x_0)\alpha_2 = 0, \end{cases} \tag{4.4.43}$$

which has precisely for its determinant of the coefficients the determinant given in the left-hand side of (4.4.42).   Thus, by Theorem II (with $n = 2$), the system (4.4.43) has a nontrivial solution, say

$$\alpha_1 = c_1 \quad \text{and} \quad \alpha_2 = c_2, \qquad [(c_1, c_2) \neq (0, 0)]. \qquad (4.4.44)$$

Let us define a function g by

$$g(x) = c_1 y_1(x) + c_2 y_2(x) \qquad (4.4.45)$$

for every x in I.   Following the same analysis as given in the proof of part (b) of Theorem 4.4.2, namely, the material after (4.4.23), we find that $g(x) = 0$ for every x in I.   Thus, (4.4.45) becomes

$$c_1 y_1(x) + c_2 y_2(x) = 0 \qquad (4.4.46)$$

for every x in I.   By (4.4.44), c_1 and c_2 are not both zero.   Thus, (4.4.46) asserts that y_1 and y_2 are linearly dependent on I.   [*See* Definition 4.3.5.] This, however, contradicts the fact that y_1 and y_2 are linearly independent on I.   Thus, the assumption that there exists a point x_0 in I such that $W(y_1, y_2; x_0) = 0$ is false.   Consequently, $W(y_1, y_2; x) \neq 0$ for every x in I, and the theorem is established.

In order to ascertain whether two solutions of $L[y] = 0$ are linearly independent on I, all we need is that the Wronskian be different from zero at one point in I.   We establish this result in the following theorem.

Theorem 4.4.4.   *Let* $y_1 = y_1(x)$ *and* $y_2 = y_2(x)$ *be two solutions of* $L[y] = 0$ *on the interval* I: $a \leq x \leq b$, *and let* x_0 *be any fixed point in* I.   *A necessary and sufficient condition that* y_1 *and* y_2 *be linearly independent on* I *is that* $W(y_1, y_2; x_0) \neq 0$.

PROOF.   *Necessity.*   If y_1 and y_2 are linearly independent on I, then by Theorem 4.4.3, $W(y_1, y_2; x) \neq 0$ for every x in I.   Since x_0 is also a point in I, $W(y_1, y_2; x_0) \neq 0$.

Sufficiency.   Suppose that $W(y_1, y_2; x_0) \neq 0$, where x_0 is any fixed point in I.   To show that y_1 and y_2 are linearly independent on I, we must show that the only constants c_1 and c_2 such that

$$c_1 y_1(x) + c_2 y_2(x) = 0 \qquad (4.4.47)$$

for every x in I are the constants $c_1 = 0$ and $c_2 = 0$.   Differentiating (4.4.47), we get

$$c_1 y_1'(x) + c_2 y_2'(x) = 0 \qquad (4.4.48)$$

for every x in I.   Now, following the analysis given in the proof of Theorem 4.4.3, namely, the material after (4.4.39), we have that y_1 and y_2 are indeed linearly independent on I, and thus the theorem is established.

Example 4.4.14. One may verify that $y_1(x) = x^{-2} \cos(3 \ln x)$ and $y_2(x) = x^{-2} \sin(3 \ln x)$ are solutions of [*see* Exercise 4.10.1]

$$x^2 y'' + 5xy' + 13y = 0, \tag{4.4.49}$$

on the interval I: $0 < x < \infty$. Now,

$$W(y_1, y_2; x)$$

$$= \begin{vmatrix} \dfrac{\cos(3\ln x)}{x^2} & \dfrac{\sin(3\ln x)}{x^2} \\[3mm] -\dfrac{3x\sin(3\ln x) + 2x\cos(3\ln x)}{x^4} & \dfrac{3x\cos(3\ln x) - 2x\sin(3\ln x)}{x^4} \end{vmatrix} .$$

A very convenient point on I to evaluate the above Wronskian is at $x_0 = 1$. Thus,

$$W(y_1, y_2; 1) = \begin{vmatrix} 1 & 0 \\ -2 & 3 \end{vmatrix} = 3.$$

Since $W(y_1, y_2; 1) \neq 0$, it follows from Theorem 4.4.4 that y_1 and y_2 are linearly independent on I. Thus, the general solution of Equation (4.4.49) on I is given by

$$y(x) = c_1 \frac{\cos(3\ln x)}{x^2} + c_2 \frac{\sin(3\ln x)}{x^2},$$

where c_1 and c_2 are arbitrary constants.

Remark 4.4.6. The following theorem follows from Theorems 4.4.3 and 4.4.4. However, we shall give an alternative proof which will lead us to develop a formula for $W(y_1, y_2; x)$.

Theorem 4.4.5. *Let* $y_1 = y_1(x)$ *and* $y_2 = y_2(x)$ *be two solutions of* $L[y] = 0$ *on the interval* I: $a \leqq x \leqq b$. *Then, either*

$$W(y_1, y_2; x) = 0, \tag{4.4.50}$$

for every x in I, or

$$W(y_1, y_2; x) \neq 0 \tag{4.4.51}$$

for every x in I.

PROOF.

$$L[y] = P(x)y'' + Q(x)y' + R(x)y = 0, \tag{4.4.1}$$

where P, Q, and R are continuous functions on I and $P(x) \neq 0$ for every x in I. Since y_1 and y_2 are solutions of $L[y] = 0$, we have

$$P(x)y_1'' + Q(x)y_1' + R(x)y_1 = 0, \tag{4.4.52}$$

and

$$P(x)y_2'' + Q(x)y_2' + R(x)y_2 = 0. \tag{4.4.53}$$

Multiplying Equations (4.4.52) and (4.4.53) by $-y_2$ and y_1, respectively, and then adding, we get

$$P(x)(y_1 y_2'' - y_2 y_1'') + Q(x)(y_1 y_2' - y_2 y_1') = 0. \qquad (4.4.54)$$

Since $P(x) \neq 0$ for every x in I, Equation (4.4.54) may be written equivalently as

$$(y_1 y_2'' - y_2 y_1'') + \frac{Q(x)}{P(x)}(y_1 y_2' - y_2 y_1') = 0. \qquad (4.4.55)$$

Since $W(y_1, y_2; x) = y_1 y_2' - y_2 y_1'$, we see that

$$\frac{dW}{dx} = y_1 y_2'' + y_1' y_2' - y_2 y_1'' - y_2' y_1' = y_1 y_2'' - y_2 y_1''.$$

Thus, we may write Equation (4.4.55) as

$$\frac{dW}{dx} + \frac{Q(x)}{P(x)} W = 0, \qquad (4.4.56)$$

which is a linear differential equation of the first order. Thus, the general solution is given by [*see* Theorem 2.6.1]

$$W(y_1, y_2; x) = c \exp\left[-\int \frac{Q(x)}{P(x)}\, dx \right], \qquad (4.4.57)$$

where c is an arbitrary constant of integration. Since $\exp\left[-\int \dfrac{Q(x)}{P(x)}\, dx \right] \neq 0$ for every x in I, it follows from (4.4.57) that if $c = 0$, then $W(y_1, y_2; x) = 0$ for every x in I; and if $c \neq 0$, then $W(y_1, y_2; x) \neq 0$ for every x in I. Thus, the theorem is established.

Remark 4.4.7. We leave it as an exercise for the reader [*see* Exercise 4.4.11] to show that

$$W(y_1, y_2; x) = W(y_1, y_2; x_0) \exp\left[-\int_{x_0}^{x} \frac{Q(\xi)}{P(\xi)}\, d\xi \right], \qquad (4.4.58)$$

where x_0 is any fixed point in I. This result is known as *Abel's identity*.

Exercises 4.4

1. Let the functions y_1 and y_2 be defined on the interval $I: -\infty < x < \infty$ by $y_1(x) = x^4$ and $y_2(x) = x^3 |x|$.
    (a) Show that the functions y_1 and y_2 are linearly independent on I.
    (b) Show that $W(y_1, y_2; x) = 0$ for every x in I.

(c) Do the results in parts (a) and (b) contradict Theorem 4.4.3?

(d) Are the functions y_1 and y_2 differentiable in I?

2. Let y_1 and y_2 be two differentiable functions in an interval I.

(a) Show that if the functions y_1 and y_2 are linearly dependent on I, then $W(y_1, y_2; x) = 0$ for every x in I.

[*Note:* The converse of this result is not necessarily true in view of Exercise 1.] However, we have the following result.

(b) Show that if $W(y_1, y_2; x) = 0$ for every x in I and if $y_1(x) \neq 0$ for every x in I, then the functions y_1 and y_2 are linearly dependent on I.

(c) Show that if $W(y_1, y_2; x_0) \neq 0$ for some point x_0 in I, then the functions y_1 and y_2 are linearly independent on I.   (This result gives us a convenient test for the linear independence of two differentiable functions in I.)

3. On the interval I: $-\infty < x < \infty$, verify for each problem given below that the functions y_1 and y_2 are linearly independent solutions of the given differential equation.   Also write the general solution for each problem.   The letters α and β denote constants, $\alpha \neq \beta$.

(a) $y'' = 0.$ $y_1(x) = 1,$ $y_2(x) = x.$

(b) $y'' - \alpha^2 y = 0.$ $y_1(x) = e^{\alpha x},$ $y_2(x) = e^{-\alpha x}.$

(c) $y'' + \beta^2 y = 0.$ $y_1(x) = \cos \beta x,$ $y_2(x) = \sin \beta x.$

(d) $y'' - 2\alpha y' + \alpha^2 y = 0.$ $y_1(x) = e^{\alpha x},$ $y_2(x) = x e^{\alpha x}.$

(e) $y'' - (\alpha + \beta)y' + \alpha\beta y = 0.$ $y_1(x) = e^{\alpha x},$ $y_2(x) = e^{\beta x}.$

(f) $y'' - 2\alpha y' + (\alpha^2 + \beta^2)y = 0.$ $y_1(x) = e^{\alpha x} \cos \beta x,$ $y_2(x) = e^{\alpha x} \sin \beta x.$

4. On an interval I not containing the origin, repeat Exercise 3 for each of the following problems.

(a) $x^2 y'' - 7xy' + 15y = 0.$ $y_1(x) = x^3,$ $y_2(x) = x^5.$

(b) $x^2 y'' - (x^2 + 2x)y' + (2 + x)y = 0.$ $y_1(x) = x,$ $y_2(x) = x e^x.$

(c) $x^2 y'' - 2xy' + (2 - x^2)y = 0.$ $y_1(x) = x e^x,$ $y_2(x) = x e^{-x}.$

5. Let y_1 and y_2 be two differentiable functions in an interval I.   Suppose that y_1 and y_2 are linearly independent on I.   Let the functions f and g be defined by $f(x) = a_1 y_1(x) + a_2 y_2(x)$ and $g(x) = b_1 y_1(x) + b_2 y_2(x)$ for every x in I, where $a_1, a_2, b_1,$ and b_2 are constants.   Let $\Delta = a_1 b_2 - a_2 b_1$.

(a) Show that $W(f, g; x) = \Delta \, W(y_1, y_2; x)$ for every x in I.

(b) Show that if $\Delta \neq 0$, then the functions f and g are linearly independent on I.

6. Determine whether the functions f and g defined below are linearly independent on I: $-\infty < x < \infty$.

(a) $f(x) = 2x + 3e^{2x},$ $g(x) = 4x + 5e^{2x}.$

(b) $f(x) = \frac{5}{2}e^{3x} - \frac{3}{2}e^{5x},$ $g(x) = -\frac{1}{2}e^{3x} + \frac{1}{2}e^{5x}.$

(c) $f(x) = \frac{1}{2}x^2 + \frac{2}{3}x,$ $g(x) = 15x^2 + 20x.$

(d) $f(x) = \cos 2x + 3 \sin 2x,$ $g(x) = 3 \cos 2x + \sin 2x.$

7. The functions y_1 and y_2 defined by $y_1(x) = \frac{5}{2}e^{3x} - \frac{3}{2}e^{5x}$ and $y_2(x) = -\frac{1}{2}e^{3x} + \frac{1}{2}e^{5x}$ are basic solutions at $x_0 = 0$ of the differential equation

$$y'' - 8y' + 15y = 0$$

on the interval I: $-\infty < x < \infty$.   Solve each initial-value problem.

(a) $\begin{cases} y'' - 8y' + 15y = 0, \\ y(0) = 6, \quad y'(0) = 8. \end{cases}$ (b) $\begin{cases} y'' - 8y' + 15y = 0, \\ y(0) = -2, \quad y'(0) = 2. \end{cases}$

8. On the interval I: $0 < x < \infty$, verify that the functions y_1 and y_2 defined by $y_1(x) = \cos(2 \ln x)$ and $y_2(x) = \sin(2 \ln x)$ are two linearly independent solutions of the differential equation [*see* Exercise 4.10.2]

$$x^2 y'' + x y' + 4y = 0.$$

Solve the initial-value problem

$$\begin{cases} x^2 y'' + xy' + 4y = 0, \\ y(1) = 5, \quad y'(1) = 4. \end{cases}$$

9. Let x_0 be any fixed point in an interval I.   Show that $y_1 = y_1(x)$ and $y_2 = y_2(x)$ cannot be basic solutions at x_0 of $L[y] = 0$ on I if either of the following conditions hold.
   (i) $y_1(x_0) = 0$ and $y_2(x_0) = 0$;
   (ii) $y_1(x)$ and $y_2(x)$ have a maximum or a minimum at x_0.
10. (a) Let the functions y_1 and y_2 be continuous on the interval I: $a \leqq x \leqq b$. The *Gramian* of the functions y_1 and y_2 is defined on I by

$$G = \begin{vmatrix} \displaystyle\int_a^b y_1^{\,2}(x)\,dx & \displaystyle\int_a^b y_1(x)y_2(x)\,dx \\[2ex] \displaystyle\int_a^b y_2(x)y_1(x)\,dx & \displaystyle\int_a^b y_2^{\,2}(x)\,dx \end{vmatrix}.$$

Prove that *a necessary and sufficient condition that the functions y_1 and y_2 be linearly dependent on I is that $G = 0$.*
   (b) Use the above result to show that the functions y_1 and y_2 defined by $y_1(x) = e^x$ and $y_2(x) = e^{x+2}$ are linearly dependent on I: $0 \leqq x \leqq 1$.   [*Note:* This result may be readily obtained by observing that $y_2(x) = ky_1(x)$, where $k = e^2$.]
11. (a) Establish Abel's identity given in (4.4.58).
   (b) Refer back to Example 4.4.14.   Use Abel's identity to find $W(y_1, y_2; x)$.
   (c) Let $u_1(x), u_2(x)$ and $v_1(x), v_2(x)$ be two fundamental systems of solutions of the equation $L[y] = P(x)y'' + Q(x)y' + R(x)y = 0$ on the interval I: $a \leqq x \leqq b$.   Show that $W(u_1 u_2; x) = kW(v_1, v_2; x)$, where the constant $k \neq 0$.
12. Let the functions y_1 and y_2 be defined by $y_1(x) = a_0 x^n + a_2 x^{n-2} + a_4 x^{n-4} + \cdots + a_{n-2} x^2$ and $y_2(x) = (a_0 x^{n-1} + a_2 x^{n-3} + a_4 x^{n-5} + \cdots + a_{n-2} x)|x|$ on the interval I: $-\infty < x < \infty$, where $a_0, a_2, \ldots, a_{n-2}$ are constants and n is a positive even integer.
   (a) Show that the functions y_1 and y_2 are linearly independent on I.
   (b) Show that $W(y_1, y_2; x) = 0$ for every x in I.
   (c) What is the geometric significance of the graphs of the functions y_1 and y_2 on I?
   [*Note:* Exercise 1 is a special case of the above exercise.]

13. Let the operator L be defined by

(1) $$L[y] = P(x)y'' + Q(x)y' + R(x)y,$$

where P, Q, and R are continuous functions on a common interval I: $a \leq x \leq b$, and $P(x) \neq 0$ for every x in I. Let the functions u and v be twice differentiable in I. As in the discussion of the proof of Theorem 4.4.5, we see that

$$uL[v] - vL[u] = u[P(x)v'' + Q(x)v' + R(x)v] - v[P(x)u'' + Q(x)u' + R(x)u]$$

$$= P(x)[uv'' - vu''] + Q(x)[uv' - vu'].$$

Thus,

(2) $$uL[v] - vL[u] = P(x)\frac{d}{dx} W(u, v; x) + Q(x)W(u, v; x).$$

The formula given in (2) is known as *Lagrange's formula*. The operator L is said to be *self-adjoint* if

(3) $$L[y] = [P(x)y']' + R(x)y.$$

Thus, we see that the operator L is self-adjoint if $Q(x) = P'(x)$ in Equation (1).
(a) Utilize Equation (2) to show that if the operator L is self-adjoint, then

(4) $$\int_a^b \{uL[v] - vL[u]\}dx = P(x)W(u, v; x)\Big|_a^b.$$

The formula given in (4) is known as *Green's formula*, and plays a very basic role in the theory and applications of differential equations.
(b) Utilize Equation (2) to show that if the operator L is self-adjoint and $u(x)$ and $v(x)$ are solutions of

(5) $$L[y] = P(x)y'' + Q(x)y' + R(x)y = 0$$

on the interval I, then

(6) $$P(x)W(u, v; x) = c,$$

for every x in I, where c is a constant.
(c) Suppose that $y_1(x)$ and $y_2(x)$ are solutions of the equation

(7) $$(\ln x)\, y'' + \frac{1}{x} y' + x^2 y = 0$$

on the interval $0 < x < \infty$. Show that

$$W(y_1, y_2; x) = \frac{c}{\ln x}.$$

4.5. Reduction of the Order of a Differential Equation

Suppose that we know a nontrivial solution of

$$L[y] = P(x)y'' + Q(x)y' + R(x)y = 0, \tag{4.5.1}$$

where as usual, the functions P, Q, and R are assumed continuous on the interval I: $a \leq x \leq b$ and $P(x) \neq 0$ for all x in I. By means of a suitable transformation which reduces $L[y] = 0$ into a first-order homogeneous linear differential equation, we are then able to find a second linearly independent solution of $L[y] = 0$ on I. Thus, a fundamental set of solutions of $L[y] = 0$ on I can be determined, and the general solution of $L[y] = 0$ on I will then be available.

Theorem 4.5.1. *Suppose that $y_1(x)$ is a solution of $L[y] = 0$ on the interval I: $a \leq x \leq b$, and $y_1(x) \neq 0$ for every x in I.*

(a) *The transformation*

$$y(x) = v(x)y_1(x) \tag{4.5.2}$$

reduces $L[y] = 0$ on I to a first-order homogeneous linear differential equation

$$P(x)y_1(x)u'(x) + [2P(x)y_1'(x) + Q(x)y_1(x)]u(x) = 0 \tag{4.5.3}$$

where $u = v'$.

(b) *Another solution $y_2(x)$ of $L[y] = 0$ on I is given by*

$$y_2(x) = y_1(x) \int \frac{\exp\left[-\int \dfrac{Q(x)}{P(x)}\,dx\right]}{[y_1(x)]^2}\,dx. \tag{4.5.4}$$

(c) *The functions y_1 and y_2 are linearly independent on I.*
(d) *The general solution $y(x)$ of $L[y] = 0$ on I is given by the linear combination of these two linearly independent solutions*

$$y(x) = c_1 y_1(x) + c_2 y_2(x), \tag{4.5.5}$$

where c_1 and c_2 are arbitrary constants.

PROOF. *Concerning* (a). Differentiating (4.5.2) two times in succession, we obtain

$$y' = vy_1' + v'y_1, \tag{4.5.6}$$

and

$$y'' = vy_1'' + 2v'y_1' + v''y_1. \tag{4.5.7}$$

Substituting y, y', and y'' of (4.5.2), (4.5.6), and (4.5.7) in (4.5.1), we obtain upon collecting terms

$$Py_1v'' + (2Py_1' + Qy_1)v' + (Py_1'' + Qy_1' + Ry_1)v = 0. \tag{4.5.8}$$

Since $y_1(x)$ is a solution of $L[y] = 0$, $Py_1'' + Qy_1' + Ry_1 = 0$ and (4.5.8) reduces to

$$Py_1v'' + (2Py_1' + Qy_1)v' = 0. \tag{4.5.9}$$

If we let $u = v'$, then (4.5.9) becomes

$$Py_1 u' + (2Py_1' + Qy_1)u = 0, \qquad (4.5.10)$$

and (4.5.3) is thus established.

Concerning (b).   The general solution of (4.5.10) is given by

$$u(x) = c \exp\left[-\int \frac{(2Py_1' + Qy_1)}{Py_1}\, dx\right] = c \exp\left[-\int \frac{2y_1'}{y_1}\, dx\right] \cdot \exp\left[-\int \frac{Q}{P}\, dx\right]$$

$$= c \exp\left[-2 \ln y_1\right] \cdot \exp\left[-\int \frac{Q}{P}\, dx\right] = \frac{c}{y_1{}^2} \exp\left[-\int \frac{Q}{P}\, dx\right], \qquad (4.5.11)$$

where c is an arbitrary constant.   Since $u = v'$, integrating (4.5.11), we obtain

$$v(x) = c \int \frac{\exp\left[-\int \frac{Q}{P}\, dx\right]}{y_1{}^2}\, dx + k, \qquad (4.5.12)$$

where k is a constant of integration.   Utilizing (4.5.2) and (4.5.12), we obtain

$$y(x) = cy_1(x) \int \frac{\exp\left[-\int \frac{Q}{P}\, dx\right]}{y_1{}^2}\, dx + ky_1(x), \qquad (4.5.13)$$

which is another solution of $L[y] = 0$ on I.   Since $y_1(x)$ is a solution of $L[y] = 0$, so is $ky_1(x)$ a solution.   [*See* Remark 4.4.1.]   Thus, in searching for another solution of $L[y] = 0$, we may omit the solution $ky_1(x)$ from (4.5.13).   Also, from Remark 4.4.1, if $cg(x)$, where c is a nonzero constant, is a solution of $L[y] = 0$, so is $g(x)$.   Thus the function y_2 defined by

$$y_2(x) = y_1(x) \int \frac{\exp\left[-\int \frac{Q(x)}{P(x)}\, dx\right]}{[y_1(x)]^2}\, dx \qquad (4.5.14)$$

may be considered as the other solution of $L[y] = 0$ on I.   Hence, part (b) of the theorem is now established.

Concerning (c).   Since the indefinite integral of the right-hand member of (4.5.14) is not a constant on I, it follows from Exercise 4.3.1 that the functions y_1 and y_2 are linearly independent on I.   [*See* Exercise 4.5.1 for an alternative proof.]   Thus, part (c) of the theorem is established.

Concerning (d).   Since the linearly independent functions y_1 and y_2 are solutions of $L[y] = 0$ on I, it follows from Definition 4.4.1, that the general solution $y(x)$ of $L[y] = 0$ on I is

$$y(x) = c_1 y_1(x) + c_2 y_2(x),$$

where c_1 and c_2 are arbitrary constants.   The proof of the theorem is now completed.

Remark 4.5.1.   Again, we would like to stress the fact that the general solution of $L[y] = 0$ obtained by the method given above is based on knowing a nonvanishing solution of $L[y] = 0$. Thus, if methods are available for finding one solution of $L[y] = 0$, then the method of reduction of order may be used to find another solution.

Example 4.5.1.   On any interval I not containing the origin, a solution of the equation

$$xy'' - (2x + 1)y' + (x + 1)y = 0 \qquad (4.5.15)$$

is $y_1(x) = e^x$.   Let us find another solution by the method of reduction of order.

SOLUTION.   Let

$$y(x) = e^x v(x). \qquad (4.5.16)$$

Then

$$y' = e^x v + e^x v' \quad \text{and} \quad y'' = e^x v + 2e^x v' + e^x v''.$$

Substituting y, y', and y'' in Equation (4.5.15), we obtain

$$xe^x v'' - e^x v' = 0,$$

or

$$v'' - \frac{1}{x} v' = 0. \qquad (4.5.17)$$

Let $u = v'$.   Thus, (4.5.17) may be written as

$$\frac{du}{u} = \frac{dx}{x}.$$

Integrating the above equation, we obtain

$$u = cx.$$

Thus,

$$dv = cx \, dx,$$

and an integration gives

$$v = \frac{c}{2} x^2. \qquad (4.5.18)$$

Utilizing (4.5.16) and (4.5.18), another solution of Equation (4.5.15) is given by

$$y_2(x) = x^2 e^x. \qquad (4.5.19)$$

Let us obtain the other solution of Equation (4.5.15) by utilizing Formula (4.5.4).   From Equations (4.5.1) and (4.5.15), we see that $P(x) = x$ and $Q(x) = -(2x + 1)$.   Thus,

$$y_2(x) = e^x \int \frac{\exp\left[\int \frac{2x+1}{x}\, dx\right]}{e^{2x}}\, dx = e^x \int \frac{\exp\left[2x + \ln x\right]}{e^{2x}}\, dx = e^x \int \frac{e^{2x}x}{e^{2x}}\, dx$$

$$= e^x \int x\, dx = \frac{1}{2}x^2 e^x.$$

Suppressing the constant $\frac{1}{2}$, we obtain the same result as in (4.5.19).

Remark 4.5.2.   Again we would like to emphasize that if $kg(x)$, where $k \neq 0$ is a constant, is a solution of $L[y] = 0$, then $g(x)$ is also a solution of $L[y] = 0$. In the above example, instead of taking $y_2(x) = (c/2)\,x^2 e^x$ as another solution of Equation (4.5.15), there is no loss of generality, by taking $y_2(x)$ as (4.5.19).
  For Example 4.5.1, the general solution on I is given by

$$y(x) = c_1 e^x + c_2 x^2 e^x, \tag{4.5.20}$$

where c_1 and c_2 are arbitrary constants.   If one had used $y_2(x) = (c/2)x^2 e^x$ as

another solution of (4.5.15), then the general solution would have been

$$y(x) = c_1 e^x + \frac{c_2\, c}{2}\, x^2 e^x. \tag{4.5.21}$$

Since $c_2\, c/2$ is another arbitrary constant, we may denote it by c_3.   Thus, (4.5.21) is the same as (4.5.20).

Exercises 4.5

1. Let $y_1(x)$ be a nonvanishing solution of $L[y] = 0$ on an interval I.   Let $y_2(x) = y_1(x)v(x)$, where

$$v(x) = \int \frac{\exp\left[-\int \frac{Q(x)}{P(x)}\, dx\right]}{[y_1(x)]^2}\, dx,$$

be another solution of $L[y] = 0$ on I.   Show that

$$W(y_1, y_2; x) = \exp\left[-\int \frac{Q(x)}{P(x)}\, dx\right] \neq 0$$

for every x in I.   Thus, in view of Theorem 4.4.3, the functions y_1 and y_2 are linearly independent on I.

In Exercises 2 through 6, find a second solution of each equation on the given interval I by the method of reduction of order.

2. $x^2y'' - 2xy' + (2 - x^2)y = 0,$ $y_1 = xe^x;$ I: all x such that $x > 0.$
3. $x^2y'' - 7xy' + 15y = 0,$ $y_1 = x^3;$ I: all x such that $x > 0.$
4. $(x^2 - x)y'' + (x + 1)y' - y = 0,$ $y_1 = 1 + x;$ I: $0 < x < 1.$
5. $x(1 + 3x^2)y'' + 2y' - 6xy = 0,$ $y_1 = 1 + x^2;$ I: all x such that $x > 0.$

6. $x^2(1 - x^2)y'' - x^3y' - 2y = 0,$ $y_1 = \dfrac{\sqrt{1 - x^2}}{x};$ I: $0 < x < 1.$

7. The *Legendre equation* is given by

$$(1) \qquad (1 - x^2)y'' - 2xy' + n(n + 1)y = 0,$$

where n is a constant.   The functions satisfying Equation (1) are called *Legendre's functions of order n*.   When n is zero or a positive integer, Equation (1) is known to have polynomial solutions.   [*See* Section 7.8.]   The *Legendre polynomials* are denoted by $P_n(x)$.   Equation (1) plays a very basic role in applied mathematics, especially in problems dealing with temperature and electrostatics in a spherical region.

When $n = 0$, Equation (1) becomes

$$(2) \qquad (1 - x^2)y'' - 2xy' = 0.$$

The Legendre polynomial $P_0(x) = 1$ is clearly a solution of (2).   When $n = 1$, Equation (1) becomes

$$(3) \qquad (1 - x^2)y'' - 2xy' + 2y = 0.$$

The Legendre polynomial $P_1(x) = x$ is clearly a solution of (3).
(a) Verify that the Legendre polynomials $P_2(x) = \frac{1}{2}(3x^2 - 1)$,

$$P_3(x) = \tfrac{1}{2}(5x^3 - 3x) \quad \text{and} \quad P_4(x) = \tfrac{1}{8}(35x^4 - 30x^2 + 3)$$

are solutions of Equation (1) for $n = 2$, 3, and 4, respectively.
(b) On the interval I: $-1 < x < 1$, utilize the method of reduction of order to find another solution of Equation (1) for $n = 0, 1, 2, 3$.
(c) Are the Legendre polynomials $P_0(x)$, $P_1(x)$, $P_2(x)$, $P_3(x)$, and $P_4(x)$ linearly independent on the interval $-1 \leq x \leq 1$?
(d) Sketch the graphs of the Legendre polynomials $P_n(x)$, $n = 0, 1, 2, 3, 4$ on the interval $-1 \leq x \leq 1$.
(e) For Legendre polynomials $P_k(x)$, $k = 0, 1, 2, 3, 4$, verify that

$$(4) \qquad \int_{-1}^{1} P_m(x)P_n(x)dx = 0, \qquad m \neq n, \quad m, n = 0, 1, 2, 3, 4.$$

[*See* also Section 7.8, Formula (7.8.26).]
(f) Let the operator L be defined by

$$(5) \qquad L[y] = (1 - x^2)y'' - 2xy' + n(n + 1)y.$$

Show that the operator L is self-adjoint.   [*See* Exercise 4.4.13(3).]   [This result will be utilized in Chapter 7, Section 8, to show that Equation (4) is valid for all $m \neq n$, $m, n = 0, 1, \ldots.$ *See* Formula (7.8.26).]

(g) For Legendre polynomials $P_n(x)$, $n = 0, 1, 2, 3, 4$, verify that

$$\int_{-1}^{1} [P_n(x)]^2 dx = \frac{2}{2n+1}.$$

[*See* also Section 7.8, Formula (7.8.26).]

(h) Determine the values C_0, C_1, C_2, C_3, and C_4 so that

$$3 + 8x - 6x^2 + 40x^3 + 35x^4$$
$$= C_0 P_0(x) + C_1 P_1(x) + C_2 P_2(x) + C_3 P_3(x) + C_4 P_4(x).$$

[This result may be extended to assert that a polynomial of degree n may be expressed as a linear combination of Legendre polynomials $P_0(x), P_1(x), \ldots, P_n(x)$.]

8. The *Bessel equation* is given by

$$(1) \qquad\qquad x^2 y'' + xy' + (x^2 - n^2)y = 0,$$

where the number n may have any value, positive or negative, integral or fractional, or even complex.   Although Equation (1) is a second-order differential equation, it is commonly called *Bessel's equation of order n*.   For example, Bessel's equation of order one-half is obtained by taking $n = \frac{1}{2}$ in Equation (1). Thus, we have

$$(2) \qquad\qquad x^2 y'' + xy' + (x^2 - \tfrac{1}{4})y = 0.$$

The functions satisfying Equation (1) are usually denoted by J_n and K_n.   [*See* Section 7.7.]   Equation (1) also plays a very important role in applied mathematics, especially in problems dealing with temperature and electrostatics in a cylindrical region.

(a) Verify that the Bessel functions $J_{-1/2}$, $J_{1/2}$, and $J_{3/2}$ defined by [*see* also Example 7.7.1 and Exercise 7.7.13]

$$J_{-1/2}(x) = \sqrt{\frac{2}{\pi x}} \cos x, \qquad J_{1/2}(x) = \sqrt{\frac{2}{\pi x}} \sin x,$$

and

$$J_{3/2}(x) = \sqrt{\frac{2}{\pi x}} \left(\frac{\sin x}{x} - \cos x \right),$$

are solutions of Equation (1) for $n = -\frac{1}{2}$, $n = \frac{1}{2}$, and $n = \frac{3}{2}$, respectively.

(b) On the interval I: $0 < x < \infty$, utilize the method of reduction of order to show that the general solution of Equation (1) for $n = \frac{1}{2}$ is given by

$$y(x) = c_1 \frac{\sin x}{\sqrt{x}} + c_2 \frac{\cos x}{\sqrt{x}},$$

where c_1 and c_2 are arbitrary constants; or by

$$y(x) = AJ_{1/2}(x) + BJ_{-1/2}(x),$$

where A and B are arbitrary constants.

(c) Are the Bessel functions $J_{-1/2}$, $J_{1/2}$, and $J_{3/2}$ given in part (a) linearly independent on the interval I: $0 < x < \infty$?

4.6. Linear Homogeneous Equations with Constant Coefficients. Forms of Solutions

Consider the differential equation given by

$$L[y] = ay'' + by' + cy = 0, \tag{4.6.1}$$

where a, b, and c are real constants and $a \neq 0$. In view of Theorem 4.4.2, the equation $L[y] = 0$ has two linearly independent solutions y_1 and y_2 on the interval I: $-\infty < x < \infty$, since $P(x) = a \neq 0$, $Q(x) = b$, and $R(x) = c$ are continuous for every x in I. Thus, the general solution of $L[y] = 0$ on I is given by

$$y(x) = c_1 y_1(x) + c_2 y_2(x), \tag{4.6.2}$$

where c_1 and c_2 are arbitrary constants.

In searching for solutions of $L[y] = 0$, it appears somewhat plausible to assume that if a linear combination like (4.6.1) of a function and its derivatives cancel to zero, the derivatives must necessarily be multiples of the function. One such function possessing this property is the exponential function e^{mx}. This motivates us to try $y(x) = e^{mx}$ as a possible solution of $L[y] = 0$ for suitably chosen values of m. Thus,

$$L[e^{mx}] = a(e^{mx})'' + b(e^{mx})' + ce^{mx} = 0. \tag{4.6.3}$$

Performing the indicated differentiations, we may write Equation (4.6.3) as

$$(am^2 + bm + c)e^{mx} = 0. \tag{4.6.4}$$

Since $e^{mx} \neq 0$ for every x in I, we see that

$$am^2 + bm + c = 0. \tag{4.6.5}$$

Thus, if m is a root of the quadratic Equation (4.6.5), then $y(x) = e^{mx}$ is a solution of $L[y] = 0$ on I. The quadratic Equation (4.6.5) is usually called the *auxiliary equation* or the *characteristic equation* of $L[y] = 0$. Note that the auxiliary equation can be obtained from Equation (4.6.1) by replacing y'' by m^2, y' by m, and $y^{(0)} \equiv y$ by $m^0 \equiv 1$.

The roots of the auxiliary Equation (4.6.5), denoted by m_1 and m_2, are given by

$$m_1 = -\frac{b + \sqrt{b^2 - 4ac}}{2a} \quad \text{and} \quad m_2 = -\frac{b - \sqrt{b^2 - 4ac}}{2a}. \tag{4.6.6}$$

It is now clear that the form of the solutions of $L[y] = 0$ on I will depend upon the discriminant $D \equiv b^2 - 4ac$ of Equation (4.6.5). Hence, we are led to consider three cases, namely, $D > 0$, $D = 0$, and $D < 0$.

Case 1. $D > 0$. The roots m_1 and m_2 are real and distinct.

Case 2. $D = 0$. The roots m_1 and m_2 are real and equal $(m_1 = m_2)$.

Case 3. $D < 0$. The roots m_1 and m_2 are imaginary.

$$(4.6.7)$$

We shall give the results for Cases 1 and 2 in the following theorem. Case 3 will be considered in a separate theorem, since a few remarks are necessary before we discuss this case.

Theorem 4.6.1. *Consider the equation*

$$L[y] = ay'' + by' + cy = 0, \qquad [4.6.1]$$

where a, b, and c are real constants and $a \neq 0$. Let m_1 and m_2 be the roots of the auxiliary equation

$$am^2 + bm + c = 0. \qquad [4.6.5]$$

(a) *If m_1 and m_2 are real and distinct, then the general solution $y(x)$ of $L[y] = 0$ on I: $-\infty < x < \infty$, is given by*

$$y(x) = c_1 e^{m_1 x} + c_2 e^{m_2 x}, \qquad (4.6.8)$$

where c_1 and c_2 are arbitrary constants.

(b) *If m_1 and m_2 are real and equal, then the general solution $y(x)$ of $L[y] = 0$ on I: $-\infty < x < \infty$, is given by*

$$y(x) = (c_1 + c_2 x)e^{m_1 x}. \qquad (4.6.9)$$

PROOF. *Concerning* (a). Since m_1 and m_2 $(m_1 \neq m_2)$ are solutions of Equation (4.6.5), the functions y_1 and y_2 given by

$$y_1(x) = e^{m_1 x} \quad \text{and} \quad y_2(x) = e^{m_2 x} \qquad (4.6.10)$$

are solutions of $L[y] = 0$. By Example 4.3.4, y_1 and y_2 are linearly independent on I. Thus, the general solution of $L[y] = 0$ on I is given by (4.6.8).

Concerning (b). When $m_1 = m_2$, the linear combination

$$c_1 e^{m_1 x} + c_2 e^{m_2 x} = ce^{m_1 x},$$

where $c = c_1 + c_2$, is not a linear combination of two linearly independent solutions of $L[y] = 0$. Since this linear combination involves only one arbitrary constant, it cannot be the general solution of $L[y] = 0$ which requires two arbitrary constants. However, knowing a nonvanishing solution of

$L[y] = 0$ on I, we may obtain another solution of $L[y] = 0$ by utilizing Theorem 4.5.1. Since $m_1 = m_2$, the discriminant D of Equation (4.6.5) is equal to zero. Thus, in view of (4.6.6), we have

$$m_1 = m_2 = -\frac{b}{2a}. \tag{4.6.11}$$

Let us denote by $y_1(x)$ one of the solutions of $L[y] = 0$ on I. That is,

$$y_1(x) = e^{m_1 x}. \tag{4.6.12}$$

Let

$$y(x) = y_1(x)v(x) = e^{m_1 x}v(x). \tag{4.6.13}$$

Differentiating (4.6.13) two times in succession, we obtain

$$y' = m_1 e^{m_1 x}v + e^{m_1 x}v', \tag{4.6.14}$$

and

$$y'' = m_1^2 e^{m_1 x}\, v + 2m_1 e^{m_1 x}\, v' + e^{m_1 x}\, v''. \tag{4.6.15}$$

Utilizing (4.6.13), (4.6.14), and (4.6.15), Equation (4.6.1) becomes, after collecting terms,

$$ae^{m_1 x}\, v'' + e^{m_1 x}(2am_1 + b)v' + e^{m_1 x}(am_1^2 + bm_1 + c)v = 0. \tag{4.6.16}$$

Since m_1 is a root of Equation (4.6.5), $am_1^2 + bm_1 + c = 0$ and Equation (4.6.16) reduces to

$$ae^{m_1 x}v'' + e^{m_1 x}(2am_1 + b)v' = 0. \tag{4.6.17}$$

Since $a \neq 0$, $e^{m_1 x} \neq 0$ for every x in I, and in view of (4.6.11), $2am_1 + b = 2a\left(-\dfrac{b}{2a}\right) + b = 0$, we see that Equation (4.6.17) simplifies to

$$v'' = 0. \tag{4.6.18}$$

Integrating (4.6.18) twice, we obtain

$$v(x) = c_1 x + c_2, \tag{4.6.19}$$

where c_1 and c_2 are arbitrary constants. Thus, the other solution $y_2(x)$ of $L[y] = 0$ on I may be taken as

$$y_2(x) = xe^{m_1 x}. \tag{4.6.20}$$

Theorem 4.5.1 assures us that the functions y_1 and y_2 of (4.6.12) and (4.6.20) are linearly independent on I. Thus, the general solution of $L[y] = 0$ on I is given by (4.6.9), and the theorem is now established.

    [*See* Exercises 4.6.1 and 4.6.16 for alternative proofs of part (b) of Theorem 4.6.1.]

Example 4.6.1.   Let us solve the initial-value problem

$$\begin{cases} y'' - 8y' + 15y = 0, & (4.6.21) \\ y(0) = 5, & (4.6.22) \\ y'(0) = 3. & (4.6.23) \end{cases}$$

SOLUTION.   The auxiliary equation of (4.6.21) is

$$m^2 - 8m + 15 = 0.$$

The roots of the above equation are $m_1 = 3$ and $m_2 = 5$.   Hence, the general solution of (4.6.21) is

$$y(x) = c_1 e^{3x} + c_2 e^{5x}. \tag{4.6.24}$$

Utilizing (4.6.22) and (4.6.24), we obtain

$$5 = y(0) = c_1 e^0 + c_2 e^0 = c_1 + c_2. \tag{4.6.25}$$

Differentiating (4.6.24), we get

$$y'(x) = 3c_1 e^{3x} + 5c_2 e^{5x}. \tag{4.6.26}$$

Utilizing (4.6.23) and (4.6.26), we obtain

$$3 = y'(0) = 3c_1 e^0 + 5c_2 e^0 = 3c_1 + 5c_2. \tag{4.6.27}$$

Solving Equations (4.6.25) and (4.6.27) simultaneously, we find $c_1 = 11$ and $c_2 = -6$.   Substituting these values in (4.6.24), the solution of the given initial-value problem is

$$y(x) = 11e^{3x} - 6e^{5x}. \tag{4.6.28}$$

Example 4.6.2.   Let us solve the initial-value problem

$$\begin{cases} 4y'' - 12y' + 9y = 0, & (4.6.29) \\ y(0) = 2, & (4.6.30) \\ y'(0) = 1. & (4.6.31) \end{cases}$$

SOLUTION.   The auxiliary equation of (4.6.29) is

$$4m^2 - 12m + 9 = 0.$$

The roots of the above equation are $m_1 = \frac{3}{2}$ and $m_2 = \frac{3}{2}$.   Thus, the general solution of (4.6.29) is [*see* (4.6.9)]

$$y(x) = c_1 e^{(3/2)x} + c_2 x e^{(3/2)x}. \tag{4.6.32}$$

Differentiating (4.6.32), we get

$$y'(x) = \tfrac{3}{2}c_1 e^{(3/2)x} + \tfrac{3}{2}c_2 x e^{(3/2)x} + c_2 e^{(3/2)x}. \tag{4.6.33}$$

Utilizing (4.6.30) through (4.6.33), we obtain

$$2 = y(0) = c_1 e^0 + c_2 \cdot 0 \cdot e^0 = c_1 \tag{4.6.34}$$

and

$$1 = y'(0) = \tfrac{3}{2} c_1 e^0 + \tfrac{3}{2} c_2 \cdot 0 \cdot e^0 + c_2 e^0 = \tfrac{3}{2} c_1 + c_2. \tag{4.6.35}$$

From (4.6.34) and (4.6.35), we find $c_1 = 2$ and $c_2 = -2$. Substituting these values in (4.6.32), the solution of the given initial-value problem is

$$y(x) = 2e^{(3/2)x} - 2xe^{(3/2)x}. \tag{4.6.36}$$

Remark 4.6.1. In case the discriminant of the auxiliary Equation (4.6.5) is negative, then the roots of this equation are imaginary. Let m_1 denote one of these roots, that is,

$$m_1 = \alpha + i\beta, \tag{4.6.37}$$

where α and β are real numbers, $\beta \neq 0$, and $i^2 = -1$. Since the coefficients of the auxiliary Equation (4.6.5) are real numbers, it is known from a result in college algebra that the other root m_2 of Equation (4.6.5) is the complex conjugate of m_1, that is,

$$m_2 = \alpha - i\beta. \tag{4.6.38}$$

The general solution of Equation (4.6.1) is then given by

$$y(x) = C_1 e^{(\alpha + i\beta)x} + C_2 e^{(\alpha - i\beta)x}, \tag{4.6.39}$$

where C_1 and C_2 are arbitrary constants. Thus, the general solution of Equation (4.6.1) is given in terms of a linear combination of complex-valued functions rather than in terms of a linear combination of real-valued functions. However, we may bring the linear combination given in (4.6.39) into a linear combination of real-valued functions by appealing to a formula which is found to be very useful in the theory of functions of a complex variable. This formula is known as *Euler's formula* and is given by

$$e^{i\theta} = \cos\theta + i\sin\theta, \tag{4.6.40}$$

where θ is a real number and $i^2 = -1$. [*See* Exercise 4.6.17 for a possible motivation of Euler's formula, but not a proof of it.] If we replace θ by $-\theta$ in (4.6.40) and recall that $\cos(-\theta) = \cos\theta$ and $\sin(-\theta) = -\sin\theta$, (4.6.40) becomes

$$e^{-i\theta} = \cos\theta - i\sin\theta. \tag{4.6.41}$$

Adding (4.6.40) and (4.6.41), we obtain

$$\cos\theta = \frac{e^{i\theta} + e^{-i\theta}}{2}. \tag{4.6.42}$$

Subtracting (4.6.41) from (4.6.40), we get

$$\sin \theta = \frac{e^{i\theta} - e^{-i\theta}}{2i}. \tag{4.6.43}$$

Utilizing the trigonometric identities, we leave it as an exercise for the reader [*see* Exercise 4.6.2] to verify that the complex-valued function $e^{i\theta}$ has the following properties which are similar to the real-valued function e^x:

$$\begin{cases} e^{i\theta_1} \cdot e^{i\theta_2} = e^{i(\theta_1 + \theta_2)}, \\ (e^{i\theta})^{-1} = e^{-i\theta}, \\ \dfrac{e^{i\theta_1}}{e^{i\theta_2}} = e^{i(\theta_1 - \theta_2)}. \end{cases} \tag{4.6.44}$$

Let $z = x + iy$, where x and y are real numbers and $y \neq 0$. We now go one step further and require that

$$e^{x+iy} = e^x \cdot e^{iy} = e^x(\cos y + i \sin y). \tag{4.6.45}$$

If $z = x + iy$, $z_1 = x_1 + iy_1$, and $z_2 = x_2 + iy_2$, then utilizing Formula (4.6.45) and the trigonometric identities, we leave it as an exercise for the reader [*see* Exercise 4.6.2] to verify that

$$e^{z_1} \cdot e^{z_2} = e^{z_1 + z_2}. \tag{4.6.46}$$

From (4.6.40), $e^0 = 1$. It now follows from (4.6.46) with $z_1 = z$ and $z_2 = -z$ that $e^z \neq 0$ for all values of z.

In view of (4.6.45), we may now write (4.6.39) as

$$\begin{aligned} y(x) &= C_1 e^{(\alpha + i\beta)x} + C_2 e^{(\alpha - i\beta)x} \\ &= C_1 e^{\alpha x} \cdot e^{i\beta x} + C_2 e^{\alpha x} \cdot e^{-i\beta x} \\ &= e^{\alpha x}[C_1(\cos \beta x + i \sin \beta x) + C_2(\cos \beta x - i \sin \beta x)] \\ &= e^{\alpha x}[(C_1 + C_2)\cos \beta x + i(C_1 - C_2)\sin \beta x] \\ &= e^{\alpha x}(c_1 \cos \beta x + c_2 \sin \beta x), \end{aligned} \tag{4.6.47}$$

where $c_1 = C_1 + C_2$ and $c_2 = i(C_1 - C_2)$ are two new arbitrary constants.

In view of Exercise 4.3.3(m), the functions y_1 and y_2 defined by $y_1(x) = e^{\alpha x} \cos \beta x$ and $y_2(x) = e^{\alpha x} \sin \beta x$ are linearly independent on I: $-\infty < x < \infty$. Note that whenever α, β, x, and y are real, the constants c_1 and c_2 in (4.6.47) are also real. Also, it is interesting to note that if c_1 and c_2 are real and $c_2 \neq 0$, the system

$$\begin{cases} C_1 + C_2 = c_1, \\ iC_1 - iC_2 = c_2, \end{cases}$$

when solved for C_1 and C_2 gives $C_1 = \frac{1}{2}(c_1 - ic_2)$ and $C_2 = \frac{1}{2}(c_1 + ic_2)$. Thus, the constants C_1 and C_2 are conjugate complex numbers. However, the essential fact to remember is that the function y of (4.6.47) satisfies the differential Equation (4.6.1) when c_1 and c_2 are arbitrary (real) constants.

We summarize the central result of the above paragraphs in the following theorem.

Theorem 4.6.2. *Consider the equation*

$$L[y] = ay'' + by' + cy = 0, \qquad\qquad [4.6.1]$$

where $a, b,$ and c are real constants and $D = b^2 - 4ac < 0$.   Let $m_1 = \alpha + i\beta$ and $m_2 = \alpha - i\beta$ be the imaginary roots of the auxiliary equation

$$am^2 + bm + c = 0. \qquad\qquad [4.6.5]$$

Then the general solution $y(x)$ of $L[y] = 0$ on I: $-\infty < x < \infty$ is given by

$$y(x) = e^{\alpha x}(c_1 \cos \beta x + c_2 \sin \beta x), \qquad\qquad [4.6.47]$$

where c_1 and c_2 are arbitrary constants.

Example 4.6.3.   Let us solve the initial-value problem

$$\begin{cases} y'' - 4y' + 13y = 0, & (4.6.48) \\[4pt] y(0) = 4, & (4.6.49) \\[4pt] y'(0) = -1. & (4.6.50) \end{cases}$$

SOLUTION.   The auxiliary equation of (4.6.48) is

$$m^2 - 4m + 13 = 0.$$

The roots of the above equation are $m_1 = 2 + i3$ and $m_2 = 2 - i3$.   Thus, with $\alpha = 2$ and $\beta = 3$, in view of (4.6.47), the general solution of (4.6.48) is

$$y(x) = e^{2x}(c_1 \cos 3x + c_2 \sin 3x). \qquad\qquad (4.6.51)$$

Differentiating (4.6.51), we obtain

$$y'(x) = e^{2x}[(2c_1 + 3c_2) \cos 3x + (2c_2 - 3c_1) \sin 3x]. \qquad\qquad (4.6.52)$$

Utilizing (4.6.49) through (4.6.52), we obtain

$$4 = y(0) = e^0(c_1 \cos 0 + c_2 \sin 0) = c_1, \qquad\qquad (4.6.53)$$

and

$$-1 = y'(0) = e^0[(2c_1 + 3c_2) \cos 0 + (2c_2 - 3c_1) \sin 0] = 2c_1 + 3c_2.$$

$$(4.6.54)$$

From (4.6.53) and (4.6.54), we find $c_1 = 4$ and $c_2 = -3$. Substituting these values in (4.6.51), the solution of the given initial-value problem is

$$y(x) = e^{2x}(4 \cos 3x - 3 \sin 3x). \qquad\qquad (4.6.55)$$

Remark 4.6.2.   For graphical or other practical purposes, we may also write (4.6.47) in the following forms [*see* Remark 3.4.4]:

$$y(x) = Ce^{\alpha t} \cos (\beta x - \phi), \tag{4.6.56}$$

where $C = \sqrt{c_1^2 + c_2^2} > 0$ and the angle ϕ is determined by the equations

$$\sin \phi = \frac{c_2}{C} \quad \text{and} \quad \cos \phi = \frac{c_1}{C}; \tag{4.6.57}$$

or

$$y(x) = Ce^{\alpha t} \sin (\beta x + \phi), \tag{4.6.58}$$

where $C = \sqrt{c_1^2 + c_2^2} > 0$ and the angle ϕ is now determined by the equations

$$\sin \phi = \frac{c_1}{C} \quad \text{and} \quad \cos \phi = \frac{c_2}{C}. \tag{4.6.59}$$

Exercises 4.6

1. *Another proof of Equation* (4.6.9).   Let the operator L be defined by

   (1) $L[y] = ay'' + by' + cy,$

   where a, b, and c are real constants and $a \neq 0$.   Then

   (2) $L[e^{mx}] = (am^2 + bm + c)e^{mx}.$

   The polynomial p defined by

   (3) $p(m) = am^2 + bm + c$

   is called the *characteristic polynomial* of the operator L or of $L[y] = 0$.   Thus,

   (4) $L[e^{mx}] = p(m)e^{mx}.$

   Suppose that m_1 is a repeated root of $p(m) = 0$.   Then

   (5) $p(m_1) = 0, \quad p'(m_1) = 0.$ (Why?)

   Show that

   (6) $\dfrac{\partial}{\partial m} L[e^{mx}] = L\left[\dfrac{\partial}{\partial m} e^{mx}\right].$

   Thus, differentiating (4) with respect to m, we get

   (7) $L[xe^{mx}] = [p'(m) + xp(m)]e^{mx}.$

   The functions y_1 and y_2 defined by

   (8) $y_1(x) = e^{m_1 x} \quad \text{and} \quad y_2(x) = xe^{m_1 x}$

   satisfy the equation $L[y] = 0$.   (Why?)   Formula (4.6.9) may now be deduced.

2. Verify Formulas (4.6.44) and (4.6.46).

3. By definition,

$$\cosh t = \frac{e^t + e^{-t}}{2}, \qquad \sinh t = \frac{e^t - e^{-t}}{2}.$$

(a) Show that the function y given by the linear combination

$$y(x) = c_1 e^{\beta x} + c_2 e^{-\beta x},$$

where c_1 and c_2 are arbitrary constants, can be written equivalently as

$$y(x) = A \cosh \beta x + B \sinh \beta x,$$

where A and B are (two new) arbitrary constants.
(b) Utilizing Formulas (4.6.42) and (4.6.43), show that

$$\cosh i\theta = \cos \theta \quad \text{and} \quad \sinh i\theta = i \sin \theta.$$

4. Consider the differential equation

(1) $$y'' + \lambda y = 0,$$

where λ is a parameter. Show that the basic solutions at $x_0 = 0$ of Equation (1) on the interval $I: \ -\infty < x < \infty$, are given by

(2) $$y_1(x) = \cos \sqrt{\lambda}\, x, \qquad y_2(x) = \frac{\sin \sqrt{\lambda}\, x}{\sqrt{\lambda}}.$$

Explain why it is preferable to write the general solution of Equation (1) as

(3) $$y(x) = c_1 \cos \sqrt{\lambda}\, x + c_2 \frac{\sin \sqrt{\lambda}\, x}{\sqrt{\lambda}},$$

rather than

(4) $$y(x) = c_1 \cos \sqrt{\lambda}\, x + c_2 \sin \sqrt{\lambda}\, x,$$

where c_1 and c_2 are arbitrary constants.

In Exercises 5 through 12, solve each initial-value problem.

5. $\begin{cases} y'' - y = 0, \\ y(1) = 2e^{-1}, \quad y'(1) = 0. \end{cases}$

6. $\begin{cases} 12y'' + y' - 6y = 0, \\ y(0) = 0, \quad y'(0) = 1. \end{cases}$

7. $\begin{cases} 4y'' - 4y' + y = 0, \\ y(0) = 4, \quad y'(0) = 1. \end{cases}$

8. $\begin{cases} y'' + y = 0, \\ y\left(\dfrac{\pi}{4}\right) = \dfrac{\sqrt{2}}{2}, \quad y'\left(\dfrac{\pi}{4}\right) = \sqrt{2}. \end{cases}$

9. $\begin{cases} y'' - 6y' + 25y = 0, \\ y(\pi) = e^{3\pi}, \quad y'(\pi) = -e^{3\pi}. \end{cases}$

10. $\begin{cases} y'' - 2\sqrt{3}\, y' + 7y = 0, \\ y(0) = 2\sqrt{3}, \quad y'(0) = -2. \end{cases}$

11. $\begin{cases} y'' + (\sqrt{2} - \sqrt{3})y' - \sqrt{6}y = 0, \\ y(0) = 0, \quad y'(0) = 0. \end{cases}$

12. $\begin{cases} \dfrac{W}{g}\dfrac{d^2 x}{dt^2} + K\dfrac{dx}{dt} + kx = 0, \\ x(0) = 4 \text{ in.}, \quad x'(0) = 0, \end{cases}$

when $W = 193.02$ lb, $g = 386.04$ in./sec^2, $K = 5$ lb-sec/in., and $k = 5$ lb/in.

13. Find the basic solutions at $x_0 = \pi/8$ of

$$y'' + 4y = 0$$

on the interval I: $-\infty < x < \infty$.

14. (a) If $\alpha < 0$, show that every solution of (4.6.47) tends to zero as $x \to +\infty$. (Such solutions are known as transients.)   [*See* also Remark 3.4.3.]

(b) If $\alpha > 0$, show that the magnitude of every solution of (4.6.47) assumes arbitrarily large values as $x \to +\infty$.

(c) Illustrate the results given in parts (a) and (b) by graphing the solution $y(x)$ of (4.6.47) when

(i) $y(x) = e^{-2x}(3 \cos x + 4 \sin x)$,
(ii) $y(x) = e^{2x}(3 \cos x + 4 \sin x)$.

15. (a) From (3.4.4), we have

$$LQ''(t) + RQ'(t) + \frac{1}{C} Q(t) = E(t),$$

where L, R, and C are positive constants.   Suppose that $E(t) = 0$ for all values of t on the interval I: $0 \leq t < \infty$.   Thus,

(1) $$LQ''(t) + RQ'(t) + \frac{1}{C} Q(t) = 0.$$

(2) Let $\alpha = \dfrac{R}{2L}$ and $\beta^2 = \dfrac{R^2}{4L^2} - \dfrac{1}{LC}$.

Show that the general solution $Q(t)$ of Equation (1) on I is given by:

Case 1. $Q(t) = c_1 e^{-(\alpha - \beta)t} + c_2 e^{-(\alpha + \beta)t}$,   if $\beta^2 > 0$;

Case 2. $Q(t) = (c_1 + c_2 t)e^{-\alpha t}$,   if $\beta^2 = 0$;

Case 3. $Q(t) = e^{-\alpha t}(c_1 \cos \beta t + c_2 \sin \beta t)$,   if $\beta^2 < 0$,

where in all three cases c_1 and c_2 are arbitrary constants.

(b) For each of the above three cases, show that $Q(t) \to 0$ as $t \to +\infty$.

(c) Solve the initial-value problem

$$\begin{cases} LQ''(t) + RQ'(t) + \dfrac{1}{C} Q(t) = 0, \\[2mm] Q(0) = 0.05 \text{ coulomb}, \\[2mm] Q'(0) = 0, \end{cases}$$

when $L = 0.1$ henry, $R = 10$ ohms, and $C = 0.002$ farad.

16. We shall now give another interesting proof of (4.6.9).   If m_1 and m_2 are two distinct real roots of Equation (4.6.5), then the general solution of $L[y] = ay'' + by' + cy = 0$ on I is given by (4.6.8):

(1) $$y(x) = c_1 e^{m_1 x} + c_2 e^{m_2 x},$$

where c_1 and c_2 are arbitrary constants.   Suppose that m_2 differs from m_1 by a finite quantity k, that is, $m_2 = m_1 + k$, $k \neq 0$.   Utilize the Maclaurin expansion of e^{kx} to show that the linear combination of the right-hand member of (1) can be expressed as

$$(2) \qquad c_1 e^{m_1 x} + c_2 e^{(m_1 + k)x} = e^{m_1 x}\left(A + \frac{B}{1!}\,x + \frac{Bk}{2!}\,x^2 + \frac{Bk^2}{3!}\,x^3 + \cdots\right),$$

where $A = c_1 + c_2$ and $B = c_2 k$ are two new arbitrary constants.   By letting $k \to 0$ in (2), infer (4.6.9).

17. For x a real number, we have

$$(1) \qquad e^x = \sum_{n=0}^{\infty} \frac{x^n}{n!}.$$

Suppose that in (1) it were permissible to replace x by $i\theta$, where θ is a real number and $i^2 = -1$.   Utilize the Maclaurin expansions of $\cos\theta$ and $\sin\theta$ to show that (1) may then be written as

$$e^{i\theta} = \cos\theta + i\sin\theta.$$

18. Verify: (a) $e^{i(\theta + 2k\pi)} = e^{i\theta}$, $k = 0, \pm 1, \pm 2, \ldots$.
    Evaluate:
    (b) $e^{i(\pi/6)}$,   (e) $e^{i(5/6)\pi}$,   (h) $e^{i(\pi/8)}$,
    (c) $e^{-i(\pi/3)}$,   (f) $e^{-i\pi}$,   (i) $e^{-i(3/8)\pi}$.
    (d) $e^{i(\pi/2)}$,   (g) $e^{i(3/2)\pi}$,

19. We shall now consider the general solution of the differential equation

$$(1) \qquad L[y] = ay'' + by' + cy = 0,$$

where a, b, and c are now complex constants and $a \neq 0$.   The reader may recall from college algebra that the *polar form* of a complex number $z = x + iy$, x and y are real numbers and $i^2 = -1$, is given by

$$(2) \qquad z = r\cos\theta + ir\sin\theta = re^{i\theta},$$

where $r = \sqrt{x^2 + y^2} \geq 0$, and θ is determined by the equations

$$(3) \qquad x = r\cos\theta, \quad y = r\sin\theta, \quad 0 \leq \theta < 2\pi.$$

Also, the *square roots* of z are given by

$$(4) \quad \sqrt{z} = \sqrt{r}\left[\cos\left(\frac{\theta}{2} + k\pi\right) + i\sin\left(\frac{\theta}{2} + k\pi\right)\right] = \sqrt{r}\,e^{i[(\theta/2) + k\pi]}, \qquad k = 0, 1.$$

In view of Formula (4.6.40), we have

$$(5) \qquad e^{i[(\theta/2) + \pi]} = \cos\left(\frac{\theta}{2} + \pi\right) + i\sin\left(\frac{\theta}{2} + \pi\right)$$

$$= -\cos\frac{\theta}{2} - i\sin\frac{\theta}{2} = -e^{i(\theta/2)}.$$

Thus, the square roots of z are given by

$$(6) \qquad \sqrt{z} = \pm\sqrt{r}\left(\cos\frac{\theta}{2} + i\sin\frac{\theta}{2}\right) = \pm\sqrt{r}\,e^{i(\theta/2)}.$$

For example, let us find the square roots of $1 + i\sqrt{3}$. Here, $z = x + iy = 1 + i\sqrt{3}$. Thus, $x = 1$, $y = \sqrt{3}$, $r = \sqrt{(1)^2 + (\sqrt{3})^2} = 2$, $\cos\theta = 1/2$, $\sin\theta = \sqrt{3}/2$, and consequently $\theta = \pi/3$. Hence,

$$\sqrt{1 + i\sqrt{3}} = \pm\sqrt{2}\left(\cos\frac{\pi}{6} + i\sin\frac{\pi}{6}\right) = \pm\frac{\sqrt{2}}{2}(\sqrt{3} + i).$$

The following formula

$$(7) \qquad \frac{d}{dx}e^{cx} = ce^{cx}$$

is also known to be true when c is a complex constant.

The auxiliary equation of $L[y] = 0$,

$$(8) \qquad am^2 + bm + c = 0,$$

where a, b, and c are now complex numbers, no longer enjoys the property that if $m_1 = \alpha + i\beta$, $\beta \neq 0$, is an imaginary root of Equation (8), then the other root of (8) is the complex conjugate of m_1, unless a, b, and c are real constants. (Recall that the real numbers are a proper subset of the complex numbers.) Reviewing the contents of the proof of Theorem 4.6.1 reveals that the results given in this theorem remain valid when a, b, and c are complex constants and $a \neq 0$.

In Exercises (a) through (h), find the general solution of each differential equation.

(a) $y'' - 2y' - \sqrt{3}\,i\,y = 0$.   (b) $y'' + 2iy' - (1 + i)y = 0$.
(c) $y'' - 2iy' - y = 0$.   (d) $y'' - 2iy' - iy = 0$.
(e) $y'' - 2iy' - (2 + i)y = 0$.
(f) $y'' - 2(2 + \sqrt{3}\,i)y' + (1 + 4\sqrt{3}\,i)y = 0$.
(g) $y'' - (\sqrt{3} + 1 + 2i)y' + [\sqrt{3} - 1 + (\sqrt{3} + 1)i]y = 0$.
(h) $(1 + i)y'' - (1 - i)y = 0$.
(i) Show that the general solution of $L[y] = 0$ can be expressed as

$$y(x) = u(x) + iv(x),$$

where the real-valued functions u and v are called, respectively, the *real part* and the *imaginary part* of the function y. Note that y is a complex-valued function of the real variable x. Do the functions u and v individually satisfy the equation $L[y] = 0$?

4.7. Nonhomogeneous Linear Differential Equations

Utilizing (4.3.5), we may write Equation (4.2.1) as

$$L[y] = P(x)y'' + Q(x)y' + R(x)y = F(x), \tag{4.7.1}$$

where the functions P, Q, R, and F are continuous real-valued functions defined on a common interval I: $\quad a \le x \le b$, $P(x) \ne 0$ for all x in I and $F(x)$ is not identically zero for all x in I.   In view of Definition 4.2.2, Equation (4.7.1) is a nonhomogeneous linear differential equation.   The related homogeneous differential equation of (4.7.1) is

$$L[y] = P(x)y'' + Q(x)y' + R(x)y = 0. \tag{4.7.2}$$

Theorem 4.7.1.   *Let $u(x)$ and $v(x)$ be two solutions of $L[y] = F(x)$ on the interval I: $\quad a \le x \le b$.   Then $u(x) - v(x)$ is a solution of $L[y] = 0$ on I.*

PROOF.   We are given that $L[u] = F(x)$ and $L[v] = F(x)$.   Since the operator L is linear, we obtain

$$L[u - v] = L[u] - L[v] = F(x) - F(x) = 0,$$

and thus $u(x) - v(x)$ is a solution of $L[y] = 0$ on I.

Theorem 4.7.2.   *Let y_1 and y_2 be two linearly independent solutions of $L[y] = 0$ on the interval I: $\quad a \le x \le b$, and let $v(x)$ be a solution of $L[y] = F(x)$ on I.   Then every solution of $L[y] = F(x)$ on I can be expressed in the form*

$$c_1 y_1(x) + c_2 y_2(x) + v(x),$$

for suitably chosen constants c_1 and c_2.

PROOF.   Let $u(x)$ be an arbitrary solution of $L[y] = F(x)$ on I.   Since we are given that $v(x)$ is a solution of $L[y] = F(x)$ on I, it follows from Theorem 4.7.1 that $u(x) - v(x)$ is a solution of $L[y] = 0$ on I.   Since, by hypothesis, y_1 and y_2 are two linearly independent solutions of $L[y] = 0$ on I, it follows from part (b) of Theorem 4.4.2 that $u(x) - v(x)$ can be expressed as a linear combination $c_1 y_1(x) + c_2 y_2(x)$ for suitably chosen constants c_1 and c_2.   From this fact, one may infer the conclusion of the theorem.

Definition 4.7.1.   Let $y_1 = y_1(x)$ and $y_2 = y_2(x)$ be two linearly independent solutions of

$$L[y] = P(x)y'' + Q(x)y' + R(x)y = 0 \tag{4.7.2}$$

on the interval I: $\quad a \le x \le b$, and let $y_p = y_p(x)$ be a solution of

$$L[y] = P(x)y'' + Q(x)y' + R(x)y = F(x) \tag{4.7.1}$$

on I. The *general solution* $y(x)$ of Equation (4.7.1) on I is defined by

$$y(x) = y_c(x) + y_p(x), \tag{4.7.3}$$

where

$$y_c(x) = c_1 y_1(x) + c_2 y_2(x), \tag{4.7.4}$$

and c_1 and c_2 are arbitrary constants.

Definition 4.7.2. The function y_c given in Equation (4.7.4) is called the *complementary function* of $L[y] = F(x)$. The function y_p is called a *particular solution* (or *particular integral*) of $L[y] = F(x)$.

Remark 4.7.1. The subscript c in y_c indicates the complementary function while the subscript p in y_p indicates a particular solution. Note that the complementary function involves two arbitrary constants c_1 and c_2, while the particular solution does not involve any arbitrary constant. Also, observe that in order to find the general solution of the nonhomogeneous equation $L[y] = F(x)$, we first find the general solution of its related homogeneous equation $L[y] = 0$; secondly we find any particular solution of $L[y] = F(x)$ and then add these solutions.

Example 4.7.1. Let us find the general solution of

$$y'' - 2y' + y = e^{3x}. \tag{4.7.5}$$

SOLUTION.   The complementary function is obtained by solving the equation

$$y'' - 2y' + y = 0.$$

Utilizing the method given in Section 4.6, we obtain

$$y_c(x) = (c_1 + c_2 x)e^x.$$

One may easily verify that

$$y_p(x) = \tfrac{1}{4}e^{3x}$$

is a solution of (4.7.5).   Thus, the general solution of (4.7.5) is

$$y(x) = y_c(x) + y_p(x) = (c_1 + c_2 x)e^x + \tfrac{1}{4}e^{3x}.$$

Note that this solution is valid for all values of x in the interval I: $-\infty < x < \infty$.

In the ensuing sections, we shall develop methods for finding a particular solution of $L[y] = F(x)$. The following theorem will be found useful in obtaining a particular solution of $L[y] = F(x)$ when F is expressed as a finite sum of functions.   [*See* also Exercise 4.7.1.]

Theorem 4.7.3. *Let $u_i(x)$ be a solution of*

$$L[y] = P(x)y'' + Q(x)y' + R(x)y = F_i(x) \tag{4.7.6}$$

for $i = 1, 2, \ldots, n$, on the interval I: $\quad a \le x \le b$. Then the function u given by the linear combination of these n solutions

$$u(x) = c_1 u_1(x) + c_2 u_2(x) + \cdots + c_n u_n(x), \tag{4.7.7}$$

where $c_1, c_2, \ldots, c_n$ are arbitrary constants, is a solution of

$$L[y] = c_1 F_1(x) + c_2 F_2(x) + \cdots + c_n F_n(x)$$

on I.

PROOF.   By hypothesis, we have

$$L[u_1] = F_1(x), \qquad L[u_2] = F_2(x), \ldots, L[u_n] = F_n(x).$$

Since the operator L is linear, we obtain

$$\begin{aligned}
L[u] &= L[c_1 u_1 + c_2 u_2 + \cdots + c_n u_n] \\
&= c_1 L[u_1] + c_2 L[u_2] + \cdots + c_n L[u_n] \\
&= c_1 F_1(x) + c_2 F_2(x) + \cdots + c_n F_n(x),
\end{aligned}$$

and thus the theorem is established.

Remark 4.7.2.   The result given in Theorem 4.7.3 is known as the *principle of superposition*.

Exercises 4.7

1. *Principle of superposition for particular solutions of $L[y] = F(x)$ on the interval I: $\quad a \le x \le b$.*   Show that if $y_{p_i}(x)$ are particular solutions of $L[y] = F_i(x)$ for $i = 1, 2, \ldots, n$, on I, then

$$y_p(x) = y_{p_1}(x) + y_{p_2}(x) + \cdots + y_{p_n}(x)$$

   is a particular solution of

$$L[y] = F_1(x) + F_2(x) + \cdots + F_n(x)$$

   on I.
2. Show that if $u(x)$ is a solution of $L[y] = F(x)$ on an interval I, and $v(x)$ is a solution of $L[y] = 0$ on I, then $u(x) + v(x)$ is a solution of $L[y] = F(x)$ on I.
3. Show that Theorem 4.4.1 and Remark 4.4.2 are special cases of Theorem 4.7.3.
4. Consider the following differential equation

   (1) $\qquad\qquad L[y] = y'' - y = 4x^2 - 10 \sin 2x + 9xe^{2x}.$

   Suppose that

$$y_{p_1}(x) = -4x^2 - 8 \text{ is a solution of } L[y] = 4x^2,$$
$$y_{p_2}(x) = 2 \sin 2x \text{ is a solution of } L[y] = -10 \sin 2x,$$
$$y_{p_3}(x) = (-4 + 3x)e^{2x} \text{ is a solution of } L[y] = 9xe^{2x}.$$

   Find the general solution of Equation (1).

5. Let y_1 and y_2 be two (distinct) solutions of $L[y] = F(x)$ on an interval I. Determine, in terms of y_1 and y_2, a family of solutions of $L[y] = F(x)$ on I depending on an arbitrary constant.

6. A *complex-valued function* of *a real variable* x is a function of the type

$$f(x) = f_1(x) + if_2(x), \qquad i^2 = -1,$$

where f_1 and f_2 are real-valued functions. Let us denote by $\mathscr{R}[f(x)]$ the *real part* of $f(x)$, that is, $\mathscr{R}[f(x)] = f_1(x)$, and denote by $\mathscr{I}[f(x)]$ the *imaginary part* of $f(x)$, that is, $\mathscr{I}[f(x)] = f_2(x)$. The functions f and g defined by

$$f(x) = f_1(x) + if_2(x) \quad \text{and} \quad g(x) = g_1(x) + ig_2(x)$$

are said to be *equal* if and only if $\mathscr{R}[f(x)] = \mathscr{R}[g(x)]$ and $\mathscr{I}[f(x)] = \mathscr{I}[g(x)]$. The usual properties encountered in calculus extend to complex-valued functions. For example,

$$\frac{d}{dx}[c_1 f_1(x) + ic_2 f_2(x)] = c_1 \frac{d}{dx} f_1(x) + ic_2 \frac{d}{dx} f_2(x),$$

$$\int [c_1 f_1(x) + ic_2 f_2(x)]dx = c_1 \int f_1(x)dx + ic_2 \int f_2(x)dx,$$

where c_1 and c_2 are complex constants.

(a) Suppose that

$$y_p(x) = u(x) + iv(x)$$

is a solution of

$$L[y] = P(x)y'' + Q(x)y' + R(x)y = U(x) + iV(x),$$

where P, Q, and R are real-valued functions. Show that $\mathscr{R}[y_p(x)]$ and $\mathscr{I}[y_p(x)]$ are solutions of $L[y] = U(x)$ and $L[y] = V(x)$, respectively.

(b) Determine whether the function y_p defined by $y_p(x) = xe^x + i \sin 2x$ is a solution of

$$L[y] = x^2 y'' - 5xy' + 6y = (x^3 - 3x^2 + x)e^x - i[(4x^2 - 6)\sin 2x + 10x \cos 2x]$$

on an interval I not containing the origin.

4.8. The Method of Undetermined Coefficients

We shall now give a method for finding a particular solution of the non-homogeneous linear differential equation with constant coefficients

$$L[y] = ay'' + by' + cy = F(x), \tag{4.8.1}$$

where a, b, c are constants, $a \neq 0$, and $F(x)$ is not identically zero in the interval I: $x_1 \leq x \leq x_2$.

Suppose that $F(x)$ contains only terms which have a finite number of linearly independent derivatives. Then it can be shown that $F(x)$ must only contain terms such as x^n, $e^{\alpha x}$, $\sin \beta x$, and $\cos \beta x$, or (finite) products of these.

Here n is a positive integer (or zero), α and β are any real constants (including zero). For such a function F, the method of undetermined coefficients is applicable. Basically, this method assumes a particular solution $y_p(x)$ of the form similar to that of the given function F and containing undetermined coefficients for each term. Then $y_p(x)$, $y_p'(x)$, and $y_p''(x)$ are substituted in the left-hand member of Equation (4.8.1). This results in an identity in the independent variable. Thus, the coefficients of like terms can be equated, and the values of the undetermined coefficients can be determined from the resulting system of linear equations. Note that this method involves no integrations, but only differentiations. The principal deficiency of the method of undetermined coefficients is that the function F must be of the form described above. Also, the coefficients a, b, and c must be constants. In spite of these restrictions, this method is very useful, since the function F of the form given above appears quite often in applications. A more general method, known as the method of the variation of parameters, will be discussed in the next section. This method will entail integrations, and thus may involve difficult computations. Some examples of functions that have more than a finite number of linearly independent derivatives are $\tan x$, $\sec x$, $\ln x$, $1/x$, and $e^{1/x}$. When the function F in Equation (4.8.1) involves any of these functions, while the method of undetermined coefficients is not applicable, the method of variation of parameters is applicable.

The following cases will assist us in determining the form of a particular solution $y_p(x)$ of $L[y] = F(x)$.

Case 1. *Suppose that*

$$L[y] = ay'' + by' + cy = P_n(x), \qquad a \neq 0, \tag{4.8.2}$$

where

$$P_n(x) = A_0 x^n + A_1 x^{n-1} + \cdots + A_n, \qquad A_0 \neq 0. \tag{4.8.3}$$

The characteristic polynomial of $L[y] = 0$ *is given by*

$$p(m) = am^2 + bm + c. \tag{4.8.4}$$

Let

$$Q_n(x) = B_0 x^n + B_1 x^{n-1} + \cdots + B_n, \qquad B_0 \neq 0. \tag{4.8.5}$$

A particular solution $y_p(x)$ *of Equation (4.8.2) has the form*

(i) $y_p(x) = Q_n(x), \quad \text{if } p(0) = c \neq 0;$ (4.8.6)

(ii) $y_p(x) = xQ_n(x), \quad \text{if } p(0) = c = 0, \quad p'(0) = b \neq 0;$ (4.8.7)

(iii) $y_p(x) = x^2 Q_n(x), \quad \text{if } p(0) = c = 0, \quad p'(0) = b = 0.$ (4.8.8)

To see that the above solutions are appropriate forms of particular solutions of $L[y] = P_n(x)$, we proceed as follows. Let us suppose that

$$y_p(x) = Q_n(x) = B_0 x^n + B_1 x^{n-1} + \cdots + B_{n-2} x^2 + B_{n-1} x + B_n. \qquad (4.8.9)$$

Differentiating (4.8.9) two times in succession, we obtain

$$y_p'(x) = Q_n'(x) = nB_0 x^{n-1} + (n-1)B_1 x^{n-2} + \cdots + 2B_{n-2} x + B_{n-1},$$

$$y_p''(x) = Q_n''(x) = n(n-1)B_0 x^{n-2} + (n-1)(n-2)B_1 x^{n-3} + \cdots + 2B_{n-2}.$$

Substituting y_p, y_p', and y_p'' in (4.8.2), we obtain upon collecting terms

$$cB_0 x^n + (cB_1 + nbB_0)x^{n-1} + \cdots + cB_n + bB_{n-1} + 2aB_{n-2}$$
$$= A_0 x^n + A_1 x^{n-1} + \cdots + A_n. \qquad (4.8.10)$$

Equation (4.8.10) will be an identity in x if the coefficients of like powers of x on each side of the equal sign have the same value. Thus,

$$\begin{cases} cB_0 = A_0, \\ cB_1 + nbB_0 = A_1, \\ \quad\vdots \\ cB_n + bB_{n-1} + 2aB_{n-2} = A_n. \end{cases} \qquad (4.8.11)$$

Concerning (i). If $p(0) = c \neq 0$, then from the first equation in (4.8.11) we obtain $B_0 = A_0/c$. Substituting this value in the second equation of (4.8.11), we determine B_1. Continuing in this fashion, we determine the remaining B_k, $k = 2, 3, \ldots, n$. Thus, (4.8.9) is now completely determined.

Concerning (ii). If $p(0) = c = 0$ and $p'(0) = b \neq 0$, then the polynomial in the left-hand side of Equation (4.8.10) is of degree $n - 1$. In order to render it a polynomial of degree n, we take

$$y_p(x) = xQ_n(x) \qquad (4.8.12)$$

and then proceed as in Case 1 (i). Note that the form of a particular solution as given in (4.8.12) does not contain a constant term. However, since $c = 0$, a constant term is a solution of the related homogeneous equation

$$ay'' + by' = 0,$$

and thus we need not incorporate it in a particular solution of the form given in (4.8.12).

Concerning (iii). If $p(0) = c = 0$ and $p'(0) = b = 0$, then the polynomial in the left-hand side of Equation (4.8.10) is of degree $n - 2$. In order to render it a polynomial of degree n, we take

$$y_p(x) = x^2 Q_n(x), \qquad (4.8.13)$$

and then proceed as in Case 1 (i).   Note that the form of a particular solution as given in (4.8.13) does not contain the linear term, say $c_1 + c_2 x$, where c_1 and c_2 are constants.   However, since $c = 0$ and $b = 0$, the linear term $c_1 + c_2 x$ is a solution of the related homogeneous equation

$$ay'' = 0,$$

and thus we need not incorporate it in a particular solution of the form given in (4.8.13).

Case 2.   *Suppose that*

$$L[y] = ay'' + by' + cy = e^{\alpha x}P_n(x), \tag{4.8.14}$$

where $P_n(x)$ is given in (4.8.3).

   A particular solution $y_p(x)$ of Equation (4.8.14) has the form

(i) $y_p(x) = e^{\alpha x}Q_n(x), \quad$ *if $p(\alpha) \neq 0$;* $\tag{4.8.15}$

(ii) $y_p(x) = xe^{\alpha x}Q_n(x), \quad$ *if $p(\alpha) = 0, \quad p'(\alpha) \neq 0$;* $\tag{4.8.16}$

(iii) $y_p(x) = x^2 e^{\alpha x}Q_n(x), \quad$ *if $p(\alpha) = 0, \quad p'(\alpha) = 0$,* $\tag{4.8.17}$

where $p(m)$ and $Q_n(x)$ are given by (4.8.4) and (4.8.5), respectively.

   Case 2 can be reduced to Case 1.   For example, let

$$y(x) = e^{\alpha x}v(x). \tag{4.8.18}$$

Then

$$y' = e^{\alpha x}(v' + \alpha v),$$
$$y'' = e^{\alpha x}(v'' + 2\alpha v' + \alpha^2 v).$$

Substituting y, y', and y'' in Equation (4.8.14), dividing by $e^{\alpha x}$ and collecting terms, we find

$$av'' + (2a\alpha + b)v' + (a\alpha^2 + b\alpha + c)v = A_0 x^n + A_1 x^{n-1} + \cdots + A_n.$$

$$\tag{4.8.19}$$

Let

$$q(m) = am^2 + (2a\alpha + b)m + (a\alpha^2 + b\alpha + c) \tag{4.8.20}$$

be the characteristic polynomial of the related homogeneous differential equation of (4.8.19).   From Case 1, a particular solution $v_p(x)$ of Equation (4.8.19) has the form

(i)′ $v_p(x) = Q_n(x), \quad$ if $q(0) = a\alpha^2 + b\alpha + c \neq 0$; $\tag{4.8.21}$

(ii)′ $v_p(x) = xQ_n(x), \quad$ if $q(0) = 0, \quad q'(0) = 2a\alpha + b \neq 0$; $\tag{4.8.22}$

(iii)′ $v_p(x) = x^2 Q_n(x), \quad$ if $q(0) = 0, \quad q'(0) = 0.$ $\tag{4.8.23}$

Utilizing (4.8.18) and (4.8.21) through (4.8.23), we may infer the results of Case 2. Note that the conditions given in (4.8.21) through (4.8.23): $q(0) = a\alpha^2 + b\alpha + c \neq 0$; $q(0) = 0$, $q'(0) = 2a\alpha + b \neq 0$; $q(0) = 0$, $q'(0) = 0$; are, respectively, equivalent to the conditions given in (4.8.15) through (4.8.17): $p(\alpha) \neq 0$; $p(\alpha) = 0$, $p'(\alpha) \neq 0$; $p(\alpha) = 0$, $p'(\alpha) = 0$.

Case 3.   *Suppose that*

$$L[y] = ay'' + by' + cy = e^{\alpha x}[P_n(x) \cos \beta x + S_j(x) \sin \beta x], \tag{4.8.24}$$

where

$$P_n(x) = A_0 x^n + A_1 x^{n-1} + \cdots + A_n,$$
$$S_j(x) = D_0 x^j + D_1 x^{j-1} + \cdots + D_j,$$

are polynomials of possibly different degree (one might even equal zero), and $(A_0, D_0) \neq (0, 0)$.   *A particular solution of Equation (4.8.24) has the form*

(i) $y_p(x) = e^{\alpha x}[Q_k(x) \cos \beta x + R_k(x) \sin \beta x]$,   *if* $p(\alpha + i\beta) \neq 0$,

(ii) $y_p(x) = xe^{\alpha x}[Q_k(x) \cos \beta x + R_k(x) \sin \beta x]$,   *if* $p(\alpha + i\beta) = 0$,
(4.8.25)

where $k = \max(n, j)$, $p(m) = am^2 + bm + c$, *and*

$$Q_k(x) = B_0 x^k + B_1 x^{k-1} + \cdots + B_k,$$
$$R_k(x) = C_0 x^k + C_1 x^{k-1} + \cdots + C_k,$$

$(B_0, C_0) \neq (0, 0)$.

We first obtain particular solutions of Equation (4.8.24) for the separate cases when $P_n(x) \equiv 0$ and when $S_j(x) \equiv 0$, and then arrive at the general statement of Case 3 by the principle of superposition.

First, suppose that $S_j(x) \equiv 0$ in Equation (4.8.24), whence

$$L[y] = e^{\alpha x}P_n(x) \cos \beta x. \tag{4.8.26}$$

Equation (4.8.26) can be reduced to Case 2 (i) and (ii) by utilizing Formulas (4.6.42) and (4.6.45).   We consider the form of a particular solution $y_p(x)$ of Equation (4.8.26).   Since $\cos \beta x = (e^{i\beta x} + e^{-i\beta x})/2$, we have

$$L[y] = e^{\alpha x}P_n(x) \frac{e^{i\beta x} + e^{-i\beta x}}{2} = P_n(x) \frac{e^{(\alpha + i\beta)x} + e^{(\alpha - i\beta)x}}{2}.$$

Thus, by Case 2, we may take y_p of the form

$$y_p(x) = e^{(\alpha + i\beta)x}q_n(x) + e^{(\alpha - i\beta)x}r_n(x), \quad \text{if } p(\alpha + i\beta) \neq 0,$$

where

$$q_n(x) = b_0 x^n + b_1 x^{n-1} + \cdots + b_n, \qquad r_n(x) = c_0 x^n + c_1 x^{n-1} + \cdots + c_n,$$

$(b_0, c_0 \neq (0, 0).$

Utilizing (4.6.45), we may write $y_p(x)$ above in the following equivalent form.

$$y_p(x) = e^{\alpha x}[\cos \beta x + i \sin \beta x]q_n(x) + e^{\alpha x}[\cos \beta x - i \sin \beta x]r_n(x)$$
$$= e^{\alpha x}[q_n(x) + r_n(x)] \cos \beta x + e^{\alpha x}[iq_n(x) - ir_n(x)] \sin \beta x$$
$$= e^{\alpha x}[(b_0 + c_0)x^n + (b_1 + c_1)x^{n-1} + \cdots + (b_n + c_n)] \cos \beta x$$
$$+ e^{\alpha x}[i(b_0 - c_0)x^n + i(b_1 - c_1)x^{n-1} + \cdots + i(b_n - c_n)] \sin \beta x.$$

Letting $B_l = (b_l + c_l)$, $C_l = i(b_l - c_l)$, $l = 0, 1, \ldots, n$, be new constants, and

$$Q_n(x) = B_0 x^n + B_1 x^{n-1} + \cdots + B_n, \qquad R_n(x) = C_0 x^n + C_1 x^{n-1} + \cdots + C_n,$$

we may write the above expression for $y_p(x)$ as follows.

(i) $\quad y_p(x) = e^{\alpha x}[Q_n(x) \cos \beta x + R_n(x) \sin \beta x], \quad$ if $p(\alpha + i\beta) \neq 0.$ $\qquad$ (4.8.27)

In a similar fashion, we have

(ii) $\quad y_p(x) = xe^{\alpha x}[Q_n(x) \cos \beta x + R_n(x) \sin \beta x], \quad$ if $p(\alpha + i\beta) = 0.$ $\qquad$ (4.8.27′)

We can likewise obtain a particular solution of Equation (4.8.24) when $P_n(x) \equiv 0$. Thus, Equation (4.8.24) becomes

$$L[y] = e^{\alpha x}S_j(x) \sin \beta x. \qquad (4.8.28)$$

Utilizing the formula $\sin \beta x = (e^{i\beta x} - e^{-i\beta x})/2i$ and proceeding as above, we leave it as an exercise for the reader [*see* Exercise 4.8.30] to show that

$$\begin{aligned}
&\text{(i)} \quad y_p(x) = e^{\alpha x}[Q_j(x) \cos \beta x + R_j(x) \sin \beta x], \quad \text{if } p(\alpha + i\beta) \neq 0, \\
&\text{(ii)} \quad y_p(x) = xe^{\alpha x}[Q_j(x) \cos \beta x + R_j(x) \sin \beta x], \quad \text{if } p(\alpha + i\beta) = 0.
\end{aligned} \qquad (4.8.29)$$

At this point, it is well to summarize briefly. If $p(\alpha + i\beta) \neq 0$, then Equations (4.8.26) and (4.8.28) have as respective particular solutions

$$y_1(x) = e^{\alpha x}[a_n(x) \cos \beta x + b_n(x) \sin \beta x]$$

and

$$y_2(x) = e^{\alpha x}[c_j(x) \cos \beta x + d_j(x) \sin \beta x],$$

where $a_n(x)$, $b_n(x)$, $c_j(x)$, and $d_j(x)$ are simply polynomials of the specified degree, which we need not define further. Thus, using the principle of superposition, we take as our particular solution of Equation (4.8.24),

$$y_p(x) = y_1(x) + y_2(x) = e^{\alpha x}\{[a_n(x) + c_j(x)] \cos \beta x + [b_n(x) + d_j(x)] \sin \beta x\}.$$

Now, it is obvious that the sum of two polynomials of possibly different degree is itself a polynomial whose degree is the maximum of the two. Thus, since $k = \max(n, j)$, we can write

$$Q_k(x) = [a_n(x) + c_j(x)] \quad \text{and} \quad R_k(x) = [b_n(x) + d_j(x)].$$

Hence,

$$y_p(x) = e^{\alpha x}[Q_k(x) \cos \beta x + R_k(x) \sin \beta x], \quad \text{if } p(\alpha + i\beta) \neq 0.$$

Similarly, we have that

$$y_p(x) = xe^{\alpha x}[Q_k(x)\cos \beta x + R_k(x)\sin \beta x], \quad \text{if } p(\alpha + i\beta) = 0,$$

and Case 3 is thus verified.

Remark 4.8.1. Consider the equation $L[y] = F(x)$. Special cases of Case 3 occur when $F(x)$ is of the form $e^{\alpha x}\cos \beta x$ or $e^{\alpha x}\sin \beta x$. [*See* Example 4.8.2.] The most general case is covered in Example 4.8.4. Case 3 must be used with care, for example, if $F(x)$ is of the form $e^{\alpha_1 x}\cos \beta x + e^{\alpha_2 x}\sin \beta x$, where $\alpha_1 \neq \alpha_2$. Then a double application of Case 3 will be required.

Remark 4.8.2. The principle of superposition [*see* Exercise 4.7.1] is found useful when $L[y] = F(x)$ is of the form

$$L[y] = ay'' + by' + cy = F_1(x) + F_2(x) + \cdots + F_n(x), \tag{4.8.30}$$

where each $F_i(x)$, $i = 1, 2, \ldots, n$ is of the type discussed in Cases 1, 2, and 3.

Example 4.8.1. Utilizing the method of undetermined coefficients, let us find a particular solution of the differential equation

$$L[y] = y'' - 5y' + 6y = 12x^2 - 20x + 4 + 3e^{2x}. \tag{4.8.31}$$

SOLUTION. We shall utilize the principle of superposition in obtaining a particular solution of (4.8.31). Let $y_{p_1}(x)$ and $y_{p_2}(x)$ denote, respectively, particular solutions of

$$L[y] = y'' - 5y' + 6y = 12x^2 - 20x + 4, \tag{4.8.32}$$

and

$$L[y] = y'' - 5y' + 6y = 3e^{2x}. \tag{4.8.33}$$

The characteristic polynomial of $L[y] = 0$ is

$$p(m) = m^2 - 5m + 6.$$

Note that $p(0) = 6 \neq 0$. In (4.8.32), $P_2(x) = 12x^2 - 20x + 4$. A particular solution $y_{p_1}(x)$ of (4.8.32) is of the form [*see* Case 1 (i)]

$$y_{p_1}(x) = B_0 x^2 + B_1 x + B_2. \tag{4.8.34}$$

Thus,

$$y'_{p_1}(x) = 2B_0 x + B_1 \quad \text{and} \quad y''_{p_1}(x) = 2B_0.$$

Substituting y_{p_1}, y'_{p_1}, and y''_{p_1} in (4.8.32), we find upon collecting terms

$$6B_0 x^2 + (6B_1 - 10B_0)x + 6B_2 - 5B_1 + 2B_0 = 12x^2 - 20x + 4. \tag{4.8.35}$$

Equation (4.8.35) will be an identity in x provided that

$$\begin{cases} 6B_0 = 12, \\ 6B_1 - 10B_0 = -20, \\ 6B_2 - 5B_1 + 2B_0 = 4. \end{cases} \tag{4.8.36}$$

Solving the system (4.8.36), we find $B_0 = 2$, $B_1 = 0$, and $B_2 = 0$. Substituting these values in (4.8.34), we obtain

$$y_{p_1}(x) = 2x^2. \tag{4.8.37}$$

Comparing the right-hand member of (4.8.33) with that of (4.8.14), we see that $\alpha = 2$ and $P_0(x) = 3$. Since $p(m) = m^2 - 5m + 6$, upon differentiation (with respect to m) we obtain $p'(m) = 2m - 5$. Thus, $p(\alpha) = p(2) = 0$ and $p'(\alpha) = p'(2) = -1 \neq 0$. Hence, a particular solution $y_{p_2}(x)$ of (4.8.33) has the form [*see* Case 2 (ii)]

$$y_{p_2}(x) = xe^{2x}Q_0(x) = B_0\, xe^{2x}. \tag{4.8.38}$$

Differentiating (4.8.38) two times in succession, we obtain

$$y'_{p_2}(x) = B_0(2x + 1)e^{2x} \quad \text{and} \quad y''_{p_2}(x) = 4B_0(x + 1)e^{2x}.$$

Substituting y_{p_2}, y'_{p_2}, and y''_{p_2} in (4.8.33), we obtain $B_0 = -3$, and (4.8.38) thus becomes

$$y_{p_2}(x) = -3xe^{2x}. \tag{4.8.39}$$

From (4.8.37), (4.8.39), and the principle of superposition, a particular solution $y_p(x)$ of (4.8.31) is

$$y_p(x) = y_{p_1}(x) + y_{p_2}(x) = 2x^2 - 3xe^{2x}. \tag{4.8.40}$$

The reader should verify that the particular solution in (4.8.40) is indeed a solution of (4.8.31).

Example 4.8.2.   Let us find a particular solution of the differential equation

$$L[y] = y'' + 2y' + 4y = 111e^{2x} \cos 3x. \tag{4.8.41}$$

SOLUTION.   We shall utilize Case 3. Comparing the right-hand member of (4.8.41) with that of (4.8.24), we see that

$$\alpha = 2, \quad \beta = 3, \quad P_0(x) = 111, \quad \text{and} \quad S_j(x) \equiv 0.$$

The characteristic polynomial of $L[y] = 0$ is

$$p(m) = m^2 + 2m + 4.$$

Note that $p(\alpha + i\beta) = p(2 + 3i) = (2 + 3i)^2 + 2(2 + 3i) + 4 = 3 + 18i \neq 0$. In view of Case 3 (i), a particular solution $y_p(x)$ of (4.8.41) has the form

$$y_p(x) = e^{2x}B_0 \cos 3x + e^{2x}C_0 \sin 3x = e^{2x}(B_0 \cos 3x + C_0 \sin 3x). \tag{4.8.42}$$

Differentiating (4.8.42) two times in succession, we obtain upon collecting terms

$$y'_p(x) = e^{2x}[(2B_0 + 3C_0) \cos 3x - (3B_0 - 2C_0) \sin 3x],$$
$$y''_p(x) = e^{2x}[(-5B_0 + 12C_0) \cos 3x - (12B_0 + 5C_0) \sin 3x].$$

Substituting these values of y_p, y_p', and y_p'' in (4.8.41), dividing by e^{2x}, and collecting terms, we find

$$(3B_0 + 18C_0) \cos 3x - (18B_0 - 3C_0) \sin 3x = 111 \cos 3x. \qquad (4.8.43)$$

Since the functions y_1 and y_2 defined by $y_1(x) = \cos 3x$ and $y_2(x) = \sin 3x$ are linearly independent on I: $-\infty < x < \infty$, Equation (4.8.43) will be an identity provided that

$$\begin{cases} 3B_0 + 18C_0 = 111, \\ 18B_0 - 3C_0 = 0. \end{cases} \qquad (4.8.44)$$

Solving the system (4.8.44), we find $B_0 = 1$ and $C_0 = 6$. Substituting these values in (4.8.42), we have

$$y_p(x) = e^{2x}(\cos 3x + 6 \sin 3x). \qquad (4.8.45)$$

The reader should verify that the particular solution in (4.8.45) is indeed a solution of (4.8.41).

Example 4.8.3. Let us solve the initial-value problem

$$\begin{cases} y'' - 4y' + 4y = 6xe^{2x}, & (4.8.46) \\ y(0) = 1, & (4.8.47) \\ y'(0) = 4. & (4.8.48) \end{cases}$$

SOLUTION. From Definition 4.7.1, the general solution $y(x)$ of (4.8.46) is given by

$$y(x) = y_c(x) + y_p(x), \qquad (4.8.49)$$

where $y_c(x)$ is the general solution of the related homogeneous differential equation

$$y'' - 4y' + 4y = 0, \qquad (4.8.50)$$

and $y_p(x)$ is a particular solution of (4.8.46). The auxiliary equation of (4.8.50) is

$$m^2 - 4m + 4 = 0.$$

The roots of the above equation are $m_1 = 2$ and $m_2 = 2$. In view of Theorem 4.6.1(b), the complementary function y_c is

$$y_c(x) = c_1 e^{2x} + c_2 x e^{2x}, \qquad (4.8.51)$$

where c_1 and c_2 are arbitrary constants.

We shall utilize Case 2 to find a particular solution $y_p(x)$ of (4.8.46). Comparing the right-hand member of (4.8.46) with that of (4.8.14), we see that $\alpha = 2$ and $P_1(x) = 6x$. The characteristic polynomial of Equation

(4.8.50) is $p(m) = m^2 - 4m + 4$.   Thus, $p'(m) = 2m - 4$.   Note that $p(\alpha) = p(2) = 4 - 8 + 4 = 0$ and $p'(\alpha) = p'(2) = 4 - 4 = 0$.   In view of Case 2 (iii), a particular solution $y_p(x)$ of (4.8.46) has the form

$$y_p(x) = x^2 e^{2x} Q_1(x) = x^2 e^{2x}(B_0 x + B_1) = e^{2x}(B_0 x^3 + B_1 x^2). \quad (4.8.52)$$

Differentiating (4.8.52) two times in succession, we obtain upon collecting terms

$$y_p'(x) = e^{2x}[2B_0 x^3 + (3B_0 + 2B_1)x^2 + 2B_1 x],$$

$$y_p''(x) = e^{2x}[4B_0 x^3 + (12B_0 + 4B_1)x^2 + (6B_0 + 8B_1)x + 2B_1].$$

Substituting y_p, y_p', and y_p'' in (4.8.46), dividing by e^{2x}, and collecting terms, we find

$$6B_0 x + 2B_1 = 6x. \qquad\qquad (4.8.53)$$

Equation (4.8.53) will be an identity in x provided that

$$6B_0 = 6 \quad \text{and} \quad 2B_1 = 0.$$

Thus, $B_0 = 1$ and $B_1 = 0$, and (4.8.52) becomes

$$y_p(x) = x^3 e^{2x}. \qquad\qquad (4.8.54)$$

In view of (4.8.51) and (4.8.54), we may write (4.8.49) as

$$y(x) = c_1 e^{2x} + c_2 xe^{2x} + x^3 e^{2x}, \qquad\qquad (4.8.55)$$

which is the general solution of (4.8.46).   Differentiating (4.8.55), we obtain

$$y'(x) = 2c_1 e^{2x} + c_2 e^{2x} + 2c_2 xe^{2x} + 3x^2 e^{2x} + 2x^3 e^{2x}. \qquad (4.8.56)$$

Utilizing (4.8.47), (4.8.48), (4.8.55), and (4.8.56), we have

$$1 = y(0) = c_1 e^0 + c_2 \cdot 0 \cdot e^0 + 0 \cdot e^0 = c_1,$$

and

$$4 = y'(0) = 2c_1 e^0 + c_2 e^0 + 2c_2 \cdot 0 \cdot e^0 + 3 \cdot 0 \cdot e^0 + 2 \cdot 0 \cdot e^0 = 2c_1 + c_2.$$

Solving these equations simultaneously, we find $c_1 = 1$ and $c_2 = 2$, and thus (4.8.55) becomes

$$y(x) = e^{2x} + 2xe^{2x} + x^3 e^{2x}, \qquad\qquad (4.8.57)$$

which is the solution of our initial-value problem.

As an exercise, the reader should verify that the function y given by Equation (4.8.57) is indeed the solution of the given initial-value problem.

Example 4.8.4.   Let us find a particular solution of the differential equation

$$L[y] = y'' + 4y = (10x - 1)e^x \cos x + 4e^x \sin x. \qquad (4.8.58)$$

SOLUTION.   Here $\alpha = 1$ and $\beta = 1$.   The characteristic polynomial of $L[y] = 0$ is $p(m) = m^2 + 4$.   Note that $p(\alpha + i\beta) = p(1 + i) = (1 + i)^2 + 4 \neq 0$.   Thus, since the maximum degree of the two coefficient polynomials in Equation (4.8.58) is one, a particular solution $y_p(x)$ of (4.8.58) may be taken of the form [*see* Case 3 (i)]

$$y_p(x) = (B_0 x + B_1)e^x \cos x + (C_0 x + C_1)e^x \sin x. \qquad (4.8.59)$$

Differentiating (4.8.59) two times in succession, we obtain upon collecting terms

$$y_p'(x) = (B_0 x + C_0 x + B_0 + B_1 + C_1)e^x \cos x + (-B_0 x + C_0 x$$
$$- B_1 + C_0 + C_1)e^x \sin x,$$

$$y_p''(x) = (2C_0 x + 2B_0 + 2C_0 + 2C_1)e^x \cos x + (-2B_0 x - 2B_0 - 2B_1$$
$$+ 2C_0)e^x \sin x.$$

Substituting y_p and y_p'' in (4.8.58), dividing by e^x, and collecting terms, we find

$$(4B_0 + 2C_0)x \cos x + (2B_0 + 4B_1 + 2C_0 + 2C_1) \cos x - (2B_0 - 4C_0)x \sin x$$
$$- (2B_0 + 2B_1 - 2C_0 - 4C_1) \sin x = 10x \cos x - \cos x + 4 \sin x.$$

$$(4.8.60)$$

Equation (4.8.60) will be an identity provided that

$$\begin{cases} 4B_0 + 2C_0 = 10, \\ 2B_0 + 4B_1 + 2C_0 + 2C_1 = -1, \\ 2B_0 - 4C_0 = 0, \\ -2B_0 - 2B_1 + 2C_0 + 4C_1 = 4. \end{cases} \qquad (4.8.61)$$

Solving the system (4.8.61), we find $B_0 = 2$, $B_1 = -2$, $C_0 = 1$, and $C_1 = \frac{1}{2}$. Substituting these values in (4.8.59), we have

$$y_p(x) = (2x - 2)e^x \cos x + (x + \tfrac{1}{2})e^x \sin x. \qquad (4.8.62)$$

The reader should verify that the particular solution in (4.8.62) is indeed a solution of (4.8.58).
   Suppose that

$$L[y] = y'' - 2y' + 2y = (10x - 1)e^x \cos x + 4e^x \sin x. \qquad (4.8.63)$$

The characteristic polynomial of $L[y] = 0$ is $p(m) = m^2 - 2m + 2$.   Note that $p(\alpha + i\beta) = p(1 + i) = (1 + i)^2 - 2(1 + i) + 2 = 0$.   Thus, a particular solution $y_p(x)$ of (4.8.63) may be taken of the form [*see* Case 3 (ii)]

$$y_p(x) = x(B_0 x + B_1)e^x \cos x + x(C_0 x + C_1)e^x \sin x. \qquad (4.8.64)$$

One may now proceed as in the above example to determine a particular solution of (4.8.63).

Remark 4.8.3.   Suppose that

$$L[y] = ay'' + by' + cy = ke^{\alpha x}, \tag{4.8.65}$$

where a, b, c, and k are real constants and $a \neq 0$.   The characteristic polynomial of $L[y] = 0$ is

$$p(m) = am^2 + bm + c. \tag{4.8.66}$$

Utilizing Case 2 with $P_0(x) = k$, we leave it as an exercise for the reader [*see* Exercise 4.8.1] to show that a *particular solution* $y_p(x)$ of Equation (4.8.65) is given by

$$y_p(x) = \frac{ke^{\alpha x}}{p(\alpha)}, \quad \text{if } p(\alpha) \neq 0; \tag{4.8.67}$$

$$y_p(x) = \frac{kxe^{\alpha x}}{p'(\alpha)}, \quad \text{if } p(\alpha) = 0, \quad p'(\alpha) \neq 0; \tag{4.8.68}$$

$$y_p(x) = \frac{kx^2 e^{\alpha x}}{p''(\alpha)}, \quad \text{if } p(\alpha) = 0, \quad p'(\alpha) = 0. \tag{4.8.69}$$

This is a very useful result.

Remark 4.8.4.   (Review Exercise 4.7.6.)   From Euler's formula (4.6.40) we have

$$e^{i\beta x} = \cos \beta x + i \sin \beta x, \tag{4.8.70}$$

where β is a real constant and x is a real number.   Thus,

$$\mathscr{R}(e^{i\beta x}) = \cos \beta x \quad \text{and} \quad \mathscr{I}(e^{i\beta x}) = \sin \beta x. \tag{4.8.71}$$

We may also find a particular solution $y_p(x)$ of

$$L[y] = ay'' + by' + cy = k \cos \beta x, \tag{4.8.72}$$

and of

$$L[y] = ay'' + by' + cy = k \sin \beta x, \tag{4.8.73}$$

where a, b, c, k, and β are real constants and $a \neq 0$, by the following method.[1] We shall carry out the analysis for Equation (4.8.72).   Consider the differential equation

$$L[w] = aw'' + bw' + cw = ke^{i\beta x}, \tag{4.8.74}$$

where

$$w = y + iv.$$

---

[1] By the principle of superposition [*see* Exercise 4.7.1], we are able to find a particular solution $y_p(x)$ of

$$L[y] = ay'' + by' + cy = k_1 \cos \beta x + k_2 \sin \beta x.$$

Note that

$$\mathscr{R}\{L[w]\} = k \cos \beta x \quad \text{and} \quad \mathscr{R}(w) = y.$$

Let

$$w_p = y_p + iv_p$$

be a particular solution of (4.8.74). Note that $\mathscr{R}(w_p) = y_p$. The characteristic polynomial of $L[w] = 0$ is

$$p(m) = am^2 + bm + c.$$

Utilizing the results given in Remark 4.8.3, we then have

$$w_p = \frac{ke^{i\beta x}}{p(i\beta)}, \quad \text{if } p(i\beta) \neq 0;$$

$$w_p = \frac{kxe^{i\beta x}}{p'(i\beta)}, \quad \text{if } p(i\beta) = 0.$$

Thus, a *particular solution* of

$$L[y] = ay'' + by' + cy = k \cos \beta x \qquad\qquad [4.8.72]$$

is given by

$$y_p(x) = \mathscr{R}\left[\frac{ke^{i\beta x}}{p(i\beta)}\right], \quad \text{if } p(i\beta) \neq 0; \qquad\qquad (4.8.75)$$

$$y_p(x) = \mathscr{R}\left[\frac{kxe^{i\beta x}}{p'(i\beta)}\right], \quad \text{if } p(i\beta) = 0. \qquad\qquad (4.8.76)$$

Similarly, if we let

$$w = u + iy, \quad \text{and} \quad w_p = u_p + iy_p,$$

then a *particular solution* of

$$L[y] = ay'' + by' + cy = k \sin \beta x \qquad\qquad [4.8.73]$$

is given by [*see* Exercise 4.8.2]

$$y_p(x) = \mathscr{I}\left[\frac{ke^{i\beta x}}{p(i\beta)}\right], \quad \text{if } p(i\beta) \neq 0; \qquad\qquad (4.8.77)$$

$$y_p(x) = \mathscr{I}\left[\frac{kxe^{i\beta x}}{p'(i\beta)}\right], \quad \text{if } p(i\beta) = 0. \qquad\qquad (4.8.78)$$

Example 4.8.5. Let us find a particular solution $y_p(x)$ of the differential equation

$$L[y] = y'' + 2y' + 24y = 8 \cos 4x. \qquad\qquad (4.8.79)$$

SOLUTION. Consider the differential equation

$$L[w] = w'' + 2w' + 24w = 8e^{i4x}.$$

The characteristic polynomial of $L[w] = 0$ is $p(m) = m^2 + 2m + 24$. Thus, $p(i\beta) = p(i4) = -16 + i8 + 24 = 8 + i8 \neq 0$. Utilizing (4.8.75), we then have that

$$y_p(x) = \mathscr{R}\left[\frac{8e^{i4x}}{8 + i8}\right] = \mathscr{R}\left[\frac{e^{i4x}}{1 + i}\right]. \tag{4.8.80}$$

The polar form of the complex number $1 + i = \sqrt{2}\, e^{i(\pi/4)}$. [*See* Exercise 4.6.19.] Thus,

$$y_p(x) = \mathscr{R}\left[\frac{e^{i4x}}{\sqrt{2}\, e^{i(\pi/4)}}\right] = \mathscr{R}\left[\frac{e^{i(4x-\pi/4)}}{\sqrt{2}}\right]$$

$$= \mathscr{R}\left[\frac{1}{\sqrt{2}}\cos\left(4x - \frac{\pi}{4}\right) + \frac{i}{\sqrt{2}}\sin\left(4x - \frac{\pi}{4}\right)\right]$$

$$= \frac{1}{\sqrt{2}}\cos\left(4x - \frac{\pi}{4}\right). \tag{4.8.81}$$

We may also write the particular solution $y_p(x)$ in (4.8.81) as

$$y_p(x) = \frac{1}{\sqrt{2}}\cos 4x \cos\frac{\pi}{4} + \frac{1}{\sqrt{2}}\sin 4x \sin\frac{\pi}{4} = \frac{1}{2}(\cos 4x + \sin 4x). \tag{4.8.82}$$

The particular solution $y_p(x)$ in (4.8.82) may be obtained directly as follows.

$$y_p(x) = \mathscr{R}\left[\frac{e^{i4x}}{1 + i}\right] = \mathscr{R}\left[\frac{1 - i}{(1 + i)(1 - i)}e^{i4x}\right]$$

$$= \mathscr{R}\left[\frac{1 - i}{2}(\cos 4x + i\sin 4x)\right]$$

$$= \mathscr{R}\left[\frac{1}{2}(\cos 4x + \sin 4x) + \frac{i}{2}(-\cos 4x + \sin 4x)\right]$$

$$= \frac{1}{2}(\cos 4x + \sin 4x).$$

Remark 4.8.5. We shall state below some additional basic results which will enable one to find more readily a particular solution of certain types of differential equations which we have been considering. We shall leave the proofs of these results as exercises for the reader [*see* Exercises 4.8.3 through 4.8.5]. In the ensuing results,

$$p(m) = am^2 + bm + c \tag{4.8.83}$$

will denote, as usual, the characteristic polynomial of the differential equation

$$L[y] = ay'' + by' + cy = 0, \tag{4.8.84}$$

where a, b, and c are real constants and $a \neq 0$. Also the letters k, α, β, and λ will denote real constants.

A *particular solution* $y_p(x)$ of the differential equation

$$L[y] = ay'' + by' + cy = ke^{\alpha x} \cos \beta x \tag{4.8.85}$$

is given by

$$y_p(x) = \mathscr{R}\left[\frac{ke^{(\alpha + i\beta)x}}{p(\alpha + i\beta)}\right], \quad \text{if } p(\alpha + i\beta) \neq 0; \tag{4.8.86}$$

$$y_p(x) = \mathscr{R}\left[\frac{kxe^{(\alpha + i\beta)x}}{p'(\alpha + i\beta)}\right], \quad \text{if } p(\alpha + i\beta) = 0. \tag{4.8.87}$$

A *particular solution* $y_p(x)$ of the differential equation

$$L[y] = ay'' + by' + cy = ke^{\alpha x} \sin \beta x \tag{4.8.88}$$

is given by

$$y_p(x) = \mathscr{I}\left[\frac{ke^{(\alpha + i\beta)x}}{p(\alpha + i\beta)}\right], \quad \text{if } p(\alpha + i\beta) \neq 0; \tag{4.8.89}$$

$$y_p(x) = \mathscr{I}\left[\frac{kxe^{(\alpha + i\beta)x}}{p'(\alpha + i\beta)}\right], \quad \text{if } p(\alpha + i\beta) = 0. \tag{4.8.90}$$

A *particular solution* $y_p(x)$ of the differential equation

$$y'' + \lambda^2 y = k \cos \beta x \tag{4.8.91}$$

is given by

$$y_p(x) = \frac{k}{\lambda^2 - \beta^2} \cos \beta x, \quad \text{if } \beta \neq \pm \lambda; \tag{4.8.92}$$

$$y_p(x) = \frac{kx}{2\lambda} \sin \lambda x, \quad \text{if } \beta = \lambda. \tag{4.8.93}$$

A *particular solution* $y_p(x)$ of the differential equation

$$y'' + \lambda^2 y = k \sin \beta x \tag{4.8.94}$$

is given by

$$y_p(x) = \frac{k}{\lambda^2 - \beta^2} \sin \beta x, \quad \text{if } \beta \neq \pm \lambda; \tag{4.8.95}$$

$$y_p(x) = -\frac{kx}{2\lambda} \cos \lambda x, \quad \text{if } \beta = \lambda. \tag{4.8.96}$$

Example 4.8.6. Let us find a particular solution $y_p(x)$ of the differential equation

$$L[y] = y'' - 2y' + 2y = 4e^{2x} + 2e^x \cos x. \tag{4.8.97}$$

SOLUTION. We shall utilize the principle of superposition in obtaining a particular solution $y_p(x)$ of (4.8.97). Let y_{p_1} and y_{p_2} denote, respectively, particular solutions of

$$L[y] = y'' - 2y' + 2y = 4e^{2x}, \tag{4.8.98}$$

and

$$L[y] = y'' - 2y' + 2y = 2e^x \cos x. \tag{4.8.99}$$

The characteristic polynomial of $L[y] = 0$ is $p(m) = m^2 - 2m + 2$. Note that $p(2) = 4 - 4 + 2 = 2$. Thus, to find $y_{p_1}(x)$ we utilize Formula (4.8.67) with $k = 4$, $\alpha = 2$, and $p(\alpha) = p(2)$. Hence,

$$y_{p_1}(x) = \frac{4e^{2x}}{2} = 2e^{2x}. \tag{4.8.100}$$

Comparing the right-hand member of (4.8.99) with that of (4.8.85), we see that $k = 2$, $\alpha = 1$, and $\beta = 1$. Note that $p(\alpha + i\beta) = p(1 + i) = (1 + i)^2 - 2(1 + i) + 2 = 0$. Thus, to find $y_{p_2}(x)$ we use Formula (4.8.87). Hence,

$$y_{p_2}(x) = \mathscr{R}\left[\frac{2xe^{(1+i)x}}{2i}\right] = \mathscr{R}\left[\frac{x}{i} e^x e^{ix}\right] = \mathscr{R}\left[\frac{x}{i} e^x(\cos x + i \sin x)\right]$$

$$= \mathscr{R}[xe^x(\sin x - i \cos x)] = xe^x \sin x. \tag{4.8.101}$$

In view of (4.8.100), (4.8.101), and the principle of superposition, a particular solution $y_p(x)$ of (4.8.97) is

$$y_p(x) = 2e^{2x} + xe^x \sin x. \tag{4.8.102}$$

Exercises 4.8

1. Prove Formulas (4.8.67), (4.8.68), and (4.8.69).
2. Prove Formulas (4.8.77) and (4.8.78).
3. Prove Formulas (4.8.86) and (4.8.87).
4. Prove Formulas (4.8.89) and (4.8.90).
5. Prove Formulas (4.8.92), (4.8.93), (4.8.95), and (4.8.96).

In Exercises 6 through 20, find the general solution of each differential equation.

6. $y'' - 5y' + 6y = 12x + 8$.   7. $y'' - 4y = 8x + 6e^x$.
8. $y'' + 4y' + 3 = y6 \cos x + 12 \sin x + 9x^2 - 32$.
9. $y'' - 3y' = -9x^2 + 18xe^{3x}$.   10. $y'' + y = 4x \cos x$.

11. $y'' + 4y = 8 \sin 2x.$
12. $y'' - 9y = 37e^{3x} \sin x.$
33. $y'' + 4y = 16x \sin 2x.$
14. $y'' + 16y = 8 \cos 4x + 16 \sin 4x.$
15. $y'' - 4y' + 5y = 2e^{2x} \sin^2 x.$
16. $y'' - 4y' + 5y = 20 \cosh 2x \cos x.$
17. $y'' - 2y' + 2y = 4xe^x \sin x.$
18. $y'' + 4y = 8 \cos 2x + 20e^{2x} \sin 2x.$
19. $y'' + y' = 4x^3 + 15x^2 + 10x - e^{-x}.$
20. $y'' + 2y' + y = x^4 + 8x^3 + 13x^2 + 5x + 3 + (12x^2 - 6x + 2)e^{-x}.$

In Exercises 21 through 26, find the solution of each initial-value problem.

21. $\begin{cases} y'' + 2y' + 2y = -30 \cos 2x, \\ y(0) = 0, \quad y'(0) = -5. \end{cases}$

22. $\begin{cases} y'' - 7y' + 10y = 100x, \\ y(0) = 0, \quad y'(0) = 5. \end{cases}$

23. $\begin{cases} y'' - y = -39e^x \sin 3x, \\ y(0) = 4, \quad y'(0) = 1. \end{cases}$

24. $\begin{cases} y'' - 2y' + y = 24x^2 e^x, \\ y(0) = 2, \quad y'(0) = 4. \end{cases}$

25. $\begin{cases} y'' + 4y = 16x \cos 2x, \\ y(\pi/2) = \pi/2, \quad y'(\pi/2) = -\pi^2. \end{cases}$

26. $\begin{cases} y'' + 2y' + 2y = 4xe^{-x} \cos x, \\ y(0) = 1, \quad y'(0) = 2. \end{cases}$

27. (a) For the solution $y(x)$ of the initial-value problem of Exercise 21, show that as $x \to \infty$, the values of x for which $y'(x) = 0$ are approximately equal to the roots of the equation

$$\tan 2x = -2.$$

[*Note:* This result is independent of the initial conditions imposed on the general solution $y(x)$ of Exercise 21.]

(b) For the solution $y(x)$ of the initial-value problem of Exercise 22, show that $y''(x) = 0$ when $x = \frac{1}{3} \ln \frac{8}{15}$.

28. (a) Solve the boundary-value problem

$$\begin{cases} y'' + 9y = 27 \cos 6x, \\ \\ y(0) = 1, \quad y\left(\dfrac{\pi}{6}\right) = 1. \end{cases}$$

(b) For the solution $y(x)$ of part (a), show that the maximum and minimum values of $y(x)$ are given, respectively, by

$$x = \frac{2}{3} n\pi \pm \frac{\pi}{9} \quad \text{and} \quad x = \frac{1}{3} n\pi,$$

where $n = \pm 1, \pm 2, \ldots$.

29. (a) Show that the solution of the initial-value problem

$$\begin{cases} y'' + 2y' + 5y = 4e^{-x}, \\ y(0) = 0, \quad y'(0) = 0, \end{cases}$$

is given by

(1) $$y(x) = 2e^{-x} \sin^2 x.$$

(b) Show that the maximum values of the function y given in (1) of part (a) are determined by the roots of the equation

$$\tan x = 2.$$

Also, sketch the graph of the function y on the interval I: $0 \leqq x \leqq 2\pi$. Note the decaying feature of this function.

30.   Establish the formulas given in (4.8.29).

4.9. The Method of Variation of Parameters

We shall now give a method for determining a particular solution of the non-homogeneous linear differential equation

$$L[y] = P(x)y'' + Q(x)y' + R(x)y = F(x), \tag{4.9.1}$$

where P, Q, R, and F are continuous real-valued functions defined on an interval I: $a \leqq x \leqq b$, $P(x) \neq 0$ for every x in I and $F(x)$ is not identically zero in I.   The related homogeneous differential equation of $L[y] = F(x)$ is

$$L[y] = P(x)y'' + Q(x)y' + R(x)y = 0. \tag{4.9.2}$$

While the method of undetermined coefficients restricted the functions P, Q, and R to be constants and F involved only terms of the form x^n, $e^{\alpha x}$, $\sin \beta x$, and $\cos \beta x$, the method of variation of parameters is applicable when these restrictions are removed.   This method depends on knowing two linearly independent solutions of $L[y] = 0$.   Also, this method involves integrations rather than differentiations, and thus the task of obtaining a particular solution of $L[y] = F(x)$ free of integrals may be very difficult and even impossible. However, particular solutions involving integrals are useful in applications. We shall now proceed to illustrate this method.

Suppose that y_1 and y_2 are two linearly independent solutions of $L[y] = 0$ on I.   Then, the general solution $y_c(x)$ of $L[y] = 0$ on I is given by

$$y_c(x) = c_1 y_1(x) + c_2 y_2(x), \tag{4.9.3}$$

where c_1 and c_2 are arbitrary constants (or parameters).   Basically, the method of variation of parameters consists of replacing the constants c_1 and c_2 in the complementary function y_c given in (4.9.3) by undetermined functions of the independent variable x.   We then determine these functions so that when the modified complementary function is substituted in the left-hand side of Equation (4.9.1) it reduces to $F(x)$.   This imposes only one condition on the two arbitrary functions.   Hence, we have another condition at our disposal, which we shall use to our advantage. Thus, let us assume that a particular solution $y_p(x)$ of $L[y] = F(x)$ on I has the form

$$y_p(x) = v_1(x)y_1(x) + v_2(x)y_2(x). \tag{4.9.4}$$

Differentiating (4.9.4), we obtain

$$y_p' = v_1 y_1' + v_2 y_2' + v_1' y_1 + v_2' y_2. \tag{4.9.5}$$

To simplify the determination of the functions v_1 and v_2, we require that v_1 and v_2 satisfy[2]

$$v_1' y_1 + v_2' y_2 = 0. \tag{4.9.6}$$

Thus,

$$y_p' = v_1 y_1' + v_2 y_2'. \tag{4.9.7}$$

Note that the requirement of (4.9.6) will avoid the appearance of v_1'' and v_2'' in y_p''. Differentiating (4.9.7), we get

$$y_p'' = v_1' y_1' + v_2' y_2' + v_1 y_1'' + v_2 y_2''. \tag{4.9.8}$$

Substituting y_p, y_p', and y_p'' of (4.9.4), (4.9.7), and (4.9.8) in (4.9.1), we obtain upon collecting terms

$$(Py_1'' + Qy_1' + Ry_1)v_1 + (Py_2'' + Qy_2' + Ry_2)v_2 + P(v_1'y_1' + v_2' y_2') = F(x). \tag{4.9.9}$$

Since y_1 and y_2 are solutions of $L[y] = 0$, the expressions in the first two parentheses of (4.9.9) are equal to zero. Thus, (4.9.9) becomes

$$P(v_1'y_1' + v_2' y_2') = F(x). \tag{4.9.10}$$

Since $P(x) \neq 0$ for every x in I, (4.9.10) may be equivalently written as

$$v_1'y_1' + v_2' y_2' = \frac{F(x)}{P(x)}. \tag{4.9.11}$$

The Equations (4.9.6) and (4.9.11) can be solved for v_1' and v_2' provided that

$$\begin{vmatrix} y_1 & y_2 \\ y_1' & y_2' \end{vmatrix} = y_1 y_2' - y_2 y_1' \neq 0. \tag{4.9.12}$$

The determinant in (4.9.12) is the Wronskian of the functions y_1 and y_2. [*See* (4.4.37).] Since y_1 and y_2 are two linearly independent solutions of $L[y] = 0$ on I, it follows from Theorem 4.4.3 that $W(y_1, y_2; x) = y_1 y_2' - y_2 y_1' \neq 0$ for every x in I. Thus, Equations (4.9.6) and (4.9.11) can be solved for v_1 and v_2. We have

---

[2] See Exercise 4.9.19 for a possible justification of the requirement made in (4.9.6).

$$v_1'(x) = \frac{\begin{vmatrix} 0 & y_2 \\ \dfrac{F(x)}{P(x)} & y_2' \\ y_1 & y_2 \\ y_1' & y_2' \end{vmatrix}}{\begin{vmatrix} y_1 & y_2 \\ y_1' & y_2' \end{vmatrix}} = -\frac{y_2(x)F(x)}{P(x)W(y_1, y_2; x)},$$

$$\tag{4.9.13}$$

$$v_2'(x) = \frac{\begin{vmatrix} y_1 & 0 \\ y_1' & \dfrac{F(x)}{P(x)} \\ y_1 & y_2 \\ y_1' & y_2' \end{vmatrix}}{\begin{vmatrix} y_1 & y_2 \\ y_1' & y_2' \end{vmatrix}} = \frac{y_1(x)F(x)}{P(x)W(y_1, y_2; x)}.$$

Integrating (4.9.13), we obtain, suppressing the constants of integration,

$$v_1(x) = -\int \frac{y_2(x)F(x)}{P(x)W(y_1, y_2; x)}\, dx,$$

$$v_2(x) = \int \frac{y_1(x)F(x)}{P(x)W(y_1, y_2; x)}\, dx.$$

$$\tag{4.9.14}$$

In view of (4.9.4) and (4.9.14), *a particular solution* $y_p(x)$ *of* $L[y] = F(x)$ *is given by*

$$y_p(x) = -y_1(x)\int \frac{y_2(x)F(x)}{P(x)W(y_1, y_2; x)}\, dx + y_2(x)\int \frac{y_1(x)F(x)}{P(x)W(y_1, y_2; x)}\, dx.$$

$$\tag{4.9.15}$$

Thus, the general solution $y(x)$ *of*

$$L[y] = P(x)y'' + Q(x)y' + R(x)y = F(x) \tag{4.9.1}$$

on I is given by

$$y(x) = y_c(x) + y_p(x), \tag{4.9.16}$$

where the complementary function y_c is given by (4.9.3) *and a particular solution is given by* (4.9.15).

Remark 4.9.1.   Note that if we had incorporated the constants of integration (k_1 and k_2) in (4.9.14), then we would have to add the terms $k_1 y_1(x)$ and $k_2 y_2(x)$ to (4.9.15).   However, these terms may then be absorbed, in the usual manner, into the complementary function y_c.

Example 4.9.1. Let us find the general solution of the differential equation

$$L[y] = y'' + y = \tan x \tag{4.9.17}$$

on the interval I: $0 < x < \pi/2$.

SOLUTION.   The complementary function y_c of $L[y] = \tan x$ on I is [*see* Theorem 4.6.2]

$$y_c(x) = c_1 \cos x + c_2 \sin x. \tag{4.9.18}$$

Replacing c_1 and c_2, respectively, by the functions v_1 and v_2, we endeavor to find a particular solution y_p of $L[y] = \tan x$ on I of the form

$$y_p(x) = v_1(x) \cos x + v_2(x) \sin x. \tag{4.9.19}$$

Differentiating (4.9.19), we obtain

$$y_p' = -v_1 \sin x + v_2 \cos x + v_1' \cos x + v_2' \sin x. \tag{4.9.20}$$

We now require that

$$v_1' \cos x + v_2' \sin x = 0. \tag{4.9.21}$$

Thus, (4.9.20) becomes

$$y_p' = -v_1 \sin x + v_2 \cos x. \tag{4.9.22}$$

Differentiating (4.9.22), we get

$$y_p'' = -v_1' \sin x + v_2' \cos x - v_1 \cos x - v_2 \sin x. \tag{4.9.23}$$

Substituting y_p and y_p'' of (4.9.19) and (4.9.23) in (4.9.17), we find

$$-v_1' \sin x + v_2' \cos x = \tan x. \tag{4.9.24}$$

Solving (4.9.21) and (4.9.24) simultaneously for v_1' and v_2', we find

$$v_1' = -\sin x \tan x, \qquad v_2' = \cos x \tan x. \tag{4.9.25}$$

Integrating (4.9.25), we obtain

$$v_1(x) = -\int \sin x \tan x \, dx = -\int \frac{\sin^2 x}{\cos x} \, dx = -\int \frac{1 - \cos^2 x}{\cos x} \, dx$$

$$= -\int \sec x \, dx + \int \cos x \, dx \tag{4.9.26}$$

$$= -\ln(\sec x + \tan x) + \sin x,$$

$$v_2(x) = \int \cos x \tan x \, dx = \int \sin x \, dx = -\cos x.$$

In view of (4.9.19) and (4.9.26), we have

$$y_p(x) = -\cos x \ln(\sec x + \tan x). \tag{4.9.27}$$

Thus, in view of (4.9.18) and (4.9.27), the general solution $y(x)$ of $L[y] = \tan x$ on I is

$$y(x) = c_1 \cos x + c_2 \sin x - \cos x \ln (\sec x + \tan x). \qquad (4.9.28)$$

Alternatively, we may determine the functions v_1 and v_2 by utilizing (4.9.14). Here, $y_1(x) = \cos x$, $\quad y_2(x) = \sin x$, $\quad F(x) = \tan x$, $\quad P(x) = 1$, and

$$W(y_1, y_2'; x) = \begin{vmatrix} \cos x & \sin x \\ -\sin x & \cos x \end{vmatrix} = 1.$$

Thus,

$$v_1(x) = -\int \sin x \tan x \, dx \quad \text{and} \quad v_2(x) = \int \cos x \tan x \, dx.$$

Example 4.9.2. Let us solve the initial-value problem

$$\begin{cases} L[y] = y'' - y = \dfrac{e^{2x}}{(e^x + 1)^2}, & (4.9.29) \\[2mm] y(0) = \ln 2, & (4.9.30) \\[2mm] y'(0) = 7/2 - \ln 2, & (4.9.31) \end{cases}$$

on the interval I: $\quad -\infty < x < \infty$.

SOLUTION. The complementary function y_c of $L[y] = e^{2x}/(e^x + 1)^2$ on I is

$$y_c(x) = c_1 e^x + c_2 e^{-x}. \qquad (4.9.32)$$

We shall find a particular solution y_p of $L[y] = e^{2x}/(e^x + 1)^2$ on I, by utilizing (4.9.15). Here, $\quad y_1(x) = e^x$, $\quad y_2(x) = e^{-x}$, $\quad F(x) = e^{2x}/(e^x + 1)^2$, $\quad P(x) = 1$, and

$$W(y_1, y_2; x) = \begin{vmatrix} e^x & e^{-x} \\ e^x & -e^{-x} \end{vmatrix} = -2.$$

Thus, (4.9.15) becomes

$$y_p(x) = \frac{e^x}{2} \int \frac{e^x}{(e^x + 1)^2} \, dx - \frac{e^{-x}}{2} \int \frac{e^{3x}}{(e^x + 1)^2} \, dx. \qquad (4.9.33)$$

Let us now evaluate the integrals in (4.9.33).

$$\int \frac{e^x}{(e^x + 1)^2} \, dx = \int (e^x + 1)^{-2} e^x \, dx = -\frac{1}{e^x + 1}. \qquad (4.9.34)$$

Concerning the integral

$$\int \frac{e^{3x}}{(e^x + 1)^2} \, dx,$$

let $e^x = z$, then $dx = z^{-1}dz$.   Thus,

$$\int \frac{e^{3x}}{(e^x + 1)^2}\, dx = \int \frac{z^2\, dz}{(z + 1)^2} = \int dz - 2\int \frac{dz}{z + 1} + \int \frac{dz}{(z + 1)^2}$$

$$= z - 2\ln(z + 1) - \frac{1}{z + 1} = e^x - 2\ln(e^x + 1) - \frac{1}{e^x + 1}.$$

$$(4.9.35)$$

Substituting the evaluations of the integrals found in (4.9.34) and (4.9.35) in (4.9.33) and simplifying the result, we obtain

$$y_p(x) = -1 + \tfrac{1}{2}e^{-x} + e^{-x}\ln(e^x + 1). \tag{4.9.36}$$

In view of (4.9.32) and (4.9.36), the general solution $y(x)$ of (4.9.29) is

$$y(x) = c_1 e^x + c_2 e^{-x} - 1 + e^{-x}\ln(e^x + 1). \tag{4.9.37}$$

[*Note:* The term $\tfrac{1}{2}e^{-x}$ in (4.9.36) was absorbed in the term $c_2 e^{-x}$ of the complementary function.]   Differentiating (4.9.37), we get

$$y'(x) = c_1 e^x - c_2 e^{-x} - e^{-x}\ln(e^x + 1) + \frac{1}{e^x + 1}. \tag{4.9.38}$$

Utilizing (4.9.30), (4.9.31), (4.9.37), and (4.9.38), we have

$$\begin{cases} \ln 2 = y(0) = c_1 + c_2 - 1 + \ln 2, \\ \tfrac{7}{2} - \ln 2 = y'(0) = c_1 - c_2 - \ln 2 + \tfrac{1}{2}. \end{cases} \tag{4.9.39}$$

Solving the system (4.9.39) for c_1 and c_2, we find $c_1 = 2$ and $c_2 = -1$.   Substituting these values in (4.9.37), the solution of our initial-value problem is

$$y(x) = 2e^x - 1 - e^{-x}[1 - \ln(e^x + 1)]. \tag{4.9.40}$$

Remark 4.9.2.   If the functions $f_1, \ldots, f_n$ which appear in the particular solution y_p of $L[y] = F(x)$ also appear in the complementary function y_c of $L[y] = F(x)$, we shall always absorb these functions in the complementary function y_c when forming the general solution of $L[y] = F(x)$.

Exercises 4.9

In Exercises 1 through 10, find the general solution of each differential equation.

1. $y'' + y = \sec x$, $0 < x < \dfrac{\pi}{2}$.

2. $y'' - y = \cos^2 x$.

3. $y'' - 4y' + 4y = e^{2x}\sec^2 x$, $0 < x < \dfrac{\pi}{2}$.

4. $y'' - 4y' + 4y = x^n e^{2x}$, $x > 0$.

5. $y'' + 2y' + y = e^{-x}\ln x$, $x > 0$.

6. $y'' + 4y' + 3y = \cos e^x$.

7. $y'' - y = \operatorname{sech} x$.

8. $y'' + 2y' + y = \dfrac{e^{-x}}{x^2}$, $x > 0$.

9. $y'' - y = \cosh 2x + e^{-2x} \sin e^{-x}$.

10. $y'' - 2y' + 2y = e^x \tan^2 x$, $0 < x < \dfrac{\pi}{2}$.

11. Solve the initial-value problem

$$\begin{cases} y'' - 3y' + 2y = \dfrac{e^{3x}}{e^{2x} + 1}, \\[2mm] y(0) = \dfrac{\pi}{4}, \quad y'(0) = \dfrac{\pi}{2}. \end{cases}$$

12. Suppose that

$$L[y] = (x^2 + 1)y'' - 2xy' + 2y = (x^2 + 1)^2$$

and $y_1(x) = x^2 - 1$, $y_2(x) = x$ are solutions of $L[y] = 0$. Find the general solution of $L[y] = (x^2 + 1)^2$ on the interval $|x| < 1$.

13. Suppose that

$$L[y] = xy'' - (2x + 1)y' + (x + 1)y = 2x^2 e^x \ln x, \qquad x > 0,$$

and $y_1(x) = x^2 e^x$, $y_2(x) = e^x$ are solutions of $L[y] = 0$. Find the general solution of $L[y] = 2x^2 e^x \ln x$.

14. (a) Show that the solution of the initial-value problem

$$\begin{cases} y'' + \lambda y = F(x), \\ y(0) = 0, \quad y'(0) = 0, \end{cases}$$

where λ is a parameter, is given by

$$y(x) = \int_0^x \frac{\sin \sqrt{\lambda}(x - t)}{\sqrt{\lambda}} F(t)\,dt.$$

(b) Use the result in part (a) to solve the initial-value problem

$$\begin{cases} y'' = e^x, \\ y(0) = 0, \quad y'(0) = 0. \end{cases}$$

15. (a) Consider the differential equation

$$L[y] = P(x)y'' + Q(x)y' + R(x)y = F(x).$$

Let $y_1 = y_1(x)$ and $y_2 = y_2(x)$ be two linearly independent solutions of $L[y] = 0$ on the interval I: $a \leq x \leq b$, containing the point x_0. Show that the solution of the initial-value problem

$$\begin{cases} L[y] = F(x), \\ y(x_0) = 0, \quad y'(x_0) = 0, \end{cases}$$

is given by

$$y(x) = \int_{x_0}^{x} \frac{y_1(t)y_2(x) - y_1(x)y_2(t)}{P(t)W(y_1, y_2 \,;\, t)} F(t)\,dt,$$

where

$$W(y_1, y_2 \,;\, x) = y_1(x)y_2'(x) - y_2(x)y_1'(x).$$

[*Note:* If we know one nonvanishing solution of $L[y] = 0$ on I, then another solution of $L[y] = 0$ may be obtained by using Formula (4.5.4). Thus, all that is necessary to know in advance is one solution of $L[y] = 0$.]

   (b) Use the result in part (a) to solve the initial-value problem

$$\begin{cases} L[y] = xy'' - (2x + 1)y' + (x + 1)y = 2x^2 e^x \ln x, & x > 0, \\ y(1) = 0, \quad y'(1) = 0, \end{cases}$$

given that $y_1(x) = x^2 e^x$ and $y_2(x) = e^x$ are solutions of $L[y] = 0$.

16. (a) Consider the differential equation

$$L[y] = P(x)y'' + Q(x)y' + R(x)y = F(x).$$

Suppose that $y_1 = y_1(x)$ and $y_2 = y_2(x)$ are two linearly independent solutions of $L[y] = 0$ on the interval I: $\ a \leq x \leq b$ containing the point x_0. Show that the solution of the initial-value problem

$$\begin{cases} L[y] = F(x), \\ y(x_0) = A, \quad y'(x_0) = B, \end{cases}$$

where A and B are constants, is given by

$$y(x) = \int_{x_0}^{x} \frac{y_1(t)y_2(x) - y_1(x)y_2(t)}{P(t)W(y_1, y_2 \,;\, t)} F(t)\,dt$$

$$+ \frac{A[y_2'(x_0)y_1(x) - y_1'(x_0)y_2(x)] - B[y_2(x_0)y_1(x) - y_1(x_0)y_2(x)]}{W(y_1, y_2 \,;\, x_0)}.$$

   (b) Use the result in part (a) to solve the initial-value problem

$$\begin{cases} L[y] = xy'' - (2x + 1)y' + (x + 1)y = 2x^2 e^x \ln x, & x > 0, \\ y(1) = 2, \quad y'(1) = 4, \end{cases}$$

given that $y_1(x) = x^2 e^x$ and $y_2(x) = e^x$ are solutions of $L[y] = 0$.

17. Solve the initial-value problem

$$\begin{cases} L[y] = x(1 + 3x^2)y'' + 2y' - 6xy = (1 + 3x^2)^2, & x > 0, \\ y(1) = 2, \quad y'(1) = 4, \end{cases}$$

given that $y_1(x) = \dfrac{1}{x}$ is a solution of $L[y] = 0$.

18. Solve the initial-value problem

$$\begin{cases} L[y] = xy'' - 2(x+1)y' + (x+2)y = 3x^3 e^{2x}, & x > 0, \\ y(1) = 3e^2, \quad y'(1) = 6e^2, \end{cases}$$

given that $y_1(x) = e^x$ is a solution of $L[y] = 0$.

19. Differentiating (4.9.5), we obtain

$$(1) \qquad y_p'' = v_1 y_1'' + v_1' y_1' + v_2 y_2'' + v_2' y_2' + (v_1' y_1 + v_2' y_2)'.$$

Substituting y_p, y_p', and y_p'' of (4.9.4), (4.9.5), and (1) into Equation (4.9.1), show that we obtain

$$(2) \quad v_1(Py_1'' + Qy_1' + Ry_1) + v_2(Py_2'' + Qy_2' + Ry_2) + Q(v_1' y_1 + v_2' y_2)$$
$$+ P(v_1' y_1 + v_2' y_2)' + P(v_1' y_1' + v_2' y_2') = F(x).$$

Since $L[y_1] = 0$, $L[y_2] = 0$, Equation (2) reduces to

$$(3) \qquad Q(v_1' y_1 + v_2' y_2) + P(v_1' y_1 + v_2' y_2)' + P(v_1' y_1' + v_2' y_2') = F(x).$$

Thus, y_p will be a solution of Equation (4.9.1) if we choose

$$(4) \qquad\qquad\qquad v_1' y_1 + v_2' y_2 = 0,$$

and

$$(5) \qquad\qquad\qquad v_1' y_1' + v_2' y_2' = \frac{F(x)}{P(x)}.$$

Equations (4) and (5) are the same as Equations (4.9.6) and (4.9.11).

4.10. The Cauchy-Euler Equation

The differential equation

$$L[y] = ax^2 y'' + bxy' + cy = F(x), \qquad\qquad (4.10.1)$$

where a, b, c are constants, $a \neq 0$, and the function F is continuous on some interval I, is called a *Cauchy-Euler linear differential equation of order two*. Observe that, in general, in order for a solution $y(x)$ of $L[y] = F(x)$ to exist, the interval I should not include the origin. Equation (4.10.1) occurs in various fields of applications, for instance, in heat and potential theory. This equation is also known as the *equidimensional equation*. The differential equation $L[y] = 0$ arises in connection with solving a certain class of partial differential equations by the "method of separation of variables." Equation (4.10.1) is one of the few linear differential equations with variable coefficients that can be solved without resorting to integration by series. The following theorem is very useful in connection with solving the equation $L[y] = F(x)$.

Theorem 4.10.1. *The transformation* $x = e^u$ *(or* $x = -e^u$*) reduces Equation* (4.10.1) *to a linear differential equation with constant coefficients.*

PROOF.   Suppose that $x > 0$ in Equation (4.10.1).   Let

$$x = e^u \quad \text{or} \quad u = \ln x, \qquad x > 0. \tag{4.10.2}$$

We then have from calculus that

$$y' = \frac{dy}{dx} = \frac{dy}{du} \cdot \frac{du}{dx} = \frac{dy}{du} \cdot \frac{1}{x} = e^{-u}\frac{dy}{du}, \tag{4.10.3}$$

and

$$y'' = \frac{d^2 y}{dx^2} = \frac{d}{dx}\left(\frac{dy}{dx}\right) = \frac{d}{du}\left(\frac{dy}{dx}\right)\frac{du}{dx} = \frac{d}{du}\left(e^{-u}\frac{dy}{du}\right) \cdot \frac{1}{x}$$

$$= e^{-u}\left(e^{-u}\frac{d^2 y}{du^2} - e^{-u}\frac{dy}{du}\right) = e^{-2u}\left(\frac{d^2 y}{du^2} - \frac{dy}{du}\right). \tag{4.10.4}$$

Substituting x, y', and y'' of (4.10.2) through (4.10.4) in (4.10.1), we obtain

$$a\left(\frac{d^2 y}{du^2} - \frac{dy}{du}\right) + b\frac{dy}{du} + cy = F(e^u),$$

or

$$a\frac{d^2 y}{du^2} + (b - a)\frac{dy}{du} + cy = F(e^u). \tag{4.10.5}$$

Equation (4.10.5) is a linear differential equation with constant coefficients. Similarly, if $x < 0$ in (4.10.1), then the transformation

$$x = -e^u \quad \text{or} \quad u = \ln(-x), \quad x < 0, \tag{4.10.6}$$

reduces (4.10.1) to a linear differential equation with constant coefficients, namely,

$$a\frac{d^2 y}{du^2} + (b - a)\frac{dy}{du} + cy = F(-e^u). \tag{4.10.7}$$

Example 4.10.1.   Let us find the general solution of the equation

$$x^2 y'' - xy' - 3y = 4x^3 \tag{4.10.8}$$

on the interval I: $0 < x < \infty$.

SOLUTION.   Comparing (4.10.8) with (4.10.1), we see that

$$a = 1, \quad b = -1, \quad c = -3, \quad \text{and} \quad F(x) = 4x^3. \tag{4.10.9}$$

Thus, under the transformation $x = e^u$, (4.10.8) reduces to [*see* (4.10.5)]

$$\frac{d^2 y}{du^2} - 2\frac{dy}{du} - 3y = 4e^{3u}. \tag{4.10.10}$$

The complementary function y_c of Equation (4.10.10) is

$$y_c(u) = c_1 e^{-u} + c_2 e^{3u}. \qquad (4.10.11)$$

The characteristic polynomial of the related homogeneous differential equation of (4.10.10) is $p(m) = m^2 - 2m - 3$, and $p'(m) = 2m - 2$. Thus, $p(\alpha) = p(3) = 9 - 6 - 3 = 0$ and $p'(\alpha) = p'(3) = 6 - 2 = 4$. Hence, by Formula (4.8.68), a particular solution y_p of (4.10.10) is

$$y_p(u) = \frac{4ue^{3u}}{4} = ue^{3u}. \qquad (4.10.12)$$

In view of (4.10.11) and (4.10.12), the general solution of (4.10.10) is

$$y = c_1 e^{-u} + c_2 e^{3u} + ue^{3u}. \qquad (4.10.13)$$

Since $x = e^u$ or $u = \ln x$, we may write (4.10.13) as

$$y(x) = \frac{c_1}{x} + (c_2 + \ln x)x^3, \qquad (4.10.14)$$

which is the general solution of (4.10.8) on the interval I.

Example 4.10.2. Let us solve the initial-value problem

$$x^2 y'' + 2xy' - 6y = 50\,\frac{\ln x}{x^3}, \quad x > 0, \qquad (4.10.15)$$

$$y(1) = 1, \quad y'(1) = 5. \qquad (4.10.16)$$

SOLUTION. Comparing (4.10.15) with (4.10.1), we see that

$$a = 1, \quad b = 2, \quad c = -6, \quad \text{and} \quad F(x) = 50\,\frac{\ln x}{x^3}.$$

Thus, under the transformation $x = e^u$, (4.10.15) reduces to [*see* (4.10.5)]

$$\frac{d^2 y}{du^2} + \frac{dy}{du} - 6y = 50\,ue^{-3u}. \qquad (4.10.17)$$

The complementary function y_c of Equation (4.10.17) is

$$y_c(u) = c_1 e^{2u} + c_2 e^{-3u}. \qquad (4.10.18)$$

Also, $p(m) = m^2 + m - 6$ and $p'(m) = 2m + 1$. Thus, $p(\alpha) = p(-3) = 9 - 3 - 6 = 0$ and $p'(\alpha) = p'(-3) = -6 + 1 = -5$. From Case 2 (ii) of Section 4.8, a particular solution y_p of (4.10.17) has the form

$$y_p(u) = ue^{-3u}(B_0 u + B_1) = e^{-3u}(B_0 u^2 + B_1 u). \qquad (4.10.19)$$

Differentiating (4.10.19) two times in succession, we obtain upon collecting terms

$$y_p'(u) = e^{-3u}(-3B_0 u^2 + 2B_0 u - 3B_1 u + B_1),$$
$$y_p''(u) = e^{-3u}(9B_0 u^2 - 12B_0 u + 9B_1 u + 2B_0 - 6B_1).$$

Substituting y_p, y_p', and y_p'' in (4.10.17), we obtain upon simplification

$$-10B_0 u + 2B_0 - 5B_1 = 50u. \tag{4.10.20}$$

Equation (4.10.20) will be an identity in u provided that

$$\begin{cases} -10B_0 = 50, \\ 2B_0 - 5B_1 = 0. \end{cases} \tag{4.10.21}$$

Solving the system (4.10.21), we find $B_0 = -5$ and $B_1 = -2$, and (4.10.19) thus becomes

$$y_p(u) = -e^{-3u}(5u^2 + 2u). \tag{4.10.22}$$

In view of (4.10.18) and (4.10.22), the general solution of (4.10.17) is

$$y = c_1 e^{2u} + c_2 e^{-3u} - e^{-3u}(5u^2 + 2u). \tag{4.10.23}$$

Since $x = e^u$ or $u = \ln x$, the general solution of (4.10.15) is

$$y(x) = c_1 x^2 + c_2 x^{-3} - x^{-3}(5 \ln^2 x + 2 \ln x), \quad x > 0. \tag{4.10.24}$$

Differentiating (4.10.24), we get

$$y'(x) = 2c_1 x - 3c_2 x^{-4} + 15x^{-4} \ln^2 x - 4x^{-4} \ln x - 2x^{-4}, \quad x > 0. \tag{4.10.25}$$

Utilizing (4.10.16), (4.10.24), and (4.10.25), we have

$$\begin{cases} 1 = y(1) = c_1 + c_2, \\ 5 = y'(1) = 2c_1 - 3c_2 - 2. \end{cases} \tag{4.10.26}$$

Solving the system (4.10.26), we find $c_1 = 2$ and $c_2 = -1$. Substituting these values in (4.10.24), the solution of our initial-value problem is

$$y(x) = 2x^2 - x^{-3}(5 \ln^2 x + 2 \ln x + 1), \quad x > 0. \tag{4.10.27}$$

Exercises 4.10

In Exercises 1 through 10, find the general solution of each differential equation on the interval I: $0 < x < \infty$.

1. $x^2 y'' + 5xy' + 13y = 0.$
2. $x^2 y'' + xy' + 4y = 0.$
3. $4x^2 y'' - 8xy' + 9y = 0.$
4. $x^2 y'' - 3xy' = 8x.$
5. $x^2 y'' - 4xy' + 6y = 12 - x^2.$
6. $x^2 y'' + 7xy' + 9y = 4/x^3.$

7. $x^2y'' - xy' + y = \ln^2 x - 3 \ln x + 1$.
8. $2x^2y'' - 4xy' + 4y = x^2 \ln^2 x$.

9. $x^2y'' + xy' - y = \dfrac{1}{1+x^2}$.

10. $x^2y'' - xy' + y = \dfrac{x}{1+\ln^2 x}$.

11. Consider the differential equation

    (1) $$a(\alpha + \beta x)^2 y'' + b(\alpha + \beta x)y' + cy = F(x),$$

    where a, b, c, α, β are constants and $a \neq 0$, $\beta \neq 0$.   Show that the transformation

    $$\alpha + \beta x = e^u, \qquad x > -\frac{\alpha}{\beta},$$

    reduces (1) into a linear differential equation with constant coefficients given by

    $$a\beta^2 \frac{d^2y}{du^2} + (b\beta - a\beta^2)\frac{dy}{du} + cy = F\left(\frac{e^u - \alpha}{\beta}\right).$$

12. Find the general solution of the differential equation

    $$(2 + 3x)^2 y'' - 3(2 + 3x)y' + 9y = 81x, \qquad x > -\tfrac{2}{3}.$$

13. Solve the initial-value problem

    $$\begin{cases} (1 + x)^2 y'' + (1 + x)y' - y = \ln (1 + 2x + x^2), \\ y(0) = 1, \quad y'(0) = 3, \end{cases}$$

    on the interval I: $-1 < x < \infty$.

14. Solve the initial-value problem

    $$\begin{cases} x^2 y'' + xy' + y = \cos (\ln x), \\ y(1) = 1, \quad y'(1) = \tfrac{1}{2}, \end{cases}$$

    on the interval I: $0 < x < \infty$.

15. The differential equation

    (1) $$P(x)y'' + Q(x)y' + R(x)y = F(x)$$

    is said to be *exact* if it can be expressed as

    (2) $$\frac{d}{dx}[M(x)y' + N(x)y] = F(x).$$

    (a) Show that if Equation (1) is exact, then the functions M and N in (2) are given by

    (3) $$M(x) = P(x), \quad N(x) = Q(x) - P'(x).$$

    Substituting the functions M and N given in (3) into (2), the general solution of an exact equation may then be obtained by integrations.

(b) Show that a necessary condition for (1) to be exact is that

$$(4) \qquad\qquad P''(x) - Q'(x) + R(x) = 0.$$

It can also be shown that the condition given in (4) is a sufficient condition for exactness.

Determine whether each equation is exact; if exact, find the general solution of each equation.

(c) $(x^2 + 2x)y'' + 4(x + 1)y' + 2y = \sin x + \ln x, \qquad x > 0.$

(d) $x^3 y'' + 4x^2 y' + 2xy = 6x^2 + 1, \qquad x > 0.$

(e) $(x^2 - x)y'' - 4(x - 1)y' + 2y = x^2 e^x, \qquad 0 < x < 1.$

Suggested Readings

Boyce and Di Prima [9]

Coddington [12]

Ritger and Rose [39]

Ross [40]

Tenenbaum and Pollard [46]

chapter 5

APPLICATIONS OF SECOND-ORDER LINEAR DIFFERENTIAL EQUATIONS

5.1. Introduction

In this chapter we shall give some applications of second-order linear differential equations with constant coefficients. These differential equations arise in pure and applied mathematics, engineering, physics, astronomy, and in other fields of science. We shall be interested primarily in linear systems (mechanical vibrations). We shall see that linear, electrical, and even torsional systems are governed by the same differential equation. Thus, except for nomenclature and units, if one knows how to solve say, linear systems, then he will also know how to solve electrical and torsional systems. One may then apply the results of Chapter 4 to obtain the solution of the given problem.

5.2. The Differential Equation of the Motion of a Mass on a Spring

Let us consider a spring attached to a support and hanging downwards. [*See* Figure 5.2.1(a).] Within certain limits of elasticity, the spring will obey the following law.

Hooke's Law: *If the spring is stretched or compressed, the change in length is directly proportional to the force exerted upon the spring.*

Moreover, when that force is removed, the spring will return to its original length and leave its previous physical properties unchanged. Thus, with each spring, there is associated a numerical constant, namely the proportionality constant, which is defined as the ratio of the force exerted to the displacement produced by that force. For example, if a force of magnitude f pounds stretches the spring s feet, the relation

$$f = ks \tag{5.2.1}$$

defines the *spring constant* k in pounds per foot. Clearly, $k > 0$.

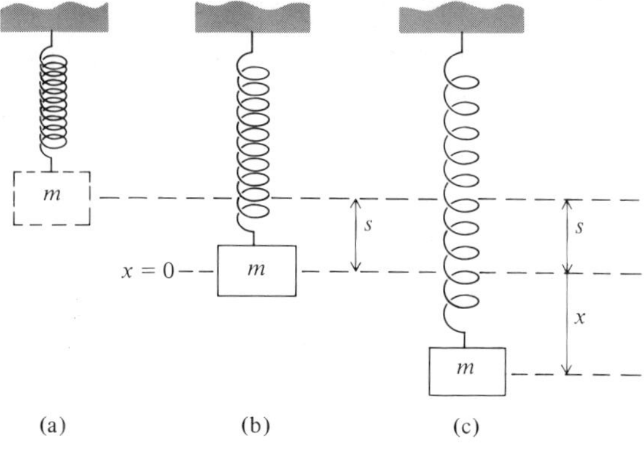

Figure 5.2.1.
(a) Unrestrained Spring   (b) Mass in Equilibrium   (c) Mass in General Position

Let us now attach a mass m to the lower end of the spring and allow it to come to rest. Suppose that the spring has stretched a distance s. [*See* Figures 5.2.1(a) and (b).]   This new position of the spring with the mass attached and at rest is called the *equilibrium position*.   In the ensuing discussion, we shall assume that the mass of the spring is negligible.

The gravitational force, which is equal to the weight of the mass, is mg. Thus, in view of (5.2.1), when the mass is at its equilibrium position, we see that

$$mg = ks. \tag{5.2.2}$$

Hence, the spring force is balanced by the weight of the mass.

Now, we may set the mass in motion by giving it an initial vertical velocity from its equilibrium position, or by first displacing the mass vertically from its equilibrium position and then either releasing it from rest or giving it an initial vertical velocity.   Thus, it is assumed that the mass is constrained to move in a vertical line passing through the point of support.   Also, for the present, we shall ignore air resistance and other effects.   Our task is to find a differential equation describing the motion of the mass attached to the lower end of the spring.

Let us take the position at which the mass is in equilibrium as the origin: $x = 0$.   (The origin is actually the center of gravity of the mass.)   Let x, measured positive downward and negative upward, represent the displacement of the mass from its equilibrium position at time t.   [*See* Figure 5.2.1(c).] The amount the spring is stretched at time t is therefore $s + x$, no matter whether the mass is above or below the equilibrium position.   Thus, the spring force is $k(s + x)$.   The spring force tends to restore the spring to its unstretched position [*see* Figure 5.2.1(a)], and thus acts in the direction opposite to the gravitational force mg, which acts in the positive x direction (that is, downward).   Consequently, the spring force is $-k(s + x)$ along the positive x direction.   Since the gravitational and spring forces are the only forces acting on the mass m, the total (net) force acting on the mass at time t along the positive x direction is $mg - k(s + x)$.   In view of (5.2.2), this net force reduces to $-kx$.   Utilizing Newton's Second Law of Motion [*see* Section 3.5] or using Formula (3.5.3), *the differential equation of the motion of a mass on a spring at time t is given by*

$$m \frac{d^2x}{dt^2} = -kx, \tag{5.2.3}$$

or

$$m \frac{d^2x}{dt^2} + kx = 0. \tag{5.2.4}$$

Equation (5.2.4) is a second-order linear homogeneous differential equation with constant coefficients.   Equation (5.2.3) asserts that the net force acting on the mass is proportional and opposite in sign to the displacement.   The mass will thus execute a simple harmonic motion about its equilibrium position, $x = 0$.   [*See* Exercise 3.5.9.]   The mechanism consisting of the spring and the mass is called a *harmonic oscillator*, and (5.2.4) is known as the equation of the harmonic oscillator.   We shall discuss the general solution of Equation (5.2.4) in Section 5.3.

We shall next develop the differential equation of the motion of a mass on a spring when the mass is also acted upon by retarding and other external forces.   Since the vibration of the mass on the spring takes place in mediums offering resistance, for example, air, water, and liquids of various densities, it is reasonable to expect that these will retard the motion of the mass.   In many instances, it has been found experimentally that for small velocities, the retarding or *damping force* may be approximated by a quantity proportional to the instantaneous velocity of the mass and oppositely directed.   We shall be interested primarily in such damping forces.   Thus, if the mass is moving in the positive x direction with instantaneous velocity dx/dt, the damping force, F_K, is given by

$$F_K = -K \frac{dx}{dt}, \tag{5.2.5}$$

where $K > 0$ is called the *damping constant*.   This constant is determined experimentally for the medium in which the motion takes place.   (Some common damping forces, such as one proportional to the cube of the velocity, will give rise to nonlinear differential equations.)

In addition to a damping force, we may have *impressed forces* acting vertically upon the mass.   One such force may be realized if the mass m is given a small shock in the vertical direction every time the mass reaches its lowest position.   Another such force may be realized by giving to the support holding the spring a small vertical prescribed motion, such as a periodic motion.   Still another such force may be due to the presence of a magnetic field.   We shall be interested primarily in impressed forces that are functions only of the independent variable t, and we shall denote such forces by $F(t)$.

If the displaced mass in the positive x direction is now acted upon by the gravitational force, the spring force, the damping force, and the impressed force, then in view of Newton's Second Law of Motion, *the differential equation of the motion of the mass on a spring at time t is given by*

$$m \frac{d^2x}{dt^2} = -K \frac{dx}{dt} - kx + F(t), \tag{5.2.6}$$

or

$$m \frac{d^2x}{dt^2} + K \frac{dx}{dt} + kx = F(t). \tag{5.2.7}$$

Equation (5.2.7) is called the equation of *forced vibrations* (or *forced motion*).   A mechanical system governed by Equation (5.2.7) is said to be *vibrating freely* if $F(t) = 0$ for all t.   In this case the mechanical system is not subjected to any impressed forces.   If $K = 0$ in Equation (5.2.7), the vibration (or motion) is called *undamped*; otherwise it is called *damped*.

5.3. Free Undamped Vibrations (or Motion)

Note that Equation (5.2.7) reduces to Equation (5.2.4) if $K = 0$ (no damping force) and $F(t) = 0$ for all t (no external force).   We shall now determine the equation of the displacement x of the mass m from its equilibrium position as a function of time, when the system is subjected to no damping force and no external force.   To this end, suppose the mass is initially displaced a distance x_0 from its equilibrium position and then released from this point with an initial velocity v_0.   We are thus led to consider the following initial-value problem.

$$m \frac{d^2x}{dt^2} + kx = 0, \tag{5.3.1}$$

$$x(0) = x_0, \tag{5.3.2}$$

$$x'(0) = v_0. \tag{5.3.3}$$

The general solution of Equation (5.3.1) is given by [*see* Theorem 4.6.2]

$$x = c_1 \cos \omega t + c_2 \sin \omega t, \tag{5.3.4}$$

where c_1 and c_2 are arbitrary constants and

$$\omega = \sqrt{\frac{k}{m}}. \tag{5.3.5}$$

Clearly, $\omega > 0$.   In view of Remark 3.4.4, we may write (5.3.4) in the following equivalent form

$$x = C \cos (\omega t - \phi), \tag{5.3.6}$$

where $C = \sqrt{c_1^2 + c_2^2} > 0$ and the angle ϕ is determined by the equations

$$\sin \phi = \frac{c_2}{C} \quad \text{and} \quad \cos \phi = \frac{c_1}{C}. \tag{5.3.7}$$

We shall now express the constants C and ϕ in terms of x_0 and v_0.   Differentiating (5.3.6) with respect to t, we obtain

$$x' = -C\omega \sin (\omega t - \phi). \tag{5.3.8}$$

Utilizing (5.3.2), (5.3.3), (5.3.6), and (5.3.8), we have

$$x_0 = x(0) = C \cos (-\phi) = C \cos \phi, \tag{5.3.9}$$

and

$$v_0 = x'(0) = -C\omega \sin (-\phi) = C\omega \sin \phi. \tag{5.3.10}$$

Eliminating ϕ from (5.3.9) and (5.3.10), we find that

$$C = \sqrt{x_0^2 + \frac{v_0^2}{\omega^2}}. \tag{5.3.11}$$

Clearly, $C > 0$.   The angle ϕ is now determined by the equations

$$\sin \phi = \frac{v_0}{\omega C} \quad \text{and} \quad \cos \phi = \frac{x_0}{C}. \tag{5.3.12}$$

SUMMARY.   *The displacement x of the mass m from the equilibrium position as a function of the time t is given by*

$$x = C \cos (\omega t - \phi), \tag{5.3.13}$$

where the constants ω, C, and ϕ are given by (5.3.5), (5.3.11), and (5.3.12), respectively.

The constant C is called the *amplitude of the vibration* (or *the amplitude of the motion*).   It represents the maximum value of the displacement x of the mass m measured from the position of equilibrium.   Since the motion is periodic, clearly the mass oscillates between $x = C$ and $x = -C$.   In view of Equation (5.3.13), $x = C$ if and only if $\cos (\omega t - \phi) = 1$, $t > 0$.   Thus, $\omega t - \phi = \pm 2n\pi$, $n = 0, 1, 2, \ldots, t > 0$.   Hence, the maximum (positive) displacements occur if and only if

$$t = \frac{1}{\omega} (\phi \pm 2n\pi) > 0, \quad n = 0, 1, 2, \ldots. \tag{5.3.14}$$

Similarly, when $x = -C$, we see that the maximum (negative) displacements occur if and only if

$$t = \frac{1}{\omega}[\phi \pm (2n + 1)\pi] > 0, \quad n = 0, 1, 2, \ldots . \qquad (5.3.15)$$

We may define the *period of vibration* (or *of motion*) to be the time interval between two successive maxima (or minima). Denoting the period of motion by T, we then have in view of (5.3.14) [or (5.3.15)] and (5.3.5) that

$$T = \frac{2\pi}{\omega} = 2\pi\sqrt{\frac{m}{k}}. \qquad (5.3.16)$$

Note that the period T is the time required (for the motion of the mass) to complete one oscillation (cycle). We define the *frequency* v to be the reciprocal of the period T. Thus,

$$v = \frac{1}{T} = \frac{\omega}{2\pi} = \frac{1}{2\pi}\sqrt{\frac{k}{m}}. \qquad (5.3.17)$$

Note that the frequency v is the number of complete oscillations (cycles) per second. The frequency v of a free undamped motion is also called the *natural frequency* of the system. The constant ω is called the *angular frequency* (or *circular angle*). Also, the constant ω may be interpreted as the angular velocity of a particle moving on the circumference of a circle. Thus, ω is expressed in radians per second. We shall also call the constant ω the natural (undamped) frequency of the motion. The value of ϕ [*see* (5.3.12)] in the interval $-\pi < \phi \leq \pi$ is called the *phase angle* (or *phase constant*) of the motion. The graph of the function x given by (5.3.13) is shown in Figure 5.3.1.

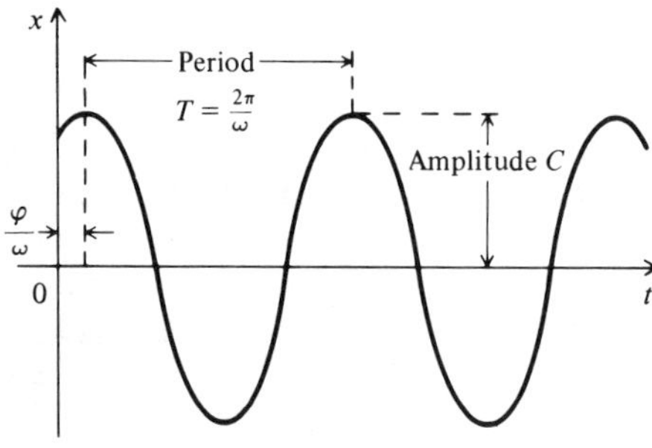

Figure 5.3.1.

Remark 5.3.1.   In view of (5.3.16), we see that the period of a simple harmonic motion is independent of the amplitude C. If the motion is not simple harmonic, the period may depend upon its amplitude. [*See* Exercise 5.3.11.]

Remark 5.3.2.   From Formula (3.5.4), we have

$$W = mg \quad \text{or} \quad m = \frac{W}{g}, \tag{5.3.18}$$

where W is the weight and the acceleration constant g, due to gravity, has the approximate values given in Remark 3.5.3. For computational convenience, we shall take $g = 32$ ft/sec^2, or $g = 980$ cm/sec^2. This convention will also be followed in the exercises. When Equation (5.3.1) is written in the form

$$\frac{W}{g}\frac{d^2x}{dt^2} + kx = 0, \tag{5.3.19}$$

the constant ω given in (5.3.5) becomes

$$\omega = \sqrt{\frac{kg}{W}}. \tag{5.3.20}$$

Remark 5.3.3.   We shall make the following observations concerning some of the possible initial conditions that may be associated with the differential Equation (5.3.1). If the mass m (or weight W) is initially displaced downward a distance x_0 from its equilibrium position and released from that point $(t = 0)$ with an initial velocity v_0 directed upward, then the initial conditions are given by

$$x(0) = x_0, \qquad x'(0) = v_0 < 0. \tag{5.3.21}$$

If in the above statement, we replace "displaced downward a distance x_0" by "displaced upward a distance x_0," then the initial conditions are given by

$$x(0) = x_0 < 0, \qquad x'(0) = v_0 < 0. \tag{5.3.22}$$

The reader should formulate other possible initial conditions that may be associated with Equation (5.3.1). Also, it should be mentioned that the units of the initial conditions should be consistent with the units in the given differential equations. This statement also applies when determining the spring constant k.

Remark 5.3.4.   In the examples and exercises of this section and Sections 5.4–5.6, unless explicitly stated to the contrary, we shall assume that the spring hangs in its medium in a vertical position with its upper end rigidly attached to a beam or support. Also, in the examples and exercises of this section, when a mass (or weight) is attached to the lower end of the spring, the medium offers no resistance to the mass, and there are no external forces acting on the mass.

Example 5.3.1.   A weight of 16 lb stretches a spring 3.84 inches.   After reaching the equilibrium position, the weight is pulled downward 6 inches and released at time $t = 0$ with an initial velocity of 2 ft/sec, directed downward.   (a) Find the resulting equation of the motion at time t.   (b) Find the amplitude, period, and frequency of the motion.   (c) Find the position, velocity, and acceleration of the weight 1/2 sec after it has been released. (d) Sketch the graph of the function describing the motion of the weight.

SOLUTION.   The differential equation of the motion of the weight is given by

$$\frac{W}{g}\frac{d^2x}{dt^2} + kx = 0. \tag{5.3.23}$$

To find the spring constant k, we utilize Hooke's Law: $f = ks$.   Since a weight of 16 lb stretches the spring 3.84 inches or 0.32 feet, we have

$$16 = 0.32\, k \quad \text{or} \quad k = 50 \text{ lb/ft.}$$

Since $W = 16$ lb and $g = 32$ ft/sec^2, $W/g = 1/2$ slug.   Thus, (5.3.23) becomes, upon substituting these values,

$$\frac{d^2x}{dt^2} + 100x = 0. \tag{5.3.24}$$

The initial conditions associated with Equation (5.3.24) are

$$x(0) = \tfrac{1}{2} \text{ ft,} \qquad x'(0) = 2 \text{ ft/sec.} \tag{5.3.25}$$

The solution of the above initial-value problem is given by (5.3.13).   Thus,

$$x = C \cos(\omega t - \phi), \tag{5.3.26}$$

where the constants ω, C, and ϕ are determined as follows.

$$\omega = \sqrt{\frac{kg}{W}} = \sqrt{\frac{(50)(32)}{16}} = 10 \text{ rad/sec,}$$

$$C = \sqrt{x_0^2 + \frac{v_0^2}{\omega^2}} = \sqrt{\left(\frac{1}{2}\right)^2 + \frac{(2)^2}{(10)^2}} = \frac{\sqrt{29}}{10} \approx 0.539 \text{ ft,}$$

$$\sin \phi = \frac{v_0}{\omega C} = \frac{2}{\sqrt{29}} \quad \text{and} \quad \cos \phi = \frac{x_0}{C} = \frac{5}{\sqrt{29}},$$

from which we obtain $\phi \approx 21.8° \approx 0.38$ radian.   Substituting these values into (5.3.26), we see that:

(a) The equation of the motion at time t is given approximately by

$$x = 0.539 \cos (10t - 0.38).\qquad\qquad (5.3.27)$$

(b) The amplitude: $C = \dfrac{\sqrt{29}}{10} \approx 0.539$ ft,

the period: $T = \dfrac{2\pi}{\omega} = \dfrac{2\pi}{10} = \dfrac{\pi}{5}$ sec,

the frequency: $v = \dfrac{1}{T} = \dfrac{5}{\pi}$ oscillations/sec.

(c) 1 radian $= 180°/\pi \approx 57.296°$. Since $\omega = 10$ rad/sec, at $t = 1/2$ sec, $\omega t - 0.38 = 4.62$ radians. Thus, 4.62 radians $\approx 264.7°$. Hence, $\cos 264.7° = -\cos 84.7° \approx -0.0924$ and $\sin 264.7° = -\sin 84.7° \approx -0.9957$. The position, velocity, and acceleration of the weight at $t = 1/2$ sec are thus given approximately by

position: $x = 0.539 \cos (10t - 0.38) = 0.539(-0.0924) = -0.05$ ft,

velocity: $x' = -5.39 \sin (10t - 0.38) = -5.39(-0.9957) = 5.37$ ft/sec,

acceleration: $x'' = -53.9 \cos (10t - 0.38) = -53.9(-0.0924)$
$$= 4.98 \text{ ft/sec}^2.$$

(d) The graph of the displacement versus time is shown in Figure 5.3.2.

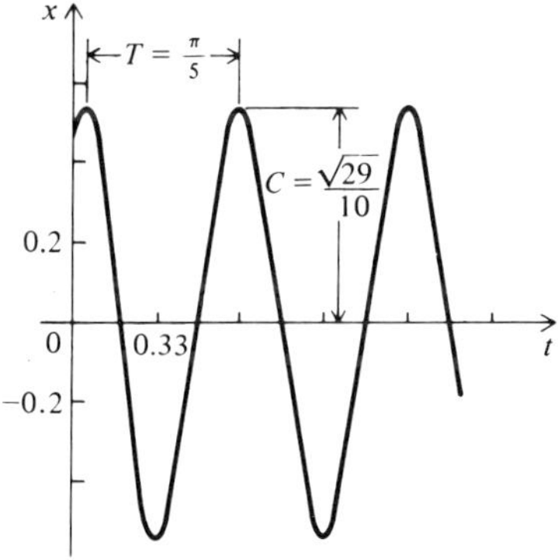

Figure 5.3.2.

Exercises 5.3

1. A weight of 8 lb stretches a spring 6 inches.   After reaching the equilibrium position, the weight is pulled down 6 inches and released from rest at time $t = 0$. (a) Find the equation of motion at time t.   (b) Find the amplitude, period, and frequency of the motion.   (c) Find the position, velocity, and accel-

eration of the weight 1 sec after it has been released.   (d) Graph the function describing the motion.

2. A force of 20 poundals [*see* Table 3.5.1] stretches a spring 4 inches.   A mass of 15 lb is attached to the lower end of the spring.   After reaching the equilibrium position, the mass is pulled up a distance of 6 inches and released from rest at time $t = 0$.   (a) Find the equation of the motion at time t.   (b) Find the amplitude, period, and frequency of the motion.   (c) Find the times when the velocity and acceleration of the mass are equal to zero.   (d) Sketch the graph of the displacement versus time.

3. A force of 200 dynes [*see* Table 3.5.1] stretches a spring 4 cm.   A mass of 50 g is attached to the lower end of the spring.   After reaching the equilibrium position, the mass is struck so as to give it a velocity of 30 cm/sec, directed downward.   (a) Find the velocity and acceleration when the mass is 10 cm below the equilibrium position and moving upward.   (b) Find the times when the mass is 15 cm below the equilibrium position and moving downward.

4. A weight W attached to the lower end of a spring having a constant $k = 50$ lb/ft is oscillating with simple harmonic motion.   If the period of the motion is $\pi/4$ sec, determine the weight W.

5. A weight of W lb stretches a spring 6 inches.   After reaching the equilibrium position, the weight is pulled down a distance of 2 ft and released at time $t = 0$ with an initial velocity of 2 ft/sec, directed downward.   Show that the maximum speed (of the weight) is given by

$$2\sqrt{2g + 1} \text{ ft/sec.}$$

6. A weight W is supported by two springs S_1 and S_2 in parallel.   [*See* Figure 5.3.3.]   Let k_1 and k_2 denote, respectively, the spring constants of S_1 and S_2. It is known that the two springs act as one, with an effective spring constant $k = k_1 + k_2$.   A weight of 20 lb stretches the springs S_1, 11 inches and S_2, 13.2 inches.   After the weight has reached the equilibrium position, it is pulled up a distance of $\sqrt{3}/2$ ft, and released at time $t = 0$ with an initial velocity of 4 ft/sec, directed downward.   (a) Find the equation of motion at time t. (b) Find the amplitude, period, and frequency of the motion.   (c) Find the times when the weight is 6 inches above the equilibrium position and moving upward.   (d) Sketch the graph of the displacement versus time.

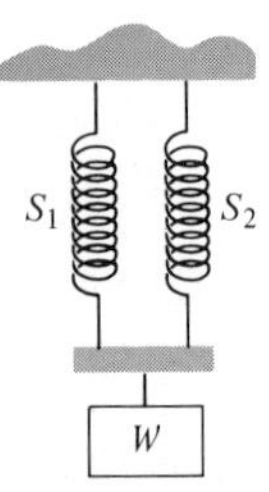

Figure 5.3.3.

7. A weight W is supported by two springs S_1 and S_2 in series.   [*See* Figure 5.3.4.] Let k_1 and k_2 denote, respectively, the spring constants of S_1 and S_2.   It is known that the two springs act as one, with an effective spring constant k given by $1/k = 1/k_1 + 1/k_2$.   Suppose that $k_1 = 30$ g/cm and $k_2 = (240/7)$ g/cm.   A weight of 245 g is attached to the lower end of the spring S_2.   After the weight has reached the equilibrium position, the weight is struck so as to give it a velocity of 80 cm/sec, directed upward.   Find the first time when the weight is 5 cm below the equilibrium position.

8. Two springs S_1 and S_2 with weights W_1 lb and W_2 lb are hung from a vertical beam as shown in Figure 5.3.5.   The weights W_1 and W_2 stretch the springs S_1, 3 inches and S_2, 12 inches, respectively.   After the weights have reached their respective equilibrium positions, they are both pulled down a distance of x_0 ft, and released from rest at time $t = 0$.   Determine the first two times $(t > 0)$ when their velocities will be equal.

9. A weight of W lb stretches a spring 6 inches.   After reaching the equilibrium position, the weight is pulled down x_0 ft.   The weight is then struck (at time $t = 0$) so as to give it an initial velocity of v_0 ft/sec, directed upward, and causing it to reach its uppermost position 4 inches above its equilibrium position in 0.35 sec. Determine the initial (a) displacement, (b) velocity, and (c) acceleration.

10. *The motion of a simple pendulum.*   A simple pendulum consists of a mass m supported by a weightless wire of length L pivoted at the top.   Let θ, in radians, be the angular displacement of the wire from its equilibrium position OQ $(\theta = 0)$ at time t.   [*See* Figure 5.3.6.] θ is positive if the weight $W = mg$ is to the right of OQ; negative if to the left of OQ.   The motion of the bob (the weight hanging from the wire) takes place along the circular arc $\overgroup{QP}$.   Acting on the motion of the bob are two forces, the tension T in the wire and the weight mg of the bob. The component $mg \cos \theta$ is equal in magnitude, but opposite in direction, to T. The component of mg tangent to the path is $mg \sin \theta$.   [*Note:* $F = -mg \sin \theta$ is that force tending to bring back the bob to its equilibrium position.]   If s denotes the length of the arc $\overgroup{QP}$, then $s = L\theta$, and the acceleration along the arc is

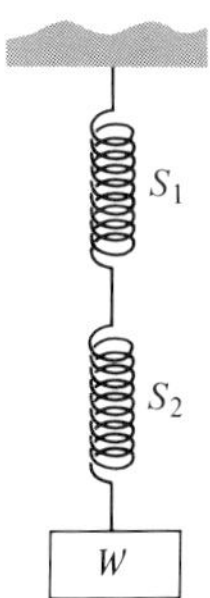

Figure 5.3.4.

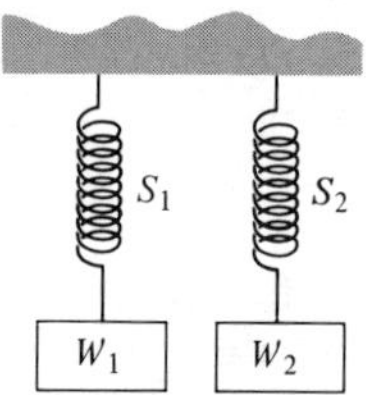

Figure 5.3.5.

$$\frac{d^2 s}{dt^2} = L\,\frac{d^2\theta}{dt^2}.$$

Neglecting air resistance and utilizing Newton's Second Law of Motion, the differential equation of motion is

$$\text{(1)} \qquad\qquad m\,\frac{d^2 s}{dt^2} = -mg\,\sin\,\theta,$$

or

$$\text{(2)} \qquad\qquad L\,\frac{d^2\theta}{dt^2} + g\,\sin\,\theta = 0.$$

Equation (2) is a second-order nonlinear homogeneous differential equation. In view of Equation (1), the mass does not execute a simple harmonic motion. (Why?)

For $0 \le \theta \le 0.08725$ or $0° \le \theta \le 5°$, $\quad |\sin\,\theta - \theta| \le 0.0001$. If we restrict θ to this range, we may replace $\sin\,\theta$ by θ in Equation (2), and thus the motion is very closely represented by the equation

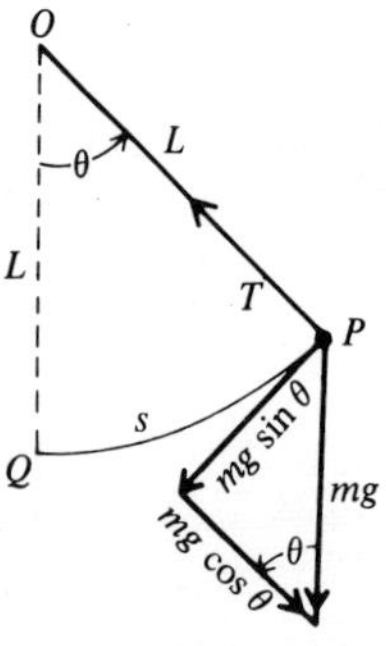

Figure 5.3.6.

(3)
$$\frac{d^2\theta}{dt^2} + \frac{g}{L}\theta = 0.$$

Equation (3) is the differential equation of a mass executing simple harmonic motion.   If we denote by ω the *angular velocity*, that is,

(4)
$$\omega = \frac{d\theta}{dt},$$

then, since $s = L\theta$, the *linear velocity* v or simply the *velocity* v, is given by

(5)
$$v = \frac{ds}{dt} = \frac{d}{dt}(L\theta) = L\frac{d\theta}{dt} = L\omega.$$

The angular velocity is positive when the weight W moves counterclockwise; negative if it moves clockwise.

(a) If the simple pendulum of length L is displaced θ_0 rad from its equilibrium position ($\theta = 0$) and released at time $t = 0$ with an angular velocity of ω_0 rad/sec directed toward ($\theta = 0$), show that the equation of motion at time t is given by

(6)
$$\theta = C\cos\left(\sqrt{\frac{g}{L}}\,t - \phi\right),$$

where the amplitude C is given by

(7)
$$C = \sqrt{\theta_0^2 + \frac{L}{g}\omega_0^2}\ \text{rad},$$

the phase angle ϕ (radians) is determined by the equations

(8)
$$\sin\phi = \frac{\omega_0\sqrt{\dfrac{L}{g}}}{C},\qquad \cos\phi = \frac{\theta_0}{C},\qquad -\pi < \phi \leqq \pi,$$

and the period T and the frequency ν of the motion are given by

(9)
$$T = 2\pi\sqrt{\frac{L}{g}}\ \text{sec},\qquad \nu = \frac{1}{2\pi}\sqrt{\frac{g}{L}}\ \text{oscillation(s)/sec}.$$

(b) A simple pendulum of length 2 ft is displaced $\theta_0 = \frac{1}{16}$ rad from the equilibrium position ($\theta = 0$), and released at time $t = 0$ with an angular velocity of $\frac{1}{4}$ rad/sec directed toward the equilibrium position.   Find the equation of motion at time t, amplitude, period, frequency, and phase angle.

(c) A simple pendulum of length 4 ft swings with amplitude $\frac{1}{16}$ rad.   Find the period of the pendulum, and its angular velocity as it crosses the equilibrium position ($\theta = 0$).

(d) A simple pendulum whose period is 1.25 sec is displaced $4°$ from the equilibrium position ($\theta = 0$) and released from rest at time $t = 0$.   Determine its linear velocity when it crosses the equilibrium position.

(e) A simple pendulum 2 ft long is released from the position $\theta = \frac{1}{16}$ rad at time $t = 0$. Find θ when $t = \pi/12$ sec, and also find the times when the pendulum crosses the equilibrium position ($\theta = 0$).

(f) A clock has a 6-inch pendulum. The clock ticks once for each time the pendulum completes a swing, returning to its original position. How many times does the clock tick in 70 sec?

(g) A simple pendulum of length L is displaced $\theta_0 = 0.02$ rad from the equilibrium position ($\theta = 0$) and released from rest at time $t = 0$. Determine what fraction of its period has elapsed when $\theta = 0.01732$ rad.

11. *The period of motion which is not simple harmonic motion.*   The differential equation of motion of a simple pendulum is given by Equation (2) of Exercise 10, namely,

$$\text{(1)} \qquad \frac{d^2\theta}{dt^2} + \frac{g}{L} \sin \theta = 0.$$

Multiplying (1) by $2(d\theta/dt)$ and integrating, show that

$$\text{(2)} \qquad \left(\frac{d\theta}{dt}\right)^2 = \frac{2g}{L} \cos \theta + A,$$

where A is a constant of integration which we shall presently determine. Suppose that the simple pendulum is displaced α (radians) from its equilibrium position ($\theta = 0$) and released from rest at time $t = 0$.   Then

$$\text{(3)} \qquad \begin{cases} \theta = \alpha & \text{when } t = 0, \\[2mm] \dfrac{d\theta}{dt} = 0 & \text{when } t = 0. \end{cases}$$

Note that α is the amplitude of vibration. With the initial conditions given in (3), we find $A = -(2g/L) \cos \alpha$, and (2) becomes

$$\text{(4)} \qquad \left(\frac{d\theta}{dt}\right)^2 = \frac{2g}{L} (\cos \theta - \cos \alpha).$$

The angular velocity is given by $\omega = d\theta/dt$; and since the linear velocity $v = L(d\theta/dt)$, the velocity in the path at the lowest point is

$$\text{(5)} \qquad \sqrt{\frac{2g}{L} (\cos \theta - \cos \alpha)} \cdot L \Bigg|_{\theta = 0} = \sqrt{2gL(1 - \cos \alpha)}.$$

This is the same velocity that would have been acquired if the bob had fallen freely under the force of gravity through the same difference in level, for $v = \sqrt{2gh}$ and $h = L(1 - \cos \alpha)$.   Integrate (4) and show that

$$\text{(6)} \qquad t = \sqrt{\frac{L}{2g}} \int_{\theta_0}^{\theta_1} \frac{d\theta}{\sqrt{\cos \theta - \cos \alpha}},$$

which gives the formula for determining the time required for the bob to move from the initial position to any other. If the lowest position of the bob is chosen as the initial position, then $\theta = 0$ when $t = 0$, and Equation (6) becomes

$$(7) \qquad t = \sqrt{\frac{L}{2g}} \int_0^{\theta_1} \frac{d\theta}{\sqrt{\cos\theta - \cos\alpha}} \, ,$$

where $0 \leqq \theta_1 \leqq \alpha$. Since $\cos\theta = 1 - 2\sin^2(\theta/2)$, show that (7) may be written as

$$(8) \qquad t = \sqrt{\frac{L}{2g}} \int_0^{\theta_1} \frac{d\theta}{\sqrt{2\left(\sin^2\dfrac{\alpha}{2} - \sin^2\dfrac{\theta}{2}\right)}} \, .$$

Letting

$$(9) \qquad \sin\frac{\theta}{2} = \sin\frac{\alpha}{2}\sin\phi,$$

where ϕ is the new variable, show that

$$(10) \qquad d\theta = \frac{2\sin\dfrac{\alpha}{2}\cos\phi \, d\phi}{\sqrt{1 - \sin^2\dfrac{\alpha}{2}\sin^2\phi}} \, .$$

Show that (8) may now be written as

$$(11) \qquad t = \sqrt{\frac{L}{g}} \int_0^{\phi_1} \frac{d\phi}{\sqrt{1 - \sin^2\dfrac{\alpha}{2}\sin^2\phi}} \, .$$

If the time involved is the time required for the completion of one-quarter of the vibration, then $\theta_1 = \alpha$ and hence $\phi_1 = \pi/2$. (Why?) The entire period T is then given by

$$(12) \qquad T = 4\sqrt{\frac{L}{g}} \int_0^{\pi/2} \frac{d\phi}{\sqrt{1 - k^2\sin^2\phi}} \, , \qquad 0 < k < 1,$$

where

$$(13) \qquad k^2 = \sin^2\frac{\alpha}{2}.$$

The integral in the right-hand member of (12) is called an *elliptic integral of the first kind* and it is usually denoted by $K(k)$. Thus,

$$(14) \qquad K(k) = \int_0^{\pi/2} \frac{d\phi}{\sqrt{1 - k^2\sin^2\phi}} \, , \qquad 0 < k < 1,$$

and (12) becomes

$$(15) \qquad T = 4\sqrt{\frac{L}{g}} \, K(k).$$

The elliptic integral $K(k)$ has been tabulated for certain values of k.   [*See* [32], p. 133.]   Some values of $K(k)$ are

$$K\left(\frac{1}{2}\right) = 1.6858, \quad K\left(\frac{\sqrt{2}}{2}\right) = 1.8541, \quad \text{and} \quad K\left(\frac{\sqrt{3}}{2}\right) = 2.1565.$$

Expanding $(1 - k^2 \sin^2 \phi)^{-1/2}$ by the binomial theorem, show that (12) may be written as

$$(16) \qquad T = 4\sqrt{\frac{L}{g}} \int_0^{\pi/2} \left(1 + \frac{1}{2} k^2 \sin^2 \phi + \frac{3}{8} k^4 \sin^4 \phi + \cdots\right) d\phi.$$

Utilizing a formula due to Wallis, namely [*see* Exercise 7.6.12],

$$(17) \qquad \int_0^{\pi/2} \sin^{2n} \phi \, d\phi = \frac{(2n - 1)(2n - 3) \cdots 1}{(2n)(2n - 2) \cdots 2} \cdot \frac{\pi}{2}, \qquad n = 1, 2, \ldots,$$

and assuming that term by term integration is valid, show that (16) becomes

$$(18) \qquad T = 2\pi \sqrt{\frac{L}{g}} \left(1 + \frac{1}{4} k^2 + \frac{9}{64} k^4 + \cdots\right).$$

It may be noted that the period T is a function of the amplitude, which was not the case of simple harmonic motion.   Also, when k (or α) is sufficiently small $(0 \leq \alpha \leq 0.08725)$, Equation (18) may be replaced by

$$(19) \qquad T = 2\pi \sqrt{\frac{L}{g}},$$

which is now the period of a simple harmonic motion.   [*See* (9) of Exercise 10.]

12. A seconds pendulum is one taking 1 sec to swing from one end of its arc to the other, so that its period is 2 sec.   (a) Find the length of a seconds pendulum that swings through an angle of 120°.   (b) Find the length of a seconds pendulum that swings through an angle of 5°.

5.4. Free Damped Vibrations

If a damping (or friction) force is present and $F(t) = 0$ for all t, then Equation (5.2.7) becomes

$$m \frac{d^2x}{dt^2} + K \frac{dx}{dt} + kx = 0, \tag{5.4.1}$$

where $K > 0$ is the damping (or friction) constant and $k > 0$ is the spring constant.   Equation (5.4.1) describes the motion of free damped vibrations. For convenience we shall write Equation (5.4.1) as

$$\frac{d^2x}{dt^2} + 2\alpha \frac{dx}{dt} + \beta^2 x = 0, \tag{5.4.2}$$

where

$$2\alpha = \frac{K}{m} > 0, \qquad \beta^2 = \frac{k}{m} > 0. \tag{5.4.3}$$

The auxiliary equation of (5.4.2) is

$$m^2 + 2\alpha m + \beta^2 = 0. \tag{5.4.4}$$

[Clearly, in Equation (5.4.4) the letter m does not represent mass. (*See* Section 4.6.)]   The roots of Equation (5.4.4) are

$$m_1 = -\alpha + \sqrt{\alpha^2 - \beta^2}, \qquad m_2 = -\alpha - \sqrt{\alpha^2 - \beta^2}. \tag{5.4.5}$$

The form of the solutions of Equation (5.4.2) will depend on the discriminant

$$D \equiv 4(\alpha^2 - \beta^2) \tag{5.4.6}$$

of Equation (5.4.4).   If $D > 0$, then m_1 and m_2 are real and $m_1 \neq m_2$.   If $D = 0$, then m_1 and m_2 are real and $m_1 = m_2$.   If $D < 0$, then m_1 and m_2 are imaginary and m_2 is the complex conjugate of m_1.

Case 1. $D > 0$ *or* $\alpha^2 > \beta^2$.   *Overdamped case.*   For this case, the general solution of Equation (5.4.2) is given by [*see* Theorem 4.6.1(a)]

$$x = c_1 e^{m_1 t} + c_2 e^{m_2 t}, \tag{5.4.7}$$

where c_1 and c_2 are arbitrary constants and m_1 and m_2 are given in (5.4.5). Since $\sqrt{\alpha^2 - \beta^2} > 0$, we see that $m_1 < 0$ and $m_2 < 0$.   Note that $e^{m_1 t} > 0$ and $e^{m_2 t} > 0$ for all t, also $e^{m_1 t} \to 0$ and $e^{m_2 t} \to 0$ as $t \to \infty$.   Observe that these functions are strictly decreasing.   That is, if $0 \leq t_1 < t_2$, then $e^{m_1 t_1} > e^{m_1 t_2}$ and $e^{m_2 t_1} > e^{m_2 t_2}$.   Now, if c_1, c_2 have the same sign and $c_1 \neq 0$, $c_2 \neq 0$, then the graph of the function x given by Equation (5.4.7) cannot cross the t axis. However, if c_1, c_2 have opposite signs and $c_1 \neq 0$, $c_2 \neq 0$, setting $x = 0$ in (5.4.7) and then solving for t, we obtain

$$t = \frac{1}{m_1 - m_2} \ln\left(\frac{-c_2}{c_1}\right) = \frac{1}{2\sqrt{\alpha^2 - \beta^2}} \ln\left(\frac{-c_2}{c_1}\right). \tag{5.4.8}$$

Thus, there exists at most one value of $t > 0$ where the graph of the motion may cross the t axis.   Differentiating (5.4.7), we get

$$\frac{dx}{dt} = m_1 c_1 e^{m_1 t} + m_2 c_2 e^{m_2 t}. \tag{5.4.9}$$

Proceeding as before, we see that there exists at most one value of $t > 0$ such that the function x given in (5.4.7) may attain a relative minimum or a

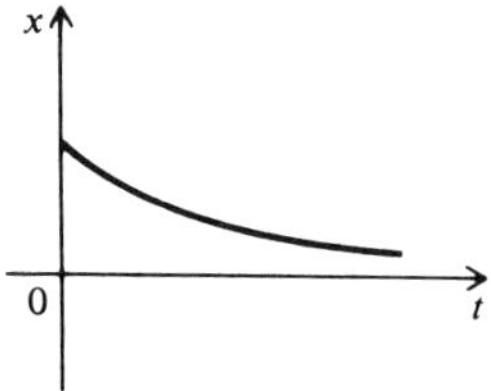

Figure 5.4.1.
$c_1 c_2 > 0$

relative maximum. Hence, the motion given by Equation (5.4.7) does not possess an oscillatory property. Clearly, the motion dies out with increasing t; that is, x is a decaying function of time. Thus, the motion governed by Equation (5.4.7) is transient.

In view of (5.4.3) and (5.4.6) when $D > 0$, Case 1 arises if

$$K > 2\sqrt{mk}. \tag{5.4.10}$$

For example, if a weight W is immersed in a medium possessing a high viscosity, or there is a shock absorber attached to the weight, then, if K is large enough, that is, if

$$K > 2\sqrt{mk} \quad \text{or} \quad K > 2\sqrt{\frac{Wk}{g}}, \tag{5.4.11}$$

the weight W will not oscillate. It will slowly approach its equilibrium position $(x = 0)$. The system in question is overdamped, thus preventing oscillations.

Figures 5.4.1 and 5.4.2 depict two possible motions that may arise for Case 1. The reader should furnish additional possible graphs of the function x given by Equation (5.4.7).

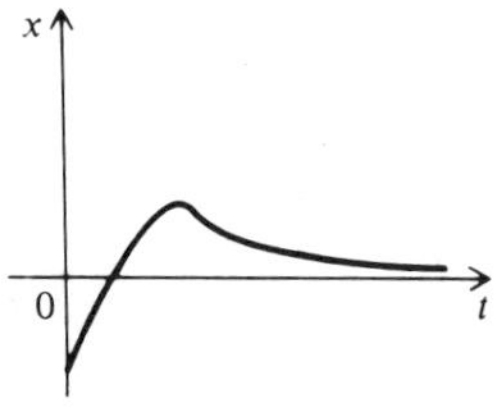

Figure 5.4.2.
$c_1 c_2 < 0$

Case 2. $D = 0$ *or* $\alpha^2 = \beta^2$. *Critically damped case.* For this case, the general solution of Equation 5.4.2 is given by [*see* Theorem 4.6.1(b)]

$$x = (c_1 + c_2 t)e^{-\alpha t}, \tag{5.4.12}$$

where c_1 and c_2 are arbitrary constants and $\alpha = K/2m > 0$. Note that $e^{-\alpha t} > 0$ and $te^{-\alpha t} > 0$ for all $t > 0$. Also, $e^{-\alpha t} \to 0$ and $te^{-\alpha t} \to 0$ as $t \to \infty$. As in Case 1, we see that there exists at most one value of $t > 0$ where the graph of the motion may cross the t axis, and where the function x given by (5.4.12) may attain a relative minimum or a relative maximum. Thus, the motion is nonoscillatory, and dies out as t becomes sufficiently large. Hence, the motion governed by Equation (5.4.12) is transient, and its graphical representation is similar to Case 1. [*See* Figure 5.4.3 for an example.]

In view of (5.4.3) and (5.4.6) when $D = 0$, Case 2 arises if

$$K = 2\sqrt{mk}. \tag{5.4.13}$$

In this case the damping (or resisting) force is just great enough to prevent oscillations. As we shall see in the forthcoming Case 3, if $K < 2\sqrt{mk}$, then the motion will possess an oscillatory property. For this reason, Case 2 is known as the critically damped case.

Case 3. $D < 0$ *or* $\alpha^2 < \beta^2$. *Oscillatory case.* For this case, the general solution of Equation (5.4.2) is given by [*see* Theorem 4.6.2]

$$x = e^{-\alpha t}(c_1 \cos \sqrt{\beta^2 - \alpha^2}\, t + c_2 \sin \sqrt{\beta^2 - \alpha^2}\, t), \tag{5.4.14}$$

where c_1 and c_2 are arbitrary constants. Equation (5.4.14) can also be written in the following equivalent form [*see* Remark 3.4.4]

$$x = Ce^{-\alpha t} \cos (\sqrt{\beta^2 - \alpha^2}\, t - \phi), \tag{5.4.15}$$

where $C = \sqrt{c_1^2 + c_2^2} > 0$, and the phase angle ϕ is determined by the equations

$$\sin \phi = \frac{c_2}{C}, \quad \cos \phi = \frac{c_1}{C}. \tag{5.4.16}$$

Since Equation (5.4.15) involves the cosine term, the motion is clearly oscillatory. The term $Ce^{-\alpha t}$ in (5.4.15) is called the *damped amplitude* of

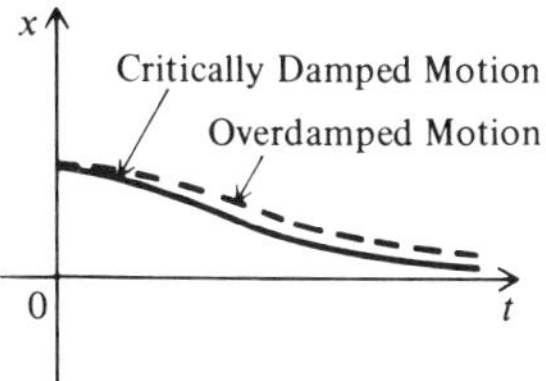

Figure 5.4.3.

motion.   It is also known as the *time-varying amplitude*.   Since $\alpha = K/2m > 0$, we see that $Ce^{-\alpha t} \to 0$ as $t \to \infty$.   Thus, with increasing time, the function x given in (5.4.15) approaches zero, since $|\cos(\sqrt{\beta^2 - \alpha^2}\, t - \phi)| \leqq 1$ for all t. Thus, the mass vibrates with smaller and smaller oscillations about its equilibrium position.   Hence, the motion of the mass is not periodic.   Also, the oscillating mass passes through the equilibrium position ($x = 0$) whenever the cosine term in (5.4.15) is equal to zero.

Even though the motion given in (5.4.15) is no longer periodic, but is oscillatory, we may define the *damped period* of vibration as being the time interval between two successive maximum (or minimum) displacements.   The damped period T is given by

$$T = \frac{2\pi}{\sqrt{\beta^2 - \alpha^2}} = \frac{4\pi m}{\sqrt{4mk - K^2}} \text{ sec.} \tag{5.4.17}$$

The *damped frequency* v of the motion is defined to be the reciprocal of the damped period.   Thus,

$$v = \frac{\sqrt{\beta^2 - \alpha^2}}{2\pi} = \frac{\sqrt{4mk - K^2}}{4\pi m} \text{ cycles/sec.} \tag{5.4.18}$$

The damped period and damped frequency are also known as the *quasi period* and *quasi frequency*.   The exponential term $e^{-\alpha t}$ occurring in (5.4.15) is known as the *damping factor*.   Clearly, it is the damping factor that causes the motion to die out with increasing time.

In view of (5.4.3) and (5.4.6) when $D < 0$, Case 3 occurs if

$$K < 2\sqrt{mk}. \tag{5.4.19}$$

For this case, the damping (or resisting) force is not strong enough to prevent oscillations.   The system in question is said to be underdamped.   The graph of the function x given in (5.4.15) is shown in Figure 5.4.4.

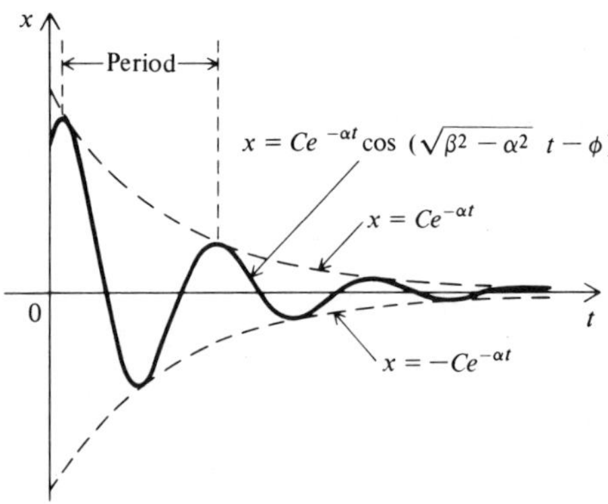

Figure 5.4.4.

Remark 5.4.1. We shall now give an interesting property concerning the motion of the mass given by (5.4.15). We shall show that the ratio of two successive maximum (or minimum) displacements is a constant. For convenience, let us write (5.4.15) as

$$x = Ce^{-\alpha t} \cos (\lambda t - \phi), \tag{5.4.20}$$

where

$$\alpha = \frac{K}{2m} > 0, \quad \beta = \sqrt{\frac{k}{m}} > 0 \quad \text{and} \quad \lambda = \sqrt{\beta^2 - \alpha^2} = \frac{\sqrt{4mk - K^2}}{2m} > 0. \tag{5.4.21}$$

Differentiating (5.4.20) and then setting dx/dt equal to zero, we obtain

$$\alpha \cos (\lambda t - \phi) + \lambda \sin (\lambda t - \phi) = 0 \quad \text{or} \quad \tan (\lambda t - \phi) = \frac{-\alpha}{\lambda}. \tag{5.4.22}$$

Thus, the values of times, t_n, which render the above function x relative maxima or relative minima, are given by

$$t_n = A + \frac{n\pi}{\lambda}, \qquad n = 0, 1, 2, \ldots, \tag{5.4.23}$$

where

$$A = \frac{1}{\lambda} \left[\phi + \arctan \left(\frac{-\alpha}{\lambda} \right) \right]. \tag{5.4.24}$$

Denoting by x_n the relative maximum or relative minimum displacements, we then have

$$x_n = Ce^{-\alpha t_n} \cos (\lambda t_n - \phi) = Ce^{-\alpha t_n} \cos \left[\lambda \left(A + \frac{n\pi}{\lambda} \right) - \phi \right]$$

$$= Ce^{-\alpha t_n} \cos [n\pi + (\lambda A - \phi)] = (-1)^n Ce^{-\alpha t_n} \cos (\lambda A - \phi)$$

$$= (-1)^n Ce^{-\alpha t_n} \cos \left[\arctan \left(\frac{-\alpha}{\lambda} \right) \right] = (-1)^n Ce^{-\alpha t_n} \frac{\lambda}{\sqrt{\alpha^2 + \lambda^2}},$$

$n = 0, 1, 2, \ldots$. Utilizing (5.4.21), we may write the above expression as

$$x_n = (-1)^n Ce^{-\alpha t_n} \sqrt{1 - \frac{K^2}{4mk}}, \qquad n = 0, 1, 2, \ldots . \tag{5.4.25}$$

The relative maximum (positive) displacements are thus given by

$$x_{2n} = Ce^{-\alpha t_{2n}} \sqrt{1 - \frac{K^2}{4mk}}, \tag{5.4.26}$$

where in view of (5.4.23)

$$t_{2n} = A + \frac{2n\pi}{\lambda}, \qquad n = 0, 1, 2, \ldots. \tag{5.4.27}$$

The relative minimum (negative) displacements are given by

$$x_{2n+1} = -Ce^{-\alpha t_{2n+1}}\sqrt{1 - \frac{K^2}{4mk}}, \tag{5.4.28}$$

where in view of (5.4.23)

$$t_{2n+1} = A + \frac{(2n+1)\pi}{\lambda}, \qquad n = 0, 1, 2, \ldots. \tag{5.4.29}$$

From the above equations, we see that

$$\frac{x_{2n+2}}{x_{2n}} = \frac{e^{-\alpha t_{2n+2}}}{e^{-\alpha t_{2n}}} = e^{-\alpha(t_{2n+2}-t_{2n})} = \exp\left(-\frac{2\pi\alpha}{\lambda}\right) = \exp\left(-\frac{2\pi K}{\sqrt{4mk - K^2}}\right),$$

$$\tag{5.4.30}$$

and

$$\frac{x_{2n+3}}{x_{2n+1}} = \exp\left(-\frac{2\pi K}{\sqrt{4mk - K^2}}\right), \tag{5.4.31}$$

$n = 0, 1, 2, \ldots.$ Thus, the ratio of two consecutive maximum or minimum displacements is a constant.   Note that

$$-\frac{2\pi K}{\sqrt{4mk - K^2}} = \ln\left(\frac{x_{2n+2}}{x_{2n}}\right), \qquad -\frac{2\pi K}{\sqrt{4mk - K^2}} = \ln\left(\frac{x_{2n+3}}{x_{2n+1}}\right),$$

$$n = 0, 1, 2, \ldots. \tag{5.4.32}$$

The quantity $2\pi K/\sqrt{4mk - K^2}$ is known as the *logarithmic decrement*.   In view of (5.4.17), we see that the logarithmic decrement is the decrease in the logarithm of the damped amplitude

$$Ce^{-\alpha t} = Ce^{-(K/2m)t}$$

over a time interval of one period.

Remark 5.4.2.   From (5.4.20) we see that $x = Ce^{-\alpha t}$ only when $\cos(\lambda t - \phi) = 1$.   That is, the times, τ_n, at which the function x given in (5.4.20) intersects the curve $x = Ce^{-\alpha t}$ occur only when

$$\tau_n = \frac{1}{\lambda}(\phi + 2n\pi), \qquad n = 0, 1, 2, \ldots. \tag{5.4.33}$$

From the discussion of Remark 5.4.1, the times, t_n, when the function x attains maximum (positive) displacements occur only when [*see* (5.4.27) and (5.4.24)]

$$t_{2n} = A + \frac{2n\pi}{\lambda} = \frac{1}{\lambda}(\phi + 2n\pi) + \frac{1}{\lambda} \arctan\left(\frac{-\alpha}{\lambda}\right), \qquad n = 0, 1, 2, \ldots .$$

$$(5.4.34)$$

Thus,

$$t_{2n} - \tau_n = \frac{1}{\lambda}\arctan\left(\frac{-\alpha}{\lambda}\right), \qquad n = 0, 1, 2, \ldots . \qquad (5.4.35)$$

Hence, the relative maxima of the function x occur shortly before the curve becomes tangent to the bounding curve $x = Ce^{-\alpha t}$. Similarly, $x = -Ce^{-\alpha t}$ only when $\cos(\lambda t - \phi) = -1$. That is, the times, σ_n, at which the function x given in (5.4.20) intersects the curve $x = -Ce^{-\alpha t}$ occur only when

$$\sigma_n = \frac{\phi + (2n+1)\pi}{\lambda}, \qquad n = 0, 1, 2, \ldots . \qquad (5.4.36)$$

We leave it for the reader [*see* Exercise 5.4.9] to show that

$$t_{2n+1} - \sigma_n = t_{2n} - \tau_n. \qquad (5.4.37)$$

What is the physical interpretation of the result given in (5.4.37)?

Example 5.4.1. A weight of 4 lb stretches a spring 4/5 ft. After reaching the equilibrium position, the weight is pulled down 4 in and released from rest at time $t = 0$. Assume that the weight is acted upon by a damping force which in pounds is numerically equal to 0.5 dx/dt, where dx/dt is the instantaneous velocity in ft/sec.

(a) Find the equation of motion at time t.
(b) Find the damped: amplitude, period, and frequency of the motion.
(c) Sketch the graph of the function describing the motion of the weight.

SOLUTION. The differential equation of the motion of the weight is given by

$$\frac{W}{g}\frac{d^2x}{dt^2} + K\frac{dx}{dt} + kx = 0. \qquad (5.4.38)$$

To determine the spring constant k, we utilize Hooke's Law: $f = ks$. Since a weight of 4 lb stretches a spring 4/5 ft, we have

$$4 = \tfrac{4}{5}k \quad \text{or} \quad k = 5 \text{ lb/ft.}$$

The damping constant $K = 0.5$. Since $W = 4$ lb and $g = 32$ ft/sec^2, $W/g = 1/8$ slug. Substituting these values in Equation (5.4.38), we obtain upon simplification

$$\frac{d^2x}{dt^2} + 4\frac{dx}{dt} + 40x = 0. \tag{5.4.39}$$

Comparing (5.4.39) with (5.4.2), we see that

$$2\alpha = 4 \quad \text{and} \quad \beta^2 = 40. \tag{5.4.40}$$

Thus, $\alpha^2 < \beta^2$ and Equation (5.4.39) gives rise to the oscillatory case. Hence, the general solution of Equation (5.4.39) is given by [*see* (5.4.14)]

$$x = e^{-2t}(c_1 \cos 6t + c_2 \sin 6t), \tag{5.4.41}$$

where c_1 and c_2 are arbitrary constants. Differentiating (5.4.41), we get

$$\frac{dx}{dt} = -2e^{-2t}[(c_1 - 3c_2) \cos 6t + (3c_1 + c_2) \sin 6t]. \tag{5.4.42}$$

To determine c_1 and c_2 we shall utilize the initial conditions of our problem, namely,

$$x(0) = \tfrac{1}{3}, \qquad x'(0) = 0. \tag{5.4.43}$$

Thus, in view of (5.4.41) through (5.4.43), we have

$$\tfrac{1}{3} = x(0) = e^0(c_1 \cos 0 + c_2 \sin 0) = c_1 \tag{5.4.44}$$

and

$$0 = x'(0) = -2e^0[(c_1 - 3c_2) \cos 0 + (3c_1 + c_2) \sin 0] = -2(c_1 - 3c_2).$$
$$\tag{5.4.45}$$

From (5.4.44) and (5.4.45), we find $c_1 = \tfrac{1}{3}$ and $c_2 = \tfrac{1}{9}$. Substituting these values in (5.4.41), we see that:

(a) The equation of motion at time t is

$$x = \tfrac{1}{9}e^{-2t}(3 \cos 6t + \sin 6t). \tag{5.4.46}$$

We may also write Equation (5.4.46) in the following equivalent form,

$$x = \frac{\sqrt{10}}{9}e^{-2t} \cos (6t - \phi), \tag{5.4.47}$$

where the phase angle ϕ is determined by the equations

$$\sin \phi = \frac{1}{\sqrt{10}}, \qquad \cos \phi = \frac{3}{\sqrt{10}}. \tag{5.4.48}$$

Thus, $\phi \approx 18.43° \approx 0.322$ radian.   Hence, the equation of motion at time t is given approximately by

$$x = 0.35e^{-2t}\cos(6t - 0.322). \qquad (5.4.49)$$

(b) The damped: amplitude $Ce^{-\alpha t} = \dfrac{\sqrt{10}}{9}e^{-2t} \approx 0.35e^{-2t}$ ft,

$$\text{period } T = \frac{2\pi}{\sqrt{\beta^2 - \alpha^2}} = \frac{2\pi}{6} = \frac{\pi}{3} \text{ sec,}$$

$$\text{frequency } v = \frac{1}{T} = \frac{3}{\pi} \text{ oscillation/sec.}$$

(c) The graph of the displacement versus time is shown in Figure 5.4.5.

Remark 5.4.3.   We leave it as an exercise for the reader to show that (5.4.49) may also be obtained by utilizing (5.4.15) and (5.4.43).

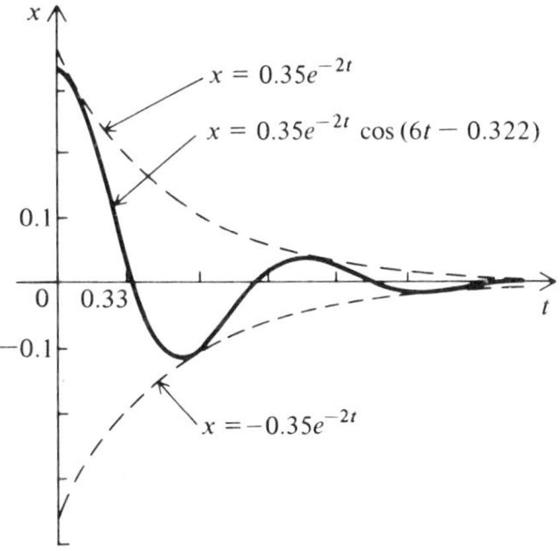

Figure 5.4.5.

Exercises 5.4

1. The damped motion of a weight suspended from a spring has a period of 1 sec.   The damped amplitude of motion decreases 50 percent in 5 sec.   Find the acceleration of the weight when it is 4 inches below its equilibrium position and is moving upwards with a speed of $\frac{1}{4}$ ft/sec.
2. A 10 lb weight, of specific gravity 2, stretches a spring 4 inches when immersed in water.   Because of resistance proportional to the velocity, the damped period of vibration is 0.02 sec longer than it would be without damping.   By what percentage will the damping factor be decreased in 1 sec?

In Exercises 3 through 7, dx/dt denotes the instantaneous velocity in ft/sec.

3. A weight of 10 lb stretches a spring 1.5 inches.   After reaching the equilibrium position, the weight is pulled down 3 inches and released from rest at time $t = 0$. Assume that the weight is acted upon by a damping force which in pounds is numerically equal to $5\sqrt{2}\, dx/dt$.   (a) Find the equation of motion at time t. (b) Find the damped amplitude, damped period, and damped frequency of motion.   (c) Find the position, velocity, and acceleration of the weight $\pi/16\sqrt{2}$ sec after it has been released.   (d) Find the time when the damped amplitude (and, hence, the damping factor) is 50 percent of its initial value. (e) Sketch the graph of the function describing the motion of the weight.

4. A weight of 8 lb stretches a spring $\frac{1}{2}$ ft.   After reaching the equilibrium position, the weight is pulled down $\frac{1}{4}$ ft and released at time $t = 0$ with an initial velocity of 2 ft/sec directed downward.   Assume that the weight is acted upon by a damping force which in pounds is numerically equal to $4\, dx/dt$.   (a) Is the motion critically damped?   (b) Determine the equation of motion at time t. (c) Find the maximum displacement of the weight from the equilibrium position. (d) Find the time required to displace the weight one-half of its maximum displacement.   (e) Find the time $(0 < t < \infty)$ when the acceleration of the weight is zero.   (f) Sketch the graph of the function describing the motion of the weight.

5. A mass of one slug is attached to the lower end of a spring whose constant is 3 lb/ft.   After reaching the equilibrium position, the mass is pulled down 6 inches and released at time $t = 0$ with an initial velocity of 2 ft/sec directed downward. Assume that the mass is acted upon by a damping force which in pounds is numerically equal to $4\, dx/dt$.   (a) Is the motion overdamped?   (b) Determine the equation of motion at time t.   (c) Find the maximum displacement of the mass from the equilibrium position.   (d) Find the time $(0 < t < \infty)$ when the acceleration of the mass is zero.   (e) Find the times when the mass is 0.6 ft below the equilibrium position.   (f) Sketch the graph of the function describing the motion of the mass.

6. A mass of one slug is attached to the lower end of a spring whose constant is 6 lb/ft.   After reaching the equilibrium position, the mass is pulled down 1 ft and released at time $t = 0$ with an initial velocity of v_0 ft/sec directed upward. Assume that the mass is acted upon by a damping force which in pounds is numerically equal to $5\, dx/dt$.   (a) Find the value of v_0 which will bring the mass to the equilibrium position in $\frac{1}{2}$ sec.   (b) Find the algebraically smallest value of v_0 which will prevent the mass from reaching its equilibrium position in any finite time.

7. A weight of 8 lb stretches a spring 4 inches.   The weight is set vibrating and is acted upon by a damping force which is proportional to dx/dt.   If the resulting motion has a damped period of $\pi/4$ sec, find the time required for the damping factor to decrease 80 percent of its initial value.

8. A 4 lb weight attached to the lower end of a spring is made to execute damped vibrations in which the resistance is proportional to the velocity.   If the damping factor is to decrease 80 percent in half a cycle, and if the damped period is to be 0.15 sec more than it would be if there were no resistance, find the amount by

which the weight should stretch the spring. Also, find the damping constant K of the system.

9. Establish (5.4.37).

5.5. Forced Motion with Damping

The differential equation governing forced motion with damping is given by Equation (5.2.7), namely,

$$m\frac{d^2x}{dt^2} + K\frac{dx}{dt} + kx = F(t), \tag{5.5.1}$$

where $K > 0$ is the damping constant, $k > 0$ is the spring constant, and $F(t)$ is the external force. The forcing function F is usually called, in engineering fields, the *input* of the system, and the solution $x(t)$ of Equation (5.5.1) is called the *output* of the system. The forcing functions F defined by $F(t) = F_0 \cos \omega t$ and $F(t) = F_0 \sin \omega t$, where the amplitude F_0 and the frequency ω are constants, play an important role in applications.

Let us write Equation (5.5.1) with $F(t) = F_0 \cos \omega t$ in the following form

$$\frac{d^2x}{dt^2} + 2\alpha\frac{dx}{dt} + \beta^2 x = E_0 \cos \omega t, \tag{5.5.2}$$

where

$$2\alpha = \frac{K}{m} > 0, \qquad \beta^2 = \frac{k}{m} > 0, \qquad \text{and} \qquad E_0 = \frac{F_0}{m} > 0. \tag{5.5.3}$$

The characteristic polynomial $p(m)$ of the related homogeneous differential equation,

$$\frac{d^2x}{dt^2} + 2\alpha\frac{dx}{dt} + \beta^2 x = 0, \tag{5.5.4}$$

is given by

$$p(m) = m^2 + 2\alpha m + \beta^2. \tag{5.5.5}$$

Note that

$$p(i\omega) = (i\omega)^2 + 2\alpha(i\omega) + \beta^2 = \beta^2 - \omega^2 + i2\alpha\omega, \qquad i^2 = -1, \tag{5.5.6}$$

and

$$|p(i\omega)|^2 = p(i\omega) \cdot \overline{p(i\omega)} = (\beta^2 - \omega^2 + i2\alpha\omega)(\beta^2 - \omega^2 - i2\alpha\omega)$$
$$= (\beta^2 - \omega^2)^2 + 4\alpha^2\omega^2. \tag{5.5.7}$$

Also, note that $p(i\omega) = 0$ if and only if $\beta^2 - \omega^2 = 0$ and $2\alpha\omega = 0$. Thus, in view of Equation (5.5.3), we see that $p(i\omega) \neq 0$ and $|p(i\omega)| > 0$. The complex-valued quantity $p(i\omega)$ given in (5.5.6) is called the *complex impedance* of the system.

The general solution $x(t)$ of Equation (5.5.2) is given by

$$x(t) = x_c(t) + x_p(t), \tag{5.5.8}$$

where x_c and x_p are, respectively, the complementary function and a particular solution of Equation (5.5.2).   [*See* Definitions 4.7.1 and 4.7.2.]   The three possible complementary functions x_c are the same as those given in Cases 1, 2, and 3 of Section 5.4.   A particular solution x_p of Equation (5.5.2) is given by [*see* (4.8.75)]

$$x_p = \mathscr{R}\left[\frac{E_0\, e^{i\omega t}}{p(i\omega)}\right], \quad \text{if} \quad p(i\omega) \neq 0. \tag{5.5.9}$$

Thus,

$$
\begin{aligned}
x_p &= \mathscr{R}\left[\frac{E_0(\cos \omega t + i \sin \omega t)}{\beta^2 - \omega^2 + i2\alpha\omega}\right] \\[2mm]
&= \mathscr{R}\left[\frac{E_0\{(\beta^2 - \omega^2) - i2\alpha\omega\}\{\cos \omega t + i \sin \omega t\}}{(\beta^2 - \omega^2)^2 + 4\alpha^2\omega^2}\right] \\[2mm]
&= \mathscr{R}\left[\frac{E_0}{|p(i\omega)|^2}\{(\beta^2 - \omega^2)\cos \omega t + 2\alpha\omega \sin \omega t\}\right. \\[2mm]
&\quad \left. + i\,\frac{E_0}{|p(i\omega)|^2}\{(\beta^2 - \omega^2)\sin \omega t - 2\alpha\omega \cos \omega t\}\right]. \tag{5.5.9'}
\end{aligned}
$$

Hence,

$$x_p = \frac{E_0}{|p(i\omega)|^2}\,[(\beta^2 - \omega^2)\cos \omega t + 2\alpha\omega \sin \omega t]. \tag{5.5.10}$$

We shall write Equation (5.5.10) in a more convenient form.   [*See* Remark 3.4.4.]

$$x_p = \frac{E_0}{|p(i\omega)|}\left[\frac{(\beta^2 - \omega^2)}{|p(i\omega)|}\cos \omega t + \frac{2\alpha\omega}{|p(i\omega)|}\sin \omega t\right].$$

Thus,

$$x_p = \frac{E_0}{|p(i\omega)|}\cos(\omega t - \phi), \tag{5.5.11}$$

where the phase angle ϕ is determined by the equations

$$\cos \phi = \frac{\beta^2 - \omega^2}{|p(i\omega)|}, \qquad \sin \phi = \frac{2\alpha\omega}{|p(i\omega)|}, \tag{5.5.12}$$

and

$$|p(i\omega)| = \sqrt{(\beta^2 - \omega^2)^2 + 4\alpha^2\omega^2}. \tag{5.5.13}$$

Hence, the general solution $x(t)$ of Equation (5.5.2) is given by

$$x(t) = x_c(t) + \frac{E_0}{\sqrt{(\beta^2 - \omega^2)^2 + 4\alpha^2\omega^2}} \cos(\omega t - \phi), \qquad (5.5.14)$$

where x_c is the general solution of Equation (5.5.4).

Since [*see* (5.5.9′)]

$$\mathscr{I}\left[\frac{E_0\, e^{i\omega t}}{p(i\omega)}\right] = \frac{E_0}{|p(i\omega)|^2}\,[(\beta^2 - \omega^2)\sin \omega t - 2\alpha\omega \cos \omega t],$$

then in view of (4.8.77), the general solution $x(t)$ of

$$\frac{d^2x}{dt^2} + 2\alpha\frac{dx}{dt} + \beta^2 x = E_0 \sin \omega t \qquad (5.5.15)$$

is given by

$$x(t) = x_c(t) + x_p(t) = x_c(t) + \frac{E_0}{|p(i\omega)|^2}\,[(\beta^2 - \omega^2)\sin \omega t - 2\alpha\omega \cos \omega t],$$

$$(5.5.16)$$

where x_c is the general solution of Equation (5.5.4). Equation (5.5.16) can also be written as

$$x(t) = x_c(t) + \frac{E_0}{\sqrt{(\beta^2 - \omega^2)^2 + 4\alpha^2\omega^2}} \sin(\omega t - \phi), \qquad (5.5.17)$$

where the phase angle ϕ is determined by the equations given in (5.5.12).

Remark 5.5.1.　Observe that the above particular solutions x_p represent a simple harmonic motion with period $2\pi/\omega$, and with amplitude of vibration equal to $E_0/|p(i\omega)|$. Also, the particular solutions x_p do not die out as t increases. However, as we have seen in Cases 1, 2, and 3 of Section 5.4, the complementary function x_c does die out as t increases. In fact, for many practical problems, x_c becomes exceedingly small in a very short time. Thus, with increasing time, the gencral solution $x(t)$ given in (5.5.14) or (5.5.17) will essentially consist of only the particular solutions x_p. The functions x_c and x_p of (5.5.14) or (5.5.17) are called, respectively, the *transient solution* (or *transient term*) and the *steady-state solution* (or *steady-state term*) of the equation of motion. The steady-state term is often denoted by $x_s(t)$.

Example 5.5.1.　A weight of 24 lb is hanging at rest on a vertical spring whose constant is 30 lb/ft. At time $t = 0$, an external force given by $F(t) = 78 \cos 4t$ lb is applied to the system. The medium offers a resistance in pounds numerically equal to $3\, dx/dt$, where dx/dt is the instantaneous velocity in ft/sec. (a) Find the equation of motion at time t. (b) Sketch the graph of the function describing the motion of the weight.

SOLUTION.   The differential equation of the motion of the weight is given by

$$\frac{24}{32}\frac{d^2x}{dt^2} + 3\frac{dx}{dt} + 30x = 78\cos 4t,$$

or

$$\frac{d^2x}{dt^2} + 4\frac{dx}{dt} + 40x = 104\cos 4t. \tag{5.5.18}$$

The initial conditions associated with Equation (5.5.18) are

$$x(0) = 0, \qquad x'(0) = 0. \tag{5.5.19}$$

Comparing (5.5.18) with (5.5.2), we see that

$$2\alpha = 4, \quad \beta^2 = 40, \quad E_0 = 104, \quad \text{and} \quad \omega = 4. \tag{5.5.20}$$

The complementary function x_c of Equation (5.5.18) is [*see* (5.4.14)]

$$x_c(t) = e^{-2t}(c_1 \cos 6t + c_2 \sin 6t), \tag{5.5.21}$$

where c_1 and c_2 are arbitrary constants.   A particular solution x_p of Equation (5.5.18) is given by Formula (5.5.10).   Thus, in view of (5.5.20), we have that

$$x_p(t) = 3\cos 4t + 2\sin 4t. \tag{5.5.22}$$

Hence, in view of (5.5.21) and (5.5.22), the general solution x of Equation (5.5.18) is given by

$$x(t) = e^{-2t}(c_1 \cos 6t + c_2 \sin 6t) + 3\cos 4t + 2\sin 4t. \tag{5.5.23}$$

Differentiating (5.5.23), we obtain

$$x'(t) = -2e^{-2t}[(c_1 - 3c_2)\cos 6t + (3c_1 + c_2)\sin 6t] - 12\sin 4t + 8\cos 4t.$$

$$\tag{5.5.24}$$

Utilizing (5.5.19), (5.5.23), and (5.5.24), we have that

$$0 = x(0) = c_1 + 3,$$

and

$$0 = x'(0) = -2(c_1 - 3c_2) + 8.$$

Thus, $c_1 = -3$ and $c_2 = -\frac{7}{3}$.   Substituting these values into (5.5.23), we see that:

(a) The equation of motion at time t is given by

$$x(t) = -\tfrac{1}{3}e^{-2t}(9\cos 6t + 7\sin 6t) + 3\cos 4t + 2\sin 4t. \tag{5.5.25}$$

Equation (5.5.25) can also be written as

$$x(t) = -\frac{\sqrt{130}}{3} e^{-2t} \cos(6t - \phi_1) + \sqrt{13} \cos(4t - \phi_2), \qquad (5.5.26)$$

where the phase angles ϕ_1 and ϕ_2 are determined, respectively, by the equations

$$\cos\phi_1 = \frac{9}{\sqrt{130}}, \quad \sin\phi_1 = \frac{7}{\sqrt{130}} \quad \text{and} \quad \cos\phi_2 = \frac{3}{\sqrt{13}}, \quad \sin\phi_2 = \frac{2}{\sqrt{13}}.$$

$$(5.5.27)$$

From (5.5.27), we find $\phi_1 \approx 37.87° \approx 0.661$ radian and $\phi_2 \approx 33.69° \approx 0.588$ radian. Thus, the equation of motion at time t is given approximately by

$$x(t) = -3.80e^{-2t} \cos(6t - 0.661) + 3.605 \cos(4t - 0.588). \qquad (5.5.28)$$

(b) The graph of the transient solution

$$x_c(t) = -\frac{\sqrt{130}}{3} e^{-2t} \cos(6t - \phi_1) = -3.80e^{-2t} \cos(6t - 0.661),$$

$$(5.5.29)$$

representing a damped oscillatory motion, is shown in Figure 5.5.1. Note that it becomes negligible in a short time. The graph of the steady-state solution

$$x_p(t) = \sqrt{13} \cos(4t - \phi_2) = 3.605 \cos(4t - 0.588) \qquad (5.5.30)$$

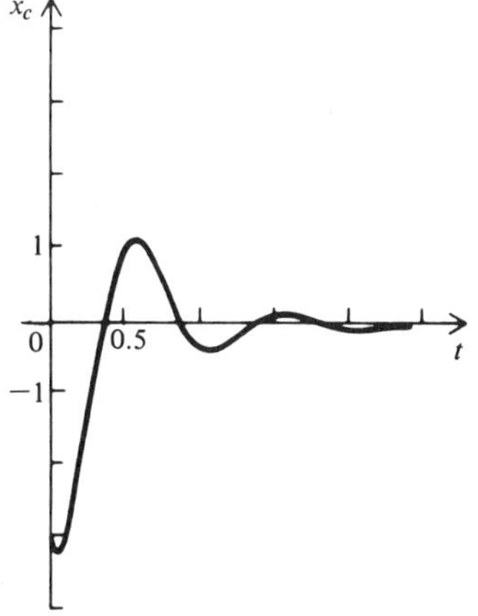

Figure 5.5.1.

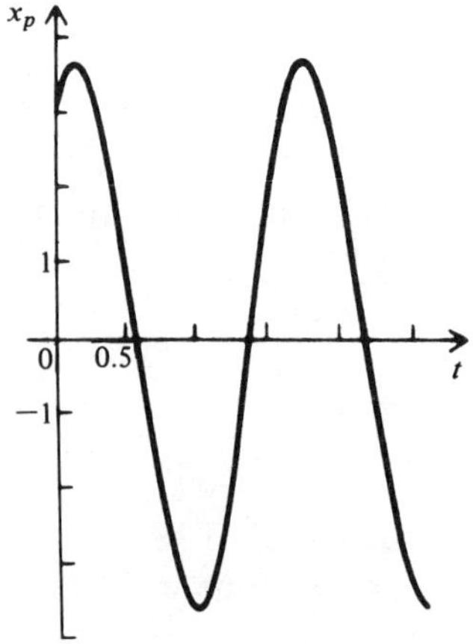

Figure 5.5.2.

is shown in Figure 5.5.2.   Observe that x_p represents a simple harmonic motion of amplitude $\sqrt{13} = 3.605$ and period $\pi/2$.   The graph of the function x given in (5.5.28) is shown in Figure 5.5.3.   It is obtained by adding the corresponding ordinates of the graphs of the functions x_c and x_p.   Figure 5.5.4 shows the graphs of the functions x_c, x_p, and x.

Remark 5.5.2.   We leave it as an exercise for the reader to show that the function x in (5.5.28) may be also obtained by utilizing (5.5.14) and (5.5.19).

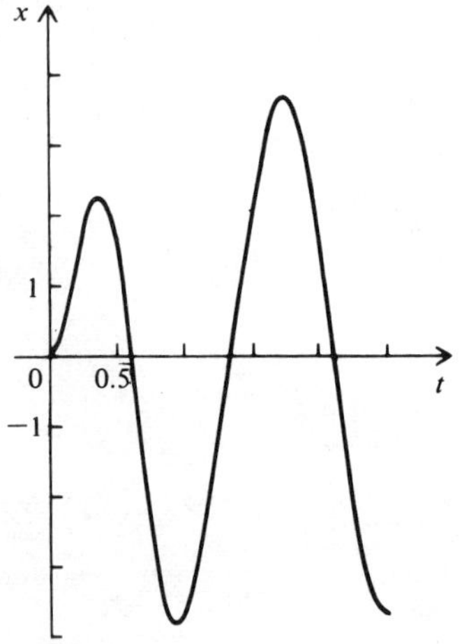

Figure 5.5.3.

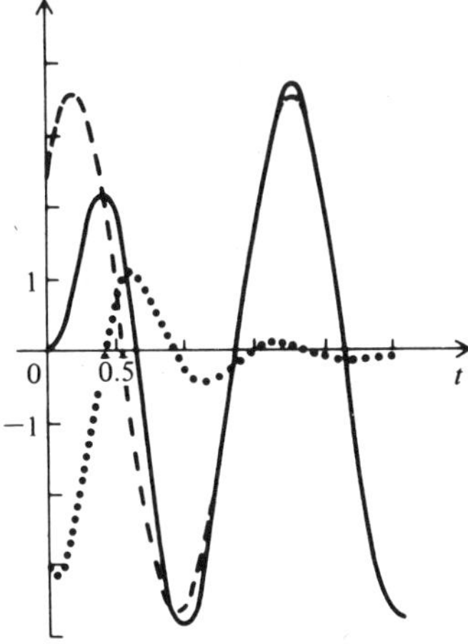

Figure 5.5.4.

Exercises 5.5

In Exercises 1 through 6, dx/dt denotes the instantaneous velocity in ft/sec.

1. A mass of one slug is attached to the lower end of a spring whose constant is 3 lb/ft. After reaching the equilibrium position, the mass is pulled down 6 inches and released from rest at time $t = 0$. At this instant, the support of the spring begins a vertical oscillation such that the distance (ft) from its initial position is given by $\frac{5}{3} \sin t$, $t \geq 0$. Assume that the mass is acted upon by a damping force which in pounds is numerically equal to $4\, dx/dt$.

   (a) Show that the differential equation for the displacement of the mass from its equilibrium position is given by

   $$\frac{32}{32}\frac{d^2x}{dt^2} = -4\frac{dx}{dt} - 3(x - y), \quad \text{where } y = \tfrac{5}{3}\sin t,$$

   which can also be written as

   $$\frac{d^2x}{dt^2} + 4\frac{dx}{dt} + 3x = 5\sin t.$$

   (b) Solve the differential equation obtained in part (a). Utilize the appropriate initial conditions to obtain the equation of motion at time t.

   (c) Sketch the graph of the function describing the motion of the mass.

2. Solve Exercise 1 above when $y = \tfrac{5}{3}\cos 2t$ ft.

3. A mass of one slug is hanging at rest on a spring whose constant is 6 lb/ft. At time $t = 0$, the support holding the spring is suddenly raised a distance $1/2$ ft and held fixed. Assume that the mass is acted upon by a damping force which in pounds is numerically equal to $5\, dx/dt$. Determine the equation of motion at time t.

4. A weight of 16 lb is hanging at rest on a spring whose constant is $2g$ lb/ft ($g = 32$). At time $t = 0$, the support of the spring begins a vertical oscillation such that the distance (ft) from its initial position is given by $2 \sin 2\sqrt{g}\, t$. Assume that the weight is acted upon by a damping force which in pounds is numerically equal to $2\sqrt{g}\, dx/dt$.   (a) Determine the equation of motion at time t.   (b) Determine the times $t > 0$ when the steady-state term of the solution in part (a) attains its maximum (positive) displacements.   (c) Find the value of the transient term of the solution in part (a) when $t = (1/2\sqrt{g})$ sec.

5. A weight of 8 lb is hanging at rest on a spring whose constant is 2 lb/ft.   At time $t = 0$, the upper end of the spring is given a vertical displacement such that the distance (ft) from its initial position is given by $\cos 4t$.   Assume that the weight is acted upon by a damping force which in pounds is numerically equal to $1\, dx/dt$.   Find the equation of motion at time t.

6. A mass of one slug is attached to the lower end of a spring whose constant is 16 lb/ft.   After reaching the equilibrium position, the mass is displaced x_0 ft and released from rest at time $t = 0$.   At this instant, the support holding the spring begins a vertical oscillation such that the distance (ft) from its initial position is given by $\sin 4t$.   Assume that the mass is acted upon by a damping force which in pounds is numerically equal to $10\, dx/dt$.   Determine the value of x_0 if the mass is to execute a simple harmonic motion.

5.6. Resonance

At times it is desirable to investigate what happens to the solution of the differential equation (5.5.2) [or (5.5.15)] when we fix the parameters α (the resistance of the medium) and β (the natural frequency of the system, that is, the frequency of a free undamped motion expressed in rad/sec), but allow the frequency ω (rad/sec) of the forcing function F defined by $F(t) = F_0 \cos \omega t$ [or $F(t) = F_0 \sin \omega t$] to vary when applied to the given system.   Thus, we may then write the general solution given in (5.5.14) [or (5.5.17)] as

$$x(t) = x_c(t) + A(\omega) \cos (\omega t - \phi), \tag{5.6.1}$$

[or as

$$x(t) = x_c(t) + A(\omega) \sin (\omega t - \phi)], \tag{5.6.2}$$

where the amplitude of vibration $A(\omega)$, of the steady-state solution, is given by

$$A(\omega) = \frac{E_0}{\sqrt{(\beta^2 - \omega^2)^2 + 4\alpha^2\omega^2}}. \tag{5.6.3}$$

Note that $A(\omega) \to 0$ as $\omega \to \infty$, and $A(\omega) \to E_0/\beta^2$ as $\omega \to 0$.   In particular, when $\omega = \beta$, we have

$$A(\beta) = \frac{E_0}{2\alpha\beta}. \tag{5.6.4}$$

Thus, if the given medium offers very little resistance, the amplitude of vibration may be dangerously large. Observe that the maximum amplitude occurs for those values of ω which render the denominator of (5.6.3) a relative minimum. Thus,

$$\frac{d}{d\omega}[(\beta^2 - \omega^2)^2 + 4\alpha^2\omega^2] = 0. \tag{5.6.5}$$

One readily finds that the value of ω given by

$$\omega = \sqrt{\beta^2 - 2\alpha^2}, \qquad \beta^2 > 2\alpha^2, \tag{5.6.6}$$

renders the function A a relative maximum. Substituting this value of ω in (5.6.3), we find

$$A_{\max} = \frac{E_0}{2\alpha\sqrt{\beta^2 - \alpha^2}}. \tag{5.6.7}$$

In view of (5.5.3), we may write (5.6.7) as

$$A_{\max} = \frac{F_0}{K\sqrt{\dfrac{k}{m} - \left(\dfrac{K}{2m}\right)^2}}. \tag{5.6.8}$$

A forcing function F defined by $F(t) = F_0 \cos \omega t$ [or $F(t) = F_0 \sin \omega t$] having the frequency ω given by (5.6.6) is said to be in *resonance* with the system. The value

$$\frac{\omega}{2\pi} = \frac{\sqrt{\beta^2 - 2\alpha^2}}{2\pi} = \frac{\sqrt{\dfrac{k}{m} - \dfrac{1}{2}\left(\dfrac{K}{m}\right)^2}}{2\pi} \text{ cycles/sec} \tag{5.6.9}$$

is called the *resonance frequency* of the system. Comparing (5.6.9) with (5.4.18), we see that the resonance frequency is less than that of the corresponding free damped vibration.

In view of (5.5.3), we may write (5.6.3) as

$$A(\omega) = \frac{\dfrac{F_0}{m}}{\sqrt{\left(\dfrac{k}{m} - \omega^2\right)^2 + \left(\dfrac{K}{m}\right)^2 \omega^2}}. \tag{5.6.10}$$

The graph of the function A given by (5.6.10) is called the *resonance curve* of the system. For a given system, if we fix m, k, and F_0, there will be a resonance curve associated with each value of the damping constant $K \geq 0$. For example, if we take $m = 1$, $k = 2$, and $F_0 = 1.5$, then (5.6.10) becomes

$$A(\omega) = \frac{1.5}{\sqrt{(2 - \omega^2)^2 + K^2\omega^2}}. \tag{5.6.11}$$

The resonance curves for the values of $K = 0.25$, 0.75, 1.25, and 1.75 are shown in Figure 5.6.1.   Note that the resonance frequency is given by

$$\frac{1}{2\pi}\sqrt{2 - \frac{K^2}{2}} \text{ cycle/sec.}$$

Thus, in this example, resonance will occur only when $K < 2$.   Also,

$$A_{\max} = \frac{1.5}{K\sqrt{2 - \dfrac{K^2}{4}}}. \tag{5.6.12}$$

It is evident from (5.6.12) and (5.6.11) that for the limiting case $K = 0$, $A_{\max}$ is not defined and $A(\omega)$ is given by

$$A(\omega) = \frac{1.5}{2 - \omega^2}. \tag{5.6.13}$$

Thus, $A(\omega)$ has an infinite discontinuity at $\omega = \sqrt{2}$, and the solution no longer exists.   [*See* Figure 5.6.1.]   This limiting case gives rise to *undamped resonance*, which we shall next consider.

The differential equation governing forced motion with no damping is obtained by setting $K = 0$ in Equation (5.5.1).   Thus, we have

$$m\frac{d^2x}{dt^2} + kx = F(t). \tag{5.6.14}$$

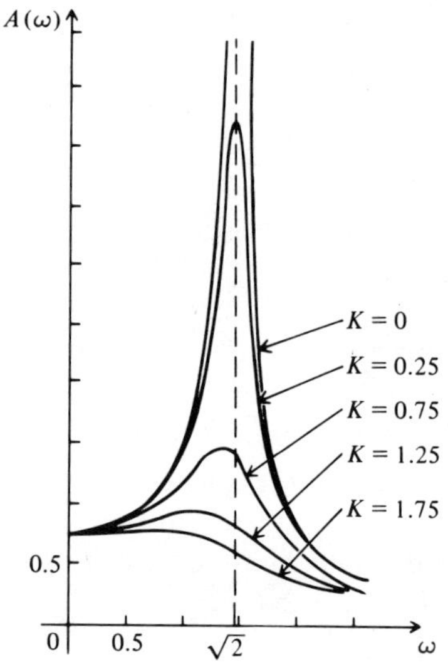

Figure 5.6.1.

With $F(t) = F_0 \cos \omega t$, we may write Equation (5.6.14) as

$$\frac{d^2x}{dt^2} + \beta^2 x = E_0 \cos \omega t, \qquad (5.6.15)$$

where

$$\beta^2 = \frac{k}{m} > 0 \quad \text{and} \quad E_0 = \frac{F_0}{m} > 0. \qquad (5.6.16)$$

The phenomenon of undamped resonance occurs when the impressed frequency ω is equal to the natural frequency β of the system. Thus [*see* (5.6.6) with $\alpha = 0$], with $\omega = \beta$, Equation (5.6.15) becomes

$$\frac{d^2x}{dt^2} + \beta^2 x = E_0 \cos \beta t. \qquad (5.6.17)$$

The complementary function x_c of Equation (5.6.17) is given by

$$x_c(t) = c_1 \cos \beta t + c_2 \sin \beta t,$$

or

$$x_c(t) = c \cos (\beta t - \phi), \qquad (5.6.18)$$

which describes a simple harmonic motion. A particular solution x_p of Equation (5.6.17) is given by [*see* (4.8.93)]

$$x_p(t) = \frac{E_0}{2\beta} t \sin \beta t. \qquad (5.6.19)$$

Note that the amplitude of x_p, namely $(E_0/2\beta)t$, increases with time. The graph of x_p is shown in Figure 5.6.2 for the case when $E_0/2\beta = \frac{1}{2}$ and $\beta = 4$. It is seen from the graph that the oscillations build up without limit.

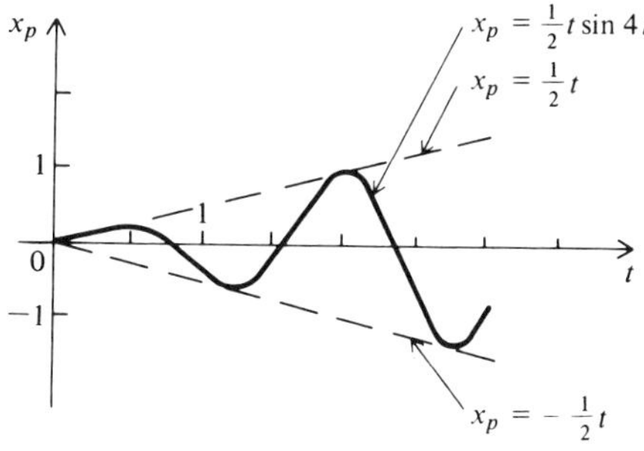

Figure 5.6.2.

Clearly, the spring will soon break.   In view of (5.6.18) and (5.6.19), the general solution x of Equation (5.6.17) is given by

$$x(t) = c \cos (\beta t - \phi) + \frac{E_0}{2\beta} t \sin \beta t, \tag{5.6.20}$$

where β and E_0 are given in (5.6.16).

Remark 5.6.1.   The phenomenon of resonance, like that of friction, is of importance in the field of engineering.   It is both a detrimental and a beneficial phenomenon.   The collapse of a certain building was attributed to the rhythmic swaying of dancing couples, who happened to strike the natural frequency of the beam supporting the structure.   It is also a known fact that soldiers are commanded to break step in crossing a bridge for fear that they may strike the note of the cables.   On the other hand, resonance is desirable in certain acoustical and radio-circuit problems.

Let us now consider the initial-value problem

$$\begin{cases} \dfrac{d^2x}{dt^2} + \beta^2 x = E_0 \cos \omega t, & [5.6.15] \\[2mm] x(0) = 0, \quad x'(0) = 0, & (5.6.21) \end{cases}$$

where β^2 and E_0 are given in (5.6.16).   We shall investigate the solution of this initial-value problem when the impressed frequency ω is close, but not equal to, the natural frequency β.   Utilizing (4.8.92) to obtain a particular solution of (5.6.15), the general solution of (5.6.15) is then given by

$$x(t) = c_1 \cos \beta t + c_2 \sin \beta t + \frac{E_0}{\beta^2 - \omega^2} \cos \omega t, \tag{5.6.22}$$

where c_1 and c_2 are arbitrary constants.   Differentiating (5.6.22), we obtain

$$x'(t) = -c_1 \beta \sin \beta t + c_2 \beta \cos \beta t - \frac{E_0 \omega}{\beta^2 - \omega^2} \sin \omega t. \tag{5.6.23}$$

Substituting the initial conditions given in (5.6.21) into (5.6.22) and (5.6.23), we find $c_1 = -E_0/(\beta^2 - \omega^2)$ and $c_2 = 0$.   Thus, the solution of the given initial-value problem is

$$x(t) = \frac{E_0}{\beta^2 - \omega^2} (\cos \omega t - \cos \beta t). \tag{5.6.24}$$

From trigonometry, we have

$$\cos \omega t - \cos \beta t = -2 \sin \left[\tfrac{1}{2}(\omega - \beta)t\right] \cdot \sin \left[\tfrac{1}{2}(\omega + \beta)t\right]. \tag{5.6.25}$$

By assumption, $\omega \approx \beta$ and thus, $\frac{1}{2}(\omega + \beta) \approx \omega$. Let $\delta = \frac{1}{2}(\omega - \beta)$. Then $\beta^2 - \omega^2 = (\beta - \omega)(\beta + \omega) \approx -4\delta\omega$. Hence, (5.6.25) becomes

$$\cos \omega t - \cos \beta t \approx -2 \sin \delta t \sin \omega t,$$

and in view of (5.6.24), the equation of motion at time t is given approximately by

$$x(t) = \frac{E_0}{2\delta\omega} \sin \delta t \sin \omega t. \tag{5.6.26}$$

Since δ is small compared with ω, we may consider the term $\sin \delta t$ as a slowly varying amplitude for the principal vibration $\sin \omega t$. Such motion illustrates the phenomenon of *beats*. [*See* Figure 5.6.3 for the case when $E_0 = 1$, $\delta = 0.25$, and $\omega = 4$.]

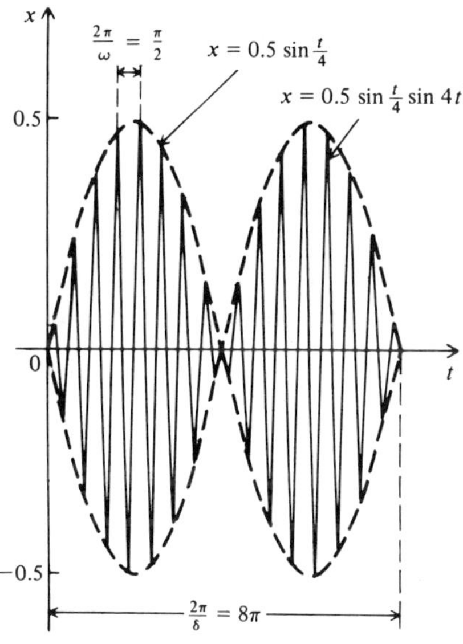

Figure 5.6.3.

Exercises 5.6

1. A weight of 16 lb is hanging at rest on a spring whose constant is 4 lb/ft. At time $t = 0$, the support of the spring begins a vertical oscillation such that the distance (ft) from its initial position is given by $\frac{1}{4} \cos 2.5\,t$. Assume that there is no damping force acting on the system.
   (a) Determine the equation of motion at time t.

   (b) Determine the approximate solution as given by Equation (5.6.26).
   (c) Sketch the graph of the function describing the motion of the weight determined in part (b).

2. A mass of one slug is hanging at rest on a spring whose constant is 12 lb/ft. Beginning at time $t = 0$, an external force given by $F(t) = 16 \sin \omega t$ lb is applied to the system.

   (a) Assume that the mass is acted upon by a damping force which in pounds is numerically equal to 4 dx/dt, where dx/dt is the instantaneous velocity in feet per second. Find the equation of motion at time t if the forcing function is in resonance with the system.
   (b) Assume that there is no damping force acting on the system. Determine the value of ω which gives rise to undamped resonance. With this value of ω, find the equation of motion of the mass at time t.

3. A mass of one slug is hanging at rest on a spring whose constant is 8 lb/ft. Beginning at time $t = 0$, an external force given by $F(t) = 2 \sin \omega t$ lb is applied to the system. Assume that the mass is acted upon by a damping force which in pounds is numerically equal to $K \, dx/dt$, where $K \geq 0$, and dx/dt is the instantaneous velocity in feet per second. Thus, the differential equation of motion of the mass is

$$\frac{d^2 x}{dt^2} + K \frac{dx}{dt} + 8x = 2 \sin \omega t.$$

   (a) Find the values of K which give rise to resonance.
   (b) Graph the resonance curves of the system for $K = 0, 1, 2,$ and 3.
   (c) If $K = 0$, find the value of ω which gives rise to undamped resonance.
   (d) If $K = 2$, find the value of ω which renders the forcing function in resonance with the system.
   (e) If $K = 1$, find the resonance frequency of the system.
   (f) If $K = 2$, find the equation of motion of the mass at time t when the forcing function is in resonance with the system.

5.7. Electric Circuits

(Review Section 3.4.)   In Section 3.4, we developed the differential equation governing the *R-L-C* series circuit, namely,

$$LQ''(t) + RQ'(t) + \frac{1}{C} Q(t) = E(t). \tag{5.7.1}$$

In Equation (5.7.1), $Q(t)$ is the charge on the capacitor at time t. The positive constants $R, L,$ and C denote, respectively, the resistance of the resistor or conductor, the inductance of the inductor, and the capacitance of the capacitor or condenser. Also, $E(t)$ is the impressed voltage or emf applied to the network by a generator or battery. [*See* Figure 3.4.1.] The reciprocal of the capacitance, namely $1/C$, is also known as the *elastance* of the network.

The relation between $Q(t)$ and $I(t)$, the current in the circuit, is given by

$$Q'(t) = I(t). \tag{5.7.2}$$

A current $I(t)$ of electricity exists in a conductor whenever an electric charge $Q(t)$ is being transferred from one point to another in that conductor. The current $I(t)$ is the rate of change of the charge $Q(t)$.

If E is a differentiable function of t, then differentiating (5.7.1) with respect to t and utilizing (5.7.2), we obtain

$$LI''(t) + RI'(t) + \frac{1}{C} I(t) = E'(t). \tag{5.7.3}$$

Equation (5.7.3) may be used to obtain directly the current $I(t)$ at time t in the circuit instead of solving Equation (5.7.1) and then using (5.7.2).

Except for nomenclature and units, we see that the differential Equations (5.7.1) and (5.7.3) are of the same form as the differential Equation (5.2.7), namely,

$$m\frac{d^2x}{dt^2} + K\frac{dx}{dt} + kx = F(t), \tag{5.7.4}$$

where $x(t)$ is the displacement of the mass m from its equilibrium position at time t, K and k are positive constants denoting, respectively, the damping constant and the spring constant, and $F(t)$ is the external force acting on the mechanical system.

Remark 5.7.1. Another important system which is governed by the differential equation of the same form as (5.7.4) is the *torsional system*. [*See* Figure 5.7.1.] Briefly, this system consists of a disc of moment of inertia I attached to a shaft of torsional stiffness K. If the disc is twisted through an angle θ (radians) and released, it will undergo torsional oscillations. Suppose that the shaft torque is proportional to the angular displacement θ,

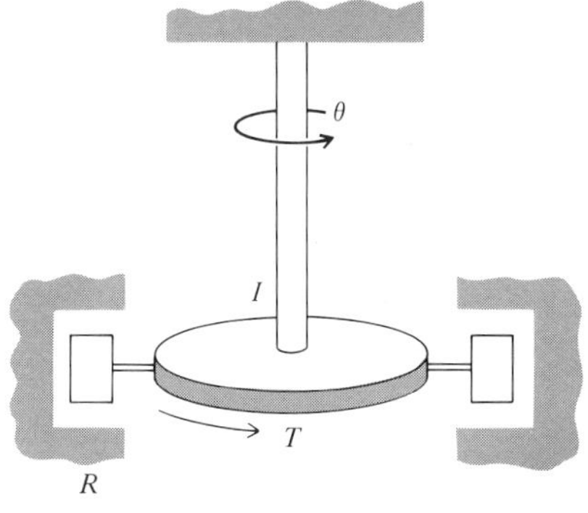

Figure 5.7.1.

the damping torque is proportional to the angular velocity $d\theta/dt$ and the external torque applied to the system is given by $T(t)$. From physics, we have that

$$\text{Torque} = I\frac{d^2\theta}{dt^2}. \tag{5.7.5}$$

Utilizing Newton's Law for Angular Motion, we have

$$I\frac{d^2\theta}{dt^2} = -R\frac{d\theta}{dt} - K\theta + T(t),$$

or

$$I\frac{d^2\theta}{dt^2} + R\frac{d\theta}{dt} + K\theta = T(t), \tag{5.7.6}$$

where $R > 0$ is the damping constant and $K > 0$ is the torsional (stiffness) constant. For example, in the English gravitational system I: slug ft^2 per rad, R: lb ft sec per rad, K: lb ft per rad, and T: lb ft.

The analogy of the above three systems is exhibited in Table 5.7.1 below.

Table 5.7.1.

Mechanical System		Electrical System		Torsional System	
Mass	m	Inductance	L	Moment of inertia	I
Spring constant	k	Elastance	$\dfrac{1}{C}$	Torsional constant	K
Damping constant	K	Resistance	R	Torsional damping	R
Impressed force	$F(t)$	Impressed voltage	$E(t)$	Impressed torque	$T(t)$
Displacement	$x(t)$	Capacitor charge	$Q(t)$	Angular displacement	$\theta(t)$
Velocity	$v(t) = \dfrac{dx}{dt}$	Current	$I(t) = \dfrac{dQ}{dt}$	Angular velocity	$\omega(t) = \dfrac{d\theta}{dt}$

Remark 5.7.2. Since the mathematical solutions of the above three systems are the same, we may solve electrical systems and torsional systems by referring to the solutions of the corresponding mechanical systems developed in Sections 5.3 through 5.6. Also, it may be more advantageous at times to solve a mechanical system via a corresponding electrical system, if it is more expedient (both financially and construction-wise) to set up the electrical system rather than the mechanical system.

Let us determine the current $I(t)$ at time t in an R-L-C series circuit when the voltage source is an ordinary alternating-current generator. Then the impressed voltage may be expressed as

$$E(t) = E_0 \sin(\omega t - \phi), \tag{5.7.7}$$

where E_0 is the amplitude ω (rad/sec) is the frequency, and ϕ is the phase angle of the impressed voltage $E(t)$.   Differentiating (5.7.7) with respect to t, we find

$$E'(t) = E_0\,\omega\,\cos\,(\omega t - \phi). \tag{5.7.8}$$

Substituting $E'(t)$ of (5.7.8) into (5.7.3), we obtain

$$I''(t) + 2\gamma I'(t) + \delta^2 I(t) = E_1 \cos\,(\omega t - \phi), \tag{5.7.9}$$

where

$$2\gamma = \frac{R}{L} > 0, \qquad \delta^2 = \frac{1}{CL} > 0, \qquad \text{and} \qquad E_1 = \frac{E_0\,\omega}{L} > 0. \tag{5.7.10}$$

The related homogeneous differential equation of (5.7.9) is

$$I''(t) + 2\gamma I'(t) + \delta^2 I(t) = 0. \tag{5.7.11}$$

The characteristic equation of (5.7.11) is

$$m^2 + 2\gamma m + \delta^2 = 0. \tag{5.7.12}$$

The discriminant D of Equation (5.7.12) is

$$D = 4\gamma^2 - 4\delta^2. \tag{5.7.13}$$

We shall be interested primarily in the case when $D < 0$; that is, when the roots of (5.7.12) are imaginary.   Note that in view of (5.7.10), the condition $D < 0$ implies that

$$R < 2\sqrt{\frac{L}{C}}, \tag{5.7.14}$$

which is the same as the inequality given in (5.4.19) when K, m, and k are replaced, respectively, by R, L, and $1/C$.   Thus, the current $I(t)$ will be oscillatory in nature.   Since in a given electrical network, the capacitance C (in farads) is usually very small, the above inequality will usually be satisfied.

The complementary function I_c of Equation (5.7.9) is given by

$$I_c(t) = e^{-\gamma t}(c_1 \cos \sqrt{\delta^2 - \gamma^2}\,t + c_2 \sin \sqrt{\delta^2 - \gamma^2}\,t), \tag{5.7.15}$$

or equivalently by

$$I_c(t) = Be^{-(R/2L)t} \cos\left(\sqrt{\frac{1}{LC} - \left(\frac{R}{2L}\right)^2}\,t - \phi_1\right), \tag{5.7.16}$$

where $B = \sqrt{c_1^2 + c_2^2} > 0$, and the phase angle ϕ_1 is determined by the equations

$$\cos \phi_1 = \frac{c_1}{B}, \qquad \sin \phi_1 = \frac{c_2}{B}. \tag{5.7.17}$$

Observe that the complementary function I_c represents a damped harmonic motion, and thus dies out with increasing time.   [*Note:* (5.7.15) and (5.7.16) are the same as (5.4.14) and (5.4.15) when, in the latter, α and β^2 are replaced, respectively, by $\gamma = R/2L$ and $\delta^2 = 1/LC$.]

Assume a particular solution I_s of Equation (5.7.9) of the form

$$I_s(t) = A_0 \cos(\omega t - \phi) + B_0 \sin(\omega t - \phi). \qquad (5.7.18)$$

Utilizing the method of undetermined coefficients [*see* Section 4.8] and also using (5.7.10), we leave it as an exercise for the reader to show that [*see* Exercise 5.7.1],

$$A_0 = \frac{E_0 X}{Z^2}, \qquad B_0 = \frac{E_0 R}{Z^2}, \qquad (5.7.19)$$

where

$$X = \frac{1}{C\omega} - L\omega, \qquad Z = \sqrt{R^2 + X^2}. \qquad (5.7.20)$$

The quantities X and Z are called, respectively, the *reactance* and the *impedance* of the circuit, and are expressed in ohms.   In view of (5.7.19), we may write (5.7.18) as

$$I_s(t) = \frac{E_0}{Z^2} [X \cos(\omega t - \phi) + R \sin(\omega t - \phi)]. \qquad (5.7.21)$$

Equation (5.7.21) may also be written in the following equivalent form

$$I_s(t) = \frac{E_0}{Z} \left[\frac{X}{Z} \cos(\omega t - \phi) + \frac{R}{Z} \sin(\omega t - \phi) \right]$$

$$= \frac{E_0}{Z} [\cos \phi_2 \cos(\omega t - \phi) + \sin \phi_2 \sin(\omega t - \phi)]$$

$$= \frac{E_0}{Z} \cos(\omega t - \phi - \phi_2), \qquad (5.7.22)$$

where the phase angle ϕ_2 is determined by the equations

$$\cos \phi_2 = \frac{X}{Z}, \qquad \sin \phi_2 = \frac{R}{Z}. \qquad (5.7.23)$$

Note that the function I_s represents a simple harmonic motion having the same frequency as the impressed voltage $E(t)$, namely, ω (rad/sec).   In view of (5.7.16) and (5.7.22), the current $I(t)$ in the circuit at time t is

$$I(t) = Be^{-(R/2L)t} \cos\left(\sqrt{\frac{1}{LC} - \left(\frac{R}{2L}\right)^2}\, t - \phi_1 \right) + \frac{E_0}{Z} \cos(\omega t - \phi - \phi_2).$$

$$(5.7.24)$$

The functions I_c and I_s of (5.7.16) and (5.7.22) are called, respectively, the *transient current* and the *steady-state current*. Clearly, with increasing time, $I(t) \approx I_s(t)$. $I_c(t)$ becomes negligible in a short period of time, since $R/2L$ is usually large, and hence $e^{-(R/2L)t}$ becomes small even for very small values of t sec. The function E of (5.7.7) or E' of (5.7.8) is called the *input* of the system; the solution given by (5.7.24) is called the *output* of the system.

Among other similarities with mechanical vibrations, electrical circuits have the property of resonance. The amplitude of the steady-state current is

$$A \equiv \frac{E_0}{Z} = \frac{E_0}{\sqrt{R^2 + X^2}} = \frac{E_0}{\sqrt{R^2 + \left(\dfrac{1}{C\omega} - L\omega\right)^2}}. \tag{5.7.25}$$

If we fix the parameters R, C, L, and E_0, and allow the frequency ω of the impressed voltage $E(t)$ to vary, then A may be considered as a function of ω. Clearly, $A(\omega)$ attains a relative maximum if $Z(\omega)$ attains a relative minimum. Thus, solving the following equation for ω,

$$0 = \frac{dZ}{d\omega} = -\frac{\left(\dfrac{1}{C\omega} - L\omega\right)\left(\dfrac{1}{C\omega^2} + L\right)}{Z}, \tag{5.7.26}$$

we find that

$$\omega = \sqrt{\frac{1}{LC}} \tag{5.7.27}$$

does indeed render the function Z a relative minimum, and hence renders the function A a relative maximum. Note that the value of ω given in (5.7.27) is obtained by setting the reactance $X = \dfrac{1}{C\omega} - L\omega$ equal to zero, and then solving for ω. Observe that in (5.7.26) when we set the term $\dfrac{1}{C\omega^2} + L$ equal to zero and solve for ω, the resulting values of ω are pure imaginary numbers, and not being in the realm of real numbers, do not render the function Z an extremum. Substituting the value of ω given in (5.7.27) into (5.7.25), we see that

$$A_{\max} = \frac{E_0}{R}. \tag{5.7.28}$$

An impressed voltage $E(t)$ having the frequency ω (rad/sec) given in (5.7.27) is said to be in *resonance* with the circuit. The value

$$\frac{\omega}{2\pi} = \frac{1}{2\pi}\sqrt{\frac{1}{LC}} \text{ cycles/sec,} \tag{5.7.29}$$

is called the *resonance frequency* of the circuit.

In view of (5.7.28), we see that if the resistance R of the conductor is small, A_{max} may become dangerously large, and there may be a breakdown in the circuit. Clearly, for a circuit in which R is almost negligible, $A_{max} \to \infty$, and there will surely be a breakdown in the circuit.

It should be noted that we may also vary the amplitude A of the steady-state current in other important ways. For example, if we fix the parameters $R, L, \omega,$ and E_0, and allow the capacitance C of the capacitor to vary, then we may consider A as a function of C. $A(C)$ has a relative maximum when

$$C = \frac{1}{\omega^2 L}, \tag{5.7.30}$$

and A_{max} is again given by Formula (5.7.28). In view of (5.7.27) and (5.7.30), we see that for fixed values of $R, L, \omega,$ and E_0, A_{max} may be obtained by *varying C* in such a manner that $1/\sqrt{LC}$ is equal to the frequency ω of the impressed voltage $E(t)$. In radio sets, tuning may be effected by changing the capacitance by means of a variable condenser. In tuning radio circuits, the object is to adjust C as given above, thereby maximizing the amplitude for a given signal (that is, frequency). Also, in radio sets, tuning may be effected by changing the inductance by means of a variable inductor.

Example 5.7.1. In an R-L-C series circuit, $R = 20$ ohms, $L = 0.5$ henry, and $C = 0.001$ farad. At time $t = 0$, the switch S is closed and an impressed voltage $E(t) = 150 \sin 120 \, \pi t$ volts is applied to the network by means of a generator. [*See* Figure 5.7.2.]
(a) Find the steady-state current at time t.
(b) Find the steady-state current when $t = 0.01$ sec.
(c) Find the value of the capacitance C which maximizes the amplitude of the steady-state current.

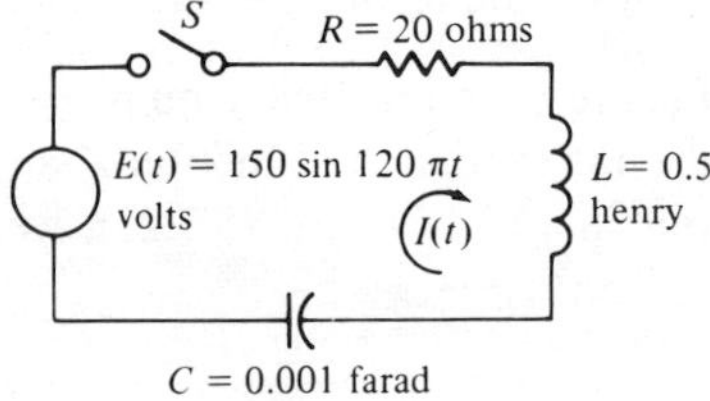

Figure 5.7.2.

SOLUTION.   (a) Comparing $E(t) = 150 \sin \pi t$ with (5.7.7), we see that $\varphi = 0$. Thus, with $\varphi = 0$ in (5.7.22), the steady-state current $I_s(t)$ is given by

$$I_s(t) = \frac{E_0}{Z} \cos(\omega t - \phi_2). \tag{5.7.31}$$

We are given that $R = 20$ ohms, $L = 0.5$ henry, $C = 0.001$ farad, $E_0 = 150$ volts, and $\omega = 120\pi$ rad/sec.   Utilizing (5.7.20) and (5.7.23), we obtain

$$Z = \sqrt{R^2 + \left(\frac{1}{C\omega} - L\omega\right)^2} = \sqrt{400 + \left(\frac{1000}{120\pi} - 60\pi\right)^2} = 186.917 \text{ ohms},$$

$$\cos \phi_2 = \frac{X}{Z} = \frac{\dfrac{1}{C\omega} - L\omega}{Z} = \frac{185.8434}{186.917} = 0.99426,$$

$$\sin \phi_2 = \frac{R}{Z} = \frac{20}{186.917} = 0.10699.$$

Hence, $\phi_2 \approx 0.107$ rad.   Substituting the above values into (5.7.31), the steady-state current $I_s(t)$ is given approximately by

$$I_s(t) = 0.8025 \cos(120\pi t - 0.107) \text{ ampere}, \tag{5.7.32}$$

which is the answer to part (a).
   (b) When $t = 0.01$ sec, we have approximately

$$I_s(0.01) = 0.8025 \cos(1.2\pi - 0.107) = 0.8025 \cos(3.663)$$
$$= (0.8025)(-0.8672) = -0.696 \text{ ampere}.$$

(c) Since the frequency ω of the steady-state current $I_s(t)$ is the same as the frequency of the impressed voltage $E(t)$, namely, $\omega = 120\pi$ rad/sec, the value of the capacitance C which maximizes the amplitude of the steady-state current is given by (5.7.30).   Thus, the value of C is given approximately by

$$C = \frac{1}{\omega^2 L} = \frac{1}{(120\pi)^2 \cdot \frac{1}{2}} = 14.07 \times 10^{-6} \text{ farad} = 14.07 \text{ microfarads}.$$

[*Note:* 1 microfarad $= 10^{-6}$ farad.]

Exercises 5.7

1. Verify (5.7.19).
2. Show that if we fix the parameters R, C, ω, and E_0, in (5.7.25) but allow the inductance L of the network to vary, then $A_{\max}$ is still given by (5.7.28).
3. Show that in an $R\text{-}C$ series circuit, the transient current cannot be oscillatory.

4. In a torsional system, the moment of inertia $I = 2$ slug ft^2 per rad, the torsional damping $R = 2$ lb ft sec per rad, the torsional constant $K = 1$ lb ft per rad, and the impressed torque $T(t) = 0$ for $t \geq 0$.   (a) Find the angular displacement $\theta(t)$ (rad) when subject to the initial conditions $\theta(0) = \pi/2$ (rad) and $\theta'(0) = 0$. (b) Find $\theta(t)$ when $t = \pi/4$ sec.

5. In an R-L-C series circuit, $R = 10$ ohms, $L = 1$ henry, and $C = 0.04$ farad. If initially $I(0) = 5$ amps and $Q(0) = 0$,   (a) find the time required for the charge on the capacitor to reach its maximum value.   (b) Find its maximum value.

6. An L-C series circuit with an emf $E(t)$ is known as a *harmonic oscillator*. [*See* Figure 5.7.3.]   Suppose that $E(t) = E_0 \sin \omega t$ volts, and initially $I(0) = I_0$ amperes and $Q(0) = Q_0$ coulombs.

(a) Find the charge $Q(t)$ and current $I(t)$ at time t when $\omega \neq \sqrt{\dfrac{1}{LC}}$.

(b) Find the charge $Q(t)$ and current $I(t)$ at time t when $\omega = \sqrt{\dfrac{1}{LC}}$.

(c) In part (b), what happens to the network as t becomes sufficiently large?

(d) In part (b), find the charge and current when $t = 0.1$ sec, if $I(0) = 20$ amperes, $Q(0) = 6$ coulombs, $L = 2$ henries, $C = 0.02$ farad, and $E_0 = 100$ volts.

7. In an R-L-C series circuit, $R = 10$ ohms, $L = 1$ henry, and $C = 0.02$ farad.   At time $t = 0$, the switch S of the circuit is closed and a constant emf of 300 volts is applied to the circuit by means of a generator. If initially $I(0) = 0$ and $Q(0) = 0$,

(a) find the charge $Q(t)$ and current $I(t)$ at time t,

(b) sketch the graph of the function I,

(c) find the maximum value of $Q(t)$ and of $I(t)$.

8. Suppose that we are given an R-L-C series circuit.   At time $t = 0$, the switch S of the circuit is closed and an emf, $E(t) = \sum_{k=1}^{n} E_k \sin \omega_k t$ volts, is applied to the circuit by means of a generator, where $E_k > 0$ are constants and ω_k (rad/sec) are distinct frequencies for $k = 1, 2, \ldots, n$.   Utilize the principle of superposition [*see* Exercise 4.7.1] to show that the steady-state current I_s is given by

$$I_s(t) = \sum_{k=1}^{n} \frac{E_k}{Z_k} \cos (\omega_k t - \phi_k),$$

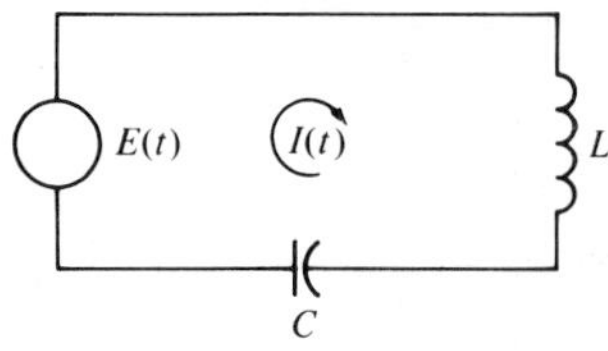

Figure 5.7.3.

where

$$X_k = \frac{1}{C\omega_k} - L\omega_k, \qquad Z_k = \sqrt{R^2 + X_k^2},$$

and the phase angles ϕ_k are determined by the equations

$$\cos \phi_k = \frac{X_k}{Z_k}, \qquad \sin \phi_k = \frac{R}{Z_k}, \qquad k = 1, 2, \ldots, n.$$

[*Note:* For a fixed value of k, the amplitude A_k of the steady-state current due to the input $E_k \sin \omega_k t$ is

$$A_k = \frac{E_k}{Z_k}.\Bigg]$$

9. In an *R-L-C* series circuit, $R = 20$ ohms, $L = 0.5$ henry, and $C = 0.001$ farad. At time $t = 0$, the switch S of the circuit is closed and an emf, $E(t) = (100 \sin 50t + 200 \sin 100t)$ volts, is applied to the circuit by means of a generator.
   (a) Find the steady-state current $I_s(t)$ at time t.
   (b) Sketch the graph of the function I_s.
   (c) Find $I_s(t)$ when $t = 0.01$ sec.
10. In an *R-L-C* series circuit, $R = 40$ ohms, $L = 0.1$ henry, $C = 0.0002$ farad. At time $t = 0$, the switch S of the circuit is closed and an emf, $E(t) = 110 \sin 120\pi t$ volts, is applied to the circuit by means of a generator. If initially $I(0) = 0$ and $Q(0) = 0$:
   (a) Find the current $I(t)$ at time t.
   (b) Sketch the graph of the function I_s representing the steady-state current in part (a).
   (c) Determine the frequency of the input $E(t)$ in order that it will be in resonance with the system.
   (d) For the resonant frequency in part (c), determine the maximum amplitude of the steady-state current.
   (e) Determine the value of the capacitance C which maximizes the amplitude of the steady-state current.
11. In an *R-L-C* series circuit, $R = 200$ ohms, $L = 1$ henry, and $C = 20$ microfarads. At time $t = 0$, the switch S of the circuit is closed and an emf, $E(t) = (110 \sin 120\pi t + 55 \sin 60 \pi t)$ volts, is applied to the circuit by means of an alternating-current generator. If initially $I(0) = 0$ and $Q(0) = 0$, find the current $I(t)$ at time t.

Suggested Readings

Reddick and Kibbey [36]   Ross [40]
Ritger and Rose [39]   Spiegel [43]

chapter 6 HIGHER-ORDER LINEAR DIFFERENTIAL EQUATIONS

6.1. Introduction

The theory developed in Chapter 4 for second-order linear differential equations extends, in a very natural fashion, to linear differential equations of order greater than two.   Thus, instead of proving all of the results of Chapter 4 for linear differential equations of order greater than two, we shall merely state these results and give proofs for just some of them.   The remaining results will be left as exercises for the reader.

6.2. Definitions.   The Fundamental Existence Theorem

Definition 6.2.1.   *A linear differential equation of order n* is an equation that can be expressed in the form

$$a_0(x)\frac{d^n y}{dx^n} + a_1(x)\frac{d^{n-1}y}{dx^{n-1}} + \cdots + a_{n-1}(x)\frac{dy}{dx} + a_n(x)y = F(x).[1] \quad (6.2.1)$$

Remark 6.2.1.   In the ensuing discussion, unless stated to the contrary, we shall assume that $a_0, a_1, \ldots, a_n$, and F, given in (6.2.1), are continuous real-valued functions defined on a common interval I: $a \leqq x \leqq b$, and $a_0(x)$ is not identically zero in I.

Definition 6.2.2.   If $F(x)$ is not identically zero, (6.2.1) is called a *nonhomogeneous* (or *inhomogeneous*) linear differential equation of order n.

Definition 6.2.3.   If $F(x)$ is identically zero in (6.2.1), the resulting equation

$$a_0(x)\frac{d^n y}{dx^n} + a_1(x)\frac{d^{n-1}y}{dx^{n-1}} + \cdots + a_{n-1}(x)\frac{dy}{dx} + a_n(x)y = 0 \quad (6.2.2)$$

is called a *homogeneous* (or *reduced*) linear differential equation of order n.

---

[1] We shall also write Equation (6.2.1) as
$$a_0(x)y^{(n)} + a_1(x)y^{(n-1)} + \cdots + a_{n-1}(x)y' + a_n(x)y = F(x).$$

If $F(x)$ is not identically zero in (6.2.1), then Equation (6.2.2) is called the *related homogeneous differential equation*.

[*Note:* When $n = 2$, Definitions 6.2.1–6.2.3 are the same as Definitions 4.2.1–4.2.3.]

We shall need the following fundamental existence theorem which is a generalization of Theorem 4.2.1. The proof of this theorem will be established in Appendix 2. [*See* Theorem 10.]

Theorem 6.2.1. *Let the functions $a_0, a_1, \ldots, a_n$, and F be continuous on the interval I: $a \leq x \leq b$, with $a_0(x) \neq 0$ for every x in I. Let x_0 be any fixed point in I, and let $y_0, y_0', \ldots, y_0^{(n-1)}$ be n arbitrary real constants. Then there exists a unique solution $y = y(x)$ satisfying the differential equation (6.2.1) for every x in I and the initial conditions*

$$y(x_0) = y_0, \qquad y'(x_0) = y_0', \ldots, y^{(n-1)}(x_0) = y_0^{(n-1)}. \tag{6.2.3}$$

Example 6.2.1. Let us show that the initial-value problem

$$\begin{cases} x^3 y''' - 3x^2 y'' + 6xy' - 6y = 2x^4, \\ y(1) = 0, \qquad y'(1) = 1, \qquad \text{and} \qquad y''(1) = -2, \end{cases}$$

has a unique solution on the interval I: $0 < x < \infty$.

SOLUTION. Here $a_0(x) = x^3 \neq 0$ for every x in I. Also, $a_0(x) = x^3$, $a_1(x) = -3x^2$, $a_2(x) = 6x$, $a_3(x) = -6$, and $F(x) = 2x^4$ are each continuous for all x in I. The point $x_0 = 1$ belongs to I. The three arbitrary constants are $y_0 = 0$, $y_0' = 1$, and $y_0'' = -2$. In view of Theorem 6.2.1, our initial-value problem has a unique solution for all x in I: $a \leq x \leq b$, where $a > 0$ and $b < \infty$. The unique solution is given by

$$y(x) = -\frac{10}{3}x + 6x^2 - 3x^3 + \frac{1}{3}x^4 \tag{6.2.4}$$

[*see* Example 6.10.1]. Note that the unique solution of our initial-value problem is actually continuous for all x: $-\infty < x < \infty$.

The following theorem is a generalization of Theorem 4.2.2, and the proof will be left as an exercise for the reader. [*See* Exercise 6.3.1.]

Theorem 6.2.2. *Let the functions $a_0, a_1, \ldots, a_n$ be continuous on the interval I: $a \leq x \leq b$, with $a_0(x) \neq 0$ for every x in I. Let x_0 be any fixed point in I. Let $y = y(x)$ be a solution of the differential equation (6.2.2) such that $y(x_0) = 0$, $y'(x_0) = 0, \ldots, y^{(n-1)}(x_0) = 0$. Then $y(x) = 0$ for every x in I.*

Example 6.2.2. The solution of the initial-value problem

$$\begin{cases} y^{(4)} - y = 0, \tag{6.2.5} \\ y(0) = 0, \qquad y'(0) = 0, \qquad y''(0) = 0, \qquad y'''(0) = 0, \tag{6.2.6} \end{cases}$$

is the trivial solution, namely, $y(x) = 0$ for all x.

The reader may verify directly the above result without utilizing Theorem 6.2.2 as follows. The general solution of Equation (6.2.5) is given by

$$y(x) = c_1 e^x + c_2 e^{-x} + c_3 \cos x + c_4 \sin x \qquad (6.2.7)$$

[*see* Example 6.6.4]. Differentiating (6.2.7) three times in succession, and then using (6.2.6), one may show that $c_1 = c_2 = c_3 = c_4 = 0$.

6.3. Linear Differential Operator of Order n

The following theorem is a generalization of Theorem 4.3.1.

Theorem 6.3.1. *Let the functions $a_0, a_1, \ldots, a_n$ be continuous on the interval I: $a \leq x \leq b$. Let the function y be n times differentiable in I. Then the operator L defined by*

$$L[y] = a_0(x)y^{(n)} + a_1(x)y^{(n-1)} + \cdots + a_{n-1}(x)y' + a_n(x)y \qquad (6.3.1)$$

is a linear operator.

The proof of this theorem follows from Exercise 4.3.8.

Remark 6.3.1. The operator L of Theorem 6.3.1 is known as a *linear differential operator of order n.*

Remark 6.3.2. For the definitions of a linear combination, linear dependence, and linear independence of n functions, see Definitions 4.3.4, 4.3.5, and 4.3.6. Definitions 4.3.4–4.3.6 may also be extended to complex-valued functions. [*See* Exercise 6.4.9.] The reader is urged to review Exercises 4.3.

Exercises 6.3

1. Prove Theorem 6.2.2.
2. The functions $u_1, u_2, \ldots, u_n$ defined below exist on the interval I: $-\infty < x < \infty$. The constants $\alpha_1, \alpha_2, \ldots, \alpha_n$ are distinct. Also, α and β are constants and n is any nonnegative integer. Show that each set of functions is linearly independent on I.
    (a) $u_1(x) = 1$, $u_2(x) = x, \ldots, u_n(x) = x^{n-1}$.
    (b) $u_1(x) = e^{\alpha_1 x}$, $u_2(x) = e^{\alpha_2 x}, \ldots, u_n(x) = e^{\alpha_n x}$.
    (c) $u_1(x) = e^{\alpha x}$, $u_2(x) = xe^{\alpha x}, \ldots, u_n(x) = x^{n-1}e^{\alpha x}$.
    (d) $u_1(x) = \sin \beta x$, $u_2(x) = x \sin \beta x, \ldots, u_n(x) = x^{n-1} \sin \beta x$.
    (e) $u_1(x) = e^{\alpha x} \sin \beta x$, $u_2(x) = e^{\alpha x}x \sin \beta x, \ldots, u_n(x) = e^{\alpha x}x^{n-1} \sin \beta x$.
3. Show that the functions $u_1, u_2, \ldots, u_n$; $v_1, v_2, \ldots, v_n$; $\ldots$; $w_1, w_2, \ldots, w_n$ defined by

$$u_1(x) = e^{m_1 x},\ u_2(x) = xe^{m_1 x}, \ldots, u_n(x) = x^{n_1}e^{m_1 x};\ v_1(x) = e^{m_2 x},\ v_2(x) = xe^{m_2 x}, \ldots,$$
$$v_n(x) = x^{n_2}e^{m_2 x}; \ldots;\ w_1(x) = e^{m_i x},\ w_2(x) = xe^{m_i x}, \ldots,\ w_n(x) = x^{n_i}e^{m_i x},$$

where $n_1, n_2, \ldots, n_i$ are any nonnegative integers and $m_1, m_2, \ldots, m_i$ are any distinct real or complex numbers, are linearly independent on the interval I: $-\infty < x < \infty$.

6.4. Some Basic Results on Homogeneous Linear Differential Equations of Order n

In view of (6.3.1), we shall often write Equation (6.2.2) as

$$L[y] = a_0(x)y^{(n)} + a_1(x)y^{(n-1)} + \cdots + a_{n-1}(x)y' + a_n(x)y = 0, \quad (6.4.1)$$

where $a_0, \ldots, a_n$ are continuous functions on the interval I: $a \leq x \leq b$, and $a_0(x) \neq 0$ for every x in I.   The following theorem is a generalization of Theorem 4.4.1, and the proof will be left as an exercise for the reader.   [*See* Exercise 6.4.1.]   In this section, $L[y] = 0$ *will always refer to the differential equation* (6.4.1).

Theorem 6.4.1.   *Let $y_1(x)$, $y_2(x)$, $\ldots$, $y_n(x)$ be n solutions of $L[y] = 0$.   Then the linear combination of these n solutions*

$$c_1 y_1(x) + c_2 y_2(x) + \cdots + c_n y_n(x),$$

where $c_1, c_2, \ldots, c_n$ are arbitrary constants, is also a solution of $L[y] = 0$.

Example 6.4.1.   One may easily verify that x, x^2, and x^3 are solutions of the differential equation [*see* Example 6.10.1]

$$x^3 y''' - 3x^2 y'' + 6xy' - 6y = 0, \qquad x > 0. \tag{6.4.2}$$

Theorem 6.4.1 asserts that $c_1 x + c_2 x^2 + c_3 x^3$, where c_1, c_2, and c_3 are arbitrary constants, is also a solution of the above differential equation.

The following theorem is a generalization of Theorem 4.4.2.

Theorem 6.4.2.   *Let the functions $a_0, a_1, \ldots, a_n$ be continuous on the interval I: $a \leq x \leq b$, and $a_0(x) \neq 0$ for every x in I.*

(a) *The equation $L[y] = 0$ always possesses n linearly independent solutions on I.*

(b) *If $y_1(x)$, $y_2(x)$, $\ldots$, $y_n(x)$ are n linearly independent solutions of $L[y] = 0$ on I, then every solution of $L[y] = 0$ can be expressed as a linear combination*

$$c_1 y_1(x) + c_2 y_2(x) + \cdots + c_n y_n(x)$$

of these n linearly independent solutions, for suitably chosen constants $c_1, c_2, \ldots, c_n$.

PROOF.   The proof of this theorem is very similar to that given for Theorem 4.4.2.

Concerning (a).   We shall actually exhibit the n linearly independent solutions of $L[y] = 0$ on I by the following useful technique.

Denote by $y_1(x), y_2(x), \ldots, y_n(x)$, n solutions of $L[y] = 0$ on I satisfying, respectively, the set of initial conditions

$$(A) \begin{cases} \begin{aligned} y_1(x_0) &= 1, \\ y_1'(x_0) &= 0, \\ &\cdots \\ y_1^{(n-1)}(x_0) &= 0, \end{aligned} \quad \begin{aligned} y_2(x_0) &= 0, \\ y_2'(x_0) &= 1, \\ &\cdots \\ y_2^{(n-1)}(x_0) &= 0, \end{aligned} \quad \cdots \quad \begin{aligned} y_n(x_0) &= 0, \\ y_n'(x_0) &= 0, \\ &\cdots \\ y_n^{(n-1)}(x_0) &= 1, \end{aligned} \end{cases} \quad (6.4.3)$$

where x_0 is a point in I.

In view of the hypothesis of our theorem, it follows from the existence Theorem 6.2.1 that $y_1(x)$ not only exists but it is also the unique solution of $L[y] = 0$ on I satisfying the first set of the initial conditions given in (6.4.3). A similar statement holds for the solutions $y_2(x), \ldots, y_n(x)$ subject to their respective initial conditions given in (6.4.3).

We shall next show that the functions $y_1, y_2, \ldots, y_n$ as given in (A) are linearly independent on I. We shall establish this fact via a contradiction. That is, suppose that $y_1, y_2, \ldots, y_n$ are linearly dependent on I. Then, in view of Definition 4.3.5, there exist constants $c_1, c_2, \ldots, c_n$, not all zero, such that

$$c_1 y_1(x) + c_2 y_2(x) + \cdots + c_n y_n(x) = 0 \tag{6.4.4}$$

for every x in I. Differentiating (6.4.4) $n - 1$ times in succession, we obtain

$$\begin{cases} c_1 y_1'(x) + c_2 y_2'(x) + \cdots + c_n y_n'(x) = 0, \\ \cdots \\ c_1 y_1^{(n-1)}(x) + c_2 y_2^{(n-1)}(x) + \cdots + c_n y_n^{(n-1)}(x) = 0, \end{cases} \tag{6.4.5}$$

for every x in I. When $x = x_0$ is any fixed point in I, Equations (6.4.4) and (6.4.5) become

$$\begin{cases} c_1 y_1(x_0) + c_2 y_2(x_0) + \cdots + c_n y_n(x_0) = 0, \\ c_1 y_1'(x_0) + c_2 y_2'(x_0) + \cdots + c_n y_n'(x_0) = 0, \\ \cdots \\ c_1 y_1^{(n-1)}(x_0) + c_2 y_2^{(n-1)}(x_0) + \cdots + c_n y_n^{(n-1)}(x_0) = 0. \end{cases} \tag{6.4.6}$$

Utilizing the initial conditions given in (6.4.3), the system of equations in (6.4.6) becomes

$$\begin{cases} c_1 \cdot 1 + c_2 \cdot 0 + \cdots + c_n \cdot 0 = 0, \\ c_1 \cdot 0 + c_2 \cdot 1 + \cdots + c_n \cdot 0 = 0, \\ \cdots \\ c_1 \cdot 0 + c_2 \cdot 0 + \cdots + c_n \cdot 1 = 0. \end{cases} \tag{6.4.7}$$

Thus, $c_1 = c_2 = \cdots = c_n = 0$, which is a contradiction. Hence, the functions $y_1, y_2, \ldots, y_n$ are linearly independent on I.

Concerning (b).   We shall now show that if $y_1(x)$, $y_2(x)$, ..., $y_n(x)$ are n linearly independent solutions of $L[y] = 0$ on I, then any other solution $y_{n+1}(x)$ of $L[y] = 0$ on I can be expressed as a linear combination of $y_1(x)$, $y_2(x)$, ..., $y_n(x)$.   [*See* Exercise 4.3.2.]   To this end, let us consider the system of n homogeneous linear equations in the $n + 1$ unknowns $\alpha_1, \alpha_2, \ldots, \alpha_{n+1}$:

$$\begin{cases} y_1(x_0)\alpha_1 + y_2(x_0)\alpha_2 + \cdots + y_n(x_0)\alpha_n + y_{n+1}(x_0)\alpha_{n+1} = 0, \\[4pt] y_1'(x_0)\alpha_1 + y_2'(x_0)\alpha_2 + \cdots + y_n'(x_0)\alpha_n + y_{n+1}'(x_0)\alpha_{n+1} = 0, \\ \qquad\qquad \cdots \\ y_1^{(n-1)}(x_0)\alpha_1 + y_2^{(n-1)}(x_0)\alpha_2 + \cdots + y_n^{(n-1)}(x_0)\alpha_n + y_{n+1}^{(n-1)}(x_0)\alpha_{n+1} = 0, \end{cases} \quad (6.4.8)$$

where x_0 is any fixed point in I.   From Theorem III of Section 4.4, the system (6.4.8) has a nontrivial solution, say

$$\alpha_1 = c_1, \quad \alpha_2 = c_2, \ldots, \quad \alpha_n = c_n \quad \alpha_{n+1} = c_{n+1} \qquad (6.4.9)$$

$[(c_1, c_2, \ldots, c_n, c_{n+1}) \neq (0, 0, \ldots, 0, 0)]$.   Let us define a function g by

$$g(x) = c_1 y_1(x) + c_2 y_2(x) + \cdots + c_n y_n(x) + c_{n+1} y_{n+1}(x) \qquad (6.4.10)$$

for every x in I.   We shall now show that $g(x) = 0$ for every x in I.   This will then prove that the functions $y_1, y_2, \ldots, y_n, y_{n+1}$ are linearly dependent on I.   [*See* Definition 4.3.5.]   Since $y_1(x)$, $y_2(x)$, ..., $y_n(x)$, $y_{n+1}(x)$ are solutions of $L[y] = 0$ on I, it follows from Theorem 6.4.1 that $g(x)$ is also a solution of $L[y] = 0$ on I.   Differentiating (6.4.10) $n - 1$ times in succession, we obtain

$$\begin{cases} g'(x) = c_1 y_1'(x) + c_2 y_2'(x) + \cdots + c_n y_n'(x) + c_{n+1} y_{n+1}'(x), \\ \qquad\qquad \cdots \\ g^{(n-1)}(x) = c_1 y_1^{(n-1)}(x) + c_2 y_2^{(n-1)}(x) + \cdots + c_n y_n^{(n-1)}(x) + c_{n+1} y_{n+1}^{(n-1)}(x), \end{cases}$$

$$(6.4.11)$$

for every x in I.   When $x = x_0$ (our fixed point in I), (6.4.10) and (6.4.11) become

$$\begin{cases} g(x_0) = c_1 y_1(x_0) + c_2 y_2(x_0) + \cdots + c_n y_n(x_0) + c_{n+1} y_{n+1}(x_0), \\ g'(x_0) = c_1 y_1'(x_0) + c_2 y_2'(x_0) + \cdots + c_n y_n'(x_0) + c_{n+1} y_{n+1}'(x_0), \\ \qquad\qquad \cdots \\ g^{(n-1)}(x_0) = c_1 y_1^{(n-1)}(x_0) + c_2 y_2^{(n-1)}(x_0) + \cdots \\ \qquad\qquad\qquad + c_n y_n^{(n-1)}(x_0) + c_{n+1} y_{n+1}^{(n-1)}(x_0). \end{cases} \quad (6.4.12)$$

From (6.4.9), we know that $c_1, c_2, \ldots, c_n, c_{n+1}$ satisfy the system (6.4.8). Thus, the right-hand members of the system (6.4.12) are each equal to zero. Hence,

$$g(x_0) = 0, \quad g'(x_0) = 0, \ldots, g^{(n-1)}(x_0) = 0. \qquad (6.4.13)$$

Since $g(x)$ is a solution of $L[y] = 0$ on I and also satisfies the initial conditions (6.4.13), it follows from Theorem 6.2.2 that $g(x) = 0$ for every x in I. Thus, part (b) is established and the proof of the theorem is now completed.

Since the general solution of a differential equation of order n should possess n essential constants (or parameters), the results given in the above theorems motivate us to make the following definition.

Definition 6.4.1. Let $y_1(x)$, $y_2(x)$, $\ldots$, $y_n(x)$ be n linearly independent solutions of the equation $L[y] = 0$ on the interval I: $a \leq x \leq b$. The *general solution* $y(x)$ of $L[y] = 0$ on I is defined by

$$y(x) = c_1 y_1(x) + c_2 y_2(x) + \cdots + c_n y_n(x), \tag{6.4.14}$$

where $c_1, c_2, \ldots, c_n$ are arbitrary constants.

Definition 6.4.2. Let x_0 be any fixed point in the interval I: $a \leq x \leq b$. The set of n linearly independent solutions $y_1(x)$, $y_2(x)$, $\ldots$, $y_n(x)$ of $L[y] = 0$ on I which satisfies the set of initial conditions

$$\begin{cases} y_1(x_0) = 1, \\ y_1'(x_0) = 0, \\ \cdots \\ y_1^{(n-1)}(x_0) = 0, \end{cases} \quad \begin{cases} y_2(x_0) = 0, \\ y_2'(x_0) = 1, \\ \cdots \\ y_2^{(n-1)}(x_0) = 0, \end{cases} \quad \cdots \quad \begin{cases} y_n(x_0) = 0, \\ y_n'(x_0) = 0, \\ \cdots \\ y_n^{(n-1)}(x_0) = 1, \end{cases}$$

is called the *basic set of solutions at* x_0 of $L[y] = 0$ on I.

Definition 6.4.3. Any set of n linearly independent solutions of $L[y] = 0$ on the interval I: $a \leq x \leq b$ is said to form a *fundamental set* or *fundamental system* of $L[y] = 0$ on I.

[*Note:* When $n = 2$, Definitions 6.4.1, 6.4.2, and 6.4.3 are the same as Definitions 4.4.1, 4.4.2, and 4.4.3.]

Example 6.4.2. One may easily verify that the functions y_1, y_2, and y_3 defined by $y_1(x) = e^{-x}$, $y_2(x) = e^{2x} \cos 3x$, and $y_3(x) = e^{2x} \sin 3x$ are solutions of the differential equation

$$y''' - 3y'' + 9y' + 13y = 0.$$

Also, one may verify that the functions y_1, y_2, and y_3 are linearly independent on the interval I: $a \leq x \leq b$, where $a > -\infty$ and $b < \infty$. Thus, the general solution of the above differential equation on I is given by

$$y(x) = c_1 e^{-x} + c_2 e^{2x} \cos 3x + c_3 e^{2x} \sin 3x,$$

where c_1, c_2, and c_3 are arbitrary constants.

Definition 6.4.4. Let each of the n functions $f_1, f_2, \ldots, f_n$ be $n - 1$ times differentiable in the interval I: $a \leq x \leq b$. The determinant

$$W(f_1, f_2, \ldots, f_n) = \begin{vmatrix} f_1 & f_2 & \cdots & f_n \\ f_1' & f_2' & \cdots & f_n' \\ \cdots & \cdots & \cdots & \cdots \\ f_1^{(n-1)} & f_2^{(n-1)} & \cdots & f_n^{(n-1)} \end{vmatrix} \qquad (6.4.15)$$

is called the *Wronskian* of $f_1, f_2, \ldots, f_n$. Its value at any point x in I will be denoted by $W(f_1, f_2, \ldots, f_n ; x)$. Thus,

$$W(f_1, f_2, \ldots, f_n ; x) = \begin{vmatrix} f_1(x) & f_2(x) & \cdots & f_n(x) \\ f_1'(x) & f_2'(x) & \cdots & f_n'(x) \\ \cdots & \cdots & \cdots & \cdots \\ f_1^{(n-1)}(x) & f_2^{(n-1)}(x) & \cdots & f_n^{(n-1)}(x) \end{vmatrix}.$$

$$(6.4.16)$$

Example 6.4.3. If $f_1(x) = x, f_2(x) = x^2$, and $f_3(x) = x^3$ in I: $-\infty < x < \infty$, then

$$W(f_1, f_2, f_3 ; x) = \begin{vmatrix} x & x^2 & x^3 \\ 1 & 2x & 3x^2 \\ 0 & 2 & 6x \end{vmatrix} = 2x^3.$$

Example 6.4.4. If $f_1(x) = \sin x$, $f_2(x) = \cos x$, $f_3(x) = \sinh x$, and $f_4(x) = \cosh x$ in I: $-\infty < x < \infty$, then

$$W(f_1, f_2, f_3, f_4 ; x) = \begin{vmatrix} \sin x & \cos x & \sinh x & \cosh x \\ \cos x & -\sin x & \cosh x & \sinh x \\ -\sin x & -\cos x & \sinh x & \cosh x \\ -\cos x & \sin x & \cosh x & \sinh x \end{vmatrix}$$

$$= \begin{vmatrix} \sin x & \cos x & 0 & 0 \\ \cos x & -\sin x & 0 & 0 \\ 0 & 0 & 2\sinh x & 2\cosh x \\ 0 & 0 & 2\cosh x & 2\sinh x \end{vmatrix}$$

$$= 4 \begin{vmatrix} \sin x & \cos x \\ \cos x & -\sin x \end{vmatrix} \cdot \begin{vmatrix} \sinh x & \cosh x \\ \cosh x & \sinh x \end{vmatrix} = 4.$$

Theorems 6.4.3 and 6.4.4 are generalizations of Theorems 4.4.3 and 4.4.4. The proofs of these theorems will be left as exercises for the reader. [*See* Exercises 6.4.2 and 6.4.3.]

Theorem 6.4.3. *Let $y_1 = y_1(x)$, $y_2 = y_2(x)$, $\ldots$, $y_n = y_n(x)$ be n solutions of $L[y] = 0$ on the interval I: $a \le x \le b$. A necessary and sufficient condition that $y_1, y_2, \ldots, y_n$ be linearly independent on I is that $W(y_1, y_2, \ldots, y_n ; x) \ne 0$ for every x in I.*

Theorem 6.4.4. *Let $y_1 = y_1(x)$, $y_2 = y_2(x)$, $\ldots$, $y_n = y_n(x)$ be n solutions of $L[y] = 0$ on the interval I: $a \le x \le b$, and let x_0 be any fixed point in I. A* *necessary and sufficient condition that $y_1, y_2, \ldots, y_n$ be linearly independent on I is that $W(y_1, y_2, \ldots, y_n; x_0) \neq 0$.*

Example 6.4.5. One may easily verify that the functions y_1, y_2, y_3, and y_4 defined by $y_1(x) = \sin x$, $y_2(x) = \cos x$, $y_3(x) = \sinh x$, and $y_4(x) = \cosh x$ are solutions of the differential equation

$$y^{(4)} - y = 0.$$

By Example 6.4.4, $W(y_1, y_2, y_3, y_4; x) = 4$ for all x in I: $-\infty < x < \infty$. Thus, by Theorem 6.4.3, the functions y_1, y_2, y_3, and y_4 are linearly independent on I. In view of Definition 6.4.1, the general solution of the above differential equation on I is given by

$$y(x) = c_1 \sin x + c_2 \cos x + c_3 \sinh x + c_4 \cosh x,$$

where c_1, c_2, c_3, and c_4 are arbitrary constants.

Remark 6.4.1. In the theory of determinants, the following result is established. If the elements a_{ij}, $i, j = 1, 2, \ldots, n$ of a determinant A are differentiable functions of the variable x, then

$$\frac{d}{dx}(\det A) = \frac{d}{dx}
\begin{vmatrix}
a_{11} & a_{12} & \cdots & a_{1n} \\
a_{21} & a_{22} & \cdots & a_{2n} \\
\cdots & \cdots & \cdots & \cdots \\
a_{n1} & a_{n2} & \cdots & a_{nn}
\end{vmatrix}
=
\begin{vmatrix}
a'_{11} & a'_{12} & \cdots & a'_{1n} \\
a_{21} & a_{22} & \cdots & a_{2n} \\
\cdots & \cdots & \cdots & \cdots \\
a_{n1} & a_{n2} & \cdots & a_{nn}
\end{vmatrix}$$

$$+
\begin{vmatrix}
a_{11} & a_{12} & \cdots & a_{1n} \\
a'_{21} & a'_{22} & \cdots & a'_{2n} \\
\cdots & \cdots & \cdots & \cdots \\
a_{n1} & a_{n2} & \cdots & a_{nn}
\end{vmatrix}
+ \cdots +
\begin{vmatrix}
a_{11} & a_{12} & \cdots & a_{1n} \\
a_{21} & a_{22} & \cdots & a_{2n} \\
\cdots & \cdots & \cdots & \cdots \\
a'_{n1} & a'_{n2} & \cdots & a'_{nn}
\end{vmatrix},$$

$$\tag{6.4.17}$$

where $a'_{ij} = (d/dx)a_{ij}(x)$. That is, the derivative of a determinant is the sum of n determinants, each the same as $\det A$ except that, in the first, the elements of the first row have been replaced by their derivatives with respect to x; in the second, the elements of the second row have been replaced by their derivatives with respect to x; and so on. (A similar result holds for columns.)

Example 6.4.6.

$$\frac{d}{dx}
\begin{vmatrix}
x & e^x & \sin x \\
x^2 & e^{2x} & \cosh x \\
x^3 & e^{3x} & \sinh x
\end{vmatrix}
=
\begin{vmatrix}
1 & e^x & \cos x \\
x^2 & e^{2x} & \cosh x \\
x^3 & e^{3x} & \sinh x
\end{vmatrix}
+
\begin{vmatrix}
x & e^x & \sin x \\
2x & 2e^{2x} & \sinh x \\
x^3 & e^{3x} & \sinh x
\end{vmatrix}$$

$$+
\begin{vmatrix}
x & e^x & \sin x \\
x^2 & e^{2x} & \cosh x \\
3x^2 & 3e^{3x} & \cosh x
\end{vmatrix}.$$

Also,

$$\frac{d}{dx}\begin{vmatrix} x & e^x & \sin x \\ x^2 & e^{2x} & \cosh x \\ x^3 & e^{3x} & \sinh x \end{vmatrix} = \begin{vmatrix} 1 & e^x & \sin x \\ 2x & e^{2x} & \cosh x \\ 3x^2 & e^{3x} & \sinh x \end{vmatrix} + \begin{vmatrix} x & e^x & \sin x \\ x^2 & 2e^{2x} & \cosh x \\ x^3 & 3e^{3x} & \sinh x \end{vmatrix}$$

$$+ \begin{vmatrix} x & e^x & \cos x \\ x^2 & e^{2x} & \sinh x \\ x^3 & e^{3x} & \cosh x \end{vmatrix}.$$

Example 6.4.7.   Let each of the n functions $f_1, f_2, \ldots, f_n$ be n-times differentiable in the interval I: $a \le x \le b$.   Let us show that

$$W'(f_1, f_2, \ldots, f_n \, ; x) = \begin{vmatrix} f_1(x) & f_2(x) & \cdots & f_n(x) \\ f_1'(x) & f_2'(x) & \cdots & f_n'(x) \\ \cdots & \cdots & \cdots & \cdots \\ f_1^{(n-2)}(x) & f_2^{(n-2)}(x) & \cdots & f_n^{(n-2)}(x) \\ f_1^{(n)}(x) & f_2^{(n)}(x) & \cdots & f_n^{(n)}(x) \end{vmatrix}, \qquad (6.4.18)$$

in I.

PROOF.   Utilizing Formula (6.4.17) to differentiate the Wronskian, we obtain

$$W'(f_1, f_2, \ldots, f_n; x) = \begin{vmatrix} f_1'(x) & f_2'(x) & \cdots & f_n'(x) \\ f_1'(x) & f_2'(x) & \cdots & f_n'(x) \\ f_1''(x) & f_2''(x) & \cdots & f_n''(x) \\ \cdots & \cdots & \cdots & \cdots \\ f_1^{(n-2)}(x) & f_2^{(n-2)}(x) & \cdots & f_n^{(n-2)}(x) \\ f_1^{(n-1)}(x) & f_2^{(n-1)}(x) & \cdots & f_n^{(n-1)}(x) \end{vmatrix}$$

$$+ \begin{vmatrix} f_1(x) & f_2(x) & \cdots & f_n(x) \\ f_1''(x) & f_2''(x) & \cdots & f_n''(x) \\ f_1''(x) & f_2''(x) & \cdots & f_n''(x) \\ \cdots & \cdots & \cdots & \cdots \\ f_1^{(n-2)}(x) & f_2^{(n-2)}(x) & \cdots & f_n^{(n-2)}(x) \\ f_1^{(n-1)}(x) & f_2^{(n-1)}(x) & \cdots & f_n^{(n-1)}(x) \end{vmatrix}$$

$$+ \begin{vmatrix} f_1(x) & f_2(x) & \cdots & f_n(x) \\ f_1'(x) & f_2'(x) & \cdots & f_n'(x) \\ f_1'''(x) & f_2'''(x) & \cdots & f_n'''(x) \\ \cdots & \cdots & \cdots & \cdots \\ f_1^{(n-2)}(x) & f_2^{(n-2)}(x) & \cdots & f_n^{(n-2)}(x) \\ f_1^{(n-1)}(x) & f_2^{(n-1)}(x) & \cdots & f_n^{(n-1)}(x) \end{vmatrix}$$

$$+ \cdots + \begin{vmatrix} f_1(x) & f_2(x) & \cdots & f_n(x) \\ f_1'(x) & f_2'(x) & \cdots & f_n'(x) \\ f_1''(x) & f_2''(x) & \cdots & f_n''(x) \\ \cdots & \cdots & \cdots & \cdots \\ f_1^{(n-1)}(x) & f_2^{(n-1)}(x) & \cdots & f_n^{(n-1)}(x) \\ f_1^{(n-1)}(x) & f_2^{(n-1)}(x) & \cdots & f_n^{(n-1)}(x) \end{vmatrix}$$

$$+ \begin{vmatrix} f_1(x) & f_2(x) & \cdots & f_n(x) \\ f_1'(x) & f_2'(x) & \cdots & f_n'(x) \\ f_1''(x) & f_2''(x) & \cdots & f_n''(x) \\ \cdots & \cdots & \cdots & \cdots \\ f_1^{(n-2)}(x) & f_2^{(n-2)}(x) & \cdots & f_n^{(n-2)}(x) \\ f_1^{(n)}(x) & f_2^{(n)}(x) & \cdots & f_n^{(n)}(x) \end{vmatrix} \qquad (6.4.19)$$

Since each determinant in (6.4.19), except the last one, has two identical rows, each of these determinants with identical rows is equal to zero. Thus, Formula (6.4.18) is established.

The following theorem follows from Theorems 6.4.3 and 6.4.4. However, we shall give an alternative proof which will lead us to develop a formula for $W(y_1, y_2, \ldots ; x)$.

Theorem 6.4.5. *Let* $y_1 = y_1(x)$, $y_2 = y_2(x)$, $\ldots$, $y_n = y_n(x)$ *be* n *solutions of* $L[y] = 0$ *on an interval* I: $a \leq x \leq b$. *Then either*

$$W(y_1, y_2, \ldots, y_n ; x) = 0 \qquad (6.4.20)$$

for every x *in* I, *or*

$$W(y_1, y_2, \ldots, y_n ; x) \neq 0 \qquad (6.4.21)$$

for every x *in* I.

PROOF. $L[y] = a_0(x)y^{(n)} + a_1(x)y^{(n-1)} + \cdots + a_{n-1}(x)y' + a_n(x)y = 0$, where $a_0, a_1, \ldots, a_n$ are continuous functions on I and $a_0(x) \neq 0$ for every x in I. Since $y_1, y_2, \ldots, y_n$ are solutions of $L[y] = 0$, we have

$$y_1^{(n)}(x) = - \frac{a_1(x)}{a_0(x)} y_1^{(n-1)}(x) - \frac{a_2(x)}{a_0(x)} y_1^{(n-2)}(x) - \cdots$$

$$- \frac{a_{n-1}(x)}{a_0(x)} y_1'(x) - \frac{a_n(x)}{a_0(x)} y_1(x) = - \sum_{k=1}^{n} \frac{a_k(x)}{a_0(x)} y_1^{(n-k)}(x),$$

and similarly

$$y_2^{(n)}(x) = - \sum_{k=1}^{n} \frac{a_k(x)}{a_0(x)} y_2^{(n-k)}(x), \ldots, y_n^{(n)}(x) = - \sum_{k=1}^{n} \frac{a_k(x)}{a_0(x)} y_n^{(n-k)}(x),$$

where $y_i^{(0)}(x) = y_i(x)$, $i = 1, 2, \ldots, n$.   Utilizing (6.4.18) with the functions f_i and their derivatives replaced by the functions y_i and their derivatives, and then substituting the above values of $y_1^{(n)}(x), y_2^{(n)}(x), \ldots, y_n^{(n)}(x)$ into the last row of the determinant, we obtain

$$W'(y_1, y_2, \ldots, y_n; x)$$

$$= \begin{vmatrix} y_1(x) & y_2(x) & \cdots & y_n(x) \\ y_1'(x) & y_2'(x) & \cdots & y_n'(x) \\ y_1''(x) & y_2''(x) & \cdots & y_n''(x) \\ \cdots & \cdots & \cdots & \cdots \\ y_1^{(n-2)}(x) & y_2^{(n-2)}(x) & \cdots & y_n^{(n-2)}(x) \\ -\sum_{k=1}^{n} \frac{a_k(x)}{a_0(x)} y_1^{(n-k)}(x) & -\sum_{k=1}^{n} \frac{a_k(x)}{a_0(x)} y_2^{(n-k)}(x) & \cdots & -\sum_{k=1}^{n} \frac{a_k(x)}{a_0(x)} y_n^{(n-k)}(x) \end{vmatrix}.$$

$$(6.4.22)$$

If we now multiply the first row, the second row, $\ldots$, the $(n-1)$ row of the determinant in (6.4.22), respectively, by

$$\frac{a_n(x)}{a_0(x)}, \frac{a_{n-1}(x)}{a_0(x)}, \ldots, \frac{a_2(x)}{a_0(x)},$$

and then add this result to the last row, we obtain

$$W'(y_1, y_2, \ldots, y_n; x)$$

$$= \begin{vmatrix} y_1(x) & y_2(x) & \cdots & y_n(x) \\ y_1'(x) & y_2'(x) & \cdots & y_n'(x) \\ y_1''(x) & y_2''(x) & \cdots & y_n''(x) \\ \cdots & \cdots & \cdots & \cdots \\ y_1^{(n-2)}(x) & y_2^{(n-2)}(x) & \cdots & y_n^{(n-2)}(x) \\ -\frac{a_1(x)}{a_0(x)} y_1^{(n-1)}(x) & -\frac{a_1(x)}{a_0(x)} y_2^{(n-1)}(x) & \cdots & -\frac{a_1(x)}{a_0(x)} y_n^{(n-1)}(x) \end{vmatrix}.$$

$$(6.4.23)$$

Thus, in view of (6.4.16), the above expression becomes, after we factor out

$$-\frac{a_1(x)}{a_0(x)}$$

from the last row of the determinant,

$$W'(y_1, y_2, \ldots, y_n; x) = -\frac{a_1(x)}{a_0(x)} W(y_1, y_2, \ldots, y_n; x). \qquad (6.4.24)$$

The general solution of the first-order differential equation (6.4.24) is

$$W(y_1, y_2, \ldots, y_n \,; x) = c\,\exp\left[-\int \frac{a_1(x)}{a_0(x)}\, dx\right], \qquad (6.4.25)$$

where c is an arbitrary constant of integration.   Since

$$\exp\left[-\int \frac{a_1(x)}{a_0(x)}\, dx\right] \neq 0$$

for every x in I, it follows from (6.4.25) that if $c = 0$, then $W(y_1, y_2, \ldots, y_n \,; x) = 0$ for every x in I; and if $c \neq 0$, then $W(y_1, y_2, \ldots, y_n \,; x) \neq 0$ for every x in I.   Thus, the theorem is established.

Remark 6.4.2.   We may also write (6.4.25) as

$$W(y_1, y_2, \ldots, y_n; x) = C\,\exp\left[-\int_{x_0}^{x} \frac{a_1(\xi)}{a_0(\xi)}\, d\xi\right], \qquad (6.4.26)$$

where x_0 is a fixed point in I.   Thus, $W(y_1, y_2, \ldots, y_n \,; x_0) = Ce^0 = C$, and (6.4.26) becomes

$$W(y_1, y_2, \ldots, y_n; x) = W(y_1, y_2, \ldots, y_n; x_0)\,\exp\left[-\int_{x_0}^{x} \frac{a_1(\xi)}{a_0(\xi)}\, d\xi\right]. \qquad (6.4.27)$$

This result is known as *Abel's identity* or *Liouville's formula*, and is very useful.

Exercises 6.4

1. Prove Theorem 6.4.1.
2. Prove Theorem 6.4.3.
3. Prove Theorem 6.4.4.
4. Let each of the n functions $y_1, y_2, \ldots, y_n$ be $n-1$ times differentiable in an interval I.
   (a) Show that if $y_1, y_2, \ldots, y_n$ are linearly dependent on I, then $W(y_1, y_2, \ldots, y_n \,; x) = 0$ for every x in I.
   (b) Show that if $W(y_1, y_2, \ldots, y_n \,; x_0) \neq 0$ for some point x_0 in I, then the functions $y_1, y_2, \ldots, y_n$ are linearly independent on I.
5. Establish the following result by mathematical induction:

$$(1) \qquad \begin{vmatrix} 1 & 1 & \cdots & 1 & 1 \\ x_n & x_{n-1} & \cdots & x_2 & x_1 \\ x_n^2 & x_{n-1}^2 & \cdots & x_2^2 & x_1^2 \\ \cdots & \cdots & \cdots & \cdots & \cdots \\ x_n^{n-2} & x_{n-1}^{n-2} & \cdots & x_2^{n-2} & x_1^{n-2} \\ x_n^{n-1} & x_{n-1}^{n-1} & \cdots & x_2^{n-1} & x_1^{n-1} \end{vmatrix} = \prod_{1 \leq i < j \leq n} (x_i - x_j),$$

where

$$(2) \quad \prod_{1 \le i < j \le n} (x_i - x_j) = [(x_1 - x_2)(x_1 - x_3)(x_1 - x_4) \cdots (x_1 - x_n)]$$
$$[(x_2 - x_3)(x_2 - x_4) \cdots (x_2 - x_n)] \ldots [(x_{n-1} - x_n)].$$

Example:

$$\begin{vmatrix} 1 & 1 & 1 & 1 \\ x_4 & x_3 & x_2 & x_1 \\ x_4^2 & x_3^2 & x_2^2 & x_1^2 \\ x_4^3 & x_3^3 & x_2^3 & x_1^3 \end{vmatrix} = (x_1 - x_2)(x_1 - x_3)(x_1 - x_4)(x_2 - x_3)(x_2 - x_4)(x_3 - x_4).$$

[*Note:* In view of (2), the determinant in (1) vanishes if and only if $x_i = x_j$ for at least one pair of integers i and j such that $1 \le i < j \le n$.]   The determinant in (1) is known as the *Vandermonde determinant*.

   Utilize Exercise 4(b) and the above result to show that the functions $y_1, y_2, \ldots, y_n$ defined by $y_1(x) = e^{\alpha_1 x}$, $y_2(x) = e^{\alpha_2 x}, \ldots, y_n(x) = e^{\alpha_n x}$, where $\alpha_1, \alpha_2, \ldots, \alpha_n$ are distinct constants, are linearly independent on the interval $I: \ -\infty < x < \infty$.

6. Let the set of n functions $y_1, y_2, \ldots, y_n$ be defined by $y_1(x) = 1$, $y_2(x) = x, \ldots, y_n(x) = x^{n-1}$.   Show that

$$W(y_1, y_2, \ldots, y_n ; x) = 1!1!2!3! \cdots (n-1)! = \prod_{k=0}^{n-1} k!$$

[By definition $0! = 1$.]   Thus, the given set of functions is linearly independent on $I: \ -\infty < x < \infty$.

7. Let the set of n functions $y_1, y_2, \ldots, y_n$ be defined by $y_1(x) = e^{\alpha x}$, $y_2(x) = xe^{\alpha x}, \ldots, y_n(x) = x^{n-1} e^{\alpha x}$, where α is a constant.   Show that

$$W(y_1, y_2, \ldots, y_n ; x) = e^{n \alpha x} \prod_{k=0}^{n-1} k!$$

Thus, the given set of functions is linearly independent on $I: \ -\infty < x < \infty$. [*Note:* The constant α may be complex.]

8. Generalize Exercise 4.4.10(a).

9. One may extend the Definitions 4.3.4 through 4.3.6 to complex-valued functions of the real variable x.   For example, the set of n complex-valued functions $f_1, f_2, \ldots, f_n$ of the real variable x is said to be *linearly dependent* on the interval $I: \ a \le x \le b$, if there exist complex constants (real, imaginary, or both) $c_1, c_2, \ldots, c_n$, not all zero, such that

$$c_1 f_1(x) + c_2 f_2(x) + \cdots + c_n f_n(x) = 0$$

for every x in I.

   Show that the complex-valued functions f_1, f_2, f_3, f_4 defined below are linearly dependent on the interval $I: \ -\infty < x < \infty$.

(a) $f_1(x) = \cos x + i \sin x$, $f_2(x) = \cos x - i \sin x$, $f_3(x) = 2i \cos x$.

(b) $f_1(x) = x + ix^2$, $f_2(x) = x^3 - ix$, $f_3(x) = x^2 + ie^x$, $f_4(x) = -e^x + ix^3$.

10. We shall now extend Exercise 4.4.10(a) to complex-valued functions of the real variable x. Let the function f be defined by $f(x) = u(x) + iv(x)$, where u and v are continuous real-valued functions on the interval I: $a \leq x \leq b$, and $i^2 = -1$. We then say that the function f is *continuous* on I.   The *complex conjugate* of f, denoted by $\overline{f}$, is defined by $\overline{f}(x) = u(x) - iv(x)$.   The *absolute value* of f, denoted by $|f|$, is defined by $|f(x)| = \sqrt{f(x)\overline{f}(x)} = \sqrt{u^2(x) + v^2(x)} \geq 0$.   Clearly, $|f(x)|^2 = f(x)\overline{f}(x) = u^2(x) + v^2(x)$.   Note that $|f(x)| = 0$ if and only if $f(x) = 0$ for every x in I; that is, if and only if $u(x) = 0$ and $v(x) = 0$ for every x in I.   By definition

$$\int_a^b f(x)dx = \int_a^b u(x)dx + i\int_a^b v(x)dx.$$

(a) Let the functions f_j be defined by $f_j(x) = u_j(x) + iv_j(x)$, $j = 1, 2, \ldots, n$, where u_j and v_j are continuous real-valued functions on the interval I: $a \leq x \leq b$.   The *Gramian* of the functions f_j, $j = 1, 2, \ldots, n$, is defined on I by

$$G = \begin{vmatrix} \int_a^b f_1(x)\overline{f_1(x)}dx & \int_a^b f_1(x)\overline{f_2(x)}dx & \cdots & \int_a^b f_1(x)\overline{f_n(x)}dx \\ \int_a^b f_2(x)\overline{f_1(x)}dx & \int_a^b f_2(x)\overline{f_2(x)}dx & \cdots & \int_a^b f_2(x)\overline{f_n(x)}dx \\ \cdots & \cdots & \cdots & \cdots \\ \int_a^b f_n(x)\overline{f_1(x)}dx & \int_a^b f_n(x)\overline{f_2(x)}dx & \cdots & \int_a^b f_n(x)\overline{f_n(x)}dx \end{vmatrix}.$$

Prove that *a necessary and sufficient condition that the functions $f_1, f_2, \ldots, f_n$ be linearly dependent on I is that $G = 0$.*
Show that the complex-valued functions f_1, f_2, f_3 defined below are linearly dependent on the indicated interval.

(b) $f_1(x) = \cos x + i \sin x$, $f_2(x) = \cos x - i \sin x$, $f_3(x) = 2i \cos x$,
$0 \leq x \leq 2\pi$.

(c) $f_1(x) = x + ix^2$, $f_2(x) = -x^2 + ix$, $f_3(x) = -3x^2 + ix$, $0 \leq x \leq 1$.

11. [Review Formulas (4.6.40) through (4.6.44).]   Since $e^{ikx} = \cos kx + i \sin kx$, we see that

$$\int e^{ikx}dx = \int \cos kx\, dx + i \int \sin kx\, dx = \frac{1}{k}(\sin kx - i \cos kx)$$

$$= -\frac{i}{k}(\cos kx + i \sin kx) = -\frac{i}{k}e^{ikx}, \qquad k \neq 0.$$

Show that the complex-valued functions f_k, $k = 1, 2, \ldots, n$, defined by

$$f_k(x) = e^{ikx} = \cos kx + i \sin kx,$$

are linearly independent on the interval I: $0 \leq x \leq 2\pi$.

6.5. Reduction of the Order of a Differential Equation

Before establishing Theorem 6.5.1 which is a generalization of Theorem 4.5.1, we would like to make the following remark.

Remark 6.5.1.   Let $y = uv$, where the functions u and v are each n times differentiable at a point x.   By *Leibnitz's rule, the nth derivative of a product of two functions is given by*

$$y^{(n)} = \sum_{k=0}^{n} C_{n,k}\, u^{(n-k)} v^{(k)}, \tag{6.5.1}$$

where $C_{n,k}$ is the binomial coefficient,

$$C_{n,k} = \frac{n!}{k!\,(n-k)!} = \frac{n(n-1)\cdots(n-k+1)}{k!}, \tag{6.5.2}$$

$u^{(0)} = u$, $v^{(0)} = v$, *and* $0! = 1$.   Formula (6.5.1) may be established by mathematical induction, and by utilizing the following relation encountered in college algebra,

$$C_{n,k} + C_{n,k+1} = C_{n+1,k+1}. \tag{6.5.3}$$

Example 6.5.1.   Suppose that $y = uv$, where $u = e^{2x}$ and $v = \sin 3x$.   Let us find $y^{(4)}$.

SOLUTION.   In view of Formula (6.5.1), with $n = 4$, we have

$$y^{(4)} = \sum_{k=0}^{4} C_{4,k}\, u^{(4-k)} v^{(k)}$$

$$= C_{4,0}\, u^{(4)} v + C_{4,1}\, u''' v' + C_{4,2}\, u'' v'' + C_{4,3}\, u' v''' + C_{4,4}\, u v^{(4)}. \tag{6.5.4}$$

Utilizing Formula (6.5.2), we find that

$$C_{4,0} = 1, \quad C_{4,1} = 4, \quad C_{4,2} = 6, \quad C_{4,3} = 4, \quad \text{and} \quad C_{4,4} = 1.$$

Also,

$$u' = 2e^{2x}, \quad u'' = 4e^{2x}, \quad u''' = 8e^{2x}, \quad u^{(4)} = 16e^{2x}, \quad v' = 3\cos 3x,$$

$$v'' = -9\sin 3x, \quad v''' = -27\cos 3x, \quad \text{and} \quad v^{(4)} = 81\sin 3x.$$

Substituting these values in (6.5.4), we obtain

$$y^{(4)} = \frac{d^4}{dx^4}(e^{2x}\sin 3x) = 16e^{2x}\sin 3x + 96e^{2x}\cos 3x - 216e^{2x}\sin 3x$$

$$- 216e^{2x}\cos 3x + 81e^{2x}\sin 3x.$$

Theorem 6.5.1.   *Consider the equation $L[y] = 0$ given in (6.4.1).   Let $y_1(x)$ be a nontrivial solution of $L[y] = 0$ on the interval I: $a \leq x \leq b$.   Then the transformation*

$$y(x) = y_1(x)v(x) \tag{6.5.5}$$

reduces $L[y] = 0$ on I to a homogeneous differential equation of order $n - 1$

$$b_0(x)u^{(n-1)} + b_1(x)u^{(n-2)} + \cdots + b_{n-1}(x)u = 0, \tag{6.5.6}$$

where

$$b_0(x) = a_0(x)y_1, \quad b_1(x) = na_0(x)y_1' + a_1(x)y_1, \ldots,$$

$$b_{n-1}(x) = na_0(x)y_1^{(n-1)} + (n-1)a_1(x)y_1^{(n-2)} + \cdots + a_{n-1}(x)y_1, \tag{6.5.7}$$

and

$$u = v'.$$

PROOF. To compute the nth derivative, the $(n-1)$st derivative, $\ldots$, the first derivative of the function y given in (6.5.5), we use Formula (6.5.1). Thus,

$$y^{(n)} = \sum_{k=0}^{n} C_{n,k}\, y_1^{(n-k)} v^{(k)},$$

$$y^{(n-1)} = \sum_{k=0}^{n-1} C_{n-1,k}\, y_1^{(n-1-k)} v^{(k)}, \ldots, \quad y' = y_1'v + y_1 v'.$$

Substituting these derivatives and the function y given in (6.5.5) in Equation (6.4.1), we have

$$a_0(x)[y_1^{(n)}v + ny_1^{(n-1)}v' + \cdots + ny_1'v^{(n-1)} + y_1 v^{(n)}]$$
$$+ a_1(x)[y_1^{(n-1)}v + (n-1)y_1^{(n-2)}v' + \cdots + (n-1)y_1'v^{(n-2)} + y_1 v^{(n-1)}]$$
$$+ \cdots + a_{n-1}(x)[y_1'v + y_1 v'] + a_n(x)y_1 v = 0.$$

This equation may also be written as

$$a_0(x)y_1 v^{(n)} + [na_0(x)y_1' + a_1(x)y_1]v^{(n-1)} + \cdots + [na_0(x)y_1^{(n-1)}$$
$$+ (n-1)a_1(x)y_1^{(n-2)} + \cdots + a_{n-1}(x)y_1]v' + [a_0(x)y_1^{(n)} + a_1(x)y_1^{(n-1)}$$
$$+ \cdots + a_{n-1}(x)y_1' + a_n(x)y_1]v = 0. \tag{6.5.8}$$

Since y_1 is a solution of $L[y] = 0$, the expression in the last bracket of Equation (6.5.8) is equal to zero. Thus, letting $u = v'$ and using the notation given in (6.5.7), Equation (6.5.8), with $L[y_1] = 0$, now reduces to Equation (6.5.6), and the theorem is established.

Example 6.5.2. On an interval I not containing the origin, one (nonvanishing) solution of the differential equation

$$x^3 y''' - 3x^2 y'' + 6xy' - 6y = 0, \tag{6.5.9}$$

is $y_1 = x^3$. Let us find the other two linearly independent solutions on I.

SOLUTION.   Let $y(x) = y_1(x)v(x) = x^3 v(x)$.   Then,

$$y''' = \sum_{k=0}^{3} C_{3,k} \, y_1^{(3-k)} v^{(k)} = C_{3,0} \, y_1''' v + C_{3,1} \, y_1'' v' + C_{3,2} \, y_1' v'' + C_{3,3} \, y_1 v'''$$

$$= 6v + 18xv' + 9x^2 v'' + x^3 v'''.$$

$$y'' = \sum_{k=0}^{2} C_{2,k} \, y_1^{(2-k)} v^{(k)} = C_{2,0} \, y_1'' v + C_{2,1} \, y_1' v' + C_{2,2} \, y_1 v''$$

$$= 6xv + 6x^2 v' + x^3 v'',$$

$$y' = 3x^2 v + x^3 v'.$$

Substituting these derivatives and $y = x^3 v$ in Equation (6.5.9), we obtain upon simplification

$$x^2 v''' + 6xv'' + 6v' = 0.$$

Letting $u = v'$, the above equation becomes

$$x^2 u'' + 6xu' + 6u = 0. \tag{6.5.10}$$

[*Note:* Equation (6.5.10) may be obtained directly by utilizing Equations (6.5.6) and (6.5.7), with $n = 3$.]   Equation (6.5.10) is a Cauchy-Euler equation. Using the method developed in Section 4.10, the two linearly independent solutions on I of Equation (6.5.10) are

$$u_1 = \frac{1}{x^3} \quad \text{and} \quad u_2 = \frac{1}{x^2}.$$

Since $v' = u$, we see that

$$v_1 = \int u_1 \, dx = \int \frac{dx}{x^3} = -\frac{1}{2} \frac{1}{x^2} \quad \text{and} \quad v_2 = \int u_2 \, dx = \int \frac{dx}{x^2} = -\frac{1}{x}.$$

As usual, we may take

$$v_1 = \frac{1}{x^2} \quad \text{and} \quad v_2 = \frac{1}{x}.$$

Since $y(x) = x^3 v(x)$, the other two linearly independent solutions on I of Equation (6.5.9) are

$$y_2 = x^3 v_1 = \frac{x^3}{x^2} = x \quad \text{and} \quad y_3 = x^3 v_2 = \frac{x^3}{x} = x^2.$$

Exercises 6.5

1. On the interval I: $-\infty < x < \infty$, $y_1 = e^x$ is one solution of the differential equation

$$y''' - 5y'' + 7y' - 3y = 0.$$

Find the other two linearly independent solutions on I.

2. On the interval I: $0 < x < \infty$, $y_1 = x^4$ is one solution of the differential equation

$$x^3 y''' - 6x^2 y'' + 18xy' - 24y = 0.$$

Find the other two linearly independent solutions on I.

6.6. Linear Homogeneous Equations of Order n with Constant Coefficients

Consider the differential equation given by

$$L[y] = a_0 y^{(n)} + a_1 y^{(n-1)} + \cdots + a_{n-1} y' + a_n y = 0, \qquad (6.6.1)$$

where a_0, a_1, ..., a_n are real constants and $a_0 \neq 0$. In view of Theorem 6.4.2, the equation $L[y] = 0$ has n linearly independent solutions $y_1(x)$, $y_2(x), \ldots, y_n(x)$ on the interval I: $-\infty < x < \infty$. Thus, the general solution $y(x)$ of $L[y] = 0$ on I is given by

$$y(x) = c_1 y_1(x) + c_2 y_2(x) + \cdots + c_n y_n(x), \qquad (6.6.2)$$

where $c_1, \ldots, c_n$ are arbitrary constants.

The results of Section 4.6 extend to nth-order linear differential equations with constant coefficients. Thus, as in the case of second-order differential equations, we try $y = e^{mx}$ as a possible solution of $L[y] = 0$ for suitably chosen values of m. Substituting $y = e^{mx}$ in Equation (6.6.1), we obtain

$$L[y] = (a_0 m^n + a_1 m^{n-1} + \cdots + a_{n-1} m + a_n)e^{mx} = 0.$$

Since $e^{mx} \neq 0$ for every x in I, we see that

$$a_0 m^n + a_1 m^{n-1} + \cdots + a_{n-1} m + a_n = 0. \qquad (6.6.3)$$

Thus, if m is a root of Equation (6.6.3), then $y = e^{mx}$ is a solution of $L[y] = 0$ on I. The polynomial p defined by

$$p(m) = a_0 m^n + a_1 m^{n-1} + \cdots + a_{n-1} m + a_n \qquad (6.6.4)$$

is called the *characteristic polynomial* or *auxiliary polynomial* of the operator L or of $L[y] = 0$. The equation $p(m) = 0$, that is, Equation (6.6.3), is called the *characteristic equation* or *auxiliary equation* of $L[y] = 0$. Observe that the characteristic equation can be obtained from Equation (6.6.1) by replacing $y^{(n)}$ by m^n, $y^{(n-1)}$ by $m^{n-1}, \ldots, y'$ by m and $y^{(0)} \equiv y$ by $m^0 \equiv 1$.

Remark 6.6.1. The form of the solutions of the equation $L[y] = 0$ given in (6.6.1), will depend on the nature of the roots of the characteristic equation $p(m) = 0$ given in (6.6.3). When the degree of the equation $p(m) = 0$ is greater than or equal to three, the task of finding its roots is usually very difficult. The higher-order differential equations that will occur in this section and elsewhere, will be selected with care, and thus are not typical of application problems. We shall consider the following three cases.

Case 1. All the roots of $p(m) = 0$ are real and simple. [*see* Remark 6.6.2.]

Case 2. All the roots of $p(m) = 0$ are real but some of the roots are of multiplicity greater than one. [*see* Remark 6.6.2.]

Case 3. All the roots of $p(m) = 0$ are imaginary but some of the roots are of multiplicity greater than one.

As we shall see, the other possible cases that may occur do not really require additional involvements.

Concerning Case 1. For this case, we have the following result.

Theorem 6.6.1. *Consider the equation $L[y] = 0$ given in (6.6.1). Let $m_1, m_2, \ldots, m_n$ be the real and simple roots of the characteristic Equation (6.6.3). Then, the general solution $y(x)$ of $L[y] = 0$ on the interval I: $-\infty < x < \infty$, is given by*

$$y(x) = c_1 e^{m_1 x} + c_2 e^{m_2 x} + \cdots + c_n e^{m_n x}, \tag{6.6.5}$$

where $c_1, c_2, \ldots, c_n$ are arbitrary constants.

PROOF. Since $m_1, m_2, \ldots, m_n$ are solutions of Equation (6.6.3), the functions $y_1, y_2, \ldots, y_n$ defined by

$$y_1(x) = e^{m_1 x}, \quad y_2(x) = e^{m_2 x}, \quad \ldots, \quad y_n(x) = e^{m_n x}, \tag{6.6.6}$$

are solutions of $L[y] = 0$. By Exercise 6.4.5, the functions $y_1, y_2, \ldots, y_n$ in (6.6.6) are linearly independent on the interval I: $-\infty < x < \infty$. Thus, in view of Definition 6.4.1, the general solution $y(x)$ of $L[y] = 0$ on I is given by (6.6.5) and the theorem is established.

Remark 6.6.2. Before considering Case 2, we give a few basic results on the theory of equations. Let

$$f(x) = a_0 x^n + a_1 x^{n-1} + \cdots + a_{n-1} x + a_n, \tag{6.6.7}$$

where $a_0, a_1, \ldots, a_n$ are constants and $a_0 \neq 0$. Denote the roots of $f(x) = 0$ by $r_1, r_2, \ldots, r_n$. Then, from algebra, the polynomial $f(x)$ can be expressed uniquely in the form

$$f(x) = a_0(x - r_1)(x - r_2) \cdots (x - r_n), \tag{6.6.8}$$

where a_0 is the leading coefficient of $f(x)$. The n linear factors, $x - r_1$, $x - r_2, \ldots, x - r_n$, in (6.6.8) need not be distinct.

A number r is a root of $f(x) = 0$ of *multiplicity* k if the factor $x - r$ occurs exactly k times in the factorization (6.6.8). The number r is also called a *k-fold root* of $f(x) = 0$. Roots of multiplicity 1, 2, 3, and 4 are called, respectively, *simple, double, triple,* and *quadruple* roots.

Example 6.6.1. In the equation

$$f(x) = (x - 2)(x + 3)(x + 3)(x - 4)(x - 4)(x - 4) = 0$$

of degree six, the number 2 is a simple root, -3 is a double root, and 4 is a triple root.

The following result is established in the theory of equations. [*See* [52], p. 47.]

Theorem I. *Let* $f(x) = a_0 x^n + a_1 x^{n-1} + \cdots + a_{n-1}x + a_n$, *where* $a_0, a_1, \ldots,$ a_n *are constants and* $a_0 \neq 0$. *Let r be a root of the equation* $f(x) = 0$. *Then, a necessary and sufficient condition that r be a k-fold root of this equation is that it be a* $(k - 1)$-*fold root of the equation* $f'(x) = 0$.

Example 6.6.2. The equation

$$f(x) = x^4 - 7x^3 + 18x^2 - 20x + 8 = 0 \tag{6.6.9}$$

is satisfied by $x = 2$. Let us show that 2 is a 3-fold root of $f(x) = 0$.

SOLUTION. Differentiating Equation (6.6.9), we obtain

$$f'(x) = 4x^3 - 21x^2 + 36x - 20 = 0. \tag{6.6.10}$$

To determine the roots of $f'(x) = 0$, we utilize the process of synthetic division.

$$
\begin{array}{rl}
4 - 21 + 36 - 20 & \underline{\,|2} \\
\underline{+\; 8 - 26 + 20} & \\
4 - 13 + 10 +\; 0, & \\[6pt]
4 - 13 + 10 & \underline{\,|2} \\
\underline{+\; 8 - 10} & \\
4 -\; 5 +\; 0. &
\end{array}
$$

Thus,

$$f'(x) = 4(x - \tfrac{5}{4})(x - 2)(x - 2) = 0, \tag{6.6.11}$$

and 2 is a 2-fold root of $f'(x) = 0$. By Theorem I, 2 is a 3-fold root of $f(x) = 0$.

Note in the above example that $f(2) = 0, f'(2) = 0, f''(2) = 0$, and $f'''(2) \neq 0$. It follows from Theorem I that if *r is a root of multiplicity k of the equation* $f(x) = 0$, *then*

$$f(r) = 0, \quad f'(r) = 0, \quad \ldots, \quad f^{(k-1)}(r) = 0, \quad \text{and} \quad f^{(k)}(r) \neq 0. \tag{6.6.12}$$

Concerning Case 2. Let the operator L be defined by

$$L[y] = a_0 y^{(n)} + a_1 y^{(n-1)} + \cdots + a_{n-1}y' + a_n y, \tag{6.6.13}$$

where $a_0, a_1, \ldots, a_n$ are real constants and $a_0 \neq 0$. Then

$$L[e^{mx}] = (a_0 m^n + a_1 m^{n-1} + \cdots + a_{n-1} m + a_n)e^{mx}. \qquad (6.6.14)$$

In view of (6.6.4), we may write (6.6.14) as

$$L[e^{mx}] = p(m)e^{mx}. \qquad (6.6.15)$$

Suppose now that m_{k+1} is a root of multiplicity s_{k+1} of $p(m) = 0$. Then, in view of (6.6.12), we have

$$p(m_{k+1}) = 0, \quad p'(m_{k+1}) = 0, \quad \ldots, \quad p^{(s_{k+1}-1)}(m_{k+1}) = 0,$$
$$\text{and} \quad p^{(s_{k+1})}(m_{k+1}) \neq 0. \qquad (6.6.16)$$

Differentiating (6.6.15) j times with respect to m, we get

$$\frac{\partial^j}{\partial m^j}\left[p(m)e^{mx}\right] = \frac{\partial^j}{\partial m^j} L[e^{mx}] = L\left[\frac{\partial^j}{\partial m^j} e^{mx}\right] = L[x^j e^{mx}]. \qquad (6.6.17)$$

Utilizing Leibnitz's formula (6.5.1) to find $\partial^j[p(m)e^{mx}]/\partial m^j$, we obtain

$$L[x^j e^{mx}] = [p^{(j)}(m) + \frac{j}{1!} p^{(j-1)}(m)x + \frac{j(j-1)}{2!} p^{(j-2)}(m)x^2 + \cdots + p(m)x^j]e^{mx}. \qquad (6.6.18)$$

In view of (6.6.16) and (6.6.18), we see that for $j = 0, 1, \ldots, s_{k+1} - 1$, the functions $u_1, u_2, \ldots, u_{s_{k+1}}$ defined by

$$u_1(x) = e^{m_{k+1}x}, \quad u_2(x) = xe^{m_{k+1}x}, \quad \ldots, \quad u_{s_{k+1}}(x) = x^{s_{k+1}-1}e^{m_{k+1}x} \qquad (6.6.19)$$

are solutions of $L[y] = 0$. Similarly, if $m_{k+2}, \ldots, m_l$ are roots of $p(m) = 0$ of multiplicity $s_{k+2}, \ldots, s_l$, then the functions $v_1, v_2, \ldots, v_{s_{k+2}}; \ldots;$ $w_1, w_2, \ldots, w_{s_l}$ defined by

$$v_1(x) = e^{m_{k+2}x}, \quad v_2(x) = xe^{m_{k+2}x}, \quad \ldots, \quad v_{s_{k+2}}(x) = x^{s_{k+2}-1}e^{m_{k+2}x}; \ldots;$$
$$w_1(x) = e^{m_l x}, \quad w_2(x) = xe^{m_l x}, \quad \ldots, \quad w_{s_l} = x^{m_l-1}e^{m_l x}, \qquad (6.6.20)$$

are also solutions of $L[y] = 0$. From Exercise 6.3.3, the functions $y_1, y_2, \ldots,$ $y_n; u_1, u_2, \ldots, u_{s_{k+1}}; \ldots; w_1, w_2, \ldots, w_{s_l}$ defined in (6.6.6), (6.6.19), and (6.6.20) are linearly independent on the interval I: $-\infty < x < \infty$. [*Note:* In view of this exercise, these functions are also linearly independent on I even when the exponents of e, which are the roots of $p(m) = 0$, are complex numbers.]

For Case 2, we thus have the following result.

Theorem 6.6.2. *Consider the equation $L[y] = 0$ given in (6.6.1). Let $m_1, m_2,$ $\ldots, m_k$ be the real and simple roots of the characteristic equation $p(m) = 0$ given in (6.6.3), and let $m_{k+1}, m_{k+2}, \ldots, m_l$ be real roots of $p(m) = 0$ of multiplicity $s_{k+1}, s_{k+2}, \ldots, s_l$, respectively, and $s_{k+1} + s_{k+2} + \cdots + s_l = n - k$. Then* the general solution $y(x)$ of $L[y] = 0$ on the interval I: $-\infty < x < \infty$, is given by

$$y(x) = b_1 e^{m_1 x} + b_2 e^{m_2 x} + \cdots + b_k e^{m_k x}$$
$$+ (c_1 + c_2 x + \cdots + c_{s_{k+1}} x^{s_{k+1}-1}) e^{m_{k+1} x}$$
$$+ (d_1 + d_2 x + \cdots + d_{s_{k+2}} x^{s_{k+2}-1}) e^{m_{k+2} x}$$
$$+ \cdots + (e_1 + e_2 x + \cdots + e_{s_l} x^{s_l-1}) e^{m_l x}, \tag{6.6.21}$$

where $b_1, \ldots, b_k, c_1, \ldots, c_{s_{k+1}}, d_1, \ldots, d_{s_{k+2}}, \ldots, e_1, \ldots, e_{s_l}$ are arbitrary constants.

Example 6.6.3. Let us find the general solution $y(x)$ on I: $-\infty < x < \infty$ of

$$y^{(7)} + y^{(6)} - 6y^{(5)} - 6y^{(4)} + 9y''' + 9y'' - 4y' - 4y = 0. \tag{6.6.22}$$

SOLUTION. The characteristic equation $p(m) = 0$ of (6.2.22) is

$$m^7 + m^6 - 6m^5 - 6m^4 + 9m^3 + 9m^2 - 4m - 4 = 0. \tag{6.6.23}$$

Utilizing the process of synthetic division, the roots of $p(m) = 0$ are $-2, 2, 1$, $1, -1, -1, -1$. Thus, 1 and -1 are roots of $p(m) = 0$ of multiplicity 2 and 3, respectively; -2 and 2 are simple roots. In view of Formula (6.6.21), the general solution $y(x)$ of Equation (6.6.22) on I is

$$y(x) = b_1 e^{-2x} + b_2 e^{2x} + (c_1 + c_2 x)e^x + (d_1 + d_2 x + d_3 x^2)e^{-x}. \tag{6.6.24}$$

The above solution is usually written as

$$y(x) = c_1 e^{-2x} + c_2 e^{2x} + (c_3 + c_4 x)e^x + (c_5 + c_6 x + c_7 x^2)e^{-x},$$

where $c_1, \ldots, c_7$ are arbitrary constants.

Concerning Case 3. Recall from algebra that if all the roots of the characteristic equation $p(m) = 0$ given in (6.6.3) are imaginary, then n, the degree of the equation $p(m) = 0$, is a positive even integer. Let $m_{2j-1} = \alpha_j + i\beta_j$, $m_{2j} = \alpha_j - i\beta_j$, $i^2 = -1, j = 1, 2, \ldots, k$ be the simple imaginary roots of $p(m) = 0$. To obtain the general solution of $L[y] = 0$ arising from these roots, for each j, proceed in a similar fashion as was used to establish Theorem 4.6.2, and then take a linear combination of the resulting solutions of each j. For the remaining imaginary roots of $p(m) = 0$ with certain multiplicities, follow the method given to establish Theorem 6.6.2. The details of the following theorem are left as an exercise for the reader. [*See* Exercise 6.6.18.]

Theorem 6.6.3. *Consider the equation $L[y] = 0$ given in (6.6.1). Let $m_{2j-1} = \alpha_j + i\beta_j, m_{2j} = \alpha_j - i\beta_j, j = 1, 2, \ldots, k$ be the simple imaginary roots of the characteristic equation $p(m) = 0$ given in (6.6.3) and $m_{2j-1} = \alpha_j + i\beta_j$, $m_{2j} = \alpha_j - i\beta_j, j = k + 1, k + 2, \ldots, l$ be the imaginary roots of $p(m) = 0$ of* *multiplicity* $s_{k+1}, s_{k+2}, \ldots, s_l$, *respectively, and* $s_{k+1} + s_{k+2} + \cdots + s_l = n - 2k$. *Then the general solution* $y(x)$ *of* $L[y] = 0$ *on the interval I:* $-\infty < x < \infty$, *is given by*

$$
\begin{aligned}
y(x) = {}& e^{\alpha_1 x}(b_1 \cos \beta_1 x + b_2 \sin \beta_1 x) + e^{\alpha_2 x}(b_3 \cos \beta_2 x + b_4 \sin \beta_2 x) + \cdots \\
& + e^{\alpha_k x}(b_{2k-1} \cos \beta_k x + b_{2k} \sin \beta_k x) \\
& + e^{\alpha_{k+1} x}[(c_1 + c_2 x + \cdots + c_{s_{k+1}} x^{s_{k+1}-1}) \cos \beta_{k+1} x \\
& + (d_1 + d_2 x + \cdots + d_{s_{k+1}} x^{s_{k+1}-1}) \sin \beta_{k+1} x] \\
& + e^{\alpha_{k+2} x}[(e_1 + e_2 x + \cdots + e_{s_{k+2}} x^{s_{k+2}-1}) \cos \beta_{k+2} x \\
& + (f_1 + f_2 x + \cdots + f_{s_{k+2}} x^{s_{k+2}-1}) \sin \beta_{k+2} x] \\
& + \cdots + e^{\alpha_l x}[(g_1 + g_2 x + \cdots + g_{s_l} x^{s_l-1}) \cos \beta_l x \\
& + (h_1 + h_2 x + \cdots + h_{s_l} x^{s_l-1}) \sin \beta_l x],
\end{aligned}
$$

$$(6.6.25)$$

where $b_1, \ldots, b_{2k}$, $c_1, \ldots, c_{s_{k+1}}$, $d_1, \ldots, d_{s_{k+1}}$, $e_1, \ldots, e_{s_{k+2}}$, $f_1, \ldots,$ $f_{s_{k+2}}, \ldots,$ $g_1, \ldots, g_{s_l}$, $h_1, \ldots, h_{s_l}$ *are arbitrary constants.*

Remark 6.6.3.   It should now be evident how one would handle other possible cases.   For example, one may utilize the results of Theorems 6.6.2 and 6.6.3 to find the general solution of the equation $L[y] = 0$ given in (6.6.1) when the roots of the characteristic equation $p(m) = 0$ given in (6.6.3) are real and imaginary, with some of these roots occurring of a certain multiplicity.

Example 6.6.4.   Let us find the general solution $y(x)$ on $I:$ $-\infty < x < \infty$ of

$$y^{(4)} - y = 0. \tag{6.6.26}$$

SOLUTION.   The characteristic equation $p(m) = 0$ of Equation (6.6.26) is

$$m^4 - 1 = 0.$$

The roots of $p(m) = 0$ are 1, -1, i, and $-i$.   In view of Theorem 6.6.1, one part of the general solution arising from the roots 1 and -1 is

$$c_1 e^x + c_2 e^{-x}.$$

From Theorem 6.6.3, with $j = 1$, $\alpha_1 = 0$, and $\beta_1 = 1$, the other part of the general solution arising from the roots i and $-i$ is

$$b_1 \cos x + b_2 \sin x.$$

Hence, the general solution $y(x)$ of Equation (6.6.26) on I is

$$y(x) = c_1 e^x + c_2 e^{-x} + b_1 \cos x + b_2 \sin x.$$

The above solution is usually written as

$$y(x) = c_1 e^x + c_2 e^{-x} + c_3 \cos x + c_4 \sin x, \tag{6.6.27}$$

where $c_1, \ldots, c_4$ are arbitrary constants.

Example 6.6.5.   Suppose that the characteristic equation $p(m) = 0$ of the equation $L[y] = 0$ has the following roots

$$1, -2, -3, -3, 4, 4, 4, -2 + 3i, -2 - 3i, 3 + 4i,$$
$$3 - 4i, 3 + 4i, 3 - 4i, 3 + 4i, 3 - 4i.$$

Then, in view of Remark 6.6.3, the general solution $y(x)$ of $L[y] = 0$ on I: $-\infty < x < \infty$, is given by

$$\begin{aligned}
y(x) &= c_1 e^x + c_2 e^{-2x} + (c_3 + c_4 x)e^{-3x} + (c_5 + c_6 x + c_7 x^2)e^{4x} \\
&\quad + e^{-2x}(c_8 \cos 3x + c_9 \sin 3x) + e^{3x}[(c_{10} + c_{11}x + c_{12}x^2) \cos 4x \\
&\quad + (c_{13} + c_{14}x + c_{15}x^2) \sin 4x],
\end{aligned}$$

where $c_1, c_2, \ldots, c_{15}$ are arbitrary constants.

Example 6.6.6.   Let us solve the initial-value problem

$$\begin{cases} y''' - 2y'' - y' + 2y = 0, & \text{(6.6.28)} \\ y(0) = 1, \quad y'(0) = 2, \quad y''(0) = -2. & \text{(6.6.29)} \end{cases}$$

SOLUTION.   The characteristic equation $p(m) = 0$ of Equation (6.6.28) is

$$m^3 - 2m^2 - m + 2 = 0. \tag{6.6.30}$$

One may utilize the process of synthetic division to determine whether the equation $p(m) = 0$ has any rational roots.   However, for the present situation, we may find the roots of $p(m) = 0$ by simply factoring it.   We have

$$m^2(m - 2) - (m - 2) = 0 \quad \text{or} \quad (m^2 - 1)(m - 2) = 0.$$

Thus, the roots of $p(m) = 0$ are -1, 1, and 2.   Hence, the general solution $y(x)$ of Equation (6.6.28) is

$$y(x) = c_1 e^{-x} + c_2 e^x + c_3 e^{2x}, \tag{6.6.31}$$

where c_1, c_2, and c_3 are arbitrary constants.   Differentiating (6.6.31) two times in succession, we get

$$y'(x) = -c_1 e^{-x} + c_2 e^x + 2c_3 e^{2x}, \tag{6.6.32}$$

$$y''(x) = c_1 e^{-x} + c_2 e^x + 4c_3 e^{2x}. \tag{6.6.33}$$

Utilizing the initial conditions (6.6.29) in (6.6.31) through (6.6.33), we obtain

$$\begin{cases} c_1 + c_2 + c_3 = 1, \\ -c_1 + c_2 + 2c_3 = 2, \\ c_1 + c_2 + 4c_3 = -2. \end{cases} \tag{6.6.34}$$

Solving the system (6.6.34), we find that $c_1 = -1, c_2 = 3$, and $c_3 = -1$. Substituting these values in (6.6.31), the solution of our initial-value problem is

$$y(x) = -e^{-x} + 3e^x - e^{2x}. \tag{6.6.35}$$

Remark 6.6.4. If the coefficients a_0, a_1, ..., a_n of the equation $L[y] = 0$ given in (6.6.1) are complex numbers instead of real numbers, then the general solution $y(x)$ of $L[y] = 0$ on I: $-\infty < x < \infty$, is a complex-valued function of the real variable x. Also, if $\alpha + i\beta$ is a complex root of $p(m) = 0$, then it is not necessarily true that its complex conjugate $\alpha - i\beta$ is also a root of $p(m) = 0$, unless the coefficients of $p(m) = 0$ are real. From Exercise 6.3.3, the functions $u_1, u_2, \ldots, u_{n_1+1}; v_1, v_2, \ldots, v_{n_2+1}; \ldots, w_1, w_2, \ldots,$ w_{n_j+1} defined by

$$u_1(x) = e^{m_1 x}, \ u_2(x) = xe^{m_1 x}, \ \ldots, \ u_{n_1+1}(x) = x^{n_1}e^{m_1 x};$$
$$v_1(x) = e^{m_2 x}, \ v_2(x) = xe^{m_2 x}, \ \ldots, \ v_{n_2+1}(x) = x^{n_2}e^{m_2 x}; \ \ldots;$$
$$w_1(x) = e^{m_j x}, \ w_2(x) = xe^{m_j x}, \ \ldots, \ w_{n_j+1}(x) = x^{n_j}e^{m_j x}, \tag{6.6.36}$$

where $n_1, n_2, \ldots, n_j$ are any nonnegative integers and $m_1, m_2, \ldots, m_j$ are any distinct real or complex numbers, are linearly independent on the interval I: $-\infty < x < \infty$. Thus, we have the following essential theorem.

Theorem 6.6.4. *Consider the equation*

$$L[y] = a_0 y^{(n)} + a_1 y^{(n-1)} + \cdots + a_{n-1}y' + a_n y = 0, \tag{6.6.37}$$

where a_0, a_1, ..., a_n are complex numbers with $a_0 \neq 0$, and its characteristic polynomial $p(m) = 0$:

$$a_0 m^n + a_1 m^{n-1} + \cdots + a_{n-1}m + a_n = 0. \tag{6.6.38}$$

Let $m_1, m_2, \ldots, m_k$ be the simple roots (not necessarily real) of $p(m) = 0$, and let $m_{k+1}, m_{k+2}, \ldots, m_l$ be roots of $p(m) = 0$ of multiplicity $s_{k+1}, s_{k+2}, \ldots, s_l$, respectively, and $s_{k+1} + s_{k+2} + \cdots + s_l = n - k$. Then the general solution $y(x)$ of $L[y] = 0$ on the interval I: $-\infty < x < \infty$, is given by

$$\begin{aligned}
y(x) = {} & A_1 e^{m_1 x} + A_2 e^{m_2 x} + \cdots + A_k e^{m_k x} \\
& + (B_1 + B_2 x + \cdots + B_{s_{k+1}}x^{s_{k+1}-1})e^{m_{k+1} x} \\
& + (C_1 + C_2 x + \cdots + C_{s_{k+2}}x^{s_{k+2}-1})e^{m_{k+2} x} + \cdots \\
& + (E_1 + E_2 x + \cdots + E_{s_l}x^{s_l-1})e^{m_l x}, \tag{6.6.39}
\end{aligned}$$

where $A_1, \ldots, A_k, B_1, \ldots, B_{s_{k+1}}, C_1, \ldots, C_{s_{k+2}}, \ldots, E_1, \ldots, E_{s_l}$ are arbitrary complex constants.

Remark 6.6.5. Since a complex number may be a real number, Theorems 6.6.1 through 6.6.3 may be obtained from Theorem 6.6.4.

Example 6.6.7.   Let us find the general solution $y(x)$ on I: $-\infty < x < \infty$ of

$$y''' - (1 + 4i)y'' - (5 - 2i)y' + (1 + 2i)y = 0. \tag{6.6.40}$$

SOLUTION.   The characteristic equation $p(m) = 0$ of Equation (6.6.40) is

$$m^3 - (1 + 4i)m^2 - (5 - 2i)m + 1 + 2i = 0.$$

Utilizing the process of synthetic division, we find that the roots of $p(m) = 0$ are $1 + 2i$, i, i.   Thus, i is a root of $p(m) = 0$ of multiplicity 2.   Hence, the general solution $y(x)$ of Equation (6.6.40) on I is

$$y(x) = C_1 e^{(1 + 2i)x} + (C_2 + C_3 x)e^{ix}, \tag{6.6.41}$$

where C_1, C_2, and C_3 are arbitrary complex constants.

Note that the function y given in (6.6.41) is a complex-valued function of the real variable x.

Exercises 6.6

In Exercises 1 through 11, find the general solution $y(x)$ of each differential equation on the interval I: $-\infty < x < \infty$.

1. $y^{(4)} - 13y'' + 36y = 0.$
2. $y''' - 5y'' + 3y' + 9y = 0.$
3. $y''' - 3y'' + 3y' - y = 0.$
4. $4y''' - 16y'' - 9y' + 36y = 0.$
5. $y''' - 4y'' + y' + 26y = 0.$
6. $y^{(4)} - 4y''' + 8y'' - 8y' + 4y = 0.$
   (One root of $p(m) = 0$ is $1 + i$.)
7. $36y^{(4)} - 12y''' - 11y'' + 2y' + y = 0.$ (Two roots of $p(m) = 0$ are $\frac{1}{2}$, $-\frac{1}{3}$.)
8. $y^{(5)} - y^{(4)} + y''' + 35y'' + 16y' - 52y = 0.$ (Two roots of $p(m) = 0$ are -2 and $2 + 3i$.)
9. $y''' - (3 + 4i)y'' - (4 - 12i)y' + 12y = 0.$ (One root of $p(m) = 0$ is 3.)
10. $y^{(4)} - (3 + i)y''' + (4 + 3i)y'' = 0.$ (Two roots of $p(m) = 0$ are 0 and $1 + 2i$.)
11. $y''' - (2 - 4i)y'' - (15 - 4i)y' + (36 - 48i)y = 0.$ (One root of $p(m) = 0$ is 3.)

In Exercises 12 through 15, solve each initial-value problem.

12. $\begin{cases} y''' - 3y'' - 4y' + 12y = 0, \\ y(0) = 1, \quad y'(0) = 5, \quad y''(0) = -1. \end{cases}$

13. $\begin{cases} y''' + 2y'' - 16y = 0, \\ y(0) = 3, \quad y'(0) = 4, \quad y''(0) = 0. \end{cases}$

14. $\begin{cases} y^{(4)} - 2y''' + 2y' - y = 0, \\ y(0) = 1, \quad y'(0) = -1, \quad y''(0) = -3, \quad y'''(0) = 3. \end{cases}$

15. $\begin{cases} y^{(5)} - y^{(4)} = 0, \\ y(1) = 3, \quad y'(1) = 2, \quad y''(1) = 1, \quad y'''(1) = -5, \quad y^{(4)}(1) = 1. \end{cases}$

16. Find the solution of the differential equation

$$y''' + 2y'' - 4y' - 8y = 0,$$

subject to the following conditions:

$$y(0) = 2, \quad y(1) = e^{-2}, \quad \text{and as } x \to \infty, y \to 0.$$

17. Find the solution of the differential equation

$$y^{(4)} + y''' - 3y'' - 5y' - 2y = 0,$$

subject to the following conditions:

$$y(0) = -1, \quad y(1) = e^{-1}, \quad y(-1) = e, \quad \text{and as } x \to \infty, y \to 0.$$

18. Carry out the details of the proof of Theorem 6.6.3.

6.7. Nonhomogeneous Linear Differential Equations of Order n

In view of (6.3.1), we shall often write Equation (6.2.1) as

$$L[y] = a_0(x)y^{(n)} + a_1(x)y^{(n-1)} + \cdots + a_{n-1}(x)y' + a_n(x)y = F(x), \quad (6.7.1)$$

where the functions $a_0, \ldots, a_n$, and F are continuous on I: $a \le x \le b$, $a_0(x) \ne 0$ for every x in I, and $F(x)$ is not identically zero for all x in I. The related homogeneous differential equation of (6.7.1) is

$$L[y] = a_0(x)y^{(n)} + a_1(x)y^{(n-1)} + \cdots + a_{n-1}(x)y' + a_n(x)y = 0. \quad (6.7.2)$$

The following two theorems are extensions of Theorems 4.7.1 and 4.7.2 and the proofs will be left as exercises for the reader. [*See* Exercise 6.7.1.]

Theorem 6.7.1. *Let $u(x)$ and $v(x)$ be two solutions of $L[y] = F(x)$ on the interval I: $a \le x \le b$. Then $u(x) - v(x)$ is a solution of $L[y] = 0$ on I.*

Theorem 6.7.2. *Let $y_1, y_2, \ldots, y_n$ be n linearly independent solutions of $L[y] = 0$ on the interval I: $a \le x \le b$, and let $v(x)$ be a solution of $L[y] = F(x)$ on I. Then every solution of $L[y] = F(x)$ on I can be expressed in the form*

$$c_1 y_1(x) + c_2 y_2(x) + \cdots + c_n y_n(x) + v(x)$$

for suitably chosen constants $c_1, c_2, \ldots, c_n$.

Definition 6.7.1. Let $y_1 = y_1(x), \quad y_2 = y_2(x), \quad \ldots, \quad y_n = y_n(x)$ be n linearly independent solutions of $L[y] = 0$ on the interval I: $a \le x \le b$, and let $y_p = y_p(x)$ be a solution of $L[y] = F(x)$ on I. The *general solution* $y(x)$ of Equation (6.7.1) on I is defined by

$$y(x) = y_c(x) + y_p(x), \quad\quad\quad (6.7.3)$$

where

$$y_c(x) = c_1 y_1(x) + c_2 y_2(x) + \cdots + c_n y_n(x), \quad\quad\quad (6.7.4)$$

and $c_1, c_2, \ldots, c_n$ are arbitrary constants.

Definition 6.7.2.   The function y_c given in (6.7.4) is called the *complementary function* of $L[y] = F(x)$.   The function y_p is called a *particular solution* (or *particular integral*) of $L[y] = F(x)$.

The following theorem is an extension of Theorem 4.7.3, and its proof will be left as an exercise for the reader.   [*See* Exercise 6.7.2.]

Theorem 6.7.3.   (*Principle of Superposition.*)   *Let* $u_i(x)$ *be a solution of*

$$L[y] = a_0(x)y^{(n)} + a_1(x)y^{(n-1)} + \cdots + a_{n-1}(x)y' + a_n(x)y = F_i(x), \quad (6.7.5)$$

for $i = 1, 2, \ldots, n$, *on the interval* I: $\ a \leq x \leq b$.   *Then the function* u *given by the linear combination of these* n *solutions*

$$u(x) = c_1 u_1(x) + c_2 u_2(x) + \cdots + c_n u_n(x), \tag{6.7.6}$$

where $c_1, c_2, \ldots, c_n$ *are arbitrary constants, is a solution of*

$$L[y] = c_1 F_1(x) + c_2 F_2(x) + \cdots + c_n F_n(x) \tag{6.7.7}$$

on I.

Exercises 6.7

1. Prove Theorems 6.7.1 and 6.7.2.
2. Prove Theorem 6.7.3.

6.8. The Method of Undetermined Coefficients

The method of undetermined coefficients given in Section 4.8 extends to differential equations of order n.   Consider the equation

$$L[y] = a_0 y^{(n)} + a_1 y^{(n-1)} + \cdots + a_{n-1}y' + a_n y = F(x), \tag{6.8.1}$$

where $a_0, a_1, \ldots, a_n$ are constants, $a_0 \neq 0$, and $F(x)$ is not identically zero in the interval I: $\ a \leq x \leq b$.   Also, we suppose that $F(x)$ contains only terms which have a finite number of linearly independent derivatives in I.   The following cases will assist us in determining the form of a particular solution $y_p(x)$ of $L[y] = F(x)$.   We shall simply state these cases, which are extensions of the cases given in Section 4.8.

Case 1.   *Suppose that*

$$L[y] = P_k(x), \tag{6.8.2}$$

where the polynomial $P_k(x)$ *is given by*

$$P_k(x) = A_0 x^k + A_1 x^{k-1} + \cdots + A_{k-1}x + A_k, \qquad A_0 \neq 0. \tag{6.8.3}$$

The characteristic polynomial $p(m)$ of the equation $L[y] = 0$ is given by

$$p(m) = a_0 m^n + a_1 m^{n-1} + \cdots + a_{n-1}m + a_n, \qquad a_0 \neq 0. \qquad (6.8.4)$$

Let the polynomial $Q_k(x)$ be given by

$$Q_k(x) = B_0 x^k + B_1 x^{k-1} + \cdots + B_{k-1}x + B_k, \qquad B_0 \neq 0. \qquad (6.8.5)$$

A particular solution $y_p(x)$ of Equation (6.8.2) has the form

(i) $y_p(x) = Q_k(x), \quad$ *if $p(0) \neq 0$;* $\qquad\qquad\qquad\qquad\qquad\qquad$ (6.8.6)

(ii) $y_p(x) = xQ_k(x), \quad$ *if $p(0) = 0, \quad p'(0) \neq 0$;* $\qquad\qquad\qquad$ (6.8.7)

(iii) $y_p(x) = x^2 Q_k(x), \quad$ *if $p(0) = 0, \quad p'(0) = 0, \quad p''(0) \neq 0$;* $\qquad$ (6.8.8)

$$\vdots$$

(iv) $y_p(x) = x^n Q_k(x), \quad$ *if $p(0) = 0, \quad p'(0) = 0, \ldots, \quad p^{(n-1)}(0) = 0.$* $\quad$ (6.8.9)

Note that

$$p(0) = a_n, \quad p'(0) = a_{n-1}, \ldots, p^{(n-1)}(0) = (n-1)!\, a_1,$$

$$\text{and} \quad p^{(n)}(0) = n!\, a_0 \neq 0. \qquad (6.8.10)$$

Thus, if 0 is a root of multiplicity j of the equation $p(m) = 0$, then the form of a particular solution that we may take is $y_p(x) = x^j Q_k(x)$.

Case 2.   *Suppose that*

$$L[y] = e^{\alpha x}P_k(x), \qquad (6.8.11)$$

where $P_k(x)$ is given in (6.8.3).

A particular solution $y_p(x)$ of Equation (6.8.11) has the form

(i) $y_p(x) = e^{\alpha x}Q_k(x), \quad$ *if $p(\alpha) \neq 0$;* $\qquad\qquad\qquad\qquad\qquad$ (6.8.12)

(ii) $y_p(x) = xe^{\alpha x}Q_k(x), \quad$ *if $p(\alpha) = 0, \quad p'(\alpha) \neq 0$;* $\qquad\qquad$ (6.8.13)

(iii) $y_p(x) = x^2 e^{\alpha x}Q_k(x), \quad$ *if $p(\alpha) = 0, \quad p'(\alpha) = 0, \quad p''(\alpha) \neq 0$;* $\quad$ (6.8.14)

$$\vdots$$

(iv) $y_p(x) = x^n e^{\alpha x}Q_k(x), \quad$ *if $p(\alpha) = 0, \quad p'(\alpha) = 0, \ldots, \quad p^{(n-1)}(\alpha) = 0,$* $\quad$ (6.8.15)

where $p(m)$ and $Q_k(x)$ are given by (6.8.4) and (6.8.5), respectively.

Thus, if α is a root of multiplicity j of the equation $p(m) = 0$, then the form of a particular solution that we may take is $y_p(x) = x^j e^{\alpha x}Q_k(x)$.

Case 3.   *Suppose that*

$$L[y] = e^{\alpha x}[P_k(x) \cos \beta x + S_j(x) \sin \beta x], \qquad (6.8.16)$$

where

$$P_k(x) = A_0 x^k + A_1 x^{k-1} + \cdots + A_{k-1}x + A_k,$$
$$S_j(x) = D_0 x^j + D_1 x^{j-1} + \cdots + D_{j-1}x + D_j,$$

and $(A_0, D_0) \neq (0, 0)$.   Let $v = \max(k, j)$.   Then a particular solution $y_p(x)$ of Equation (6.8.16) has the form

(i) $y_p(x) = e^{\alpha x}[Q_v(x) \cos \beta x + R_v(x) \sin \beta x], \quad$ *if $p(\alpha + i\beta) \neq 0$;* $\qquad (6.8.17)$

(ii) $y_p(x) = xe^{\alpha x}[Q_v(x) \cos \beta x + R_v(x) \sin \beta x], \quad$ *if $p(\alpha + i\beta) = 0$,*

$$p'(\alpha + i\beta) \neq 0; \quad (6.8.18)$$

(iii) $y_p(x) = x^2 e^{\alpha x}[Q_v(x) \cos \beta x + R_v(x) \sin \beta x], \quad$ *if $p(\alpha + i\beta) = 0$,*

$$p'(\alpha + i\beta) = 0, \quad p''(\alpha + i\beta) \neq 0; \quad (6.8.19)$$

$$\vdots$$

(iv) $y_p(x) = x^l e^{\alpha x}[Q_v(x) \cos \beta x + R_v(x) \sin \beta x], \quad$ *if $p(\alpha + i\beta) = 0$,*

$$p'(\alpha + i\beta) = 0, \quad \ldots, \quad p^{(l-1)}(\alpha + i\beta) = 0, \quad p^{(l)}(\alpha + i\beta) \neq 0, \quad (6.8.20)$$

where

$$Q_v(x) = B_0 x^v + B_1 x^{v-1} + \cdots + B_{v-1}x + B_v,$$
$$R_v(x) = C_0 x^v + C_1 x^{v-1} + \cdots + C_{v-1}x + C_v,$$
$$(B_0, C_0) \neq (0, 0) \qquad (6.8.21)$$

and $p(m)$ is given by (6.8.4).

Thus, if $\alpha + i\beta$ is a root of multiplicity s of the equation $p(m) = 0$, then the form of a particular solution that we may take is

$$y_p(x) = x^s e^{\alpha x}[Q_v(x) \cos \beta x + R_v(x) \sin \beta x].$$

Example 6.8.1.   Let us find a particular solution of the differential equation

$$L[y] = y^{(4)} + 2y'' + y = 8 \cos x. \qquad (6.8.22)$$

SOLUTION.   Comparing the right-hand member of Equation (6.8.22) with that of Equation (6.8.16), we see that $\alpha = 0$, $\beta = 1$, $P_0(x) = 8$, and $S_j(x) \equiv 0$. The characteristic polynomial $p(m)$ of $L[y] = 0$ is

$$p(m) = m^4 + 2m^2 + 1 = (m^2 + 1)^2.$$

Thus,

$$p'(m) = 4m(m^2 + 1) \quad \text{and} \quad p''(m) = 8m^2 + 4(m^2 + 1).$$

Since $\alpha + i\beta = 0 + i \cdot 1 = i$, we see that $p(i) = 0$, $p'(i) = 0$, and $p''(i) = -8$. In view of Case 3 (iii), a particular solution $y_p(x)$ of Equation (6.8.22) has the form $y_p(x) = B_0 x^2 \cos x + C_0 x^2 \sin x$, or if we wish to avoid subscripts, we have

$$y_p(x) = Bx^2 \cos x + Cx^2 \sin x. \tag{6.8.23}$$

Differentiating (6.8.23) four times in succession and collecting terms, we obtain

$$y_p'(x) = (Cx^2 + 2Bx) \cos x + (-Bx^2 + 2Cx) \sin x,$$

$$y_p''(x) = (-Bx^2 + 4Cx + 2B) \cos x + (-Cx^2 - 4Bx + 2C) \sin x,$$

$$y_p'''(x) = (-Cx^2 - 6Bx + 6C) \cos x + (Bx^2 - 6Cx - 6B) \sin x,$$

$$y_p^{(4)}(x) = (Bx^2 - 8Cx - 12B) \cos x + (Cx^2 + 8Bx - 12C) \sin x.$$

Substituting y_p, y_p'', and $y_p^{(4)}$ into (6.8.22), we obtain upon simplification

$$-B \cos x - C \sin x = \cos x. \tag{6.8.24}$$

Equation (6.8.24) will be an identity provided that $C = 0$ and $B = -1$. Substituting these values into (6.8.23), a particular solution $y_p(x)$ of (6.8.22) is

$$y_p(x) = -x^2 \cos x. \tag{6.8.25}$$

Remark 6.8.1.   The results given in Remarks 4.8.3 through 4.8.5 extend to differential equations of order n. These results will be left as exercises for the reader. [*See* Exercises 6.8.1 through 6.8.5.] One may readily obtain a particular solution of Equation (6.8.22) by utilizing Exercise 6.8.2(iii). For,

$$y_p(x) = \mathscr{R}\left[\frac{8x^2 e^{ix}}{p''(i)}\right] = \mathscr{R}\left[\frac{8x^2}{-8} e^{ix}\right] = -x^2 \mathscr{R}(e^{ix})$$

$$= -x^2 \mathscr{R}[\cos x + i \sin x] = -x^2 \cos x.$$

Exercises 6.8

In Exercises 1 through 5, the operator L is defined by

$$L[y] = a_0 y^{(n)} + a_1 y^{(n-1)} + \cdots + a_{n-1} y' + a_n y,$$

where $a_0, a_1, \ldots, a_n$ are constants, $a_0 \neq 0$, and the characteristic polynomial $p(m)$ of the operator L is

$$p(m) = a_0 m^n + a_1 m^{n-1} + \cdots + a_{n-1} m + a_n.$$

Also, k, α, and β are real constants.

1. Suppose that $L[y] = ke^{\alpha x}$. Show that a *particular solution* $y_p(x)$ of this equation is given by

   (i) $y_p(x) = \dfrac{ke^{\alpha x}}{p(\alpha)}$,   if $p(\alpha) \neq 0$;

   (ii) $y_p(x) = \dfrac{kxe^{\alpha x}}{p'(\alpha)}$,   if $p(\alpha) = 0$, $p'(\alpha) \neq 0$;

   (iii) $y_p(x) = \dfrac{kx^2 e^{\alpha x}}{p''(\alpha)}$,   if $p(\alpha) = 0$, $p'(\alpha) = 0$, $p''(\alpha) \neq 0$;

   $\vdots$

   (iv) $y_p(x) = \dfrac{kx^n e^{\alpha x}}{p^{(n)}(\alpha)}$,   if $p(\alpha) = 0$, $p'(\alpha) = 0$, $\ldots$, $p^{(n-1)}(\alpha) = 0$.

2. Suppose that $L[y] = k \cos \beta x$. Show that a *particular solution* $y_p(x)$ of this equation is given by

   (i) $y_p(x) = \mathscr{R}\left[\dfrac{ke^{i\beta x}}{p(i\beta)}\right]$,   if $p(i\beta) \neq 0$;

   (ii) $y_p(x) = \mathscr{R}\left[\dfrac{kxe^{i\beta x}}{p'(i\beta)}\right]$,   if $p(i\beta) = 0$, $p'(i\beta) \neq 0$;

   (iii) $y_p(x) = \mathscr{R}\left[\dfrac{kx^2 e^{i\beta x}}{p''(i\beta)}\right]$,   if $p(i\beta) = 0$, $p'(i\beta) = 0$, $p''(i\beta) \neq 0$;

   $\vdots$

   (iv) $y_p(x) = \mathscr{R}\left[\dfrac{kx^l e^{i\beta x}}{p^{(l)}(i\beta)}\right]$,   if $p(i\beta) = 0$, $p'(i\beta) = 0$, $\ldots$,

   $$p^{(l-1)}(i\beta) = 0, \quad p^{(l)}(i\beta) \neq 0.$$

3. Suppose that $L[y] = k \sin \beta x$. Show that a *particular solution* $y_p(x)$ of this equation is given by the result of Exercise 2 with $\mathscr{R}$ replaced by $\mathscr{I}$.

4. Suppose that $L[y] = ke^{\alpha x} \cos \beta x$. Show that a *particular solution* $y_p(x)$ of this equation is given by

   (i) $y_p(x) = \mathscr{R}\left[\dfrac{ke^{(\alpha + i\beta)x}}{p(\alpha + i\beta)}\right]$,   if $p(\alpha + i\beta) \neq 0$;

   (ii) $y_p(x) = \mathscr{R}\left[\dfrac{kxe^{(\alpha + i\beta)x}}{p'(\alpha + i\beta)}\right]$,   if $p(\alpha + i\beta) = 0$, $p'(\alpha + i\beta) \neq 0$;

   (iii) $y_p(x) = \mathscr{R}\left[\dfrac{kx^2 e^{(\alpha + i\beta)x}}{p''(\alpha + i\beta)}\right]$,   if $p(\alpha + i\beta) = 0$, $p'(\alpha + i\beta) = 0$, $p''(\alpha + i\beta) \neq 0$;

   $\vdots$

   (iv) $y_p(x) = \mathscr{R}\left[\dfrac{kx^l e^{(\alpha + i\beta)x}}{p^{(l)}(\alpha + i\beta)}\right]$,   if $p(\alpha + i\beta) = 0$,

   $$p'(\alpha + i\beta) = 0, \quad \ldots, \quad p^{(l-1)}(\alpha + i\beta) = 0, \quad p^{(l)}(\alpha + i\beta) \neq 0.$$

5. Suppose that $L[y] = ke^{\alpha x} \sin \beta x$. Show that a *particular solution* $y_p(x)$ of this equation is given by the result of Exercise 4 with $\mathscr{R}$ replaced by $\mathscr{I}$.

In Exercises 6 through 15, find the general solution of each differential equation.

6. $y^{(5)} - y''' = 210x^4 - 240x^3 - 600x^2 + 480x - 1920$.

7. $y^{(4)} - 6y''' + 13y'' - 12y' + 4y = 2e^x - 4e^{2x}$.

8. $y''' - 6y'' + 21y' - 26y = 36e^{2x} \sin 3x$.

9. $y^{(4)} + 6y''' + 13y'' + 12y' + 4y = 200 \cos 2x + 400 \sin 2x$.

10. $y^{(4)} + 6y''' - 7y'' - 96y' - 144y = (240x^2 - 64x + 118)e^x$.

11. $y^{(5)} + 9y^{(4)} + 11y''' - 117y'' - 432y' - 432y = 84e^{-3x}$.

12. $y^{(6)} - 12y^{(5)} + 63y^{(4)} - 184y''' + 315y'' - 300y' + 125y$
$$= e^{2x}(48 \cos x + 96 \sin x).$$

[$2 + i$ is a root of $p(m) = 0$ of multiplicity 3.]

13. $y''' + y'' - y' - y = (2x^2 + 4x + 8) \cos x + (6x^2 + 8x + 12) \sin x$.

14. $y^{(4)} + 4y'' = 24x^2 - 6x + 14 + 32 \cos 2x$.

15. $y^{(4)} - 3y''' + 3y'' - y' = 6x - 20 - 120x^2 e^x$.

In Exercises 16 through 20, find the solution of each initial-value problem.

16. $\begin{cases} y''' - y'' + y' - y = 2e^x, \\ y(0) = 1, \quad y'(0) = 3, \quad y''(0) = -3. \end{cases}$

17. $\begin{cases} y''' + 2y'' + 5y' = 20e^{-x} \cos 2x, \\ y(0) = 1, \quad y'(0) = 2, \quad y''(0) = -3. \end{cases}$

18. $\begin{cases} y''' - 2y'' - y' + 2y = 2x^2 + 4x - 9, \\ y(0) = 2, \quad y'(0) = -4, \quad y''(0) = -1. \end{cases}$

19. $\begin{cases} y^{(4)} - 5y'' + 4y = 10 \cos x - 20 \sin x, \\ y(0) = 1, \quad y'(0) = 10, \quad y''(0) = 2, \quad y'''(0) = -4. \end{cases}$

20. $\begin{cases} y^{(4)} + 5y'' + 4y = 40 \cos 3x, \\ y\!\left(\dfrac{\pi}{2}\right) = 2, \quad y'\!\left(\dfrac{\pi}{2}\right) = 2, \quad y''\!\left(\dfrac{\pi}{2}\right) = 1, \quad y'''\!\left(\dfrac{\pi}{2}\right) = 4. \end{cases}$

6.9. The Method of Variation of Parameters

The method of variation of parameters given in Section 4.9 extends to differential equations of order n. Consider the equation

$$L[y] = a_0(x)y^{(n)} + a_1(x)y^{(n-1)} + \cdots + a_{n-1}(x)y' + a_n(x)y = F(x), \quad (6.9.1)$$

where the functions $a_0, a_1, \ldots, a_n$, and F are continuous on the interval I: $a \leqq x \leqq b$, $a_0(x) \neq 0$ for every x in I, and $F(x)$ is not identically zero in I. The related homogeneous differential equation of $L[y] = F(x)$ is

$$L[y] = a_0(x)y^{(n)} + a_1(x)y^{(n-1)} + \cdots + a_{n-1}(x)y' + a_n(x)y = 0. \quad (6.9.2)$$

Suppose that the functions $y_1, y_2, \ldots, y_n$ are n linearly independent solutions of $L[y] = 0$ on I. Then, the general solution $y_c(x)$ of $L[y] = 0$ on I is given by [*see* Definition 6.4.1]

$$y_c(x) = c_1 y_1(x) + c_2 y_2(x) + \cdots + c_n y_n(x), \qquad (6.9.3)$$

where $c_1, c_2, \ldots, c_n$ are arbitrary constants. Basically, the method of variation of parameters consists of replacing the constants $c_1, c_2, \ldots, c_n$ in the function y_c given in (6.9.3) by undetermined functions of the independent variable x. We then determine those functions so that when the modified function y_c is substituted in the left-hand side of Equation (6.9.1) it reduces to $F(x)$. This imposes only one condition on the n arbitrary functions. Hence, we have $n - 1$ other conditions at our disposal, which we shall use to our advantage. Let us assume that a particular solution $y_p(x)$ of $L[y] = F(x)$ on I has the form

$$y_p(x) = y_1(x)v_1(x) + y_2(x)v_2(x) + \cdots + y_n(x)v_n(x). \qquad (6.9.4)$$

Our task is to determine the functions $v_1, v_2, \ldots, v_n$ so that the particular solution $y_p(x)$ in (6.9.4) satisfies the equation $L[y] = F(x)$ given in (6.9.1).

Differentiating (6.9.4), we get

$$y_p' = y_1'v_1 + y_2'v_2 + \cdots + y_n'v_n + y_1v_1' + y_2v_2' + \cdots + y_nv_n'. \qquad (6.9.5)$$

To simplify the determination of the functions $v_1, v_2, \ldots, v_n$, as a first condition, we require that these functions satisfy

$$y_1v_1' + y_2v_2' + \cdots + y_nv_n' = 0. \qquad (6.9.6)$$

Thus, (6.9.5) reduces to

$$y_p' = y_1'v_1 + y_2'v_2 + \cdots + y_n'v_n. \qquad (6.9.7)$$

Differentiating (6.9.7), we obtain

$$y_p'' = y_1''v_1 + y_2''v_2 + \cdots + y_n''v_n + y_1'v_1' + y_2'v_2' + \cdots + y_n'v_n'. \qquad (6.9.8)$$

As a second condition on the functions $v_1, v_2, \ldots, v_n$, we also require that they satisfy

$$y_1'v_1' + y_2'v_2' + \cdots + y_n'v_n' = 0. \qquad (6.9.9)$$

Equation (6.9.8) now becomes

$$y_p'' = y_1''v_1 + y_2''v_2 + \cdots + y_n''v_n. \qquad (6.9.10)$$

Continuing in this manner through $(n - 1)$ derivatives of $y_p(x)$, we obtain

$$y_p^{(k)} = y_1^{(k)}v_1 + y_2^{(k)}v_2 + \cdots + y_n^{(k)}v_n, \qquad k = 3, 4, \ldots, n - 1, \qquad (6.9.11)$$

and the following additional conditions that the functions $v_1, v_2, \ldots, v_n$ are required to satisfy

$$y_1^{(k-1)}v_1' + y_2^{(k-1)}v_2' + \cdots + y_n^{(k-1)}v_n' = 0, \qquad k = 3, 4, \ldots, n-1. \quad (6.9.12)$$

In view of (6.9.11), we see that the nth derivative of $y_p(x)$ is (with accessory conditions)

$$y_p^{(n)} = y_1^{(n)}v_1 + y_2^{(n)}v_2 + \cdots + y_n^{(n)}v_n + y_1^{(n-1)}v_1'$$
$$+ y_2^{(n-1)}v_2' + \cdots + y_n^{(n-1)}v_n'. \quad (6.9.13)$$

Substituting $y_p, y_p', y_p'', \ldots, y_p^{(n)}$ of (6.9.4), (6.9.7), (6.9.10), $\ldots$, (6.9.13) into (6.9.1), we obtain upon collecting terms

$$[a_0(x)y_1^{(n)} + a_1(x)y_1^{(n-1)} + \cdots + a_n(x)y_1]v_1$$
$$+ [a_0(x)y_2^{(n)} + a_1(x)y_2^{(n-1)} + \cdots + a_n(x)y_2]v_2 + \cdots$$
$$+ [a_0(x)y_n^{(n)} + a_1(x)y_n^{(n-1)} + \cdots + a_n(x)y_n]v_n$$
$$+ a_0(x)[y_1^{(n-1)}v_1' + y_2^{(n-1)}v_2' + \cdots + y_n^{(n-1)}v_n'] = F(x). \quad (6.9.14)$$

Since, by assumption, $y_1, y_2, \ldots, y_n$ are solutions of $L[y] = 0$, the coefficients of $v_1, v_2, \ldots, v_n$ in (6.9.14) are each equal to zero. Since $a_0(x) \neq 0$ for every x in I, Equation (6.9.14) now becomes

$$y_1^{(n-1)}v_1' + y_2^{(n-1)}v_2' + \cdots + y_n^{(n-1)}v_n' = \frac{F(x)}{a_0(x)}. \quad (6.9.15)$$

Equations (6.9.6), (6.9.9), (6.9.12), and (6.9.15) represent a nonhomogeneous system of n linear (algebraic) equations in the unknowns $v_1', v_2', \ldots, v_n'$:

$$\begin{cases} y_1 v_1' + y_2 v_2' + \cdots + y_n v_n' = 0, \\ y_1' v_1' + y_2' v_2' + \cdots + y_n' v_n' = 0, \\ y_1'' v_1' + y_2'' v_2' + \cdots + y_n'' v_n' = 0, \\ \qquad \cdots \\ y_1^{(n-1)}v_1' + y_2^{(n-1)}v_2' + \cdots + y_n^{(n-1)}v_n' = \dfrac{F(x)}{a_0(x)}. \end{cases} \quad (6.9.16)^2$$

Note that the determinant of the coefficients of the variables $v_1', v_2', \ldots, v_n'$ of the system (6.9.16) is precisely the Wronskian of the functions $y_1, y_2, \ldots, y_n$. [See Equation (6.4.16).] Since the functions $y_1, y_2, \ldots, y_n$ are linearly independent solutions of $L[y] = 0$ on I, it follows from Theorem 6.4.3 that $W(y_1, y_2, \ldots, y_n; x) \neq 0$ for every x in I. Thus, by Theorem I of Section

---

[2] The system (6.9.16) may be obtained directly by following a method similar to that given in Exercise 4.9.19.

4.4, the system (6.9.16) has a unique solution. By Cramer's rule, we then have

$$
v_j'(x) = \frac{\begin{vmatrix} y_1 & y_2 & \cdots & y_{j-1} & 0 & y_{j+1} & \cdots & y_n \\ y_1' & y_2' & \cdots & y_{j-1}' & 0 & y_{j+1}' & \cdots & y_n' \\ \vdots & \vdots & & \vdots & \vdots & \vdots & & \vdots \\ y_1^{(n-1)} & y_2^{(n-1)} & \cdots & y_{j-1}^{(n-1)} & \dfrac{F(x)}{a_0(x)} & y_{j+1}^{(n-1)} & \cdots & y_n^{(n-1)} \end{vmatrix}}{W(y_1, y_2, \ldots, y_n; x)},
$$

$$j = 1, 2, \ldots, n. \quad (6.9.17)$$

If we denote by $W_j(x)$ the determinant obtained from $W(y_1, y_2, \ldots, y_n; x)$ by replacing the jth column by the column $(0, 0, \ldots, 1)$, we may then write (6.9.17) as

$$v_j'(x) = \frac{F(x)}{a_0(x)} \frac{W_j(x)}{W(y_1, y_2, \ldots, y_n; x)}, \qquad j = 1, 2, \ldots, n. \qquad (6.9.18)$$

Integrating the above expression, we obtain

$$v_j(x) = \int \frac{F(x) W_j(x)}{a_0(x) W(y_1, y_2, \ldots, y_n; x)} \, dx, \qquad j = 1, 2, \ldots, n. \quad (6.9.19)$$

In view of (6.9.4) and (6.9.19), *a particular solution y_p of the equation $L[y] = F(x)$ given in (6.9.1) is*

$$y_p(x) = \sum_{j=1}^{n} y_j(x) \int \frac{F(x) W_j(x)}{a_0(x) W(y_1, y_2, \ldots, y_n; x)} \, dx. \qquad (6.9.20)$$

Thus, the general solution $y(x)$ of the equation $L[y] = F(x)$ given in (6.9.1) is

$$y(x) = y_c(x) + y_p(x), \qquad (6.9.21)$$

where the complementary function y_c is given in (6.9.3) and a particular solution y_p is given in (6.9.20).   [This solution is valid for all x in I.]

Remark 6.9.1.   When utilizing the method of variation of parameters to find a particular solution y_p of $L[y] = F(x)$, one may save some time evaluating $W(y_1, y_2, \ldots, y_n; x)$ by using Abel's identity given in (6.4.27).

Example 6.9.1.   Let us find the general solution of the differential equation

$$L[y] = y''' - 3y'' + 4y' - 2y = e^x \sec x, \qquad (6.9.22)$$

on the interval $I: 0 < x < \pi/2$.

SOLUTION.   The characteristic equation $p(m) = 0$ of $L[y] = 0$ is

$$m^3 - 3m^2 + 4m - 2 = 0.$$

Utilizing the process of synthetic division, we find that the roots of the above equation are 1, $1 + i$, $1 - i$. Thus, the complementary function y_c of Equation (6.9.22) is

$$y_c(x) = c_1 e^x + e^x(c_2 \cos x + c_3 \sin x). \qquad (6.9.23)$$

Replacing c_1, c_2, and c_3 in (6.9.23) by the functions v_1, v_2, and v_3, respectively, we endeavor to find a particular solution y_p of Equation (6.9.22) on I of the form

$$y_p(x) = v_1(x)e^x + v_2(x)e^x \cos x + v_3(x)e^x \sin x. \qquad (6.9.24)$$

We shall utilize Formula (6.9.19) to find v_1, v_2, and v_3. First, observe that

$$W(y_1, y_2, y_3 ; x) = W(e^x, e^x \cos x, e^x \sin x; x)$$

$$= \begin{vmatrix} e^x & e^x \cos x & e^x \sin x \\ e^x & e^x(\cos x - \sin x) & e^x(\sin x + \cos x) \\ e^x & -2e^x \sin x & 2e^x \cos x \end{vmatrix}. \qquad (6.9.25)$$

One may save some time evaluating the above determinant by utilizing Abel's Formula (6.4.27) with $a_0(x) = 1$, $a_1(x) = -3$, and $x_0 = 0$. We find

$$W(y_1, y_2, y_3 ; x) = W(y_1, y_2, y_3 ; 0) \exp\left[-\int_0^x - 3 \, d\xi \right] = W(y_1, y_2, y_3 ; 0)e^{3x}$$

$$= e^{3x} \begin{vmatrix} 1 & 1 & 0 \\ 1 & 1 & 1 \\ 1 & 0 & 2 \end{vmatrix} = e^{3x} \cdot 1 = e^{3x}. \qquad (6.9.26)$$

We shall next compute $W_1(x)$, $W_2(x)$, and $W_3(x)$. Recall that $W_1(x)$ is obtained by replacing the first column in the determinant given in (6.9.25) by the column $(0, 0, 1)$. Thus,

$$W_1(x) = \begin{vmatrix} 0 & e^x \cos x & e^x \sin x \\ 0 & e^x(\cos x - \sin x) & e^x(\sin x + \cos x) \\ 1 & -2e^x \sin x & 2e^x \cos x \end{vmatrix}$$

$$= \begin{vmatrix} e^x \cos x & e^x \sin x \\ e^x(\cos x - \sin x) & e^x(\sin x + \cos x) \end{vmatrix} = e^{2x}.$$

Similarly, we have

$$W_2(x) = \begin{vmatrix} e^x & 0 & e^x \sin x \\ e^x & 0 & e^x(\sin x + \cos x) \\ e^x & 1 & 2e^x \cos x \end{vmatrix} = -e^{2x} \cos x,$$

$$W_3(x) = \begin{vmatrix} e^x & e^x \cos x & 0 \\ e^x & e^x(\cos x - \sin x) & 0 \\ e^x & -2e^x \sin x & 1 \end{vmatrix} = -e^{2x} \sin x.$$

Utilizing Formula (6.9.19) with $F(x) = e^x \sec x$, $a_0(x) = 1$, $W(y_1, y_2, y_3 ; x) = e^{3x}$, and the $W_j(x)$, $j = 1, 2, 3$, determined above, we find

$$v_1(x) = \int \frac{e^{3x} \sec x}{e^{3x}} \, dx = \int \sec x \, dx = \ln (\sec x + \tan x),$$

$$v_2(x) = - \int \frac{e^{3x} \sec x \cos x}{e^{3x}} \, dx = - \int dx = -x,$$

$$v_3(x) = - \int \frac{e^{3x} \sec x \sin x}{e^{3x}} \, dx = - \int \tan x \, dx = \ln \cos x.$$

Substituting these $v_j(x)$ in (6.9.24), we obtain

$$y_p(x) = e^x \ln (\sec x + \tan x) - e^x x \cos x + e^x \sin x \ln \cos x. \quad (6.9.27)$$

In view of (6.9.23) and (6.9.27), the general solution $y(x)$ of Equation (6.9.22) on the interval I is

$$y(x) = e^x[c_1 + \ln (\sec x + \tan x) + (c_2 - x) \cos x + (c_3 + \ln \cos x) \sin x].$$

$$(6.9.28)$$

Exercises 6.9

In Exercises 1 through 8, find the general solution of each differential equation.

1. $y''' + y' = \tan x, \quad 0 < x < \dfrac{\pi}{2}$.

2. $y''' + y' = 3x^2 + 4x + \sec x, \quad 0 < x < \dfrac{\pi}{2}$.

3. $y''' + y' = 2 \sin x - \csc x, \quad 0 < x < \dfrac{\pi}{2}$.

4. $y''' + 3y'' + 3y' + y = e^{-x} \ln x, \quad x > 0$.

5. $y''' - 3y'' + 3y' - y = \dfrac{e^x \ln x}{x}, \quad x > 0$.

6. $y''' - 6y'' + 11y' - 6y = \dfrac{e^{3x}}{e^{2x} + 1}$.

7. $y^{(4)} - 4y''' + 6y'' - 4y' + y = \dfrac{e^x}{x^n}, \quad x > 0$.

8. $y^{(4)} + 2y'' + y = \sec^3 x, \quad 0 < x < \dfrac{\pi}{2}$.

In Exercises 9 and 10, find the solution of each initial-value problem.

9.
$$\begin{cases} y''' - 3y'' + 2y' = 4x - 8 + \dfrac{2e^{2x}}{e^x + 1}, \\[2mm] y(0) = 1, \quad y'(0) = -1, \quad y''(0) = 2. \end{cases}$$

10.
$$\begin{cases} y^{(5)} = \dfrac{288}{x}, \quad x > 0, \\[2mm] y(1) = 7, \quad y'(1) = 2, \quad y''(1) = 0, \quad y'''(1) = 0, \quad y^{(4)}(1) = 0. \end{cases}$$

11. Consider the function I_n defined by

$$(1) \qquad I_n(x) = \int_{x_0}^{x} (x - t)^{n-1} f(t)\, dt,$$

where n is a positive integer and x_0 is a constant.   Differentiating (1) with respect to x, we obtain [*see* Exercise 1.3.8 with the role of x and t being interchanged]

$$(2) \qquad I_n'(x) = \frac{d}{dx} \int_{x_0}^{x} (x - t)^{n-1} f(t)\, dt$$

$$= (x - t)^{n-1} f(t)\big|_{t=x} + (n - 1) \int_{x_0}^{x} (x - t)^{n-2} f(t)\, dt.$$

Thus, for $n > 1$, we have

$$(3) \qquad I_n'(x) = (n - 1) \int_{x_0}^{x} (x - t)^{n-2} f(t)\, dt = (n - 1) I_{n-1}(x), \qquad n > 1,$$

and for $n = 1$, we have

$$(4) \qquad I_1'(x) = f(x).$$

Show that repeated use of Formula (3) gives the following general relation

$$(5) \qquad I_n^{(k)}(x) = (n - 1)(n - 2)\ldots(n - k) I_{n-k}(x), \qquad n > k.$$

In particular, when $k = n - 1$, Formula (5) becomes

$$(6) \qquad I_n^{(n-1)}(x) = (n - 1)! I_1(x).$$

Differentiating (6) and utilizing (4), we get

$$(7) \qquad I_n^{(n)}(x) = (n - 1)! f(x).$$

Observe that $I_n(x_0) = 0$ for $n \geq 1$.   Thus, in view of (5) and (6), we see that $I_n'(x_0) = 0$, $I_n''(x_0) = 0, \ldots, I_n^{(n-1)}(x_0) = 0$.   Utilizing Formula (7), show that

$$(8) \qquad \begin{aligned} I_1(x) &= \int_{x_0}^{x} f(t)\, dt = \int_{x_0}^{x} f(x_1)\, dx_1, \\[2mm] I_2(x) &= \int_{x_0}^{x} I_1(x_2)\, dx_2 = \int_{x_0}^{x} \int_{x_0}^{x_2} f(x_1)\, dx_1 dx_2, \end{aligned}$$

and, in the general case

(9) $$I_n(x) = (n-1)! \int_{x_0}^{x} \int_{x_0}^{x_n} \cdots \int_{x_0}^{x_3} \int_{x_0}^{x_2} f(x_1)dx_1\, dx_2 \ldots dx_{n-1}\, dx_n .$$

Hence, in view of (1), we have the following important formula

(10)
$$\int_{x_0}^{x} \int_{x_0}^{x_n} \cdots \int_{x_0}^{x_3} \int_{x_0}^{x_2} f(x_1)dx_1\, dx_2 \ldots dx_{n-1}\, dx_n$$
$$= \frac{1}{(n-1)!} \int_{x_0}^{x} (x-t)^{n-1} f(t)dt .$$

One may interpret the left-hand member of (10) as the result of first integrating f from x_0 to x and then integrating this operation $n-1$ additional times. It is often represented symbolically in the form

$$\overbrace{\int_{x_0}^{x} \cdots \int_{x_0}^{x}}^{n\ \text{times}} f(x)dx \cdots dx \quad \text{or} \quad \overbrace{\int_{x_0}^{x}}^{n\ \text{times}} f(x)dx^n .$$

Example:

$$\int_{1}^{x} \frac{288}{x}\, dx^5 = \frac{1}{4!} \int_{1}^{x} (x-t)^4 \frac{288}{t}\, dt$$

$$= 12 \left[x^4 \int_{1}^{x} \frac{dt}{t} - 4x^3 \int_{1}^{x} dt + 6x^2 \int_{1}^{x} t\, dt - 4x \int_{1}^{x} t^2\, dt + \int_{1}^{x} t^3\, dt \right]$$

$$= 12x^4 \ln x - 25x^4 + 48x^3 - 36x^2 + 16x - 3 .$$

12. Consider the differential equation

(1)
$$\frac{d^n y}{dx^n} = f(x),$$

where the function f is continuous on the interval I: $a \leq x \leq b$. Show that the general solution $y(x)$ of Equation (1) on I is given by

(2) $$y(x) = c_0 + c_1 x + c_2 x^2 + \cdots + c_{n-1}x^{n-1} + \frac{1}{(n-1)!} \int^{x} (x-t)^{n-1} f(t)dt,$$

where $c_0, c_1, \ldots, c_{n-1}$ are arbitrary constants. [*Note:* If

$$F(x,t) = \int (x-t)^{n-1} f(t)dt, \text{ then } \int^{x} (x-t)^{n-1} f(t)dt = F(x,t)\Big|_{t=x} = F(x,x). \Big]$$

13. Find the solution of the initial-value problem

$$\begin{cases} y^{(4)} = 6 \ln x, \quad x > 0, \\[2mm] y(1) = -\dfrac{25}{48}, \quad y'(1) = -\dfrac{5}{6}, \quad y''(1) = \dfrac{3}{2}, \quad y'''(1) = 6. \end{cases}$$

14. Consider the initial-value problem

$$(1) \qquad y^{(n)} + a_1(x)y^{(n-1)} + \cdots + a_{n-1}(x)y' + a_n(x)y = g(x),$$

$$(2) \qquad y(0) = c_0, \quad y'(0) = c_1, \ \ldots, \quad y^{(n-2)}(0) = c_{n-2}, \quad y^{(n-1)}(0) = c_{n-1},$$

where the functions $a_1, \ldots, a_n$, and g are continuous on the interval I: $\ 0 \le x \le b$. Let u be a function continuous on I, and consider the transformation

$$(3) \qquad y^{(n)} = u(x).$$

Utilizing (2) and integrating (3) n times from 0 to k, show that

$$(4) \quad \begin{cases} y^{(n-1)} = \displaystyle\int_0^x u(x)dx + c_{n-1}, \\[2em] y^{(n-2)} = \displaystyle\int_0^x u(x)dx^2 + c_{n-1}x + c_{n-2}, \\[2em] y^{(n-3)} = \displaystyle\int_0^x u(x)dx^3 + c_{n-1}\dfrac{x^2}{2!} + c_{n-2}x + c_{n-3}, \\[1em] \qquad \vdots \\[1em] y = \displaystyle\int_0^x u(x)dx^n + c_{n-1}\dfrac{x^{n-1}}{(n-1)!} + c_{n-2}\dfrac{x^{n-2}}{(n-2)!} + \cdots + c_1 x + c_0. \end{cases}$$

Utilizing (3), (4), and (10) of Exercise 11, show that Equation (1) is transformed into

$$(5) \qquad u(x) + \int_0^x K(x, t)u(t)dt = f(x),$$

where

$$(6) \qquad K(x, t) = \sum_{k=1}^{n} a_k(x)\frac{(x-t)^{k-1}}{(k-1)!},$$

and

$$(7) \qquad f(x) = g(x) - c_{n-1}a_1(x) - (c_{n-1}x + c_{n-2})a_2(x)$$

$$- \cdots - \left(c_{n-1}\frac{x^{n-1}}{(n-1)!} + c_{n-2}\frac{x^{n-2}}{(n-2)!} + \cdots + c_1 x + c_0\right)a_n(x).$$

Conversely, solving (5) for $u(x)$ with $K(x, t)$ and $f(x)$ given in (6) and (7), and then substituting this value of $u(x)$ into the last equation of (4), we obtain the (unique) solution of (1) which satisfies the initial conditions (2).

In Equation (5) the unknown function u also occurs under the integral sign. Equation (5) is called a *Volterra integral equation of the second kind with kernel $K(x, t)$*. If the leading coefficient in (1) is not unity but $a_0(x) \ne 0$ for every x in I, then in (5) we replace $u(x)$ by $a_0(x) u(x)$; the functions K and f are still given by (6) and (7).

As an example, let us form the integral equation corresponding to the initial-value problem

(8) $$y'' + xy = e^x,$$

(9) $$y(0) = 0, \quad y'(0) = 1.$$

Comparing (8) with (1), we see that $n = 2$, $a_1(x) = 0$, $a_2(x) = x$, and $g(x) = e^x$. Also, comparing (9) with (2), we see that $c_0 = 0$ and $c_1 = 1$. Utilizing (6), we obtain

(10)
$$K(x, t) = \sum_{k=1}^{2} a_k(x) \frac{(x - t)^{k-1}}{(k - 1)!}$$

$$= a_1(x) + a_2(x)(x - t) = 0 + x(x - t) = x(x - t).$$

Using (7) with $n = 2$, we obtain

(11)
$$f(x) = g(x) - c_1 a_1(x) - (c_1 x + c_0) a_2(x)$$

$$= e^x - 1 \cdot 0 - (1 \cdot x + 0)x = e^x - x^2.$$

Substituting $K(x, t)$ and $f(x)$ found in (10) and (11) into (5), we obtain the desired integral equation,

(12)
$$u(x) + \int_0^x x(x - t)u(t)dt = e^x - x^2.$$

15. Form the integral equation corresponding to each of the following initial-value problems.

(a)
$$\begin{cases} y''' - 3y'' + 3y' - y = \cos^2 x, \\ y(0) = 1, \quad y'(0) = 1, \quad y''(0) = 1. \end{cases}$$

(b)
$$\begin{cases} y^{(4)} - y = g(x), \\ y(0) = 1, \quad y'(0) = -1, \quad y''(0) = 2, \quad y'''(0) = -6. \end{cases}$$

(c)
$$\begin{cases} y''' + x^2 y'' - xy' + 2y = g(x), \\ y(0) = 1, \quad y'(0) = 0, \quad y''(0) = 2. \end{cases}$$

16. Consider the integral equation

(1)
$$u(x) = 2x + \int_0^x (t - x)u(t)dt.$$

[*Note:* $u(0) = 0$.]   Differentiating Equation (1) with respect to x, we get

$$u'(x) = 2 + (t - x)u(t)\Big|_{t=x} + \int_0^x -u(t)dt,$$

or

$$u'(x) = 2 - \int_0^x u(t)dt.$$

[*Note:* $u'(0) = 2.$]   Differentiating the above equation with respect to x, we obtain

$$u''(x) = -u(x).$$

Thus, we have the following initial-value problem,

$$\begin{cases} u'' + u = 0, \\ u(0) = 0, \quad u'(0) = 2. \end{cases}$$

Show that the solution $u(x)$ of Equation (1) is given by $u(x) = 2 \sin x$.

In Exercises 17 through 20, utilize the technique given in Exercise 16 to find the solution of each integral equation.

17. $u(x) = e^x - \displaystyle\int_0^x (x - t)u(t)dt.$

18. $u(x) = e^x - 2 \sin x + x - 1 - \displaystyle\int_0^x (x - t + 2)u(t)dt.$

19. $u(x) = x^3 - \frac{3}{2}x^2 + x + \displaystyle\int_0^x [3 - 3(x - t) + \frac{1}{2}(x - t)^2]u(t)dt.$

20. $u(x) = \cos x + \frac{1}{6} \displaystyle\int_0^x (x - t)^3 u(t)dt.$

6.10. The Cauchy-Euler Equation

The differential equation

$$a_0 x^n y^{(n)} + a_1 x^{n-1} y^{(n-1)} + \cdots + a_{n-1}xy' + a_n y = F(x), \qquad (6.10.1)$$

where $a_0, a_1, \ldots, a_{n-1}$ are constants, $a_0 \neq 0$, and the function F is continuous on some interval I, is called a *Cauchy-Euler linear differential equation of order n.* This equation is also known as the *equidimensional equation.* Before establishing Theorem 6.10.1, we shall first establish a useful differentiation formula.

Consider the transformation

$$x = e^u \quad \text{or} \quad u = \ln x, \quad x > 0. \qquad (6.10.2)$$

Utilizing the chain rule and the above equation, we have

$$\frac{dy}{dx} = \frac{dy}{du} \cdot \frac{du}{dx} = \frac{dy}{du} \cdot \frac{1}{x} = e^{-u} \frac{dy}{du}. \qquad (6.10.3)$$

$$\frac{d^2 y}{dx^2} = \frac{d}{dx}\left(\frac{dy}{dx}\right) = \frac{d}{du}\left(\frac{dy}{dx}\right) \cdot \frac{du}{dx} = \frac{d}{du}\left(e^{-u}\frac{dy}{du}\right) \cdot \frac{1}{x}$$

$$= e^{-u}\left[e^{-u}\frac{d^2 y}{du^2} - e^{-u}\frac{dy}{du}\right] = e^{-2u}\left(\frac{d^2 y}{du^2} - \frac{dy}{du}\right). \qquad (6.10.4)$$

Utilizing the principle of mathematical induction, we shall now establish the following useful differentiation formula.

$$\frac{d^r}{dx^r} = e^{-ru}\frac{d}{du}\left(\frac{d}{du} - 1\right)\left(\frac{d}{du} - 2\right)\cdots\left(\frac{d}{du} - r + 1\right). \qquad (6.10.5)$$

In view of (6.10.3), Formula (6.10.5) has already been verified for $r = 1$. Suppose that Formula (6.10.5) is true for $r = k$, that is,

$$\frac{d^k}{dx^k} = e^{-ku}\frac{d}{du}\left(\frac{d}{du} - 1\right)\left(\frac{d}{du} - 2\right)\cdots\left(\frac{d}{du} - k + 1\right). \qquad (6.10.6)$$

If we now can show that Formula (6.10.5) is also true for $r = k + 1$, then Formula (6.10.5) will be true for all positive integral values of r. Utilizing the chain rule, (6.10.2), and (6.10.6), we have

$$\frac{d^{k+1}}{dx^{k+1}} = \frac{d}{dx}\left(\frac{d^k}{dx^k}\right) = \frac{d}{du}\left(\frac{d^k}{dx^k}\right)\cdot\frac{du}{dx}$$

$$= \frac{d}{du}\left[e^{-ku}\frac{d}{du}\left(\frac{d}{du} - 1\right)\left(\frac{d}{du} - 2\right)\cdots\left(\frac{d}{du} - k + 1\right)\right]\cdot\frac{1}{x}$$

$$= e^{-u}\left\{e^{-ku}\frac{d}{du}\left[\frac{d}{du}\left(\frac{d}{du} - 1\right)\left(\frac{d}{du} - 2\right)\cdots\left(\frac{d}{du} - k + 1\right)\right]\right.$$

$$\left. - ke^{-ku}\left[\frac{d}{du}\left(\frac{d}{du} - 1\right)\left(\frac{d}{du} - 2\right)\cdots\left(\frac{d}{du} - k + 1\right)\right]\right\}$$

$$= e^{-(k+1)u}\left[\left(\frac{d}{du} - k\right)\frac{d}{du}\left(\frac{d}{du} - 1\right)\left(\frac{d}{du} - 2\right)\cdots\left(\frac{d}{du} - k + 1\right)\right]$$

$$= e^{-(k+1)u}\frac{d}{du}\left(\frac{d}{du} - 1\right)\left(\frac{d}{du} - 2\right)\cdots\left(\frac{d}{du} - k + 1\right)\left(\frac{d}{du} - k\right),$$

which is the same as (6.10.5) when r is replaced by $k + 1$. Hence, Formula (6.10.5) is valid for all positive integral values of r.

Theorem 6.10.1. *The transformation* $x = e^u$ *(or* $x = e^{-u}$*) reduces Equation* (6.10.1) *to a linear differential equation of order n with constant coefficients.*

PROOF. Suppose that $x > 0$ in Equation (6.10.1). Let

$$x = e^u \quad \text{or} \quad u = \ln x, \quad x > 0. \qquad [6.10.2]$$

Substituting in (6.10.1) $d^r y/dx^r$, $r = 1, 2, \ldots, n$, according to Formula (6.10.5), and also substituting in (6.10.1) for x^r the expression e^{ru}, $r = 1, 2, \ldots, n$, obtained from (6.10.2), we get upon simplification

$$b_0\frac{d^n y}{du^n} + b_1\frac{d^{n-1}y}{du^{n-1}} + \cdots + b_{n-1}\frac{dy}{du} + b_n y = F(e^u), \qquad (6.10.7)$$

where $b_0, b_1, \ldots, b_n$ are constants and $b_0 \neq 0$. Thus, the theorem is established for the case when $x > 0$ in Equation (6.10.1).

Similarly, if $x < 0$ in Equation (6.10.1), then the transformation

$$x = e^{-u} \quad \text{or} \quad u = \ln(-x) \qquad x < 0, \tag{6.10.8}$$

reduces Equation (6.10.1) into a linear differential equation of order n with constant coefficients, namely,

$$b_0 \frac{d^n y}{du^n} + b_1 \frac{d^{n-1} y}{du^{n-1}} + \cdots + b_{n-1} \frac{dy}{du} + b_n y = F(e^{-u}). \tag{6.10.9}$$

Example 6.10.1. Let us solve the initial-value problem

$$\begin{cases} x^3 \dfrac{d^3 y}{dx^3} - 3x^2 \dfrac{d^2 y}{dx^2} + 6x \dfrac{dy}{dx} - 6y = 2x^4, & x > 0, \tag{6.10.10} \\[2mm] y(1) = 0, \quad y'(1) = 1, \quad y''(1) = -2. & \tag{6.10.11} \end{cases}$$

SOLUTION. Let

$$x = e^u \quad \text{or} \quad u = \ln x, \qquad x > 0. \tag{6.10.2}$$

From (6.10.3), (6.10.4), and (6.10.5), we have

$$\frac{dy}{dx} = e^{-u} \frac{dy}{du},$$

$$\frac{d^2 y}{dx^2} = e^{-2u} \left(\frac{d^2 y}{du^2} - \frac{dy}{du} \right),$$

$$\frac{d^3 y}{dx^3} = e^{-3u} \left(\frac{d^3 y}{du^3} - 3 \frac{d^2 y}{du^2} + 2 \frac{dy}{du} \right).$$

Substituting these derivatives in (6.10.10) and using (6.10.2), we obtain upon simplification·

$$\frac{d^3 y}{du^3} - 6 \frac{d^2 y}{du^2} + 11 \frac{dy}{du} - 6y = 2e^{4u}. \tag{6.10.12}$$

The characteristic equation $p(m) = 0$ of the related homogeneous differential equation of Equation (6.10.12) is

$$m^3 - 6m^2 + 11m - 6 = 0.$$

Utilizing synthetic division, we find that the roots of the above equation are 1, 2, and 3. Thus, the complementary function y_c of Equation (6.10.12) is

$$y_c(u) = c_1 e^u + c_2 e^{2u} + c_3 e^{3u}, \tag{6.10.13}$$

where c_1, c_2, and c_3 are arbitrary constants. Using Exercises 6.8.1(i), we find that a particular solution y_p of Equation (6.10.12) is

$$y_p(u) = \tfrac{1}{3} e^{4u}. \tag{6.10.14}$$

In view of (6.10.13) and (6.10.14), the general solution y of Equation (6.10.12) is

$$y = c_1 e^u + c_2 e^{2u} + c_3 e^{3u} + \tfrac{1}{3} e^{4u}. \tag{6.10.15}$$

Since $x = e^u$, we may write (6.10.15) as

$$y(x) = c_1 x + c_2 x^2 + c_3 x^3 + \tfrac{1}{3} x^4, \tag{6.10.16}$$

which is the general solution of Equation (6.10.10).   Differentiating (6.10.16) two times in succession, we obtain

$$y'(x) = c_1 + 2c_2 x + 3c_3 x^2 + \tfrac{4}{3} x^3, \tag{6.10.17}$$

$$y''(x) = 2c_2 + 6c_3 x + 4x^2. \tag{6.10.18}$$

Substituting the initial conditions given in (6.10.11) into (6.10.16) through (6.10.18), we have

$$\begin{cases} c_1 + c_2 + c_3 = -\tfrac{1}{3}, \\ c_1 + 2c_2 + 3c_3 = -\tfrac{1}{3}, \\ 2c_2 + 6c_3 = -6. \end{cases} \tag{6.10.19}$$

Solving the system (6.10.19), we find $c_1 = -\tfrac{10}{3}$, $c_2 = 6$, and $c_3 = -3$.   Substituting these values in (6.10.16), we find that the solution of the given initial-value problem is

$$y(x) = -\tfrac{10}{3} x + 6x^2 - 3x^3 + \tfrac{1}{3} x^4. \tag{6.10.20}$$

Exercises 6.10

1. Show that under the transformation $x = e^u$ or $u = \ln x$, $x > 0$, the Cauchy-Euler equations

$$a_0 x^3 \frac{d^3 y}{dx^3} + a_1 x^2 \frac{d^2 y}{dx^2} + a_2 x \frac{dy}{dx} + a_3 y = F(x)$$

and

$$a_0 x^4 \frac{d^4 y}{dx^4} + a_1 x^3 \frac{d^3 y}{dx^3} + a_2 x^2 \frac{d^2 y}{dx^2} + a_3 x \frac{dy}{dx} + a_4 y = F(x),$$

are transformed, respectively, into

$$a_0 \frac{d^3 y}{du^3} + (a_1 - 3a_0) \frac{d^2 y}{du^2} + (a_2 - a_1 + 2a_0) \frac{dy}{du} + a_3 y = F(e^u)$$

and

$$a_0 \frac{d^4 y}{du^4} + (a_1 - 6a_0) \frac{d^3 y}{du^3} + (a_2 - 3a_1 + 11a_0) \frac{d^2 y}{du^2}$$

$$+ (a_3 - a_2 + 2a_1 - 6a_0) \frac{dy}{du} + a_4 y = F(e^u).$$

In Exercises 2 through 9, find the general solution of each differential equation on the interval I: $0 < x < \infty$.

2. $x^3 y''' - 6x^2 y'' + 18xy' - 24y = 48 - 2x^2 + 4x^4$.

3. $x^4 y^{(4)} + 12x^3 y''' + 38x^2 y'' + 32xy' + 4y = \dfrac{2}{x} + \dfrac{4}{x^2}$.

4. $x^4 y^{(4)} + 4x^3 y''' + 2x^2 y'' + 2xy' - 2y = 2\ln^4 x - 38\ln^2 x + 56$.

5. $x^3 y''' = \dfrac{2x^2}{x+1}$.

6. $x^4 y^{(4)} + 6x^3 y''' + 9x^2 y'' + 3xy' + y = 8\cos \ln x$.

7. $x^5 y^{(5)} + 10x^4 y^{(4)} + 25x^3 y''' + 15x^2 y'' + xy' = \dfrac{288}{\ln x}$.

8. $x^4 y^{(4)} - 2x^3 y''' + 2xy' + 16y = 68x^4 + 600\sin^2 \ln x$.

9. $x^4 y^{(4)} + 6x^3 y''' + 7x^2 y'' + xy' - y$
$$= x(20\ln^4 x + 136\ln^3 x + 336\ln^2 x + 264\ln x).$$

10. Consider the differential equation
$$a_0(\alpha + \beta x)^n y^{(n)} + a_1(\alpha + \beta x)^{n-1} y^{(n-1)} + \cdots + a_{n-1}(\alpha + \beta x)y' + a_n y = f(x),$$

where $a_0, a_1, \ldots, a_n$, α, β are constants and $a_0 \neq 0$, $\beta \neq 0$. Show that the transformation

$$\alpha + \beta x = e^u, \qquad x > -\frac{\alpha}{\beta},$$

reduces the above differential equation to a linear differential equation of order n with constant coefficients.

11. Show that under the transformation
$$\alpha + \beta x = e^u \quad \text{or} \quad u = \ln(\alpha + \beta x), \qquad x > -\alpha/\beta,$$

the differential equation
$$a_0(\alpha + \beta x)^4 y^{(4)} + a_1(\alpha + \beta x)^3 y''' + a_2(\alpha + \beta x)^2 y'' + a_3(\alpha + \beta x)y' + a_4 y = F(x),$$

is transformed into

$$a_0 \beta^4 \frac{d^4 y}{du^4} + (a_1\beta^3 - 6a_0\beta^4)\frac{d^3 y}{du^3} + (a_2\beta^2 - 3a_1\beta^3 + 11a_0\beta^4)\frac{d^2 y}{du^2}$$

$$+ (a_3\beta - a_2\beta^2 + 2a_1\beta^3 - 6a_0\beta^4)\frac{dy}{du} + a_4 y = F\left(\frac{e^u - \alpha}{\beta}\right).$$

12. Find the general solution of the differential equation
$$(4+x)^4 y^{(4)} + 6(4+x)^3 y''' + 5(4+x)^2 y'' - (4+x)y' + y$$
$$= 9x^2 + \ln(16 + 8x + x^2), \qquad x > -4.$$

In Exercises 13 through 15, use the technique employed in Exercise 6.9.16 to find the solution of each integral equation.

13. $y(x) = \ln x + \displaystyle\int_1^x \frac{x}{t^2} y(t)dt,$ $x > 0.$ [*Note:* $y(1) = 0,$ $y'(1) = 1.$]

14. $y(x) = \ln x + (x - 1)^3 + \displaystyle\int_1^x \frac{x^2}{t^3} y(t)dt,$ $x > 0.$

15. $y(x) = \ln x + 1 - x - 3 \displaystyle\int_1^x \frac{(x-t)^2}{t^3} y(t)dt,$ $x > 0.$

16. Consider the integral equation

$$(1) \qquad y(x) = \ln x + (x - 1)^n + \int_1^x \frac{(x-t)^{n-1}}{t^n} y(t)dt, \qquad x > 0,$$

where n is a positive integer.   Show that by differentiating Equation (1) n times with respect to x, we obtain the following initial-value problem

$$\begin{cases} x^n y^{(n)} - (n-1)!y = n!x^n + (-1)^{n-1}(n-1)!, \\ y(1) = 0, \quad y'(1) = 1, \quad y''(1) = -1, \quad y'''(1) = 2, \quad y^{(4)}(1) = -6, \ldots, \\ y^{(n-1)}(1) = (-1)^n(n-2)! \quad \text{for } n > 2. \end{cases}$$

Suggested Readings

Coddington [12]   Tricomi [48]
Goursat [17]   Weisner [52]
Ince [23]

chapter 7   SERIES SOLUTIONS OF LINEAR DIFFERENTIAL EQUATIONS

7.1. Introduction

There is a large class of ordinary differential equations that cannot be solved in closed form. That is, their solutions cannot be expressed in terms of elementary functions.[1] However, over certain intervals, these differential equations may possess solutions expressible in terms of power series or "Frobenius series." We shall be interested primarily in second-order homogeneous linear differential equations with variable coefficients which are polynomials. These equations occur most frequently in theoretical and applied mathematics and other branches of science, in particular mathematical physics. A complete systematic development of solving differential equations with variable coefficients via infinite series is based on the theory of functions of a complex variable. Thus, our present discussion shall be by no means complete. However, in some respects, it shall be self-contained.

7.2. Some Basic Definitions and Results on Power Series

We shall state without proofs (with a few exceptions) some of the basic results of power series which will give the reader a better understanding of the forthcoming material. Some of these proofs have been established in calculus.

---

[1] The *elementary operations* on functions $f = f(x)$ and $g = g(x)$ are those that yield any of the following: $f(x) \pm g(x)$, $f(x) \cdot g(x)$, $f(x)/g(x)$, $[f(x)]^\alpha$, $[\alpha]^{f(x)}$, $\ln [f(x)]$; where α is a constant. The *elementary functions* are those generated by constants and the independent variable by means of a finite number of elementary operations. Examples of elementary functions: $e^{5 \ln x}$, $\ln [3 + \sqrt{2 + \cos x}]$, $(5x^2 + 7x + 2)$ arc sin $(\ln \cos x)$, $\sinh^{-1}[x + \sqrt[4]{3x^2 + 1}] + 3 \cosh^{-1}[\ln \sqrt[3]{e^x + 1}] + \sinh x + 5x^2 \cos 2x$.

Definition 7.2.1. An expression of the form

$$a_0 + a_1(x - x_0) + a_2(x - x_0)^2 + \cdots + a_n(x - x_0)^n + \cdots = \sum_{n=0}^{\infty} a_n(x - x_0)^n$$

$$(7.2.1)$$

is called a *power series* in powers of $(x - x_0)$, and its value is the limit

$$\lim_{N \to \infty} \sum_{n=0}^{N} a_n(x - x_0)^n \tag{7.2.2}$$

for those values of x for which the limit exists.   For such values of x the series is said to *converge*.   The numbers a_n, $n = 0, 1, \ldots$, are called the *coefficients* of the power series.   The series in (7.2.1) is said to be *divergent*, or to *diverge*, if the limit given in (7.2.2) either does not exist or is infinite.

Definition 7.2.2.   A power series $\sum_{n=0}^{\infty} a_n(x - x_0)^n$ is said to *converge absolutely* or be *absolutely convergent* if the series $\sum_{n=0}^{\infty} |a_n(x - x_0)^n|$ converges.

Theorem 7.2.1.   *Every power series $\sum_{n=0}^{\infty} a_n(x - x_0)^n$ has a "radius of convergence" R such that when $0 < R < \infty$, the series converges absolutely for $|x - x_0| < R$ and diverges for $|x - x_0| > R$.   When $R = 0$, the series converges only for $x = x_0$.   When $R = \infty$, the series converges for all x.   The number R is given by*

$$R = \lim_{n \to \infty} \left| \frac{a_n}{a_{n+1}} \right|, \quad \text{if the limit exists,} \tag{7.2.3}$$

$$R = \frac{1}{\lim\limits_{n \to \infty} \sqrt[n]{|a_n|}}, \quad \text{if the limit exists,} \tag{7.2.4}$$

and in any case by the formula

$$R = \frac{1}{\overline{\lim}\limits_{n \to \infty} \sqrt[n]{|a_n|}}. \tag{7.2.5}$$

[*Note:*   The formula for R given in (7.2.5) involves the concept of the upper limit of a sequence $\{a_n\}$ of real numbers.]

Remark 7.2.1.   When $0 < R < \infty$, the interval $|x - x_0| < R$ or $x_0 - R < x < x_0 + R$ is called the *interval of convergence* of the power series

$$\sum_{n=0}^{\infty} a_n(x - x_0)^n.$$

The series may or may not converge at the end points $x = x_0 \pm R$.   It may converge at both, at just one, or at neither.   When $R = \infty$, the interval of convergence is the entire x axis.   In Figure 7.2.1, the interval of convergence is depicted by the heavy line.

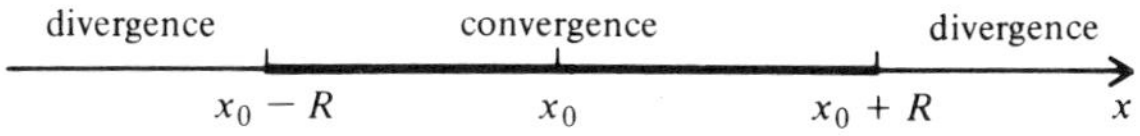

<table>
<tr><td>divergence</td><td>convergence</td><td>divergence</td><td></td></tr>
</table>

$$x_0 - R \qquad\qquad x_0 \qquad\qquad x_0 + R \qquad\qquad x$$

Figure 7.2.1.
Interval of Convergence

Example 7.2.1. Let us determine the radius of convergence R of the power series

$$\sum_{n=0}^{\infty} \left[\frac{1 \cdot 3 \cdot 5 \cdots (2n + 1)}{2 \cdot 4 \cdot 6 \cdots (2n + 2)} \right] \frac{(x - \frac{3}{2})^n}{n + 1}. \tag{7.2.6}$$

SOLUTION.　Here we have

$$a_n = \frac{1 \cdot 3 \cdot 5 \cdots (2n + 1)}{2 \cdot 4 \cdot 6 \cdots (2n + 2)} \cdot \frac{1}{n + 1} \quad \text{and} \quad a_{n+1} = \frac{1 \cdot 3 \cdot 5 \cdots (2n + 3)}{2 \cdot 4 \cdot 6 \cdots (2n + 4)} \cdot \frac{1}{n + 2}.$$

Thus,

$$\lim_{n \to \infty} \left| \frac{a_n}{a_{n+1}} \right| = \lim_{n \to \infty} \left| \frac{(2n + 4)(n + 2)}{(2n + 3)(n + 1)} \right| = 1.$$

Hence, $R = 1$. By Theorem 7.2.1, we are assured that the series (7.2.6) converges absolutely for all x such that $\left| x - \frac{3}{2} \right| < 1$ or $\frac{1}{2} < x < \frac{5}{2}$. It may be shown that the series (7.2.6) converges at $x = \frac{1}{2}$ and $x = \frac{5}{2}$. Thus, the interval of convergence of the series (7.2.6) is $\frac{1}{2} \leq x \leq \frac{5}{2}$.

Theorem 7.2.2. *If the power series $\sum_{n=0}^{\infty} a_n(x - x_0)^n$ converges absolutely, then $\sum_{n=0}^{\infty} a_n(x - x_0)^n$ converges.*

Note that the converse of the above theorem is false. We need only to observe that $\sum_{n=0}^{\infty} (-1)^n/(n + 1)$ is convergent, but not absolutely.

Another useful theorem is the following.

Theorem 7.2.3. *(Comparison Test for Convergence.) Suppose that the constants A_n are nonnegative and $|a_n| \leq A_n$, for $n = 0, 1, 2, \ldots$. If the power series $\sum_{n=0}^{\infty} A_n(x - x_0)^n$ converges for $|x - x_0| < r$, then the power series $\sum_{n=0}^{\infty} a_n(x - x_0)^n$ also converges at least for $|x - x_0| < r$.*

In the ensuing results, unless stated to the contrary, we shall assume that the radius of convergence $R > 0$.

Remark 7.2.2. Suppose that a series $\sum_{n=0}^{\infty} a_n(x - x_0)^n$ converges to a function f at each point x in a set S. We then say that the series represents a function f in S, and we write

$$f(x) = \sum_{n=0}^{\infty} a_n(x - x_0)^n \tag{7.2.7}$$

for x in S.

Theorem 7.2.4.   *A power series represents a continuous function at every point interior to its interval of convergence I:* $|x - x_0| < R.$ *That is,*

$$f(x) = \sum_{n=0}^{\infty} a_n(x - x_0)^n, \qquad |x - x_0| < R. \tag{7.2.8}$$

Theorem 7.2.5.   *A power series can be integrated term by term at every point interior to its interval of convergence I:* $|x - x_0| < R.$ *That is, if*

$$f(x) = \sum_{n=0}^{\infty} a_n(x - x_0)^n, \qquad |x - x_0| < R,$$

then

$$\int_a^b f(x)dx = \sum_{n=0}^{\infty} \int_a^b a_n(x - x_0)^n \, dx$$

$$= \sum_{n=0}^{\infty} \frac{a_n}{n+1} [(b - x_0)^{n+1} - (a - x_0)^{n+1}], \qquad |x - x_0| < R, \tag{7.2.9}$$

where a and b are any two points interior to I and a < b.

Theorem 7.2.6.   *A power series can be differentiated term by term at every point interior to its interval of convergence I:* $|x - x_0| < R.$ *That is, if*

$$f(x) = \sum_{n=0}^{\infty} a_n(x - x_0)^n, \qquad |x - x_0| < R,$$

then

$$f'(x) = \sum_{n=1}^{\infty} na_n(x - x_0)^{n-1}, \qquad |x - x_0| < R. \tag{7.2.10}$$

Also, we have

$$f^{(k)}(x) = \sum_{n=k}^{\infty} \frac{n!}{(n-k)!} a_n(x - x_0)^{n-k}, \quad k = 0, 1, 2, \ldots, \qquad |x - x_0| < R.$$

$$\tag{7.2.11}$$

[*Note:*   When $k = 0$, (7.2.11) becomes $f^{(0)}(x) \equiv f(x) = \sum_{n=0}^{\infty} a_n(x - x_0)^n.$]
From (7.2.11), we see that

$$f^{(n)}(x) = n!a_n + (n + 1)n(n - 1) \cdots 2a_{n+1}(x - x_0)$$

$$+ (n + 2)(n + 1)n \cdots 3a_{n+2}(x - x_0)^2 + \cdots,$$

$$n = 0, 1, 2, \ldots, \qquad |x - x_0| < R. \tag{7.2.12}$$

Setting $x = x_0$ in (7.2.12), we obtain

$$f^{(n)}(x_0) = n!a_n \quad \text{or} \quad a_n = \frac{f^{(n)}(x_0)}{n!}, \qquad n = 0, 1, 2, \ldots. \tag{7.2.13}$$

[*Note:* $0! \equiv 1.$]   Thus,

$$f(x) = \sum_{n=0}^{\infty} \frac{f^{(n)}(x_0)}{n!} (x - x_0)^n, \qquad |x - x_0| < R. \tag{7.2.14}$$

Observe that the relationship between the function f and the power series representing it is an intimate one. The coefficients a_n are completely determined by f in a neighborhood of x_0.

Definition 7.2.3. The series in (7.2.14) is called the *Taylor series* expansion of f in powers of $(x - x_0)$, or in a neighborhood of x_0.

It is clear that not every function possesses a Taylor series expansion, since, in order that (7.2.14) be well defined, all derivatives of f must exist at $x = x_0$. For example, the function f defined by $f(x) = 1/(x - 1)$ does not possess a Taylor series expansion in powers of $(x - 1)$. However, it does possess one in powers of $(x - x_0)$ for $x_0 \neq 1$.

In particular, when $x_0 = 0$, (7.2.14) reduces to

$$f(x) = \sum_{n=0}^{\infty} \frac{f^{(n)}(0)}{n!} x^n, \qquad |x| < R. \tag{7.2.15}$$

Definition 7.2.4. The series in (7.2.15) is called the *Maclaurin series* expansion of f in powers of x, or in a neighborhood of zero.

Example 7.2.2. Let us determine the Taylor series expansion about the point $x_0 = 2$ for the function f defined by $f(x) = 1/x^2$. Also, let us determine the radius of convergence of the series.

SOLUTION. Differentiating $f(x)$ n times, we obtain

$$f^{(n)}(x) = \frac{(-1)^n(n + 1)!}{x^{n+2}}, \qquad n = 0, 1, 2, \ldots .$$

Thus,

$$f^{(n)}(2) = \frac{(-1)^n(n + 1)!}{2^{n+2}}, \qquad n = 0, 1, 2, \ldots .$$

Hence,

$$\frac{1}{x^2} = \sum_{n=0}^{\infty} \frac{f^{(n)}(2)}{n!} (x - 2)^n = \sum_{n=0}^{\infty} \frac{(-1)^n(n + 1)}{2^{n+2}} (x - 2)^n. \tag{7.2.16}$$

Concerning the radius of convergence of the series in (7.2.16), we have

$$\lim_{n \to \infty} \left| \frac{a_n}{a_{n+1}} \right| = \lim_{n \to \infty} \left| \frac{(-1)^n 2^{n+3} (n + 1)}{(-1)^{n+1} 2^{n+2}(n + 2)} \right| = \lim_{n \to \infty} \left| -2 \cdot \frac{n + 1}{n + 2} \right| = 2.$$

Hence, $R = 2$. Theorem 7.2.1 assures us that the series in (7.2.16) converges absolutely for all x such that $|x - 2| < 2$. The reader should determine whether the series converges at the end points of the interval of convergence.

Definition 7.2.5.   A function f is said to be *analytic* at a point $x = x_0$ if it can be expanded in a Taylor series in powers of $(x - x_0)$ valid[2] for every x in a neighborhood of x_0.

Definition 7.2.6.   A function f is said to be *analytic on an interval I* if it is analytic at every point in I.

Example 7.2.3.   The following are analytic functions in the indicated interval.

$$(1) \ \ e^x = \sum_{n=0}^{\infty} \frac{x^n}{n!}, \qquad -\infty < x < \infty,$$

$$(2) \ \ \sin x = \sum_{n=0}^{\infty} \frac{(-1)^n}{(2n+1)!} x^{2n+1}, \qquad -\infty < x < \infty,$$

$$(3) \ \ \cosh x = \sum_{n=0}^{\infty} \frac{x^{2n}}{(2n)!}, \qquad -\infty < x < \infty,$$

$$(4) \ \ \frac{1}{1+x} = \sum_{n=0}^{\infty} (-1)^n x^n, \qquad -1 < x < 1,$$

$$(5) \ \ \ln(1+x) = \sum_{n=1}^{\infty} \frac{(-1)^{n-1}}{n} x^n, \qquad -1 < x \leq 1.$$

A polynomial P given by

$$(6) \ \ P(x) = a_0 + a_1 x + a_2 x^2 + \cdots + a_n x^n$$

being a finite series, is analytic for all (finite) values of x.   Any rational function P/Q, where $P(x)$ and $Q(x)$ are polynomials, is analytic for all (finite) values of x, except for those values x for which $Q(x) = 0$.

Theorem 7.2.7.   *If two power series $\sum_{n=0}^{\infty} a_n(x - x_0)^n$ and $\sum_{n=0}^{\infty} b_n(x - x_0)^n$ both converge to the same function f in some neighborhood N: $\ |x - x_0| < r$ of x_0, then the series are identical.   That is, $a_n = b_n$, $n = 0, 1, 2, \ldots$.*
    In particular, if a power series $\sum_{n=0}^{\infty} c_n(x - x_0)^n$ converges to zero at every point in some neighborhood of the point x_0, then $c_n = 0$, $n = 0, 1, 2, \ldots$.
    We shall often make use of the above theorem, especially the latter result.

Theorem 7.2.8.   *If two power series*

$$f(x) = \sum_{n=0}^{\infty} a_n(x - x_0)^n \quad and \quad g(x) = \sum_{n=0}^{\infty} b_n(x - x_0)^n$$

---

² By "valid" we mean that for every value of x in a neighborhood of x_0, the series expansion of the function converges to the value of the function.

have nonzero convergence radii r and R, respectively, $r \leq R$, then

$$f(x) \pm g(x) = \sum_{n=0}^{\infty} (a_n \pm b_n)(x - x_0)^n, \qquad |x - x_0| < r, \tag{7.2.17}$$

$$f(x)g(x) = \sum_{n=0}^{\infty} \sum_{k=0}^{n} a_k(x - x_0)^k b_{n-k}(x - x_0)^{n-k}$$

$$= \sum_{n=0}^{\infty} \sum_{k=0}^{n} a_k b_{n-k}(x - x_0)^n, \qquad |x - x_0| < r. \tag{7.2.18}$$

Also,

$$f(x)g(x) = \sum_{n=0}^{\infty} \sum_{k=0}^{n} a_{n-k}(x - x_0)^{n-k} b_k(x - x_0)^k$$

$$= \sum_{n=0}^{\infty} \sum_{k=0}^{n} a_{n-k} b_k(x - x_0)^n, \qquad |x - x_0| < r. \tag{7.2.19}$$

Note that

$$\sum_{k=0}^{n} a_k b_{n-k} = a_0 b_n + a_1 b_{n-1} + \cdots + a_{n-1}b_1 + a_n b_0. \tag{7.2.20}$$

The expression in (7.2.18) or (7.2.19) is known as the *Cauchy product* of the two given power series.

Example 7.2.4.   One may easily verify that the Taylor series expansions of the functions f and g defined by $f(x) = 1/(1 - 4x)$ and $g(x) = 1/(1 - 3x)$ about the point $x_0 = 0$ are

$$f(x) = \sum_{n=0}^{\infty} 4^n x^n, \quad |x| < \tfrac{1}{4}, \qquad g(x) = \sum_{n=0}^{\infty} 3^n x^n, \quad |x| < \tfrac{1}{3}.$$

Thus,

$$f(x)g(x) = \sum_{n=0}^{\infty} \sum_{k=0}^{n} 4^k x^k 3^{n-k} x^{n-k} = \sum_{n=0}^{\infty} \sum_{k=0}^{n} 4^k 3^{n-k} x^n, \qquad |x| < \tfrac{1}{4}. \tag{7.2.21}$$

We may simplify the result in (7.2.21).   Observe that $\sum_{k=0}^{n} 4^k 3^{n-k}$ forms a geometric progression with ratio $4/3$ and having $(n + 1)$ terms.   Thus, utilizing the formula for the sum of a geometric progression, we obtain

$$\sum_{k=0}^{n} 4^k 3^{n-k} = 3^n \frac{[1 - (\tfrac{4}{3})^{n+1}]}{1 - \tfrac{4}{3}} = 4^{n+1} - 3^{n+1},$$

and consequently we have that

$$\frac{1}{(1 - 4x)(1 - 3x)} = \sum_{n=0}^{\infty} (4^{n+1} - 3^{n+1})x^n, \qquad |x| < \tfrac{1}{4}. \tag{7.2.22}$$

Theorem 7.2.9.   *If two power series*

$$f(x) = \sum_{n=0}^{\infty} a_n(x - x_0)^n \quad and \quad g(x) = \sum_{n=0}^{\infty} b_n(x - x_0)^n$$

have nonzero radii of convergence r and R, respectively, and if $g(x_0) \neq 0$, then there exists a power series $\sum_{n=0}^{\infty} c_n(x - x_0)^n$ and a number $\sigma > 0$ such that

$$\frac{f(x)}{g(x)} = \sum_{n=0}^{\infty} c_n(x - x_0)^n, \qquad |x - x_0| < \sigma. \tag{7.2.23}$$

The coefficients c_n satisfy the equation

$$a_n = c_0 b_n + c_1 b_{n-1} + c_2 b_{n-2} + \cdots + c_{n-1} b_1 + c_n b_0. \tag{7.2.24}$$

Remark 7.2.3.   Let ρ be the minimum distance between x_0 and the zeros of $g(x)$.   If $g(x)$ has no zeros, let $\rho = \infty$.   Note that in (7.2.23) we may take $\sigma = \min(\rho, r, R)$.   Letting τ be the radius of convergence of the power series in (7.2.23), we have $\tau \geq \min(\rho, r, R)$.

Remark 7.2.4.   Since in Theorem 7.2.9 we are given that $g(x_0) \neq 0$, it follows that $b_0 \neq 0$ in the expansion of the function g about the point $x = x_0$.   Thus, from (7.2.24), we obtain

$$c_n = \frac{a_n - c_0 b_n - c_1 b_{n-1} - \cdots - c_{n-1} b_1}{b_0}. \tag{7.2.25}$$

This formula enables us to compute c_n when $c_0, c_1, \ldots, c_{n-1}$ have already been computed.

Example 7.2.5.   Given the power series expansion

$$f(x) = \sum_{n=0}^{\infty} 4^n x^n, \quad |x| < \tfrac{1}{4}, \qquad g(x) = \sum_{n=0}^{\infty} 3^n x^n, \quad |x| < \tfrac{1}{3},$$

let us find the first four terms of the power series expansion of f/g about the point $x_0 = 0$.

SOLUTION.   Since

$$f(x) = \sum_{n=0}^{\infty} 4^n x^n = 1 + 4x + 4^2 x^2 + 4^3 x^3 + \cdots,$$

$$g(x) = \sum_{n=0}^{\infty} 3^n x^n = 1 + 3x + 3^2 x^2 + 3^3 x^3 + \cdots,$$

we see that $a_0 = 1, a_1 = 4, a_2 = 4^2, a_3 = 4^3, \ldots, b_0 = 1, b_1 = 3, b_2 = 3^2, b_3 = 3^3, \ldots$.   Utilizing Formula (7.2.25), we find

$$n = 0, \quad c_0 = \frac{a_0}{b_0} = 1.$$

$$n = 1, \quad c_1 = \frac{a_1 - c_0 b_1}{b_0} = \frac{4 - 1 \cdot 3}{1} = 1.$$

$$n = 2, \quad c_2 = \frac{a_2 - c_0 b_2 - c_1 b_1}{b_0} = \frac{4^2 - 1 \cdot 3^2 - 1 \cdot 3}{1} = 4.$$

$$n = 3, \quad c_3 = \frac{a_3 - c_0 b_3 - c_1 b_2 - c_2 b_1}{b_0} = \frac{4^3 - 1 \cdot 3^3 - 1 \cdot 3^2 - 4 \cdot 3}{1} = 16.$$

Thus,

$$\frac{f(x)}{g(x)} = 1 + x + 4x^2 + 16x^3 + \cdots . \tag{7.2.26}$$

Theorem 7.2.10. *(Cauchy's Inequality.)* *If the power series $\sum_{n=0}^{\infty} a_n(x - x_0)^n$ has a nonzero radius of convergence R, then for every x satisfying $|x - x_0| = r < R$, there exists a constant $M = M_r > 0$ such that*

$$|a_n| \leq \frac{M}{r^n}, \qquad n = 0, 1, 2, \ldots . \tag{7.2.27}$$

PROOF. The series $\sum_{n=0}^{\infty} a_n(x - x_0)^n$ converges for $|x - x_0| = r$. Now, a necessary (but not sufficient) condition for the series $\sum_{n=0}^{\infty} a_n r^n$ to converge is that $\lim_{n \to \infty} |a_n r^n| = 0$. Thus, there exists an integer N such that for $n > N$, we have

$$|a_n| r^n \leq 1, \qquad n > N.$$

Let

$$M = \max (|a_0|, |a_1| r, \ldots, |a_N| r^N, 1).$$

It is now evident that, for this particular value of M, the inequality in (7.2.27) is indeed satisfied.

Example 7.2.6. Consider the power series

$$\sum_{n=0}^{\infty} \frac{(-1)^n(n + 1)}{2^{n+2}} (x - 2)^n, \qquad |x - 2| < 2.$$

Then, for every x such that $|x - 2| = \frac{3}{2}$, the inequality in (7.2.27) becomes

$$\left(\frac{3}{2}\right)^n \frac{(n + 1)}{2^{n+2}} \leq M, \qquad n = 0, 1, 2, \ldots .$$

We leave it to the reader to show that the above inequality is satisfied when $M = 1$.

Theorem 7.2.11. *Let the two power series $f(x) = \sum_{n=j}^{\infty} a_n(x - x_0)^n$, $g(x) = \sum_{n=k}^{\infty} b_n(x - x_0)^n$ have nonzero radii of convergence r and R, respectively, $r \leq R$. Suppose that $j < k$ and $a_j \neq 0$, $b_k \neq 0$. Then the functions f and g are linearly independent on the interval of convergence I: $x_0 - r < x < x_0 + r$.*

[*Note:* The condition $j < k$ and $a_j \neq 0$, $b_k \neq 0$, guarantees us that the series represented by the functions f and g start with a term $(x - x_0)^j$ and $(x - x_0)^k$, respectively.]

PROOF. We shall establish this result via a contradiction. Thus, suppose that the functions f and g were linearly dependent on I. Then, in view of Definition 4.3.5, there would exist constants c_1 and c_2, not both zero, such that

$$c_1 f(x) + c_2 g(x) = 0$$

for every x in I. The above expression may be written as

$$c_1 \sum_{n=j}^{\infty} a_n(x - x_0)^n + c_2 \sum_{n=k}^{\infty} b_n(x - x_0)^n = 0, \qquad (7.2.28)$$

for every x in I. Since $j < k$, let us suppose for definiteness that $k = j + l$, $l > 0$. Then utilizing Theorem 7.2.8, we may write (7.2.28) as

$$\begin{aligned}
c_1[a_j(x - x_0)^j &+ a_{j+1}(x - x_0)^{j+1} + \cdots + a_{j+l-1}(x - x_0)^{j+l-1}] \\
&+ (c_1 a_k + c_2 b_k)(x - x_0)^k + (c_1 a_{k+1} + c_2 b_{k+1})(x - x_0)^{k+1} \\
&+ \cdots = 0, \qquad |x - x_0| < r. \qquad (7.2.29)
\end{aligned}$$

In view of Theorem 7.2.7 and (7.2.29), it follows that

$$c_1 a_j = 0, \quad c_1 a_{j+1} = 0, \quad \cdots, \quad c_1 a_{j+l-1} = 0, \quad c_1 a_k + c_2 b_k = 0,$$
$$c_1 a_{k+1} + c_2 b_{k+1} = 0, \quad \ldots.$$

Since $a_j \neq 0$, $c_1 a_j = 0$ implies that $c_1 = 0$. Thus, (7.2.29) reduces to

$$c_2[b_k(x - x_0)^k + b_{k+1}(x - x_0)^{k+1} + \cdots] = 0, \qquad |x - x_0| < r.$$

Again applying Theorem 7.2.7 to the above series, we see that

$$c_2 b_k = 0, \quad c_2 b_{k+1} = 0, \quad \ldots.$$

Since $b_k \neq 0$, $c_2 b_k = 0$ implies that $c_2 = 0$. Thus, $c_1 = 0$, $c_2 = 0$ and consequently, the functions f and g are not linearly dependent on I. Hence, they are linearly independent on I. [*See* Definition 4.3.6.]

[*Note:* Two power series are dependent only if all their coefficients are proportional.]

Example 7.2.7. The functions f and g given by

$$f(x) = \sum_{n=0}^{\infty} (-1)^n (n + 1)(2n + 1)(2n + 3)x^{2n}, \qquad |x| < 1,$$

$$g(x) = \sum_{n=0}^{\infty} (-1)^n (n + 1)(n + 2)(2n + 3)x^{2n+1}, \qquad |x| < 1,$$

are linearly independent on the interval I: $-1 < x < 1$.

Remark 7.2.5.   In the forthcoming sections, we shall make use of the following concepts.   The letter used to denote the index of summation in a power series acts as a "dummy" variable.   That is,

$$\sum_{n=0}^{\infty} a_n x^n = \sum_{j=0}^{\infty} a_j x^j = \sum_{k=0}^{\infty} a_k x^k.$$

Also, a power series may be written equivalently in many different ways by changing the index of summation.   That is,

$$\sum_{n=3}^{\infty} n^2 a_n x^n = \sum_{k=0}^{\infty} (k+3)^2 a_{k+3} x^{k+3} = \sum_{n=0}^{\infty} (n+3)^2 a_{n+3} x^{n+3}.$$

Note that the second series was obtained from the first series by setting $n = k + 3$.   Then, when $n = 3$, $k = 0$; $n^2 = (k+3)^2$ and $a_n = a_{k+3}$.   As another illustration, we have

$$\sum_{n=0}^{\infty} (n+1)a_n x^n = \sum_{k=1}^{\infty} k a_{k-1} x^{k-1} = \sum_{j=2}^{\infty} (j-1)a_{j-2} x^{j-2} = \sum_{l=3}^{\infty} (l-2)a_{l-3} x^{l-3}$$

$$= \sum_{n=3}^{\infty} (n-2)a_{n-3} x^{n-3} = \sum_{n=5}^{\infty} (n-4)a_{n-5} x^{n-5}.$$

Note that the third series was obtained from the second series by setting $k = j - 1$.   Then, when $k = 1, j = 2$; $a_{k-1} = a_{j-1-1} = a_{j-2}$.

Let us verify

$$4x \sum_{n=2}^{\infty} n(n-1)a_n x^{n-2} + 2 \sum_{n=1}^{\infty} n a_n x^{n-1} - \sum_{n=0}^{\infty} a_n x^n$$

$$= -a_0 + 2a_1 + \sum_{n=2}^{\infty} [2n(2n-1)a_n - a_{n-1}]x^{n-1}. \qquad (7.2.30)$$

Our task is to make the exponents of x equal to $n - 1$, in each of the series in the left-hand member of (7.2.30), so that we may add the series.   Since $4x$ is independent of the index of summation n, we may write

$$4x \sum_{n=2}^{\infty} n(n-1)a_n x^{n-2}$$

as

$$\sum_{n=2}^{\infty} 4n(n-1)a_n x^{n-1}.$$

Summing out the first term in the second series in (7.2.30), we have

$$2a_1 + \sum_{n=2}^{\infty} 2n a_n x^{n-1}.$$

Changing the index of summation in the third series in (7.2.30) and then summing out the first term, we have

$$-\sum_{n=1}^{\infty} a_{n-1}x^{n-1} = -a_0 - \sum_{n=2}^{\infty} a_{n-1}x^{n-1}.$$

Thus, in the interior of their common interval of convergence, we obtain

$$4x\sum_{n=2}^{\infty} n(n-1)a_n x^{n-2} + 2\sum_{n=1}^{\infty} na_n x^{n-1} - \sum_{n=0}^{\infty} a_n x^n$$

$$= \sum_{n=2}^{\infty} 4n(n-1)a_n x^{n-1} + 2a_1 + \sum_{n=2}^{\infty} 2na_n x^{n-1} - a_0 - \sum_{n=2}^{\infty} a_{n-1}x^{n-1}$$

$$= -a_0 + 2a_1 + \sum_{n=2}^{\infty} [4n(n-1)a_n + 2na_n - a_{n-1}]x^{n-1}$$

$$= -a_0 + 2a_1 + \sum_{n=2}^{\infty} [2n(2n-1)a_n - a_{n-1}]x^{n-1}.$$

Remark 7.2.6.   All the theorems stated in this section are also valid when we replace the real variable x by the complex variable $z = x + iy$, where x and y are real numbers and $i^2 = -1$.   The only change is that "the interval of convergence" is replaced by "the circle of convergence," since the expression $|z - z_0| = R$ now represents a circle of radius R with center at z_0.   That is, if $z = x + iy$ and $z_0 = x_0 + iy_0$, then $z - z_0 = (x - x_0) + i(y - y_0)$, and $|z - z_0| = R$ becomes $\sqrt{(x - x_0)^2 + (y - y_0)^2} = R$ or $(x - x_0)^2 + (y - y_0)^2 = R^2$.

The function f given by

$$f(x) = \frac{1}{1 + x^2} \tag{7.2.31}$$

is defined for all real values of x.   Its Taylor series expansion about the point $x_0 = 0$ is [*see* Exercise 7.2.2(c)]

$$\frac{1}{1 + x^2} = \sum_{n=0}^{\infty} (-1)^n x^{2n}, \tag{7.2.32}$$

and its interval of convergence I is

$$I: \quad -1 < x < 1. \tag{7.2.33}$$

However, in the realm of complex numbers, the function f given by (7.2.31) is not defined for $x = \pm i$.   The Taylor series expansion of $f(z) = 1/(1 + z^2)$ about the point $z = 0$ is

$$f(z) = \sum_{n=0}^{\infty} (-1)^n z^{2n}. \tag{7.2.34}$$

Utilizing Formula (7.2.3), the radius of convergence $R = 1$.   Theorem 7.2.1 assures us that the power series in (7.2.34) converges absolutely at every point

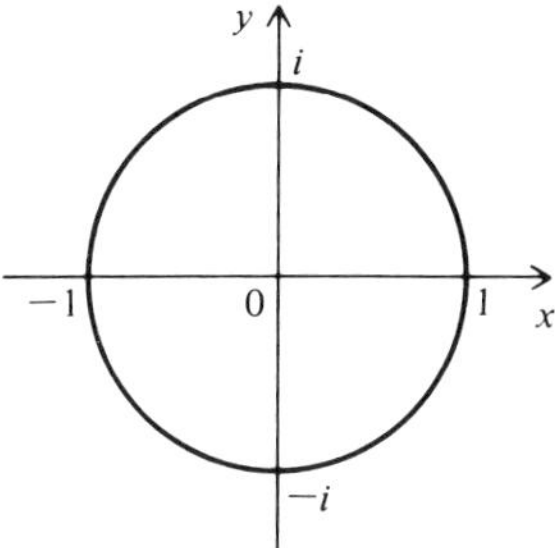

Figure 7.2.2.

in the interior of its circle of convergence: $|z - 0| = |z| = 1$, and diverges at every point exterior to its circle of convergence. On the circle of convergence, it may converge at some points and diverge at others. The power series in (7.2.34) diverges at all points on the circle of convergence. In particular, it diverges when $z = x = 1$ and $z = x = -1$. In the complex plane, $|z| = 1$ denotes the circle $x^2 + y^2 = 1$. [*See* Figure 7.2.2.] Now, the interval of convergence I in (7.2.33) is merely a cross section of the circle of convergence $|z| = 1$ in the complex plane.

The above analysis explains the reason why the interval of convergence of the power series in (7.2.32) is I: $-1 < x < 1$, and not, as one might have conjectured, the interval: $-\infty < x < \infty$.

In the complex plane, the distance r between two complex numbers $z_1 = x_1 + iy_1$, $z_2 = x_2 + iy_2$ is given by [*see* Figure 7.2.3]

$$r = |z_2 - z_1| = \sqrt{(x_2 - x_1)^2 + (y_2 - y_1)^2}. \tag{7.2.35}$$

Note that the distance from $z = 0$ to the nearest zero of $z^2 + 1 = 0$, namely $z = i$ or $z = -i$, is

$$r = |i - 0| = \sqrt{(0 - 0)^2 + (1 - 0)^2} = 1.$$

The power series (7.2.32) will converge at least for all values of x in the interval $|x - 0| < r$ or $|x| < 1$. At the end points of the interval it diverges.

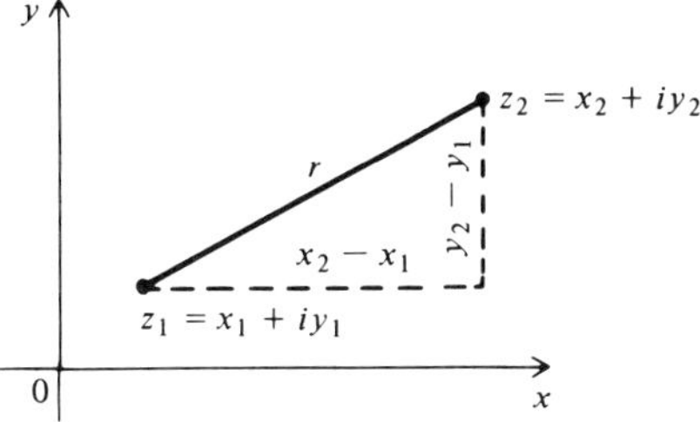

Figure 7.2.3.

Exercises 7.2

1. Determine the radius of convergence of the following power series.

(a) $\displaystyle\sum_{n=0}^{\infty} \frac{(x+1)^n}{(n+1)^2}$,

(b) $\displaystyle\sum_{n=0}^{\infty} 2^n (x-3)^n$,

(c) $\displaystyle\sum_{n=0}^{\infty} \frac{(x-1)^n}{n!}$,

(d) $\displaystyle\sum_{n=0}^{\infty} (n^2+4)(x+4)^n$,

(e) $\displaystyle\sum_{n=0}^{\infty} \frac{(-1)^{n+1}2^n}{4n^2-1} x^{2n+1}$,

(f) $\displaystyle\sum_{n=0}^{\infty} \frac{(-1)^n(n+1)}{1\cdot 3\cdot 5\cdots(2n+1)} x^{2n+1}$,

(g) $\displaystyle\sum_{n=1}^{\infty} \frac{(-1)^n (x-2)^{3n+1}}{3^n n![4\cdot 7\cdot 10\cdots(3n+1)]}$,

(h) $\displaystyle\sum_{n=1}^{\infty} \frac{1\cdot 5\cdot 9\cdots(4n-3)}{n!(4n^2-1)} x^{2n+1}$,

(i) $\displaystyle\sum_{n=1}^{\infty} \frac{n!}{n^n} x^n$,

(j) $\displaystyle\sum_{n=1}^{\infty} \frac{\ln n^n}{n!} x^n$,

(k) $\displaystyle\sum_{n=0}^{\infty} (-1)^n \frac{(n+1)(2n+1)(2n+3)}{3} x^{2n}$,

(l) $\displaystyle\sum_{n=0}^{\infty} (-1)^n \frac{(n+1)(n+2)(2n+3)}{6} x^{2n+1}$.

2. Expand the following functions in a Taylor series about the indicated point x_0. Also determine the interval of convergence of the series.

(a) $\dfrac{1}{1-x}$, $x_0=0$,

(b) $\dfrac{1}{2x+3}$, $x_0=0$,

(c) $\dfrac{1}{1+x^2}$, $x_0=0$,

(d) e^{x-1}, $x_0=2$,

(e) $\cosh x$, $x_0=0$,

(f) $\dfrac{1}{x+3}$, $x_0=-2$,

(g) $\dfrac{x}{1-x}$, $x_0=0$,

(h) $\dfrac{1}{x^2}$, $x_0=1$,

(i) $\dfrac{1}{(x-3)(x-4)}$, $x_0=0$,

(j) $\sin x$, $x_0=\dfrac{\pi}{2}$,

(k) $\ln(x+1)$, $x_0=0$,

(l) $\dfrac{1}{(1-x)^k}$, $x_0=0$, $k=1,2,3,\ldots$.

3. Utilize Theorem 7.2.8 to establish the following results. (The power series expansions of the functions are taken about $x_0 = 0$.)

(a) $\dfrac{1}{(1-x)(1-2x)} = \sum_{n=0}^{\infty} \sum_{k=0}^{n} 2^{n-k} x^n = \sum_{n=0}^{\infty} (2^{n+1} - 1)x^n, \qquad -\dfrac{1}{2} < x < \dfrac{1}{2}.$

(b) $e^{x^2} \cos x = \sum_{n=0}^{\infty} \sum_{k=0}^{n} \dfrac{(-1)^k}{(2k)!(n-k)!} x^{2n}, \qquad -\infty < x < \infty.$

(c) $\dfrac{\ln(x+1)}{1-x} = \sum_{n=0}^{\infty} \sum_{k=0}^{n} \dfrac{(-1)^k}{k+1} x^{n+1} = \sum_{n=1}^{\infty} \sum_{k=0}^{n-1} \dfrac{(-1)^k}{k+1} x^n$

$$= \sum_{n=1}^{\infty} \sum_{k=1}^{n} \dfrac{(-1)^{k-1}}{k} x^n, \qquad -1 < x < 1.$$

(d) $\ln^2(x+1) = \sum_{n=0}^{\infty} \sum_{k=0}^{n} \dfrac{(-1)^n}{(k+1)(n-k+1)} x^{n+2}, \qquad -1 < x \leq 1.$

(e) $e^{-x^2} \sinh x = \sum_{n=0}^{\infty} \sum_{k=0}^{n} \dfrac{(-1)^k}{k!(2n-2k+1)!} x^{2n+1}, \qquad -\infty < x < \infty.$

4. Suppose that a power series has only even powers: $\sum_{n=0}^{\infty} a_n x^{2n}$. Suppose that another power series contains all nonnegative powers: $\sum_{n=0}^{\infty} b_n x^n$, $b_n \neq 0$, $n = 0, 1, \ldots$. Show that

$$\left(\sum_{n=0}^{\infty} a_n x^{2n} \right) \left(\sum_{n=0}^{\infty} b_n x^n \right) = \sum_{n=0}^{\infty} \sum_{k=0}^{[n/2]} a_k b_{n-2k} x^n.$$

Here $[p]$ means the greatest integer less than or equal to the real number p. That is, $[p]$ is that integer m which satisfies the inequalities

$$m \leq p < m + 1.$$

Thus, we have

$$p - 1 < [p] \leq p.$$

For example, $[0] = 0$, $[0.5] = 0$, $[2.33] = 2$, $[7.5] = 7$, $[9] = 9$, $[-0.5] = -1$, $[-1.25] = -2$, and $[-3.33] = -4$.

If a power series has only odd powers, $\sum_{n=0}^{\infty} a_n x^{2n+1}$, while another power series contains all nonnegative powers, then

$$\left(\sum_{n=0}^{\infty} a_n x^{2n+1} \right) \left(\sum_{n=0}^{\infty} b_n x^n \right) = x \left(\sum_{n=0}^{\infty} a_n x^{2n} \right) \left(\sum_{n=0}^{\infty} b_n x^n \right) = \sum_{n=0}^{\infty} \sum_{k=0}^{[n/2]} a_k b_{n-2k} x^{n+1}.$$

5. Utilize the results of Exercise 4 to find the Maclaurin series expansions of the following functions.

(a) $e^x \cosh x$,

(b) $\dfrac{\sinh x}{1-x}$,

(c) $\dfrac{e^{-x}}{1-x^2}$,

(d) $e^x \sin 2x$.

6. Given: $\sin x = \sum\limits_{n=0}^{\infty} \dfrac{(-1)^n x^{2n+1}}{(2n+1)!}$ and $\cos x = \sum\limits_{n=0}^{\infty} \dfrac{(-1)^n x^{2n}}{(2n)!}$. Utilize Formula (7.2.25) to determine the first four nonzero terms of the Maclaurin series expansions of $\tan x$ and $\sec x$.

7. Verify

$$x^2 \sum_{n=2}^{\infty} n(n-1)a_n x^{n-2} + (x^2+x) \sum_{n=1}^{\infty} na_n x^{n-1} - \sum_{n=0}^{\infty} a_n x^n$$

$$= -a_0 + \sum_{n=2}^{\infty} [(n^2-1)a_n + (n-1)a_{n-1}]x^n.$$

7.3. Power Series Solutions about an Ordinary Point

Definition 7.3.1. A point $x = x_0$ is called an *ordinary point* of the linear differential equation

$$y^{(n)} + a_1(x)y^{(n-1)} + a_2(x)y^{(n-2)} + \cdots + a_{n-1}(x)y' + a_n(x)y = F(x), \quad (7.3.1)$$

if each function $a_1, a_2, \ldots, a_n$, and F is analytic at $x = x_0$.

Definition 7.3.2. A point $x = x_0$ is called a *singular point* or a *singularity* of Equation (7.3.1) if one or more of the functions $a_1, a_2, \ldots, a_n$, and F is not analytic at $x = x_0$.

We shall be interested primarily in second-order homogeneous linear differential equations of the form

$$y'' + P(x)y' + Q(x)y = 0. \tag{7.3.2}$$

Definition 7.3.3. A point $x = x_0$ is called an *ordinary point* of Equation (7.3.2), if both of the functions P and Q are analytic at $x = x_0$.

Definition 7.3.4. A point $x = x_0$ is called a *singular point* or a *singularity* of Equation (7.3.2), if either (or both) of the functions P and Q is not analytic at $x = x_0$.

Clearly, Definitions 7.3.3 and 7.3.4 are special cases of Definitions 7.3.1 and 7.3.2. For the present, we shall be interested in ordinary and singular points of (7.3.2) which are finite. In Section 7.9, we shall also discuss ordinary and singular points of (7.3.2) at infinity.

Example 7.3.1. Consider the differential equation

$$x(x+2)y'' + (x+1)y' - 3xy = 0. \tag{7.3.3}$$

Let us write (7.3.3) in the form given in (7.3.2). Thus,

$$y'' + \frac{x+1}{x(x+2)} y' - \frac{3}{x+2} y = 0. \tag{7.3.4}$$

Comparing (7.3.4) with (7.3.2), we see that

$$P(x) = \frac{x + 1}{x(x + 2)} \quad \text{and} \quad Q(x) = -\frac{3}{x + 2}. \tag{7.3.5}$$

Since the function P does not possess Taylor series expansions about $x = 0$ and $x = -2$, these points are singular points of (7.3.4) and hence of (7.3.3). Except for the points $x = 0$ and $x = -2$, all other (finite) points are ordinary points of (7.3.3).

Example 7.3.2.   Consider the differential equation

$$(1 + x^2)y'' + 10xy' + 20y = 0. \tag{7.3.6}$$

Writing (7.3.6) in the form given in (7.3.2), we have

$$y'' + \frac{10x}{1 + x^2} y' + \frac{20}{1 + x^2} y = 0. \tag{7.3.7}$$

Thus, the points $x = \pm i$ are the only (finite) singularities of Equation (7.3.6).

Remark 7.3.1.   Before establishing Theorem 7.3.1, which will give us a sufficient condition for the existence of power series solutions of the differential equation (7.3.2) subject to the initial conditions $y(x_0) = a$ and $y'(x_0) = b$, we shall first give an instructive example.   In this example, we shall manipulate formally, that is, we shall not concern ourselves with convergence properties of power series.   We shall then show that this formal procedure does indeed lead us to a bona fide power series solution of the given initial-value problem.

Example 7.3.3.   Let us find the power series solution of the initial-value problem

$$\begin{cases} (1 + x^2)y'' + 10xy' + 20y = 0, & \tag{7.3.8} \\ y(0) = 2, \qquad y'(0) = 3, & \tag{7.3.9} \end{cases}$$

in powers of x.

SOLUTION.   From Example 7.3.2, we know that $x = 0$ is an ordinary point of (7.3.8).   In view of Theorem 4.2.1, we are guaranteed a unique solution $y(x)$ of the given initial-value problem in a small neighborhood of the point $x = 0$.   Let us assume that there exists a power series solution of (7.3.8) of the form

$$y(x) = \sum_{n=0}^{\infty} a_n x^n \tag{7.3.10}$$

which is valid in a small neighborhood of the origin.   Differentiating (7.3.10) twice, we obtain

$$y'(x) = \sum_{n=1}^{\infty} na_n x^{n-1} \quad \text{and} \quad y''(x) = \sum_{n=2}^{\infty} n(n-1)a_n x^{n-2}. \tag{7.3.11}$$

Substituting the series for y, y', and y'' into Equation (7.3.8), we obtain

$$(1 + x^2) \sum_{n=2}^{\infty} n(n-1)a_n x^{n-2} + 10x \sum_{n=1}^{\infty} na_n x^{n-1} + 20 \sum_{n=0}^{\infty} a_n x^n = 0. \quad (7.3.12)$$

We may also write (7.3.12) in the following form.

$$\sum_{n=2}^{\infty} n(n-1)a_n x^{n-2} + \sum_{n=2}^{\infty} n(n-1)a_n x^n + \sum_{n=1}^{\infty} 10na_n x^n + \sum_{n=0}^{\infty} 20a_n x^n = 0.$$

$$(7.3.13)$$

Rewriting the first series in (7.3.13) so that the power of x is n, that is, replacing n by $n + 2$, we obtain

$$\sum_{n=0}^{\infty} (n+2)(n+1)a_{n+2} x^n + \sum_{n=2}^{\infty} n(n-1)a_n x^n + \sum_{n=1}^{\infty} 10na_n x^n + \sum_{n=0}^{\infty} 20a_n x^n = 0.$$

$$(7.3.14)$$

In (7.3.14), summing out the first two terms in the first and last series, and the first term in the third series, we obtain

$$2a_2 + 6a_3 x + \sum_{n=2}^{\infty} (n+2)(n+1)a_{n+2} x^n + \sum_{n=2}^{\infty} n(n-1)a_n x^n + 10a_1 x$$

$$+ \sum_{n=2}^{\infty} 10na_n x^n + 20a_0 + 20a_1 x + \sum_{n=2}^{\infty} 20a_n x^n = 0. \quad (7.3.15)$$

Upon combining terms and simplifying, (7.3.15) may be written as

$$20a_0 + 2a_2 + (30a_1 + 6a_3)x$$

$$+ \sum_{n=2}^{\infty} [(n+1)(n+2)a_{n+2} + (n+4)(n+5)a_n]x^n = 0. \quad (7.3.16)$$

Since the power series (7.3.16) converges to zero in a neighborhood of the origin, in view of Theorem 7.2.7, we must have

$$20a_0 + 2a_2 = 0 \quad \text{or} \quad a_2 = -10a_0, \quad (7.3.17)$$

$$30a_1 + 6a_3 = 0 \quad \text{or} \quad a_3 = -5a_1, \quad (7.3.18)$$

and

$$(n+1)(n+2)a_{n+2} + (n+4)(n+5)a_n = 0, \quad n \geq 2. \quad (7.3.19)$$

The condition (7.3.19) is called a *recurrence relation*. It gives a_{n+2} in terms of the preceding a's. From (7.3.17), (7.3.18), and (7.3.19), we see that each a_{n+2} with an even subscript may be expressed in terms of a_0, and that each a_{n+2} with an odd subscript may be expressed in terms of a_1. More clearly, let us first write (7.3.19) as

$$a_{n+2} = -\frac{(n+4)(n+5)}{(n+1)(n+2)} a_n, \quad n \geq 2. \quad (7.3.20)$$

Observe that $a_2 = -10a_0$ and $a_3 = -5a_1$ may also be written as $a_2 = -(4 \cdot 5)/(1 \cdot 2)a_0$ and $a_3 = -(5 \cdot 6)/(2 \cdot 3)a_1$ [*Note:* These values of a_2 and a_3 would agree with those obtained if we substituted $n = 0$ and $n = 1$ in (7.3.20).] Thus, substituting $n = 2, 3, 4, \ldots$, in (7.3.20), and also writing a_2 and a_3, we get

$$a_2 = -\frac{4 \cdot 5}{1 \cdot 2}a_0, \qquad\qquad a_3 = -\frac{5 \cdot 6}{2 \cdot 3}a_1,$$

$$a_4 = -\frac{6 \cdot 7}{3 \cdot 4}a_2 = \frac{4 \cdot 5 \cdot 6 \cdot 7}{1 \cdot 2 \cdot 3 \cdot 4}a_0, \qquad a_5 = -\frac{7 \cdot 8}{4 \cdot 5}a_3 = \frac{5 \cdot 6 \cdot 7 \cdot 8}{2 \cdot 3 \cdot 4 \cdot 5}a_1,$$

$$a_6 = -\frac{8 \cdot 9}{5 \cdot 6}a_4 \qquad\qquad a_7 = -\frac{9 \cdot 10}{6 \cdot 7}a_5$$

$$= -\frac{4 \cdot 5 \cdot 6 \cdot 7 \cdot 8 \cdot 9}{1 \cdot 2 \cdot 3 \cdot 4 \cdot 5 \cdot 6}a_0, \qquad = -\frac{5 \cdot 6 \cdot 7 \cdot 8 \cdot 9 \cdot 10}{2 \cdot 3 \cdot 4 \cdot 5 \cdot 6 \cdot 7}a_1.$$

By mathematical induction, the reader may show that

$$a_{2k} = (-1)^k \frac{4 \cdot 5 \cdot 6 \cdot 7 \cdots (2k + 2)(2k + 3)}{1 \cdot 2 \cdot 3 \cdot 4 \cdots (2k - 1)(2k)}a_0, \qquad k = 1, 2, \ldots, \quad (7.3.21)$$

and

$$a_{2k+1} = (-1)^k \frac{5 \cdot 6 \cdot 7 \cdot 8 \cdots (2k + 3)(2k + 4)}{2 \cdot 3 \cdot 4 \cdot 5 \cdots (2k)(2k + 1)}a_1, \qquad k = 1, 2, \ldots. \quad (7.3.22)$$

The expressions a_{2k} and a_{2k+1} may be expressed more compactly as follows:

$$a_{2k} = (-1)^k \frac{4 \cdot 5 \cdot 6 \cdot 7 \cdots (2k - 2)(2k - 1)(2k)(2k + 1)(2k + 2)(2k + 3)}{1 \cdot 2 \cdot 3 \cdot 4 \cdots (2k - 3)(2k - 2)(2k - 1)(2k)}a_0$$

$$= (-1)^k \frac{(2k + 1)(2k + 2)(2k + 3)}{1 \cdot 2 \cdot 3}a_0,$$

or

$$a_{2k} = (-1)^k \frac{(k + 1)(2k + 1)(2k + 3)}{3}a_0, \qquad k = 1, 2, \ldots. \quad (7.3.23)$$

Similarly,

$$a_{2k+1} = (-1)^k \frac{(k + 1)(k + 2)(2k + 3)}{6}a_1, \qquad k = 1, 2, \ldots. \quad (7.3.24)$$

Thus, all the coefficients of the power series in (7.3.10) are now determined in terms of a_0 and a_1. The constants a_0 and a_1 in return can be determined from the initial conditions.   For, from (7.3.9), (7.3.10), and (7.3.11), we see that

$$2 = y(0) = a_0 + a_1 \cdot 0 + a_2 \cdot 0^2 + \cdots = a_0,$$

and   (7.3.25)

$$3 = y'(0) = a_1 + 2a_2 \cdot 0 + 3a_3 \cdot 0^2 + \cdots = a_1.$$

In view of (7.3.23) through (7.3.25), it is desirable to split up the power series in (7.3.10) into two series, one containing all the terms with even subscripts, and the other containing all the terms with odd subscripts.   Thus,

$$y(x) = a_0 + a_2 x^2 + a_4 x^4 + \cdots + a_{2k} x^{2k} + \cdots + a_1 x + a_3 x^3$$

$$+ a_5 x^5 + \cdots + a_{2k+1} x^{2k+1} + \cdots$$

$$= a_0 + \sum_{k=1}^{\infty} a_{2k} x^{2k} + a_1 x + \sum_{k=1}^{\infty} a_{2k+1} x^{2k+1}$$

$$= 2 + 2 \sum_{k=1}^{\infty} (-1)^k \frac{(k+1)(2k+1)(2k+3)}{3} x^{2k} + 3x$$

$$+ 3 \sum_{k=1}^{\infty} (-1)^k \frac{(k+1)(k+2)(2k+3)}{6} x^{2k+1}.$$

We may also write the above expression as

$$y(x) = 2 \sum_{k=0}^{\infty} (-1)^k \frac{(k+1)(2k+1)(2k+3)}{3} x^{2k}$$

$$+ 3 \sum_{k=0}^{\infty} (-1)^k \frac{(k+1)(k+2)(2k+3)}{6} x^{2k+1},   \qquad (7.3.26)$$

since when $k = 0$, the first series gives 2 and the second series gives $3x$.   Thus, in a purely formal way, we have found the power series solution of the initial-value problem, which consists of the sum of two power series.

In view of Exercises 7.2.1 (k) and (l), each power series in (7.3.26) converges for $|x| < 1$.   Thus, the formal method given above is completely justified for $|x| < 1$ by utilizing the pertinent results on power series given in Section 7.2.

In view of Example 7.2.7, the functions y_1 and y_2 represented by the power series

$$y_1(x) = \sum_{k=0}^{\infty} (-1)^k \frac{(k+1)(2k+1)(2k+3)}{3} x^{2k},   \qquad (7.3.27)$$

and

$$y_2(x) = \sum_{k=0}^{\infty} (-1)^k \frac{(k+1)(k+2)(2k+3)}{6} x^{2k+1},   \qquad (7.3.28)$$

are linearly independent on the interval I: $-1 < x < 1$.   Thus, $y_1(x)$ and $y_2(x)$ are two linearly independent solutions of Equation (7.3.8) on I.   Hence, if it is desired, the general solution of Equation (7.3.8) on I is

$$y(x) = ay_1(x) + by_2(x), \tag{7.3.29}$$

where a and b are arbitrary constants.

Remark 7.3.2.   In Example 7.3.3, we were very fortunate to obtain compact expressions for the unknown coefficients of the power series in (7.3.10) in terms of the arbitrary constants [*see* (7.3.23) and (7.3.24)].   In general, it is either very difficult or even impossible to obtain such compact expressions. In such cases, one merely computes some of the unknown coefficients in terms of the arbitrary constants, by utilizing the recurrence relation arising in the process of solving the given differential equation.   The number of coefficients to be computed depends, as a rule, on the nature of the problem that is being considered, and on the degree of accuracy that one ultimately wishes to obtain.

Remark 7.3.3.   The values of a_{2k} and a_{2k+1} of (7.3.23) and (7.3.24) may also be obtained by the following simple device.   With $a_2 = -(4 \cdot 5)/(1 \cdot 2)a_0$, $a_3 = -(5 \cdot 6)/(2 \cdot 3)a_1$, and the recurrence relation (7.3.20), we find

$$a_2 = -\frac{4 \cdot 5}{1 \cdot 2}a_0, \qquad\qquad a_3 = -\frac{5 \cdot 6}{2 \cdot 3}a_1,$$

$$a_4 = -\frac{6 \cdot 7}{3 \cdot 4}a_2, \qquad\qquad a_5 = -\frac{7 \cdot 8}{4 \cdot 5}a_3,$$

$$a_6 = -\frac{8 \cdot 9}{5 \cdot 6}a_4, \qquad\qquad a_7 = -\frac{9 \cdot 10}{6 \cdot 7}a_5,$$

$$\vdots \qquad\qquad\qquad\qquad \vdots$$

$$a_{2k} = -\frac{(2k+2)(2k+3)}{(2k-1)(2k)}a_{2k-2}, \qquad a_{2k+1} = -\frac{(2k+3)(2k+4)}{(2k)(2k+1)}a_{2k-1}.$$

Note that a_{2k} and a_{2k+1} are obtained by setting $n = 2k - 2$ and $n = 2k - 1$ in (7.3.20), respectively.   Now, performing the indicated multiplication, we have

$$a_2\, a_4\, a_6 \cdots a_{2k}$$

$$= \left(-\frac{4 \cdot 5}{1 \cdot 2}a_0\right)\left(-\frac{6 \cdot 7}{3 \cdot 4}a_2\right)\left(-\frac{8 \cdot 9}{5 \cdot 6}a_4\right) \cdots \left(-\frac{(2k+2)(2k+3)}{(2k-1)(2k)}a_{2k-2}\right).$$

Cancelling out the common factors in the above expression, namely, $a_2\, a_4\, a_6 \cdots a_{2k-2}$, we obtain

$$a_{2k} = (-1)^k \frac{4 \cdot 5 \cdot 6 \cdot 7 \cdot 8 \cdot 9 \cdots (2k+2)(2k+3)}{1 \cdot 2 \cdot 3 \cdot 4 \cdot 5 \cdot 6 \cdots (2k-1)(2k)}a_0$$

$$= (-1)^k \frac{(k+1)(2k+1)(2k+3)}{3}a_0, \qquad k = 1, 2, \dots.$$

Similarly, performing the indicated multiplication,

$$a_3 a_5 a_7 \cdots a_{2k+1}$$

$$= \left(-\frac{5\cdot6}{2\cdot3}a_1\right)\left(-\frac{7\cdot8}{4\cdot5}a_3\right)\left(-\frac{9\cdot10}{6\cdot7}a_5\right)\cdots\left(-\frac{(2k+3)(2k+4)}{(2k)(2k+1)}a_{2k-1}\right),$$

and now cancelling out the common factors, we obtain

$$a_{2k+1} = (-1)^k \frac{5\cdot6\cdot7\cdot8\cdot9\cdot10\cdots(2k+3)(2k+4)}{2\cdot3\cdot4\cdot5\cdot6\cdot7\cdots(2k)(2k+1)}a_1$$

$$= (-1)^k \frac{(k+1)(k+2)(2k+3)}{6}a_1, \qquad k = 1, 2, \ldots.$$

We shall now establish the following essential theorem.

Theorem 7.3.1. *Let $x = x_0$ be an ordinary point of the differential equation*

$$L[y] = y'' + P(x)y' + Q(x)y = 0, \qquad\qquad [7.3.2]$$

that is, the functions P and Q have power series expansions in powers of $(x - x_0)$ convergent for $|x - x_0| < \rho$, $\rho > 0$.[3] Let a and b be two arbitrary constants. Then there exists a unique solution $y(x)$ of the initial-value problem

$$\begin{cases} L[y] = 0, \\ y(x_0) = a, \qquad y'(x_0) = b, \end{cases}$$

with a power series expansion

$$y(x) = \sum_{n=0}^{\infty} a_n(x - x_0)^n \qquad\qquad (7.3.30)$$

convergent for $|x - x_0| < \rho$. Moreover, we have $a_0 = a$, $a_1 = b$, and the coefficients a_n for $n \geq 2$ may be determined in terms of a_0 and a_1 by substituting the series (7.3.30) into Equation (7.3.2).

PROOF. Let us suppose, for the moment, that there exists a power series solution $y(x)$ of the initial-value problem of the form

$$y(x) = \sum_{n=0}^{\infty} a_n(x - x_0)^n \qquad\qquad [7.3.30]$$

convergent for $|x - x_0| < \rho$, $\rho > 0$. In view of Theorem 7.2.6, we then have

$$y'(x) = \sum_{n=1}^{\infty} na_n(x - x_0)^{n-1}, \qquad |x - x_0| < \rho, \qquad (7.3.31)$$

---

[3] In general, the interval of convergence of the power series expansions in powers of $(x - x_0)$ of the functions P and Q are different. Here, we have taken that common interval in which both series expansions converge.

and

$$y''(x) = \sum_{n=2}^{\infty} n(n-1)a_n(x-x_0)^{n-2}, \qquad |x-x_0| < \rho. \qquad (7.3.32)$$

Since the solution $y(x)$ must satisfy the initial conditions $y(x_0) = a$ and $y'(x_0) = b$, we see from (7.3.30) and (7.3.31) that

$$a = y(x_0) = a_0 \quad \text{and} \quad b = y'(x_0) = a_1. \qquad (7.3.33)$$

From hypothesis, we have that

$$P(x) = \sum_{n=0}^{\infty} p_n(x-x_0)^n, \qquad Q(x) = \sum_{n=0}^{\infty} q_n(x-x_0)^n, \qquad |x-x_0| < \rho. \qquad (7.3.34)$$

Also, since the solution $y(x)$ must satisfy the equation $L[y] = 0$, substituting the series for y, y', y'', P, and Q into Equation (7.3.2), we obtain

$$\sum_{n=2}^{\infty} n(n-1)a_n(x-x_0)^{n-2} + \left(\sum_{n=0}^{\infty} p_n(x-x_0)^n\right)\left(\sum_{n=1}^{\infty} na_n(x-x_0)^{n-1}\right)$$
$$+ \left(\sum_{n=0}^{\infty} q_n(x-x_0)^n\right)\left(\sum_{n=0}^{\infty} a_n(x-x_0)^n\right) = 0. \qquad (7.3.35)$$

In (7.3.35) we wish to make all the exponents of $(x-x_0)$ equal to n. Thus, shifting the indices in the series beginning with $n = 2$ and $n = 1$, (7.3.35) can be written as

$$\sum_{n=0}^{\infty} (n+2)(n+1)a_{n+2}(x-x_0)^n + \left(\sum_{n=0}^{\infty} p_n(x-x_0)^n\right)\left(\sum_{n=0}^{\infty} (n+1)a_{n+1}(x-x_0)^n\right)$$
$$+ \left(\sum_{n=0}^{\infty} q_n(x-x_0)^n\right)\left(\sum_{n=0}^{\infty} a_n(x-x_0)^n\right) = 0. \qquad (7.3.36)$$

Utilizing Theorem 7.2.8, we may express (7.3.36) as

$$\sum_{n=0}^{\infty} (n+2)(n+1)a_{n+2}(x-x_0)^n + \sum_{n=0}^{\infty} \sum_{k=0}^{n} p_{n-k}(k+1)a_{k+1}(x-x_0)^n$$
$$+ \sum_{n=0}^{\infty} \sum_{k=0}^{n} q_{n-k} a_k(x-x_0)^n = 0. \qquad (7.3.37)$$

We may also write (7.3.37) as

$$\sum_{n=0}^{\infty} \left\{(n+1)(n+2)a_{n+2} + \sum_{k=0}^{n} [p_{n-k}(k+1)a_{k+1} + q_{n-k} a_k]\right\}(x-x_0)^n = 0,$$
$$|x-x_0| < \rho. \qquad (7.3.38)$$

Since the power series (7.3.38) converges to zero at every point in the interval $|x-x_0| < \rho$, in view of Theorem 7.2.7 it follows that

$$(n+1)(n+2)a_{n+2} + \sum_{n=0}^{n} [p_{n-k}(k+1)a_{k+1} + q_{n-k} a_k] = 0,$$

or

$$(n + 1)(n + 2)a_{n+2} = - \sum_{k=0}^{n} [p_{n-k}(k + 1)a_{k+1} + q_{n-k} a_k], \qquad n = 0, 1, 2, \ldots,$$

$$(7.3.39)$$

which gives us a *recurrence relation.* Thus, the coefficients a_n for $n \geq 2$ must satisfy the above recurrence relation.

Our task now is to show that if the coefficients a_n for $n \geq 2$ are defined by (7.3.39), then the series

$$\sum_{n=0}^{\infty} a_n(x - x_0)^n \qquad (7.3.40)$$

converges for $|x - x_0| < \rho$. (This result in conjunction with Theorem 4.2.1 will then establish the theorem.) To this end, let r be any real number satisfying $0 < r < \rho$. Since the series given in (7.3.34) are convergent for $|x - x_0| = r$, it follows from Theorem 7.2.10 that there exists a constant $M > 0$ such that

$$|p_n| \leq \frac{M}{r^n} \quad \text{and} \quad |q_n| \leq \frac{M}{r^n}, \qquad n = 0, 1, 2, \ldots. \qquad (7.3.41)$$

Utilizing (7.3.41) with n replaced by $n - k$, and the so-called generalized triangle inequality: $|\sum_{k=0}^{n} B_k| \leq \sum_{k=0}^{n} |B_k|$, upon taking the absolute value of the recurrence formula (7.3.39), we obtain

$$(n + 1)(n + 2)|a_{n+2}| \leq \frac{M}{r^n} \sum_{k=0}^{n} [(k + 1)|a_{k+1}| + |a_k|]r^k$$

$$\leq \frac{M}{r^n} \sum_{k=0}^{n} [(k + 1)|a_{k+1}| + |a_k|]r^k$$

$$+ M|a_{n+1}|r, \qquad n = 0, 1, 2, \ldots. \qquad (7.3.42)$$

The latter inequality is preferred, since it will simplify considerably the task of ultimately determining the convergence of the power series (7.3.40). (Also, it is clear that $M|a_{n+1}|r$ is a positive number whenever $a_{n+1} \neq 0$.)

Let us define

$$A_0 = |a_0|, \quad A_1 = |a_1|,$$

and for $n \geq 2$, let us define A_n recursively by

$$(n + 1)(n + 2)A_{n+2}$$

$$= \frac{M}{r^n} \sum_{k=0}^{n} [(k + 1)A_{k+1} + A_k]r^k + MA_{n+1}r, \qquad n = 0, 1, 2, \ldots. \qquad (7.3.43)$$

Comparing (7.3.42) with (7.3.43) and utilizing mathematical induction, we leave it as an exercise for the reader [*see* Exercise 7.3.1] to show that

$$0 \leqq |a_n| \leqq A_n, \qquad n = 0, 1, 2, \ldots . \tag{7.3.44}$$

Let us consider the series

$$\sum_{n=0}^{\infty} A_n(x - x_0)^n. \tag{7.3.45}$$

We shall now determine the interval of convergence of the power series (7.3.45). In doing so, we shall utilize Theorem 7.2.1. Thus, we would like to obtain an expression for the ratio A_n/A_{n+1}. This is achieved as follows. Replacing n, respectively, by $n - 1$ and $n - 2$ in (7.3.43), we obtain (for n large)

$$n(n + 1)A_{n+1} = \frac{M}{r^{n-1}} \sum_{k=0}^{n-1} [(k + 1)A_{k+1} + A_k]r^k + MA_n r, \tag{7.3.46}$$

and

$$(n - 1)nA_n = \frac{M}{r^{n-2}} \sum_{k=0}^{n-2} [(k + 1)A_{k+1} + A_k]r^k + MA_{n-1} r. \tag{7.3.47}$$

Summing out the term $k = n - 1$ in (7.3.46), we obtain

$$n(n + 1)A_{n+1} = \frac{M}{r^{n-1}} \sum_{k=0}^{n-2} [(k + 1)A_{k+1} + A_k]r^k + MnA_n + MA_{n-1} + MA_n r. \tag{7.3.48}$$

We may also write (7.3.47) as

$$\frac{M}{r^{n-1}} \sum_{k=0}^{n-2} [(k + 1)A_{k+1} + A_k]r^k = \frac{(n - 1)nA_n}{r} - MA_{n-1}. \tag{7.3.49}$$

In view of (7.3.49), we may now write (7.3.48) after simplification as

$$\frac{A_n}{A_{n+1}} = \frac{rn(n + 1)}{(n - 1)n + Mrn + Mr^2}, \tag{7.3.50}$$

which is a desired expression for the ratio A_n/A_{n+1}. Thus,

$$\lim_{n \to \infty} \frac{A_n}{A_{n+1}} = r. \tag{7.3.51}$$

Hence, by Theorem 7.2.1, the power series (7.3.45) converges (absolutely) for $|x - x_0| < r$. In view of (7.3.44) and Theorem 7.2.3, the power series (7.3.40) also converges for $|x - x_0| < r$. But r was any number satisfying $0 < r < \rho$. Consequently, the power series (7.3.40) converges for $|x - x_0| < \rho$. In view of Theorem 4.2.1, the solution $y(x)$ is necessarily unique. Thus, the theorem is established.

Remark 7.3.4.   Note, the above theorem asserts that at an ordinary point of Equation (7.3.2), every solution of the equation is analytic.   A polynomial $a_0 + a_1 x + a_2 x^2 + \cdots + a_n x^n$, being a finite series, thus converges for all (finite) values of x.   Hence, if $P(x)$ and $Q(x)$ given in (7.3.2) are polynomials, then the power series in (7.3.30) will converge for all (finite) values of x.

Remark 7.3.5.   Suppose that in Theorem 7.3.1 Equation (7.3.2) is replaced by the nonhomogeneous linear differential equation

$$y'' + P(x)y' + Q(x)y = F(x), \tag{7.3.52}$$

where the function F also has a power series expansion in powers of $(x - x_0)$ convergent for $|x - x_0| < \rho,\ \rho > 0$.   That is,

$$F(x) = \sum_{n=0}^{\infty} f_n(x - x_0)^n, \quad |x - x_0| < \rho. \tag{7.3.53}$$

Then following a similar method as given in the proof of Theorem 7.3.1 to arrive at Equation (7.3.39), we leave it as an exercise for the reader [*see* Exercise 7.3.2] to show that the coefficients a_n for $n \geq 2$, of the assumed power series solution $y(x) = \sum_{n=0}^{\infty} a_n(x - x_0)^n$ of Equation (7.3.52), must satisfy the following *recurrence relation*:

$$(n + 1)(n + 2)a_{n+2} + \sum_{k=0}^{n} [p_{n-k}(k + 1)a_{k+1} + q_{n-k} a_k] = f_n, \quad n = 0, 1, 2, \ldots. \tag{7.3.54}$$

From the above recurrence relation, we may now determine the coefficients a_n for $n \geq 2$ in terms of a_0 and a_1.

The following theorem is a generalization of Theorem 7.3.1 and its proof is parallel.

Theorem 7.3.2.   *Let $x = x_0$ be an ordinary point of the differential equation*

$$L[y] = y^{(n)} + a_1(x)y^{(n-1)} + \cdots + a_{n-1}(x)y' + a_n(x)y = 0, \qquad [7.3.1]$$

that is, the functions $a_1, a_2, \ldots, a_n$ have power series expansions in powers of $(x - x_0)$ convergent for $|x - x_0| < \rho, \rho > 0$.   Let $b_0, b_1, \ldots, b_{n-1}$ be n arbitrary constants.   Then there exists a unique solution $y(x)$ of the initial-value problem

$$\begin{cases} L[y] = 0 \\ y(x_0) = b_0, y'(x_0) = b_1, \ldots, y^{(n-1)}(x_0) = b_{n-1}, \end{cases}$$

with a power series expansion

$$y(x) = \sum_{k=0}^{\infty} c_k(x - x_0)^k \tag{7.3.55}$$

convergent for $|x - x_0| < \rho.$ *Moreover, we have*

$$c_k = \frac{b_k}{k!}, \quad k = 0, 1, \ldots, n - 1,$$

and the coefficients c_k for $k \geq n$ may be determined in terms of $c_0, c_1, \ldots, c_{n-1}$ by substituting the series (7.3.55) into Equation (7.3.1).

Remark 7.3.6.   Let $x = x_0$ be an ordinary point of Equation (7.3.1).   Let r denote the distance from $x = x_0$ to the nearest singular point of Equation (7.3.1) in the complex plane.   Then the power series in (7.3.55) will converge at least for all values of x in the interval $|x - x_0| < r$.

Remark 7.3.7.   In finding the power series solution about an ordinary point $x = x_0$, of the differential equation of the form

$$y^{(n)} + a_1(x - x_0)^{r_1} y^{(n-1)} + a_2(x - x_0)^{r_2} y^{(n-2)} + \cdots + a_n(x - x_0)^{r_n} y = 0,$$

$$(7.3.56)$$

where $a_1, a_2, \ldots, a_n$ are constants and $r_1, r_2, \ldots, r_n$ are nonnegative integers, it is desirable to first translate the axis by setting

$$x - x_0 = v. \tag{7.3.57}$$

Then,

$$\frac{dy}{dx} = \frac{dy}{dv}, \quad \frac{d^2 y}{dx^2} = \frac{d^2 y}{dv^2}, \ldots, \frac{d^n y}{dx^n} = \frac{d^n y}{dv^n}. \tag{7.3.58}$$

Using (7.3.57) and (7.3.58), Equation (7.3.56) becomes

$$\frac{d^n y}{dv^n} + a_1 v^{r_1} \frac{d^{n-1} y}{dv^{n-1}} + a_2 v^{r_2} \frac{d^{n-2} y}{dv^{n-2}} + \cdots + a_n v^{r_n} y = 0. \tag{7.3.59}$$

We now find the power series solution of Equation (7.3.59) about the ordinary point $v = 0$.   In this answer, we then substitute the value v given in (7.3.57) to obtain the desired power series solution of Equation (7.3.56) in powers of $(x - x_0)$.   Since the coefficients of $y^{(n-1)}, y^{(n-2)}, \ldots, y$ of Equation (7.3.56) are polynomials, it follows from Remark 7.3.4 that the power series solution of Equation (7.3.56) in powers of $x - x_0$ will converge for all (finite) values of x.

Example 7.3.4.   Let us find the power series solution of the differential equation

$$y'' - (x - 1)y' + (x - 1)^2 y = 0, \tag{7.3.60}$$

about the point $x = 1$.

SOLUTION.   Clearly, $x = 1$ is an ordinary point of Equation (7.3.60). Let $x - 1 = v$.   Then, in view of (7.3.58), Equation (7.3.60) becomes

$$\frac{d^2 y}{dv^2} - v\frac{dy}{dv} + v^2 y = 0. \tag{7.3.61}$$

Let

$$y = \sum_{n=0}^{\infty} a_n v^n. \tag{7.3.62}$$

Then,

$$\frac{dy}{dv} = \sum_{n=1}^{\infty} na_n v^{n-1} \quad \text{and} \quad \frac{d^2 y}{dv^2} = \sum_{n=2}^{\infty} n(n-1)a_n v^{n-2}. \tag{7.3.63}$$

Substituting the series of y, dy/dv, and $d^2 y/dv^2$ into Equation (7.3 61), we obtain

$$\sum_{n=2}^{\infty} n(n-1)a_n v^{n-2} - \sum_{n=1}^{\infty} na_n v^n + \sum_{n=0}^{\infty} a_n v^{n+2} = 0. \tag{7.3.64}$$

In the second and third series in (7.3.64), changing the exponents of v to $n - 2$, we obtain

$$\sum_{n=2}^{\infty} n(n-1)a_n v^{n-2} - \sum_{n=3}^{\infty} (n-2)a_{n-2} v^{n-2} + \sum_{n=4}^{\infty} a_{n-4} v^{n-2} = 0. \tag{7.3.65}$$

In (7.3.65), summing out the first two terms in the first series and the first term in the second series, we obtain

$$2a_2 + (6a_3 - a_1)v + \sum_{n=4}^{\infty} [n(n-1)a_n - (n-2)a_{n-2} + a_{n-4}]v^{n-2} = 0.$$

$$\tag{7.3.66}$$

In view of Theorem 7.2.7 and (7.3.66), it follows that

$$2a_2 = 0 \quad \text{or} \quad a_2 = 0,$$
$$6a_3 - a_1 = 0 \quad \text{or} \quad a_3 = \tfrac{1}{6}a_1,$$

and

$$n(n-1)a_n - (n-2)a_{n-2} + a_{n-4} = 0, \qquad n \geq 4,$$

which gives us the recurrence relation.   Thus,

$$a_n = \frac{(n-2)a_{n-2} - a_{n-4}}{n(n-1)}, \qquad n \geq 4. \tag{7.3.67}$$

The coefficients a_n for $n \geq 4$ may now be determined in terms of a_0 and a_1 by utilizing Formula (7.3.67).   We shall determine just some of these coefficients, since we cannot obtain an explicit formula for a_n. With a_0 and a_1 arbitrary, $a_2 = 0$, $a_3 = a_1/6$, we find

$$a_4 = \frac{2a_2 - a_0}{4 \cdot 3} = -\frac{a_0}{12},$$

$$a_5 = \frac{3a_3 - a_1}{5 \cdot 4} = \frac{3 \cdot \frac{1}{6}a_1 - a_1}{5 \cdot 4} = -\frac{a_1}{40},$$

$$a_6 = \frac{4a_4 - a_2}{6 \cdot 5} = \frac{4a_4}{6 \cdot 5} = -\frac{a_0}{90}, \tag{7.3.68}$$

$$a_7 = \frac{5a_5 - a_3}{7 \cdot 6} = \frac{-5 \cdot \dfrac{a_1}{40} - \dfrac{a_1}{6}}{7 \cdot 6} = -\frac{a_1}{144},$$

$$a_8 = \frac{6a_6 - a_4}{8 \cdot 7} = \frac{-6 \cdot \dfrac{a_0}{90} + \dfrac{a_0}{12}}{8 \cdot 7} = \frac{a_0}{3360}.$$

Substituting the above values of the a_n's in (7.3.62), we obtain upon collecting terms in a_0 and a_1

$$y = a_0\left(1 - \frac{1}{12} v^4 - \frac{1}{90} v^6 + \frac{1}{3360} v^8 + \cdots\right)$$

$$+ a_1\left(v + \frac{1}{6} v^3 - \frac{1}{40} v^5 - \frac{1}{144} v^7 + \cdots\right). \tag{7.3.69}$$

Substituting $v = x - 1$ in (7.3.69), we obtain

$$y(x) = a_0\left[1 - \frac{1}{12}(x - 1)^4 - \frac{1}{90}(x - 1)^6 + \frac{1}{3360}(x - 1)^8 + \cdots\right]$$

$$+ a_1\left[(x - 1) + \frac{1}{6}(x - 1)^3 - \frac{1}{40}(x - 1)^5 - \frac{1}{144}(x - 1)^7 + \cdots\right],$$

$$\tag{7.3.70}$$

which gives the solution of Equation (7.3.60) in powers of $(x - 1)$ through terms in $(x - 1)^8$. Clearly, the two series in the brackets converge for all (finite) values of x. Moreover, these two series are the power series expansions of two linearly independent solutions of Equation (7.3.60). Since a_0 and a_1 are arbitrary constants, (7.3.70) represents the general solution of Equation (7.3.60) in powers of $(x - 1)$ [through terms in $(x - 1)^8$].

Remark 7.3.8.   In the following example, we shall illustrate another method of finding the power series solution of an initial-value problem.   The method involves the Taylor or the Maclaurin series expansion, and it is applicable even when the given differential equation is not linear. It works nicely when the successive derivatives, which are obtained from the given differential equation, do not require laborious computations. However, if we wish to find only a few terms of the power series solution, this method should definitely be considered.

Example 7.3.5. Let us find the power series solution of the initial-value problem

$$\begin{cases} y'' + xy' - y = 0, & (7.3.71) \\ y(0) = 1, \quad y'(0) = -1, & (7.3.72) \end{cases}$$

in powers of x.

SOLUTION.   Clearly $x = 0$ is an ordinary point of Equation (7.3.71). Comparing (7.3.71) with (7.3.2), we have $P(x) = x$ and $Q(x) = -1$. In view of Theorem 7.3.1 and Remark 7.3.4, the sought power series solution of the initial-value problem is unique and converges for all (finite) values of x. We now seek a solution in the form of a Maclaurin series (7.2.15). Thus,

$$y(x) = \sum_{n=0}^{\infty} \frac{y^{(n)}(0)}{n!} x^n. \tag{7.3.73}$$

Our task is to find the values of the derivatives $y^{(n)}(0)$ for $n = 2, 3, \ldots$. [*Note:* $y(0)$ and $y'(0)$ are already given in (7.3.72).]   From (7.3.71), we have

$$y'' = -xy' + y. \tag{7.3.74}$$

Utilizing the initial conditions given in (7.3.72) in Equation (7.3.74), we get

$$y''(0) = -0 \cdot y'(0) + y(0) = -0 \cdot 1 + 1 = 1. \tag{7.3.75}$$

Differentiating Equation (7.3.74), we obtain

$$y''' = -xy'' - y' + y' = -xy''. \tag{7.3.76}$$

Thus,

$$y'''(0) = -0 \cdot y''(0) = -0 \cdot 1 = 0. \tag{7.3.77}$$

Differentiating successively Equation (7.3.76) and evaluating its derivatives at $x = 0$, we find

$$\begin{aligned}
y^{(4)} &= -xy''' - y'', & y^{(4)}(0) &= -y''(0) = -1, \\
y^{(5)} &= -xy^{(4)} - 2y''', & y^{(5)}(0) &= -2y'''(0) = 0, \\
y^{(6)} &= -xy^{(5)} - 3y^{(4)}, & y^{(6)}(0) &= -3y^{(4)}(0) = 1 \cdot 3, \\
y^{(7)} &= -xy^{(6)} - 4y^{(5)}, & y^{(7)}(0) &= -4y^{(5)}(0) = 0, \\
y^{(8)} &= -xy^{(7)} - 5y^{(6)}, & y^{(8)}(0) &= -5y^{(6)}(0) = -1 \cdot 3 \cdot 5.
\end{aligned}$$

By mathematical induction, one may easily establish the following results.

$$y^{(2n-1)} = -xy^{(2n-2)} - (2n-4)y^{(2n-3)},$$

$$y^{(2n-1)}(0) = -(2n-4)y^{(2n-3)}(0) = 0, \ n \geq 2, \qquad (7.3.78)$$

and

$$y^{(2n)} = -xy^{(2n-1)} - (2n-3)y^{(2n-2)},$$

$$y^{2n}(0) = -(2n-3)y^{(2n-2)}(0)$$

$$= (-1)^{n+1}[1 \cdot 3 \cdot 5 \cdots (2n-3)], \qquad n \geq 2. \qquad (7.3.79)$$

Let us write (7.3.73) in the following form.

$$y(x) = y(0) + \frac{y'(0)}{1!} x + \frac{y''(0)}{2!} x^2 + \sum_{n=2}^{\infty} \frac{y^{(2n-1)}(0)}{(2n-1)!} x^{2n-1} + \sum_{n=2}^{\infty} \frac{y^{(2n)}(0)}{(2n)!} x^{2n}.$$

$$(7.3.80)$$

Using (7.3.72), (7.3.75), (7.3.78), and (7.3.79), we may write the above series as

$$y(x) = 1 - x + \frac{x^2}{2} + \sum_{n=2}^{\infty} (-1)^{n+1} \frac{[1 \cdot 3 \cdot 5 \cdots (2n-3)]}{(2n)!} x^{2n}, \quad (7.3.81)$$

which is the desired power series solution of our initial-value problem.

Exercises 7.3

1. Use mathematical induction to establish the result in (7.3.44).
2. Establish the recurrence Formula (7.3.54).

Find the power series solutions in powers of x of each of the differential equations in Exercises 3 through 12. Also state the region of convergence of the power series solutions.

3. $y'' - xy' - 2y = 0.$
4. $y'' + xy = 0.$
5. $y'' - x^2y' - xy = 0.$
6. $(1 + x^2)y'' + 2xy' - 2y = 0.$
7. $(1 - x^2)y'' - 2xy' + 6y = 0.$
8. $(x^3 + 1)y'' - 6xy = 0.$
9. $y''' - x^2y'' - 4xy' - 2y = 0.$
10. $y^{(4)} - y = 0.$
11. $(2 - x^3)y'' - 5x^2y' - 4xy = 0.$
12. $y'' + x^2y' + 2xy = 12x^2 + 6x + 2.$

Find the power series solution of each of the initial-value problems in Exercises 13 through 20. Sum in closed form if possible. Also state the region of convergence of the power series solutions.

13. $\begin{cases} y'' - xy = 0, \\ y(0) = 1, \quad y'(0) = -1. \end{cases}$

14. $\begin{cases} y'' + xy = x^2, \\ y(0) = 0, \quad y'(0) = 2. \end{cases}$

15. $\begin{cases} y'' - (x+1)y' - y = 0, \\ y(0) = 1, \quad y'(0) = 1. \end{cases}$

16. $\begin{cases} y'' - xy' - (x+1)y = 0, \\ y(0) = 2, \quad y'(0) = -1. \end{cases}$

17. $\begin{cases} (3 + x^2)y'' + xy' - (x + 4)y = 0, \\ y(0) = 3, \quad y'(0) = 5. \end{cases}$ 18. $\begin{cases} (x + 2)y'' + (x + 1)y' - y = 0, \\ y(0) = 2, \quad y'(0) = -2. \end{cases}$

19. $\begin{cases} y'' - xy - y = \sin x, \\ y(0) = 2, \quad y'(0) = 1. \end{cases}$ 20. $\begin{cases} y''' - xy'' - xy' - y = 6 + 10x^3, \\ y(0) = 6, \quad y'(0) = 4, \quad y''(0) = 4. \end{cases}$

Find the power series solutions about the point indicated of each of the differential equations in Exercises 21 through 26.   Also state the region of convergence of the power series solutions.

21. $y'' + (1 - x)y' - y = 0$.   About the point $x = 1$.
22. $y'' + (x - 1)y' + y = 0$.   About the point $x = 2$.
23. $x^2y'' + (x + 1)y' - 4y = 0$.   About the point $x = -1$.
24. $y'' + (x - 4)y' + (x - 2)y = 0$.   About the point $x = 3$.
25. $y'' - e^x y = 0$.   About the point $x = 0$.
26. $y'' - (\cos x)y = e^x$.   About the point $x = 0$.
27. (a) Consider the differential equation of Exercise 6, namely,

$$(1) \qquad\qquad (1 + x^2)y'' + 2xy' - 2y = 0.$$

We may also utilize Leibnitz's rule [*see* Remark 6.5.1] to find the general solution of (1) in terms of a Maclaurin series expansion.   Let $y = y(x)$ be the solution of (1).   Taking the nth derivative of (1), we obtain

$$(2) \qquad\qquad [(1 + x^2)y'']^{(n)} + 2(xy')^{(n)} - 2y^{(n)} = 0.$$

In view of Formulas (6.5.1) and (6.5.2), Equation (2) becomes

$$(3) \; (1 + x^2)y^{(n+2)} + 2nxy^{(n+1)} + n(n - 1)y^{(n)} + 2xy^{(n+1)} + 2ny^{(n)} - 2y^{(n)} = 0.$$

Setting $x = 0$ in (3), we obtain

$$y^{(n+2)}(0) + n(n - 1)y^{(n)}(0) + 2ny^{(n)}(0) - 2y^{(n)}(0) = 0,$$

or

$$(4) \qquad y^{(n+2)}(0) + (n - 1)(n + 2)y^{(n)}(0) = 0, \quad n = 0, 1, 2, \ldots .$$

Setting $n = 0, 1, 2, \ldots$, successively in (4), we obtain

$$y''(0) = -(-1 \cdot 2)y^{(0)}(0) = -(-1 \cdot 2)y(0), \quad y'''(0) = 0,$$

$$y^{(4)}(0) = -(1 \cdot 4)y^{(2)}(0) = (-1 \cdot 2)(1 \cdot 4)y(0);$$

and by mathematical induction, we leave it for the reader to show that

$$y^{(2k)}(0) = (-1)^k\{(-1 \cdot 2)(1 \cdot 4)(3 \cdot 6) \cdots [(2k - 3)(2k)]\}y(0), \quad y^{(2k+1)}(0) = 0,$$

$$k = 1, 2, 3, \ldots .$$

[*Note:* $y(0)$ and $y'(0)$ are arbitrary.]   Denoting $y(0)$ and $y'(0)$ by a_0 and a_1, respectively, we have

$$y(x) = \sum_{k=0}^{\infty} \frac{y^{(k)}(0)}{k!} x^k$$

$$= y(0) + \frac{y'(0)}{1!} x + \sum_{k=1}^{\infty} \frac{y^{(2k)}(0)}{(2k)!} x^{2k} + \sum_{k=1}^{\infty} \frac{y^{(2k+1)}(0)}{(2k + 1)!} x^{2k+1}$$

$$= a_0 + a_1 x + a_0 \sum_{k=1}^{\infty} \frac{(-1)^k \{(-1 \cdot 2)(1 \cdot 4)(3 \cdot 6) \cdots [(2k-3)(2k)]\}}{(2k)!} x^{2k}$$

$$= a_0 + a_1 x + a_0 \sum_{k=1}^{\infty} \frac{(-1)^{k+1}}{2k-1} x^{2k} = a_1 x + a_0 \left[1 + \sum_{k=1}^{\infty} \frac{(-1)^{k-1}}{2k-1} x^{2k} \right],$$

which is the same answer as found in Exercise 6.

(b) Utilize the method given in part (a) to solve Exercises 3, 5, and 8.

28. For $x < 0$, we define

(1) $$\ln x = \ln(-x) + i\pi, \qquad i^2 = -1.$$

If r is a complex number and $x \neq 0$, we define

(2) $$x^r = e^{r \ln x}.$$

Show that:

(3) $$x^r = e^{i\pi r} |x|^r, \qquad x < 0;$$

(4) $$\frac{d}{dx} |x|^r = \frac{r|x|^r}{x}, \qquad x \neq 0;$$

(5) $$\frac{\partial}{\partial r} |x|^r = |x|^r \ln |x|, \qquad x \neq 0.$$

7.4. Solutions about a Regular Singular Point.   Examples

Review Definitions 7.3.2 and 7.3.4 for the meaning of a singular point.

Definition 7.4.1.   A point $x = x_0$ is called a *regular singular point* of the linear differential equation

$$y^{(n)} + a_1(x)y^{(n-1)} + a_2(x)y^{(n-2)} + \cdots + a_{n-1}(x)y' + a_n(x)y = 0, \quad (7.4.1)$$

if it is a singular point and if the functions defined by the products

$$(x - x_0)a_1(x), \quad (x - x_0)^2 a_2(x), \ldots, (x - x_0)^{n-1}a_{n-1}(x), \quad (x - x_0)^n a_n(x)$$

$$(7.4.2)$$

are each analytic at $x = x_0$.

Definition 7.4.2.   A point $x = x_0$ is called an *irregular singular point* of Equation (7.4.1), if one or more of the functions defined by the products (7.4.2) is not analytic at $x = x_0$.

We shall primarily be interested in second-order homogeneous linear differential equations of the form

$$y'' + P(x)y' + Q(x)y = 0. \tag{7.4.3}$$

Definition 7.4.3.   A point $x = x_0$ is called a *regular singular point* of Equation (7.4.3), if it is a singular point and if the functions defined by the products

$$(x - x_0)P(x) \quad \text{and} \quad (x - x_0)^2 Q(x) \tag{7.4.4}$$

are both analytic at $x = x_0$.

Definition 7.4.4.   A point $x = x_0$ is called an *irregular singular point* of Equation (7.4.3), if either (or both) of the functions defined by the products (7.4.4) is not analytic at $x = x_0$.

Example 7.4.1.   Let us classify the singular points for finite values of x of the differential equation

$$x^2(x - 2)y'' + (x + 1)y' - 3xy = 0. \tag{7.4.5}$$

SOLUTION.   Writing (7.4.5) in the form given in (7.4.3), we have

$$y'' + \frac{x + 1}{x^2(x - 2)} y' - \frac{3}{x(x - 2)} y = 0.$$

Thus,

$$P(x) = \frac{x + 1}{x^2(x - 2)} \quad \text{and} \quad Q(x) = - \frac{3}{x(x - 2)}.$$

Clearly, the singular points of (7.4.5) are $x = 0$ and $x = 2$.

Concerning $x = 0$:   form the functions defined by the products

$$xP(x) = \frac{x + 1}{x(x - 2)} \quad \text{and} \quad x^2 Q(x) = - \frac{3x}{x - 2}.$$

The product function defined by $xP(x)$ is not analytic at $x = 0$.   Thus, $x = 0$ is an irregular singular point of (7.4.5).   [*Note:* The product function defined by $x^2 Q(x)$ is analytic at $x = 0$.]

Concerning $x = 2$:   form the functions defined by the products

$$(x - 2)P(x) = \frac{x + 1}{x^2} \quad \text{and} \quad (x - 2)^2 Q(x) = - \frac{3(x - 2)}{x}.$$

Both of the product functions thus defined are analytic at $x = 2$.   Hence, $x = 2$ is a regular singular point of (7.4.5).

Example 7.4.2.   Let us classify the singular points for finite values of x of the differential equation

$$(x^2 + 1)(x^2 - 1)^2 y'' + (x^3 - 1)y' + (x + 1)y = 0. \tag{7.4.6}$$

SOLUTION.   Writing (7.4.6) in the form given in (7.4.3), we have upon simplification

$$y'' + \frac{x^2 + x + 1}{(x^2 + 1)(x - 1)(x + 1)^2} y' + \frac{1}{(x^2 + 1)(x - 1)^2(x + 1)} y = 0.$$

Thus,

$$P(x) = \frac{x^2 + x + 1}{(x - i)(x + i)(x - 1)(x + 1)^2}$$

and

$$Q(x) = \frac{1}{(x - i)(x + i)(x - 1)^2(x + 1)}.$$

Clearly, the singular points of (7.4.6) are $x = 1$, $x = -1$, $x = i$, and $x = -i$. Form the functions defined by the products

$$(x - 1)P(x) = \frac{x^2 + x + 1}{(x^2 + 1)(x + 1)^2}, \quad (x - 1)^2 Q(x) = \frac{1}{(x^2 + 1)(x + 1)}, \quad (7.4.7)$$

$$(x + 1)P(x) = \frac{x^2 + x + 1}{(x^2 + 1)(x - 1)(x + 1)}, \quad (x + 1)^2 Q(x) = \frac{x + 1}{(x^2 + 1)(x - 1)^2},$$

$$(7.4.8)$$

$$(x - i)P(x) = \frac{x^2 + x + 1}{(x + i)(x - 1)(x + 1)^2},$$

$$(x - i)^2 Q(x) = \frac{x - i}{(x + i)(x - 1)^2(x + 1)}, \quad (7.4.9)$$

$$(x + i)P(x) = \frac{x^2 + x + 1}{(x - i)(x - 1)(x + 1)^2},$$

$$(x + i)^2 Q(x) = \frac{x + i}{(x - i)(x - 1)^2(x + 1)}. \quad (7.4.10)$$

In view of (7.4.7), both product functions thus defined are analytic at $x = 1$. Hence, $x = 1$ is a regular singular point of (7.4.6).   From (7.4.8), the product function defined by $(x + 1)P(x)$ is not analytic at $x = -1$.   Thus, $x = -1$ is an irregular singular point of (7.4.6).   From (7.4.9), both product functions thus defined are analytic at $x = i$.   Hence, $x = i$ is a regular singular point of (7.4.6).   Similarly, in view of (7.4.10), $x = -i$ is a regular singular point of (7.4.6).

Remark 7.4.1.   Let us denote by p and q the functions defined, respectively, by the products in (7.4.4).   That is,

$$p(x) = (x - x_0)P(x) \quad \text{and} \quad q(x) = (x - x_0)^2 Q(x). \quad (7.4.11)$$

Then we may write Equation (7.4.3) in the following form:

$$(x - x_0)^2 y'' + (x - x_0)p(x)y' + q(x)y = 0. \quad (7.4.12)$$

Suppose that $x = x_0$ is a singular point of Equation (7.4.3). Then $x = x_0$ is a *regular singular point* of Equation (7.4.3), if it can be written in the form given by Equation (7.4.12), where p and q are analytic functions at $x = x_0$. The point $x = x_0$ is an *irregular singular point* of Equation (7.4.3), if either (or both) functions p and q in (7.4.12) is not analytic at $x = x_0$.

In Example 7.4.1, we may write Equation (7.4.5) in the following forms.

$$x^2 y'' + x\left(\frac{x+1}{x(x-2)}\right) y' - \frac{3x}{x-2} y = 0, \tag{7.4.13}$$

and

$$(x-2)^2 y'' + (x-2)\left(\frac{x+1}{x^2}\right) y' - \frac{3(x-2)}{x} y = 0. \tag{7.4.14}$$

Comparing (7.4.13) with (7.4.12) with $x_0 = 0$, we see that $p(x) = (x+1)/x(x-2)$. Thus, the function p is not analytic at $x = 0$. Hence, $x = 0$ is an irregular singular point of (7.4.5). Comparing (7.4.14) with (7.4.12) with $x_0 = 2$, we see that $p(x) = (x+1)/x^2$ and $q(x) = -3(x-2)/x$. Thus, the functions p and q are both analytic at $x = 2$. Hence, $x = 2$ is a regular singular point of (7.4.5).

Remark 7.4.2. Let $x_0 \neq 0$ be a regular singular point of Equation (7.4.3). Then Equation (7.4.3) may be written in the form

$$(x - x_0)^2 y'' + (x - x_0)p(x)y' + q(x)y = 0, \tag{7.4.12}$$

where p and q are analytic functions at $x = x_0$. Thus,

$$p(x) = \sum_{n=0}^{\infty} p_n(x - x_0)^n, \qquad q(x) = \sum_{n=0}^{\infty} q_n(x - x_0)^n, \tag{7.4.15}$$

and the power series converge in some interval $|x - x_0| < \rho$, $\rho > 0$. To simplify the task of finding a series solution of Equation (7.4.12), it is desirable to change this equation into an equivalent equation with a regular point at the origin. This is accomplished by a translation of the axes. Thus, let $t = x - x_0$, then when $x = x_0$, $t = 0$. Let us set

$$\bar{p}(t) = p(t + x_0), \qquad \bar{q}(t) = q(t + x_0). \tag{7.4.16}$$

Then the functions $\bar{p}$ and $\bar{q}$ are analytic at $t = 0$, since the functions p and q are analytic at $x = x_0$. Also, let us set

$$\bar{y}(t) = y(t + x_0). \tag{7.4.17}$$

Since $t = x - x_0$, $dx/dt = 1$. Utilizing the chain rule from calculus, we have

$$\frac{d\bar{y}}{dt} = \frac{d\bar{y}}{dx}\frac{dx}{dt} = \frac{d}{dx}y(t + x_0), \qquad \frac{d^2\bar{y}}{dt^2} = \frac{d}{dt}\left(\frac{d\bar{y}}{dt}\right) = \frac{d}{dx}\left(\frac{dy}{dx}\right)\frac{dx}{dt} = \frac{d^2}{dx^2}y(t + x_0).$$

$$\tag{7.4.18}$$

Using (7.4.16), (7.4.17), (7.4.18), and the fact that $t = x - x_0$, we may write Equation (7.4.12) as

$$t^2 \frac{d^2 \bar{y}}{dt^2} + t\bar{p}(t)\frac{d\bar{y}}{dt} + \bar{q}(t)\bar{y} = 0. \tag{7.4.19}$$

Since the functions $\bar{p}$ and $\bar{q}$ are analytic at $t = 0$, it follows that $t = 0$ is a regular singular point of Equation (7.4.19).   Conversely, if $\bar{y}(t)$ is a solution of Equation (7.4.19), the function y given by $y(x) = \bar{y}(x - x_0)$ is a solution of Equation (7.4.12).   In this sense, Equation (7.4.19) is equivalent to Equation (7.4.12).

In the ensuing discussion, we shall assume that such a change of variable has been done, and we shall concern ourselves with the differential equation

$$L[y] = x^2 y'' + xp(x)y' + q(x)y = 0, \tag{7.4.20}$$

where the functions p and q are now both analytic at $x = 0$, and thus have power series expansions

$$p(x) = \sum_{n=0}^{\infty} p_n x^n, \qquad q(x) = \sum_{n=0}^{\infty} q_n x^n \tag{7.4.21}$$

which are convergent in some interval $|x| < \rho$, $\rho > 0$, and such that the coefficients p_0, q_0, and q_1 are not all zero.   If p_0, q_0, and q_1 were all equal to zero, then $x = 0$ would be an ordinary point of Equation (7.4.20).

Remark 7.4.3.   If the functions p and q in (7.4.20) are both constants, then we obtain a Cauchy-Euler homogeneous differential equation.   [*See* Equation (4.10.1) with $F(x) = 0$.]   In view of Theorem 4.10.1 and the results of Section 4.6, [*see* also Exercise 1.3.9] it may be seen that at least one solution of Equation (4.10.1) with $F(x) = 0$, is of the form $|x|^r$, $x \neq 0$, where r is a real number or a complex number.   Thus, if the functions p and q in (7.4.20) are not both constants, then it is somewhat plausible that at least one solution of (7.4.20) is of the form

$$y(x) = |x|^r \sum_{n=0}^{\infty} a_n x^n, \qquad a_0 \neq 0, \tag{7.4.22}$$

where the constants r and a_n may be determined by substituting the series in Equation (7.4.20), and the power series $\sum_{n=0}^{\infty} a_n x^n$ converges in some neighborhood of the origin.   The series $|x|^r \sum_{n=0}^{\infty} a_n x^n$ is known as a *Frobenius series*.   Clearly, a Frobenius series is a generalization of a power series.   It reduces to a power series when r is a nonnegative integer.

In preparation for the general theory to be given in Section 7.5, the following examples will serve a useful purpose in discovering what types of solutions are possible when solving a differential equation having a regular singular point at $x = 0$.

Example 7.4.3.   Let us find two linearly independent solutions valid near the origin of the differential equation

$$x^2 y'' + \tfrac{3}{2}x(x - 1)y' + y = 0. \tag{7.4.23}$$

SOLUTION.   Here $p(x) = \tfrac{3}{2}(x - 1)$ and $q(x) = 1$.   The functions p and q are both analytic at $x = 0$.   Thus, $x = 0$ is a regular singular point of Equation (7.4.23).   Let us restrict our attention to $x > 0$.   In a formal (nonrigorous) manner, we shall endeavor to find Frobenius series solutions of Equation (7.4.23).   Thus, let us suppose that a solution $y(x)$ of Equation (7.4.23) has the form

$$y(x) = x^r \sum_{n=0}^{\infty} a_n x^n = \sum_{n=0}^{\infty} a_n x^{r+n}, \qquad a_0 \neq 0. \tag{7.4.24}$$

Differentiating (7.4.24) twice, we obtain

$$y'(x) = \sum_{n=0}^{\infty} (r + n)a_n x^{r+n-1}, \qquad y''(x) = \sum_{n=0}^{\infty} (r + n)(r + n - 1)a_n x^{r+n-2}. \tag{7.4.25}$$

Substituting the series for y, y', and y'' into Equation (7.4.23), we obtain upon collecting terms with the same exponent of x

$$\sum_{n=0}^{\infty} [(r + n)(r + n - 1) - \tfrac{3}{2}(r + n) + 1]a_n x^{r+n} + \sum_{n=0}^{\infty} \tfrac{3}{2}(r + n)a_n x^{r+n+1} = 0,$$

which can also be written upon shifting the index in the second series as

$$\sum_{n=0}^{\infty} [(r + n)(r + n - 1) - \tfrac{3}{2}(r + n) + 1]a_n x^{r+n} + \sum_{n=1}^{\infty} \tfrac{3}{2}(r + n - 1)a_{n-1} x^{r+n} = 0. \tag{7.4.26}$$

Summing out the first term in the first series, (7.4.26) may be expressed as

$$[r(r - 1) - \tfrac{3}{2}r + 1]a_0 x^r + \sum_{n=1}^{\infty} \{[(r + n)(r + n - 1) - \tfrac{3}{2}(r + n) + 1]a_n$$

$$+ \tfrac{3}{2}(r + n - 1)a_{n-1}\}x^{r+n} = 0. \tag{7.4.27}$$

For convenience of notation, we let

$$f(r) = r(r - 1) - \tfrac{3}{2}r + 1 = (r - 2)(r - \tfrac{1}{2}). \tag{7.4.28}$$

Then

$$f(r + n) = (r + n)(r + n - 1) - \tfrac{3}{2}(r + n) + 1, \tag{7.4.29}$$

and we may now write (7.4.27) as

$$f(r)a_0 x^r + \sum_{n=1}^{\infty} [f(r + n)a_n + \tfrac{3}{2}(r + n - 1)a_{n-1}]x^{r+n} = 0. \tag{7.4.30}$$

Thus, if the function y given by (7.4.24) is to satisfy Equation (7.4.23), then all the coefficients of x must vanish in (7.4.30).   Since we are assuming that $a_0 \neq 0$, we must have

$$f(r) = (r - 2)(r - \tfrac{1}{2}) = 0, \tag{7.4.31}$$

and the recurrence relation

$$f(r + n)a_n + \tfrac{3}{2}(r + n - 1)a_{n-1} = 0, \qquad n \geq 1. \tag{7.4.32}$$

The quadratic equation (7.4.31) is called the *indicial equation* of (7.4.23), and its roots $r_1 = 2$ and $r_2 = 1/2$ are called the *indices* or the *exponents* of (7.4.23). The roots of $f(r) = 0$ are also called the *indicial roots*.   The indicial equation is obtained by setting equal to zero the coefficient of the lowest power of x $(n = 0)$ in (7.4.26).   The indicial roots are the only allowable values of r which will give rise to possible Frobenius series solutions of Equation (7.4.23). The expression $f(r)$ in (7.4.28) is called the *indicial polynomial* of (7.4.23).

Equation (7.4.32) enables us to express $a_1, a_2, \ldots$ in terms of a_0 and r.   If $f(r + n) \neq 0$ for $n \geq 1$, then we may write (7.4.32) as

$$a_n = -\frac{3}{2}\frac{(r + n - 1)}{f(r + n)} a_{n-1}, \qquad n \geq 1. \tag{7.4.33}$$

Thus,

$$a_1 = -\frac{3}{2}\frac{r}{f(r + 1)} a_0, \quad a_2 = -\frac{3}{2}\frac{(r + 1)}{f(r + 2)} a_1 = \left(\frac{3}{2}\right)^2 \frac{(r + 1)r}{f(r + 2)f(r + 1)} a_0,$$

$$a_3 = -\frac{3}{2}\frac{(r + 2)}{f(r + 3)} a_2 = -\left(\frac{3}{2}\right)^3 \frac{(r + 2)(r + 1)r}{f(r + 3)f(r + 2)f(r + 1)} a_0,$$

and by mathematical induction, we have

$$a_n = (-1)^n \left(\frac{3}{2}\right)^n \frac{(r + n - 1)(r + n - 2) \cdots (r + 2)(r + 1)r}{f(r + n)f(r + n - 1) \cdots f(r + 3)f(r + 2)f(r + 1)} a_0, \qquad n \geq 1. \tag{7.4.34}$$

Since $f(r) = (r - 2)(r - \tfrac{1}{2})$ is equal to zero only when $r = r_1 = 2$ or $r = r_2 = \tfrac{1}{2}$, we see that $f(r_1 + n) \neq 0$ and $f(r_2 + n) \neq 0$ for $n \geq 1$.   Consider now the series

$$\sum_{n=0}^{\infty} a_n x^n, \tag{7.4.35}$$

where the a_n for $n \geq 1$ are given by (7.4.34).   Utilizing Theorem 7.2.1, we obtain

$$R = \lim_{n \to \infty} \left| \frac{a_n}{a_{n+1}} \right|$$

$$= \lim_{n \to \infty} \left| (-1)^n (\tfrac{3}{2})^n \frac{(r+n-1)\cdots r}{f(r+n)\cdots f(r+1)} \right.$$

$$\left. \cdot \frac{f(r+n+1)f(r+n)\cdots f(r+1)}{(-1)^{n+1}(\tfrac{3}{2})^{n+1}\,(r+n)(r+n-1)\cdots r} \right|$$

$$= \lim_{n \to \infty} \left| \frac{2}{3} \cdot \frac{f(r+n+1)}{r+n} \right|.$$

Since $f(r+n+1)$ is a quadratic expression in n, namely,

$$f(r+n+1) = (r+n-1)(r+n+\tfrac{1}{2}),$$

we see that

$$R \to \infty \quad \text{as} \quad n \to \infty.$$

Thus, the series (7.4.35) converges for $|x| < \infty$. In view of the relevant results on power series of Section 7.2, all the above formal manipulations are now completely justified.

Using $a_0 = 1$, the a_n from (7.4.34), and the pertinent values of r, $r_1 = 2$ and $r_2 = \tfrac{1}{2}$, the solutions of Equation (7.4.23) for $x > 0$ are $y = y_1(x)$ and $y = y_2(x)$, where

$$y_1(x) = x^2 + x^2 \sum_{n=1}^{\infty} (-1)^n \left(\frac{3}{2}\right)^n \frac{(n+1)(n)\cdots 3 \cdot 2}{f(n+2)f(n+1)\cdots f(4)f(3)} x^n, \qquad (7.4.36)$$

and

$$y_2(x) = x^{1/2} + x^{1/2} \sum_{n=1}^{\infty} (-1)^n \left(\frac{3}{2}\right)^n \frac{(n-\tfrac{1}{2})(n-\tfrac{3}{2})\cdots \tfrac{3}{2}\cdot\tfrac{1}{2}}{f(n+\tfrac{1}{2})f(n-\tfrac{1}{2})\cdots f(\tfrac{5}{2})f(\tfrac{3}{2})} x^n. \qquad (7.4.37)$$

Utilizing (7.4.28), we see that

$$f(n+2)f(n+1)\cdots f(4)f(3) = \left[n\left(n+\frac{3}{2}\right)\right]\left[(n-1)\left(n+\frac{1}{2}\right)\right] \cdots \left[2\cdot\frac{7}{2}\right]\left[1\cdot\frac{5}{2}\right]$$

$$= [n(n-1)\cdots 2 \cdot 1]\left[\frac{(2n+3)(2n+1)\cdots 7 \cdot 5}{2^n}\right]$$

$$= n!\,\frac{[5\cdot 7 \cdots (2n+3)]}{2^n}$$

and

$$f\left(n+\frac{1}{2}\right)f\left(n-\frac{1}{2}\right)\cdots f\left(\frac{5}{2}\right)f\left(\frac{3}{2}\right)$$

$$= \left[\left(n-\frac{3}{2}\right)n\right]\left[\left(n-\frac{5}{2}\right)(n-1)\right] \cdots \left[\frac{1}{2}\cdot 2\right]\left[-\frac{1}{2}\cdot 1\right]$$

$$= \left[\left(n - \frac{3}{2}\right)\left(n - \frac{5}{2}\right) \cdots \left(\frac{1}{2}\right)\left(-\frac{1}{2}\right)\right][n(n - 1) \cdots 2 \cdot 1]$$

$$= \left[\frac{(2n - 3)(2n - 5) \cdots 1}{2^{n-1}} \cdot -\frac{1}{2}\right] n!.$$

Thus, $y_1(x)$ and $y_2(x)$ may be expressed as

$$y_1(x) = x^2 + x^2 \sum_{n=1}^{\infty} \frac{(-1)^n 3^n (n + 1)}{5 \cdot 7 \cdots (2n + 3)} x^n = x^2 \sum_{n=0}^{\infty} \frac{(-1)^n 3^{n+1}(n + 1)}{3 \cdot 5 \cdot 7 \cdots (2n + 3)} x^n$$

$$(7.4.38)$$

and

$$y_2(x) = x^{1/2} + x^{1/2} \sum_{n=1}^{\infty} \frac{(-1)^{n+1} 3^n (2n - 1)}{2^n n!} x^n = x^{1/2} \sum_{n=0}^{\infty} \frac{(-1)^{n+1} 3^n (2n - 1)}{2^n n!} x^n,$$

$$(7.4.39)$$

which are the desired Frobenius series solutions of Equation (7.4.23) for $x > 0$.

To obtain the solutions for $x < 0$, we observe [*see* also Exercise 7.3.28] that all the above computations go through when x^r is replaced everywhere by $|x|^r$, where

$$|x|^r = e^{r \ln |x|}. \tag{7.4.40}$$

Thus, for $x \neq 0$, the desired Frobenius series solutions of Equation (7.4.23) are

$$y_1(x) = x^2 \sum_{n=0}^{\infty} \frac{(-1)^n 3^{n+1}(n + 1)}{3 \cdot 5 \cdot 7 \cdots (2n + 3)} x^n \tag{7.4.41}$$

and

$$y_2(x) = |x|^{1/2} \sum_{n=0}^{\infty} \frac{(-1)^{n+1} 3^n (2n - 1)}{2^n n!} x^n. \tag{7.4.42}$$

[*Note:* $x^2 = |x|^2$. Also, in the series of (7.4.41), we may cancel out the common factor 3 when $n = 1, 2, \ldots$.]

By a modification of Theorem 7.2.11 (so that it is applicable also for the above infinite series) we leave it as an exercise for the reader to show that the solutions y_1 and y_2 of (7.4.41) and (7.4.42) are linearly independent on the intervals I_1: $-\infty < x < 0$ and I_2: $0 < x < \infty$, and hence on any interval not containing $x = 0$. Thus, on any interval not containing the origin, the general solution $y(x)$ of Equation (7.4.23) is

$$y(x) = c_1 y_1(x) + c_2 y_2(x), \tag{7.4.43}$$

where c_1 and c_2 are arbitrary constants.

Remark 7.4.4.   Observe that even though the differential equation (7.4.23) has no meaning at the regular singular point $x = 0$, the solutions y_1 and y_2 given by (7.4.41) and (7.4.42) are well defined at $x = 0$.   Also, in the above example, the indicial roots were distinct, and $r_1 - r_2 = 2 - \frac{1}{2} = \frac{3}{2}$ is not an integer.   As will be seen in Section 7.5 [*see* Theorem 7.5.1], whenever the indicial equation of (7.4.20) has roots r_1, r_2 such that $r_1 - r_2$ is not zero or a positive integer, Equation (7.4.20) will always possess for $0 < |x| < \rho$, $\rho > 0$, two linearly independent solutions y_1 and y_2 of the form

$$y_1(x) = |x|^{r_1} \sum_{n=0}^{\infty} a_n x^n, \qquad a_0 = 1, \tag{7.4.44}$$

and

$$y_2(x) = |x|^{r_2} \sum_{n=0}^{\infty} b_n x^n, \qquad b_0 = 1. \tag{7.4.45}$$

The coefficients a_n and b_n may be obtained by substituting the solutions into Equation (7.4.20).

Remark 7.4.5.   When the indicial equation of (7.4.20) has roots r_1, r_2 such that $r_1 = r_2$ or $r_1 - r_2 = m$, where m is a positive integer, these are known as *exceptional cases*.   The procedure for finding two linearly independent solutions, for these exceptional cases, becomes more involved than the one given in the above example.   Example 7.4.4 below illustrates a procedure for finding two linearly independent solutions of the given differential equation when its indicial equation has equal roots.

Example 7.4.4.   Let us find two linearly independent solutions valid near the origin of the differential equation

$$x^2 \frac{d^2 y}{dx^2} - x \frac{dy}{dx} - (2x^2 - 1)y = 0. \tag{7.4.46}$$

SOLUTION.   Here $p(x) = -1$ and $q(x) = -2x^2 + 1$.   Thus, the functions p and q are both analytic at $x = 0$.   Hence, $x = 0$ is a regular singular point of Equation (7.4.46).   Let us restrict our attention to some interval $0 < x < \rho$, $\rho > 0$.   As usual, we first manipulate in a formal manner, and later justify these formal manipulations.   Thus, let us suppose that a solution $y(x)$ of Equation (7.4.46) is of the form

$$y(x) = x^r \sum_{n=0}^{\infty} a_n x^n = \sum_{n=0}^{\infty} a_n x^{r+n}, \qquad a_0 \neq 0. \tag{7.4.47}$$

Differentiating (7.4.47) twice, we get

$$y'(x) = \sum_{n=0}^{\infty} (r + n)a_n x^{r+n-1}, \quad y''(x) = \sum_{n=0}^{\infty} (r + n)(r + n - 1)a_n x^{r+n-2}.$$

$$\tag{7.4.48}$$

Substituting the series for y, y', and y'' into Equation (7.4.46), we obtain upon collecting terms with the same exponent of x

$$\sum_{n=0}^{\infty} [(r+n)(r+n-1) - (r+n) + 1]a_n x^{r+n} - \sum_{n=0}^{\infty} 2a_n x^{r+n+2} = 0.$$

The above expression can also be written upon shifting the index in the second series as

$$\sum_{n=0}^{\infty} (r+n-1)^2 a_n x^{r+n} - \sum_{n=2}^{\infty} 2a_{n-2} x^{r+n} = 0. \tag{7.4.49}$$

Summing out the first two terms in the first series, (7.4.49) may be expressed as

$$(r-1)^2 a_0 x^r + r^2 a_1 x^{r+1} + \sum_{n=2}^{\infty} [(r+n-1)^2 a_n - 2a_{n-2}]x^{r+n} = 0. \tag{7.4.50}$$

Let

$$f(r) = (r-1)^2. \tag{7.4.51}$$

Then,

$$f(r+n) = (r+n-1)^2. \tag{7.4.52}$$

In view of (7.4.51) and (7.4.52), we may write (7.4.50) as

$$f(r)a_0 x^r + r^2 a_1 x^{r+1} + \sum_{n=2}^{\infty} [f(r+n)a_n - 2a_{n-2}]x^{r+n} = 0. \tag{7.4.53}$$

Thus, if the function y given in (7.4.47) is to satisfy Equation (7.4.46), then all the coefficients of x in (7.4.53) must vanish. Since by assumption $a_0 \neq 0$, we must have

$$f(r) = (r-1)^2 = 0. \tag{7.4.54}$$

Hence, the indicial equation $f(r) = 0$ has two equal roots, namely, $r_1 = r_2 = 1$. Thus, for any possible solution of the form given in (7.4.47), we must choose $r = 1$. However, we then can only obtain one such Frobenius series solution. To find another solution of Equation (7.4.46), though not necessarily of the same form, we shall utilize a method similar to that used in Exercise 4.6.1.

Let us denote the left-hand member of Equation (7.4.46) by $L[y]$, that is

$$L[y] = x^2 \frac{d^2 y}{dx^2} - x \frac{dy}{dx} - (2x^2 - 1)y. \tag{7.4.55}$$

Substituting the series for y, y', and y'' into $L[y]$, we obtain

$$L[y] = f(r)a_0 x^r + r^2 a_1 x^{r+1} + \sum_{n=2}^{\infty} [f(r+n)a_n - 2a_{n-2}]x^{n+r}, \tag{7.4.56}$$

where $f(r)$ and $f(r + n)$ are given by (7.4.51) and (7.4.52).   We now proceed differently from the method given in Example 7.4.3.   Since the indicial equation arises by setting to zero the coefficient of the lowest exponent of x (namely, when $n = 0$), we deliberately, for the present, avoid trying to make this term vanish.   However, by choosing the a's and leaving r as a parameter, we can make every term except the first one in $L[y]$ vanish.   Thus, setting each coefficient of x in (7.4.56), equal to zero, except the first one, we have

$$r^2 a_1 = 0 \quad \text{or} \quad a_1 = 0, \tag{7.4.57}$$

and the recurrence relation

$$f(r + n)a_n - 2a_{n-2} = 0, \qquad n \geq 2. \tag{7.4.58}$$

We shall now obtain an expression for the a_n, $n \geq 2$, in terms of a_0 and r.   In view of (7.4.57) and (7.4.58), we have

$$a_2 = \frac{2a_0}{f(r + 2)}, \qquad a_3 = \frac{2a_1}{f(r + 3)} = 0, \qquad a_4 = \frac{2a_2}{f(r + 4)} = \frac{2^2 a_0}{f(r + 4)f(r + 2)},$$

$$a_5 = \frac{2a_3}{f(r + 5)} = 0, \qquad a_6 = \frac{2a_4}{f(r + 6)} = \frac{2^3 a_0}{f(r + 6)f(r + 4)f(r + 2)},$$

$$a_7 = \frac{2a_5}{f(r + 7)} = 0;$$

and by mathematical induction, we see that

$$a_{2n+1} = 0, \qquad n \geq 1, \tag{7.4.59}$$

and

$$a_{2n} = \frac{2^n a_0}{f(r + 2n)f(r + 2n - 2) \cdots f(r + 4)f(r + 2)}, \qquad n \geq 1. \tag{7.4.60}$$

Let $a_0 = 1$.   Using the a's determined in (7.4.59) and (7.4.60), we may write a function y which is dependent upon both x and r, namely,

$$y(x, r) = x^r + \sum_{n=1}^{\infty} a_{2n}(r)x^{r + 2n}, \tag{7.4.61}$$

where

$$a_{2n}(r) = \frac{2^n}{f(r + 2n)f(r + 2n - 2) \cdots f(r + 4)f(r + 2)}, \qquad n \geq 1. \tag{7.4.62}$$

Clearly, the function y given by (7.4.61) has been determined so that when it is substituted in (7.4.56), $L[y(x, r)]$ reduces to the first term.   That is, with $a_0 = 1$, we obtain

$$L[y(x, r)] = f(r)x^r = (r - 1)^2 x^r. \tag{7.4.63}$$

Clearly, a solution of Equation (7.4.46) is a function y such that $L[y] = 0$. When $r = 1$, we obtain one solution, namely, $y(x, 1)$, since $L[y(x, 1)] = 0 \cdot x = 0$. This fact was already known from our earlier work. Since $r = 1$ is a root of multiplicity 2 of $f(r) = 0$, it follows that $f'(r) = 0$ when $r = 1$. [See Equation (6.6.12) for the case when $k = 2$.] Thus, to obtain another solution of Equation (7.4.46), we differentiate (7.4.63) with respect to r and then set $r = 1$. That is,

$$\frac{\partial L[y(x, r)]}{\partial r} = 2(r - 1)x^r + (r - 1)^2 x^r \ln x,$$

and inverting the operation $\partial/\partial r$ and L, we have

$$\frac{\partial L[y(x, r)]}{\partial r} = L\left[\frac{\partial y(x, r)}{\partial r}\right] = 2(r - 1)x^r + (r - 1)^2 x^r \ln x. \tag{7.4.64}$$

Finally, setting $r = 1$ in (7.4.64), we get

$$\frac{\partial L[y(x, r)]}{\partial r}\bigg|_{r=1} = L\left[\frac{\partial y(x, r)}{\partial r}\bigg|_{r=1}\right] = 0. \tag{7.4.65}$$

Hence, the two formal solutions of Equation (7.4.46) are given by $y = y_1(x)$ and $y = y_2(x)$, where

$$y_1(x) = [y(x, r)]_{r=1} = y(x, 1), \tag{7.4.66}$$

and

$$y_2(x) = \left[\frac{\partial y(x, r)}{\partial r}\right]_{r=1}. \tag{7.4.67}$$

To obtain the solution $y_1(x)$ we substitute $r = 1$ in (7.4.61), which is a Frobenius series. To obtain the solution $y_2(x)$ we first differentiate (7.4.61) with respect to r, and we get

$$\frac{\partial y(x, r)}{\partial x^r} = x^r \ln x + \sum_{n=1}^{\infty} a_{2n}(r)x^{r+2n} \ln x + \sum_{n=1}^{\infty} a'_{2n}(r)x^{r+2n}. \tag{7.4.68}$$

Thus,

$$y_1(x) = x + \sum_{n=1}^{\infty} a_{2n}(1)x^{2n+1}, \tag{7.4.69}$$

and

$$y_2(x) = x \ln x + \ln x \sum_{n=1}^{\infty} a_{2n}(1)x^{2n+1} + \sum_{n=1}^{\infty} a'_{2n}(1)x^{2n+1}. \tag{7.4.70}$$

We shall now evaluate $a_{2n}(r)$ and $a'_{2n}(r)$ at $r = 1$.   From (7.4.51) and (7.4.62), we have

$$a_{2n}(r) = \frac{2^n}{f(r + 2n)f(r + 2n - 2) \,\cdots\, f(r + 4)f(r + 2)}$$

$$= \frac{2^n}{(r + 2n - 1)^2(r + 2n - 3)^2 \cdots (r + 3)^2(r + 1)^2}. \qquad (7.4.71)$$

To find $a'_{2n}(r)$ we proceed as follows.

$$\ln a_{2n}(r) = \ln 2^n$$
$$- 2\,[\ln\,(r + 2n - 1) + \ln\,(r + 2n - 3)$$
$$+ \cdots + \ln\,(r + 3) + \ln\,(r + 1)].$$

Differentiating the above expression with respect to r, we obtain

$$\frac{a'_{2n}(r)}{a_{2n}(r)} = -2\left(\frac{1}{r + 2n - 1} + \frac{1}{r + 2n - 3} + \cdots + \frac{1}{r + 3} + \frac{1}{r + 1}\right). \qquad (7.4.72)$$

Substituting $r = 1$ in (7.4.71), we find

$$a_{2n}(1) = \frac{2^n}{(2n)^2(2n - 2)^2 \cdots 4^2 \cdot 2^2} = \frac{2^n}{2^{2n}[n(n - 1) \cdots 2 \cdot 1]^2} = \frac{1}{2^n(n!)^2}.$$

$$(7.4.73)$$

Substituting $r = 1$ in (7.4.72) and using the above value of $a_{2n}(1)$, we find

$$a'_{2n}(1) = -\frac{2}{2^n(n!)^2}\left(\frac{1}{2n} + \frac{1}{2n - 2} + \cdots + \frac{1}{4} + \frac{1}{2}\right)$$

$$= -\frac{1}{2^n(n!)^2}\left(1 + \frac{1}{2} + \frac{1}{3} + \cdots + \frac{1}{n}\right). \qquad (7.4.74)$$

A notation often used to express the *partial sum* of the above harmonic series is H_n, that is

$$H_n = 1 + \frac{1}{2} + \frac{1}{3} + \cdots + \frac{1}{n} = \sum_{k=1}^{n} \frac{1}{k}. \qquad (7.4.75)$$

Thus, we may write (7.4.74) compactly as

$$a'_{2n}(1) = -\frac{H_n}{2^n(n!)^2}. \qquad (7.4.76)$$

In view of (7.4.73) and (7.4.76), we may write (7.4.69) and (7.4.70) as follows:

$$y_1(x) = x + x \sum_{n=1}^{\infty} \frac{x^{2n}}{2^n(n!)^2} = x \sum_{n=0}^{\infty} \frac{x^{2n}}{2^n(n!)^2} \qquad (7.4.77)$$

and

$$y_2(x) = x \ln x + x \ln x \sum_{n=1}^{\infty} \frac{x^{2n}}{2^n (n!)^2} - x \sum_{n=1}^{\infty} \frac{H_n}{2^n (n!)^2} x^{2n}. \qquad (7.4.78)$$

Using (7.4.77), we may also write $y_2(x)$ as

$$y_2(x) = y_1(x) \ln x - x \sum_{n=1}^{\infty} \frac{H_n}{2^n (n!)^2} x^{2n}. \qquad [7.4.78]$$

Utilizing Theorem 7.2.1, it is not too difficult to show that the series

$$\sum_{n=0}^{\infty} \frac{x^{2n}}{2^n (n!)^2} \quad \text{and} \quad \sum_{n=1}^{\infty} \frac{H_n}{2^n (n!)^2} x^{2n}$$

both converge absolutely for all $|x| < \infty$. Thus, in view of the relevant results on power series of Section 7.2, all the formal manipulations used to arrive at the solutions y_1 and y_2 of (7.4.77) and (7.4.78) are now completely justified for $x > 0$.

The solutions for $x < 0$ may be obtained by replacing x and $\ln x$ everywhere by $|x|$ and $\ln |x|$. Thus, for $x \neq 0$, the solutions $y_1(x)$ and $y_2(x)$ of Equation (7.4.46) are

$$y_1(x) = |x| \sum_{n=0}^{\infty} \frac{x^{2n}}{2^n (n!)^2}, \qquad (7.4.79)$$

and

$$y_2(x) = y_1(x) \ln |x| - |x| \sum_{n=1}^{\infty} \frac{H_n}{2^n (n!)^2} x^{2n}. \qquad (7.4.80)$$

In view of (7.4.80), the functions y_1 and y_2 are linearly independent on the intervals I_1: $-\infty < x < 0$ and I_2: $0 < x < \infty$, and hence on any interval not containing $x = 0$. Thus, on any interval not containing the origin, the general solution $y(x)$ of Equation (7.4.46) is

$$y(x) = c_1 y_1(x) + c_2 y_2(x), \qquad (7.4.81)$$

where c_1 and c_2 are arbitrary constants.

Note that we may also write (7.4.80) as

$$y_2(x) = y_1(x) \ln |x| - |x|^3 \sum_{n=0}^{\infty} \frac{H_{n+1}}{2^{n+1}[(n+1)!]^2} x^{2n}.$$

Clearly, the expansion of $y_2(x)$ is not a Frobenius series.

Remark 7.4.6. As will be seen in Section 7.5 [*see* Theorem 7.5.2], whenever the indicial equation of (7.4.20) has two equal roots $(r_1 = r_2)$, Equation

(7.4.20) will always possess for $0 < |x| < \rho,\ \rho > 0$, two linearly independent solutions y_1 and y_2 of the form

$$y_1(x) = |x|^{r_1} \sum_{n=0}^{\infty} a_n x^n, \qquad a_0 = 1, \tag{7.4.82}$$

and

$$y_2(x) = y_1(x) \ln |x| + |x|^{r_1+1} \sum_{n=0}^{\infty} b_n x^n. \tag{7.4.83}$$

The coefficients a_n and b_n may be determined by substituting the solutions into Equation (7.4.20). However, in some cases, it is more convenient to determine the coefficients a_n and b_n by the method given in the above example. Note that $a_n = a_n(r_1)$ and $b_n = a_n'(r_1)$.

Whenever the indicial equation has roots whose difference is a positive integer, the general solution of the given differential equation may or may not involve a logarithmic term. Examples 7.4.5 and 7.4.6 below illustrate these cases.

Example 7.4.5. Let us find two linearly independent solutions valid near the origin of the differential equation

$$xy'' + (x^3 - 1)y' + x^2 y = 0. \tag{7.4.84}$$

SOLUTION. Clearly, $x = 0$ is a regular singular point of Equation (7.4.84). Let us restrict our attention to the interval $0 < x < \rho,\ \rho > 0$. As usual, we first manipulate in a formal manner. Suppose that a solution $y(x)$ of Equation (7.4.84) is of the form

$$y(x) = x^r \sum_{n=0}^{\infty} a_n x^n = \sum_{n=0}^{\infty} a_n x^{r+n}, \qquad a_0 \neq 0. \tag{7.4.85}$$

Differentiating (7.4.85) twice, we get

$$y'(x) = \sum_{n=0}^{\infty} (r+n)a_n x^{r+n-1}, \qquad y''(x) = \sum_{n=0}^{\infty} (r+n)(r+n-1)a_n x^{n+r-2}. \tag{7.4.86}$$

Substituting the series for y, y', and y'' into Equation (7.4.84), we obtain upon collecting terms with the same exponent of x

$$\sum_{n=0}^{\infty} (r+n)(r+n-2)a_n x^{r+n-1} + \sum_{n=0}^{\infty} (r+n+1)a_n x^{r+n+2} = 0$$

or

$$\sum_{n=0}^{\infty} (r+n)(r+n-2)a_n x^{r+n-1} + \sum_{n=3}^{\infty} (r+n-2)a_{n-3} x^{r+n-1} = 0. \tag{7.4.87}$$

Equating to zero the coefficient of the lowest power of x ($n = 0$) in (7.4.87), we obtain the indicial equation

$$r(r - 2) = 0. \tag{7.4.88}$$

The indicial roots are $r_1 = 2$ and $r_2 = 0$. Let $m = r_1 - r_2 = 2 - 0 = 2$. Thus, the difference of the indicial roots is a positive integer $m = 2$. Summing out the first three terms in the first series in (7.4.87), we have

$$r(r - 2)a_0\, x^{r-1} + (r + 1)(r - 1)a_1 x^r + (r + 2)ra_2\, x^{r+1}$$

$$+ \sum_{n=3}^{\infty} [(r + n)(r + n - 2)a_n + (r + n - 2)a_{n-3}]x^{r+n-1} = 0. \tag{7.4.89}$$

If the function y given by (7.4.85) is to satisfy (7.4.84), we must have

$$r(r - 2)a_0 = 0, \tag{7.4.90}$$

$$(r + 1)(r - 1)a_1 = 0, \tag{7.4.91}$$

$$(r + 2)ra_2 = 0, \tag{7.4.92}$$

and the recurrence formula

$$(r + n)(r + n - 2)a_n + (r + n - 2)a_{n-3} = 0$$

or

$$a_n = -\frac{a_{n-3}}{r + n}, \qquad n \geq 3. \tag{7.4.93}$$

From the above equations, it is now apparent that when we substitute the larger root $r = r_1 = 2$, we will obtain a solution $y = y_1(x)$ of Equation (7.4.84) of the form

$$y_1(x) = x^2 \sum_{n=0}^{\infty} a_n x^n, \tag{7.4.94}$$

where a_0 is arbitrary, $a_1 = a_2 = 0$ and each a_n for $n \geq 3$ is expressible in terms of a_0. If we substitute the smaller root $r = r_2 = 0$ in the above equations, we see that a_0 is arbitrary, $a_1 = 0$, and $a_m = a_2$ is also arbitrary. In this case, in view of (7.4.93), the other solution $y = y_2(x)$ of Equation (7.4.84) will involve two arbitrary constants a_0 and a_2. As we shall see, the solution $y_2(x)$ actually contains the solution $y_1(x)$. Thus, if we use the smaller root to begin with, we are then able to obtain at once the formal general solution of Equation (7.4.84).

If, when we substituted the smaller root $r = r_2 = 0$ in the above equations, the coefficient $a_m = a_2$ could not be determined when imposing the condition that $a_0 \neq 0$, then the solution $y_2(x)$ would involve a logarithmic term, as in the case when we had equal indicial roots. Example 7.4.6 will further clarify this point.   Thus, when the indicial equation has roots r_1, r_2 such that $r_1 - r_2 = m$, where m is a positive integer, the given differential equation will always have at least one Frobenius series solution of the form

$$y_1(x) = x^{r_1} \sum_{n=0}^{\infty} a_n x^n, \qquad a_0 \neq 0. \tag{7.4.95}$$

[*See* also Theorem 7.5.2.]   One may then, if he wishes, utilize the method of reduction of order [*see* Section 4.5] to find the other solution of the given differential equation.   However, in general, the application of this method often involves laborious computations.

Substituting $r = r_1 = 2$ in Equations (7.4.90) through (7.4.93), we have that a_0 is arbitrary, $a_1 = 0$, $a_2 = 0$ and

$$a_n = -\frac{a_{n-3}}{n+2}, \qquad n \geq 3.$$

Thus,

$$a_3 = -\frac{a_0}{5}, \qquad\qquad a_4 = -\frac{a_1}{6} = 0, \qquad a_5 = -\frac{a_2}{7} = 0,$$

$$a_6 = -\frac{a_3}{8} = \frac{a_0}{5 \cdot 8}, \qquad a_7 = -\frac{a_4}{9} = 0, \qquad a_8 = -\frac{a_5}{10} = 0,$$

$$a_9 = -\frac{a_6}{11} = -\frac{a_0}{5 \cdot 8 \cdot 11}, \qquad a_{10} = -\frac{a_7}{12} = 0, \qquad a_{11} = -\frac{a_8}{13} = 0;$$

and by mathematical induction, we have

$$a_{3n} = \frac{(-1)^n a_0}{5 \cdot 8 \cdot 11 \cdots (3n+2)}, \qquad a_{3n+1} = 0, \quad a_{3n+2} = 0, \qquad n \geq 1. \tag{7.4.96}$$

Let us write $y_1(x)$ of (7.4.94) as

$$y_1(x) = x^2[a_0 + a_3 x^3 + a_6 x^6 + a_9 x^9 + \cdots + a_1 x + a_4 x^4$$
$$+ a_7 x^7 + a_{10}x^{10} + \cdots + a_2 x^2 + a_5 x^5 + a_8 x^8 + a_{11}x^{11} + \cdots]. \tag{7.4.97}$$

We may also write (7.4.97) as

$$y_1(x) = x^2\left[a_0 + \sum_{n=1}^{\infty} a_{3n} x^{3n} + a_1 x + \sum_{n=1}^{\infty} a_{3n+1} x^{3n+1} \right.$$

$$\left. + a_2 x^2 + \sum_{n=1}^{\infty} a_{3n+2} x^{3n+2} \right]. \tag{7.4.98}$$

Since $a_1 = 0$, $a_2 = 0$, $a_{3n+1} = 0$, and $a_{3n+2} = 0$ for $n \geq 1$, taking $a_0 = 1$ and using the value of a_{3n} from (7.4.96), the formal solution $y_1(x)$ of (7.4.84) is

$$y_1(x) = x^2\left[1 + \sum_{n=1}^{\infty} \frac{(-1)^n x^{3n}}{5 \cdot 8 \cdot 11 \cdots (3n+2)}\right]. \tag{7.4.99}$$

If we now substitute the smaller root $r = r_2 = 0$ in Equations (7.4.90) through (7.4.93), we have that a_0 is arbitrary, $a_1 = 0$, a_2 is also arbitrary and

$$a_n = -a_{n-3}/n, \qquad n \geq 3.$$

Thus,

$$a_3 = -\frac{a_0}{3}, \qquad\qquad a_4 = -\frac{a_1}{4} = 0, \qquad a_5 = -\frac{a_2}{5},$$

$$a_6 = -\frac{a_3}{6} = \frac{a_0}{3 \cdot 6}, \qquad a_7 = -\frac{a_4}{7} = 0, \qquad a_8 = -\frac{a_5}{8} = \frac{a_2}{5 \cdot 8},$$

$$a_9 = -\frac{a_6}{9} = -\frac{a_0}{3 \cdot 6 \cdot 9}, \qquad a_{10} = -\frac{a_7}{10} = 0, \qquad a_{11} = -\frac{a_8}{11} = -\frac{a_2}{5 \cdot 8 \cdot 11};$$

and by mathematical induction, we have

$$a_{3n} = \frac{(-1)^n a_0}{3 \cdot 6 \cdot 9 \cdots 3n} = \frac{(-1)^n a_0}{3^n n!}, \qquad a_{3n+1} = 0,$$

$$a_{3n+2} = \frac{(-1)^n a_2}{5 \cdot 8 \cdot 11 \cdots (3n+2)}, \qquad n \geq 1. \tag{7.4.100}$$

With $r = r_2 = 0$, $a_1 = 0$, and the values of the a_n's from (7.4.100), another formal solution of Equation (7.4.84) for $x > 0$ is $y = y_2(x)$, where

$$y_2(x) = x^0 \sum_{n=0}^{\infty} a_n x^n$$

$$= a_0 + \sum_{n=1}^{\infty} a_{3n} x^{3n} + a_1 x + \sum_{n=1}^{\infty} a_{3n+1} x^{3n+1} + a_2 x^2 + \sum_{n=1}^{\infty} a_{3n+2} x^{3n+2}$$

$$= a_0 + \sum_{n=1}^{\infty} \frac{(-1)^n a_0 x^{3n}}{3^n n!} + a_2 x^2 + \sum_{n=1}^{\infty} \frac{(-1)^n a_2 x^{3n+2}}{5 \cdot 8 \cdot 11 \cdots (3n+2)}$$

$$= a_0 \sum_{n=0}^{\infty} \frac{(-1)^n x^{3n}}{3^n n!} + a_2\left[x^2 + x^2 \sum_{n=1}^{\infty} \frac{(-1)^n x^{3n}}{5 \cdot 8 \cdot 11 \cdots (3n+2)}\right]. \tag{7.4.101}$$

In particular, when $a_2 = 1$, the solution $y_2(x)$ of (7.4.101) contains the solution $y_1(x)$ already found.   [*See* (7.4.99).]   With a_0 and a_2 as arbitrary constants, the solution $y_2(x)$ of (7.4.101) may be taken as the formal general solution of Equation (7.4.84).   The series in the right-hand member of (7.4.101) both converge for all $|x| < \infty$. Hence, all the formal manipulations used to arrive at the solution $y_2(x)$ are now completely justified for $x > 0$. For $x < 0$, the solution $y_2(x)$ is also given by (7.4.101). (Why?) The functions y_{21} and y_{22}, represented by

$$y_{21}(x) = \sum_{n=0}^{\infty} \frac{(-1)^n x^{3n}}{3^n n!}, \qquad y_{22}(x) = x^2 + x^2 \sum_{n=1}^{\infty} \frac{(-1)^n x^{3n}}{5 \cdot 8 \cdots (3n + 2)}, \qquad (7.4.102)$$

are linearly independent on the intervals I_1: $-\infty < x < 0$ and I_2: $0 < x < \infty$, and hence on any interval not containing $x = 0$. Thus, on any interval not containing the origin, the solution $y_2(x)$ of (7.4.101) may be taken as the general solution of Equation (7.4.84).

Example 7.4.6. Let us find two linearly independent solutions valid near the origin of the Bessel equation of order one. [*See* Exercise 4.5.8.]

$$x^2 y'' + xy' + (x^2 - 1)y = 0. \qquad (7.4.103)$$

SOLUTION. Clearly, $x = 0$ is a regular singular point of Equation (7.4.103). Let us restrict our attention to the interval $0 < x < \rho, \rho > 0$. Suppose that a trial solution of Equation (7.4.103) is of the form

$$y(x) = x^r \sum_{n=0}^{\infty} a_n x^n = \sum_{n=0}^{\infty} a_n x^{r+n}, \qquad a_0 \neq 0. \qquad (7.4.104)$$

Differentiating (7.4.104) twice, we get

$$y'(x) = \sum_{n=0}^{\infty} (r + n)a_n x^{r+n-1}, \qquad y''(x) = \sum_{n=0}^{\infty} (r + n)(r + n - 1)a_n x^{r+n-2}. \qquad (7.4.105)$$

Substituting the series for y, y', and y'' into Equation (7.4.103), we obtain

$$\sum_{n=0}^{\infty} [(r + n)(r + n - 1) + r + n - 1]a_n x^{r+n} + \sum_{n=0}^{\infty} a_n x^{r+n+2} = 0$$

or

$$\sum_{n=0}^{\infty} [(r + n - 1)(r + n + 1)]a_n x^{r+n} + \sum_{n=2}^{\infty} a_{n-2} x^{r+n} = 0 \qquad (7.4.106)$$

Equating to zero the coefficient of the lowest power of x in (7.4.106), we obtain the indicial equation $r^2 - 1 = 0$. The indicial roots are $r_1 = 1$ and $r_2 = -1$. Let $m = r_1 - r_2 = 1 + 1 = 2$. Thus, the difference of the indicial roots is a positive integer $m = 2$. Summing out the first two terms in the first series in (7.4.106), we have

$$(r - 1)(r + 1)a_0 x^r + r(r + 2)a_1 x^{r+1}$$

$$+ \sum_{n=2}^{\infty} [(r + n - 1)(r + n + 1)a_n + a_{n-2}]x^{r+n} = 0. \qquad (7.4.107)$$

If the function y given in (7.4.104) is to satisfy (7.4.103), we must have

$$(r - 1)(r + 1)a_0 = 0, \tag{7.4.108}$$

$$r(r + 2)a_1 = 0, \tag{7.4.109}$$

and the recurrence relation

$$(r + n - 1)(r + n + 1)a_n + a_{n-2} = 0, \qquad n \geq 2. \tag{7.4.110}$$

From Equations (7.4.108) through (7.4.110), it is now apparent that when we substitute the larger root $r = r_1 = 1$, we will obtain a solution $y = y_1(x)$ of Equation (7.4.103) of the form

$$y_1(x) = x \sum_{n=0}^{\infty} a_n x^n, \tag{7.4.111}$$

where a_0 is arbitrary, $a_1 = 0$, and each a_n for $n \geq 2$ is expressible in terms of a_0. If we substitute the smaller root $r = r_2 = -1$, we see that a_0 is arbitrary, $a_1 = 0$, and $a_m = a_2$ cannot be determined; since, when $n = 2$, Equation (7.4.110) becomes

$$0 \cdot a_2 + a_0 = 0 \quad \text{or} \quad 0 \cdot a_2 = -a_0 .$$

But, by assumption $a_0 \neq 0$. Hence, a_2 cannot be determined. To circumvent this difficulty, we shall now proceed in a similar fashion as in the case when we had equal indicial roots. [*See* Example 7.4.4.]

Let us denote the left-hand member of Equation (7.4.103) by $L[y]$, that is,

$$L[y] = x^2 y'' + xy' + (x^2 - 1)y. \tag{7.4.112}$$

Substituting the series for y, y', and y'' into $L[y]$, we have

$$L[y] = (r - 1)(r + 1)a_0 x^r + r(r + 2)a_1 x^{r+1}$$

$$+ \sum_{n=2}^{\infty} [(r + n - 1)(r + n + 1)a_n + a_{n-2}]x^{r+n}. \tag{7.4.113}$$

By choosing the a's and leaving r as a parameter, we can make every term, except the first one, in $L[y]$ vanish. Thus, in (7.4.113), setting each coefficient of x except the first one equal to zero, we have $a_1 = 0$, and the recurrence relation,

$$(r + n - 1)(r + n + 1)a_n + a_{n-2} = 0$$

or

$$\frac{a_n = -a_{n-2}}{(r + n - 1)(r + n + 1)}, \qquad n \geq 2. \tag{7.4.114}$$

Utilizing (7.4.114) and mathematical induction, one may easily establish the following results:

$$a_{2n} = \frac{(-1)^n a_0}{[(r+1)(r+3)(r+5)\cdots(r+2n-1)][(r+3)(r+5)\cdots(r+2n+1)]},$$

$$a_{2n+1} = 0, \qquad n \geq 1. \tag{7.4.115}$$

With the values of a_{2n} and a_{2n+1} from (7.4.115) and $a_1 = 0$, $L[y]$ of (7.4.113) reduces to the first term, namely,

$$L[y] = (r-1)(r+1)a_0 x^r. \tag{7.4.116}$$

For the larger root $r = r_1 = 1$ only one solution can be obtained.   (Why?) For the smaller root $r = r_1 = -1$, two solutions would be available provided that the factor $(r+1)$ were $(r+1)^2$.   However, since a_0 is still arbitrary, we may take $a_0 = (r+1)$ in order to get the desired square factor in the right-hand member of (7.4.116).   This is the key point of the present analysis.

With $a_0 = (r+1)$, $a_1 = 0$, and using the values of a_{2n} and a_{2n+1} from (7.4.115), we may write a function y which is dependent upon both x and r, namely,

$$y(x, r) = (r+1)x^r$$

$$+ x^r \sum_{n=1}^{\infty} \frac{(-1)^n (r+1)x^{2n}}{[(r+1)(r+3)(r+5)\cdots(r+2n-1)][(r+3)(r+5)\cdots(r+2n+1)]}. \tag{7.4.117}$$

Clearly, the function y given by (7.4.117) has been so determined that when it is substituted in (7.4.113), $L[y(x, r)]$ reduces to the first term.   That is, with $a_0 = (r+1)$, we have

$$L[y(x, r)] = (r-1)(r+1)^2 x^r. \tag{7.4.118}$$

As in the case of equal indicial roots, the two formal linearly independent solutions being sought may be obtained as

$$y_1(x) = y(x, -1) \tag{7.4.119}$$

and

$$y_2(x) = \left[\frac{\partial y(x, r)}{\partial r}\right]_{r=-1}. \tag{7.4.120}$$

Summing out the first term in the series in (7.4.117) and then cancelling the common factor $(r+1)$, we have

$$y(x, r) = (r + 1)x^r + x^r \left[-\frac{x^2}{r + 3} \right.$$

$$+ \sum_{n=2}^{\infty} \frac{(-1)^n x^{2n}}{[(r + 3)(r + 5) \cdots (r + 2n - 1)][(r + 3)(r + 5) \cdots (r + 2n + 1)]} \left. \right].$$

$$(7.4.121)$$

Differentiating (7.4.121) with respect to r, we get

$$\frac{\partial y(x, r)}{\partial r} = (r + 1)x^r \ln x$$

$$+ x^r \ln x \left[-\frac{x^2}{r + 3} \right.$$

$$+ \sum_{n=2}^{\infty} \frac{(-1)^n x^{2n}}{[(r + 3)(r + 5) \cdots (r + 2n - 1)][(r + 3)(r + 5) \cdots (r + 2n + 1)]} \left. \right]$$

$$+ x^r + x^r \left[\frac{x^2}{(r + 3)^2} \right.$$

$$- \sum_{n=2}^{\infty} \frac{(-1)^n}{[(r + 3)(r + 5) \cdots (r + 2n - 1)][(r + 3)(r + 5) \cdots (r + 2n + 1)]}$$

$$\times \left(\frac{1}{r + 3} + \frac{1}{r + 5} + \cdots + \frac{1}{r + 2n - 1} \right.$$

$$+ \frac{1}{r + 3} + \frac{1}{r + 5} + \cdots + \frac{1}{r + 2n + 1} \left. \right) x^{2n} \left. \right].$$

$$(7.4.122)$$

Substituting $r = r_2 = -1$ in (7.4.121), we find

$$y_1(x) = y_1(x, -1) = x^{-1} \left[-\frac{x^2}{2} + \sum_{n=2}^{\infty} \frac{(-1)^n x^{2n}}{[2 \cdot 4 \cdots (2n - 2)][2 \cdot 4 \cdots 2n]} \right]$$

$$= x^{-1} \left[-\frac{x^2}{2} + \sum_{n=2}^{\infty} \frac{(-1)^n x^{2n}}{2^{2n-1}(n - 1)! \, n!} \right]$$

or

$$y_1(x) = x^{-1} \sum_{n=1}^{\infty} \frac{(-1)^n x^{2n}}{2^{2n-1}(n - 1)! \, n!}. \qquad (7.4.123)$$

Substituting $r = r_2 = -1$ in (7.4.122) and then using (7.4.123), we obtain

$$y_2(x) = \left[\frac{\partial y(x, r)}{\partial r}\right]_{r=-1}$$

$$= x^{-1} \ln x \left[-\frac{x^2}{2} + \sum_{n=2}^{\infty} \frac{(-1)^n x^{2n}}{[2 \cdot 4 \cdots (2n-2)][2 \cdot 4 \cdots 2n]}\right] + x^{-1}$$

$$+ x^{-1}\left[\frac{x^2}{2^2} - \sum_{n=2}^{\infty} \frac{(-1)^n}{[2 \cdot 4 \cdots (2n-2)][2 \cdot 4 \cdots 2n]}\right.$$

$$\left. \times \left(\frac{1}{2} + \frac{1}{4} + \cdots + \frac{1}{2n-2} + \frac{1}{2} + \frac{1}{4} + \cdots + \frac{1}{2n}\right)x^{2n}\right]$$

$$= y_1(x) \ln x + x^{-1} - x^{-1}\left[-\frac{x^2}{2^2} + \sum_{n=2}^{\infty} \frac{(-1)^n}{2^{2n-1}(n-1)!\, n!}\right.$$

$$\left. \times \frac{1}{2}\left(1 + \frac{1}{2} + \cdots + \frac{1}{n-1} + 1 + \frac{1}{2} + \cdots + \frac{1}{n}\right)x^{2n}\right].$$

Let us denote the partial sums of the harmonic series in the parenthesis of the above expression by H_{n-1} and H_n, that is,

$$H_{n-1} = 1 + \frac{1}{2} + \cdots + \frac{1}{n-1}$$

and

$$H_n = 1 + \frac{1}{2} + \cdots + \frac{1}{n},$$

(7.4.124)

and, by definition, $H_0 = 0$. We may then write $y_2(x)$ compactly as

$$y_2(x) = y_1(x) \ln x + x^{-1} - x^{-1} \sum_{n=1}^{\infty} \frac{(-1)^n (H_{n-1} + H_n) x^{2n}}{2^{2n}(n-1)!\, n!}. \tag{7.4.125}$$

[*Note:* When $n = 1$, the first term in the series becomes

$$-x^{-1}\left[\frac{(-1)^1 (H_0 + H_1)}{2^2 0!\, 1!} x^2\right] = -x^{-1}\left[-\frac{(0+1)}{2^2} x^2\right] = -x^{-1}\left(-\frac{x^2}{2^2}\right).]$$

Utilizing Theorem 7.2.1, it is not difficult to show that the series

$$\sum_{n=1}^{\infty} \frac{(-1)^n x^{2n}}{2^{2n-1}(n-1)!\, n!} \quad \text{and} \quad \sum_{n=1}^{\infty} \frac{(-1)^n (H_{n-1} + H_n) x^{2n}}{2^{2n}(n-1)!\, n!}$$

both converge absolutely for all $|x| < \infty$. Thus, in view of the relevant results on power series of Section 7.2, all the formal manipulations used to arrive at the solutions $y_1(x)$ and $y_2(x)$ of (7.4.123) and (7.4.125) are now completely justified for $x > 0$.

The solutions for $x < 0$ may be obtained by replacing x and $\ln x$ by $|x|$ and $\ln |x|$ everywhere.   Thus, for $x \neq 0$, the solutions $y_1(x)$ and $y_2(x)$ of Equation (7.4.103) are

$$y_1(x) = |x|^{-1} \sum_{n=1}^{\infty} \frac{(-1)^n x^{2n}}{2^{2n-1}(n-1)! \, n!} \tag{7.4.126}$$

and

$$y_2(x) = y_1(x) \ln |x| + |x|^{-1} \left[1 - \sum_{n=1}^{\infty} \frac{(-1)^n (H_{n-1} + H_n) x^{2n}}{2^{2n}(n-1)! \, n!} \right]. \tag{7.4.127}$$

In view of (7.4.127), the functions y_1 and y_2 are linearly independent on the intervals $I_1: \ -\infty < x < 0$, and $I_2: \ 0 < x < \infty$ and hence on any interval not containing $x = 0$.   Thus, on any interval not containing the origin, the general solution $y(x)$ of Equation (7.4.103) is

$$y(x) = c_1 y_1(x) + c_2 y_2(x), \tag{7.4.128}$$

where c_1 and c_2 are arbitrary constants.   The reader may show that if we used the larger root $r = r_1 = 1$ in (7.4.110) and chose $a_0 = -\tfrac{1}{2}$, the solution $y = y_1(x)$ of Equation (7.4.103) would be

$$y_1(x) = |x| \sum_{n=0}^{\infty} \frac{(-1)^{n+1} x^{2n}}{2^{2n+1} n! \, (n+1)!} \tag{7.4.129}$$

which is equivalent to the solution of (7.4.126).

Remark 7.4.7.   From Examples 7.4.5 and 7.4.6, we see that if the given differential equation has an indicial equation with roots r_1, r_2 such that $r_1 - r_2 = m$, where m is a positive integer, then the solution $y_1(x)$ associated with the larger root r_1 always exists and is of the form

$$y_1(x) = |x|^{r_1} \sum_{n=0}^{\infty} a_n x^n, \qquad a_0 = 1,$$

while the other solution $y_2(x)$, associated with the smaller root r_2, may or may not be of the same form as $y_1(x)$.   As we shall see in Section 7.5 [*see* Theorem 7.5.2], that whenever the indicial equation of (7.4.20) has roots r_1, r_2 such that $r_1 - r_2 = m$, where m is a positive integer, Equation (7.4.20) will always possess for $0 < |x| < \rho, \rho > 0$, two linearly independent solutions y_1, y_2 of the form

$$y_1(x) = |x|^{r_1} \sum_{n=0}^{\infty} a_n x^n, \qquad a_0 \neq 0, \tag{7.4.130}$$

and

$$y_2(x) = |x|^{r_2} \sum_{n=0}^{\infty} b_n x^n + c y_1(x) \ln |x|, \tag{7.4.131}$$

where the constant c may be zero or it may not.   Moreover, the coefficients a_n and b_n may be determined by substituting the solutions in Equation (7.4.20).

Remark 7.4.8.   Even though all the examples given in this section involved indicial roots which were real numbers, the above results are also true when the indicial roots are complex numbers.   The condition $r_1 \geq r_2$ is now replaced by $\mathscr{R}(r_1) \geq \mathscr{R}(r_2)$, where $\mathscr{R}(r)$ denotes the real part of the complex root $r = \alpha + i\beta$, that is, $\mathscr{R}(r) = \alpha$.

Exercises 7.4

In Exercises 1 through 9, locate and classify the singular points for finite values of x of each differential equation.

1. $x^3(x^2 - 1)^2 y'' + (x^3 + x^2)y' + y = 0.$
2. $(x + 1)^3 y'' + (x^2 - 1)^2 y' + (x^3 + 1)y = 0.$
3. $x^2(x^2 + 5x + 6)^2 y'' + (x + 2)y' + e^x y = 0.$
4. $(x^4 - 3x^3)y'' + x^2 y' - e^{-x} y = 0.$
5. $(x^4 - 4x^2)^2 y'' + (x^4 + 2x^3)y' + (x^2 + x)y = 0.$
6. $(x^2 - 1)^3 y'' + (x^4 - 2x^2 + 1)y' + (x^3 + x^2 + x + 1)y = 0.$
7. $x^3 y'' + x^2 y' + (\sin x)y = 0.$
8. $x^3 y'' + (1 - e^{x^2})y' + (1 - \cos x)y = 0.$
9. $(\cos x)x^3 y'' + (\sin x)x^2 y' + (x + 1)y = 0.$

In Exercises 10 through 36, utilize the techniques given in the examples of this section to obtain two linearly independent solutions of each differential equation near $x = 0$.   Also, determine the intervals on which these solutions are linearly independent.

10. $4xy'' + 2(1 + x)y' + y = 0.$ 11. $2x^2 y'' + x(1 + 2x)y' + 2xy = 0.$
12. $6x^2 y'' + xy' + (1 + x^2)y = 0.$ 13. $2x^2 y'' + 3xy' - (1 - 2x)y = 0.$
14. $2(x^2 + x^3)y'' - (x - 3x^2)y' + y = 0.$
15. $9(-2x^2 + x^3)y'' - 18xy' + 2(1 + x)y = 0.$
16. $6(x^2 - x^3)y'' + xy' + y = 0.$ 17. $(2x^2 + x^4)y'' + 3xy' - 6x^2 y = 0.$
18. $xy'' + y' + xy = 0.$ (This is known as Bessel's equation of order zero.)   [*See* Exercise 4.5.8.]
19. $xy'' + y' - 4xy = 0.$ 20. $x^2 y'' - 3xy' + 4(1 - x^2)y = 0.$
21. $x^2 y'' - (3x + x^3)y' + (4 - x^2)y = 0.$ 22. $4x^2 y'' + (8x + x^2)y' + (1 + x)y = 0.$
23. $16x^2 y'' + 4(6x + x^2)y' + (1 + 4x)y = 0.$
24. $4x^2 y'' + 2(8x + x^3)y' + (9 + 4x^2)y = 0.$
25. $(x^2 + x^4)y'' + (9x + 5x^3)y' + 4(4 + x^2)y = 0.$
26. $x^2 y'' - (3x - x^2)y' + (3 - 2x)y = 0.$
27. $(x^2 + x^4)y'' - 2xy' - 2(2 + x^2)y = 0.$
28. $4x^2 y'' + 2(2x - x^2)y' - (25 + 3x)y = 0.$
29. $xy'' + (6 - x^2)y' + \left(\dfrac{4}{x} - 3x\right)y = 0.$ 30. $xy'' + y = 0.$

31. $x^2y'' - 5xy' + 4(2 - x^2)y = 0.$ 32. $x^2y'' + (4x - x^2)y' + (2 - x)y = 0.$
33. $xy'' + (3 - 4x)y' - 16y = 0.$ 34. $x^2y'' - xy' - (3 + 16x^4)y = 0.$
35. $9x^2y'' + 3(2x + x^2)y' - 2(1 - 2x)y = 0.$
36. $(x^2 - x^3)y'' + 4(x - x^2)y' + 2y = 0.$

Find two linearly independent solutions near the indicated singular point of each of the differential equations in Exercises 37 through 40.

37. $2(x - 1)^2y'' - (x - 1)y' + xy = 0.$ The point $x = 1.$
38. $3(x + 2)y'' + 4y' + xy = 0.$ The point $x = -2.$
39. $x(x - 2)^2y'' - (x + 1)(x - 2)y' + (4 - 3x + x^2)y = 0.$ The point $x = 2.$
40. $(x - 1)y'' + (x - 2)(x^2 - x + 1)y' + x(x - 1)^2y = 0.$ The point $x = 1.$

7.5. Solutions about a Regular Singular Point. General Theory

Let us consider the differential equation

$$x^2y'' + xp(x)y' + q(x)y = 0, \tag{7.5.1}$$

where the functions p and q are analytic at $x = 0$ and have power series expansions which are convergent for $|x| < \rho,\ \rho > 0.$ That is,

$$p(x) = \sum_{n=0}^{\infty} p_n x^n \quad \text{and} \quad q(x) = \sum_{n=0}^{\infty} q_n x^n. \tag{7.5.2}$$

[Thus, $x = 0$ is a regular singular point of (7.5.1).]   Let us restrict our attention to $x > 0.$ Let us suppose that we have a solution of (7.5.1) of the form

$$y(x) = x^r \sum_{n=0}^{\infty} a_n x^n, \qquad a_0 \neq 0, \tag{7.5.3}$$

which is convergent for $0 < x < \rho.$ Differentiating (7.5.3) twice, we obtain

$$y'(x) = x^{r-1} \sum_{n=0}^{\infty} (n + r)a_n x^n, \qquad y''(x) = x^{r-2} \sum_{n=0}^{\infty} (n + r)(n + r - 1)a_n x^n. \tag{7.5.4}$$

Substituting the series for p, q, y, y', and y'' into Equation (7.5.1), we obtain

$$x^r \sum_{n=0}^{\infty} (n + r)(n + r - 1)a_n x^n + x^r \left[\sum_{n=0}^{\infty} p_n x^n \right] \left[\sum_{n=0}^{\infty} (n + r)a_n x^n \right]$$

$$+ x^r \left[\sum_{n=0}^{\infty} q_n x^n \right] \left[\sum_{n=0}^{\infty} a_n x^n \right] = 0. \tag{7.5.5}$$

Utilizing Theorem 7.2.8, we may write (7.5.5) as

$$x^r \sum_{n=0}^{\infty} (n + r)(n + r - 1)a_n x^n + x^r \sum_{n=0}^{\infty} \sum_{k=0}^{n} p_{n-k}(k + r)a_k x^n$$

$$+ x^r \sum_{n=0}^{\infty} \sum_{k=0}^{n} q_{n-k} a_k x^n = 0$$

or as

$$x^r \sum_{n=0}^{\infty} \left\{ (n+r)(n+r-1)a_n + \sum_{k=0}^{n} [p_{n-k}(k+r) + q_{n-k}]a_k \right\} x^n = 0. \quad (7.5.6)$$

If the function y given by (7.5.3) is to satisfy Equation (7.5.1), then in view of (7.5.6) and Theorem 7.2.7, we must have

$$(n+r)(n+r-1)a_n + \sum_{k=0}^{n} [p_{n-k}(k+r) + q_{n-k}]a_k = 0, \qquad n = 0, 1, 2, \dots .$$

$$(7.5.7)$$

Summing out the term $k = n$, Equation (7.5.7) may be written as

$$[(n+r)(n+r-1) + p_0(n+r) + q_0]a_n$$

$$+ \sum_{k=0}^{n-1} [p_{n-k}(k+r) + q_{n-k}]a_k = 0, \qquad n = 0, 1, 2, \dots . \quad (7.5.8)$$

Since $a_0 \neq 0$, for $n = 0$, we must have

$$r(r-1) + p_0 r + q_0 = 0, \qquad (7.5.9)$$

which is called the *indicial equation* of (7.5.1). Note that the indicial equation is a quadratic equation in r. The roots of this equation are called the *indicial roots*, and they are the only allowable values of r which will give rise to possible Frobenius series solutions of Equation (7.5.1). The polynomial f defined by

$$f(r) = r(r-1) + p_0 r + q_0 \qquad (7.5.10)$$

is called the *indicial polynomial* of (7.5.1). Note that

$$f(r+n) = (r+n)(r+n-1) + p_0(r+n) + q_0 .$$

Thus, Equation (7.5.8) asserts that

$$f(r) = 0 \qquad (7.5.11)$$

and

$$f(r+n)a_n + \sum_{k=0}^{n-1} [(k+r)p_{n-k} + q_{n-k}]a_k = 0, \qquad n = 1, 2, \dots, \quad (7.5.12)$$

which is known as a *recurrence relation*. Letting

$$F_n = \sum_{k=0}^{n-1} [(k+r)p_{n-k} + q_{n-k}]a_k , \qquad (7.5.13)$$

we may now write Equation (7.5.12) as

$$f(r+n)a_n = -F_n, \qquad n = 1, 2, \dots . \quad (7.5.14)$$

From (7.5.13) we have

$$F_1 = (rp_1 + q_1)a_0,$$
$$F_2 = (rp_2 + q_2)a_0 + [(1 + r)p_1 + q_1]a_1,$$
$$F_3 = (rp_3 + q_3)a_0 + [(1 + r)p_2 + q_2]a_1 + [(2 + r)p_1 + q_1]a_2,$$
$$\vdots$$
$$F_n = (rp_n + q_n)a_0 + [(1 + r)p_{n-1} + q_{n-1}]a_1$$
$$+ [(2 + r)p_{n-2} + q_{n-2}]a_2 + \cdots + [(n - 1 + r)p_1 + q_1]a_{n-1}.$$

Clearly, F_n is a linear combination of a_0, a_1, ..., a_{n-1} with coefficients involving r and the known functions p and q given by (7.5.2). If, for the present, we leave r and a_0 indeterminate, in view of Equation (7.5.14), we see that we can solve for $a_1, a_2, \ldots$ in terms of r and a_0 provided that $f(r + n) \neq 0$, $n = 1, 2, \ldots$. Let us denote these solutions by $a_n(r)$, $n = 1, 2, \ldots$. With F_n and a_k replaced by $F_n(r)$ and $a_k(r)$, we then have, from (7.5.14), that

$$a_n(r) = - \frac{F_n(r)}{f(r + n)}, \tag{7.5.15}$$

provided that $f(r + n) \neq 0$, $n = 1, 2, \ldots$. Thus,

$$a_1(r) = - \frac{F_1(r)}{f(r + 1)} = - \frac{(rp_1 + q_1)a_0}{f(r + 1)},$$

$$a_2(r) = - \frac{F_2(r)}{f(r + 2)} = - \frac{\{(rp_2 + q_2)a_0 + [(1 + r)p_1 + q_1]a_1(r)\}}{f(r + 2)}$$

$$= - \frac{\left\{(rp_2 + q_2)a_0 + [(1 + r)p_1 + q_1]\left[- \dfrac{(rp_1 + q_1)}{f(r + 1)} a_0\right]\right\}}{f(r + 2)}$$

$$= \frac{\{-f(r + 1)(rp_2 + q_2) + [(1 + r)p_1 + q_1](rp_1 + q_1)\}a_0}{f(r + 2)f(r + 1)},$$

and continuing in this fashion, we see that

$$a_n(r) = - \frac{F_n(r)}{f(r + n)}$$

$$= \frac{P_n(r)a_0}{f(r + n)f(r + n - 1) \cdots f(r + 2)f(r + 1)}, \qquad n = 1, 2, \ldots, \tag{7.5.16}$$

where $P_n(r)$ is a polynomial in r (not necessarily of degree n). Thus, a_n is a rational function of r, that is, a quotient of two polynomials in r.

Let us define a function y by

$$y(x, r) = a_0 x^r + x^r \sum_{n=1}^{\infty} a_n(r)x^n, \tag{7.5.17}$$

where the a_n are given recursively by $f(r + n)a_n(r) = -F_n(r)$, $n = 1, 2, \ldots$. Let us denote the left-hand member of Equation (7.5.1) by $L[y]$, that is,

$$L[y] = x^2 y'' + xp(x)y' + q(x)y. \qquad (7.5.18)$$

In view of the above results, we see that when the function y given by (7.5.17) is substituted in (7.5.18), $L[y(x, r)]$ reduces to a single term, namely,

$$L[y(x, r)] = a_0[r(r - 1) + p_0 r + q_0]x^r = a_0 f(r)x^r. \qquad (7.5.19)$$

Thus, it is now evident that if the function y given by (7.5.3) is a solution of (7.5.1), then r is a root of the indicial equation $f(r) = 0$, and the coefficients a_n for $n \geq 1$ are determined uniquely in terms of a_0 and r, provided that $f(r + n) \neq 0$, $n = 1, 2, \ldots$. [See Equation (7.5.16).]   Conversely, if r is a root of the indicial equation $f(r) = 0$, and if $f(r + n) \neq 0$, $n = 1, 2, \ldots$, so that the $a_n(r)$ can be determined uniquely, then the function y given by $y(x) = y(x, r)$ is a solution of (7.5.1) with a_0 arbitrary, provided that the series in (7.5.17) can be shown to be convergent.

Let r_1 and r_2 be the roots of the indicial equation $f(r) = 0$.   For definiteness, suppose that $\mathscr{R}(r_1) \geq \mathscr{R}(r_2)$.   [Remember that the roots of the indicial equation may be complex.   Also, if $r = \alpha + i\beta$, then $\mathscr{R}(r) = \alpha$.]   We may now write the indicial polynomial as

$$f(r) = (r - r_1)(r - r_2). \qquad (7.5.20)$$

Then

$$f(r_1 + n) = (r_1 + n - r_1)(r_1 + n - r_2) = n(n + r_1 - r_2), \qquad (7.5.21)$$

and

$$f(r_2 + n) = (r_2 + n - r_1)(r_2 + n - r_2) = n(n + r_2 - r_1). \qquad (7.5.22)$$

Since $\mathscr{R}(r_1) \geq \mathscr{R}(r_2)$, in view of (7.5.21), we always have that $f(r_1 + n) \neq 0$, $n = 1, 2, \ldots$.   If r_1 and r_2 do not differ by a positive integer, that is, if $r_1 - r_2 \neq m$, where m is a positive integer, then in view of (7.5.22), we also have that $f(r_2 + n) \neq 0$, $n = 1, 2, \ldots$.   Hence, whenever the roots of the indicial equations are distinct and do not differ by a positive integer, the Frobenius series solutions of Equation (7.5.1) for $x > 0$ are $y = y_1(x)$ and $y = y_2(x)$, where

$$y_1(x) = x^{r_1} \sum_{n=0}^{\infty} a_n(r_1)x^n, \qquad a_0(r_1) = 1, \qquad (7.5.23)$$

and

$$y_2(x) = x^{r_2} \sum_{n=0}^{\infty} a_n(r_2)x^n, \qquad a_0(r_2) = 1, \qquad (7.5.24)$$

provided that we can show that the above series are convergent.   For $x < 0$, we replace x^{r_1} and x^{r_2} by $|x|^{r_1}$ and $|x|^{r_2}$.

We shall now proceed to establish the convergence of the above series.   We shall carry out the proof for the series in (7.5.23).   Also, the proof that we shall presently give is similar to the proof of Theorem 7.3.1.   Actually, the proof given below contains the proof of Theorem 7.3.1 as a special case.

Let us consider the series

$$\sum_{n=0}^{\infty} a_n(r)x^n, \tag{7.5.25}$$

where $a_0(r) = 1$ and the $a_n(r)$ are given recursively by $f(r + n)a_n(r) = -F_n(r)$, that is,

$$f(r + n)a_n(r) = -\sum_{k=0}^{n-1} [(k + r)p_{n-k} + q_{n-k}]a_k(r), \qquad n = 1, 2, \ldots . \tag{7.5.26}$$

Applying the inequality $|a + b| \geq |a| - |b|$ to (7.5.21) and (7.5.22), we see that

$$|f(r_1 + n)| = n|n + (r_1 - r_2)| \geq n(n - |r_1 - r_2|) \tag{7.5.27}$$

and

$$|f(r_2 + n)| = n|n + (r_2 - r_1)| \geq n(n - |r_2 - r_1|). \tag{7.5.28}$$

Let μ be any real number satisfying $0 < \mu < \rho$.   Since the series in (7.5.2) are convergent for $|x| = \mu$, it follows from Theorem 7.2.10 (with $x_0 = 0$) that there exists a constant $M > 0$ such that

$$|p_n| \leq \frac{M}{\mu^n} \quad \text{and} \quad |q_n| \leq \frac{M}{\mu^n}, \qquad n = 0, 1, 2, \ldots . \tag{7.5.29}$$

Using (7.5.29) with n replaced by $n - k$ and (7.5.27), and taking the absolute value of the recurrence relation in (7.5.26) with $r = r_1$, we obtain

$$n(n - |r_1 - r_2|)|a_n(r_1)| \leq \sum_{k=0}^{n-1} \left(\frac{M|k + r_1|}{\mu^{n-k}} + \frac{M}{\mu^{n-k}} \right)|a_k(r_1)|$$

$$\leq M \sum_{k=0}^{n-1} (k + |r_1| + 1)\mu^{k-n}|a_k(r_1)|,$$

$$n = 1, 2, \ldots . \tag{7.5.30}$$

Let N be an integer such that

$$N - 1 \leq |r_1 - r_2| < N. \tag{7.5.31}$$

Let us now define $A_0, A_1, \cdots, A_{N-1}$ by

$$A_0 = a_0(r_1) = 1, \quad A_1 = |a_1(r_1)|, \quad \ldots, \quad A_{n-1} = |a_{N-1}(r_1)|, \tag{7.5.32}$$

and

$$n(n - |r_1 - r_2|)A_n = M \sum_{k=0}^{n-1} (k + |r_1| + 1)\mu^{k-n}A_k, \quad n = N, N + 1, \ldots . $$

$$\tag{7.5.33}$$

Comparing (7.5.30) with (7.5.33) and utilizing mathematical induction, we see that

$$0 \leq |a_n(r_1)| \leq A_n, \qquad n = 0, 1, 2, \ldots. \qquad (7.5.34)$$

Let us consider the series

$$\sum_{n=0}^{\infty} A_n x^n. \qquad (7.5.35)$$

We shall now show that the series (7.5.35) converges for $|x| < \mu$. Replacing n by $n + 1$ in (7.5.33), we obtain

$$(n + 1)(n + 1 - |r_1 - r_2|)A_{n+1}$$

$$= M \sum_{k=0}^{n} (k + |r_1| + 1)\mu^{k-n-1}A_k, \qquad n \geq N. \qquad (7.5.36)$$

Summing out the term $k = n$ in (7.5.36), we obtain

$$(n + 1)(n + 1 - |r_1 - r_2|)A_{n+1}$$

$$= M \sum_{k=0}^{n-1} (k + |r_1| + 1)\mu^{k-n-1}A_k + M(n + |r_1| + 1)\mu^{-1}A_n$$

or

$$\mu(n + 1)(n + 1 - |r_1 - r_2|)A_{n+1}$$

$$= M \sum_{k=0}^{n-1} (k + |r_1| + 1)\mu^{k-n}A_k + M(n + |r_1| + 1)A_n. \qquad (7.5.37)$$

Using (7.5.33), we may write (7.5.37) as

$$\mu(n + 1)(n + 1 - |r_1 - r_2|)A_{n+1} = n(n - |r_1 - r_2|)A_n + M(n + |r_1| + 1)A_n.$$

Thus,

$$\frac{A_n}{A_{n+1}} = \frac{\mu(n + 1)(n + 1 - |r_1 - r_2|)}{n(n - |r_1 - r_2|) + M(n + |r_1| + 1)}. \qquad (7.5.38)$$

Hence,

$$\lim_{n \to \infty} \frac{A_n}{A_{n+1}} = \mu.$$

By Theorem 7.2.1, the series (7.5.35) converges for $|x| < \mu$. In view of (7.5.34) and Theorem 7.2.3, the series (7.5.25) with $r = r_1$ also converges for $|x| < \mu$. But μ was any real number satisfying $0 < \mu < \rho$. Consequently, the series

$$\sum_{n=0}^{\infty} a_n(r_1)x^n, \qquad a_0(r_1) = 1, \qquad (7.5.39)$$

converges for $|x| < \rho$.

The same argument as given above with r_1 replaced by r_2 everywhere shows that the series

$$\sum_{n=0}^{\infty} a_n(r_2)x^n, \qquad a_0(r_2) = 1, \tag{7.5.40}$$

converges for $|x| < \rho$.

The functions y_1 and y_2 given by

$$y_1(x) = |x|^{r_1} \sum_{n=0}^{\infty} a_n(r_1)x^n, \qquad a_0(r_1) = 1, \tag{7.5.41}$$

and

$$y_2(x) = |x|^{r_2} \sum_{n=0}^{\infty} a_n(r_2)x^n, \qquad a_0(r_2) = 1, \tag{7.5.42}$$

are linearly independent for $0 < |x| < \rho$.   (Why?)

In view of (7.5.2), we see that $p(0) = p_0$ and $q(0) = q_0$.   Thus, we may write the *indicial equation* (7.5.9) as

$$r(r - 1) + p(0)r + q(0) = 0. \tag{7.5.43}$$

Combining the above results, we have established the following essential theorem.

Theorem 7.5.1.   *Consider the differential equation*

$$x^2 y'' + xp(x)y' + q(x)y = 0, \tag{7.5.1}$$

where p and q are analytic at $x = 0$ and have power series expansions

$$p(x) = \sum_{n=0}^{\infty} p_n x^n, \qquad q(x) = \sum_{n=0}^{\infty} q_n x^n$$

which converge for $|x| < \rho$, $\rho > 0$.   Let r_1 and r_2 $(\mathscr{R}(r_1) \geqq \mathscr{R}(r_2))$ be the roots of the indicial equation

$$f(r) = r(r - 1) + p(0)r + q(0) = 0.$$

For $0 < |x| < \rho$, there is a solution y_1 of the form

$$y_1(x) = |x|^{r_1} \sum_{n=0}^{\infty} a_n x^n, \qquad a_0 = 1,$$

where the power series converges for $|x| < \rho$.   If $r_1 - r_2$ is not zero or a positive integer, then there is a second linearly independent solution y_2 for $0 < |x| < \rho$ of the form

$$y_2(x) = |x|^{r_2} \sum_{n=0}^{\infty} b_n x^n, \qquad b_0 = 1,$$

where the power series converges for $|x| < \rho$.

The coefficients a_n and b_n can be determined by substituting the solutions into the differential equation (7.5.1).

Note that whenever the $\mathscr{R}(r_1) > \mathscr{R}(r_2)$, the above theorem always guarantees us a Frobenius series solution of Equation (7.5.1) for the index r_1. However, in order to obtain a Frobenius series solution of Equation (7.5.1) for the index r_2, it required also that $r_1 - r_2$ was not equal to a positive integer. We now ask, what is the form of the second linearly independent solution of Equation (7.5.1) if $r_1 = r_2$ or if $r_1 - r_2$ is a positive integer? These are known as *exceptional cases*, and we shall now consider them.

Case 1. $r_1 = r_2$. We restrict our attention to $x > 0$. We shall first manipulate in a formal (nonrigorous) manner. We retain our previous work [from (7.5.1)] until we reach Equation (7.5.19)

$$L[y(x, r)] = a_0 f(r) x^r, \qquad a_0 \neq 0,$$

where $y(x, r)$ is given by (7.5.17). We now proceed in a similar fashion as in Example 7.4.4. [*See* (7.4.63), ...] Since $r = r_1$ is a root of multiplicity two of $f(r) = 0$, it follows that $f(r_1) = 0$ and $f'(r_1) = 0$. Differentiating (7.5.19) with respect to r, we obtain

$$\frac{\partial}{\partial r} L[y(x, r)] = L\left[\frac{\partial}{\partial r} y(x, r)\right] = a_0[f'(r) + f(r) \ln x] x^r.$$

Thus,

$$\frac{\partial}{\partial r} L[y(x, r)]\bigg|_{r=r_1} = L\left[\frac{\partial}{\partial r} y(x, r)\bigg|_{r=r_1}\right] = a_0[f'(r_1) + f(r_1) \ln x] x^{r_1} = 0.$$

Hence,

$$y_2(x) = \frac{\partial}{\partial r} y(x, r)\bigg|_{r=r_1}$$

is a solution of Equation (7.5.1) provided the series involved converge. Utilizing (7.5.17) we see that

$$\frac{\partial}{\partial r} y(x, r) = \frac{\partial}{\partial r}\left[x^r \sum_{n=0}^{\infty} a_n(r)x^n\right] = x^r \ln x \sum_{n=0}^{\infty} a_n(r)x^n + x^r \sum_{n=0}^{\infty} a_n'(r)x^n.$$

Hence,

$$y_2(x) = \frac{\partial}{\partial r} y(x, r)\bigg|_{r=r_1} = x^{r_1} \ln x \sum_{n=0}^{\infty} a_n(r_1)x^n + x^{r_1} \sum_{n=0}^{\infty} a_n'(r_1)x^n.$$

Taking $a_0(r_1) = 1$ and using the solution $y_1(x)$ already obtained, namely

$$y_1(x) = x^{r_1} \sum_{n=0}^{\infty} a_n(r_1)x^n, \qquad a_0(r_1) = 1,$$

we may write $y_2(x)$ in the form

$$y_2(x) = y_1(x) \ln x + x^{r_1} \sum_{n=0}^{\infty} a_n'(r_1)x^n. \tag{7.5.44}$$

For $x < 0$, we replace x^{r_1} and $\ln x$ by $|x|^{r_1}$ and $\ln |x|$.

Since $a_0(r_1) = 1$, it follows that $a_0'(r_1) = 0$. Thus, the series $\sum_{n=0}^{\infty} a_n'(r_1)x^n$ begins with a term x. In view of (7.5.16), the coefficient a_n is a rational function of r. Also, from (7.5.21) with $r_2 = r_1$, we see that $f(r_1 + n) \neq 0$, $n = 1, 2, \ldots$. Hence, $a_n'(r_1)$ exists for $n = 1, 2, \ldots$. Consequently, the series $\sum_{n=0}^{\infty} a_n'(r_1)x^n$ is well defined. These considerations suggest that in the case $r_1 = r_2$, Equation (7.5.1) has a second linearly independent solution y_2 of the form

$$y_2(x) = y_1(x) \ln |x| + |x|^{r_1+1} \sum_{n=0}^{\infty} b_n x^n \qquad (7.5.45)$$

valid for $0 < |x| < \rho$, $\rho > 0$, whose coefficients b_n may be determined by substituting $y_2(x)$ into Equation (7.5.1).

Remark 7.5.1. Once we have discovered, in a formal manner, the form of the second linearly independent solution $y_2(x)$ of (7.5.1), a method of justifying this solution is the following. We substitute $y_2(x)$ into Equation (7.5.1) and also utilize (7.5.2). This will enable us to determine the coefficients of the series involved. We may then give a proof of the convergence of this series fashioned after the convergence proof of series (7.5.39). However, we shall not give this proof.

Case 2. $r_1 - r_2 = m$, *where m is a positive integer.* As usual we first restrict our attention to $x > 0$ and manipulate in a formal manner. In view of (7.5.22), we see that if $r_1 - r_2 = m$, then

$$f(r_2 + n) = n(n - m). \qquad (7.5.46)$$

Thus, for $n = 1, 2, \ldots, m - 1$, $f(r_2 + n) \neq 0$. From (7.5.16), it is apparent that if a_0 is given, then the coefficients $a_n(r_2)$ will certainly exist for $n = 1$, $2, \ldots, m - 1$. However, when $n = m$, $f(r_2 + m) = 0$, and consequently we run into difficulty in computing $a_m(r_2)$. As we shall presently see, this difficulty can be circumvented.

Since $r_1 = r_2 + m$, using (7.5.20) we obtain

$$\begin{aligned}
f(r + m) &= (r + m - r_1)(r + m - r_2) \\
&= (r + m - r_2 - m)(r + m - r_2) = (r - r_2)(r + m - r_2). \qquad (7.5.47)
\end{aligned}$$

If the polynomial $P_m(r)$ in (7.5.16) vanishes when $r = r_2$, that is, $P_m(r)$ has $r - r_2$ as a factor, then using (7.5.47) we see that (for $r \neq r_2$)

$$\begin{aligned}
a_m(r) &= \frac{P_m(r)a_0}{f(r + m)f(r + m - 1) \cdots f(r + 2)f(r + 1)} \\
&= \frac{(r - r_2)Q_m(r)a_0}{(r - r_2)(r + m - r_2)f(r + m - 1) \cdots f(r + 2)f(r + 1)}.
\end{aligned}$$

Thus,

$$a_m(r) = \frac{Q_m(r)a_0}{(r + m - r_2)f(r + m - 1) \cdots f(r + 2)f(r + 1)}. \qquad (7.5.48)$$

Clearly, when $r = r_2$, $a_m(r_2)$ is a finite number. In view of (7.5.22), $f(r_2 + n) \neq 0$ for $n = m + 1, m + 2, \ldots$. Consequently, we may now continue to compute $a_{m+1}(r_2), a_{m+2}(r_2), \ldots$, which are all finite numbers. Hence, if $P_m(r)$ has $r - r_2$ as a factor, then a second solution y_2 of (7.5.1) for $x > 0$ has the form

$$y_2(x) = x^{r_2} \sum_{n=0}^{\infty} a_n(r_2)x^n, \qquad a_0(r_2) = 1, \qquad (7.5.49)$$

which is also a Frobenius series. For $x < 0$, we replace x^{r_2} by $|x|^{r_2}$. In view of Remark 7.5.1, we have for this rather special situation that a second solution y_2 of (7.5.1) has the form

$$y_2(x) = |x|^{r_2} \sum_{n=0}^{\infty} b_n x^n, \qquad b_0 = 1, \qquad (7.5.50)$$

valid for $0 < |x| < \rho$, $\rho > 0$, whose coefficients b_n may be determined by substituting the solution $y_2(x)$ in Equation (7.5.1). The reader may verify that the functions y_1 and y_2 given by

$$y_1(x) = |x|^{r_1} \sum_{n=0}^{\infty} a_n x^n \quad \text{and} \quad y_2(x) = |x|^{r_2} \sum_{n=0}^{\infty} b_n x^n \qquad (7.5.51)$$

are linearly independent for $0 < |x| < \rho$. Thus, we have obtained two linearly independent solutions of Equation (7.5.1).

If the polynomial $P_m(r)$ in (7.5.16) does not vanish when $r = r_2$, and this is the usual case in practice, it is still possible to determine $a_m(r_2)$. In fact, as we shall see, we can always determine $a_m(r_2)$. Utilizing (7.5.16) and (7.5.47) we have (for $r \neq r_2$)

$$a_m(r) = \frac{P_m(r)a_0}{f(r + m)f(r + m - 1) \cdots f(r + 2)f(r + 1)}$$

$$= \frac{P_m(r)a_0}{(r - r_2)(r + m - r_2)f(r + m - 1) \cdots f(r + 2)f(r + 1)}. \qquad (7.5.52)$$

Now, since a_0 is arbitrary, we may take $a_0(r) = r - r_2$. (This is the key point involved in solving Case 2 when $P_m(r_2) \neq 0$.) Substituting this value of a_0 in (7.5.52) and then cancelling out the common factor $r - r_2$, we see now that $a_m(r_2)$ is indeed a finite number. Let us define

$$y(x, r) = x^r \sum_{n=0}^{\infty} a_n(r)x^n, \qquad a_0(r) = r - r_2. \qquad (7.5.53)$$

Proceeding formally as in Case 1, we find that [*see* Equation (7.5.19) with a_0 replaced by $r - r_2$]

$$L[y(x, r)] = (r - r_2)f(r)x^r. \tag{7.5.54}$$

Substituting $r = r_2$ in (7.5.54), we obtain a solution y given by

$$y(x) = y(x, r_2). \tag{7.5.55}$$

If we now examine closely the expression

$$a_n(r) = -\frac{F_n(r)}{f(r+n)} = \frac{P_n(r)(r - r_2)}{f(r+n)f(r+n-1)\cdots f(r+2)f(r+1)}, \tag{7.5.56}$$

since $f(r_2 + n) \neq 0$ for $n = 1, 2, \ldots, m - 1$, we see that

$$a_1(r_2) = a_2(r_2) = \cdots = a_{m-1}(r_2) = 0. \tag{7.5.57}$$

Thus,

$$y(x) = y(x, r_2) = x^{r_2}\sum_{n=0}^{\infty} a_n(r_2)x^n = x^{r_2}\sum_{n=m}^{\infty} a_n(r_2)x^n.$$

Since $r_1 = r_2 + m$, we may write the above solution as

$$y(x) = x^{r_2}[a_m(r_2)x^m + a_{m+1}(r_2)x^{m+1} + a_{m+2}(r_2)x^{m+2} + \cdots]$$

$$= x^{r_2+m}[a_m(r_2) + a_{m+1}(r_2)x + a_{m+2}(r_2)x^2 + \cdots]$$

$$= x^{r_1}\sum_{n=0}^{\infty} b_n(r_2)x^n, \tag{7.5.58}$$

where

$$b_n(r_2) = a_{m+n}(r_2), \qquad n = 0, 1, 2, \ldots. \tag{7.5.59}$$

In view of the solution $y_1(x)$ already known, namely,

$$y_1(x) = x^{r_1}\sum_{n=0}^{\infty} a_n(r_1)x^n, \qquad a_0(r_1) = 1,$$

we leave it as an exercise for the reader to show that the solution y found in (7.5.58) is a constant multiple of $y_1(x)$. That is,

$$y(x) = cy_1(x) \tag{7.5.60}$$

for some constant c. Thus, even though our present discussion did not lead us to a second linearly independent solution of (7.5.1), the result obtained in (7.5.60) will be utilized below.

To obtain a second linearly independent solution y_2 of (7.5.1) for $x > 0$, we proceed in a formal manner as follows. Differentiating (7.5.54) with respect to r, we obtain

$$\frac{\partial}{\partial r} L[y(x, r)] = L\left[\frac{\partial}{\partial r} y(x, r)\right] = f(r)x^r + (r - r_2)[f'(r) + f(r)\ln x]x^r. \tag{7.5.61}$$

Substituting $r = r_2$ in (7.5.61), a solution y_2 of Equation (7.5.1) is given by

$$y_2(x) = \frac{\partial}{\partial r} y(x, r)\Big|_{r=r_2}. \tag{7.5.62}$$

Differentiating (7.5.53) with respect to r, we obtain

$$\frac{\partial}{\partial r} y(x, r) = x^r \ln x \sum_{n=0}^{\infty} a_n(r)x^n + x^r \sum_{n=0}^{\infty} a_n'(r)x^n, \qquad a_0(r) = r - r_2. \tag{7.5.63}$$

Thus, in view of (7.5.62) and (7.5.63), we see that

$$y_2(x) = x^{r_2} \ln x \sum_{n=0}^{\infty} a_n(r_2)x^n + x^{r_2} \sum_{n=0}^{\infty} a_n'(r_2)x^n. \tag{7.5.64}$$

From (7.5.57), $a_1(r_2) = a_2(r_2) = \cdots = a_{m-1}(r_2) = 0$. Utilizing the result of (7.5.60), we may write $y_2(x)$ of (7.5.64) as

$$y_2(x) = cy_1(x) \ln x + x^{r_2} \sum_{n=0}^{\infty} a_n'(r_2)x^n. \tag{7.5.65}$$

For $x < 0$, we replace x^{r_1}, x^{r_2}, and $\ln x$ by $|x|^{r_1}$, $|x|^{r_2}$, and $\ln |x|$. In view of Remark 7.5.1, a second solution y_2 of (7.5.1) has the form

$$y_2(x) = cy_1(x) \ln |x| + |x|^{r_2} \sum_{n=0}^{\infty} b_n x^n, \tag{7.5.66}$$

valid for $0 < |x| < \rho$, $\rho > 0$, where c is a constant (possibly zero) and the coefficients b_n may be determined by substituting the solution $y_2(x)$ into Equation (7.5.1). The reader may verify that the functions y_1 and y_2 given by

$$y_1(x) = |x|^{r_1} \sum_{n=0}^{\infty} a_n x^n \quad \text{and} \quad y_2(x) = cy_1(x) \ln |x| + |x|^{r_2} \sum_{n=0}^{\infty} b_n x^n \tag{7.5.67}$$

are linearly independent for $0 < |x| < \rho$. Thus, if $P_m(r_2) \neq 0$, we can always obtain two linearly independent solutions of (7.5.1). If when we substitute the solution $y_2(x)$ given by (7.5.67) into Equation (7.5.1), it turns out that $c = 0$, then the two linearly independent solutions of (7.5.1) will be of the form given by (7.5.51).

Let us summarize the results of Cases 1 and 2 in the following theorem.

Theorem 7.5.2. *Consider the differential equation*

$$x^2 y'' + xp(x)y' + q(x)y = 0, \tag{7.5.1}$$

where p and q are analytic at $x = 0$ and have power series expansions

$$p(x) = \sum_{n=0}^{\infty} p_n x^n, \qquad q(x) = \sum_{n=0}^{\infty} q_n x^n$$

which converge for $|x| < \rho$, $\rho > 0$. Let r_1 and r_2 ($\mathscr{R}(r_1) \geqq \mathscr{R}(r_2)$) be the roots of the indicial equation

$$f(r) = r(r-1) + p(0)r + q(0).$$

If $r_1 = r_2$, then there are two linearly independent solutions y_1, y_2 for $0 < |x| < \rho$ of the form

$$y_1(x) = |x|^{r_1} \sum_{n=0}^{\infty} a_n x^n, \qquad a_0 \neq 0,$$

$$y_2(x) = |x|^{r_1+1} \sum_{n=0}^{\infty} b_n x^n + y_1(x) \ln |x|,$$

where the power series converge for $|x| < \rho$.

If $r_1 - r_2$ is a positive integer m, then there are two linearly independent solutions y_1, y_2 for $0 < |x| < \rho$ of the form

$$y_1(x) = |x|^{r_1} \sum_{n=0}^{\infty} a_n x^n, \qquad a_0 \neq 0,$$

$$y_2(x) = |x|^{r_2} \sum_{n=0}^{\infty} b_n x^n + cy_1(x) \ln |x|, \qquad b_0 \neq 0,$$

where the power series converge for $|x| < \rho$, and c is a constant which may be zero or it may not.

The coefficients a_n and b_n can be determined by substituting the solutions into the differential equation (7.5.1).

Remark 7.5.2. Note that the roots of the indicial equation may be obtained in advance by utilizing Formula (7.5.43):

$$r(r-1) + p(0)r + q(0) = 0.$$

As an illustration, let us refer back to Example 7.4.3. Here $p(x) = -\frac{3}{2} + \frac{3}{2}x$ and $q(x) = 1$. Thus, $p(0) = -\frac{3}{2}$ and $q(0) = 1$. Substituting these values in (7.5.43), we obtain

$$2r^2 - 5r + 2 = 0.$$

The roots of this equation are $r = 2$ and $r = \frac{1}{2}$, which are the same as those given by Equation (7.4.31).

As an exercise, the reader should compute again the indicial roots found in Examples 7.4.4 through 7.4.6 by utilizing Formula (7.5.43).

7.6. The Gamma Function

For $n > 0$, the *gamma function* is defined by the convergent improper integral

$$\Gamma(n) = \int_0^\infty x^{n-1} e^{-x} \, dx. \tag{7.6.1}$$

The gamma function was first studied by Euler, and it is sometimes called the *second Eulerian integral.* This function is used in evaluating certain definite integrals which play an important role in theory and application. In Section 7.7, we shall study Bessel functions and we shall make use of the gamma function.

When $n = 1$, we obtain

$$\Gamma(1) = \int_0^\infty e^{-x} \, dx = -e^{-x} \Big|_0^\infty = 1. \tag{7.6.2}$$

For $n > 0$, we see by successive applications of L'Hospital's rule that

$$\lim_{x \to \infty} x^n e^{-x} = 0.$$

Thus, integrating by parts the following integral, we have for $n > 0$

$$\int_0^\infty x^n e^{-x} \, dx = -x^n e^{-x} \Big|_0^\infty + n \int_0^\infty x^{n-1} e^{-x} \, dx = n \int_0^\infty x^{n-1} e^{-x} \, dx. \tag{7.6.3}$$

In view of (7.6.1), we may write (7.6.3) as

$$\Gamma(n + 1) = n\Gamma(n), \qquad n > 0, \tag{7.6.4}$$

which may be regarded as a sort of reduction formula or recurrence formula.

The gamma function has been tabulated for values of $1 \le n \le 2$. [*See* [1], p. 267.] Using (7.6.4), we may now determine the values of $\Gamma(n)$ for $2 < n \le 3$, from which the values of $\Gamma(n)$ for $3 < n \le 4$ are determined, and so on. Thus, the values of $\Gamma(n)$ are determined for all $n \ge 1$.

From (7.6.4), we have for $n > 0$

$$\Gamma(n) = \frac{\Gamma(n + 1)}{n}. \tag{7.6.5}$$

Knowing the values of $\Gamma(n)$ for $1 \le n \le 2$, we may now use (7.6.5) to determine the values of $\Gamma(n)$ for $0 < n < 1$. Thus, the values of $\Gamma(n)$ are determined for all $n > 0$.

For $n \le 0$, the integral (7.6.1) diverges. Thus, $\Gamma(n)$ is not defined by (7.6.1) for $n \le 0$. However, if k is any positive integer, $\Gamma(n + k)$ is well defined for $n > -k$. In view of (7.6.4), we have for $n > -k$,

$$\Gamma(n + k) = (n + k - 1)\Gamma(n + k - 1) = (n + k - 1)(n + k - 2)\Gamma(n + k - 2)$$

$$= \cdots = (n + k - 1)(n + k - 2) \cdots (n + 1)n\Gamma(n).$$

Thus, we may define $\Gamma(n)$ for $-k < n < 0$, except for $n = 0, -1, -2, \ldots, -k$, by

$$\Gamma(n) = \frac{\Gamma(n + k)}{n(n + 1) \cdots (n + k - 1)}. \tag{7.6.6}$$

Now, if we repeat this process for $k = 1, 2, \ldots$, we may define $\Gamma(n)$ for all $n < 0$, except for $n = 0, -1, -2, \ldots$. Hence, the values of $\Gamma(n)$ are now determined for all n, except for $n = 0, -1, -2, \ldots$.

Example 7.6.1. Let us find the (approximate) value of

(1) $\Gamma(4.5)$,   (2) $\Gamma(0.4)$,   (3) $\Gamma(-0.6)$,   and   (4) $\Gamma(-3.8)$.

SOLUTION. From the table, $\Gamma(1.2) = 0.9182$, $\Gamma(1.4) = 0.8873$, and $\Gamma(1.5) = 0.8862$.

(1) $\Gamma(4.5) = (3.5)\Gamma(3.5) = (3.5)(2.5)\Gamma(2.5) = (3.5)(2.5)(1.5)\Gamma(1.5)$
$= (3.5)(2.5)(1.5)(0.8862) = 11.6314,$

(2) $\Gamma(0.4) = \dfrac{\Gamma(1.4)}{0.4} = \dfrac{0.8873}{0.4} = 2.2182,$

(3) $\Gamma(-0.6) = \dfrac{\Gamma(0.4)}{-0.6} = \dfrac{2.2182}{-0.6} = -3.6970,$

(4) $\Gamma(-3.8) = \dfrac{\Gamma(-2.8)}{-3.8} = \dfrac{\Gamma(-1.8)}{(-3.8)(-2.8)} = \dfrac{\Gamma(-0.8)}{(-3.8)(-2.8)(-1.8)}$

$= \dfrac{\Gamma(0.2)}{(-3.8)(-2.8)(-1.8)(-0.8)} = \dfrac{\Gamma(1.2)}{(-3.8)(-2.8)(-1.8)(-0.8)(0.2)}$

$= \dfrac{0.9182}{(-3.8)(-2.8)(-1.8)(-0.8)(0.2)} = 0.2996.$

Remark 7.6.1. One may generalize the gamma function for complex values of n. The integral (7.6.1) is known to be convergent for $\mathscr{R}(n) > 0$. The above formulas are also true for complex values of n. Clearly, in any inequality involving n, one replaces n by $\mathscr{R}(n)$.

Remark 7.6.2. When n is a positive integer, then successive applications of Formula (7.6.4) give

$$\Gamma(n + 1) = n\Gamma(n) = n(n - 1)\Gamma(n - 1) = n(n - 1)(n - 2)\Gamma(n - 2)$$

$$= \cdots = n(n - 1)(n - 2) \cdots 2 \cdot 1 \cdot \Gamma(1). \tag{7.6.7}$$

In view of (7.6.2), we may write (7.6.7) as

$$\Gamma(n + 1) = n!$$

or   (7.6.8)

$$\Gamma(n) = (n - 1)!.$$

Thus, when n is a positive integer, the gamma function reduces to the *factorial function.*   Since $\Gamma(1) = 1$, we define $0! = 1$.

Figure 7.6.1 depicts the graph of the gamma function for n real.

Example 7.6.2.   Let us show that

$$\Gamma(\tfrac{1}{2}) = \sqrt{\pi}.$$ (7.6.9)

SOLUTION.   Utilizing (7.6.1) with $n = \tfrac{1}{2}$, we have

$$\Gamma(\tfrac{1}{2}) = \int_0^\infty x^{-1/2} e^{-x}\, dx.$$

Let $x = u^2$, then $dx = 2u\, du$.   Thus,

$$\Gamma(\tfrac{1}{2}) = 2 \int_0^\infty e^{-u^2}\, du = 2 \int_0^\infty e^{-v^2}\, dv.$$ (7.6.10)

The following result is known to be true.   [*See* [45], pp. 656–665.]

$$\int_0^\infty e^{-u^2}\, du \cdot \int_0^\infty e^{-v^2}\, dv = \int_0^\infty \int_0^\infty e^{-(u^2 + v^2)}\, du\, dv.$$

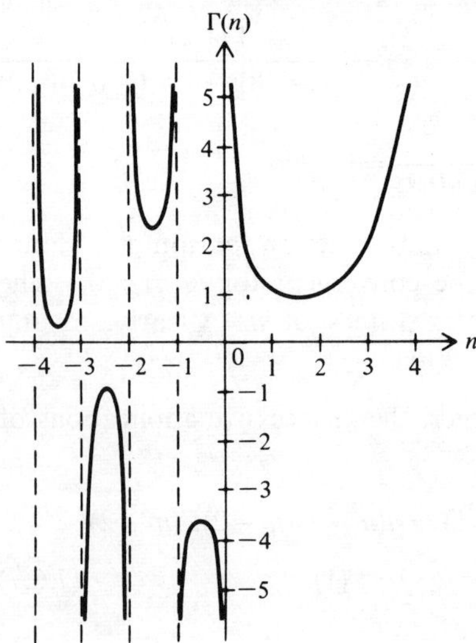

Figure 7.6.1.

Consequently, we have

$$[\Gamma(\tfrac{1}{2})]^2 = 4 \int_0^\infty \int_0^\infty e^{-(u^2+v^2)}\, du\, dv.$$

Let $u = r \cos \theta$ and $v = r \sin \theta$. We may then write the above expression as

$$[\Gamma(\tfrac{1}{2})]^2 = 4 \int_0^{\pi/2} \int_0^\infty e^{-r^2} r\, dr\, d\theta = -2 \int_0^{\pi/2} \left[e^{-r^2} \Big|_0^\infty \right] d\theta = 2 \int_0^{\pi/2} d\theta = \pi.$$

Therefore,

$$\Gamma(\tfrac{1}{2}) = \sqrt{\pi}.$$

Note that in view of (7.6.9) and (7.6.10), we have

$$\int_0^\infty e^{-x^2}\, dx = \tfrac{1}{2}\Gamma\left(\frac{1}{2}\right) = \frac{\sqrt{\pi}}{2}. \tag{7.6.11}$$

This integral plays an important part in the theory of probability.

Example 7.6.3. Let us show that when n is a positive integer,

$$\Gamma\left(n + \frac{1}{2}\right) = \frac{(2n)!}{2^{2n} n!} \sqrt{\pi}. \tag{7.6.12}$$

SOLUTION. By repeated applications of (7.6.4), we obtain

$$\Gamma\left(n + \frac{1}{2}\right) = \left(n - \frac{1}{2}\right)\left(n - \frac{3}{2}\right) \cdots \frac{3}{2} \cdot \frac{1}{2} \cdot \Gamma\left(\frac{1}{2}\right) = \frac{(2n-1)(2n-3)\cdots 3 \cdot 1}{2^n} \Gamma\left(\frac{1}{2}\right)$$

$$= \frac{2n(2n-1)(2n-2)(2n-3)\cdots 4 \cdot 3 \cdot 2 \cdot 1}{2^n[2n(2n-2)\cdots 4 \cdot 2]} \sqrt{\pi} = \frac{(2n)!}{2^{2n} n!} \sqrt{\pi}.$$

[*Note:* When $n = 0$, Formula (7.6.12) gives $\Gamma(\tfrac{1}{2}) = \sqrt{\pi}$, which agrees with (7.6.9).]

Exercises 7.6

1. From the table, $\Gamma(1.325) = 0.8940$ and $\Gamma(1.68) = 0.9050$. Find the (approximate) value of (a) $\Gamma(3.325)$ and (b) $\Gamma(-2.32)$.

2. By means of the substitution $y = \alpha x$, show that

$$\int_0^\infty x^n e^{-\alpha x}\, dx = \frac{\Gamma(n+1)}{\alpha^{n+1}}, \qquad n > -1,\ \alpha > 0.$$

3. The integral

   (a)
$$\int_0^1 x^m \left(\ln \frac{1}{x}\right)^n dx$$

is known to be finite and well defined for $m > -1$ and $n > -1$. By means of the substitution $y = -\ln x$, show that (a) becomes

(b)
$$\int_0^1 x^m \left(\ln \frac{1}{x} \right)^n dx = \frac{\Gamma(n+1)}{(m+1)^{n+1}}, \qquad m > -1, n > -1.$$

(c) Show that

$$\int_0^1 x^4 \left(\ln \frac{1}{x} \right)^3 dx = \frac{6}{625}, \qquad \int_0^1 \sqrt{x} \left(\ln \frac{1}{x} \right)^{-1/2} dx = \sqrt{\frac{2\pi}{3}}.$$

4. By means of the substitution $y = \alpha^2 x^2$, show that

$$\int_0^\infty x^n e^{-\alpha^2 x^2} dx = \frac{\Gamma\left(\dfrac{n+1}{2}\right)}{2\alpha^{n+1}}, \qquad n > -1, \alpha > 0.$$

5. Show that

$$\int_0^\infty e^{-\alpha^2 x^2} dx = \frac{\Gamma(\tfrac{1}{2})}{2\alpha} = \frac{\sqrt{\pi}}{2\alpha}, \qquad \alpha > 0.$$

This integral plays an important role in the theory of probability.

6. For $m > 0$ and $n > 0$, the *beta function* or *first Eulerian integral* is defined by the integral

(a)
$$B(m, n) = \int_0^1 x^{m-1}(1 - x)^{n-1} dx.$$

The beta function is known to be finite and well defined for $m > 0$ and $n > 0$. We shall express this important integral in terms of gamma functions.
   Let $y = 1 - x$. Show that

(b)
$$B(m, n) = B(n, m).$$

By means of the substitution $x = \dfrac{y}{1 + y}$, show that (a) becomes

(c)
$$B(m, n) = \int_0^\infty \frac{y^{m-1}}{(1 + y)^{m+n}} dy = \int_0^\infty \frac{x^{m-1}}{(1 + x)^{m+n}} dx.$$

Utilizing Exercise 2 with n replaced by $m - 1$, show that

(d)
$$\Gamma(m) = \int_0^\infty \alpha^m x^{m-1} e^{-\alpha x} dx.$$

Thus,

$$\Gamma(m)\alpha^{n-1} e^{-\alpha} = \int_0^\infty \alpha^{m+n-1} x^{m-1} e^{-\alpha(1+x)} dx.$$

The following result is known to be true.

$$\int_0^\infty \left[\int_0^\infty x^{m-1} \alpha^{m+n-1} e^{-\alpha(1+x)} \, dx \right] d\alpha = \int_0^\infty x^{m-1} \left[\int_0^\infty \alpha^{m+n-1} e^{-\alpha(1+x)} \, d\alpha \right] dx.$$

Consequently, we have

$$\Gamma(m) \int_0^\infty \alpha^{n-1} e^{-\alpha} \, d\alpha = \int_0^\infty x^{m-1} \left[\int_0^\infty \alpha^{m+n-1} e^{-\alpha(1+x)} \, d\alpha \right] dx.$$

For the integral within the bracket, utilizing Exercise 2 with x replaced by α, n replaced by $m+n-1$, and α replaced by $1+x$, show that the above expression can be written as

(e)
$$\int_0^\infty \frac{x^{m-1}}{(1+x)^{m+n}} \, dx = \frac{\Gamma(m)\Gamma(n)}{\Gamma(m+n)}, \qquad m > 0, n > 0,$$

or in view of (c),

(f)
$$B(m, n) = \int_0^1 x^{m-1}(1-x)^{n-1} \, dx = \frac{\Gamma(m)\Gamma(n)}{\Gamma(m+n)}, \qquad m > 0, n > 0.$$

Thus, the beta function is expressed in terms of gamma functions.

7. Show that

(a) $B\left(\dfrac{1}{2}, \dfrac{1}{2}\right) = \pi$,   (b) $B\left(\dfrac{3}{2}, \dfrac{1}{2}\right) = \dfrac{\pi}{2}$,   (c) $B\left(\dfrac{1}{3}, \dfrac{1}{4}\right) = \dfrac{7\Gamma(\frac{4}{3})\Gamma(\frac{5}{4})}{\Gamma(\frac{19}{12})}$.

8. Show that when n is a positive integer,

(a) $B\left(n, \dfrac{1}{2}\right) = \dfrac{2^{2n}(n!)^2}{n(2n)!}$,

(b) $B\left(n + \dfrac{1}{2}, \dfrac{1}{2}\right) = \dfrac{(2n)!}{2^{2n-1}(n!)^2} \dfrac{\pi}{2}$,

(c) $B\left(n + 1, \dfrac{1}{2}\right) = \dfrac{2^{2n+1}(n!)^2}{(2n+1)!}$,

(d) $B(n, n) = \dfrac{\sqrt{\pi}\,\Gamma(n)}{2^{2n-1}\Gamma(n + \frac{1}{2})}$.

9. Show that

$$\Gamma(m)\Gamma(1 - m) = \int_0^\infty \frac{x^{m-1}}{1+x} \, dx = \int_0^1 \frac{x^{m-1}}{(1-x)^m} \, dx, \qquad 0 < m < 1.$$

10. Show that

$$\int_0^\infty \frac{dx}{\sqrt{x(1+x)}} = \pi.$$

11. By means of the substitution $y = \sin x$, show that

(a)
$$\int_0^{\pi/2} \sin^n x \, dx = \int_0^1 y^n (1 - y^2)^{-1/2} \, dy, \qquad n > -1.$$

By means of the substitution $z = y^2$, show that

(b)
$$\int_0^1 y^n(1-y^2)^{-1/2}\, dy = \frac{1}{2} B\left(\frac{n+1}{2}, \frac{1}{2}\right) = \frac{1}{2}\frac{\Gamma\left(\frac{n+1}{2}\right)\Gamma\left(\frac{1}{2}\right)}{\Gamma\left(\frac{n}{2}+1\right)}$$

$$= \frac{\sqrt{\pi}}{2}\frac{\Gamma\left(\frac{n+1}{2}\right)}{\Gamma\left(\frac{n}{2}+1\right)}, \qquad n > -1.$$

Thus,

(c)
$$\int_0^{\pi/2} \sin^n x\, dx = \frac{\sqrt{\pi}}{2}\frac{\Gamma\left(\frac{n+1}{2}\right)}{\Gamma\left(\frac{n}{2}+1\right)}, \qquad n > -1.$$

By means of the substitution $y = \dfrac{\pi}{2} - x$, show that

(d)
$$\int_0^{\pi/2} \sin^n x\, dx = \int_0^{\pi/2} \cos^n x\, dx.$$

12. Show that when n is a positive integer,

(a)
$$\int_0^{\pi/2} \cos^{2n}x\, dx = \int_0^{\pi/2} \sin^{2n} x\, dx = \frac{(2n)!}{2^{2n}(n!)^2}\frac{\pi}{2} = \frac{1}{2} B\left(n+\frac{1}{2}, \frac{1}{2}\right),$$

(b)
$$\int_0^{\pi/2} \cos^{2n+1} x\, dx = \int_0^{\pi/2} \sin^{2n+1} x\, dx = \frac{2^{2n}(n!)^2}{(2n+1)!} = \frac{1}{2} B\left(n+1, \frac{1}{2}\right).$$

13. By means of the substitution $x = \sin y$, show that

(a)
$$\int_0^1 \frac{x^{2n}}{\sqrt{1-x^2}}\, dx = \int_0^{\pi/2} \sin^{2n} y\, dy = \frac{\sqrt{\pi}\,\Gamma(n+\frac{1}{2})}{2\,\Gamma(n+1)}, \qquad n > -\frac{1}{2},$$

(b)
$$\int_0^1 \frac{x^{2n+1}}{\sqrt{1-x^2}}\, dx = \int_0^{\pi/2} \sin^{2n+1} y\, dy = \frac{\sqrt{\pi}\,\Gamma(n+1)}{2\,\Gamma(n+\frac{3}{2})}, \qquad n > -1.$$

Thus, when n is a positive integer, we obtain

(c)
$$\int_0^1 \frac{x^{2n}}{\sqrt{1-x^2}}\, dx = \frac{1}{2} B\left(n+\frac{1}{2}, \frac{1}{2}\right),$$

(d)
$$\int_0^1 \frac{x^{2n+1}}{\sqrt{1-x^2}}\, dx = \frac{1}{2} B\left(n+1, \frac{1}{2}\right).$$

14. If $0 < x < \dfrac{\pi}{2}$ and n is a positive integer, then it follows that

(a) $$\sin^{2n+1} x < \sin^{2n} x < \sin^{2n-1} x.$$

Hence,

(b) $$\int_0^{\pi/2} \sin^{2n+1} x \, dx < \int_0^{\pi/2} \sin^{2n} x \, dx < \int_0^{\pi/2} \sin^{2n-1} x \, dx.$$

Using Exercise 12, show that (b) can be expressed as

(c) $$\sqrt{\frac{2n}{2n+1}} \sqrt{\frac{\pi}{2}} < \frac{2^{2n}(n!)^2}{(2n+1)!} \sqrt{2n+1} < \sqrt{\frac{\pi}{2}}.$$

Show that

(d) $$\lim_{n \to \infty} \frac{2^{2n}(n!)^2}{(2n+1)!} \sqrt{2n+1} = \sqrt{\frac{\pi}{2}}.$$

The expression in (d) is called *Wallis' formula*. Show that (d) can also be written as

(e) $$\lim_{n \to \infty} \left[\frac{2 \cdot 4 \cdot 6 \cdots (2n)}{1 \cdot 3 \cdot 5 \cdots (2n-1)} \right]^2 \frac{1}{2n+1} = \frac{\pi}{2}.$$

15. By means of the substitution $y = x^n$, $n > 0$, show that

(a) $$\int_0^1 x^m (1 - x^n)^p \, dx = \frac{1}{n} \frac{\Gamma(p+1)\Gamma\!\left(\dfrac{m+1}{n}\right)}{\Gamma\!\left(p + 1 + \dfrac{m+1}{n}\right)}, \qquad m > -1, p > -1.$$

Show that

(b) $$\int_0^1 x^{1/2}(1 - x^3)^{-1/2} \, dx = \frac{\pi}{3},$$

(c) $$\int_0^1 \sqrt{x}\sqrt{1 - \sqrt{x}} \, dx = \tfrac{32}{105},$$

(d) $$\int_0^1 x^5 (1 - x^2)^4 \, dx = \tfrac{1}{105}.$$

7.7. Bessel's Equation and Bessel Functions

One of the most important differential equations, one with widespread applications and theoretical interest, is

$$x^2 y'' + xy' + (x^2 - \alpha^2)y = 0, \tag{7.7.1}$$

where the number α may have any value, positive or negative, integral, fraction, or irrational; or even complex. Equation (7.7.1) is called *Bessel's equation of order (or index)* α. [*See* also Exercise 4.5.8.]

We have already solved Equation (7.7.1) when $\alpha = 0$ and $\alpha = 1$. [*See* Exercise 7.4.18 and Example 7.4.6.] Comparing (7.7.1) with (7.5.1), we see that $p(x) = 1$ and $q(x) = x^2 - \alpha^2$. Thus, the functions p and q are both analytic at $x = 0$, and in view of Remark 7.3.4, their power series expansions converge for all $|x| < \infty$. Clearly, $x = 0$ is a regular singular point of (7.7.1). Since $p(0) = 1$ and $q(0) = -\alpha^2$, the indicial equation of (7.7.1) is [*see* (7.5.43)]

$$r(r - 1) + r - \alpha^2 = 0 \quad \text{or} \quad r^2 - \alpha^2 = 0,$$

and the indicial roots are

$$r_1 = \alpha \quad \text{and} \quad r_2 = -\alpha. \tag{7.7.2}$$

We shall first find solutions of (7.7.1) when $\alpha \neq 0$ and $\mathscr{R}(\alpha) \geq 0$. Also, as usual, we shall first restrict our attention to $x > 0$. For $r_1 = \alpha$, in view of Theorem 7.5.1, there exists a solution y_1 of (7.7.1) of the form

$$y_1(x) = x^\alpha \sum_{n=0}^{\infty} a_n x^n, \qquad a_0 \neq 0, \tag{7.7.3}$$

where the coefficients a_n are determined by substituting the solution into (7.7.1). Differentiating (7.7.3) twice, we obtain

$$y_1'(x) = x^\alpha \sum_{n=0}^{\infty} (\alpha + n)a_n x^{n-1}, \qquad y_1''(x) = x^\alpha \sum_{n=0}^{\infty} (\alpha + n)(\alpha + n - 1)a_n x^{n-2}.$$

Substituting the series for y_1, y_1', and y_1'' into (7.7.1), we obtain upon the usual simplifications

$$x^\alpha \sum_{n=0}^{\infty} n(2\alpha + n)a_n x^n + x^\alpha \sum_{n=2}^{\infty} a_{n-2} x^n = 0. \tag{7.7.4}$$

Summing out the first two terms from the first series, (7.7.4) becomes

$$0 \cdot a_0 x^\alpha + (2\alpha + 1)a_1 x^{\alpha+1} + x^\alpha \sum_{n=2}^{\infty} [n(2\alpha + n)a_n + a_{n-2}]x^n = 0. \tag{7.7.5}$$

Thus, a_0 is arbitrary, $a_1 = 0$, and the recurrence formula is

$$a_n = -\frac{a_{n-2}}{n(2\alpha + n)}, \qquad n \geq 2. \tag{7.7.6}$$

Since $a_1 = 0$, it follows from (7.7.6) that

$$a_{2n+1} = 0, \qquad n = 1, 2, \dots . \tag{7.7.7}$$

That is, all the coefficients a_n with odd subscripts are equal to zero. Substituting $n = 2, 4, 6, \ldots$ in (7.7.6), we obtain

$$a_2 = -\frac{a_0}{2(2\alpha + 2)} = -\frac{a_0}{2^2(\alpha + 1)},$$

$$a_4 = -\frac{a_2}{4(2\alpha + 4)} = \frac{a_0}{2^4 2!(\alpha + 1)(\alpha + 2)},$$

$$a_6 = -\frac{a_4}{6(2\alpha + 6)} = -\frac{a_0}{2^6 3!(\alpha + 1)(\alpha + 2)(\alpha + 3)},$$

and by mathematical induction, one may easily show that

$$a_{2n} = \frac{(-1)^n a_0}{2^{2n} n!(\alpha + 1)(\alpha + 2) \cdots (\alpha + n)}, \qquad n \geq 1. \qquad (7.7.8)$$

In view of (7.7.3), (7.7.7), and (7.7.8), one solution of Bessel's equation for $x > 0$ is

$$y_1(x) = a_0 x^\alpha \left[1 + \sum_{n=1}^{\infty} \frac{(-1)^n x^{2n}}{2^{2n} n!(\alpha + 1)(\alpha + 2) \cdots (\alpha + n)} \right]. \qquad (7.7.9)$$

Since we are seeking a particular solution of (7.7.1) and a_0 is arbitrary, if we wish, we may choose $a_0 = 1$. However, it is customary to express a_0 in terms of the gamma function, namely,

$$a_0 = \frac{1}{2^\alpha \Gamma(\alpha + 1)}. \qquad (7.7.10)$$

Substituting this value of a_0 in (7.7.9), and then noticing that

$$\Gamma(n + \alpha + 1) = (n + \alpha)(n + \alpha - 1) \cdots (\alpha + 2)(\alpha + 1)\Gamma(\alpha + 1), \qquad (7.7.11)$$

we obtain

$$y_1(x) = \left(\frac{x}{2}\right)^\alpha \sum_{n=0}^{\infty} \frac{(-1)^n}{n! \, \Gamma(n + \alpha + 1)} \left(\frac{x}{2}\right)^{2n}, \qquad \mathscr{R}(\alpha) \geq 0. \qquad (7.7.12)$$

This solution of the Bessel equation is denoted by $J_\alpha(x)$. That is,

$$J_\alpha(x) = \left(\frac{x}{2}\right)^\alpha \sum_{n=0}^{\infty} \frac{(-1)^n}{n! \, \Gamma(n + \alpha + 1)} \left(\frac{x}{2}\right)^{2n}, \qquad \mathscr{R}(\alpha) \geq 0, \qquad (7.7.13)$$

and J_α is called *the Bessel function of the first kind of order* (*or index*) α.

To find the solution of (7.7.1) for $x < 0$, we replace x^α by $|x|^\alpha$ in (7.7.13).

Remark 7.7.1. From the result of Exercise 7.4.18, we have that

$$y_1(x) = \sum_{n=0}^{\infty} \frac{(-1)^n x^{2n}}{2^{2n}(n!)^2} = \sum_{n=0}^{\infty} \frac{(-1)^n}{(n!)^2} \left(\frac{x}{2}\right)^{2n}. \qquad (7.7.14)$$

Setting $\alpha = 0$ in (7.7.11), we obtain

$$\Gamma(n + 1) = n(n - 1) \cdots 2 \cdot 1 \cdot \Gamma(1) = n!.$$

Thus, setting $\alpha = 0$ in (7.7.13), we have

$$J_0(x) = \sum_{n=0}^{\infty} \frac{(-1)^n}{(n!)^2} \left(\frac{x}{2}\right)^{2n}, \tag{7.7.15}$$

which agrees with the result in (7.7.14).

We are now left with two cases to be considered. First, we consider the case when $r_1 - r_2 = 2\alpha$ is not a positive integer. In view of Theorem 7.5.1, there exists another solution of (7.7.1) of the form

$$y_2(x) = x^{-\alpha} \sum_{n=0}^{\infty} b_n x^n \qquad b_0 \neq 0. \tag{7.7.16}$$

A close observation of the calculations which led to $J_\alpha(x)$ of (7.7.13) reveals that they carry over for $r_2 = -\alpha$ with no change. Thus, if $r_1 - r_2 = 2\alpha$ is not a positive integer, a second solution of (7.7.1) is

$$J_{-\alpha}(x) = \left(\frac{x}{2}\right)^{-\alpha} \sum_{n=0}^{\infty} \frac{(-1)^n}{n! \Gamma(n - \alpha + 1)} \left(\frac{x}{2}\right)^{2n}. \tag{7.7.17}$$

Observe that if α is not a positive integer, then $\Gamma(n - \alpha + 1)$ exists for $n = 0$, $1, 2, \ldots$, and thus $J_{-\alpha}(x)$ is well defined. [*Note:* $J_\alpha(x)$ may exist even if $r_1 - r_2 = 2\alpha$ is a positive integer; for example, if 2α is equal to an odd positive integer.]

Summarizing the above findings and using Theorem 7.5.1, we have the following result.

Theorem 7.7.1. *If α is not zero or a positive integer, then for $0 < |x| < \infty$, the two linearly independent solutions J_α and $J_{-\alpha}$ of the Bessel equation (7.7.1) are given by*

$$J_\alpha(x) = \left|\frac{x}{2}\right|^\alpha \sum_{n=0}^{\infty} \frac{(-1)^n}{n! \Gamma(n + \alpha + 1)} \left(\frac{x}{2}\right)^{2n} \tag{7.7.18}$$

and

$$J_{-\alpha}(x) = \left|\frac{x}{2}\right|^{-\alpha} \sum_{n=0}^{\infty} \frac{(-1)^n}{n! \Gamma(n - \alpha + 1)} \left(\frac{x}{2}\right)^{2n}. \tag{7.7.19}$$

We shall now consider the other case when α is a positive integer, say $\alpha = m$. Let us restrict our attention to $x > 0$. In view of Theorem 7.5.2 and (7.7.13) with α replaced by m, the two solutions of Equation (7.7.1) are given by

$$y_1(x) = J_m(x) = \left(\frac{x}{2}\right)^m \sum_{k=0}^{\infty} \frac{(-1)^k}{k! \Gamma(k + m + 1)} \left(\frac{x}{2}\right)^{2k} \tag{7.7.20}$$

and

$$y_2(x) = x^{-m} \sum_{n=0}^{\infty} b_n x^n + cJ_m(x) \ln x, \qquad b_0 \neq 0. \qquad (7.7.21)$$

Our task now is to determine the coefficients b_n and the constant c. Since m is a positive integer, $\Gamma(k + m + 1) = (k + m)!$, and we may write $J_m(x)$ as

$$J_m(x) = x^m \sum_{k=0}^{\infty} c_{2k} x^{2k}, \qquad (7.7.22)$$

where

$$c_{2k} = \frac{(-1)^k}{2^{2k+m}k!(k + m)!}. \qquad (7.7.23)$$

Differentiating (7.7.21) twice, we obtain

$$y_2'(x) = x^{-m} \sum_{n=0}^{\infty} (n - m)b_n x^{n-1} + \frac{cJ_m(x)}{x} + cJ_m'(x) \ln x,$$

$$y_2''(x) = x^{-m} \sum_{n=0}^{\infty} (n - m)(n - m - 1)b_n x^{n-2} - \frac{cJ_m(x)}{x^2} + \frac{2cJ_m'(x)}{x} + cJ_m''(x) \ln x.$$

Substituting the series for y_2, y_2', and y_2'' into (7.7.1) with $\alpha^2 = m^2$, we obtain upon the usual simplifications

$$0 \cdot b_0 x^{-m} + (1 - 2m)b_1 x^{1-m} + x^{-m} \sum_{n=2}^{\infty} [n(n - 2m)b_n + b_{n-2}]x^n$$

$$+ 2cxJ_m'(x) + c[x^2 J_m''(x) + xJ_m'(x) + (x^2 - m^2)J_m(x)] \ln x = 0.$$

$$(7.7.24)$$

Since $J_m(x)$ is a solution of Equation (7.7.1), the above expression in the bracket vanishes. Thus, upon multiplying (7.7.24) by x^m, we now obtain

$$0 \cdot b_0 + (1 - 2m)b_1 x + \sum_{n=2}^{\infty} [n(n - 2m)b_n + b_{n-2}]x^n = -2cx^{m+1}J_m'(x).$$

$$(7.7.25)$$

Differentiating (7.7.22) and then substituting $J_m'(x)$ into (7.7.25), we obtain

$$0 \cdot b_0 + (1 - 2m)b_1 x + \sum_{n=2}^{\infty} [n(n - 2m)b_n + b_{n-2}]x^n$$

$$= -2c \sum_{k=0}^{\infty} (m + 2k)c_{2k} x^{2m+2k}. \qquad (7.7.26)$$

Since m is a positive integer, we see that b_0 is arbitrary, $b_1 = 0$, and for $m > 1$

$$n(n - 2m)b_n + b_{n-2} = 0, \qquad n = 2, 3, \ldots, 2m - 1. \qquad (7.7.27)$$

Since $b_1 = 0$, it follows from (7.7.27) that

$$b_3 = b_5 = \cdots = b_{2m-1} = 0. \tag{7.7.28}$$

Since b_0 is arbitrary, substituting $n = 2, 4, 6, \ldots$ in (7.7.27), we find

$$b_2 = \frac{b_0}{2(2m-2)} = \frac{b_0}{2^2(m-1)},$$

$$b_4 = \frac{b_2}{4(2m-4)} = \frac{b_0}{2^4 2!(m-1)(m-2)},$$

$$b_6 = \frac{b_4}{6(2m-6)} = \frac{b_0}{2^6 3!(m-1)(m-2)(m-3)},$$

and by mathematical induction, we see that

$$b_{2j} = \frac{b_0}{2^{2j}j!(m-1)(m-2)\cdots(m-j)}, \qquad j = 1, 2, \ldots, m-1. \tag{7.7.29}$$

We shall now determine the constant c. Comparing the coefficients of x^{2m} in (7.7.26), obtained by setting $n = 2m$ and $k = 0$, we see that

$$b_{2m-2} = -2cmc_0. \tag{7.7.30}$$

From (7.7.29) with $j = m - 1$, we obtain

$$b_{2m-2} = \frac{b_0}{2^{2m-2}(m-1)!(m-1)(m-2)\cdots 1} = \frac{b_0}{2^{2m-2}[(m-1)!]^2}. \tag{7.7.31}$$

From (7.7.23) with $k = 0$, we have

$$c_0 = \frac{1}{2^m m!}. \tag{7.7.32}$$

Substituting these values of b_{2m-2} and c_0 in (7.7.30) and then solving for c, we find

$$c = -\frac{b_0}{2^{m-1}(m-1)!}. \tag{7.7.33}$$

Since the series in the right-hand member of (7.7.26) contains only even powers of x, it follows that

$$b_{2m+1} = b_{2m+3} = \cdots = 0. \tag{7.7.34}$$

Since $b_1 = 0$, in view of (7.7.28) and (7.7.34), we see that all the coefficients b_n with odd subscripts are zero. It is apparent from (7.7.26) and (7.7.29) that the coefficient b_{2m} is undetermined. However, the remaining coefficients

$b_{2m+2}, b_{2m+4}, \ldots$ can be determined as follows. Comparing the coefficients of x^{2m+2k} in (7.7.26), obtained by setting $n = 2m + 2k$, we see that

$$2k(2m + 2k)b_{2m+2k} + b_{2m+2k-2} = -2c(m + 2k)c_{2k}, \qquad k = 1, 2, \ldots. \tag{7.7.35}$$

For $k = 1$, we have

$$b_{2m+2} = -\frac{2c(m + 2)c_2}{2(2m + 2)} - \frac{b_{2m}}{2(2m + 2)},$$

which can also be written as

$$b_{2m+2} = -\frac{cc_2}{2}\left(1 + \frac{1}{m + 1}\right) - \frac{b_{2m}}{4(m + 1)}. \tag{7.7.36}$$

It is convenient to choose

$$b_{2m} = -\frac{cc_0}{2} H_m, \tag{7.7.37}$$

where

$$H_m = 1 + \frac{1}{2} + \cdots + \frac{1}{m}. \tag{7.7.38}$$

With this choice of b_{2m}, (7.7.36) becomes

$$b_{2m+2} = -\frac{cc_2}{2}\left(1 + \frac{1}{m + 1}\right) + \frac{cc_0}{2}\frac{H_m}{4(m + 1)}. \tag{7.7.39}$$

From (7.7.23) and (7.7.32), we have that

$$c_2 = -\frac{1}{2^{2+m}(m + 1)!} = -\frac{1}{2^2(m + 1)2^m m!} = -\frac{c_0}{4(m + 1)},$$

or

$$c_0 = -4(m + 1)c_2. \tag{7.7.40}$$

Substituting this value of c_0 in (7.7.39), we obtain

$$b_{2m+2} = -\frac{cc_2}{2}\left(1 + \frac{1}{m + 1} + H_m\right) = -\frac{cc_2}{2}(1 + H_{m+1}). \tag{7.7.41}$$

Note:

$$H_{m+1} = 1 + \frac{1}{2} + \cdots + \frac{1}{m} + \frac{1}{m + 1}.$$

Substituting $k = 2$ in (7.7.35), and also utilizing (7.7.41), we get

$$b_{2m+4} = -\frac{2c(m+4)c_4}{4(2m+4)} - \frac{b_{2m+2}}{4(2m+4)} = -\frac{cc_4}{2}\left(\frac{1}{2} + \frac{1}{m+2}\right) - \frac{b_{2m+2}}{4(2m+4)}$$

$$= -\frac{cc_4}{2}\left(\frac{1}{2} + \frac{1}{m+2}\right) + \frac{cc_2}{2}\frac{(1 + H_{m+1})}{2^2 2!(m+2)}. \tag{7.7.42}$$

From (7.7.23), we have

$$c_4 = \frac{1}{2^{4+m}2!(2+m)!} = \frac{1}{2^2 2!(m+2)[2^{2+m}(m+1)!]} = -\frac{c_2}{2^2 2!(m+2)},$$

or

$$c_2 = -2^2 2!(m+2)c_4. \tag{7.7.43}$$

Substituting this value of c_2 in (7.7.42), we obtain

$$b_{2m+4} = -\frac{cc_4}{2}\left(\frac{1}{2} + \frac{1}{m+2}\right) - \frac{cc_4}{2}(1 + H_{m+1})$$

$$= -\frac{cc_4}{2}\left(1 + \frac{1}{2} + H_{m+1} + \frac{1}{m+2}\right)$$

$$= -\frac{cc_4}{2}\left(1 + \frac{1}{2} + H_{m+2}\right). \tag{7.7.44}$$

By mathematical induction, we leave it as an exercise for the reader [*see* Exercise 7.7.1] to show that

$$b_{2m+2k} = -\frac{cc_{2k}}{2}(H_k + H_{m+k}), \tag{7.7.45}$$

where

$$H_{m+k} = 1 + \frac{1}{2} + \cdots + \frac{1}{m} + \frac{1}{m+1} + \cdots + \frac{1}{m+k}, \qquad k = 1, 2, \ldots. \tag{7.7.46}$$

Utilizing (7.7.28), (7.7.29), (7.7.34), (7.7.37), and (7.7.45), we may now write the solution $y_2(x)$ of (7.7.21) in the form

$$y_2(x) = b_0 x^{-m} + x^{-m}\sum_{j=1}^{m-1} b_{2j}x^{2j} + x^{-m}b_{2m}x^{2m}$$

$$+ x^{-m}\sum_{k=1}^{\infty} b_{2m+2k}x^{2m+2k} + cJ_m(x)\ln x$$

$$= b_0 x^{-m} + b_0 x^{-m}\sum_{j=1}^{m-1}\frac{x^{2j}}{2^{2j}j!(m-1)(m-2)\cdots(m-j)} - \frac{cc_0}{2}H_m x^m$$

$$- \frac{c}{2}\sum_{k=1}^{\infty} c_{2k}(H_k + H_{m+k})x^{m+2k} + cJ_m(x)\ln x, \tag{7.7.47}$$

where the coefficients c_{2k}, the constant c, the partial harmonic series H_m, and H_{m+k} are given, respectively, by (7.7.23), (7.7.33), (7.7.38), and (7.7.46).

For the particular choice of $c = 1$, the function y_2 in (7.7.47) is customarily denoted by K_m. In view of (7.7.33), we then have that

$$b_0 = -2^{m-1}(m-1)!. \tag{7.7.48}$$

Substituting $c = 1$, the values of c_{2k}, c_0, and b_0 of (7.7.23), (7.7.32), and (7.7.48) into (7.7.47), and then simplifying, we obtain

$$K_m(x) = -\frac{1}{2}\left(\frac{x}{2}\right)^{-m}\sum_{j=0}^{m-1}\frac{(m-j-1)!}{j!}\left(\frac{x}{2}\right)^{2j} - \frac{1}{2}\frac{H_m}{m!}\left(\frac{x}{2}\right)^{m}$$
$$-\frac{1}{2}\left(\frac{x}{2}\right)^{m}\sum_{k=1}^{\infty}\frac{(-1)^k(H_k + H_{m+k})}{k!(k+m)!}\left(\frac{x}{2}\right)^{2k} + J_m(x)\ln x. \tag{7.7.49}$$

The expression K_m is called the *Bessel function of the second kind of order* (or *index*) *m*.

To find the solution of (7.7.1) for $x < 0$, we replace x^{-m}, x^m, and $\ln x$ by $|x|^{-m}$, $|x|^m$, and $\ln|x|$ in (7.7.49).

Remark 7.7.2. If for $m = 0$, we define $H_0 = 0$ and the first sum in (7.7.49) to be zero, then (7.7.49) becomes

$$K_0(x) = -\frac{1}{2}\sum_{k=1}^{\infty}\frac{(-1)^k(H_k + H_k)}{k!\,k!}\left(\frac{x}{2}\right)^{2k} + J_0(x)\ln x$$
$$= -\sum_{k=1}^{\infty}\frac{(-1)^k H_k}{2^{2k}(k!)^2}x^{2k} + J_0(x)\ln x, \tag{7.7.50}$$

which agrees with the solution $y_2(x)$ of Exercise 7.4.18. [*See* also Remark 7.7.1.]

Summarizing the above findings and using Theorem 7.5.2, we have the following result.

Theorem 7.7.2. *If α is a positive integer m, then for $0 < |x| < \infty$, the two linearly independent solutions J_m and K_m of the Bessel equation (7.7.1) are given by*

$$J_m(x) = \left|\frac{x}{2}\right|^{m}\sum_{k=0}^{\infty}\frac{(-1)^k}{k!(m+k)!}\left(\frac{x}{2}\right)^{2k} \tag{7.7.51}$$

and

$$K_m(x) = -\frac{1}{2}\left|\frac{x}{2}\right|^{-m}\left\{\sum_{j=0}^{m-1}\frac{(m-j-1)!}{j!}\left(\frac{x}{2}\right)^{2j} + \frac{H_m}{m!}\left(\frac{x}{2}\right)^{2m}\right.$$
$$\left. + \left(\frac{x}{2}\right)^{2m}\sum_{k=1}^{\infty}\frac{(-1)^k(H_k + H_{m+k})}{k!(m+k)!}\left(\frac{x}{2}\right)^{2k}\right\}$$
$$+ J_m(x)\ln|x|. \tag{7.7.52}$$

If $\alpha = 0$, then for $0 < |x| < \infty$, the two linearly independent solutions J_0 and K_0 are given by

$$J_0(x) = \sum_{k=0}^{\infty} \frac{(-1)^k}{(k!)^2} \left(\frac{x}{2}\right)^{2k} \tag{7.7.53}$$

and

$$K_0(x) = -\sum_{k=1}^{\infty} \frac{(-1)^k H_k}{(k!)^2} \left(\frac{x}{2}\right)^{2k} + J_0(x) \ln |x|. \tag{7.7.54}$$

Example 7.7.1.   Let us show that

$$J_{1/2}(x) = \sqrt{\frac{2}{\pi x}} \sin x, \qquad x > 0. \tag{7.7.55}$$

SOLUTION.   With $\alpha = 1/2$, (7.7.13) becomes

$$J_{1/2}(x) = \sqrt{\frac{x}{2}} \sum_{n=0}^{\infty} \frac{(-1)^n x^{2n}}{2^{2n} n! \, \Gamma\left(\dfrac{2n+3}{2}\right)}. \tag{7.7.56}$$

By a repeated application of Formula (7.6.4), we obtain

$$\Gamma\left(\frac{2n+3}{2}\right) = \frac{2n+1}{2} \frac{2n-1}{2} \cdots \frac{1}{2} \Gamma\left(\frac{1}{2}\right).$$

From (7.6.9), $\Gamma(\tfrac{1}{2}) = \sqrt{\pi}$. Thus, (7.7.56) may be expressed as

$$J_{1/2}(x) = \sqrt{\frac{x}{2\pi}} \sum_{n=0}^{\infty} \frac{2(-1)^n x^{2n}}{2^n n! \, [(2n+1)(2n-1)\cdots 1]}.$$

But, $2^n n! [(2n+1)(2n-1) \cdots 1] = (2n+1)!$. Hence,

$$J_{1/2}(x) = \sqrt{\frac{x}{2\pi}} \sum_{n=0}^{\infty} \frac{2(-1)^n x^{2n}}{(2n+1)!} = \sqrt{\frac{2}{\pi x}} \sum_{n=0}^{\infty} \frac{(-1)^n x^{2n+1}}{(2n+1)!} = \sqrt{\frac{2}{\pi x}} \sin x.$$

Note that $J_{1/2}$ represents a damped periodic function with period 2π.

Some of the Bessel functions of the first kind and second kind have been tabulated and graphed. [*See* [1], pp. 359, 409.] The graphs of the functions J_0 and J_1 are shown in Figure 7.7.1. Table 7.7.1 gives the first six non-negative roots of $J_0(x) = 0$ and $J_1(x) = 0$ to four decimal places, together with their differences. Also, it gives the values of $J_1(x_j)$ and $J_0(x_k)$, where x_j and x_k denote the nonnegative roots of $J_0(x) = 0$ and $J_1(x) = 0$, respectively.

From (7.7.53) and (7.7.51) with $m = 1$, we have

$$J_0(x) = \sum_{n=0}^{\infty} \frac{(-1)^n x^{2n}}{2^{2n}(n!)^2}, \qquad J_1(x) = \sum_{n=0}^{\infty} \frac{(-1)^n x^{2n+1}}{2^{2n+1} n!(n+1)!}. \tag{7.7.57}$$

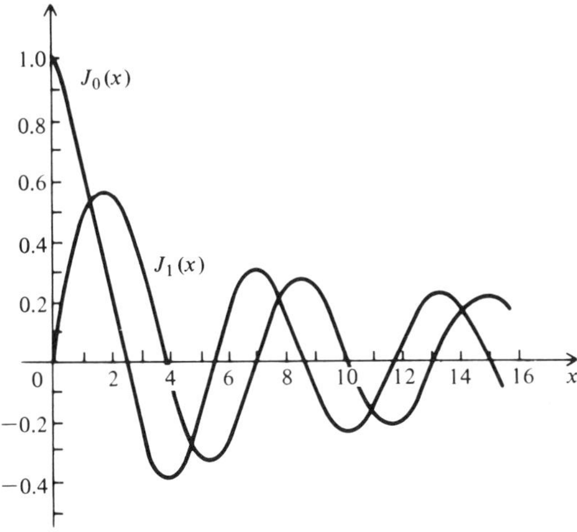

Figure 7.7.1.

Thus, J_0 is an even function, and J_1 is an odd function. Also, $J_0'(x) = -J_1(x)$. [*See* Exercise 7.7.3.] Now, from Figure 7.7.1, one might expect that the equation $J_0(x) = 0$ has

(1) an infinite number of real roots, all simple,
(2) no purely imaginary roots (ib, $b \neq 0$, $i^2 = -1$),
(3) no complex roots ($a + ib$, $a \neq 0$, $b \neq 0$),
(4) no common root with the equation $J_1(x) = 0$.

Table 7.7.1.

	$J_0(x_j)$			$J_1(x_k)$	
x_j	$x_{j+1} - x_j$	$J_1(x_j)$	x_k	$x_{k+1} - x_k$	$J_0(x_k)$
2.4048		0.5191	0.0000		1.0000
	3.1153			3.8317	
5.5201		−0.3403	3.8317		−0.4028
	3.1336			3.1839	
8.6537		0.2715	7.0156		0.3001
	3.1378			3.1579	
11.7915		−0.2325	10.1735		−0.2497
	3.1394			3.1502	
14.9309		0.2065	13.3237		0.2184
	3.1402			3.1469	
18.0711		−0.1877	16.4706		−0.1965

It turns out that these statements are true. However, to establish them requires a significant amount of interesting work and more advanced study on Bessel functions. [*See* Exercises 7.7.23 through 7.7.30.]

Also, it can be shown that

$$\lim_{j \to \infty} (x_{j+1} - x_j) = \pi, \qquad \lim_{k \to \infty} (x_{k+1} - x_k) = \pi. \qquad (7.7.58)$$

Thus, for x sufficiently large, $J_0(x)$ and $J_1(x)$ are "almost-periodic" with period very close to 2π. Since $J_0(-x) = J_0(x)$, $J_1(-x) = -J_1(x)$, and $J_0'(x) = -J_1(x)$, we see that for x sufficiently large, the similarity between $J_0(x)$ and $\cos x$ and between $J_1(x)$ and $\sin x$ becomes more striking.

Exercises 7.7

1. By mathematical induction establish Formula (7.7.45).

2. Define $\dfrac{1}{\Gamma(p)} = 0$ for $p = 0, -1, -2, \ldots.$ Show that

$$J_{-m}(x) = (-1)^m J_m(x), \qquad m = 1, 2, \ldots.$$

Thus, the Bessel functions J_{-m} and J_m are linearly dependent for $0 < |x| < \infty$.

In Exercises 3 through 12, establish the given formula.

3. $J_0'(x) = -J_1(x)$.
4. $K_0'(x) = -K_1(x)$.
5. $[xJ_1(x)]' = xJ_0(x)$.
6. $[x^{\alpha+1}J_{\alpha+1}(x)]' = x^{\alpha+1}J_\alpha(x)$.
7. $[x^{-\alpha}J_\alpha(x)]' = -x^{-\alpha}J_{\alpha+1}(x)$.

8. $J_\alpha'(x) = J_{\alpha-1}(x) - \dfrac{\alpha}{x} J_\alpha(x)$.

9. $J_\alpha'(x) = \dfrac{\alpha}{x} J_\alpha(x) - J_{\alpha+1}(x)$.

10. $J_\alpha(x) = \dfrac{x}{2\alpha} [J_{\alpha-1}(x) + J_{\alpha+1}(x)]$.

11. $[J_\alpha{}^2(x)]' = \dfrac{x}{2\alpha} [J_{\alpha-1}^2(x) - J_{\alpha+1}^2(x)]$.

12. $[xJ_0(x)J_1(x)]' = x[J_0^2(x) - J_1^2(x)]$.
13. Show that for $x > 0$,

(a) $J_{-1/2}(x) = \sqrt{\dfrac{2}{\pi x}} \cos x,$ (b) $J_{3/2}(x) = \sqrt{\dfrac{2}{\pi x}} \left(\dfrac{\sin x}{x} - \cos x \right),$

(c) $J_{5/2}(x) = \sqrt{\dfrac{2}{\pi x}} \left(\dfrac{3 \sin x}{x} - \sin x - \dfrac{3 \cos x}{x} \right).$

14. Utilize the appropriate series expansions to compute, correct to four decimal places, the values of $J_0(\tfrac{1}{2})$, $J_0(1)$, $J_0(2)$, $J_1(\tfrac{1}{2})$, $J_1(1)$, $J_1(2)$, $J_2(\tfrac{1}{2})$, $J_2(1)$, and $J_2(2)$.

15. Evaluate, correct to four decimal places, the following integrals.

    (a) $\displaystyle\int_0^1 x[J_0^2(x) - J_1{}^2(x)]dx,$ (c) $\displaystyle\int_{\pi/2}^{\pi} x[J_{-1/2}^2(x) - J_{3/2}^2(x)]dx,$

    (b) $\displaystyle\int_0^1 x^2 J_1(x)dx,$ (d) $\displaystyle\int_{1/2}^2 \left[J_1(x) - \frac{2}{x} J_2(x)\right] dx.$

16. If β is any root of the equation $J_0(x) = 0$, show that

    (a) $\displaystyle\int_0^1 J_1(\beta x)dx = \frac{1}{\beta},$ (c) $\displaystyle\int_0^{\infty} J_1(x)dx = 1.$

    (b) $\displaystyle\int_0^{\beta} J_1(x)dx = 1,$

    If $\beta\ (\neq 0)$ is a root of the equation $J_1(x) = 0$, show that

    (d) $\displaystyle\int_0^1 x J_0(\beta x)dx = 0.$

17. By eliminating $J_{\alpha+1}(x)$ between the formulas in Exercises 6 and 7, show that we obtain the Bessel equation (7.7.1) with y replaced by $J_\alpha(x)$.

18. Show that

    (a) $y = J_0(x)$ is a solution of the Bessel equation

$$xy'' + y' + xy = 0,$$

    (b) $y = J_0(\beta x)$ is a solution of the equation

$$xy'' + y' + \beta^2 xy = 0.$$

19. Show that $y = x^\alpha J_\alpha(x)$ is a solution of the equation

$$xy'' + (1 - 2\alpha)y' + xy = 0.$$

20. Show that

    (a) $y = \sqrt{x}\, J_\alpha(x)$ is a solution of the equation

$$x^2 y'' + (x^2 - \alpha^2 + \tfrac{1}{4})y = 0,$$

    (b) $y = \sqrt{x}\, J_\alpha(\beta x)$ is a solution of the equation

$$x^2 y'' + (\beta^2 x^2 - \alpha^2 + \tfrac{1}{4})y = 0, \qquad \beta > 0.$$

21. Consider the Bessel equation of order zero

    (1) $\qquad\qquad\qquad\qquad xy'' + y' + xy = 0.$

    Let $u = J_0(\beta x)$ and $v = J_0(\gamma x)$, where β and γ are constants. Then, in view of Exercise 18, u and v are solutions of the equations

    (2) $\qquad\qquad\qquad\qquad xu'' + u' + \beta^2 xu = 0,$

    (3) $\qquad\qquad\qquad\qquad xv'' + v' + \gamma^2 xv = 0.$

Multiply (2) by v and (3) by u, and then subtract (3) from (2).   Show that

$$(4) \qquad [x(u'v - uv')]' = (\gamma^2 - \beta^2)xuv.$$

Thus (suppressing the constant of integration), we have

$$(5) \qquad (\gamma^2 - \beta^2) \int xuv \, dx = x(u'v - uv').$$

Show that (5) may be expressed as

$$(6) \qquad (\gamma^2 - \beta^2) \int x J_0(\beta x) J_0(\gamma x) dx = x[\beta J_0'(\beta x) J_0(\gamma x) - \gamma J_0(\beta x) J_0'(\gamma x)].$$

Multiply (2) by $2xu'$.   Show that

$$(7) \qquad [x^2(u')^2 + \beta^2 x^2 u^2]' = 2\beta^2 x u^2.$$

Integrate (7) and use Exercise 3.   Show that

$$(8) \qquad \int x J_0^2(\beta x) dx = \tfrac{1}{2} x^2 [J_0^2(\beta x) + J_1^2(\beta x)].$$

In particular, integrating (6) and (8) between the limits 0 and 1, we obtain

$$(9) \qquad (\gamma^2 - \beta^2) \int_0^1 x J_0(\beta x) J_0(\gamma x) dx = \beta J_0'(\beta) J_0(\gamma) - \gamma J_0(\beta) J_0'(\gamma),$$

$$(10) \qquad \int_0^1 x J_0^2(\beta x) dx = \tfrac{1}{2} [J_0^2(\beta) + J_1^2(\beta)].$$

If β, γ ($\beta^2 \neq \gamma^2$) are two roots of the equation $J_0(x) = 0$, show that

$$(11) \qquad \int_0^1 x J_0(\beta x) J_0(\gamma x) dx = 0.$$

If β, γ ($\beta^2 \neq \gamma^2$) are two roots of the equation

$$x J_0'(x) + C J_0(x) = 0,$$

where C is a constant, show that

$$(12) \qquad \int_0^1 x J_0(\beta x) J_0(\gamma x) dx = 0.$$

If $\beta^2 \neq \gamma^2$, then in view of Exercise 3, we may write (9) as

$$(13) \qquad \int_0^1 x J_0(\beta x) J_0(\gamma x) dx = \frac{-\beta J_1(\beta) J_0(\gamma) + \gamma J_0(\beta) J_1(\gamma)}{\gamma^2 - \beta^2} .$$

If $\beta = \gamma$, each being a root of the equation $J_0(x) = 0$, then the right-hand member of (13) is indeterminate, namely, $\tfrac{0}{0}$.   Now, by considering β as the root of $J_0(x) = 0$ and γ as a variable approaching β, we may evaluate this indeterminate form by utilizing l'Hospital's rule.   Show that

$$\lim_{\gamma \to \beta} \left[\frac{-\beta J_1(\beta) J_0(\gamma) + \gamma J_0(\beta) J_1(\gamma)}{\gamma^2 - \beta^2} \right] = \frac{1}{2} J_1^2(\beta).$$

Thus

$$(14) \qquad \int_0^1 x J_0^2(\beta x)\,dx = \frac{1}{2} J_1^2(\beta).$$

Hence, in view of (11) and (14), we have that if β and γ are two (positive) real roots of the equation $J_0(x) = 0$, then

$$(15) \qquad \int_0^1 x J_0(\beta x) J_0(\gamma x)\,dx = \begin{cases} 0, & \text{if } \gamma \neq \beta, \\ \frac{1}{2} J_1^2(\beta), & \text{if } \gamma = \beta.^6 \end{cases}$$

This result is found useful in the Fourier series expansion of a function f, on the interval $0 < x < 1$, in terms of Bessel functions of order zero.

22. *Behavior of Bessel functions when x is (sufficiently) large.*   Consider the Bessel equation of order zero,

$$(1) \qquad xy'' + y' + xy = 0.$$

Let $u = \sqrt{x}\, y$.   Under this substitution, show that (1) takes on the form

$$(2) \qquad u'' = -\left(1 + \frac{1}{4x^2}\right)u.$$

For x sufficiently large, show that the general solution of (2) behaves like

$$(3) \qquad u(x) = C \cos{(x - \mu)},$$

where C and μ are constants.   Thus, for x sufficiently large, the general solution of (1) behaves like

$$(4) \qquad y(x) = C\, \frac{\cos{(x - \mu)}}{\sqrt{x}}.$$

Now, it can be shown [*see* [8], p. 82], that if $x > 0$,

$$(5) \qquad J_0(x) = \sqrt{\frac{2}{\pi x}} \left[\cos{\left(x - \frac{\pi}{4}\right)} + p(x)\right],$$

$$(6) \qquad K_0(x) = \sqrt{\frac{2}{\pi x}} \left[\sin{\left(x - \frac{\pi}{4}\right)} + q(x)\right],$$

where $p(x) \to 0$ and $q(x) \to 0$ as $x \to \infty$.   Thus, for x sufficiently large, any solution $f(x)$ of Equation (1) can be written in the form

$$(7) \qquad f(x) = A J_0(x) + B K_0(x),$$

where A and B are arbitrary constants.   Show that (7) can also be written in the following form:

$$(8) \qquad f(x) = \sqrt{A^2 + B^2}\, \sqrt{\frac{2}{\pi x}}\, [\cos(x - \mu) + r(x)],$$

---

[6] Equation (15) expresses the orthogonality of the Bessel function $J_0 = J_0(\beta x)$ with respect to the weight function x over the interval zero to one. [*See* Definition 7.8.1.]

where $\mu = \dfrac{\pi}{4} + \tan^{-1}\left(\dfrac{B}{A}\right)$ and $r(x) \to 0$ as $x \to \infty$.

23. Let f denote any Bessel function of order zero.
    (a) Show that for x sufficiently large, the roots of the equation $f(x) = 0$ are approximately those of $\cos(x - \mu) = 0$.   Thus, one may infer that $f(x) = 0$ has an infinite number of real roots.
    (b) Show that the sufficiently large real roots of $f(x) = 0$ are $\mu + (N - \tfrac{1}{2})\pi$, approximately, where N is any sufficiently large positive integer.
    (c) Show that the sufficiently large positive roots of the equation $J_0(x) = 0$ are $(N - \tfrac{1}{4})\pi$, approximately, and those of $K_0(x) = 0$ are $(N - \tfrac{3}{4})\pi$, approximately, where N is any sufficiently large positive integer.
24. Let f denote any Bessel function of order zero.   Show that none of the roots of the equation $f(x) = 0$ can be a repeated root.   (Thus, all the roots of the equation $f(x) = 0$ are simple roots.)
25. Show that the equation $J_0(x) = 0$ has an infinite number of simple real roots.
26. Show that the equation $J_0(x) = 0$ has no purely imaginary roots.
27. Show that the equation $J_0(x) = 0$ has no complex roots.
28. Show that the equation $J_0'(x) = 0$ has an infinite number of real roots.
29. Show that the equation $J_1(x) = 0$ has an infinite number of real roots.
30. Show that the equations $J_0(x) = 0$ and $J_1(x) = 0$ have no root in common.
31. It can be shown that for each fixed m $(m = 0, 1, 2, \ldots)$ the set of all positive roots of the equation $J_m(x) = 0$ consists of an infinite sequence $x = x_k$ $(k = 1, 2, 3, \ldots)$ such that $x_k \to \infty$ as $k \to \infty$.
    (a) For each fixed m, show that between every two consecutive positive roots of $J_m(x) = 0$, there exists a root of $J_{m+1}(x) = 0$.
    (b) For each fixed m, show that between every two consecutive positive roots of $J_{m+1}(x) = 0$, there exists a root of $J_m(x) = 0$.   (For each fixed m, the positive roots of $J_m(x) = 0$ and $J_{m+1}(x) = 0$ interlace each other.)
32. *The ber and bei functions.*   Consider the differential equation

$$(1) \qquad\qquad xy'' + y' - ixy = 0, \qquad i^2 = -1.$$

This equation arises in electrical engineering.   In view of Exercise 18(b), with $\beta^2 = -i$, we see that a solution of (1) is given by

$$(2) \qquad\qquad y = J_0(\beta x) = J_0[(-i)^{1/2}x] = J_0(i^{3/2}x).$$

From (6) of Exercise 4.6.19, we find that $\sqrt{-i} = \pm e^{-i\pi/4}$.   Thus,

$$\sqrt{-i} = -e^{-i\pi/4} = e^{i\pi}e^{-i\pi/4} = e^{i(3/4)\pi}.$$

For a particular solution of (2), let us choose $i^{3/2} = e^{i(3/4)\pi}$.   Utilizing (7.7.15) with x replaced by $i^{3/2}x$, show that

$$(3) \qquad J_0(i^{3/2}x) = 1 + \sum_{k=1}^{\infty} \frac{(-1)^k x^{4k}}{2^{4k}(2k!)^2} + i \sum_{k=1}^{\infty} \frac{(-1)^{k+1} x^{4k-2}}{2^{4k-2}[(2k-1)!]^2}.$$

Clearly, when x is real, $J_0(i^{3/2}x)$ is complex.   It is customary to denote the real part and the imaginary part of the Bessel function J_0 of (3) by ber and bei, respectively.   That is,

$$(4) \qquad \text{ber } x = 1 + \sum_{k=1}^{\infty} \frac{(-1)^k x^{4k}}{2^{4k}(2k!)^2}, \qquad \text{bei } x = \sum_{k=1}^{\infty} \frac{(-1)^{k+1} x^{4k-2}}{2^{4k-2}[(2k-1)!]^2}.$$

Thus, we may write (3) as

$$(5) \qquad\qquad J_0(i^{3/2}x) = \text{ber } x + i \text{ bei } x.$$

Show that

$$(6) \qquad\qquad \frac{d}{dx}(x \text{ ber}' x) = -x \text{ bei } x,$$

$$(7) \qquad\qquad \frac{d}{dx}(x \text{ bei}' x) = x \text{ ber } x,$$

Where "$'$" indicates differentiation with respect to x. The functions ber, ber', bei, and bei' have been tabulated. [*See* [1], p. 430.] The ber and bei functions are also known as the *Kelvin functions*.

7.8. Legendre's Equation and Legendre Polynomials

Another differential equation which plays an important role in applied mathematics and mathematical physics is

$$L[y] = (1 - x^2)y'' - 2xy' + \alpha(\alpha + 1)y = 0, \qquad\qquad (7.8.1)$$

where α is any constant. Equation (7.8.1) is called *Legendre's equation of order α*. [*See* also Exercise 4.5.7.]

In view of Definition 7.4.3, we see that $x = \pm 1$ are regular singular points of (7.8.1). All other (finite) points are ordinary points of (7.8.1). We shall be particularly interested in obtaining solutions of (7.8.1) about the ordinary point $x = 0$, since these solutions are of considerable importance in applications.

In view of Theorem 7.3.1, we are assured that the initial-value problem

$$\begin{cases} L[y] = 0, \\ y(0) = a, \qquad y'(0) = b, \end{cases} \qquad\qquad (7.8.2)$$

has a (unique) power series solution

$$y(x) = \sum_{n=0}^{\infty} a_n x^n \qquad\qquad (7.8.3)$$

which is convergent in a neighborhood of the origin, namely, at least for $|x| < 1$. Our present task is to determine the coefficients a_n.

Differentiating (7.8.3) twice, we obtain

$$y'(x) = \sum_{n=1}^{\infty} n a_n x^{n-1}, \qquad y''(x) = \sum_{n=2}^{\infty} n(n-1)a_n x^{n-2}. \qquad (7.8.4)$$

From (7.8.2), (7.8.3), and (7.8.4), we see that

$$a = y(0) = a_0 \quad \text{and} \quad b = y'(0) = a_1. \tag{7.8.5}$$

Thus, a_0 and a_1 are arbitrary. Substituting the series of y, y', and y'' into (7.8.1), we obtain

$$\sum_{n=0}^{\infty} (n+1)(n+2)a_{n+2} x^n - \sum_{n=2}^{\infty} n(n-1)a_n x^n$$

$$- \sum_{n=1}^{\infty} 2na_n x^n + \sum_{n=0}^{\infty} \alpha(\alpha+1)a_n x^n = 0.$$

In the above equation, summing out the first two terms in the first and last series, and the first term in the third series, we obtain upon simplifications

$$2a_2 + \alpha(\alpha+1)a_0 + [6a_3 + (\alpha-1)(\alpha+2)a_1]x$$

$$+ \sum_{n=2}^{\infty} [(n+1)(n+2)a_{n+2} + (\alpha-n)(\alpha+n+1)a_n]x^n = 0. \tag{7.8.6}$$

Since the power series (7.8.6) converges to zero in a neighborhood of the origin, it follows from Theorem 7.2.7 that

$$2a_2 + \alpha(\alpha+1)a_0 = 0 \quad \text{or} \quad a_2 = -\frac{\alpha(\alpha+1)}{2!} a_0, \tag{7.8.7}$$

$$6a_3 + (\alpha-1)(\alpha+2)a_1 = 0 \quad \text{or} \quad a_3 = -\frac{(\alpha-1)(\alpha+2)}{3!} a_1, \tag{7.8.8}$$

and

$$(n+1)(n+2)a_{n+2} + (\alpha-n)(\alpha+n+1)a_n = 0$$

or

$$a_{n+2} = -\frac{(\alpha-n)(\alpha+n+1)}{(n+1)(n+2)} a_n, \qquad n \geq 2, \tag{7.8.9}$$

which gives us the recurrence relation.

Utilizing (7.8.7), (7.8.8), and (7.8.9), we leave it as an exercise for the reader [*see* Exercise 7.8.1] to show, by mathematical induction, that

$$a_{2k} = (-1)^k \frac{[(\alpha-2k+2)(\alpha-2k+4)\cdots(\alpha-2)\alpha]\,[(\alpha+1)(\alpha+3)\cdots(\alpha+2k-3)(\alpha+2k-1)]}{(2k)!} a_0,$$

$$k = 1, 2, \ldots, \tag{7.8.10}$$

and

$$a_{2k+1} = (-1)^k \frac{\begin{array}{c}[(\alpha - 2k + 1)(\alpha - 2k + 3) \cdots (\alpha - 3)(\alpha - 1)] \\ [(\alpha + 2)(\alpha + 4) \cdots (\alpha + 2k - 2)(\alpha + 2k)]\end{array}}{(2k + 1)!} a_1,$$

$$k = 1, 2, \ldots. \tag{7.8.11}$$

In view of (7.8.11), (7.8.10), and (7.8.5), the unique solution of our initial-value problem on the interval $|x| < 1$ is

$$y(x) = a + \sum_{k=1}^{\infty} a_{2k} x^{2k} + bx + \sum_{k=1}^{\infty} a_{2k+1} x^{2k+1}. \tag{7.8.12}$$

Clearly, the functions y_1 and y_2 given by

$$y_1(x) = a_0 \left\{ 1 + \sum_{k=1}^{\infty} \frac{\begin{array}{c}(-1)^k[(\alpha - 2k + 2)(\alpha - 2k + 4) \cdots (\alpha - 2)\alpha] \\ [(\alpha + 1)(\alpha + 3) \cdots (\alpha + 2k - 3)(\alpha + 2k - 1)]\end{array}}{(2k)!} x^{2k} \right\},$$

$$\tag{7.8.13}$$

$$y_2(x) = a_1 \left\{ x + \sum_{k=1}^{\infty} \frac{\begin{array}{c}(-1)^k[(\alpha - 2k + 1)(\alpha - 2k + 3) \cdots (\alpha - 3)(\alpha - 1)] \\ [(\alpha + 2)(\alpha + 4) \cdots (\alpha + 2k - 2)(\alpha + 2k)]\end{array}}{(2k + 1)!} x^{2k+1} \right\},$$

$$\tag{7.8.14}$$

where a_0 and a_1 are arbitrary constants, are linearly independent solutions of the Legendre equations on the interval $|x| < 1$. Note that y_1 and y_2 are even and odd functions, respectively.

Remark 7.8.1. If α is a nonnegative even integer, say $\alpha = 2n$, $n = 0, 1, \ldots$, then the series (7.8.13) terminates when $k = n + 1$. Thus, series (7.8.13) reduces to a polynomial. If α is a positive odd integer, say $\alpha = 2n + 1$, $n = 0, 1, \ldots$, then the series (7.8.14) terminates also when $k = n + 1$, and series (7.8.14) reduces to a polynomial. Consequently, when α is zero or a positive integer, the general solution of Legendre's equation contains a polynomial solution $P_n(x)$ and an infinite series solution, say $Q_n(x)$. Since a_0 and a_1 in (7.8.13) and (7.8.14) are arbitrary constants, it is found convenient to choose them so that $P_n(1) = 1$, $n = 1, 2, \ldots$. This is accomplished by choosing

$$a_0 = (-1)^{n/2} \frac{1 \cdot 3 \cdots (n - 1)}{2 \cdot 4 \cdots n}, \qquad a_1 = (-1)^{(n-1)/2} \frac{1 \cdot 3 \cdots n}{2 \cdot 4 \cdots (n - 1)}. \tag{7.8.15}$$

With these values of a_0 and a_1 substituted in (7.8.13) and (7.8.14), and $P_0(x) \equiv 1$, we obtain the so-called *Legendre polynomials* $P_n(x)$.   That is,

$$P_n(x) = (-1)^{n/2} \frac{1 \cdot 3 \cdots (n-1)}{2 \cdot 4 \cdots n} \sum_{k=0}^{n/2} (-1)^k$$

$$\times \frac{[(n-2k+2)(n-2k+4)\cdots(n-2)n]}{} \underset{\displaystyle \frac{[(n+1)(n+3)\cdots(n+2k-3)(n+2k-1)]}{(2k)!}}{} x^{2k},$$

$$n \text{ even,} \tag{7.8.16}$$

$$P_n(x) = (-1)^{(n-1)/2} \frac{1 \cdot 3 \cdots n}{2 \cdot 4 \cdots (n-1)} \sum_{k=0}^{(n-1)/2} (-1)^k$$

$$\times \frac{[(n-2k+1)(n-2k+3)\cdots(n-3)(n-1)]}{} \underset{\displaystyle \frac{[(n+2)(n+4)\cdots(n+2k-2)(n+2k)]}{(2k+1)!}}{} x^{2k+1},$$

$$n \text{ odd.} \tag{7.8.17}$$

$P_n(x)$, being polynomials, converge for all (finite) values of x.   The Q_n, associated with the Legendre polynomials $P_n(x)$, are called the *Legendre functions of the second kind*.   It turns out that $Q_n(x)$ becomes infinite at $x = \pm 1$.   For examples, see answers to Exercise 4.5.7(b).   In the remainder of this section, we shall primarily concern ourselves with the Legendre polynomials.

We shall now endeavor to combine Formulas (7.8.16) and (7.8.17) into a single compact formula.   Writing out the sum in (7.8.16) in descending powers of x, that is, starting the summation first with $k = n/2$, then $k = n/2 - 1$, and so on, we obtain

$$P_n(x) = \frac{1 \cdot 3 \cdots (n-1)}{2 \cdot 4 \cdots n}$$

$$\times \left[\frac{2 \cdot 4 \cdots (n-2)n(n+1)(n+3)\cdots(2n-3)(2n-1)}{n!} x^n \right.$$

$$- \frac{4 \cdot 6 \cdots (n-2)n(n+1)(n+3)\cdots(2n-5)(2n-3)}{(n-2)!} x^{n-2}$$

$$\left. + \frac{6 \cdot 8 \cdots (n-2)n(n+1)(n+3)\cdots(2n-7)(2n-5)}{(n-4)!} x^{n-4} - \cdots \right]$$

$$= \frac{(2n-1)(2n-3)\cdots 3 \cdot 1}{n!}$$

$$\times \left[x^n - \frac{n(n-1)}{2(2n-1)} x^{n-2} + \frac{n(n-1)(n-2)(n-3)}{2 \cdot 4(2n-1)(2n-3)} x^{n-4} - \cdots \right].$$

$$\tag{7.8.18}$$

Similarly, we have from (7.8.17) that

$$P_n(x) = \frac{1 \cdot 3 \cdots n}{2 \cdot 4 \cdots (n-1)}$$

$$\times \left[\frac{2 \cdot 4 \cdots (n-3)(n-1)(n+2)(n+4) \cdots (2n-3)(2n-1)}{n!} x^n \right.$$

$$- \frac{4 \cdot 6 \cdots (n-3)(n-1)(n+2)(n+4) \cdots (2n-5)(2n-3)}{(n-2)!} x^{n-2}$$

$$+ \frac{6 \cdot 8 \cdots (n-3)(n-1)(n+2)(n+4) \cdots (2n-7)(2n-5)}{(n-4)!} x^{n-4} - \cdots \left. \right]$$

$$= \frac{(2n-1)(2n-3) \cdots 3 \cdot 1}{n!}$$

$$\times \left[x^n - \frac{n(n-1)}{2(2n-1)} x^{n-2} + \frac{n(n-1)(n-2)(n-3)}{2 \cdot 4(2n-1)(2n-3)} x^{n-4} - \cdots \right].$$

$$\tag{7.8.19}$$

From (7.8.18) and (7.8.19) we see that whether n is even or odd, we obtain the
same form for the Legendre polynomials. Also, from (7.8.18) or (7.8.19),
it is apparent that the coefficient of x^{n-2k} has the form

$$(-1)^k \frac{(2n-1)(2n-3) \cdots 3 \cdot 1}{n!}$$

$$\times \frac{n(n-1) \cdots (n-2k+1)}{(2 \cdot 4 \cdots 2k)[(2n-1)(2n-3) \cdots (2n-2k+1)]}.$$

$$\tag{7.8.20}$$

The expression in (7.8.20) may also be written as

$$(-1)^k \frac{(2n-1)(2n-3) \cdots (2n-2k+1)(2n-2k-1) \cdots 3 \cdot 1}{n(n-1) \cdots (n-2k+1)(n-2k)!}$$

$$\times \frac{n(n-1) \cdots (n-2k+1)}{2^k k! [(2n-1)(2n-3) \cdots (2n-2k+1)]}$$

$$= (-1)^k \frac{(2n-2k-1) \cdots 3 \cdot 1}{(n-2k)! \, 2^k k!}$$

$$= (-1)^k \frac{(2n-2k)(2n-2k-1)(2n-2k-2) \cdots 4 \cdot 3 \cdot 2 \cdot 1}{(n-2k)! \, 2^k k! [(2n-2k)(2n-2k-2) \cdots 4 \cdot 2]}$$

$$= (-1)^k \frac{(2n-2k)!}{(n-2k)! \, 2^k k! \, 2^{n-k}[(n-k)(n-k-1) \cdots 2 \cdot 1]}$$

$$= (-1)^k \frac{(2n-2k)!}{2^n k! (n-k)! (n-2k)!}.$$

Thus, we may write the *Legendre polynomials* in a single compact formula as

$$P_n(x) = \sum_{k=0}^{N} \frac{(-1)^k(2n-2k)!}{2^n k!(n-k)!(n-2k)!}\, x^{n-2k}, \tag{7.8.21}$$

where $N = n/2$ when n is even and $N = (n-1)/2$ when n is odd.

The first six Legendre polynomials are

$$P_0(x) = 1, \qquad P_1(x) = x, \qquad P_2(x) = \tfrac{1}{2}(3x^2 - 1), \qquad P_3(x) = \tfrac{1}{2}(5x^3 - 3x),$$

$$\tag{7.8.22}$$

$$P_4(x) = \tfrac{1}{8}(35x^4 - 30x^2 + 3), \qquad P_5(x) = \tfrac{1}{8}(63x^5 - 70x^3 + 15x).$$

Note that, as planned, $P_0(1) = P_1(1) = P_2(1) = P_3(1) = P_4(1) = P_5(1) = 1$. Clearly, P_0, P_2, and P_4 are even functions, and P_1, P_3, and P_5 are odd functions. Figure 7.8.1 depicts the graphs of the first five Legendre polynomials on the interval $-1 \le x \le 1$.

Rodrigues' formula. We shall develop a useful expression for the Legendre polynomials. Expanding $(x^2 - 1)^n$ by the binomial theorem, we obtain for every positive integer n

$$(x^2 - 1)^n = \sum_{k=0}^{n} \frac{(-1)^k n!}{k!(n-k)!}\, x^{2n-2k}. \tag{7.8.23}$$

Differentiating (7.8.23) n times, we obtain

$$\frac{d^n}{dx^n}(x^2 - 1)^n = \sum_{k=0}^{n} \frac{(-1)^k n!}{k!(n-k)!} \cdot (2n-2k)(2n-2k-1)\cdots(n-2k+1)x^{n-2k}$$

$$= \sum_{k=0}^{n} 2^n n! \left[\frac{(-1)^k}{2^n k!(n-k)!} \cdot (2n-2k)(2n-2k-1) \right.$$

$$\left. \cdots (n-2k+1) \cdot \frac{(n-2k)!}{(n-2k)!}\, x^{n-2k} \right]$$

$$= 2^n n! \sum_{k=0}^{n} \frac{(-1)^k}{2^n k!(n-k)!} \cdot \frac{(2n-2k)!}{(n-2k)!}\, x^{n-2k} = 2^n n!\, P_n(x).$$

Thus,

$$P_n(x) = \frac{1}{2^n n!} \frac{d^n}{dx^n}(x^2 - 1)^n. \tag{7.8.24}$$

This is *Rodrigues' formula* for the Legendre polynomials.

Definition 7.8.1. A set of functions $f_0, f_1, \ldots, f_n, \ldots$, defined on an interval I: $a \le x \le b$, is called an *orthogonal set with respect to the weight function σ on the interval I if*

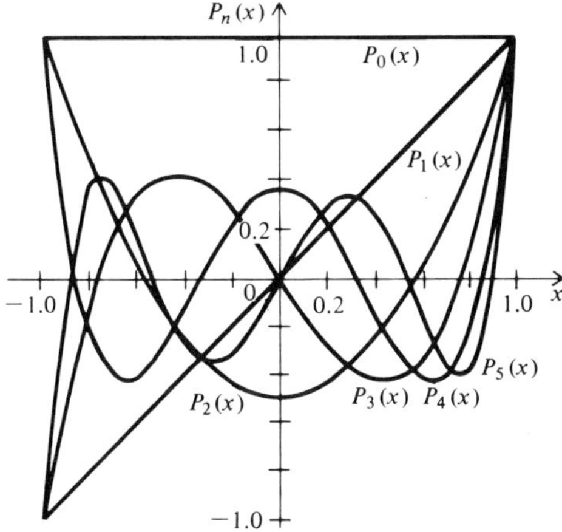

Figure 7.8.1.

$$\int_a^b \sigma(x) f_m(x) f_n(x)\,dx = 0 \quad \text{for } m \neq n,$$

$$\neq 0 \quad \text{for } m = n. \tag{7.8.25}$$

[Usually $\sigma(x) > 0$ for all x in I and consequently, $\int_a^b \sigma(x) f_n^2(x)\,dx > 0$.]

Orthogonality of the Legendre polynomials. We shall now show that the set of the Legendre polynomials $P_0, P_1, \ldots, P_n$, defined on the interval I: $-1 \leqq x \leqq 1$, is orthogonal with respect to the weight function $\sigma = 1$ on the interval I. That is, we shall show that

$$\int_{-1}^{1} P_m(x) P_n(x)\,dx = \begin{cases} 0, & m \neq n, \\ \dfrac{2}{2n+1}, & m = n. \end{cases} \tag{7.8.26}$$

Since $y = P_m(x)$ and $y = P_n(x)$ satisfy the Legendre equation (7.8.1) with $\alpha = m$ and $\alpha = n$, respectively, we have

$$(1 - x^2)P_m''(x) - 2xP_m'(x) + m(m+1)P_m(x) = 0 \tag{7.8.27}$$

and

$$(1 - x^2)P_n''(x) - 2xP_n'(x) + n(n+1)P_n(x) = 0. \tag{7.8.28}$$

Let $\lambda_m = m(m+1)$ and $\lambda_n = n(n+1)$. Equations (7.8.27) and (7.8.28) may also be written as

$$\frac{d}{dx}\left[(1 - x^2)P_m'(x)\right] + \lambda_m P_m(x) = 0 \tag{7.8.29}$$

and

$$\frac{d}{dx}[(1 - x^2)P_n'(x)] + \lambda_n P_n(x) = 0. \tag{7.8.30}$$

Suppose that $m \neq n$ so that $\lambda_m \neq \lambda_n$. Multiplying (7.8.29) and (7.8.30) by $P_n(x)$ and $P_m(x)$, respectively, and then subtracting, we obtain

$$(\lambda_m - \lambda_n)P_m(x)P_n(x) = P_m(x)\frac{d}{dx}[(1 - x^2)P_n'(x)] - P_n(x)\frac{d}{dx}[(1 - x^2)P_m'(x)].$$
$$\tag{7.8.31}$$

Integrating (7.8.31) between $x = -1$ and $x = 1$, we get

$$(\lambda_m - \lambda_n)\int_{-1}^{1} P_m(x)P_n(x)dx = \int_{-1}^{1} P_m(x)\frac{d}{dx}[(1 - x^2)P_n'(x)]dx$$

$$- \int_{-1}^{1} P_n(x)\frac{d}{dx}[(1 - x^2)P_m'(x)]dx. \tag{7.8.32}$$

Integrating by parts the integrals in the right-hand side of (7.8.32), we obtain

$$(\lambda_m - \lambda_n)\int_{-1}^{1} P_m(x)P_n(x)dx$$

$$= [P_m(x)(1 - x^2)P_n'(x)]\Big|_{-1}^{1} - [P_n(x)(1 - x^2)P_m'(x)]\Big|_{-1}^{1}$$

$$- \int_{-1}^{1} P_m'(x)(1 - x^2)P_n'(x)dx + \int_{-1}^{1} P_n'(x)(1 - x^2)P_m'(x)dx = 0.$$

Since $\lambda_m \neq \lambda_n$, it follows that

$$\int_{-1}^{1} P_m(x)P_n(x)dx = 0, \qquad m \neq n.$$

To show that

$$\int_{-1}^{1} [P_n(x)]^2\, dx = \frac{2}{2n + 1},$$

we shall utilize Rodrigues' formula. Multiplying both sides of (7.8.24) by x^n and then integrating from $x = -1$ and $x = 1$, we have

$$2^n n!\int_{-1}^{1} x^n P_n(x)dx = \int_{-1}^{1} x^n \frac{d^n}{dx^n}(x^2 - 1)^n\, dx. \tag{7.8.33}$$

By n successive integrations by parts of the integral on the right-hand side of (7.8.33), we leave it as an exercise for the reader [*see* Exercise 7.8.2] to show that

$$2^n n! \int_{-1}^{1} x^n P_n(x)dx = (-1)^n n! \int_{-1}^{1} (x^2 - 1)^n \, dx = n! \int_{-1}^{1} (1 - x^2)^n \, dx.$$

$$(7.8.34)$$

Letting $x = \sin \theta$ in the rightmost integral in (7.8.34), we obtain

$$2^n \int_{-1}^{1} x^n P_n(x)dx = \int_{-\pi/2}^{\pi/2} \cos^{2n+1}\theta \, d\theta = 2 \int_{0}^{\pi/2} \cos^{2n+1}\theta \, d\theta. \quad (7.8.35)$$

In view of Exercise 7.6.12 (b), we obtain

$$\int_{-1}^{1} x^n P_n(x)dx = \frac{2^{n+1}(n!)^2}{(2n+1)!}. \tag{7.8.36}$$

Since $\int_{-1}^{1} x^m P_n(x)dx = 0$, $m \neq n$, using (7.8.21) with $k = 0$ and (7.8.36), we obtain

$$\int_{-1}^{1} P_n(x)P_n(x)dx = \int_{-1}^{1} \frac{(2n)!}{2^n(n!)^2} x^n P_n(x)dx = \frac{(2n)!}{2^n(n!)^2} \cdot \frac{2^{n+1}(n!)^2}{(2n+1)!} = \frac{2}{2n+1}.$$

Thus, (7.8.26) is established.

Remark 7.8.2.   The results in (7.8.26) are used in the Fourier series expansion of a function f, on the interval $|x| < 1$, in terms of the Legendre polynomials.

Remark 7.8.3.   When $P_n(x)$ is a solution of the Legendre equation, one may utilize the method of reduction of order of a differential equation [*see* Section 4.5], to obtain the other linearly independent solution $Q_n(x)$ of the Legendre equation.   The *Legendre functions of the second kind*, Q_n are found to be

$$Q_n(x) = \frac{P_n(x)}{2} \ln \frac{1+x}{1-x} - \sum_{k=1}^{n} \frac{1}{k} P_{k-1}(x)P_{n-k}(x). \tag{7.8.37}$$

[*See* Exercise 4.5.7 (b) when $n = 0, 1, 2, 3$.]   Since

$$Q_0(x) = \frac{P_0(x)}{2} \ln \frac{1+x}{1-x} = \frac{1}{2} \ln \frac{1+x}{1-x}, \tag{7.8.38}$$

we may also write (7.8.37) as

$$Q_n(x) = P_n(x)Q_0(x) - \sum_{k=1}^{n} \frac{1}{k} P_{k-1}(x)P_{n-k}(x). \tag{7.8.39}$$

While the Legendre polynomials are defined for all (finite) values of x, the Legendre functions of the second kind are defined for $|x| < 1$.   However, if we write

$$Q_n^*(x) = \frac{P_n(x)}{2} \ln \frac{x+1}{x-1} - \sum_{k=1}^{n} \frac{1}{k} P_{k-1}(x)P_{n-k}(x), \tag{7.8.40}$$

we obtain a second solution which is valid for $|x| > 1$.

Utilizing (7.8.22), (7.8.38), and (7.8.39), the first six Legendre functions of the second kind are given by

$$Q_0(x) = \frac{1}{2} \ln \frac{1+x}{1-x}, \qquad Q_1(x) = xQ_0(x) - 1,$$

$$Q_2(x) = \frac{1}{2}(3x^2 - 1)Q_0(x) - \frac{3}{2}x, \qquad Q_3(x) = \frac{1}{2}(5x^3 - 3x)Q_0(x) - \frac{5}{2}x^2 + \frac{2}{3},$$

$$Q_4(x) = \frac{1}{8}(35x^4 - 30x^2 + 3)Q_0(x) - \frac{35}{8}x^3 + \frac{55}{24}x,$$

$$Q_5(x) = \frac{1}{8}(63x^5 - 70x^3 + 15x)Q_0(x) - \frac{63}{8}x^4 + \frac{49}{8}x^2 - \frac{8}{15}. \qquad (7.8.41)$$

Exercises 7.8

1. By mathematical induction, establish (7.8.10) and (7.8.11).
2. Establish (7.8.34).
3. Utilize Rodrigues' formula to show that $\int_{-1}^{1} P_m(x)P_n(x)dx = 0$ if $m \neq n$.
4. From the binomial series expansion, we have

$$(1 + x)^n = 1 + \sum_{v=1}^{\infty} \frac{n(n-1)\cdots(n-v+1)}{v!} x^v,$$

where n is any real number.   This series converges for $|x| < 1$.   In particular, when n is a nonnegative integer, we have

$$(1 + x)^n = \sum_{v=0}^{n} \frac{n!}{v!(n-v)!} x^v.$$

If $|2xr - r^2| < 1$ or $|2xr| + r^2 < 1$, show that

(1) $$[1 - (2xr - r^2)]^{-1/2} = 1 + \sum_{v=1}^{\infty} \frac{(2v)!}{2^{2v}(v!)^2}(2xr - r^2)^v.$$

For every positive integer v, show that

(2) $$(2xr - r^2)^v = \sum_{j=0}^{v} \frac{(-1)^j v!}{2^{j-v}j!(v-j)!} x^{v-j}r^{v+j}.$$

Thus, in view of (1) and (2), we have

(3) $$(1 - 2xr + r^2)^{-1/2} = 1 + \sum_{v=1}^{\infty} \sum_{j=0}^{v} \frac{(-1)^j(2v)!x^{v-j}}{2^{v+j}v!j!(v-j)!} r^{v+j}.$$

Show that (3) can be written as

(4) $$(1 - 2xr + r^2)^{-1/2} = \sum_{n=0}^{\infty} P_n(x)r^n,$$

where the P_n are the Legendre polynomials given in (7.8.21).   This relation is very useful in generating important recurrence formulas.   [*See* Exercises 5, 6, ... below.]   The expression $(1 - 2xr + r^2)^{-1/2}$ is called the *generating function* for the Legendre polynomials.

In Exercises 5 through 11, establish the given formula.   In each of these exercises, $n = 1, 2, \ldots$.

5. $(n + 1)P_{n+1}(x) - (2n + 1)xP_n(x) + nP_{n-1}(x) = 0.$

6. $nP_n(x) = xP_n'(x) - P_{n-1}'(x).$

7. $(2n + 1)P_n(x) = P_{n+1}'(x) - P_{n-1}'(x).$

8. $(n + 1)P_n(x) = P_{n+1}'(x) - xP_n'(x).$

9. $(2n + 1)(1 - x^2)P_n'(x) = n(n + 1)[P_{n-1}(x) - P_{n+1}(x)].$

10. $P_n'(x) = (2n - 1)P_{n-1}(x) + (2n - 5)P_{n-3}(x) + (2n - 9)P_{n-5}(x) + \cdots$, ending with $3P_1(x)$ if n is even and with $P_0(x)$ if n is odd.

11. $xP_n'(x) = nP_n(x) + (2n - 3)P_{n-2}(x) + (2n - 7)P_{n-4}(x) + \cdots$,   ending with $P_0(x)$ if n is even and with $3P_1(x)$ if n is odd.

[Exercises 5 through 9 can be shown to be true if n is any real number.   Also these results are true when $P_n(x)$ is replaced by $Q_n(x)$, where n is any real number.]

12. For $n = 0, 1, 2, \ldots$, show that

(a) $P_n(1) = 1,$

(b) $P_n(-1) = (-1)^n,$

(c) $P_{2n+1}(0) = 0,$

(d) $P_{2n}(0) = (-1)^n \dfrac{(2n)!}{2^{2n}(n!)^2},$

(e) $P_{2n}'(0) = 0,$

(f) $P_{2n+1}'(0) = (-1)^n \dfrac{(2n + 1)!}{2^{2n}(n!)^2}.$

13. For $n = 1, 2, \ldots$, show that

(a) $\displaystyle\int_{-1}^{1} xP_{n-1}(x)P_n(x)dx = \dfrac{2n}{4n^2 - 1},$

(b) $\displaystyle\int_{-1}^{1} (1 - x^2)[P_n'(x)]^2 dx = \dfrac{2n(n + 1)}{2n + 1},$

(c) $\displaystyle\int_{-1}^{1} P_n(x)dx = 0,$

(d) $\displaystyle\int_{x}^{1} P_n(\xi)d\xi = \dfrac{1}{2n + 1}[P_{n-1}(x) - P_{n+1}(x)].$

14. Let $x = \cos\theta$.   Utilize (7.8.22) and some of the trigonometric identities to show that

(a) $P_0(\cos\theta) = 1,$

(b) $P_1(\cos\theta) = \cos\theta,$

(c) $P_2(\cos\theta) = \tfrac{1}{4}(3\cos 2\theta + 1),$

(d) $P_3(\cos\theta) = \tfrac{1}{8}(5\cos 3\theta + 3\cos\theta),$

(e) $P_4(\cos\theta) = \tfrac{1}{64}(35\cos 4\theta + 20\cos 2\theta + 9),$

(f) $P_5(\cos\theta) = \tfrac{1}{128}(63\cos 5\theta + 35\cos 3\theta + 30\cos\theta).$

15. Consider the Legendre equation

(1) $$(1 - x^2)y'' - 2xy' + n(n+1)y = 0.$$

Differentiating (1) m times, show that

(2) $$(1 - x^2)y^{(m+2)} - 2x(m+1)y^{(m+1)} + [n(n+1) - m(m+1)]y^{(m)} = 0.$$

Let $u = (1 - x^2)^{m/2}y^{(m)}$. Show that (2) becomes

(3) $$(1 - x^2)u'' - 2xu' + \left[n(n+1) - \frac{m^2}{1 - x^2}\right]u = 0.$$

Equation (3) is called the *associated Legendre equation*. From the manner in which (3) was derived, it is evident that (3) will be satisfied by

(4) $$P_n^m(x) = (1 - x^2)^{m/2} D^m[P_n(x)]$$

and

(5) $$Q_n^m(x) = (1 - x^2)^{m/2} D^m[Q_n(x)],$$

where $D^m \equiv \dfrac{d^m}{dx^m}$, $m = 0, 1, 2, \ldots$. [Recall that $D^0 y \equiv y$, so that $P_n^0(x) = D^0[P_n(x)] = P_n(x)$.]

The expressions P_n^m and Q_n^m of (4) and (5) are called respectively the *associated Legendre functions of the first kind and of the second kind*. In view of Rodrigues' formula (7.8.24), we may also write (4) as

(6) $$P_n^m(x) = \frac{(1 - x^2)^{m/2}}{2^n n!} D^{n+m}(1 - x^2)^n.$$

16. If $s > m$ and $n > m$, show that

(a) $$\int_{-1}^{1} P_s^m(x)P_n^m(x)dx = 0, \quad \text{if } s \neq n.$$

Show that

(b) $$P_n^{m+1}(x) = (1 - x^2)^{1/2}D[P_n^m(x)] + mx(1 - x^2)^{-1/2}P_n^m(x).$$

Show that

(c) $$D\{(1 - x^2)D[P_n^m(x)]\} = -\left[n(n+1) - \frac{m^2}{1 - x^2}\right]P_n^m(x).$$

Show that

(d) $$\int_{-1}^{1} [P_n^{m+1}(x)]^2 dx = (n - m)(n + m + 1)\int_{-1}^{1} [P_n^m(x)]^2 dx.$$

[This is a useful recurrence or reduction formula.]

Show that

(e)
$$\int_{-1}^{1} [P_n^m(x)]^2 dx = \frac{(n+m)!}{(n-m)!} \int_{-1}^{1} [P_n(x)]^2 dx.$$

Thus, in view of (e) and (7.8.26), we have

(f)
$$\int_{-1}^{1} [P_n^m(x)]^2 dx = \frac{2}{2n+1} \cdot \frac{(n+m)!}{(n-m)!}.$$

From (a) and (f) we have

(g)
$$\int_{-1}^{1} P_s^m(x)P_n^m(x)dx = \begin{cases} 0, & s \neq n, \\ \dfrac{2}{2n+1} \cdot \dfrac{(n+m)!}{(n-m)!}, & s = n. \end{cases}$$

Thus, the associated Legendre functions of the first kind form an orthogonal set with respect to the weight function $\sigma = 1$ on the interval I: $-1 \leq x \leq 1$.

17. Let $x = \cos\theta$.   Show that
   (a) $P_1^1(\cos\theta) = \sin\theta$,
   (b) $P_2^1(\cos\theta) = 3\sin\theta\cos\theta$,
   (c) $P_2^2(\cos\theta) = 3\sin^2\theta$,
   (d) $P_3^1(\cos\theta) = \frac{3}{2}\sin\theta(5\cos^2\theta - 1)$,
   (e) $P_3^2(\cos\theta) = \frac{15}{2}\sin^2\theta\cos\theta$,
   (f) $P_3^3(\cos\theta) = \frac{15}{2}\sin^3\theta$.

18. Suppose that a set of functions $f_1, f_2, \ldots$, defined on an interval $I: a \leq x \leq b$, forms an orthogonal set with respect to a weight function σ defined on I.   Show that this set of functions is linearly independent on I.   (This result is important in the study of Fourier series.)

7.9. Singularities at Infinity

It is of practical and theoretical interest to investigate the behavior of the solutions of the equation

$$y'' + P(x)y' + Q(x)y = 0, \tag{7.9.1}$$

for sufficiently large values of $|x|$ (that is, in a neighborhood of infinity). This may be achieved by making the change of variable $x = 1/t$ in (7.9.1) and then investigating the solutions of the resulting differential equation in a neighborhood of $t = 0$.   Thus, let $y = y(x)$ be a solution of (7.9.1) for $|x| > R$, $R > 0$, and let

$$\bar{y}(t) = y\left(\frac{1}{t}\right), \quad \bar{P}(t) = P\left(\frac{1}{t}\right), \quad \text{and} \quad \bar{Q}(t) = Q\left(\frac{1}{t}\right). \tag{7.9.2}$$

Clearly, the functions $\bar{y}$, $\bar{P}$, and $\bar{Q}$ of (7.9.2) are well defined for $|t| < 1/R$ whenever the functions y, P, and Q are well defined for $|x| > R$.   Utilizing the chain rule and the fact that $dt/dx = -t^2$, we obtain

$$\frac{d}{dx}\, y\!\left(\frac{1}{t}\right) = -t^2 \frac{d}{dt}\, \bar{y}(t),$$

$$\frac{d^2}{dx^2}\, y\!\left(\frac{1}{t}\right) = t^4 \frac{d^2}{dt^2}\, \bar{y}(t) + 2t^3 \frac{d}{dt}\, \bar{y}(t). \tag{7.9.3}$$

From (7.9.1),

$$\frac{d^2}{dx^2}\, y\!\left(\frac{1}{t}\right) + P\!\left(\frac{1}{t}\right) \frac{d}{dx}\, y\!\left(\frac{1}{t}\right) + Q\!\left(\frac{1}{t}\right) y\!\left(\frac{1}{t}\right) = 0. \tag{7.9.4}$$

Thus, in view of (7.9.2), (7.9.3), and (7.9.4), we see that under the transformation $x = 1/t$, Equation (7.9.1) becomes

$$t^4 \frac{d^2}{dt^2}\, \bar{y}(t) + [2t^3 - t^2 \bar{P}(t)]\frac{d}{dt}\, \bar{y}(t) + \bar{Q}(t)\bar{y}(t) = 0,$$

or

$$t^4 \bar{y}'' + [2t^3 - t^2 \bar{P}(t)]\bar{y}' + \bar{Q}(t)\bar{y} = 0. \tag{7.9.5}$$

Hence, the function $\bar{y}$ satisfies the transformed differential equation

$$t^4 y'' + [2t^3 - t^2 \bar{P}(t)]y' + \bar{Q}(t)y = 0. \tag{7.9.6}$$

Conversely, if $\bar{y}(t)$ satisfies (7.9.6), then $y(x)$ satisfies (7.9.1).

Remark 7.9.1. We shall say that *infinity is an ordinary point, a regular singular point, or an irregular singular point* of Equation (7.9.1) if and only if for the transformed Equation (7.9.6), the point $t = 0$ is an ordinary point, a regular singular point, or an irregular singular point, respectively.

Remark 7.9.2. Let us write (7.9.6) as

$$t^2 y'' + \left[2 - \frac{\bar{P}(t)}{t}\right]ty' + \frac{\bar{Q}(t)}{t^2}\, y = 0. \tag{7.9.7}$$

Thus [*see* Remarks 7.4.1 and 7.4.2], $t = 0$ is a regular singular point of (7.9.6) if and only if the functions defined by $\bar{P}(t)/t$ and $\bar{Q}(t)/t^2$ are analytic at $t = 0$. Consequently, we have that

$$\bar{P}(t) = t \sum_{n=0}^{\infty} p_n t^n \quad \text{and} \quad \bar{Q}(t) = t^2 \sum_{n=0}^{\infty} q_n t^n, \tag{7.9.8}$$

where the series converge for $|t| < 1/R$, $R > 0$. Since $t = 1/x$, in view of (7.9.2) and (7.9.8), we see that

$$P(x) = \frac{1}{x} \sum_{n=0}^{\infty} \frac{p_n}{x^n} \quad \text{and} \quad Q(x) = \frac{1}{x^2} \sum_{n=0}^{\infty} \frac{q_n}{x^n}, \tag{7.9.9}$$

where the series converge for $|x| > R$.   Thus, infinity is a regular singular point of (7.9.1) if and only if (7.9.1) can be expressed as

$$x^2 y'' + p(x)xy' + q(x)y = 0, \tag{7.9.10}$$

where the functions p and q are analytic at infinity, that is,

$$p(x) = \sum_{n=0}^{\infty} \frac{p_n}{x^n} \quad \text{and} \quad q(x) = \sum_{n=0}^{\infty} \frac{q_n}{x^n} \tag{7.9.11}$$

and these series converge for $|x| > R$ for some $R > 0$.

Example 7.9.1.   Let us show that infinity is a regular singular point of the Legendre equation

$$(1 - x^2)y'' - 2xy' + \alpha(\alpha + 1)y = 0. \tag{7.9.12}$$

SOLUTION.   Let us write (7.9.12) as

$$y'' - \frac{2}{1 - x^2}\, xy' + \frac{\alpha(\alpha + 1)}{1 - x^2}\, y = 0. \tag{7.9.13}$$

Multiplying (7.9.13) by x^2, we get

$$x^2 y'' - \frac{2x^2}{1 - x^2}\, xy' + \frac{\alpha(\alpha + 1)x^2}{1 - x^2}\, y = 0. \tag{7.9.14}$$

Comparing (7.9.14) with (7.9.10), we see that

$$p(x) = -\frac{2x^2}{1 - x^2} \quad \text{and} \quad q(x) = \frac{\alpha(\alpha + 1)x^2}{1 - x^2}.$$

Now,

$$p(x) = -\frac{2x^2}{1 - x^2} = \frac{2}{1 - \dfrac{1}{x^2}} = 2 \sum_{n=0}^{\infty} \frac{1}{x^{2n}},$$

and the series converges for $|x| > 1$.   Similarly,

$$q(x) = -\alpha(\alpha + 1) \sum_{n=0}^{\infty} \frac{1}{x^{2n}},$$

and the series converges for $|x| > 1$.   Thus, infinity is a regular singular point of (7.9.12).

Exercises 7.9

1. (a) Find two linearly independent solutions of the Legendre equation
$$(1 - x^2)y'' - 2xy' + k(k + 1)y = 0$$
in the neighborhood of infinity.

(b) For k a nonnegative integer, show that one of the solutions in part (a) reduces to a polynomial.

2. A very important differential equation which has been extensively studied is

$$(1) \qquad x(1-x)y'' + [\gamma - (\alpha + \beta + 1)x]y' - \alpha\beta y = 0,$$

where α, β, and γ are constants. This equation is known as *Gauss' equation* or as the *hypergeometric equation*. For convenience, we shall denote Equation (1) by $G(\alpha, \beta, \gamma; x) = 0$.

(a) Show that $0, 1,$ and ∞ are regular singular points of $G(\alpha, \beta, \gamma; x) = 0$. Assume a solution of $G(\alpha, \beta, \gamma; x) = 0$ of the form

$$(2) \qquad y(x) = x^r \sum_{n=0}^{\infty} a_n x^n, \qquad a_0 \neq 0.$$

(b) Show that

$$a_0 r(r - 1 + \gamma)x^{r-1} + \sum_{n=0}^{\infty} [(n + r + 1)(n + r + \gamma)a_{n+1}$$

$$- (n + r + \alpha)(n + r + \beta)a_n]x^{r+n} = 0.$$

Thus, the indicial roots are $r_1 = 0$ and $r_2 = 1 - \gamma$, and the recurrence formula is

$$a_{n+1} = \frac{(n + r + \alpha)(n + r + \beta)}{(n + r + 1)(n + r + \gamma)} a_n, \qquad n = 0, 1, 2, \ldots .$$

(c) If γ is not an integer, show that for $0 < |x| < 1$, two linearly independent solutions y_1, y_2 of $G(\alpha, \beta, \gamma; x) = 0$ are

$$(3) \quad y_1(x) = 1 + \frac{\alpha\beta}{\gamma} x + \frac{\alpha(\alpha + 1)\beta(\beta + 1)}{\gamma(\gamma + 1)} \frac{x^2}{2!}$$

$$+ \cdots + \frac{\alpha(\alpha + 1)\cdots(\alpha + k - 1)\beta(\beta + 1)\cdots(\beta + k - 1)}{\gamma(\gamma + 1)\cdots(\gamma + k - 1)} \frac{x^k}{k!} + \cdots,$$

$$(4) \quad y_2(x) = |x|^{1-\gamma}\left[1 + \frac{(\alpha - \gamma + 1)(\beta - \gamma + 1)}{2 - \gamma} x \right.$$

$$+ \frac{(\alpha - \gamma + 1)(\alpha - \gamma + 2)(\beta - \gamma + 1)(\beta - \gamma + 2)}{(2 - \gamma)(3 - \gamma)} \frac{x^2}{2!} + \cdots$$

$$\left. + \frac{(\alpha - \gamma + 1)(\alpha - \gamma + 2)\cdots(\alpha - \gamma + k)(\beta - \gamma + 1)(\beta - \gamma + 2)\cdots(\beta - \gamma + k)}{(2 - \gamma)(3 - \gamma)\cdots(k + 1 - \gamma)} \frac{x^k}{k!} + \cdots \right].$$

The series in (3) is known as the *hypergeometric series*, and the function defined by it is the *hypergeometric function* usually denoted by $F(\alpha, \beta, \gamma; x)$. That is,

$$(5) \ F(\alpha, \beta, \gamma; x) = 1 + \sum_{k=1}^{\infty} \frac{\alpha(\alpha + 1)\cdots(\alpha + k - 1)\beta(\beta + 1)\cdots(\beta + k - 1)}{k!\gamma(\gamma + 1)\cdots(\gamma + k - 1)} x^k.$$

In view of (5), we may write (4) as

$$(6) \qquad y_2(x) = |x|^{1-\gamma} F(\alpha - \gamma + 1, \beta - \gamma + 1, 2 - \gamma; x).$$

Clearly, the notation $F(\alpha - \gamma + 1, \beta - \gamma + 1, 2 - \gamma; x)$ means that we replace α by $\alpha - \gamma + 1$, β by $\beta - \gamma + 1$, and γ by $2 - \gamma$ in (5).
Note that if $\gamma = 1$, then $y_1(x) = y_2(x)$; and for any other integer value of γ, one of the two series diverges.

3. Verify the following identities:

(a) $F(\alpha, \beta, \beta; x) = (1 - x)^{-\alpha}$,

(b) $xF(1, 1, 2; -x) = \ln(1 + x)$,

(c) $xF(\tfrac{1}{2}, \tfrac{1}{2}, \tfrac{3}{2}; x^2) = \arcsin x$,

(d) $xF(\tfrac{1}{2}, 1, \tfrac{3}{2}; -x^2) = \arctan x$,

(e) $2xF(\tfrac{1}{2}, 1, \tfrac{3}{2}; x^2) = \ln \dfrac{1 + x}{1 - x}$,

(f) $F'(\alpha, \beta, \gamma; x) = \dfrac{\alpha\beta}{\gamma} F(\alpha + 1, \beta + 1, \gamma + 1; x)$.

4. Consider the differential equation

$$(1) \qquad (x^2 + a_1 x + a_2)y'' + (a_3 x + a_4)y' + a_5 y = 0,$$

where $a_1, \ldots, a_5$ are constants. Suppose that the roots of the equation $x^2 + a_1 x + a_2 = 0$ are a and b, $a \neq b$. Thus, $x^2 + a_1 x + a_2 = (x - a)(x - b)$. Let

$$(2) \qquad x = a + (b - a)t.$$

Thus,

$$x - b = (x - a) - (b - a) = (b - a)t - (b - a) = (a - b)(1 - t).$$

Under the change of variable given in (2), show that Equation (1) is transformed into

$$(3) \qquad t(1 - t)\frac{d^2y}{dt^2} + (\lambda + \mu t)\frac{dy}{dt} + vy = 0,$$

where λ, μ, and v are constants. This is a hypergeometric equation in which $\gamma = \lambda$, and μ and v are defined by the equations

$$(4) \qquad \alpha + \beta + 1 = -\mu \quad \text{and} \quad \alpha\beta = -v.$$

5. Consider the Legendre equation

$$(1) \qquad (1 - x^2)y'' - 2xy' + n(n + 1)y = 0,$$

where n is a nonnegative integer. Utilize the result of Exercise 4 with $a = 1$, $b = -1$, and $x = 1 - 2t$, to show that Equation (1) can be expressed as

$$t(1 - t)\frac{d^2y}{dt^2} + (1 - 2t)\frac{dy}{dt} + n(n + 1)y = 0,$$

a hypergeometric equation with $\gamma = 1$, $\alpha = n + 1$, and $\beta = -n$. Thus, the Legendre equation (1) is equivalent to

$$G\left(n+1, -n, 1; \frac{1-x}{2}\right) = 0.$$

6. Consider the differential equation

(1) $$(1 - x^2)y'' - xy' + n^2 y = 0,$$

where n is a nonnegative integer. This equation is known as *Tschebyscheff's equation.*

(a) Utilize the result of Exercise 4 to show that Equation (1) is equivalent to

(2) $$G\left(n, -n, \frac{1}{2}; \frac{1-x}{2}\right) = 0.$$

(b) Utilize the result of Exercise 2 to show that the general solution $y_n(x)$ of Equation (1) is

(3) $$y_n(x) = c_1 F\left(n, -n, \frac{1}{2}; \frac{1-x}{2}\right)$$
$$+ c_2 \left(\frac{1-x}{2}\right)^{1/2} F\left(n+\frac{1}{2}, -n+\frac{1}{2}, \frac{3}{2}; \frac{1-x}{2}\right), \qquad |x| < 1,$$

where c_1 and c_2 are arbitrary constants.

(c) In view of (5) of Exercise 2, show that the particular solution $y_n(x) = F\left(n, -n, \frac{1}{2}; \frac{1-x}{2}\right)$, $n = 0, 1, 2, \ldots$, are polynomials. These are known as *Tschebyscheff's polynomials* and are usually denoted by $T_n(x)$. These polynomials are useful in statistics.

(d) Show that

(4) $$T_n(x) = 1 + \sum_{k=1}^{n} \frac{(-1)^k 2^k n(n+k-1)!}{(n-k)!(2k)!} (1-x)^k,$$
$$T_n(1) = 1, \quad n = 0, 1, 2, \ldots .$$

(e) Verify that

$$T_0(x) = 1, \qquad T_1(x) = x, \qquad T_2(x) = 2x^2 - 1, \qquad T_3(x) = 4x^3 - 3x,$$
$$T_4(x) = 8x^4 - 8x^2 + 1, \qquad T_5(x) = 16x^5 - 20x^3 + 5x.$$

7. Consider the differential equation

(1) $$xy'' + (\gamma - x)y' - \alpha y = 0,$$

where α and γ are constants. This equation is known as *Laguerre's equation.* Assume a solution of Equation (1) of the form

$$y(x) = x^r \sum_{n=0}^{\infty} a_n x^n, \qquad a_0 \neq 0.$$

(a) If γ is not an integer, show that for $0 < |x| < \infty$, two linearly independent solutions y_1, y_2 of Equation (1) are

$$(2) \qquad\qquad y_1(x) = F(\alpha, \gamma\,;\, x),$$

$$(3) \qquad\qquad y_2(x) = |x|^{1-\gamma} F(\alpha - \gamma + 1, 2 - \gamma;\, x),$$

where

$$(4) \qquad F(\alpha, \gamma;\, x) = 1 + \sum_{k=1}^{\infty} \frac{\alpha(\alpha+1)\cdots(\alpha+k-1)}{k!\,\gamma(\gamma+1)\cdots(\gamma+k-1)}\, x^k.$$

If n is a nonnegative integer, the polynomials defined by

$$(5) \qquad L_n^{(\mu)}(x) = \frac{(1+\mu)(2+\mu)\cdots(n+\mu)}{n!}\, F(-n, 1+\mu;\, x)$$

are known as the *generalized or the associated Laguerre's polynomials*. For the special case when $\mu = 0$, the polynomials are known as *Laguerre's polynomials*, and they are usually denoted by $L_n(x)$. That is,

$$(6) \qquad\qquad L_n(x) = F(-n, 1;\, x).$$

The associated Laguerre polynomials $L_n^{(1)}(x)$, are utilized in quantum mechanics in obtaining the wave equation of the hydrogen atom.

(b) Show that

$$(7) \qquad L_n^{(\mu)}(x) = \sum_{k=0}^{n} \frac{(-1)^k(1+\mu)(2+\mu)\cdots(n+\mu)}{k!(n-k)!(1+\mu)(2+\mu)\cdots(k+\mu)}\, x^k,$$

and

$$(8) \qquad\qquad L_n(x) = \sum_{k=0}^{n} \frac{(-1)^k n!}{(k!)^2(n-k)!}\, x^k,$$

where in (7), $(1+\mu)(2+\mu)\cdots(k+\mu)$ is defined to be 1 when $k = 0$. Clearly, $y = L_n^{(\mu)}(x)$ satisfies the differential equation (1) when $\alpha = -n$ and $\gamma = 1 + \mu$. Thus,

$$(9) \qquad x[L_n^{(\mu)}(x)]'' + (\mu + 1 - x)[L_n^{(\mu)}(x)]' + nL_n^{(\mu)}(x) = 0.$$

(c) If μ is independent of n and if $\mathscr{R}(\mu) > -1$, show that

$$(10) \qquad \int_0^{\infty} x^{\mu} e^{-x} L_m^{(\mu)}(x) L_n^{(\mu)}(x)\, dx = 0, \qquad m \neq n.$$

(d) *Rodrigues' formula* for $L_n(x)$. Show that

$$(11) \qquad\qquad L_n(x) = \frac{e^x}{n!}\, D^n(x^n e^{-x}).$$

(e) *Orthogonality of the Laguerre polynomials.* Show that

$$(12) \qquad \int_0^{\infty} e^{-x} L_m(x) L_n(x)\, dx = \begin{cases} 0, & m \neq n, \\ 1, & m = n. \end{cases}$$

Thus, the Laguerre polynomials form an orthogonal set with respect to the weight function e^{-x} on the interval $0 < x < \infty$.

(f) Show that

(13)
$$xL_n'(x) = nL_n(x) - nL_{n-1}(x),$$

(14)
$$L_n'(x) = L_{n-1}'(x) - L_{n-1}(x),$$

(15)
$$nL_n(x) = (2n - 1 - x)L_{n-1}(x) - (n - 1)L_{n-2}(x),$$

(16)
$$L_n'(x) = -\sum_{k=0}^{n-1} L_k(x).$$

[The above results are also true for $L_n^{(\mu)}(x)$.]

(g) Verify that

$$L_0(x) = 1, \qquad L_1(x) = 1 - x, \qquad L_2(x) = 1 - 2x + \tfrac{1}{2}x^2,$$
$$L_3(x) = 1 - 3x + \tfrac{3}{2}x^2 - \tfrac{1}{6}x^3, \qquad L_1^{(1)}(x) = 2 - x,$$
$$L_2^{(1)}(x) = 3 - 3x + \tfrac{1}{2}x^2, \qquad L_3^{(1)}(x) = 4 - 6x + 2x^2 - \tfrac{1}{6}x^3,$$
$$L_4^{(1)}(x) = 5 - 10x + 5x^2 - \tfrac{5}{6}x^3 + \tfrac{1}{24}x^4,$$
$$L_4^{(3)}(x) = 35 - 35x + \tfrac{21}{2}x^2 - \tfrac{7}{6}x^3 + \tfrac{1}{24}x^4.$$

8. Consider the differential equation

(1)
$$y'' - 2xy' + 2\alpha y = 0,$$

where α is a constant.   This equation is known as *Hermite's equation*.   Assume a solution of Equation (1) of the form

$$y(x) = \sum_{k=0}^{\infty} a_k x^k, \qquad a_0 \neq 0.$$

(a) Show that the general solution of Equation (1) valid for all (finite) values of x is

(2)
$$y(x) = a_0 \left[1 + \sum_{k=1}^{\infty} \frac{(-1)^k 2^k \alpha(\alpha - 2) \cdots (\alpha - 2k + 2)}{(2k)!} x^{2k} \right]$$
$$+ a_1 \left[x + \sum_{k=1}^{\infty} \frac{(-1)^k 2^k (\alpha - 1) \cdots (\alpha - 2k + 1)}{(2k + 1)!} x^{2k+1} \right],$$

where a_0 and a_1 are arbitrary constants.

(b) With $a_0 = (-1)^{n/2} \dfrac{2^{n/2} n!}{2 \cdot 4 \cdots n}$, $a_1 = (-1)^{(n-1)/2} \dfrac{2^{(n+1)/2} n!}{2 \cdot 4 \cdots (n-1)}$,   and $H_0(x) \equiv 1$, follow the method given in Section 8 [*see* Remark 7.8.1] to obtain the *Hermitian polynomials* $H_n(x)$

(3)
$$H_n(x) = \sum_{k=0}^{[n/2]} \frac{(-1)^k n!}{k!(n - 2k)!} (2x)^{n-2k},$$

where $[p]$ means the greatest integer less than or equal to the real number p. [*See* Exercise 7.2.4.]   These polynomials are useful in statistics and probability.

(c) Verify that

$$H_1(x) = 2x, \qquad H_2(x) = 4x^2 - 2, \qquad H_3(x) = 8x^3 - 12x,$$
$$H_4(x) = 16x^4 - 48x^2 + 12, \qquad H_5(x) = 32x^5 - 160x^3 + 120x,$$
$$H_6(x) = 64x^6 - 480x^4 + 720x^2 - 120.$$

(d) *A generating function for the Hermitian polynomials.*   Show that

$$(4) \qquad e^{2xt - t^2} = \sum_{n=0}^{\infty} \frac{H_n(x)}{n!} t^n.$$

(e) *Rodrigues' formula for $H_n(x)$.*   Show that

$$(5) \qquad H_n(x) = (-1)^n e^{x^2} \frac{d^n}{dx^n} (e^{-x^2}).$$

Clearly, $y = H_n(x)$ satisfies Equation (1) with $\alpha = n$.   That is,

$$(6) \qquad H_n''(x) - 2xH_n'(x) + 2nH_n(x) = 0.$$

(f) *Orthogonality of the Hermitian polynomials.*   Show that

$$(7) \qquad \int_{-\infty}^{\infty} e^{-x^2} H_m(x)H_n(x)\,dx = \begin{cases} 0, & m \neq n, \\ \sqrt{\pi}\, 2^n n!, & m = n. \end{cases}$$

Thus, the Hermitian polynomials form an orthogonal set with respect to the weight function e^{-x^2} on the interval $-\infty < x < \infty$.

(g) Show that

$$(8) \qquad H_n(x) = 2xH_{n-1}(x) - 2(n-1)H_{n-2}(x),$$

$$(9) \qquad H_n'(x) = 2nH_{n-1}(x),$$

$$(10) \qquad xH_n'(x) = nH_{n-1}'(x) + nH_n(x).$$

(h) Show that

$$(11) \qquad H_n(-x) = (-1)^n H_n(x),$$

$$(12) \qquad H_{2n}(0) = (-1)^n \frac{(2n)!}{n!}, \qquad H_{2n+1}(0) = 0,$$

$$(13) \qquad H_{2n}'(0) = 0, \qquad H_{2n+1}'(0) = (-1)^n \frac{2(2n+1)!}{n!}.$$

(i) Show that

$$(14) \qquad x^n = \sum_{k=0}^{[n/2]} \frac{n! H_{n-2k}(x)}{2^n k!(n-2k)!}.$$

Suggested Readings

Brauer and Nohel [10]   Goursat [17]
Coddington [12]   Rainville [35]

chapter 8   SYSTEMS OF LINEAR DIFFERENTIAL EQUATIONS

8.1. Introduction

In many problems occurring in mechanics, electricity, and other branches of science, one has to find a simultaneous solution of more than one differential equation with more than one unknown function; each unknown function depending upon a single independent variable. [*See* Section 8.5 for some examples.] Such situations usually arise when we are dealing with physical systems involving more than one degree of freedom. Roughly speaking, a physical system is said to have *n degrees of freedom* if the state of the system can be described by n independent conditions. In this chapter, not only shall we develop methods that will enable us to solve these problems of applications, but we shall also develop methods for solving an important class of "systems of linear differential equations." Also, in developing a method for solving a "system of linear first-order differential equations" we shall utilize some basic concepts from matrix theory.

8.2. Definitions and Existence Theorems

Definition 8.2.1. *A system of n first-order equations* is a system of the form

$$
\begin{cases}
\dfrac{dx_1}{dt} = F_1(t, x_1, x_2, \ldots, x_n), \\[2mm]
\dfrac{dx_2}{dt} = F_2(t, x_1, x_2, \ldots, x_n), \\[2mm]
\quad\vdots \\[2mm]
\dfrac{dx_n}{dt} = F_n(t, x_1, x_2, \ldots, x_n),
\end{cases}
\qquad (8.2.1)
$$

where each F_i $(i = 1, 2, \ldots, n)$ is a function of t, x_1, x_2, $\ldots$, x_n, defined in some region R of $(n + 1)$-dimensional $(t, x_1, x_2, \ldots, x_n)$ space.

Remark 8.2.1. Clearly in (8.2.1), the unknown functions are $x_1, x_2, \ldots, x_n$ and the independent variable is t. We shall often denote differentiation with respect to t by a prime, that is, dx_i/dt will be denoted by x_i'. Also, d^2x_i/dt^2, $d^3x_i/dt^3, \ldots$, will be denoted by $x_i'', x_i''', \ldots$.

Definition 8.2.2. A set of functions

$$x_1 = \phi_1(t), \quad x_2 = \phi_2(t), \ldots, x_n = \phi_n(t), \tag{8.2.2}$$

each defined on an interval I, $\alpha \leqq t \leqq \beta$, is said to be a *solution* of the system (8.2.1) if for each t in I

(i) $\phi_i'(t)$ exist, $\quad i = 1, 2, \ldots, n$,
(ii)   the point $(t, \phi_1(t), \phi_2(t), \ldots, \phi_n(t)) \subset R$,

and

(iii) $\begin{cases} \phi_1'(t) = F_1(t, \phi_1(t), \phi_2(t), \ldots, \phi_n(t)), \\ \phi_2'(t) = F_2(t, \phi_1(t), \phi_2(t), \ldots, \phi_n(t)), \\ \quad\vdots \\ \phi_n'(t) = F_n(t, \phi_1(t), \phi_2(t), \ldots, \phi_n(t)). \end{cases}$ (8.2.3)

Note that condition (iii) asserts that the set of functions in (8.2.2) simultaneously satisfies all the equations of the system (8.2.1) identically for each t in I. At times, we shall also write the set of functions in (8.2.2) as

$$x_1 = x_1(t), \quad x_2 = x_2(t), \ldots, x_n = x_n(t). \tag{8.2.4}$$

The following definition is a special case of Definition 8.2.1, and occurs when each function F_i $(i = 1, 2, \ldots, n)$ in (8.2.1) is a linear function of the dependent variables $x_1, x_2, \ldots, x_n$.

Definition 8.2.3. A *system of n linear first-order equations* is a system of the form

$$\begin{cases} x_1' = a_{11}(t)x_1 + a_{12}(t)x_2 + \cdots + a_{1n}(t)x_n + g_1(t), \\ x_2' = a_{21}(t)x_1 + a_{22}(t)x_2 + \cdots + a_{2n}(t)x_n + g_2(t), \\ \quad\vdots \\ x_n' = a_{n1}(t)x_1 + a_{n2}(t)x_2 + \cdots + a_{nn}(t)x_n + g_n(t), \end{cases} \tag{8.2.5}$$

where the a_{ij} and g_i $(i, j = 1, 2, \ldots, n)$ are functions of the independent variable t.

If each of the functions g_i $(i = 1, 2, \ldots, n)$ in (8.2.5) is identically zero in I, $\alpha \leqq t \leqq \beta$, then the system (8.2.5) is called *homogeneous*; if at least one of the functions g_i is not identically zero in I, then the system (8.2.5) is called *nonhomogeneous* (or *inhomogeneous*).

The system (8.2.5) may be written compactly as

$$x_i' = \sum_{j=1}^{n} a_{ij}(t)x_j + g_i(t), \qquad i = 1, 2, \ldots, n. \tag{8.2.6}$$

Remark 8.2.2. Let a set of initial conditions associated with the system (8.2.1) be of the form

$$x_1(t_0) = x_1^0, \quad x_2(t_0) = x_2^0, \ldots, x_n(t_0) = x_n^0, \tag{8.2.7}$$

where t_0 is some fixed point in I, $\alpha \leq t \leq \beta$, and the *n*-tuple of real numbers $(x_1^0, x_2^0, \ldots, x_n^0)$ is given in advance.

We shall now state two sufficiency existence theorems concerning the systems (8.2.1) and (8.2.5). [For outlines of proofs of these theorems, *see* Theorem 7 and the material immediately following its proof, and Theorem 9 of Appendix 2.]

Theorem 8.2.1. *Let each of the functions F_i $(i = 1, 2, \ldots, n)$ be defined in a domain D of $(n + 1)$-dimensional $(t, x_1, x_2, \ldots, x_n)$ space. Let each of the functions F_i and $\partial F_i/\partial x_j$ $(i, j = 1, 2, \ldots, n)$ be continuous in D, containing the point $(t_0, x_1^0, x_2^0, \ldots, x_n^0)$. Then an interval I_0: $|t - t_0| \leq h$ $(h > 0)$ exists on which there is a unique solution*

$$x_1 = \phi_1(t), \quad x_2 = \phi_2(t), \ldots, x_n = \phi_n(t),$$

satisfying the system (8.2.1) and the initial conditions (8.2.7).

Theorem 8.2.2. *Let each of the functions a_{ij} and g_i $(i, j = 1, 2, \ldots, n)$ be continuous on an interval I, $\alpha \leq t \leq \beta$, containing the point $t = t_0$. Then there exists a unique solution*

$$x_1 = \phi_1(t), \quad x_2 = \phi_2(t), \ldots, x_n = \phi_n(t),$$

satisfying the system (8.2.5) for each t in I and the initial conditions (8.2.7).

Remark 8.2.3. Note that in Theorem 8.2.1, the size of the interval I_0 is not specified. However, in Theorem 8.2.2, the size of the interval I on which there exists a unique solution is the same as the interval on which the given functions a_{ij} and g_i are assumed continuous.

Example 8.2.1. Let us show that the initial-value problem

$$\begin{cases} x_1' = t^2 \quad x_1 + e^t x_2 + \cos^2 t, \\ x_2' = (\sin t)x_1 - t^3 x_2 + e^{-2t} \sin 2t, \end{cases} \tag{8.2.8}$$

$$x_1(0) = 1, \quad x_2(0) = 2, \tag{8.2.9}$$

has a unique solution on the interval I, $\alpha \leq t \leq \beta$, where $\alpha > -\infty$ and $\beta < \infty$.

SOLUTION. Comparing the system (8.2.8) with the system (8.2.5) with $n = 2$, we see that $a_{11}(t) = t^2$, $a_{12}(t) = e^t$, $g_1(t) = \cos^2 t$, $a_{21}(t) = \sin t$, $a_{22}(t) = -t^3$, and $g_2(t) = e^{-2t} \sin 2t$. Clearly, each of these functions is continuous for all finite values of t. Thus, the interval I in Theorem 8.2.2 is given by $\alpha \leq t \leq \beta$, where $\alpha > -\infty$ and $\beta < \infty$. From the initial conditions (8.2.9), the point

$t = t_0 = 0$ belongs to I. Consequently, Theorem 8.2.2 assures us a unique solution $x_1 = \phi_1(t)$, $x_2 = \phi_2(t)$ satisfying the system (8.2.8) for each t in I and the initial conditions (8.2.9).

Example 8.2.2. Let us verify that the unique solution of the initial-value problem

$$\begin{cases} x_1' = -2x_1 - 4x_2 + 1 + 4t, \\ x_2' = -x_1 + x_2 + \frac{3}{2}t^2, \end{cases} \tag{8.2.10}$$

$$x_1(0) = 1, \quad x_2(0) = 4, \tag{8.2.11}$$

is given by $x_1 = \phi_1(t)$ and $x_2 = \phi_2(t)$, where

$$\phi_1(t) = -3e^{2t} + 4e^{-3t} + t + t^2 \quad \text{and} \quad \phi_2(t) = 3e^{2t} + e^{-3t} - \frac{1}{2}t^2. \tag{8.2.12}$$

The solution is valid for each t in I, $\alpha \leq t \leq \beta$, where $\alpha > -\infty$ and $\beta < \infty$.

VERIFICATION. As in Example 8.2.1, one may readily verify that the given initial-value problem has a unique solution which is valid for each t in I. We shall now verify that the set of functions $x_1 = \phi_1(t)$ and $x_2 = \phi_2(t)$ simultaneously satisfies the two equations of the system (8.2.10) identically for each t in I. Differentiating $\phi_1(t)$ and $\phi_2(t)$, we obtain

$$\phi_1'(t) = -6e^{2t} - 12e^{-3t} + 1 + 2t \quad \text{and} \quad \phi_2'(t) = 6e^{2t} - 3e^{-3t} - t. \tag{8.2.13}$$

Substituting $\phi_1(t)$, $\phi_2(t)$, $\phi_1'(t)$, and $\phi_2'(t)$ of (8.2.12) and (8.2.13) into the system (8.2.10), we find

$$\begin{aligned} -6e^{2t} - 12e^{-3t} + 1 + 2t &= -2(-3e^{2t} + 4e^{-3t} + t + t^2) \\ &\quad - 4(3e^{2t} + e^{-3t} - \tfrac{1}{2}t^2) + 1 + 4t \\ &= -6e^{2t} - 12e^{-3t} + 1 + 2t, \end{aligned}$$

and

$$\begin{aligned} 6e^{2t} - 3e^{-3t} - t &= -(-3e^{2t} + 4e^{-3t} + t + t^2) + 3e^{2t} + e^{-3t} - \tfrac{1}{2}t^2 + \tfrac{3}{2}t^2 \\ &= 6e^{2t} - 3e^{-3t} - t. \end{aligned}$$

Finally, we verify the initial conditions in (8.2.11). Substituting $t = 0$ in (8.2.12), we obtain

$$\phi_1(0) = -3e^0 + 4e^0 + 0 + 0^2 = 1 \quad \text{and} \quad \phi_2(0) = 3e^0 + e^0 - \tfrac{1}{2}0^2 = 4.$$

Remark 8.2.4. Let us consider the differential equation of order n

$$y^{(n)} = F(t, y, y', \ldots, y^{(n-1)}). \tag{8.2.14}$$

Equation (8.2.14) can be transformed into a system of n first-order equations by making the following change of variables. Let

$$x_1 = y, \quad x_2 = y', \quad x_3 = y'', \ldots, x_{n-1} = y^{(n-2)}, \ x_n = y^{(n-1)}. \tag{8.2.15}$$

Then,

$$x_1' = y' = x_2, \quad x_2' = y'' = x_3, \ldots, x_{n-1}' = y^{(n-1)} = x_n,$$

and in view of (8.2.14), $x_n' = y^{(n)} = F(t, x_1, x_2, \ldots, x_n)$. Thus, we have

$$\begin{cases} x_1' = x_2, \\ x_2' = x_3, \\ \quad \vdots \\ x_{n-1}' = x_n, \\ x_n' = F(t, x_1, x_2, \ldots, x_n), \end{cases} \tag{8.2.16}$$

which is a special case of the system (8.2.1). We leave it for the reader [*see* Exercise 8.2.1] to show that (8.2.14) and (8.2.16) are equivalent. In view of this result, we see that the theory of a differential equation or order n is a special case of the corresponding theory of a system of n first-order equations.

Example 8.2.3. Let us establish the equivalence of the following systems.

$$\begin{cases} y_1'' = a_{11}y_1 + a_{12}y_2 + a_{13}y_3 = \sum_{n=1}^{3} a_{1n}y_n, \\[2ex] y_2'' = a_{21}y_1 + a_{22}y_2 + a_{23}y_3 = \sum_{n=1}^{3} a_{2n}y_n, \\[2ex] y_3'' = a_{31}y_1 + a_{32}y_2 + a_{33}y_3 = \sum_{n=1}^{3} a_{3n}y_n, \end{cases} \tag{8.2.17}$$

where the a_{ij}, $i, j = 1, 2, 3$ are constants, and

$$\begin{cases} x_1' = x_2, \\ x_2' = a_{11}x_1 + a_{12}x_3 + a_{13}x_5 = \sum_{n=1}^{3} a_{1n}x_{2n-1}, \\ x_3' = x_4, \\ x_4' = a_{21}x_1 + a_{22}x_3 + a_{23}x_5 = \sum_{n=1}^{3} a_{2n}x_{2n-1}, \\ x_5' = x_6, \\ x_6' = a_{31}x_1 + a_{32}x_3 + a_{33}x_5 = \sum_{n=1}^{3} a_{3n}x_{2n-1}. \end{cases} \tag{8.2.18}$$

SOLUTION. Let $x_1 = y_1$, $x_2 = y_1'$, $x_3 = y_2$, $x_4 = y_2'$, $x_5 = y_3$, $x_6 = y_3'$. Now utilizing (8.2.17), we obtain (8.2.18).

Let the set of functions $y_i = \phi_i(t)$, $i = 1, 2, 3$, be a solution of the system (8.2.17) on an interval I, $\alpha \le t \le \beta$. Then,

$$\begin{cases} \phi_1'' = \displaystyle\sum_{n=1}^{3} a_{1n}\phi_n, \\[2mm] \phi_2'' = \displaystyle\sum_{n=1}^{3} a_{2n}\phi_n, \\[2mm] \phi_3'' = \displaystyle\sum_{n=1}^{3} a_{3n}\phi_n, \end{cases} \tag{8.2.19}$$

for each t in I. Setting $\psi_1(t) = \phi_1(t)$, $\psi_2(t) = \phi_1'(t)$, $\psi_3(t) = \phi_2(t)$, $\psi_4(t) = \phi_2'(t)$, $\psi_5(t) = \phi_3(t)$, $\psi_6(t) = \phi_3'(t)$, and using (8.2.19) to evaluate ϕ_1'', ϕ_2'', and ϕ_3'', we obtain

$$\begin{cases} \psi_1' = \phi_1' = \psi_2, \\[2mm] \psi_2' = \phi_1'' = \displaystyle\sum_{n=1}^{3} a_{1n}\psi_{2n-1}, \\[2mm] \psi_3' = \phi_2' = \psi_4, \\[2mm] \psi_4' = \phi_2'' = \displaystyle\sum_{n=1}^{3} a_{2n}\psi_{2n-1}, \\[2mm] \psi_5' = \phi_3' = \psi_6, \\[2mm] \psi_6' = \phi_3'' = \displaystyle\sum_{n=1}^{3} a_{3n}\psi_{2n-1}, \end{cases} \tag{8.2.20}$$

for each t in I. Thus, the set of functions $x_i = \psi_i(t)$, $i = 1, \ldots, 6$, is a solution of the system (8.2.18) for each t in I.

Conversely, suppose that the set of functions $x_i = \psi_i(t)$, $i = 1, \ldots, 6$, is a solution of the system (8.2.18) for each t in I. Setting

$$\phi_1(t) = \psi_1(t), \quad \phi_2(t) = \psi_3(t), \quad \phi_3(t) = \psi_5(t),$$

for each t in I, and utilizing the system (8.2.20) to evaluate $\psi_i'(t)$, $i = 1, \ldots, 6$, we obtain

$$\begin{cases} \phi_1' = \psi_1' = \psi_2, \\[2mm] \phi_1'' = \psi_2' = \displaystyle\sum_{n=1}^{3} a_{1n}\psi_{2n-1} = \displaystyle\sum_{n=1}^{3} a_{1n}\phi_n, \\[2mm] \phi_2' = \psi_3' = \psi_4, \\[2mm] \phi_2'' = \psi_4' = \displaystyle\sum_{n=1}^{3} a_{2n}\psi_{2n-1} = \displaystyle\sum_{n=1}^{3} a_{2n}\phi_n, \\[2mm] \phi_3' = \psi_5' = \psi_6, \\[2mm] \phi_3'' = \psi_6' = \displaystyle\sum_{n=1}^{3} a_{3n}\psi_{2n-1} = \displaystyle\sum_{n=1}^{3} a_{3n}\phi_n, \end{cases}$$

for each t in I.   From the above equations, we have

$$
\begin{cases}
\phi_1'' = \sum_{n=1}^{3} a_{1n}\phi_n, \\[2ex]
\phi_2'' = \sum_{n=1}^{3} a_{2n}\phi_n, \\[2ex]
\phi_3'' = \sum_{n=1}^{3} a_{3n}\phi_n,
\end{cases}
$$

for each t in I.   Thus, the set of functions $y_i = \phi_i(t), i = 1, 2, 3,$ is a solution of the system (8.2.17) for each t in I.   The equivalence between the systems (8.2.17) and (8.2.18) is now established.

Note in Example 8.2.3, a system of 3 second-order equations was transformed into a system of 6 first-order equations.   Systems of higher–order differential equations can also be transformed into systems of first-order equations.   For example, we leave it to the reader [*see* Exercise 8.2.2] to show the equivalence of the systems

$$
\begin{cases}
u''' = G_1(t,\, u,\, u',\, u'',\, v,\, v',\, v'',\, w,\, w',\, w''), \\
v''' = G_2(t,\, u,\, u',\, u'',\, v,\, v',\, v'',\, w,\, w',\, w''), \\
w''' = G_3(t,\, u,\, u',\, u'',\, v,\, v',\, v'',\, w,\, w',\, w''),
\end{cases}
\tag{8.2.21}
$$

and

$$
\begin{cases}
x_1' = x_2, \\
x_2' = x_3, \\
x_3' = G_1(t,\, x_1,\, x_2,\, x_3,\, x_4,\, x_5,\, x_6,\, x_7,\, x_8,\, x_9), \\
x_4' = x_5, \\
x_5' = x_6, \\
x_6' = G_2(t,\, x_1,\, x_2,\, x_3,\, x_4,\, x_5,\, x_6,\, x_7,\, x_8,\, x_9), \\
x_7' = x_8, \\
x_8' = x_9, \\
x_9' = G_3(t,\, x_1,\, x_2,\, x_3,\, x_4,\, x_5,\, x_6,\, x_7,\, x_8,\, x_9).
\end{cases}
\tag{8.2.22}
$$

Remark 8.2.5.   It should become apparent by now that a knowledge of a system of n first-order equations is not as restrictive as one might have initially anticipated.   In particular, many problems of applications lead to a system of equations which can usually be transformed into a system of linear first-order equations, whose theory is a special case of the corresponding theory of a system of first-order equations.   However, it should be mentioned that, except in very special cases, it is extremely difficult or even impossible to obtain explicitly, in terms of elementary functions, a solution for the system (8.2.1).

Exercises 8.2

1. Establish the equivalence of (8.2.14) and (8.2.16).
2. Establish the equivalence of the systems (8.2.21) and (8.2.22).
3. Consider the initial-value problem

$$y''' + a_1(t)y'' + a_2(t)y' + a_3(t)y = g(t),$$

$$y(t_0) = y_0, \quad y'(t_0) = y_0', \quad y''(t_0) = y_0''.$$

Show that this initial-value problem is equivalent to

$$\begin{cases} x_1' = x_2, \\ x_2' = x_3, \\ x_3' = -a_3(t)x_1 - a_2(t)x_2 - a_1(t)x_3 + g(t), \end{cases}$$

$$x_1(t_0) = y_0, \quad x_2(t_0) = y_0', \quad x_3(t_0) = y_0''.$$

4. Verify that the unique solution of the initial-value problem

$$\begin{cases} x_1' = 3x_1 - \tfrac{1}{2}x_2 - 3t^2 - \tfrac{1}{2}t + \tfrac{3}{2}, \\ x_2' = 2x_2 - 2t - 1, \end{cases}$$

$$x_1(0) = 2, \quad x_2(0) = 3,$$

is given by $x_1 = \phi_1(t)$ and $x_2 = \phi_2(t)$, where

$$\phi_1(t) = e^{2t} + e^{3t} + t^2 + t \quad \text{and} \quad \phi_2(t) = 2e^{2t} + 1 + t.$$

The solution is valid for each t in I, $-\infty < t < \infty$.

8.3. Solution of a Linear Algebraic System of Equations.
Linear Differential Operators

(a) *Solution of a Linear Algebraic System of Equations.* Before developing a
method in the next section for solving a system of linear (differential) equations with constant coefficients, we shall recall a few basic results from college
algebra concerning a method of solving a linear algebraic system of equations.
Consider, for example, the system

$$\begin{cases} a_{11}x_1 + a_{12}x_2 + a_{13}x_3 = b_1, \\ a_{21}x_1 + a_{22}x_2 + a_{23}x_3 = b_2, \\ a_{31}x_1 + a_{32}x_2 + a_{33}x_3 = b_3, \end{cases} \tag{8.3.1}$$

where the a_{ij}, b_i, $i, j = 1, 2, 3$, are either constants or quantities that are independent of the three unknowns x_1, x_2, and x_3. Let us denote the determinant of the coefficients of the unknowns by $\det A$, that is,

$$\det A = \begin{vmatrix} a_{11} & a_{12} & a_{13} \\ a_{21} & a_{22} & a_{23} \\ a_{31} & a_{32} & a_{33} \end{vmatrix}. \tag{8.3.2}$$

For the present, suppose that det $A \neq 0$. Then the system (8.3.1) can be transformed into an *equivalent triangular system*:

$$\begin{cases} c_{11}x_1 + c_{12}x_2 + c_{13}x_3 = d_1, \\ \qquad\quad c_{22}x_2 + c_{23}x_3 = d_2, \\ \qquad\qquad\qquad c_{33}x_3 = d_3, \end{cases} \qquad \begin{cases} c_{11}x_1 \qquad\qquad\qquad\qquad = f_1, \\ c_{21}x_1 + c_{22}x_2 \qquad\qquad = f_2, \\ c_{31}x_1 + c_{32}x_2 + c_{33}x_3 = f_3, \end{cases} \quad (8.3.3)$$

or

$$\begin{cases} e_{11}x_1 + e_{12}x_2 + e_{13}x_3 = g_1, \\ e_{21}x_1 + e_{22}x_2 \qquad\quad = g_2, \\ e_{31}x_1 \qquad\qquad\qquad = g_3, \end{cases} \qquad \begin{cases} \qquad\qquad\qquad e_{13}x_3 = h_1, \\ \qquad\quad e_{22}x_2 + e_{23}x_3 = h_2, \\ e_{31}x_1 + e_{32}x_2 + e_{33}x_3 = h_3, \end{cases} \quad (8.3.4)$$

where the c_{ij}, e_{ij}, d_i, f_i, g_i, and h_i $(i, j = 1, 2, 3)$ are either constants or quantities which are independent of the unknowns x_1, x_2, and x_3. Moreover,

$$\det C = \begin{vmatrix} c_{11} & c_{12} & c_{13} \\ 0 & c_{22} & c_{23} \\ 0 & 0 & c_{33} \end{vmatrix} = \begin{vmatrix} c_{11} & 0 & 0 \\ c_{21} & c_{22} & 0 \\ c_{31} & c_{32} & c_{33} \end{vmatrix} = c_{11}c_{22}c_{33} \neq 0,$$

and

$$\det E = \begin{vmatrix} e_{11} & e_{12} & e_{13} \\ e_{21} & e_{22} & 0 \\ e_{31} & 0 & 0 \end{vmatrix} = \begin{vmatrix} 0 & 0 & e_{13} \\ 0 & e_{22} & e_{23} \\ e_{31} & e_{32} & e_{33} \end{vmatrix} = -e_{31}e_{22}e_{13} \neq 0.$$

$$(8.3.5)$$

Thus, the elements along the diagonals in det C and det E are each different from zero. Conversely, if det $C \neq 0$ or det $E \neq 0$, so is det $A \neq 0$.

The unique solution of the system (8.3.1) may now be readily obtained from the systems (8.3.3) or (8.3.4). For example, consider the first system in (8.3.3). From the third equation, we obtain $x_3 = d_3/c_{33}$. Substituting x_3 into the second equation, we obtain $x_2 = (c_{33}d_2 - c_{23}d_3)/c_{22}c_{33}$. Finally, substituting x_3 and x_2 into the first equation we can find x_1.

If det $A = 0$, so are det $C = 0$ and det $E = 0$; and conversely, if det $C = 0$ or det $E = 0$, so is det $A = 0$. In this case the system (8.3.1) may have either infinitely many solutions or no solution. The equivalent triangular systems show up these cases very nicely. [*See* Examples 8.3.2 and 8.3.3.]

Example 8.3.1.   Solve the system

$$\begin{cases} x_1 + x_2 - 2x_3 = 1, \\ 2x_1 + 7x_2 - 9x_3 = 12, \\ 3x_1 - x_2 + 4x_3 = 7. \end{cases} \qquad (8.3.6)$$

SOLUTION.   In (8.3.6), add -2 times the first equation to the second equation; add -3 times the first equation to the third equation; and we obtain

$$\begin{cases} x_1 + x_2 - 2x_3 = 1, \\ \qquad\quad 5x_2 - 5x_3 = 10, \\ \qquad\quad -4x_2 + 10x_3 = 4. \end{cases} \tag{8.3.7}$$

In (8.3.7), multiply the second equation by $\frac{1}{5}$; multiply the third equation by $\frac{1}{2}$; and we get

$$\begin{cases} x_1 + x_2 - 2x_3 = 1, \\ \qquad\quad x_2 - x_3 = 2, \\ \qquad\quad -2x_2 + 5x_3 = 2. \end{cases} \tag{8.3.8}$$

In (8.3.8), add 2 times the second equation to the third equation; and we obtain a desired equivalent triangular system

$$\begin{cases} x_1 + x_2 - 2x_3 = 1, \\ \qquad\quad x_2 - x_3 = 2, \\ \qquad\qquad\quad 3x_3 = 6. \end{cases} \tag{8.3.9}$$

From (8.3.9), we readily find $x_3 = 2$, $x_2 = 4$, and $x_1 = 1$, which is the solution of the system (8.3.6).

For a linear algebraic system of equations

$$\sum_{j=1}^{n} a_{ij} x_j = b_i \qquad i = 1, \ldots, n, \tag{8.3.10}$$

suppose that an *equivalent triangular system* has the form

$$\sum_{j=1}^{n} c_{ij} x_j = d_i \qquad i = 1, 2, \ldots, n, \tag{8.3.11}$$

where $c_{ij} = 0$ for $i > j$. If $\det A \neq 0$ so is $\det C \neq 0$, and conversely. The unique solution of the system (8.3.10) is obtained by solving the system (8.3.11). If $\det A = 0$ so is $\det C = 0$, and conversely. In this case, the system (8.3.10) may have either infinitely many solutions or else no solution. The following examples illustrate these cases.

Example 8.3.2.   Let us determine whether the system

$$\begin{cases} x_1 - x_2 + x_3 = 2, \\ 2x_1 - 3x_2 + 4x_3 = 8, \\ 3x_1 - 4x_2 + 5x_3 = 10, \end{cases} \tag{8.3.12}$$

has no solution or infinitely many solutions.

SOLUTION.   Following the method given in Example 8.3.1, an equivalent triangular system of (8.3.12) is

$$\begin{cases} x_1 - x_2 + x_3 = 2, \\ \qquad\quad -x_2 + 2x_3 = 4, \\ \qquad\qquad\quad 0x_3 = 0. \end{cases} \tag{8.3.13}$$

From the third equation in (8.3.13), we see that x_3 is arbitrary. From the second equation in (8.3.13), we find $x_2 = 2x_3 - 4$. Substituting this value of x_2 into the first equation in (8.3.13), we find $x_1 = x_3 - 2$. Thus, the system (8.3.12) has infinitely many solutions,

$$x_1 = x_3 - 2, \qquad x_2 = 2x_3 - 4, \qquad x_3 = x_3,$$

where x_3 is arbitrary. For instance, if $x_3 = 1$, then $x_1 = -1$ and $x_2 = -2$; if $x_3 = 0$, then $x_1 = -2$ and $x_2 = -4$.

Example 8.3.3. Let us determine whether the system

$$\begin{cases} x_1 - x_2 + x_3 = 2, \\ 2x_1 - 3x_2 + 4x_3 = 8, \\ 3x_1 - 4x_2 + 5x_3 = 9, \end{cases} \tag{8.3.14}$$

has no solution or infinitely many solutions.

SOLUTION. An equivalent triangular system of (8.3.14) is

$$\begin{cases} x_1 - x_2 + x_3 = 2, \\ -x_2 + 2x_3 = 4, \\ 0x_3 = -1. \end{cases} \tag{8.3.15}$$

From the third equation in (8.3.15), we see that x_3 cannot be determined. Thus, the system (8.3.14) has no solution.

(b) *Linear Differential Operators.* In view of Exercise 4.3.8, the operator L defined by

$$L[x] = a_0(t)x^{(n)} + a_1(t)x^{(n-1)} + \cdots + a_{n-1}(t)x' + a_n(t)x, \tag{8.3.16}$$

is a *linear differential operator of order n.* In (8.3.16) x is a function of t. If we denote $d^k x/dt^k$ by $D^k x$ $(k = 1, \ldots, n)$ and $D^0 x \equiv x$, we may write (8.3.16) as

$$L[x] = a_0(t)D^n x + a_1(t)D^{n-1}x + \cdots + a_{n-1}(t)Dx + a_n(t)x. \tag{8.3.17}$$

Setting

$$P(D) = a_0(t)D^n + a_1(t)D^{n-1} + \cdots + a_{n-1}(t)D + a_n(t), \tag{8.3.18}$$

we may also write (8.3.16) as

$$L[x] = P(D)[x]. \tag{8.3.19}$$

$P(D)$ in (8.3.18) is known as a *polynomial operator of order n.*

Definition 8.3.1. Two operators L_1 and L_2 are said to be *equal* if and only if

$$L_1[x] = L_2[x] \tag{8.3.20}$$

for all functions x for which both $L_1[x]$ and $L_2[x]$ are defined.

Let

$$L_1[x] = P_1(D)[x], \quad L_2[x] = P_2(D)[x], \quad \text{and} \quad L_3[x] = P_3(D)[x], \quad (8.3.21)$$

where $P_1(D)$, $P_2(D)$, and $P_3(D)$ are polynomial operators of order m, n, and s, respectively.

Definition 8.3.2. The *sum* of two operators L_1 and L_2 is the operator defined by

$$(L_1 + L_2)[x] = L_1[x] + L_2[x], \quad (8.3.22)$$

for all functions x for which both $L_1[x]$ and $L_2[x]$ are defined.

Example 8.3.4. Let $L_1 = t^2 D^2 + \sin t D + e^t$, $L_2 = \cos t D + 4$, and $x = \sin 2t$. Let us determine $(L_1 + L_2)[x]$.

SOLUTION.

$$
\begin{aligned}
(L_1 + L_2)[x] &= (t^2 D^2 + \sin t D + e^t)[\sin 2t] + (\cos t D + 4)[\sin 2t] \\
&= t^2 D^2 \sin 2t + \sin t D \sin 2t + e^t \sin 2t + \cos t D \sin 2t \\
&\quad + 4 \sin 2t \\
&= -4t^2 \sin 2t + 2 \sin t \cos 2t + e^t \sin 2t + 2 \cos t \cos 2t \\
&\quad + 4 \sin 2t \\
&= (4 - 4t^2 + e^t) \sin 2t + 2(\sin t + \cos t)\cos 2t.
\end{aligned}
$$

Definition 8.3.3. The *product* of two operators L_1 and L_2 is the operator defined by

$$(L_1 L_2)[x] = L_1[L_2[x]], \quad (8.3.23)$$

for all functions x for which both $L_2[x]$ and $L_1[L_2[x]]$ are defined.

Example 8.3.5. Let $L_1 = e^t D + \cos t$, $L_2 = D^2 + \sin t D + t^2$, and $x = t^3 + 1$. Let us determine $(L_1 L_2)[x]$.

SOLUTION.

$$
\begin{aligned}
(L_1 L_2)[x] &= (e^t D + \cos t)\,[(D^2 + \sin t D + t^2)[t^3 + 1]] \\
&= (e^t D + \cos t)\,[D^2(t^3 + 1) + \sin t D(t^3 + 1) + t^5 + t^2] \\
&= (e^t D + \cos t)\,[6t + 3t^2 \sin t + t^5 + t^2] \\
&= e^t D\,[6t + 3t^2 \sin t + t^5 + t^2] \\
&\quad + \cos t(6t + 3t^2 \sin t + t^5 + t^2) \\
&= e^t(6 + 6t \sin t + 3t^2 \cos t + 5t^4 + 2t) \\
&\quad + \cos t(6t + 3t^2 \sin t + t^5 + t^2) \\
&= 6e^t + (2e^t + 6e^t \sin t + 6 \cos t)t + (3e^t \cos t + \cos t \\
&\quad + 3 \cos t \sin t)t^2 + 5e^t t^4 + (\cos t)t^5.
\end{aligned}
$$

For linear differential operators, the following results are known to be true.

Commutative Law. $L_1 + L_2 = L_2 + L_1.$ (8.3.24)

Associative Laws. $L_1 + (L_2 + L_3) = (L_1 + L_2) + L_3,$

$$L_1(L_2 L_3) = (L_1 L_2)L_3.$$ (8.3.25)

Distributive Law. $L_1(L_2 + L_3) = L_1 L_2 + L_1 L_3.$ (8.3.26)

For linear differential operators with constant coefficients, the following additional results are also known to be true.

Commutative Law. $L_1 L_2 = L_2 L_1.$ (8.3.27)

$$P(D) = a_0 D^n + a_1 D^{n-1} + \cdots + a_{n-1}D + a_n = a_0(D - r_1)(D - r_2)\cdots(D - r_n),$$

$a_0 \neq 0,$ *and where* $r_1, r_2, \ldots, r_n$ *are the roots of* $P(r) = 0.$ (8.3.28)

That is, polynomial operators with constant coefficients can be factored as in the case of ordinary polynomials.

Remark 8.3.1.   Notice that the linear differential operators with constant coefficients obey all the laws of ordinary algebra with respect to addition and multiplication.

Example 8.3.6.   Let $L_1 = tD$ and $L_2 = \sin tD.$ Let us show that

$$L_1 L_2 \neq L_2 L_1.$$

SOLUTION.

$$(L_1 L_2)[x] = tD\,[(\sin tD)[x]] = tD[\sin tDx] = t(\sin tD^2 x + \cos tDx)$$
$$= t \sin tD^2 x + t \cos tDx.$$
$$(L_2 L_1)[x] = \sin tD\,[(tD)[x]] = \sin tD[tDx] = \sin t(tD^2 x + Dx)$$
$$= t \sin tD^2 x + \sin tDx.$$

Hence, in view of Definition 8.3.1, $L_1 L_2 \neq L_2 L_1.$ Thus, if the linear differential operators have variable coefficients, then the commutative law is not necessarily true.

Example 8.3.7.   Let $L_1 = D - 2,$ $L_2 = D - 3,$ and $L_3 = D^2 - 5D + 6.$ Let us show

$$D^2 - 5D + 6 = (D - 2)(D - 3).$$

SOLUTION.

$$(L_1 L_2)[x] = (D - 2)\,[(D - 3)[x]] = (D - 2)\,[Dx - 3x]$$
$$= D[Dx - 3x] - 2(Dx - 3x) = D^2 x - 3Dx - 2Dx + 6x$$
$$= D^2 x - 5Dx + 6x = L_3[x].$$

Hence, in view of Definition 8.3.1, we have

$$D^2 - 5D + 6 = (D - 2)(D - 3).$$

Example 8.3.8. Let $L_1 = D^n$, $L_2 = D^m$, and $L_3 = D^{n+m}$. Let us show that

$$D^n D^m = D^{n+m}.$$

SOLUTION.

$$(L_1 L_2)[x] = D^n[D^m[x]] = D^n[D^m x] = D^{n+m} x = L_3[x],$$

from which the result now follows.

Exercises 8.3

In Exercises 1 through 4, solve each system by first transforming it into an equivalent triangular system.

1. $\begin{cases} 2x_1 + 5x_2 = 3, \\ 4x_1 - 2x_2 = 1. \end{cases}$

3. $\begin{cases} 2x_1 - 3x_2 + 4x_3 = 8, \\ 3x_1 + 4x_2 - 5x_3 = -4, \\ 4x_1 - 5x_2 + 6x_3 = 12. \end{cases}$

2. $\begin{cases} x_1 + 2x_2 + x_3 = 7, \\ x_1 + x_2 - x_3 = 2, \\ 3x_1 - x_2 + 2x_3 = 12. \end{cases}$

4. $\begin{cases} x_1 + 3x_2 - x_3 + 4x_4 = 5, \\ 2x_1 - x_2 + 2x_3 + x_4 = 4, \\ 4x_1 + 4x_2 + 3x_3 + 2x_4 = 0, \\ 3x_1 - 2x_2 + x_3 - x_4 = -5. \end{cases}$

In Exercises 5 through 7, determine whether each system has no solution or infinitely many solutions.

5. $\begin{cases} 2x_1 + 4x_2 = 3, \\ 4x_1 + 8x_2 = 1. \end{cases}$

7. $\begin{cases} x_1 + 2x_2 - x_3 + x_4 = -2, \\ -x_1 - 3x_2 + 3x_3 - 2x_4 = 6, \\ 3x_1 + 5x_2 - 3x_3 + 5x_4 = -6, \\ 3x_1 + 4x_2 - x_3 + 4x_4 = -2. \end{cases}$

6. $\begin{cases} x_1 + x_2 - x_3 = 1, \\ 2x_1 - 3x_2 + 6x_3 = -3, \\ 4x_1 - x_2 + 4x_3 = -1. \end{cases}$

8. Let $L_1 = D^2 - (r_1 + r_2)D + r_1 r_2$ and $L_2 = (D - r_1)(D - r_2)$. Show that

$$D^2 - (r_1 + r_2)D + r_1 r_2 = (D - r_1)(D - r_2).$$

9. Let $L_1 = (D - r_1)(D - r_2)$ and $L_2 = (D - r_2)(D - r_1)$. Show that

$$(D - r_1)(D - r_2) = (D - r_2)(D - r_1).$$

10. Let $L_1 = e^t D + 1$ and $L_2 = t^3 D + \cos t$. Show that

$$L_1 L_2 \neq L_2 L_1.$$

11. Let $L_1 = D^2 + 2$, $L_2 = D^2 + 3$, and $L_3 = D^4 + 5D^2 + 6$. Show that

$$D^4 + 5D^2 + 6 = (D^2 + 2)(D^2 + 3).$$

12. Let $L[x] = P(D)[x]$, where $P(D)$ is given in (8.3.18).   Show that

$$L[e^{mt}] = P(m)e^{mt}.$$

13. Let $L = (D - m)^k$ $(k = 1, \ldots)$ and $(D - m)^0[x] \equiv x$.   Show that

$$L[t^k e^{mt}] = k! e^{mt}.$$

14. Let $L = (D - \alpha)^n$.   Show that

$$L[t^k e^{\alpha t}] = 0, \text{ for } k = 0, 1, \ldots, n - 1.$$

15. Let $L = D^n$.   Show that

$$L[e^{-\alpha t} F(t)] = e^{-\alpha t}(D - \alpha)^n[F(t)],$$

for all functions F which are differentiable n times.

8.4. Solution of a System of Linear Equations

Definition 8.4.1.   The system of equations

$$\begin{cases} P_{11}(D)x_1 + P_{12}(D)x_2 + \cdots + P_{1n}(D)x_n = g_1(t), \\ P_{21}(D)x_1 + P_{22}(D)x_2 + \cdots + P_{2n}(D)x_n = g_2(t), \\ \qquad\qquad\qquad \vdots \\ P_{n1}(D)x_1 + P_{n2}(D)x_2 + \cdots + P_{nn}(D)x_n = g_n(t), \end{cases} \tag{8.4.1}$$

where each P_{ij} $(i, j = 1, \ldots, n)$ is a polynomial operator, is called a *system of n linear differential equations*.

Definition 8.4.2.   A set of functions

$$x_1 = \phi_1(t), \quad x_2 = \phi_2(t), \quad \ldots, \quad x_n = \phi_n(t), \tag{8.4.2}$$

each defined on an interval I, $\alpha \leqq t \leqq \beta$, is said to be a *solution* of the system (8.4.1), if this set of functions simultaneously satisfies all the equations of the system (8.4.1) identically for each t in I.   If a solution of the system contains the correct number of arbitrary constants, then it is called the *general solution*.   [*See* Theorems 8.4.1 and 8.4.2 for special cases.]

(a) *Reduction to an Equivalent Triangular System.*   We shall primarily be interested in the system (8.4.1) for the case when $n = 2, 3$; and the polynomial operators P_{ij} are assumed to have for their coefficients constants only. The theory of solving such systems parallels the theory of solving corresponding linear algebraic systems of equations.

Examples of systems of linear differential equations:

$$\begin{cases} (D^2 - 2D + 1)x_1 + (D^3 + D)x_2 = 1 + 3t, \\ \qquad (3D + 4)x_1 + (D^4 - 1)x_2 = e^t. \end{cases}$$

$$\begin{cases} (3D^3 + 4D^2 + D + 1)x_1 + \qquad (D^2 - 1)x_2 - \quad (3D^4 + 4D + 6)x_3 = \sin t, \\ (2D^2 - 4D + 2)x_1 + (D^3 + 2D^2 + D)x_2 + (2D^5 - 5D^3 + 2D)x_3 = 5, \\ (D^2 + 5D + 3)x_1 + (D^4 - 2D^2 + 1)x_2 + \quad (D^2 - 4D + 4)x_3 = e^t \sin t. \end{cases}$$

Consider the system of 2 linear differential equations

$$\begin{cases} P_{11}(D)x_1 + P_{12}(D)x_2 = g_1(t), \\ P_{21}(D)x_1 + P_{22}(D)x_2 = g_2(t). \end{cases} \tag{8.4.3}$$

Let us define the *determinant of the system* (8.4.3) by $\Delta(D)$:

$$\Delta(D) \equiv \begin{vmatrix} P_{11}(D) & P_{12}(D) \\ P_{21}(D) & P_{22}(D) \end{vmatrix} \equiv P_{11}(D)P_{22}(D) - P_{12}(D)P_{21}(D). \tag{8.4.4}$$

The *order* of the determinant $\Delta(D)$ is defined as the highest (order) derivative in $P_{11}(D)P_{22}(D) - P_{12}(D)P_{21}(D)$. If $\Delta(D) = 0$, then the system (8.4.3) is said to be *degenerate*.

Case 1. *Suppose that $\Delta(D) \neq 0$.* Our present task is to transform the system (8.4.3) into an "equivalent triangular system," from which the solution of the given system may then be obtained by utilizing the methods developed in Chapters 4 and 6.

Let $F(D)$ denote a polynomial operator with constant coefficients. In view of Remark 8.3.1, polynomial operators with constant coefficients obey all the laws of ordinary algebra with respect to addition and multiplication. Operating (multiplying) on the first equation in (8.4.3) by $F(D)$ and then adding it to the second equation in (8.4.3), we obtain

$$\begin{cases} P_{11}(D)x_1 + \qquad\qquad\qquad P_{12}(D)x_2 = g_1(t) \\ [F(D)P_{11}(D) + P_{21}(D)]x_1 + [F(D)P_{12}(D) + P_{22}(D)]x_2 = F(D)g_1(t) + g_2(t). \end{cases}$$

$$\tag{8.4.5}$$

The determinant of the system (8.4.5) is

$$\begin{vmatrix} P_{11}(D) & P_{12}(D) \\ F(D)P_{11}(D) + P_{21}(D) & F(D)P_{12}(D) + P_{22}(D) \end{vmatrix}$$

$$= P_{11}(D)F(D)P_{12}(D) + P_{11}(D)P_{22}(D)$$
$$\quad - P_{12}(D)F(D)P_{11}(D) - P_{12}(D)P_{21}(D)$$
$$= P_{11}(D)P_{22}(D) - P_{12}(D)P_{21}(D). \tag{8.4.6}$$

Hence, the determinant of the system (8.4.5) is the same as the determinant of the system (8.4.3). One may easily verify that a solution of the system (8.4.3) is also a solution of the system (8.4.5), and conversely. Thus, the systems (8.4.3) and (8.4.5) may be considered as being equivalent systems.

It can be shown that by operating on the equations of the system (8.4.3) and on the equations of the resulting equivalent systems, we ultimately obtain an equivalent system in which a coefficient of x_1 or x_2 is zero. For example, say, we obtained the equivalent system

$$\begin{cases} R_{11}(D)x_1 + R_{12}(D)x_2 = G_1(t), \\ \qquad\qquad R_{22}(D)x_2 = G_2(t). \end{cases} \qquad (8.4.7)$$

The reader is urged to write down the other three possible equivalent systems. A system of the type given in (8.4.7) is called an *equivalent triangular system*.

Note that the determinant of the system (8.4.7) is $R_{11}(D)R_{22}(D)$. The order of this determinant is the sum of the orders of the operators $R_{11}(D)$ and $R_{22}(D)$. [*See* Example 8.3.8 for a special case]. The second equation in (8.4.7) is a nonhomogeneous linear differential equation with constant coefficients. We may now utilize the methods developed in Chapters 4 and 6 to find the general solution $x_2 = \phi_2(t)$. The number of arbitrary constants in $\phi_2(t)$ is equal to the order of the operator $R_{22}(D)$. Now, substituting $x_2 = \phi_2(t)$ into the first equation in (8.4.7), we obtain a nonhomogeneous linear differential equation (in the variable x_1) with constant coefficients, whose order is equal to that of the operator $R_{11}(D)$. Hence, in obtaining the solution $x_1 = \phi_1(t)$, we will be introducing an additional number of arbitrary constants equal to the order of the operator $R_{11}(D)$. Thus, the solution $x_1 = \phi_1(t)$, $x_2 = \phi_2(t)$ of the system (8.4.7) will have as many arbitrary constants as the sum of the orders of the operators $R_{11}(D)$ and $R_{22}(D)$, which is precisely the order of the determinant of the system (8.4.7). But the order of the determinant of the system (8.4.7) is equal to the order of the determinant of the system (8.4.3).

Let us summarize the above result in the form of a theorem.

Theorem 8.4.1. *Consider the linear system*

$$\begin{cases} P_{11}(D)x_1 + P_{12}(D)x_2 = g_1(t), \\ P_{21}(D)x_1 + P_{22}(D)x_2 = g_2(t). \end{cases} \qquad [8.4.3]$$

Suppose that the determinant of the system (8.4.3): $\Delta(D) = P_{11}(D)P_{22}(D) - P_{12}(D)P_{21}(D) \neq 0$. *Then the number of arbitrary constants in the general solution* $x_1 = \phi_1(t)$, $x_2 = \phi_2(t)$ *of the system (8.4.3) is equal to the order of the determinant* $\Delta(D)$.

Remark 8.4.1. Suppose that the system (8.4.3) is transformed into an equivalent system in which a coefficient of x_1 or x_2 is a constant k ($\neq 0$). For example, say, we obtained the system

$$\begin{cases} R_{11}(D)x_1 + R_{12}(D)x_2 = G_1(t), \\ R_{21}(D)x_1 + \qquad kx_2 = G_2(t). \end{cases} \qquad (8.4.8)$$

Then, clearly the system (8.4.8) can be readily transformed into an equivalent triangular system by simply operating on the second equation in (8.4.8) by

$-(1/k)R_{12}(D)$ and then adding it to the first equation in (8.4.8). Thus, we obtain

$$\begin{cases} \left[R_{11}(D) - \dfrac{1}{k} R_{12}(D)R_{21}(D) \right] x_1 & = G_1(t) - \dfrac{1}{k} R_{12}(D)G_2(t), \\[2ex] \qquad\qquad R_{21}(D)x_1 + kx_2 = G_2(t). \end{cases} \tag{8.4.9}$$

Case 2. *Suppose that* $\Delta(D) = 0$. Then as in the case of a linear algebraic system of equations [*see* Section 8.3], the system (8.4.3) may have no solution or infinitely many solutions. For example, if an equivalent triangular system has either form

(a)
$$\begin{cases} R_{11}(D)x_1 + R_{12}(D)x_2 = G_1(t), \\ \qquad 0x_1 + \qquad\quad 0x_2 = 0, \end{cases}$$

or
$$\tag{8.4.10}$$

(b)
$$\begin{cases} R_{11}(D)x_1 + R_{12}(D)x_2 = G_1(t), \\ \qquad 0x_1 + \qquad\quad 0x_2 = G_2(t), \end{cases}$$

where $G_2(t) \not\equiv 0$, then the system (8.4.3) will have infinitely many solutions if it has the form (a), and it will have no solution if it has the form (b). To obtain the infinitely many solutions, we assign x_2 (or x_1) arbitrary, and solve for x_1 (or x_2) in the first equation in (8.4.10) (a). Note that, since $\Delta(D) = 0$, the solutions $x_1 = \phi_1(t)$, $x_2 = \phi_2(t)$ of the system (8.4.3) will not contain any arbitrary constants.

Remark 8.4.2. The above method extends to systems of 3 or more linear differential equations. Let us briefly consider the linear system

$$\begin{cases} P_{11}(D)x_1 + P_{12}(D)x_2 + P_{13}(D)x_3 = g_1(t), \\ P_{21}(D)x_1 + P_{22}(D)x_2 + P_{23}(D)x_3 = g_2(t), \\ P_{31}(D)x_1 + P_{32}(D)x_2 + P_{33}(D)x_3 = g_3(t). \end{cases} \tag{8.4.11}$$

The determinant of the system (8.4.11) is

$$\Delta(D) = \begin{vmatrix} P_{11}(D) & P_{12}(D) & P_{13}(D) \\ P_{21}(D) & P_{22}(D) & P_{23}(D) \\ P_{31}(D) & P_{32}(D) & P_{33}(D) \end{vmatrix}. \tag{8.4.12}$$

Case 3. *Suppose that* $\Delta(D) \neq 0$. An equivalent triangular system is

$$\begin{cases} R_{11}(D)x_1 + R_{12}(D)x_2 + R_{13}(D)x_3 = G_1(t), \\ \qquad\qquad R_{22}(D)x_2 + R_{23}(D)x_3 = G_2(t), \\ \qquad\qquad\qquad\qquad R_{33}(D)x_3 = G_3(t). \end{cases} \tag{8.4.13}$$

The determinant of the equivalent system (8.4.13) is

$$\begin{vmatrix} R_{11}(D) & R_{12}(D) & R_{13}(D) \\ 0 & R_{22}(D) & R_{23}(D) \\ 0 & 0 & R_{33}(D) \end{vmatrix} = R_{11}(D)R_{22}(D)R_{33}(D). \qquad (8.4.14)$$

From the third equation in (8.4.13), we find the solution $x_3 = \phi_3(t)$. Substituting $x_3 = \phi_3(t)$ in the second equation in (8.4.13), we find the solution $x_2 = \phi_2(t)$. Finally, substituting $x_3 = \phi_3(t)$ and $x_2 = \phi_2(t)$ in the first equation in (8.4.13), we find the solution $x_1 = \phi_1(t)$.

Theorem 8.4.2. *The number of arbitrary constants in the general solution* $x_1 = \phi_1(t)$, $x_2 = \phi_2(t)$, $x_3 = \phi_3(t)$ *of the system* (8.4.11) *is equal to the order of the determinant* $\Delta(D)$ *given by* (8.4.12).

Case 4. *Suppose that* $\Delta(D) = 0$. If an equivalent triangular system is, for example, of the form

$$\begin{cases} R_{11}(D)x_1 + R_{12}(D)x_2 + R_{13}(D)x_3 = G_1(t), \\ 0x_1 + R_{22}(D)x_2 + R_{23}(D)x_3 = G_2(t), \\ 0x_1 + 0x_2 + 0x_3 = 0, \end{cases} \qquad (8.4.15)$$

then the system (8.4.11) will have infinitely many solutions. However, if the third equation in (8.4.15) is of the form $0x_1 + 0x_2 + 0x_3 = G_3(t)\,[G_3(t) \not\equiv 0]$, then the system (8.4.11) will have no solution. To obtain the infinitely many solutions, we assign, say, $x_3 = \phi_3(t)$ arbitrary. Substituting $x_3 = \phi_3(t)$ in the second equation in (8.4.15), we find $x_2 = \phi_2(t)$. Finally, substituting $x_3 = \phi_3(t)$ and $x_2 = \phi_2(t)$ into the first equation in (8.4.15), we find $x_1 = \phi_1(t)$. These infinitely many solutions of the system (8.4.15) will not contain any arbitrary constants.

Example 8.4.1. Solve the system

$$\begin{cases} (D + 2)x_1 - Dx_2 = 0, \\ (D - 1)x_2 = t. \end{cases} \qquad (8.4.16)$$

SOLUTION.

$$\Delta(D) = \begin{vmatrix} D + 2 & -D \\ 0 & D - 1 \end{vmatrix} = (D + 2)(D - 1) = D^2 + D - 2.$$

The order of the determinant $\Delta(D)$ is two. Thus, the general solution $x_1 = x_1(t)$, $x_2 = x_2(t)$ of the system (8.4.16) should contain two arbitrary constants. The second equation in (8.4.16) is a linear differential equation of the first order. Utilizing Formula (2.6.2), we find

$$x_2(t) = c_1 e^t - t - 1.$$

Substituting $x_2(t)$ in the first equation in (8.4.16), we have

$$(D + 2)x_1(t) - D[c_1 e^t - t - 1] = 0,$$

or

$$(D + 2)x_1(t) = c_1 e^t - 1,$$

which is also a linear differential equation of the first order. Again utilizing Formula (2.6.2), we find

$$x_1(t) = \tfrac{1}{3} c_1 e^t + c_2 e^{-2t} - \tfrac{1}{2}.$$

Example 8.4.2. Solve the system

$$\begin{cases} (D + 2)x_1 + Dx_2 = 0, \\ (D - 2)x_1 - 2x_2 = e^t. \end{cases} \tag{8.4.17}$$

SOLUTION.

$$\Delta(D) = \begin{vmatrix} D + 2 & D \\ D - 2 & -2 \end{vmatrix} = -2(D + 2) - D(D - 2) = -D^2 - 4.$$

Since the order of the determinant $\Delta(D)$ is two, the general solution $x_1 = x_1(t)$, $x_2 = x_2(t)$ of the system (8.4.17) will contain two arbitrary constants. In view of Remark 8.4.1, the system (8.4.17) may be transformed into an equivalent triangular system by operating on the second equation in (8.4.17) by $\tfrac{1}{2}D$ and then adding it to the first equation. We obtain

$$\begin{cases} [(D + 2) + \tfrac{1}{2}D(D - 2)]x_1 = \tfrac{1}{2}D[e^t], \\ (D - 2)x_1 - 2x_2 = e^t, \end{cases}$$

which simplifies to

$$\begin{cases} (D^2 + 4)x_1 = e^t, \\ (D - 2)x_1 - 2x_2 = e^t. \end{cases} \tag{8.4.18}$$

To solve the first equation in (8.4.18), we utilize the methods given in Section 4.8. We find

$$x_1(t) = c_1 \cos 2t + c_2 \sin 2t + \tfrac{1}{5}e^t.$$

Substituting $x_1(t)$ in the second equation in (8.4.18), we find

$$\begin{aligned}
x_2(t) &= \tfrac{1}{2}(D - 2)[c_1 \cos 2t + c_2 \sin 2t + \tfrac{1}{5}e^t] - \tfrac{1}{2}e^t \\
&= \tfrac{1}{2}D[c_1 \cos 2t + c_2 \sin 2t + \tfrac{1}{5}e^t] - (c_1 \cos 2t + c_2 \sin 2t + \tfrac{1}{5}e^t) - \tfrac{1}{2}e^t \\
&= \tfrac{1}{2}(-2c_1 \sin 2t + 2c_2 \cos 2t + \tfrac{1}{5}e^t) - (c_1 \cos 2t + c_2 \sin 2t + \tfrac{1}{5}e^t) - \tfrac{1}{2}e^t \\
&= (c_2 - c_1) \cos 2t - (c_2 + c_1) \sin 2t - \tfrac{3}{5}e^t.
\end{aligned}$$

Example 8.4.3.   Solve the system

$$\begin{cases} (D + 2)x_1 + (D + 1)x_2 = e^{2t}, \\ (D^2 + 6D + 2)x_1 + (D^2 + 2D - 2)x_2 = 3t^2 + 6e^{2t}. \end{cases} \tag{8.4.19}$$

SOLUTION.

$$\Delta(D) = \begin{vmatrix} D + 2 & D + 1 \\ D^2 + 6D + 2 & D^2 + 2D - 2 \end{vmatrix} = -3D^2 - 6D - 6.$$

Since the order of the determinant $\Delta(D)$ is two, the general solution $x_1 = x_1(t)$, $x_2 = x_2(t)$ of the system (8.4.19) will contain two arbitrary constants. In (8.4.19), operating on the first equation by $-(D + 1)$ and then adding it to the second equation, we obtain

$$\begin{cases} (D + 2)x_1 + (D + 1)x_2 = e^{2t}, \\ [(D^2 + 6D + 2) - (D + 1)(D + 2)]x_1 + [(D^2 + 2D - 2) - (D + 1)(D + 1)]x_2 = 3t^2 \\ \qquad\qquad\qquad\qquad\qquad\qquad\qquad\qquad\qquad + 6e^{2t} - (D + 1)[e^{2t}], \end{cases}$$

which simplifies to

$$\begin{cases} (D + 2)x_1 + (D + 1)x_2 = e^{2t}, \\ Dx_1 - x_2 = t^2 + e^{2t}. \end{cases} \tag{8.4.20}$$

In (8.4.20), operating on the second equation by $(D + 1)$ and then adding it to the first equation, we obtain

$$\begin{cases} [(D + 2) + D(D + 1)]x_1 = e^{2t} + (D + 1)[t^2 + e^{2t}], \\ Dx_1 - x_2 = t^2 + e^{2t}, \end{cases}$$

or

$$\begin{cases} (D^2 + 2D + 2)x_1 = t^2 + 2t + 4e^{2t}, \\ Dx_1 - x_2 = t^2 + e^{2t}. \end{cases} \tag{8.4.21}$$

Utilizing the methods of Section 4.8 to obtain the solution $x_1 = x_1(t)$ of the first equation in (8.4.21), we find

$$x_1(t) = e^{-t}(c_1 \cos t + c_2 \sin t) + \tfrac{2}{5}e^{2t} + \tfrac{1}{2}t^2 - \tfrac{1}{2}.$$

Substituting $x_1(t)$ in the second equation in (8.4.21), we find

$$\begin{aligned} x_2(t) &= D[e^{-t}(c_1 \cos t + c_2 \sin t) + \tfrac{2}{5}e^{2t} + \tfrac{1}{2}t^2 - \tfrac{1}{2}] - t^2 - e^{2t} \\ &= e^{-t}(-c_1 \cos t - c_2 \sin t - c_1 \sin t + c_2 \cos t) + \tfrac{4}{5}e^{2t} + t - t^2 - e^{2t} \\ &= e^{-t}[(c_2 - c_1) \cos t - (c_2 + c_1) \sin t] - \tfrac{1}{5}e^{2t} - t^2 + t. \end{aligned}$$

Example 8.4.4.   Solve the system

$$\begin{cases} x_1 + (D^2 + D + 1)x_2 = \sin t, \\ (D - 1)x_1 + (D^3 - 1)x_2 = \cos t - \sin t. \end{cases} \tag{8.4.22}$$

SOLUTION.

$$\Delta(D) = \begin{vmatrix} 1 & D^2 + D + 1 \\ D - 1 & D^3 - 1 \end{vmatrix} = D^3 - 1 - (D^2 + D + 1)(D - 1) = 0.$$

Hence, the system (8.4.22) is degenerate. In (8.4.22), operating on the first equation by $-(D - 1)$ and then adding it to the second equation, we obtain

$$\begin{cases} x_1 + (D^2 + D + 1)x_2 = \sin t, \\ [(D - 1) - (D - 1)]x_1 + [(D^3 - 1) \\ \qquad - (D - 1)(D^2 + D + 1)]x_2 = \cos t - \sin t - (D - 1)[\sin t], \end{cases}$$

which simplifies to

$$\begin{cases} x_1 + (D^2 + D + 1)x_2 = \sin t, \\ 0x_1 + \qquad\qquad 0x_2 = 0. \end{cases} \tag{8.4.23}$$

Thus, in view of Case 2 [*see* the system (8.4.10)(a)], the system (8.4.22) has infinitely many solutions. If in the first equation of (8.4.23), we assign, say, x_2 arbitrary, the infinitely many solutions of the system (8.4.22) are given by

$$x_1(t) = \sin t - (D^2 + D + 1)[x_2(t)]. \tag{8.4.24}$$

For instance, if we take $x_2(t) = t^2$ in (8.4.24), then

$$x_1(t) = \sin t - (D^2 + D + 1)[t^2] = \sin t - 2 - 2t - t^2.$$

Thus, one solution of the system (8.4.22) is

$$x_1(t) = \sin t - 2 - 2t - t^2, \quad x_2(t) = t^2. \tag{8.4.25}$$

Note, as expected, the solution in (8.4.25) contains no arbitrary constants.

Example 8.4.5. Show that the system

$$\begin{cases} (D^2 + D)x_1 + (D^2 - 1)x_2 = \sin t, \\ D^2 x_1 + (D^2 - D)x_2 = \cos t, \end{cases} \tag{8.4.26}$$

has no solution.

SOLUTION.

$$\Delta(D) = \begin{vmatrix} D^2 + D & D^2 - 1 \\ D^2 & D^2 - D \end{vmatrix} = (D^2 + D)(D^2 - D) - (D^2 - 1)D^2 = 0.$$

Hence, the system (8.4.26) is degenerate. In (8.4.26), add -1 times the second equation to the first equation, and we get

$$\begin{cases} Dx_1 + (D - 1)x_2 = \sin t - \cos t, \\ D^2 x_1 + (D^2 - D)x_2 = \cos t. \end{cases} \tag{8.4.27}$$

In (8.4.27), operating on the first equation by $-D$ and then adding it to the second equation, we get

$$\begin{cases} Dx_1 + (D-1)x_2 = \sin t - \cos t, \\ 0x_1 + \qquad 0x_2 = -\sin t. \end{cases}$$

Thus, in view of Case 2 [*see* the system (8.4.10)(b)], the system (8.4.26) has no solution.

Example 8.4.6.   Solve the system

$$\begin{cases} \dfrac{dx}{dt} + x - y - z = t, \\[2mm] \dfrac{dx}{dt} + 2\dfrac{dy}{dt} + \dfrac{dz}{dt} - y = 0, \\[2mm] \dfrac{dx}{dt} - \dfrac{dy}{dt} - 4x + y = 6e^{2t} - 1. \end{cases} \qquad (8.4.28)$$

SOLUTION.   In operator notation, we may write the above system as

$$\begin{cases} (D+1)x - \qquad\quad y - \quad z = t, \\ Dx + (2D-1)y + Dz = 0, \\ (D-4)x - \quad (D-1)y \qquad = 6e^{2t} - 1. \end{cases} \qquad (8.4.29)$$

$$\Delta(D) = \begin{vmatrix} D+1 & -1 & -1 \\ D & 2D-1 & D \\ D-4 & -D+1 & 0 \end{vmatrix} = D^3 + 2D^2 - 7D + 4.$$

Since the order of the determinant $\Delta(D)$ is three, the general solution $x = x(t)$, $y = y(t)$, $z = z(t)$ of the system (8.4.29) will contain three arbitrary constants. In (8.4.29), operating on the first equation by D and then adding it to the second equation, we get

$$\begin{cases} (D+1)x - \qquad\quad y - z = t, \\ (D^2 + 2D)x + (D-1)y \qquad = 1, \\ (D-4)x - (D-1)y \qquad = 6e^{2t} - 1. \end{cases} \qquad (8.4.30)$$

In (8.4.30), adding the second equation to the third equation, we obtain a desired equivalent triangular system

$$\begin{cases} (D+1)x - \qquad\quad y - z = t, \\ (D^2 + 2D)x + (D-1)y \qquad = 1, \\ (D^2 + 3D - 4)x \qquad\qquad = 6e^{2t}. \end{cases} \qquad (8.4.31)$$

Utilizing the methods of Section 4.8 to obtain the solution $x = x(t)$ of the third equation in (8.4.31), we find

$$x(t) = c_1 e^t + c_2 e^{-4t} + e^{2t}. \tag{8.4.32}$$

Substituting $x(t)$ into the second equation in (8.4.31), we have

$$\begin{aligned}
(D-1)y(t) &= 1 - (D^2 + 2D)[c_1 e^t + c_2 e^{-4t} + e^{2t}] \\
&= 1 - D^2[c_1 e^t + c_2 e^{-4t} + e^{2t}] - 2D[c_1 e^t + c_2 e^{-4t} + e^{2t}] \\
&= 1 - (c_1 e^t + 16c_2 e^{-4t} + 4e^{2t}) - 2(c_1 e^t - 4c_2 e^{-4t} + 2e^{2t}) \\
&= 1 - (3c_1 e^t + 8c_2 e^{-4t} + 8e^{2t}),
\end{aligned}$$

which is a first-order linear differential equation. Utilizing Formula (2.6.2), we find

$$\begin{aligned}
y(t) &= e^t \left[\int e^{-t}\, dt - 3c_1 \int dt - 8c_2 \int e^{-5t}\, dt - 8 \int e^t\, dt + c_3 \right] \\
&= e^t(-e^{-t} - 3c_1 t + \tfrac{8}{5}c_2 e^{-5t} - 8e^t + c_3) \\
&= (c_3 - 3c_1 t)e^t + \tfrac{8}{5}c_2 e^{-4t} - 8e^{2t} - 1. \tag{8.4.33}
\end{aligned}$$

Substituting $x(t)$ and $y(t)$ found in (8.4.32) and (8.4.33) into the first equation in (8.4.31), we find

$$\begin{aligned}
z(t) &= (D+1)[c_1 e^t + c_2 e^{-4t} + e^{2t}] - (c_3 - 3c_1 t)e^t - \tfrac{8}{5}c_2 e^{-4t} + 8e^{2t} + 1 - t \\
&= c_1 e^t - 4c_2 e^{-4t} + 2e^{2t} + c_1 e^t + c_2 e^{-4t} \\
&\quad + e^{2t} - (c_3 - 3c_1 t)e^t - \tfrac{8}{5}c_2 e^{-4t} + 8e^{2t} + 1 - t \\
&= (2c_1 - c_3 + 3c_1 t)e^t - \tfrac{23}{5}c_2 e^{-4t} + 11e^{2t} - t + 1.
\end{aligned}$$

(b) *Solution of Linear Systems by Elimination.* We shall briefly outline the method of elimination to solve the systems (8.4.3) and (8.4.11). You will notice that this method has both advantages and disadvantages over the method of reducing the system to an equivalent triangular system. Let us rewrite the system

$$\begin{cases} P_{11}(D)x_1 + P_{12}(D)x_2 = g_1(t), \\ P_{21}(D)x_1 + P_{22}(D)x_2 = g_2(t). \end{cases} \tag{8.4.3}$$

Suppose that the determinant $\Delta(D) \neq 0$. In (8.4.3), let us first eliminate, say, the function x_1. Operating on the first equation by $P_{21}(D)$ and on the second equation by $P_{11}(D)$, we have

$$\begin{cases} P_{21}(D)P_{11}(D)x_1 + P_{21}(D)P_{12}(D)x_2 = P_{21}(D)[g_1(t)], \\ P_{11}(D)P_{21}(D)x_1 + P_{11}(D)P_{22}(D)x_2 = P_{11}(D)[g_2(t)]. \end{cases} \tag{8.4.34}$$

To eliminate x_1 in (8.4.34), we subtract the second equation from the first. We obtain

$$[P_{21}(D)P_{12}(D) - P_{11}(D)P_{22}(D)]x_2 = P_{21}(D)[g_1(t)] - P_{11}(D)[g_2(t)], \quad (8.4.35)$$

which is a nonhomogeneous linear differential equation with constant coefficients whose order is equal to the order of the determinant $\Delta(D) \neq 0$. Utilizing known methods, we now find the solution $x_2 = x_2(t)$ of Equation (8.4.35). Substitute the solution $x_2(t)$ into either equation of the system (8.4.3), preferably into the one which will give a differential equation of the lowest order. Say that we substitute $x_2(t)$ into the first equation in (8.4.3). Now solve this equation and obtain the solution $x_1 = x_1(t)$. If it turns out that the number of arbitrary constants in $x_1(t)$, $x_2(t)$ is equal to the order of the determinant $\Delta(D)$, we have then obtained the desired general solution of the system (8.4.3). However, if it turns out that the number of arbitrary constants in $x_1(t)$, $x_2(t)$ is greater than the order of the determinant $\Delta(D)$, then additional work is required to obtain the general solution of the system (8.4.3). This is the disadvantage of the method of elimination. [Observe that the number of arbitrary constants in $x_1(t)$, $x_2(t)$ cannot be less than the order of the determinant $\Delta(D)$. (Why?)] In order to obtain the relation between the extraneous constants and the arbitrary constants [equal in number to the order of $\Delta(D)$,] we substitute $x_1(t)$, $x_2(t)$ into the equation in (8.4.3) which was not used to obtain the solution $x_1(t)$. For the present case, we would substitute into the second equation in (8.4.3).

Example 8.4.7.   Solve the system

$$\begin{cases} \dfrac{dx}{dt} + \dfrac{dy}{dt} - x + y = t, \\[2mm] \dfrac{dx}{dt} + \dfrac{dy}{dt} + x - y = e^t. \end{cases} \qquad (8.4.36)$$

SOLUTION.   In operator notation, we may write the system (8.4.36) as

$$\begin{cases} (D - 1)x + (D + 1)y = t, \\ (D + 1)x + (D - 1)y = e^t. \end{cases} \qquad (8.4.37)$$

$$\Delta(D) = \begin{vmatrix} D - 1 & D + 1 \\ D + 1 & D - 1 \end{vmatrix} = (D - 1)^2 - (D + 1)^2 = -4D.$$

Since the order of the determinant $\Delta(D)$ is one, the general solution $x = x(t)$, $y = y(t)$ of the system (8.4.37) should contain one arbitrary constant. In (8.4.37), operating on the first equation by $(D + 1)$ and on the second equation by $(D - 1)$, we have

$$\begin{cases} (D + 1)(D - 1)x + (D + 1)^2 y = (D + 1)[t], \\ (D - 1)(D + 1)x + (D - 1)^2 y = (D - 1)[e^t]. \end{cases} \qquad (8.4.38)$$

In (8.4.38), subtracting the second equation from the first and then simplifying, we get

$$Dy = \frac{1}{4}(1 + t). \tag{8.4.39}$$

Integrating (8.4.39), we find

$$y(t) = \frac{1}{4}\left(t + \frac{t^2}{2}\right) + C_1, \tag{8.4.40}$$

where C_1 is an arbitrary constant of integration. Substituting $y(t)$ into the first equation in (8.4.37), we get

$$(D - 1)x = t - (D + 1)\left[\frac{1}{4}\left(t + \frac{t^2}{2}\right) + C_1\right]$$

$$= t - \frac{1}{4}(1 + t) - \frac{1}{4}\left(t + \frac{t^2}{2}\right) - C_1$$

$$= -\frac{1}{4} + \frac{1}{2}t - \frac{1}{8}t^2 - C_1.$$

Utilizing Formula (2.6.2), we find

$$x(t) = e^t\left[-\frac{1}{4}\int e^{-t}\,dt + \frac{1}{2}\int te^{-t}\,dt - \frac{1}{8}\int t^2 e^{-t}\,dt - C_1\int e^{-t}\,dt + C_2\right]$$

$$= -\frac{1}{4}t + \frac{1}{8}t^2 + C_1 + C_2 e^t, \tag{8.4.41}$$

where C_2 is an arbitrary constant of integration. Thus, $x(t)$ and $y(t)$ contain two arbitrary constants. We know that the general solution of the system (8.4.37) must contain only one arbitrary constant. Thus, substituting $y(t)$, $x(t)$ found in (8.4.40) and (8.4.41) into the second equation in (8.4.37), we obtain

$$(D + 1)\left[-\frac{1}{4}t + \frac{1}{8}t^2 + C_1 + C_2 e^t\right] + (D - 1)\left[\frac{1}{4}\left(t + \frac{t^2}{2}\right) + C_1\right] \equiv e^t,$$

which simplifies to

$$2C_2 e^t \equiv e^t.$$

Hence, $C_2 = \frac{1}{2}$. Thus, the general solution of the system (8.4.37) [and hence of (8.4.36)] is

$$x(t) = -\frac{1}{4}t + \frac{1}{8}t^2 + \frac{1}{2}e^t + C_1, \quad y(t) = \frac{1}{4}t + \frac{1}{8}t^2 + C_1. \tag{8.4.42}$$

Let us check this answer by utilizing the method of reducing the system (8.4.37) into an equivalent triangular system. In (8.4.37), add -1 times the first equation to the second equation, and we get

$$\begin{cases} (D-1)x + (D+1)y = t, \\ \quad x - \qquad\quad y = \tfrac{1}{2}(e^t - t). \end{cases} \tag{8.4.43}$$

In (8.4.43), operating on the second equation by $-(D-1)$ and then adding it to the first equation, we obtain

$$\begin{cases} \quad 2Dy = t - (D-1)[\tfrac{1}{2}(e^t - t)], \\ x - \quad y = \tfrac{1}{2}(e^t - t), \end{cases}$$

or

$$\begin{cases} \quad Dy = \tfrac{1}{4} + \tfrac{1}{4}t, \\ x - \quad y = \tfrac{1}{2}(e^t - t). \end{cases} \tag{8.4.44}$$

Integrating the first equation in (8.4.44), we find

$$y(t) = \tfrac{1}{4}t + \tfrac{1}{8}t^2 + C_1.$$

Substituting $y(t)$ into the second equation in (8.4.44), we find

$$x(t) = y(t) + \tfrac{1}{2}(e^t - t) = \tfrac{1}{4}t + \tfrac{1}{8}t^2 + C_1 + \tfrac{1}{2}e^t - \tfrac{1}{2}t = -\tfrac{1}{4}t + \tfrac{1}{8}t^2 + \tfrac{1}{2}e^t + C_1.$$

The following example illustrates the technique involved in solving a system of three linear differential equations by eliminating the functions.

Example 8.4.8. Solve the system

$$\begin{cases} (D-1)x_1 + (D-4)x_2 + Dx_3 = 0, \\ (D^2 - 1)x_1 - (D+2)x_2 + D^2 x_3 = t, \\ Dx_1 + (D-2)x_2 \qquad\qquad = e^t. \end{cases} \tag{8.4.45}$$

SOLUTION.

$$\Delta(D) = \begin{vmatrix} D-1 & D-4 & D \\ D^2-1 & -(D+2) & D^2 \\ D & D-2 & 0 \end{vmatrix} = \begin{vmatrix} D-1 & D-4 & D \\ D-1 & -(D^2-3D+2) & 0 \\ D & D-2 & 0 \end{vmatrix}$$

$$= D\begin{vmatrix} D-1 & -(D^2-3D+2) \\ D & D-2 \end{vmatrix} = D^4 - 2D^3 - D^2 + 2D.$$

Since the order of the determinant $\Delta(D)$ is four, the general solution $x_1 = x_1(t)$, $x_2 = x_2(t)$, $x_3 = x_3(t)$ of the system (8.4.45) should contain four arbitrary constants. In (8.4.45), we first eliminate x_3. Operating on the first equation by $-D$ and then adding it to the second equation, we get

$$\begin{cases} (D-1)x_1 - (D-2)(D-1)x_2 = t, \\ Dx_1 + \qquad\qquad (D-2)x_2 = e^t. \end{cases} \tag{8.4.46}$$

In (8.4.46), we now eliminate x_2. Operating on the second equation by $(D - 1)$ and then adding it to the first equation, we obtain

$$(D^2 - 1)x_1 = t. \tag{8.4.47}$$

The solution $x_1 = x_1(t)$ of Equation (8.4.47) is found to be

$$x_1(t) = c_1 e^t + c_2 e^{-t} - t. \tag{8.4.48}$$

Substituting $x_1(t)$ in the second equation of (8.4.46), we find

$$(D - 2)x_2 = -c_1 e^t + e^t + c_2 e^{-t} + 1. \tag{8.4.49}$$

Utilizing Formula (2.6.2), the solution $x_2 = x_2(t)$ of Equation (8.4.49) is

$$x_2(t) = c_1 e^t - e^t - \frac{c_2}{3} e^{-t} + c_3 e^{2t} - \frac{1}{2}. \tag{8.4.50}$$

Substituting $x_1(t)$ and $x_2(t)$ found in (8.4.48) and (8.4.50) into the first equation in (8.4.45), we get

$$Dx_3 = 3c_1 e^t - 3e^t + \tfrac{1}{3}c_2 e^{-t} + 2c_3 e^{2t} - t - 1. \tag{8.4.51}$$

Integrating Equation (8.4.51), we find

$$x_3(t) = 3c_1 e^t - 3e^t - \tfrac{1}{3}c_2 e^{-t} + c_3 e^{2t} - \tfrac{1}{2}t^2 - t + c_4. \tag{8.4.52}$$

As an exercise, the reader should check the general solution of the system (8.4.45) by substituting $x_1(t)$, $x_2(t)$, $x_3(t)$ found in (8.4.48), (8.4.50), and (8.4.52) into the second equation in (8.4.45).

Exercises 8.4

In Exercises 1 through 20, solve each system.

1. $\begin{cases} Dx_1 + (D + 1)x_2 = \cos t, \\ \qquad\qquad Dx_2 = \sin t. \end{cases}$

2. $\begin{cases} \qquad\qquad D^2 x_2 = 12t^2 + e^t, \\ Dx_1 + (D - 1)x_2 = 4t^4 + 8t^3. \end{cases}$

3. $\begin{cases} (D - 1)x_1 + (D + 1)x_2 = 1, \\ (D + 1)x_1 \qquad\qquad = e^{2t}. \end{cases}$

4. $\begin{cases} D^3 x_1 \qquad\qquad = 0, \\ Dx_1 + (D - 2)x_2 = e^{4t}. \end{cases}$

5. $\begin{cases} (D - 2)x_1 + (D + 2)x_2 = 0, \\ (D + 2)x_1 + (D - 2)x_2 = 8te^t. \end{cases}$

6. $\begin{cases} \dfrac{dx}{dt} + x - 6y = 0, \\[2mm] \dfrac{dy}{dt} - x + 2y = 0. \end{cases}$

7. $\begin{cases} \dfrac{dx}{dt} + x - y = \cos t, \\[2mm] \dfrac{dy}{dt} - 6x \quad\; = \sin t + 5e^{2t}. \end{cases}$

8. $\begin{cases} \dfrac{dx}{dt} + \dfrac{dy}{dt} - x + 8y = 6t, \\[2mm] \dfrac{dx}{dt} + \dfrac{dy}{dt} + x + 4y = 0. \end{cases}$

9. $\begin{cases} \dfrac{d^2x}{dt^2} + 2x - y = 0, \\[2mm] \dfrac{dy}{dt} - 6x + 3y = 0. \end{cases}$

10. $\begin{cases} \dfrac{d^2x}{dt^2} - y = t^2, \\[2mm] \dfrac{d^2y}{dt^2} - x = 4e^t. \end{cases}$

11. $\begin{cases} (D-3)x_1 + \qquad\qquad Dx_2 = t^2 + 1, \\[2mm] (D^2 - 4D - 12)x_1 + 2(D^2 + 2D)x_2 = 2t + 4. \end{cases}$

12. $\begin{cases} (D^2 - 1)x_1 + \quad (D+2)x_2 = 4e^{-t}, \\[2mm] (D^3 + 2D^2 - D - 2)x_1 + (5D^2 + 4)x_2 = 8. \end{cases}$

13. $\begin{cases} \dfrac{d^2x}{dt^2} + 2x - y \qquad\quad = 0, \\[3mm] -3\dfrac{dx}{dt} + \dfrac{dy}{dt} - 6x + 2y = 10e^{-t}\cos t. \end{cases}$

14. $\begin{cases} \dfrac{d^2y}{dt^2} - x + 4y \qquad\quad = 4\cos t, \\[3mm] \dfrac{d^2x}{dt^2} - 2\dfrac{d^2y}{dt^2} - 2x + 7y = 0. \end{cases}$

15. $\begin{cases} -(D+2)x + \qquad\qquad (D^3 + D^2)y = e^{2t}, \\[2mm] (D^2 - 4)x - (3D^3 - D^2 - 16D + 12)y = 50\cos t + 100\sin t. \end{cases}$

16. $\begin{cases} \dfrac{dx}{dt} - y - z = 0, \\[2mm] \dfrac{dy}{dt} - x - z = 0, \\[2mm] \dfrac{dz}{dt} - x - y = 0. \end{cases}$

17. $\begin{cases} \dfrac{dx}{dt} - x + y \qquad = 0, \\[2mm] \dfrac{dy}{dt} - x - 2y - z = 0, \\[2mm] \dfrac{dz}{dt} + 2x - y + z = 0. \end{cases}$

18. $\begin{cases} \dfrac{d^2x}{dt^2} - y - z = 0, \\[2mm] \dfrac{d^2y}{dt^2} - x - z = 0, \\[2mm] \dfrac{d^2z}{dt^2} - x - y = 0. \end{cases}$

19. $\begin{cases} (D+1)x - \qquad\quad y - \quad z = 0, \\[2mm] x - (D+1)y + \quad z = 2t, \\[2mm] (D^2 + 2D)x + \qquad 2y - Dz = 4e^t. \end{cases}$

20. $\begin{cases} \qquad\qquad Dx + \quad y - \quad z = 2t, \\ \qquad\qquad\quad x + Dy + \quad z = t^2, \\ (-D^3 + D^2 + 2D)x - \quad y + D^2 z = 2 - t^2. \end{cases}$

In Exercises 21 through 25, verify that each system is degenerate. Also, find the solutions if they exist.

21. $\begin{cases} (D+1)x_1 + \qquad\qquad x_2 = e^t, \\ (D^2 - 1)x_1 + (D-1)x_2 = 0. \end{cases}$

22. $\begin{cases} \qquad (D+2)x_1 + \qquad\qquad Dx_2 = \cos t + t^2, \\ (D^2 + 4D + 4)x_1 + (D^2 + 2D)x_2 = 2\cos t - \sin t + 2t^2 + 2t. \end{cases}$

23. $\begin{cases} (D^2 - 1)x_1 - (D^2 - 3D + 2)x_2 = 2t - t^2 + 6e^t, \\ (D+1)x_1 - \qquad\qquad (D-2)x_2 = t^2 + e^t. \end{cases}$

24. $\begin{cases} \qquad x - \quad (D+1)y + \qquad\quad Dz = t, \\ \quad -Dx + \qquad D^2 y - (D^2 - 1)z = e^t, \\ (D-1)x - (D^2 - 1)y + (D^2 - D)z = 1 - t. \end{cases}$

25. $\begin{cases} \quad (D-1)x + \qquad\qquad Dy - \qquad\quad z = \cos t, \\ \quad (D^2 - 1)x + \quad (D^2 + 2)y + (D+3)z = \sin t + e^t, \\ (D^2 + D - 2)x + (D^2 + 2D)y - (D+2)z = 2\cos t + \sin t. \end{cases}$

8.5. Applications of Systems of Linear Differential Equations

(a) *Electrical networks.* The reader is urged to review the material of Section 3.4. Let us consider the network displayed in Figure 8.5.1. A *loop* is a single closed path for the current. The network in Figure 8.5.1 consists of three loops, namely, *abcdefa*, *abefa*, and *bcdeb*. However, as we shall see in the discussion of Example 8.5.1 only two of these three loops are independent. A *node* is a point in the network at which two or more (circuit) elements meet. For example, in Figure 8.5.1, the point *b* is a node. (Are there any other nodes?) If three or more elements meet at a node, that node is called a *junction*. In the above network, the point *b* is a junction. A *branch* of a network extends from one junction to another, and may consist

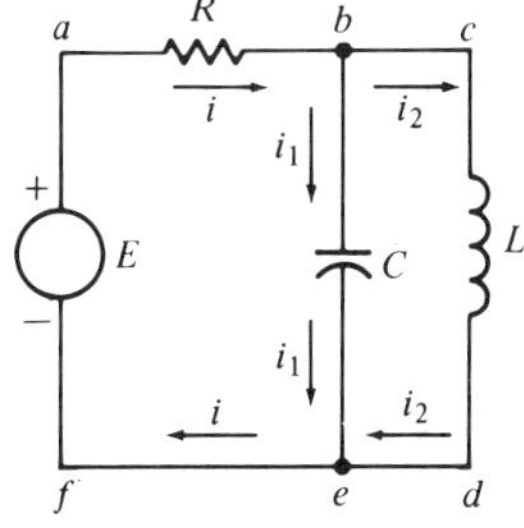

Figure 8.5.1.

of one element or several elements in series.   For example, *be*, *bcde*, and *efab* are branches.   (Are there any other branches?)   The current along a given branch is the same.   Let us arbitrarily assign the direction of the current flow by arrows as indicated in Figure 8.5.1.

In order to solve problems involving multiple loop networks, besides Kirchhoff's Voltage Law, we shall utilize the following:

Kirchhoff's Current Law.   The sum of the currents directed toward any junction in a circuit equals the sum of the currents directed away from that junction.

If the currents toward a junction are considered positive and those away from the same junction are considered negative, then this law states that:

The algebraic sum of all the currents meeting at a common junction is zero.

For example, in Figure 8.5.1, at the junction *b*, we have

$$i - i_1 - i_2 = 0,$$

or

$$i = i_1 + i_2.$$

(8.5.1)

Example 8.5.1.   In Figure 8.5.1, suppose that $E = 100$ volts, $R = 50$ ohms, $C = 2 \cdot 10^{-4}$ farad, and $L = 0.04$ henry.   Let us find the currents $i_1 = i_1(t)$ and $i_2 = i_2(t)$ of the network at time t, if $i_1(0) = 2$ amps and $i_2(0) = 0$.

SOLUTION.   Let i, i_1, and i_2 denote, respectively, the currents flowing in the branches *efab*, *be*, and *bcde*.   Utilizing Kirchhoff's Voltage Law [*see* Section 3.4 and Equations (3.4.1) and (3.4.2)] for the loop *abcdefa*, we have

$$50i + 0.04\frac{di_2}{dt} = 100. \tag{8.5.2}$$

For the loop *abefa*, we have

$$50i + \frac{1}{2 \cdot 10^{-4}} \int i_1 \, dt = 100. \tag{8.5.3}$$

For the loop *bcdeb*, we have

$$0.04\frac{di_2}{dt} - \frac{1}{2 \cdot 10^{-4}} \int i_1 \, dt = 0. \tag{8.5.4}$$

Note, in (8.5.4) the minus sign before the integral is necessary, since we are traversing the branch *eb* in the direction opposite to that of the current i_1 as we complete the loop *bcdeb*.   Also, since this loop is free of an electromotive force, the right-hand side of (8.5.4) is zero.   Further, observe that Equation (8.5.4) can be obtained by subtracting Equation (8.5.3) from Equation (8.5.2). Thus, only two of the three loops are independent.   Let us simply retain Equations (8.5.2) and (8.5.3).

Differentiating (8.5.3), we obtain

$$50\frac{di}{dt} + \frac{10^4}{2} i_1 = 0. \tag{8.5.5}$$

At the junction b, from (8.5.1) we have that $i = i_1 + i_2$. Thus, $\dfrac{di}{dt} = \dfrac{di_1}{dt} + \dfrac{di_2}{dt}$.
Substituting di/dt and i into (8.5.5) and (8.5.2), respectively, we obtain

$$\begin{cases} 50\dfrac{di_1}{dt} + 50\dfrac{di_2}{dt} + 5000i_1 = 0, \\[2mm] 0.04\dfrac{di_2}{dt} + 50\,i_1 + 50i_2 = 100. \end{cases} \tag{8.5.6}$$

In operator form, we may write the system (8.5.6) upon simplification as

$$\begin{cases} (D + 100)i_1 + \qquad\qquad Di_2 = 0, \\ \qquad 1250i_1 + (D + 1250)i_2 = 2500. \end{cases} \tag{8.5.7}$$

[*Note:*

$$\Delta(D) = \begin{vmatrix} D + 100 & D \\ 1250 & D + 1250 \end{vmatrix} = D^2 + 100D + 125{,}000.]$$

Since the determinant $\Delta(D)$ is of order two, the general solution $i_1 = i_1(t)$, $i_2 = i_2(t)$ of the system (8.5.7) will contain two arbitrary constants. These arbitrary constants can then be determined by utilizing the initial conditions on the currents, namely,

$$i_1(0) = 2, \qquad i_2(0) = 0. \tag{8.5.8}$$

In (8.5.7), operating on the second equation by $-\frac{1}{1250}(D + 100)$ and then adding it to the first equation, we obtain

$$\begin{cases} (D^2 + 100D + 125{,}000)i_2 = 250{,}000, \\ 1250\,i_1 + \qquad\qquad (D + 1250)i_2 = 2500. \end{cases} \tag{8.5.9}$$

Utilizing the methods of Section 4.8 to obtain the solution $i_2 = i_2(t)$ of the first equation in (8.5.9), we find

$$i_2(t) = e^{-50t}(c_1 \cos 350t + c_2 \sin 350t) + 2. \tag{8.5.10}$$

Substituting (8.5.10) into the second equation in (8.5.9), we find

$$i_1(t) = -\frac{e^{-50t}}{1250}[(1200c_1 + 350c_2) \cos 350t + (1200c_2 - 350c_1) \sin 350t].$$

$$\tag{8.5.11}$$

Utilizing (8.5.8), (8.5.10), and (8.5.11), we obtain

$$\begin{cases} c_1 + 2 = 0, \\ 1200c_1 + 350c_2 = -2500. \end{cases}$$

Hence, $c_1 = -2$ and $c_2 = -\frac{2}{7}$. Substituting these values into (8.5.10) and (8.5.11), the desired solution is

$$\begin{cases} i_1(t) = e^{-50t}(2 \cos 350t - \frac{2}{7} \sin 350t), \\ i_2(t) = -e^{-50t}(2 \cos 350t + \frac{2}{7} \sin 350t) + 2. \end{cases} \tag{8.5.12}$$

(Review Remark 3.4.3.) For example, when $t = 0.2$ sec, $e^{-50t} = e^{-10} \approx 0.00005$. Thus, the current i_1 is entirely transient, that is, it approaches 0 in a very short period of time. Also, in a very short period of time, the current i_2 approaches its steady-state current $i_{2,s} = 2$. As expected, from the data given in the example, the steady-state current can also be obtained by utilizing Ohm's Law

$$i = \frac{E}{R} = \frac{100}{50} = 2 \text{ amps.}$$

(b) *A two degree-of-freedom spring mass system.* The reader is urged to reread the material of Section 5.2. We shall consider a mechanical system consisting of two masses m_1, m_2, and three springs as shown in Figure 8.5.2. The springs have spring constants k_1, k_2, k_3, and they obey Hooke's Law. The masses of the springs are assumed to be negligible. At equilibrium position, the springs have their natural lengths. We assume the motion of the masses takes place in a straight line on a frictionless, horizontal surface, and that the only forces acting on the masses are those due to the restoring forces in the springs. Let $x_1 = x_1(t)$ and $x_2 = x_2(t)$ denote, respectively, the displacements of the masses m_1 and m_2 from their equilibrium positions O_1 and O_2 at time t. [*See* Figure 8.5.3.] We assume that x_1 is positive when the

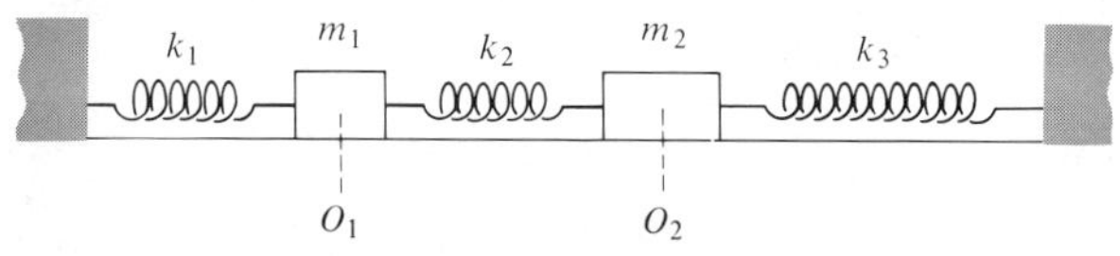

Figure 8.5.2.
System at equilibrium position

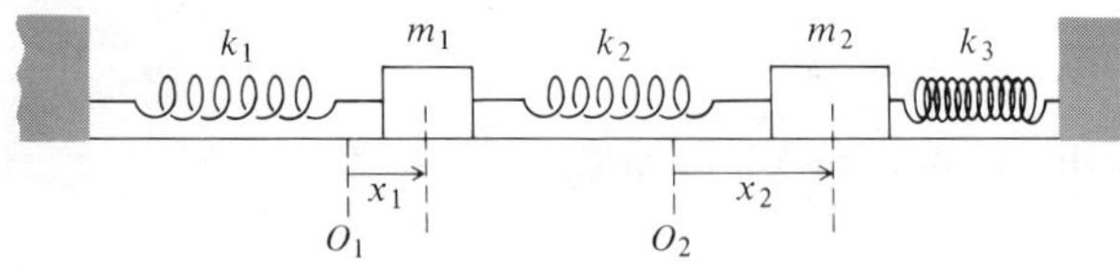

Figure 8.5.3.
System at time t

mass m_1 is to the right of O_1, and that x_2 is positive when the mass m_2 is to the right of O_2.   Following a similar argument as given in Section 5.2, we find at time $t > 0$, the net force acting on the mass m_1 is

$$-k_1 x_1 + k_2(x_2 - x_1),$$

and the net force acting on the mass m_2 is

$$-k_2(x_2 - x_1) - k_3 x_2.$$

[*See* Figure 8.5.4.]   Thus, utilizing Newton's Second Law of Motion [*see* Section 3.5], we obtain

$$\begin{cases} m_1 \dfrac{d^2 x_1}{dt^2} = -k_1 x_1 + k_2(x_2 - x_1), \\[2mm] m_2 \dfrac{d^2 x_2}{dt^2} = -k_2(x_2 - x_1) - k_3 x_2, \end{cases}$$

or

$$\begin{cases} m_1 \dfrac{d^2 x_1}{dt^2} + (k_1 + k_2)x_1 - k_2 x_2 = 0, \\[2mm] m_2 \dfrac{d^2 x_2}{dt^2} - k_2 x_1 + (k_2 + k_3)x_2 = 0. \end{cases} \tag{8.5.13}$$

A possible set of initial conditions for this problem is given by

$$x_1(0) = a, \qquad x_2(0) = b, \qquad x_1'(0) = c, \qquad x_2'(0) = d, \tag{8.5.14}$$

where a, b, c, and d are constants.

Example 8.5.2.   Let us solve the system (8.5.13) when $k_3 = k_1$ and $m_2 = m_1$.

SOLUTION.   In operator form, we may write the system (8.5.13) with $k_3 = k_1$ and $m_2 = m_1$ as

$$\begin{cases} (D^2 + \alpha^2)x_1 - \qquad\quad \beta^2 x_2 = 0, \\[2mm] -\beta^2 x_1 + (D^2 + \alpha^2)x_2 = 0, \end{cases} \tag{8.5.15}$$

where

$$\alpha^2 = \frac{k_1 + k_2}{m_1}, \qquad \beta^2 = \frac{k_2}{m_1}. \tag{8.5.16}$$

Figure 8.5.4.
A free body diagram of the masses

Note that $\alpha > \beta > 0$. Also,

$$\Delta(D) = \begin{vmatrix} D^2 + \alpha^2 & -\beta^2 \\ -\beta^2 & D^2 + \alpha^2 \end{vmatrix} = D^4 + 2\alpha^2 D^2 + \alpha^4 - \beta^4.$$

Since the determinant $\Delta(D)$ is of order four, the general solution $x_1 = x_1(t)$ and $x_2 = x_2(t)$ of the system (8.5.15) will contain four arbitrary constants. In (8.5.15), operating on the second equation by $\dfrac{1}{\beta^2}(D^2 + \alpha^2)$ and then adding it to the first equation, we obtain

$$\begin{cases} [(D^2 + \alpha^2)^2 - \beta^4]x_2 = 0, \\ -\beta^2 x_1 + \qquad (D^2 + \alpha^2)x_2 = 0. \end{cases} \tag{8.5.17}$$

The auxiliary equation of the first equation in (8.5.17) is

$$(m^2 + \alpha^2)^2 - \beta^4 = 0.$$

The roots of this equation are $\pm\sqrt{\alpha^2 - \beta^2}\, i$ and $\pm\sqrt{\alpha^2 + \beta^2}\, i$. Thus,

$$x_2(t) = c_1 \cos \sqrt{\alpha^2 - \beta^2}\, t + c_2 \sin \sqrt{\alpha^2 - \beta^2}\, t$$
$$+ c_3 \cos \sqrt{\alpha^2 + \beta^2}\, t + c_4 \sin \sqrt{\alpha^2 + \beta^2}\, t.$$

Substituting $x_2(t)$ into the second equation in (8.5.17), we obtain

$$x_1(t) = c_1 \cos \sqrt{\alpha^2 - \beta^2}\, t + c_2 \sin \sqrt{\alpha^2 - \beta^2}\, t$$
$$- c_3 \cos \sqrt{\alpha^2 + \beta^2}\, t - c_4 \sin \sqrt{\alpha^2 + \beta^2}\, t.$$

From (8.5.16), we see that

$$\sqrt{\alpha^2 - \beta^2} = \sqrt{\frac{k_1}{m_1}} \quad \text{and} \quad \sqrt{\alpha^2 + \beta^2} = \sqrt{\frac{k_1 + 2k_2}{m_1}}.$$

Let

$$\gamma_1 = \sqrt{\frac{k_1}{m_1}} \quad \text{and} \quad \gamma_2 = \sqrt{\frac{k_1 + 2k_2}{m_1}}.$$

Thus, the general solution of the system (8.5.15) is

$$\begin{cases} x_1(t) = c_1 \cos \gamma_1 t + c_2 \sin \gamma_1 t - c_3 \cos \gamma_2 t - c_4 \sin \gamma_2 t, \\ x_2(t) = c_1 \cos \gamma_1 t + c_2 \sin \gamma_1 t + c_3 \cos \gamma_2 t + c_4 \sin \gamma_2 t. \end{cases} \tag{8.5.18}$$

Note that the frequencies of the system are

$$\nu_1 = \frac{1}{2\pi}\gamma_1 = \frac{1}{2\pi}\sqrt{\frac{k_1}{m_1}} \quad \text{and} \quad \nu_2 = \frac{1}{2\pi}\gamma_2 = \frac{1}{2\pi}\sqrt{\frac{k_1 + 2k_2}{m_1}}. \tag{8.5.19}$$

Exercises 8.5

1. *Thomson's experiment.*   In the determination of the ratio e/m, in which electrons were subjected to an electric field of intensity E and a magnetic field of intensity H, Thomson discovered that the motion of the electrons is governed by

$$
(1) \qquad
\begin{cases}
m\,\dfrac{d^2x}{dt^2} + He\,\dfrac{dy}{dt} = Ee, \\[2ex]
m\,\dfrac{d^2y}{dt} - He\,\dfrac{dx}{dt} = 0,
\end{cases}
$$

where m and e are, respectively, the mass and charge of the electron.   If the initial conditions at $t = 0$ are

$$
x(0) = 0, \qquad y(0) = 0, \qquad x'(0) = 0, \qquad y'(0) = 0,
$$

show that the motion (or path) of the electrons is a cycloid.

2. A projectile of mass m is fired into the air from the ground with an initial velocity of v_0 and at an angle α with the horizontal.   Suppose that the projectile travels in a vertical plane.   If the air resists the motion with a force per unit mass of $k\ (k > 0)$ times the velocity, show that the equations of motion are (assuming, as usual, that the mass remains constant)

$$
(1) \qquad
\begin{cases}
m\,\dfrac{d^2x}{dt^2} = -\,mk\,\dfrac{dx}{dt}, \\[2ex]
m\,\dfrac{d^2y}{dt^2} = -mg - mk\,\dfrac{dy}{dt}.
\end{cases}
$$

[*See* Figure 8.5.5.]   Solve the system (1) subject to the initial conditions

$$
x(0) = 0, \qquad y(0) = 0, \qquad x'(0) = v_0 \cos \alpha, \qquad y'(0) = v_0 \sin \alpha.
$$

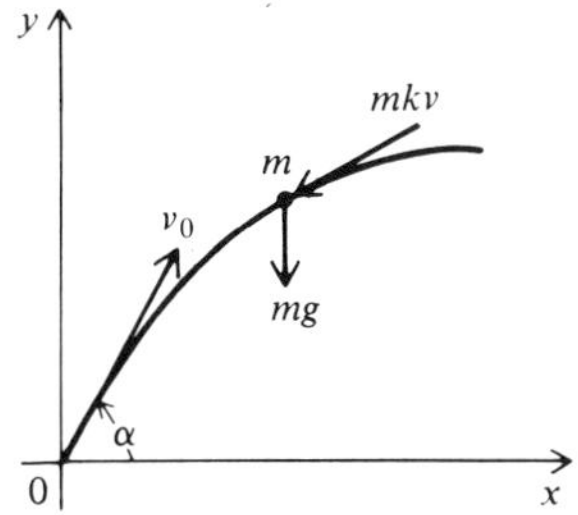

Figure 8.5.5.

3. (Review material of Section 3.2.—Mixture Problems.) Two identical tanks have a capacity of 100 gallons each. Initially both tanks are filled with water. Brine containing 1 lb of dissolved salt per gallon flows into the first tank at a rate of 3 gal/min, and the resulting uniform mixture flows into the second tank at a rate of 5 gal/min. The resulting uniform mixture in the second tank simultaneously flows back into the first tank at a rate of 2 gal/min, and flows out of the second tank at a rate of 3 gal/min.

(a) Show that the amount of salt $x_1(t)$, $x_2(t)$ in each tank at time t is given by

$$(1) \qquad \begin{cases} x_1' + \frac{1}{20}x_1 - \frac{1}{50}x_2 = 3, \\ x_2' - \frac{1}{20}x_1 + \frac{1}{20}x_2 = 0. \end{cases}$$

Also, the initial conditions are

$$(2) \qquad x_1(0) = 0, \qquad x_2(0) = 0.$$

(b) Solve the system (1) subject to the initial conditions (2).

4. Two identical tanks have a capacity of G gallons each. The first is filled initially with brine containing p pounds of salt, while the second is filled with water. Solution is pumped from each tank to the other at the same rate of r gallons per minute.

(a) Find the quantity of salt $x_1(t)$ and $x_2(t)$ in each tank at time t.

(b) Determine the amount of salt in each tank as $t \to \infty$.

5. Consider the mechanical system consisting of two masses m_1, m_2 and two springs having spring constants k_1 and k_2 as shown in Figure 8.5.6. The masses m_1, m_2 are displaced s ft downward from the equilibrium position, and then released. Assume that the motion is along a straight line and that the masses of the springs are negligible. Determine the equations of motion $x_1(t)$, $x_2(t)$ at time t.

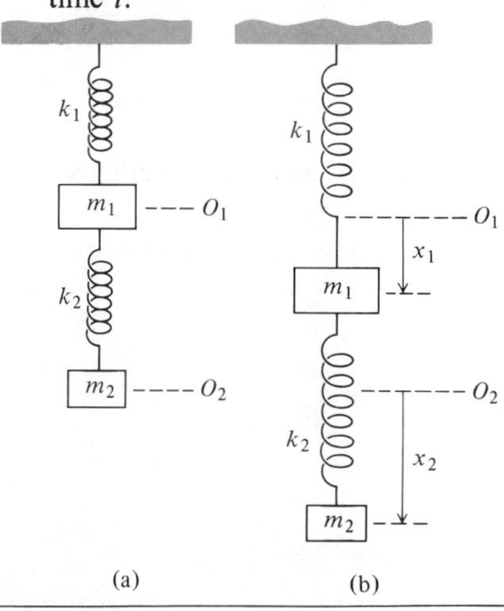

Figure 8.5.6.
(a)   System in equilibrium
(b)   System at time t

6. Consider the mechanical system consisting of two springs S_1 and S_2 as shown in Figure 8.5.7.   Suppose that $k_1 = 100$ lb/ft, $k_2 = 87.5$ lb/ft, $W_1 = 4$ lb and $W_2 = 14$ lb.   After the system has reached its equilibrium position, determine the equations of motion $x_1(t)$, $x_2(t)$ at time t, when

(a) each weight W_1 and W_2 is pulled down 1 ft and released from rest at time $t = 0$;

(b) the weight W_2 is held fast, while the weight W_1 is pulled up 6 inches and released from rest at time $t = 0$;

(c) the weight W_2 is projected downward with a speed of 2 ft/sec;

(d) the weight W_1 is pulled up 9 inches and the weight W_2 is pulled down 6 inches and released from rest at time $t = 0$;

(e)
$$\begin{cases} x_1(0) = -\tfrac{3}{4}\text{ft}, & x_2(0) = -\tfrac{1}{2}\text{ft}, \\ x_1'(0) = -1\text{ ft/sec}, & x_2'(0) = 2\text{ ft/sec}. \end{cases}$$

What is the physical interpretation of these initial conditions?

7. In the network of Figure 8.5.8, $E = 20$ volts, $R_1 = 20$ ohms, $R_2 = 40$ ohms, $L_1 = 0.01$ henry, and $L_2 = 0.02$ henry.   If initially the currents are zero, determine the currents $i_1(t)$, $i_2(t)$ at time t.

8. In the network of Figure 8.5.9, $E = 50$ volts, $R_1 = 20$ ohms, $R_2 = 10$ ohms, $C_1 = 10^{-4}$ farad, and $C_2 = 2 \cdot 10^{-4}$ farad.   If the initial drop in voltage across each capacitor C_1 and C_2 is zero, determine the currents $i_1(t)$, $i_2(t)$ at time t.

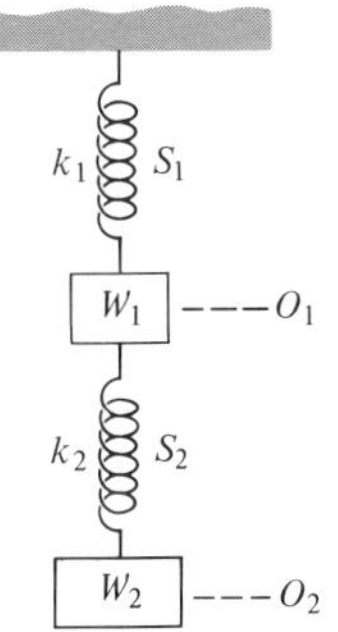

Figure 8.5.7.
System in equilibrium

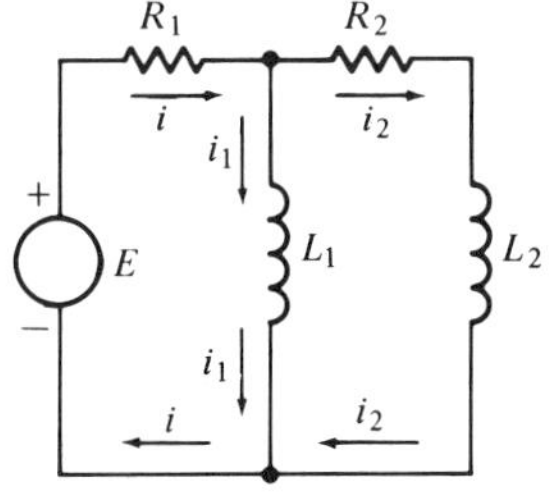

Figure 8.5.8.

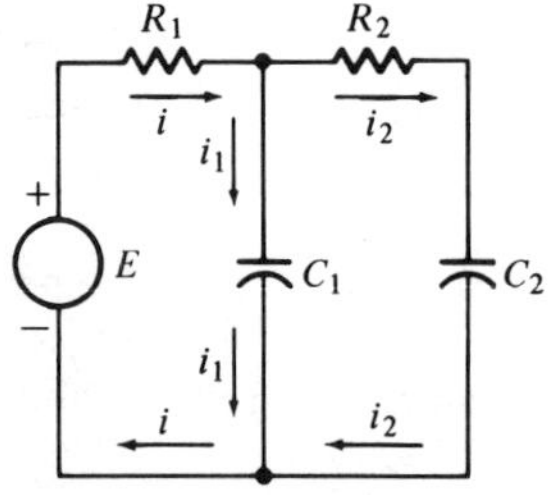

Figure 8.5.9.

9. In the network of Figure 8.5.10, $E = 25$ volts, $R = 25$ ohms, $C = 4 \cdot 10^{-4}$ farad, and $L = 0.02$ henry. If initially the currents $i_1 = 5$ amps and $i_2 = 1$ amp, determine the currents $i_1(t)$, $i_2(t)$ at time t.

10. Consider the network displayed in Figure 8.5.11 with $E = E_0 \sin \omega t$.
    (a) If initially the currents are zero, determine the currents $i_1(t)$, $i_2(t)$ at time t.
    (b) Find the currents $i_1(t)$, $i_2(t)$ of part (a) for the case when $E_0 = 150$ volts, $\omega = 120\pi$ rad, $R_1 = 50$ ohms, $R_2 = 10$ ohms, and $C = 0.002$ farad.

11. In the network of Figure 8.5.12, initially the charge on the capacitor is Q_0 coulombs, and the branch current $i_2(t)$ is zero.
    (a) Determine the charge $Q(t)$ on the capacitor and the currents in the various branches at time t.
    (b) Find $Q(t)$, $i_1(t)$, $i_2(t)$, and $i(t)$ in part (a) for the case when $Q_0 = 1$ coulomb, $E = 50$ volts, $R_1 = 25$ ohms, $R_2 = 5$ ohms, $C = 0.04$ farad, and $L = 0.8$ henry.

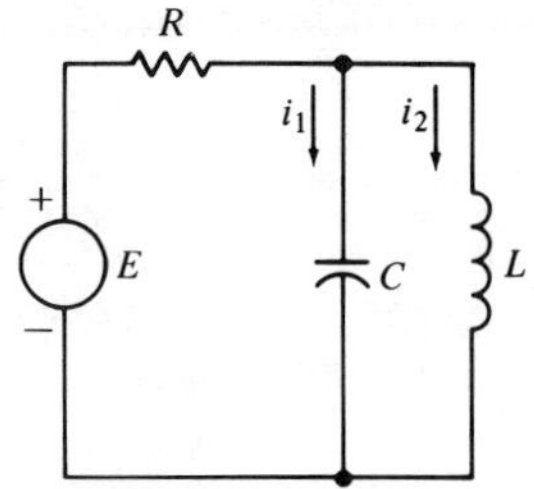

Figure 8.5.10.

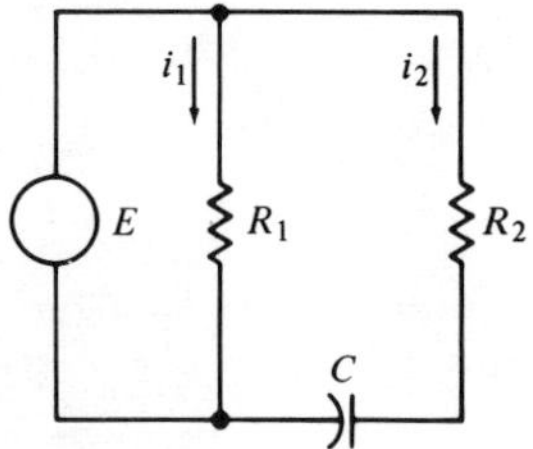

Figure 8.5.11.

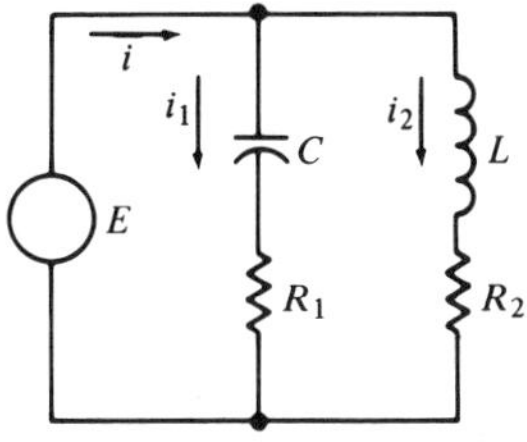

Figure 8.5.12.

12. In the network of Figure 8.5.12, initially the charge on the capacitor is Q_0 coulombs, and the branch current $i_2(t)$ is zero. If $E = E_0 \sin \omega t$, where E_0 and ω are constants, determine the charge on the capacitor and the currents $i_1(t)$, $i_2(t)$ at time t.

13. Consider the network shown in Figure 8.5.13.   Suppose that at time $t = 0$, the network is at rest and the switch S is closed; that is, there are no charges on the capacitors and no currents are flowing at $t = 0$.   Utilizing Kirchhoff's laws, show that the circuit equations for the branch currents $i_2(t)$, $i_3(t)$ at time t are given by

$$\left\{\begin{aligned}
&\left((L_1 + L_2)D^2 + (R_1 + R_2)D + \left(\frac{1}{C_1} + \frac{1}{C_2}\right)\right)i_2 + \left(L_1 D^2 + R_1 D + \frac{1}{C_1}\right)i_3 = 0, \\
&\quad - \left(L_2 D^2 + R_2 D + \frac{1}{C_2}\right)i_2 + \left(L_3 D^2 + R_3 D + \frac{1}{C_3}\right)i_3 = 0.
\end{aligned}\right.$$

14. Suppose in the network of Figure 8.5.13, the capacitors C_1, C_2, and C_3 are short circuited; and $R_1 = 20$ ohms, $R_2 = 10$ ohms, $R_3 = 15$ ohms, $L_1 = 0.04$ henry, $L_2 = 0.02$ henry, and $L_3 = 0.03$ henry.   If the network is at rest initially, find the currents $i_1(t)$, $i_2(t)$, $i_3(t)$ at time t, when voltages $E_1(t) = 60$ and $E_2(t) = 30$ are impressed in the network at time $t = 0$.

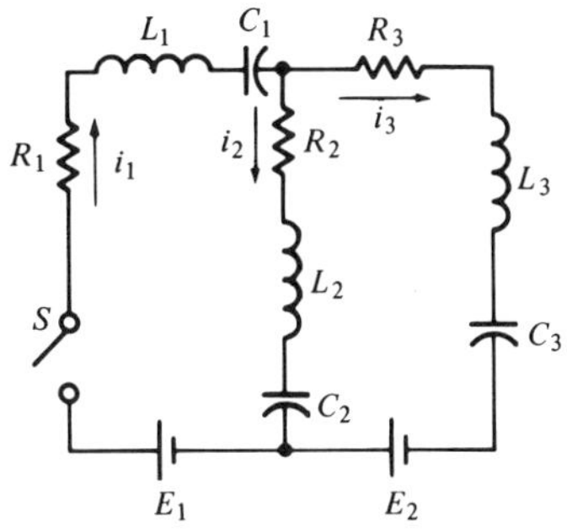

Figure 8.5.13.

15. In the network shown in Figure 8.5.14, initially there are no charges on the capacitors.

   (a) Show that the current $i(t)$ at time t is given by

$$i = \frac{E}{R}\, e^{-t/RC},$$

   where $C = C_1 + C_2 + C_3$.   [*Note: C* is an equivalent capacitance for capacitors in parallel.]

   (b) Find the current $i(t)$ in part (a) for the case when $E = 100$ volts, $R = 20$ ohms, $C_1 = 10^{-3}$ farad, $C_2 = 2 \cdot 10^{-3}$ farad, and $C_3 = 3 \cdot 10^{-3}$ farad.

16. Consider the network displayed in Figure 8.5.15.   In this figure, the elements of the circuit are coupled electromagnetically; L_1 and L_2 refer, respectively, to the inductance of the primary and secondary coils of a transformer, and L_{12} is called the *mutual inductance*. L_{12} is considered positive or negative if the magnetic fields due to L_1 and L_2 add or subtract, respectively.   Applying Kirchhoff's Voltage Law to the two closed circuits, we have

$$\left| \begin{array}{l} R_1 i_1 + L_{12}\dfrac{di_2}{dt} + L_1 \dfrac{di_1}{dt} = E_1, \\[2ex] L_2 \dfrac{di_2}{dt} + L_{12}\dfrac{di_1}{dt} + R_2 i_2 = E_2. \end{array} \right.$$

   Suppose that in the network of Figure 8.5.15, $R_1 = 1$ ohm, $R_2 = 2$ ohms, $L_1 = 1$ henry, $L_2 = 3$ henrys, and $L_{12} = -1$ henry.   If the network is at rest initially, find the currents $i_1(t)$, $i_2(t)$ at time t, when voltages $E_1(t) = 5 \cos t$ and $E_2(t) = 100$ are impressed in the network at time $t = 0$.

17. In circuit analysis, it is often convenient to consider currents flowing in meshes. A *mesh* is a loop for a current that contains no other loops.   For example, in Figure 8.5.16, *ABGFA* and *BDHGB* are meshes, while *ADKJA* is not a mesh.   A particular direction (indicated by a "curved" arrow) is chosen for each mesh current.   For example, if one wishes, he may consider a "curved" arrow in a clockwise direction as being positive.   Applying Kirchhoff's Voltage and Current Laws, we may write the differential equation for each mesh.   It is very important that all currents flowing through an element of a mesh are considered. For example, when writing the equation for the mesh containing the current i_1,

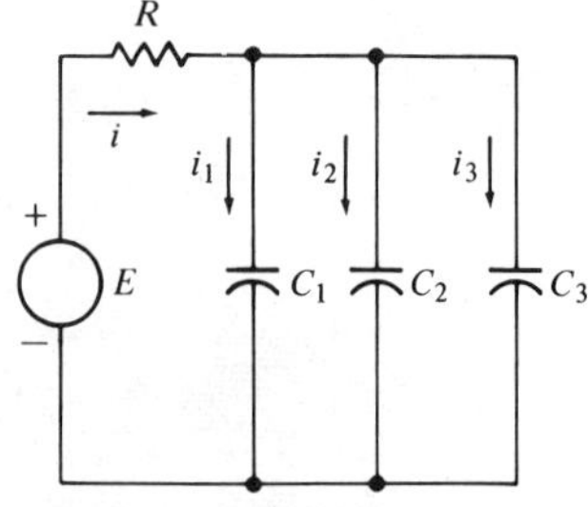

Figure 8.5.14.

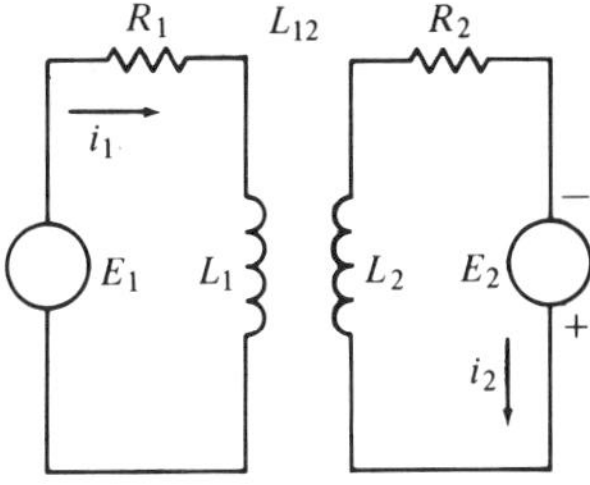

Figure 8.5.15.

the current flowing through R_3 is $i_1 - i_3$; whereas, when writing the equation for the mesh containing the current i_3, the current flowing through R_3 is now $i_3 - i_1$. To obtain a solution for a desired quantity, for example, the current flowing through the resistor R_4, it is simply a matter of adding or subtracting the proper mesh currents. For instance, in Figure 8.5.16, the current through the resistor R_4 is $i_3 - i_2$ or $i_2 - i_3$, or simply $|i_3 - i_2|$.

(a) In Figure 8.5.16, suppose that the charge on the capacitor is initially zero. Find the mesh currents $i_1(t)$, $i_2(t)$, and $i_3(t)$ at time t.

(b) Find the current flowing through R_4 in part (a) for the case when $E = 60$ volts, $R_1 = 10$ ohms, $R_2 = 10$ ohms, $R_3 = 4$ ohms, $R_4 = 2$ ohms, and $C = 0.003$ farad.

In order to facilitate calculations in Exercises 18 through 20, we have assigned convenient values to the elements of the network. These values, however, may not be typical in practical applications.

18. In Figure 8.5.17, a configuration similar to a Wheatstone Bridge has mesh currents i_1, i_2, and i_3 all initially zero. Show that the equations that result from this network are given in operator notation by

$$\begin{cases} (R_1 + R_3)i_1 - & R_1 i_2 - & R_3 i_3 = E, \\ -R_1 i_1 + [LD + (R_1 + R_2)]i_2 - LDi_3 & = 0, \\ -R_3 i_1 - & LDi_2 + [LD + (R_3 + R_4)]i_3 = 0. \end{cases}$$

Find the current flowing through the inductor L at time t, if $R_1 = R_3 = 10$ ohms, $R_2 = R_4 = 5$ ohms, $L = 1.5$ henrys, and $E = 30$ volts.

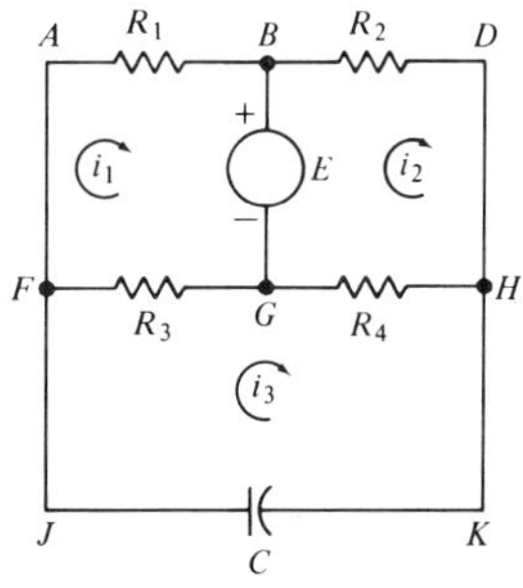

Figure 8.5.16.

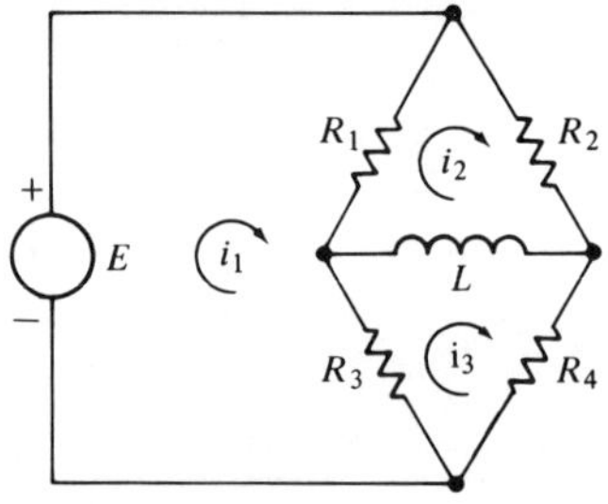

Figure 8.5.17.

19. In the network shown in Figure 8.5.18, each mesh current flows through an inductor, a resistor, and a capacitor. Initially, all of the mesh currents are zero. Show that the equations that result from this network are given by the system below when the forcing function $E = E_0 \sin \omega t$.

$$\begin{cases} \left(L_1 D^2 + R_1 D + \dfrac{1}{C_1}\right)i_1 & -\dfrac{1}{C_1}i_2 & -L_1 D^2 i_3 = \omega E_0 \cos \omega t, \\[2ex] -\dfrac{1}{C_1}i_1 + \left(L_2 D^2 + R_2 D + \dfrac{1}{C_1}\right)i_2 & -R_2 D i_3 = 0, \\[2ex] -L_1 D^2 i_1 & -R_2 D i_2 + \\[1ex] & \left(L_1 D^2 + R_2 D + \dfrac{1}{C_2}\right)i_3 = 0. \end{cases}$$

Find the currents flowing through the capacitors C_1 and C_2 at time t, if the inductors are shorted and $R_1 = 2$ ohms, $R_2 = 4$ ohms, $C_1 = C_2 = 0.5$ farad, and $E = 10 \sin t$ volts.

20. Throughout Section 8.5, when considering networks, we have dealt entirely with currents flowing through the elements of the network. Quite often, voltage drops across elements are of interest. When dealing with such problems, we must recall the basic V-i relations for circuit elements. For a resistor, $V = iR$;

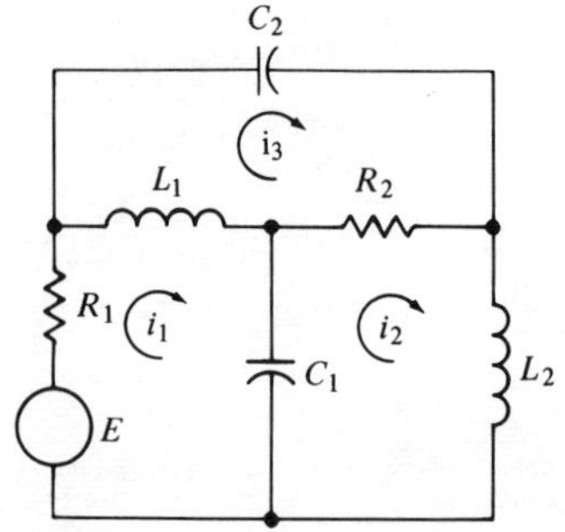

Figure 8.5.18.

for an inductor, $V = L(di/dt)$; and for a capacitor, $V = (\int i\,dt)/C$. When solving for voltage drops, these relations must be written establishing the relation of i in terms of V and the circuit element. That is, for a resistor, $i = V/R$; for an inductor, $i = (\int V\,dt)/L$; and for a capacitor, $i = C(dV/dt)$.

The equations for the network are now written at the nodes instead of around the meshes. Consider the network of Exercise 19. Using a method referred to as the "node voltage method," an arbitrary zero voltage reference is chosen and labeled O. [*See* Figure 8.5.19.] Now the remaining nodes are used to produce the proper number of equations.

From Kirchhoff's Current Law, at A, the algebraic sum of the currents entering this node must be zero. Therefore,

$$C_2 \frac{d(V_3 - V_1)}{dt} + \frac{1}{L_1} \int (V_2 - V_1)dt + \frac{(E - V_1)}{R_1} = 0,$$

which can be written in operator notation after some simplification as

$$(1) \qquad \left(C_2 D^2 + \frac{1}{R_1} D + \frac{1}{L_1} \right) V_1 - \frac{1}{L_1} V_2 - C_2 D^2 V_3 = \frac{\omega E_0 \cos \omega t}{R_1}.$$

At node B, the resulting equation is

$$\frac{1}{L_1} \int (V_1 - V_2)dt + C_1 \frac{d(O - V_2)}{dt} + \frac{(V_3 - V_2)}{R_2} = 0,$$

or in operator notation

$$(2) \qquad -\frac{1}{L_1} V_1 + \left(C_1 D^2 + \frac{1}{R_2} D + \frac{1}{L_1} \right) V_2 - \frac{1}{R_2} D V_3 = 0.$$

Finally, at the node D, the resulting equation is

$$C_2 \frac{d(V_1 - V_3)}{dt} + \frac{(V_2 - V_3)}{R_2} + \frac{1}{L_2} \int (O - V_3)dt = 0,$$

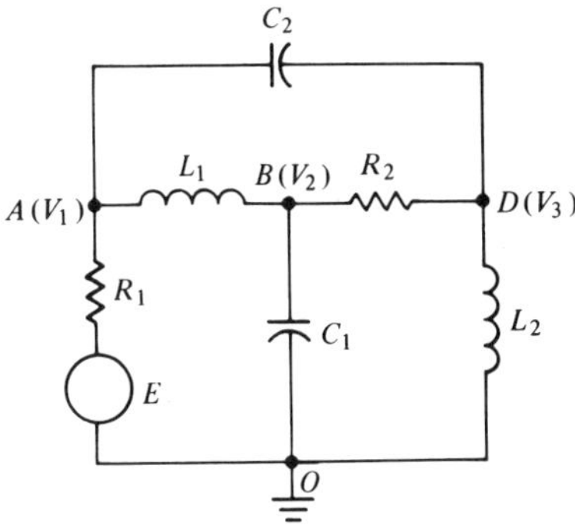

Figure 8.5.19.

or in operator notation

$$(3) \qquad -C_2 D^2 V_1 - \frac{1}{R_2} DV_2 + \left(C_2 D^2 + \frac{1}{R_2} D + \frac{1}{L_2}\right) V_3 = 0.$$

Note that O has been treated as equal to zero.

As in the mesh current method, to obtain a solution for a desired quantity, it is simply a matter of adding or subtracting the proper node voltages. For example, the voltage drop across the inductor L_1 is $|V_1 - V_2|$, while the voltage drop across the inductor L_2 is $|V_3 - O| = |V_3|$.

(a) In Figure 8.5.20, show that the equations that result from this network are given in operator notation by

$$\left[\left(\frac{1}{R_1} + \frac{1}{R_4}\right) D + \frac{1}{L_2}\right] V_1 \qquad\qquad - \frac{1}{R_1} DV_2 \quad - \frac{1}{L_2} V_3 \quad = \frac{\omega E_0}{R_4} \cos \omega t,$$

$$- \frac{1}{R_1} DV_1 + \left[\left(\frac{1}{R_1} + \frac{1}{R_2}\right) D + \frac{1}{L_1}\right] V_2 \quad - \frac{1}{R_2} DV_3 = 0,$$

$$- \frac{1}{L_2} V_1 \qquad\qquad - \frac{1}{R_2} DV_2 + $$

$$\left[\left(\frac{1}{R_2} + \frac{1}{R_3}\right) D + \frac{1}{L_2}\right] V_3 \quad = 0,$$

when the forcing function E is of the form $E_0 \sin \omega t$.

(b) If $R_1 = R_2 = R_3 = R_4 = 1$ ohm, $L_1 = L_2 = 1$ henry, and $E = \sin t$ volt, find the node voltages $V_1(t)$, $V_2(t)$, and $V_3(t)$ at time t. Assume that initially these node voltages are zero.

21. Three masses m, m, and M are connected by three identical springs, whose spring constant is equal to k. The masses are allowed to move around a frictionless, circular rod. We assume that the mass of each spring is negligible, and that the only forces acting on the masses are those due to the restoring forces in the springs. Also, each spring obeys Hooke's Law. Let $x_1(t)$, $x_2(t)$, and $x_3(t)$ denote, respectively, the displacements of the masses m, m, and M from their equilibrium positions O_1, O_2, and O_3. [*See* Figure 8.5.21.] We assume that a counterclockwise displacement of a mass is considered positive.

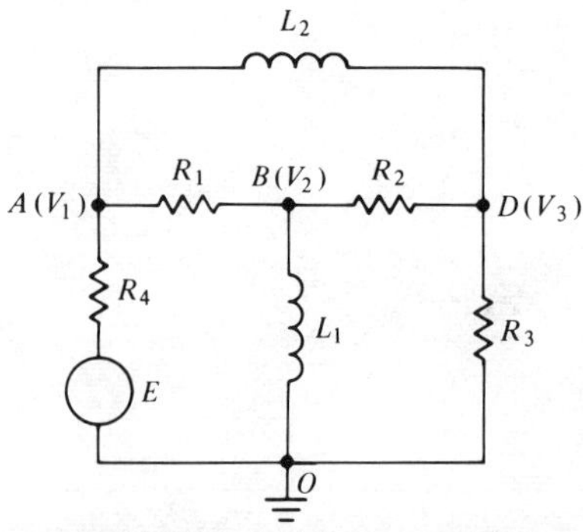

Figure 8.5.20.

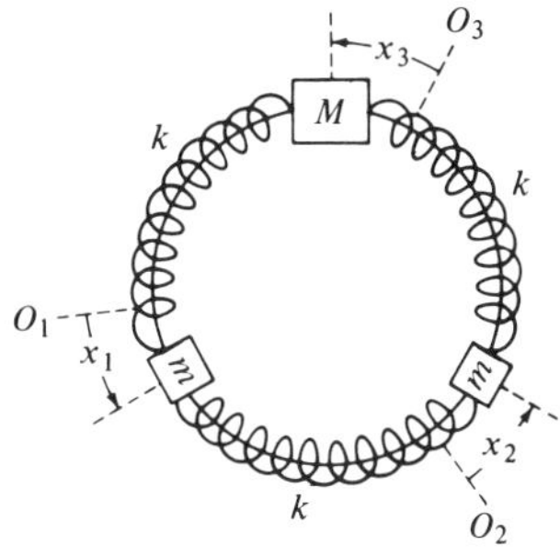

Figure 8.5.21.

(a) Show that the equations of motion of the masses m, m, and M in operator notation are given by

$$\begin{cases} (mD^2 + 2k)x_1 & - \; kx_2 & - \; kx_3 = 0, \\ - \; kx_1 + (mD^2 + 2k)x_2 & & - \; kx_3 = 0, \\ - \; kx_1 & - \; kx_2 + (MD^2 + 2k)x_3 = 0. \end{cases}$$

(b) Obtain the general solution $x_1(t)$, $x_2(t)$, and $x_3(t)$ at time t of the system of equations given in part (a).

(c) Suppose that at time $t = 0$, the identical masses (m) are at their equilibrium positions O_1 and O_2, with no initial velocity; while the mass M is displaced counterclockwise s ft from its equilibrium position O_3, and then given an initial velocity of v_0 ft/sec in a counterclockwise direction. Obtain the equations of motion $x_1(t)$, $x_2(t)$, and $x_3(t)$ at time t.

8.6. Matrix Algebra

In Section 8.11, we shall develop another method of solving a system of n linear first-order equations. In doing so, we shall need some basic concepts of matrix theory. As will be seen, this method has definite advantages over the methods developed in Section 8.4, especially when the system contains more than three equations. This method also generates interesting results and is utilized in various fields of applications.

We shall denote "matrices" by bold-faced capital letters **A**, **B**, ... An expression of the form

$$\begin{bmatrix} a_{11} & a_{12} & \cdots & a_{1n} \\ a_{21} & a_{22} & \cdots & a_{2n} \\ \vdots & \vdots & & \vdots \\ a_{m1} & a_{m2} & \cdots & a_{mn} \end{bmatrix} \tag{8.6.1}$$

is called a *rectangular array* consisting of m rows and n columns; the numbers or functions a_{ij} occurring in this array are called *elements*. We shall assume the elements a_{ij} to be real or complex-valued.   An array of the type (8.6.1), subject to certain rules of operations defined below [*see* Definitions 8.6.1 through 8.6.4], is called a *matrix*.   We shall denote this matrix by $\mathbf{A}$.

$$\mathbf{A} = \begin{bmatrix} a_{11} & a_{12} & \cdots & a_{1n} \\ a_{21} & a_{22} & \cdots & a_{2n} \\ \vdots & \vdots & & \vdots \\ a_{m1} & a_{m2} & \cdots & a_{mn} \end{bmatrix}. \tag{8.6.2}$$

The subscripts i and j of the element a_{ij} of a matrix identify, respectively, the *row* and *column* where the element a_{ij} is located.   For example, in (8.6.2), the element a_{32} occurs in the third row and second column, while the element a_{23} occurs in the second row and third column.

A matrix $\mathbf{A}$ consisting of m rows and n columns is said to be of *order* "m by n" or "$m \times n$."   A matrix $\mathbf{A}$ consisting of n rows and n columns is called a *square matrix* of order n, or an *n-squared matrix*.   In a square matrix, the elements $a_{11}, a_{22}, \ldots, a_{nn}$ are called the (*main*) *diagonal elements*.

At times, it is convenient to abbreviate the symbol (8.6.2) to the form $[a_{ij}]_{(m \times n)}$, which means "the matrix of order $m \times n$ whose elements are the a_{ij}s."   When the order of the matrix has been established, this is abbreviated further to the form $[a_{ij}]$.

Definition 8.6.1.   *Equality   of   Matrices.*   Two   matrices $\mathbf{A} = [a_{ij}]$ and $\mathbf{B} = [b_{ij}]$ are said to be equal ($\mathbf{A} = \mathbf{B}$) if and only if they have the same order, and each element of the matrix $\mathbf{A}$ is equal to the corresponding element of the matrix $\mathbf{B}$.

Thus, $\mathbf{A} = \mathbf{B}$, if and only if

$$a_{ij} = b_{ij} \tag{8.6.3}$$

for all values of i and j.   In other words, two matrices are equal if and only if they are identical.

Example 8:6.1.   If $\mathbf{A} = \begin{bmatrix} x^2 & u \\ y & v^2 \end{bmatrix}$ and $\mathbf{B} = \begin{bmatrix} 1 & 3 \\ 2 & -4 \end{bmatrix}$, then $\mathbf{A} = \mathbf{B}$ provided that $x = \pm 1$, $y = 2$, $u = 3$, and $v = \pm 2i$, $i^2 = -1$.

Definition 8.6.2.   *Addition* (*Subtraction*) *of Matrices.*   If $\mathbf{A} = [a_{ij}]$ and $\mathbf{B} = [b_{ij}]$ are two $m \times n$ matrices, their sum (difference), $\mathbf{A} \pm \mathbf{B}$, is defined to be the $m \times n$ matrix $\mathbf{C} = [c_{ij}]$, where each element of $\mathbf{C}$ is the sum (difference) of the corresponding elements of $\mathbf{A}$ and $\mathbf{B}$.   Thus,

$$[c_{ij}] = [a_{ij}] \pm [b_{ij}] = [a_{ij} \pm b_{ij}]. \tag{8.6.4}$$

Remark 8.6.1. We do not define addition (subtraction) for matrices that are not of the same order. Two matrices of the same order are said to be *conformable for addition*.

Example 8.6.2. If $\mathbf{A} = \begin{bmatrix} 1 & -1 & 0 \\ 2 & 2 & 1 \\ 3 & 3 & 2 \end{bmatrix}$ and $\mathbf{B} = \begin{bmatrix} -1 & 1 & -1 \\ -2 & 3 & 2 \\ -3 & 1 & 4 \end{bmatrix}$, then

$$\mathbf{A} + \mathbf{B} = \begin{bmatrix} 1-1 & -1+1 & 0-1 \\ 2-2 & 2+3 & 1+2 \\ 3-3 & 3+1 & 2+4 \end{bmatrix} = \begin{bmatrix} 0 & 0 & -1 \\ 0 & 5 & 3 \\ 0 & 4 & 6 \end{bmatrix}$$

and

$$\mathbf{A} - \mathbf{B} = \begin{bmatrix} 1+1 & -1-1 & 0+1 \\ 2+2 & 2-3 & 1-2 \\ 3+3 & 3-1 & 2-4 \end{bmatrix} = \begin{bmatrix} 2 & -2 & 1 \\ 4 & -1 & -1 \\ 6 & 2 & -2 \end{bmatrix}.$$

Definition 8.6.3. *Scalar Multiplication of Matrices.* The product of a scalar α and a matrix $\mathbf{A} = [a_{ij}]$ is the matrix $\mathbf{B} = [b_{ij}]$, where $b_{ij} = \alpha a_{ij}$. Thus,

$$[b_{ij}] = \alpha[a_{ij}] = [\alpha a_{ij}]. \tag{8.6.5}$$

[*Note:* Each element of the matrix $\mathbf{A}$ is multiplied by the scalar α.]

Example 8.6.3. If $\mathbf{A} = \begin{bmatrix} 2 & -1 \\ 1 & 3 \end{bmatrix}$, then $2\begin{bmatrix} 2 & -1 \\ 1 & 3 \end{bmatrix} = \begin{bmatrix} 4 & -2 \\ 2 & 6 \end{bmatrix}$.

If $\mathbf{A} = \begin{bmatrix} i & -1 \\ 3i & 2 \end{bmatrix}$, where $i^2 = -1$, then $2i\begin{bmatrix} i & -1 \\ 3i & 2 \end{bmatrix} = \begin{bmatrix} -2 & -2i \\ -6 & 4i \end{bmatrix}$.

If $\mathbf{A}$, $\mathbf{B}$, $\mathbf{C}$ are conformable for addition and α, β are scalars, we have [*see* Exercise 8.6.1]

$$\mathbf{A} + \mathbf{B} = \mathbf{B} + \mathbf{A}. \qquad \text{Addition is commutative.} \tag{8.6.6}$$

$$\mathbf{A} + (\mathbf{B} + \mathbf{C}) = (\mathbf{A} + \mathbf{B}) + \mathbf{C}. \qquad \text{Addition is associative.} \tag{8.6.7}$$

$$\alpha(\mathbf{A} + \mathbf{B}) = \alpha\mathbf{A} + \alpha\mathbf{B}, \qquad (\alpha + \beta)\mathbf{A} = \alpha\mathbf{A} + \beta\mathbf{A}. \qquad \text{Distributive laws.} \tag{8.6.8}$$

Definition 8.6.4. *Multiplication of Matrices.* The product $\mathbf{AB}$ "*in that order*" of the $m \times n$ matrix $\mathbf{A} = [a_{ij}]$ and the $n \times p$ matrix $\mathbf{B} = [b_{ij}]$ is the $m \times p$ matrix $\mathbf{C} = [c_{ij}]$, where

$$c_{ij} = a_{i1}b_{1j} + a_{i2}b_{2j} + \cdots + a_{in}b_{nj} = \sum_{k=1}^{n} a_{ik}b_{kj}, \quad i = 1, \ldots, m; \; j = 1, \ldots, p.$$

$$\tag{8.6.9}$$

Remark 8.6.2. Since the product is dependent on the order, we write $[a_{ij}][b_{ij}] = [c_{ij}]$. Note that $[b_{ij}][a_{ij}]$ is not defined unless $p = m$. In view of (8.6.9), to obtain the element in the ith row and jth column of the matrix $\mathbf{C}$, we multiply each element of the *i*th row of **A** by the corresponding element of the *j*th column of **B**, and then add the resulting products. When the number of columns of a matrix **A** is the same as the number of rows of a matrix **B**, **A** *is said to be conformable to* **B** *for multiplication.* Even when **A** is conformable to **B** for multiplication and **B** is conformable to **A** for multiplication, it is not necessarily true that **AB** = **BA**. That is, in general, for matrix multiplication, the commutative law does not hold. Thus, in general,

$$\mathbf{AB} \neq \mathbf{BA}. \tag{8.6.10}$$

[*See* Example 8.6.5.]

Example 8.6.4. Let $\mathbf{A} = \begin{bmatrix} 2 & -3 & 4 \\ 1 & 0 & -2 \\ 3 & -1 & 2 \end{bmatrix}$ and $\mathbf{B} = \begin{bmatrix} 1 & 2 \\ -2 & 4 \\ 0 & 3 \end{bmatrix}$.

Observe that the number of columns of the matrix **A** is the same as the number of rows of the matrix **B**, namely, three. Since the orders of the matrices **A** and **B** are, respectively, 3×3 and 3×2, the order of the matrix $\mathbf{C} = \mathbf{AB}$ is 3×2. In view of Definition 8.6.4, we obtain

$$\begin{bmatrix} 2 & -3 & 4 \\ 1 & 0 & -2 \\ 3 & -1 & 2 \end{bmatrix} \begin{bmatrix} 1 & 2 \\ -2 & 4 \\ 0 & 3 \end{bmatrix}$$

$$= \begin{bmatrix} 2 \cdot 1 + (-3) \cdot (-2) + 4 \cdot 0 & 2 \cdot 2 + (-3) \cdot 4 + 4 \cdot 3 \\ 1 \cdot 1 + 0 \cdot (-2) + (-2) \cdot 0 & 1 \cdot 2 + 0 \cdot 4 + (-2) \cdot 3 \\ 3 \cdot 1 + (-1) \cdot (-2) + 2 \cdot 0 & 3 \cdot 2 + (-1) \cdot 4 + 2 \cdot 3 \end{bmatrix}$$

$$= \begin{bmatrix} 8 & 4 \\ 1 & -4 \\ 5 & 8 \end{bmatrix}.$$

Example 8.6.5. If $\mathbf{A} = \begin{bmatrix} 2 & 0 & -1 \\ -1 & 2 & 0 \\ 1 & 3 & 4 \end{bmatrix}$ and $\mathbf{B} = \begin{bmatrix} 1 & 2 & 0 \\ 0 & 1 & -1 \\ 1 & 3 & -2 \end{bmatrix}$, then

$$\mathbf{AB} = \begin{bmatrix} 2 \cdot 1 + 0 \cdot 0 + (-1) \cdot 1 & 2 \cdot 2 + 0 \cdot 1 + (-1) \cdot 3 \\ -1 \cdot 1 + 2 \cdot 0 + 0 \cdot 1 & -1 \cdot 2 + 2 \cdot 1 + 0 \cdot 3 \\ 1 \cdot 1 + 3 \cdot 0 + 4 \cdot 1 & 1 \cdot 2 + 3 \cdot 1 + 4 \cdot 3 \end{bmatrix}$$

$$\begin{matrix} 2 \cdot 0 + 0 \cdot (-1) + (-1) \cdot (-2) \\ -1 \cdot 0 + 2 \cdot (-1) + 0 \cdot (-2) \\ 1 \cdot 0 + 3 \cdot (-1) + 4 \cdot (-2) \end{matrix} \Bigg] = \begin{bmatrix} 1 & 1 & 2 \\ -1 & 0 & -2 \\ 5 & 17 & -11 \end{bmatrix}.$$

Similarly, we find

$$\mathbf{BA} = \begin{bmatrix} 0 & 4 & -1 \\ -2 & -1 & -4 \\ -3 & 0 & -9 \end{bmatrix}.$$

Thus, $\mathbf{AB} \neq \mathbf{BA}$.

If $\mathbf{A}$, $\mathbf{B}$, $\mathbf{C}$ are conformable for the indicated sums and products, we have [*see* Exercise 8.6.2]

$$\mathbf{A(BC)} = \mathbf{(AB)C}. \quad \text{Multiplication is associative.} \tag{8.6.11}$$

$$\mathbf{A(B + C)} = \mathbf{AB + AC}, \quad \mathbf{(A + B)C} = \mathbf{AC + BC}.$$
$$\text{Multiplication is distributive with respect to addition.} \tag{8.6.12}$$

A matrix, each of whose elements is zero is called a *zero matrix*, and we denote it by $\mathbf{0}$. If

$$\mathbf{A} = \begin{bmatrix} a_{11} & a_{12} & a_{13} \\ a_{21} & a_{22} & a_{23} \\ a_{31} & a_{32} & a_{33} \end{bmatrix} \quad \text{and} \quad \mathbf{0} = \begin{bmatrix} 0 & 0 & 0 \\ 0 & 0 & 0 \\ 0 & 0 & 0 \end{bmatrix},$$

then

$$\mathbf{A + 0} = \begin{bmatrix} a_{11} + 0 & a_{12} + 0 & a_{13} + 0 \\ a_{21} + 0 & a_{22} + 0 & a_{23} + 0 \\ a_{31} + 0 & a_{32} + 0 & a_{33} + 0 \end{bmatrix} = \mathbf{A}.$$

Similarly, we have $\mathbf{0 + A = A}$. Thus

$$\mathbf{A + 0 = 0 + A = A}. \tag{8.6.13}$$

The zero matrix is an identity for addition.

Remark 8.6.3. Let $\mathbf{A} = \begin{bmatrix} 1 & 2 \\ 0 & 0 \end{bmatrix}$ and $\mathbf{B} = \begin{bmatrix} 0 & 2 \\ 0 & -1 \end{bmatrix}$, then $\mathbf{AB = 0}$. Thus, if $\mathbf{AB = 0}$, then it is not necessarily true that $\mathbf{A = 0}$ or $\mathbf{B = 0}$.[1] Let

$$\mathbf{A} = \begin{bmatrix} 1 & -1 \\ -1 & 1 \end{bmatrix}, \quad \mathbf{B} = \begin{bmatrix} 1 & 1 \\ 0 & 1 \end{bmatrix}, \quad \text{and} \quad \mathbf{C} = \begin{bmatrix} 0 & 1 \\ -1 & 1 \end{bmatrix}, \quad \text{then}$$

$$\mathbf{AB} = \begin{bmatrix} 1 & 0 \\ -1 & 0 \end{bmatrix} \quad \text{and} \quad \mathbf{AC} = \begin{bmatrix} 1 & 0 \\ -1 & 0 \end{bmatrix}.$$

Thus, $\mathbf{AB = AC}$ does not imply $\mathbf{B = C}$. That is, the cancellation law is not necessarily true for matrices. Let $\mathbf{A} = \begin{bmatrix} 1 & 1 \\ -1 & -1 \end{bmatrix} \neq \mathbf{0}$, then $\mathbf{A^2 = A \cdot A = 0}$.

---

[1] If $\mathbf{A}$ and $\mathbf{B}$ are nonzero n-square matrices such that $\mathbf{AB = 0}$, then $\mathbf{A}$ and $\mathbf{B}$ are called *divisors of zero*.

Thus, if $\mathbf{A} \neq \mathbf{0}$, then it is not necessarily true that $\mathbf{A}^p \neq \mathbf{0}$, where p is a positive integer. Matrices such that $\mathbf{A}^p = \mathbf{0}$, for some positive integer p, are called *nilpotent*.

A square matrix $\mathbf{A}$ whose elements $a_{ij} = 0$ for $i \neq j$ and $a_{ii} = 1$ is called an *identity matrix*, and we denote it by $\mathbf{I}$.

$$\mathbf{I} = \begin{bmatrix} 1 & 0 & \cdots & 0 \\ 0 & 1 & \cdots & 0 \\ \vdots & \vdots & & \vdots \\ 0 & 0 & \cdots & 1 \end{bmatrix}. \tag{8.6.14}$$

If

$$\mathbf{A} = \begin{bmatrix} a_{11} & a_{12} & a_{13} \\ a_{21} & a_{22} & a_{23} \\ a_{31} & a_{32} & a_{33} \end{bmatrix} \quad \text{and} \quad \mathbf{I} = \begin{bmatrix} 1 & 0 & 0 \\ 0 & 1 & 0 \\ 0 & 0 & 1 \end{bmatrix},$$

then

$$\mathbf{AI} = \begin{bmatrix} a_{11} \cdot 1 & a_{12} \cdot 1 & a_{13} \cdot 1 \\ a_{21} \cdot 1 & a_{22} \cdot 1 & a_{23} \cdot 1 \\ a_{31} \cdot 1 & a_{32} \cdot 1 & a_{33} \cdot 1 \end{bmatrix} = \mathbf{A}.$$

Similarly, we find $\mathbf{IA} = \mathbf{A}$. Thus,

$$\mathbf{AI} = \mathbf{IA} = \mathbf{A}. \tag{8.6.15}$$

The identity matrix is an identity for multiplication. The identity matrix $\mathbf{I}$ commutes with a matrix $\mathbf{A}$, when $\mathbf{A}$ is conformable to $\mathbf{I}$ for multiplication.

If $\mathbf{A} = [a_{ij}]$ is an $m \times n$ matrix, its *transpose* is defined as the $n \times m$ matrix $\mathbf{B} = [b_{ji}]$, where $b_{ji} = a_{ij}$. The transpose of the matrix $\mathbf{A}$ is usually denoted by $\mathbf{A}^T$ or $\mathbf{A}'$. Note that we form the transpose of a matrix $\mathbf{A}$ by simply interchanging the rows and columns of $\mathbf{A}$. If

$$\mathbf{A} = \begin{bmatrix} 1 & 4 & 7 \\ 2 & 5 & 8 \\ 3 & 6 & 9 \end{bmatrix}, \quad \text{then} \quad \mathbf{A}^T = \begin{bmatrix} 1 & 2 & 3 \\ 4 & 5 & 6 \\ 7 & 8 & 9 \end{bmatrix}.$$

If $\mathbf{A}, \mathbf{B}$ are conformable for the indicated sums and products, we have [*see* Exercise 8.6.3]

$$(\mathbf{A}^T)^T = \mathbf{A}. \tag{8.6.16}$$

$$(\mathbf{A} + \mathbf{B})^T = \mathbf{A}^T + \mathbf{B}^T. \tag{8.6.17}$$

$$(\mathbf{AB})^T = \mathbf{B}^T \mathbf{A}^T. \tag{8.6.18}$$

A square matrix $\mathbf{A}$ is called *symmetric* if $\mathbf{A}^T = \mathbf{A}$. Thus, a square matrix $\mathbf{A}$ is symmetric provided $a_{ij} = a_{ji}$, for all values of i and j. For example,

$$\mathbf{A} = \begin{bmatrix} 1 & 2 & 3 \\ 2 & -4 & -6 \\ 3 & -6 & 7 \end{bmatrix} \text{ is a symmetric matrix.}$$

If the elements a_{ij} of a matrix $\mathbf{A}$ are complex numbers or complex-valued functions, then the complex conjugate of a_{ij} is denoted by $\bar{a}_{ij}$. If $\mathbf{A} = [a_{ij}]$ is an $m \times n$ matrix, the *conjugate* of $\mathbf{A}$ is defined as the $m \times n$ matrix $\mathbf{B} = [b_{ij}]$, where $b_{ij} = \bar{a}_{ij}$. The conjugate of $\mathbf{A}$ is denoted by $\overline{\mathbf{A}}$. For example, if

$$\mathbf{A} = \begin{bmatrix} 1 - 2i & 0 & 2 \\ 2i & -i & 3 \\ 3 & -1 - i & 4 - i \end{bmatrix}, \quad \text{then} \quad \overline{\mathbf{A}} = \begin{bmatrix} 1 + 2i & 0 & 2 \\ -2i & i & 3 \\ 3 & -1 + i & 4 + i \end{bmatrix}.$$

A square matrix $\mathbf{A}$ is called *Hermitian* if $\overline{\mathbf{A}}^T = \mathbf{A}$. Thus, a square matrix $\mathbf{A}$ is Hermitian provided $a_{ij} = \bar{a}_{ji}$, for all values of i and j. For example,

$$\mathbf{A} = \begin{bmatrix} 3 & 2 + i & -5i \\ 2 - i & 5 & 3 - i \\ 5i & 3 + i & 4 \end{bmatrix} \text{ is a Hermitian matrix.}$$

Note, for a Hermitian matrix, the diagonal elements a_{ii} must be real elements. (Why?)

Matrix functions. Suppose that the elements a_{ij} of the $m \times n$ matrix $\mathbf{A}$ are functions of the variable t on the interval J: $\alpha \leq t \leq \beta$. Then the matrix $\mathbf{A}$ will depend upon t, and we write

$$\mathbf{A}(t) = \begin{bmatrix} a_{11}(t) & a_{12}(t) & \cdots & a_{1n}(t) \\ a_{21}(t) & a_{22}(t) & \cdots & a_{2n}(t) \\ \vdots & \vdots & & \vdots \\ a_{m1}(t) & a_{m2}(t) & \cdots & a_{mn}(t) \end{bmatrix}. \tag{8.6.19}$$

A *matrix function* $\mathbf{A}$ is a correspondence which assigns to each value t in J a unique $m \times n$ matrix $\mathbf{A}(t)$. A matrix function $\mathbf{A}$ is said to be *continuous at a point t_0 in J* or *continuous on the interval J*, if each element a_{ij} is continuous at a point t_0 in J or continuous on the interval J. A matrix function $\mathbf{A}$ is *differentiable at a point t_0 in J* or *differentiable in the interval J*, if each element a_{ij} is differentiable at a point t_0 in J or differentiable in the interval J. The *derivative* of a matrix function $\mathbf{A}$ is defined by

$$\frac{d\mathbf{A}}{dt} = \begin{bmatrix} \dfrac{d}{dt}a_{11}(t) & \dfrac{d}{dt}a_{12}(t) & \cdots & \dfrac{d}{dt}a_{1n}(t) \\ \dfrac{d}{dt}a_{21}(t) & \dfrac{d}{dt}a_{22}(t) & \cdot\cdot & \dfrac{d}{dt}a_{2n}(t) \\ \vdots & \vdots & & \vdots \\ \dfrac{d}{dt}a_{m1}(t) & \dfrac{d}{dt}a_{m2}(t) & \cdots & \dfrac{d}{dt}a_{mn}(t) \end{bmatrix} \equiv \left[\frac{d}{dt}a_{ij}(t) \right]. \tag{8.6.20}$$

The *integral* of the matrix function $\mathbf{A}$, on the interval J, is defined by

$$
\int_\alpha^\beta \mathbf{A}(t)\,dt =
\begin{bmatrix}
\int_\alpha^\beta a_{11}(t)\,dt & \int_\alpha^\beta a_{12}(t)\,dt & \cdots & \int_\alpha^\beta a_{1n}(t)\,dt \\[2mm]
\int_\alpha^\beta a_{21}(t)\,dt & \int_\alpha^\beta a_{22}(t)\,dt & \cdots & \int_\alpha^\beta a_{2n}(t)\,dt \\[2mm]
\vdots & \vdots & & \vdots \\[2mm]
\int_\alpha^\beta a_{m1}(t)\,dt & \int_\alpha^\beta a_{m2}(t)\,dt & \cdots & \int_\alpha^\beta a_{mn}(t)\,dt
\end{bmatrix}
\equiv \left[\int_\alpha^\beta a_{ij}(t)\,dt \right].
\tag{8.6.21}
$$

We leave it as an exercise for the reader [*see* Exercise 8.6.4] to show that

$$
\frac{d}{dt}\left(\mathbf{C}\mathbf{A}(t)\right) = \mathbf{C}\,\frac{d\mathbf{A}(t)}{dt}, \quad \text{where } \mathbf{C} \text{ is a constant matrix.}
\tag{8.6.22}
$$

$$
\frac{d}{dt}\left(\mathbf{A}(t) + \mathbf{B}(t)\right) = \frac{d\mathbf{A}(t)}{dt} + \frac{d\mathbf{B}(t)}{dt}.
\tag{8.6.23}
$$

$$
\frac{d}{dt}\left(\mathbf{A}(t)\mathbf{B}(t)\right) = \mathbf{A}(t)\,\frac{d\mathbf{B}(t)}{dt} + \frac{d\mathbf{A}(t)}{dt}\,\mathbf{B}(t).
\tag{8.6.24}
$$

$$
\int_\alpha^\beta \left(\mathbf{A}(t) + \mathbf{B}(t)\right)\,dt = \int_\alpha^\beta \mathbf{A}(t)\,dt + \int_\alpha^\beta \mathbf{B}(t)\,dt.
\tag{8.6.25}
$$

Note that the order of multiplication must be preserved in (8.6.22) and (8.6.24). For example, in (8.6.22), it is not necessarily true that $\mathbf{C}\dfrac{d\mathbf{A}(t)}{dt} = \dfrac{d\mathbf{A}(t)}{dt}\mathbf{C}$. In (8.6.25), the elements of $\mathbf{A}$ and $\mathbf{B}$ are assumed integrable in J.

Exercises 8.6

1. Prove (8.6.6), (8.6.7), and (8.6.8).
2. Prove (8.6.11) and (8.6.12).
3. Prove (8.6.16), (8.6.17), and (8.6.18).
4. Prove (8.6.22), (8.6.23), (8.6.24), and (8.6.25).

5. If $\mathbf{A} = \begin{bmatrix} 1 & 2 & 0 \\ -2 & 3 & 1 \\ 0 & 1 & 2 \end{bmatrix}$, $\mathbf{B} = \begin{bmatrix} 2 & -1 & 1 \\ 1 & 0 & 2 \\ 3 & 2 & -1 \end{bmatrix}$, and $\mathbf{C} = \begin{bmatrix} 0 & 1 & -2 \\ -1 & 2 & 0 \\ -2 & 1 & 2 \end{bmatrix}$,

   find

   (a) $2\mathbf{A} + 3\mathbf{B}$,   (b) $3\mathbf{A} - 2\mathbf{B} - \mathbf{C}$,   (c) $\mathbf{A}^T + \mathbf{B}^T - (2\mathbf{C})^T$,   (d) $(\mathbf{A}\mathbf{B})\mathbf{C}$,
   (e) $\mathbf{A}(\mathbf{B}\mathbf{C})$,   (f) $\mathbf{A}(\mathbf{B} + \mathbf{C})$,   (g) $\mathbf{A}\mathbf{B} + \mathbf{A}\mathbf{C}$,   (h) $(\mathbf{A}\mathbf{B})^T$,   (i) $\mathbf{B}^T\mathbf{A}^T$.

6. Perform the matrix multiplications:

(a) $\begin{bmatrix} 1 & 2 & 1 \\ 0 & -1 & 2 \\ 1 & 2 & 0 \end{bmatrix} \begin{bmatrix} -1 & 2 \\ 1 & 1 \\ 0 & -2 \end{bmatrix}$,   (b) $\begin{bmatrix} 1 & -1 & 2 & -2 \end{bmatrix} \begin{bmatrix} 3 \\ 0 \\ -1 \\ 2 \end{bmatrix}$,

(c) $\begin{bmatrix} 1 & 0 & 0 & 0 \\ 0 & 1 & 0 & 0 \\ 0 & 0 & 1 & 0 \\ 0 & 0 & 0 & 1 \end{bmatrix} \begin{bmatrix} a_1 & a_2 \\ b_1 & b_2 \\ c_1 & c_2 \\ d_1 & d_2 \end{bmatrix}$,

(d) $\begin{bmatrix} \alpha & 0 & 0 \\ 0 & \alpha & 0 \\ 0 & 0 & \alpha \end{bmatrix} \begin{bmatrix} a_1 & a_2 & a_3 \\ b_1 & b_2 & b_3 \\ c_1 & c_2 & c_3 \end{bmatrix}$,

(e) $\begin{bmatrix} -1 \\ 2 \\ 3 \\ -1 \end{bmatrix} \begin{bmatrix} 2 & -1 & -2 & -6 \end{bmatrix}$,

(f) $\begin{bmatrix} i & 0 \\ 1 & 1 \end{bmatrix} \begin{bmatrix} 1 & 1 \\ i & 0 \end{bmatrix} \begin{bmatrix} 1 & -i \\ 0 & 1 \end{bmatrix}$, $i^2 = -1$.

7. If $\mathbf{A} = \begin{bmatrix} -1 & 0 & -2 \\ 2 & 1 & 1 \end{bmatrix}$, $\mathbf{B} = \begin{bmatrix} 0 & 1 & 1 \\ 1 & 0 & -1 \\ -1 & -1 & 0 \end{bmatrix}$, $\mathbf{C} = \begin{bmatrix} 0 & 1 \\ 1 & -1 \\ -1 & 1 \end{bmatrix}$,   verify

that $(\mathbf{AB})\mathbf{C} = \mathbf{A}(\mathbf{BC})$.

8. Suppose that $\mathbf{A}$ is a square matrix. Define $\mathbf{A}^{p+1} = \mathbf{A}^p \cdot \mathbf{A}$, where p is a positive integer.

(a) If

$$\mathbf{A} = \begin{bmatrix} 0 & a & b \\ 0 & 0 & c \\ 0 & 0 & 0 \end{bmatrix},$$

show that $\mathbf{A}^3 = \mathbf{0}$.

(b) If

$$\mathbf{A} = \begin{bmatrix} i & 0 & 0 \\ 0 & i & 0 \\ 0 & 0 & i \end{bmatrix}, \quad i^2 = -1,$$

show that $\mathbf{A}^4 = \mathbf{I}$.

9. Determine the values of x for which

$$\begin{bmatrix} x & 2 \end{bmatrix} \begin{bmatrix} 1 & -1 \\ 2 & 0 \end{bmatrix} \begin{bmatrix} x \\ 1 \end{bmatrix} = \mathbf{0}.$$

10. (a) If $\mathbf{A} = \begin{bmatrix} \alpha_1 & 0 & 0 \\ 0 & \alpha_2 & 0 \\ 0 & 0 & \alpha_3 \end{bmatrix}$ and $\mathbf{B} = \begin{bmatrix} \beta_1 & 0 & 0 \\ 0 & \beta_2 & 0 \\ 0 & 0 & \beta_3 \end{bmatrix}$, does $\mathbf{AB} = \mathbf{BA}$?

(Such matrices are called diagonal matrices.)[2]

(b) Compute $\mathbf{A}^p$, where p is a positive integer.

11. Explain why, in general,

$$(\mathbf{A} \pm \mathbf{B})^2 \neq \mathbf{A}^2 \pm 2\mathbf{AB} + \mathbf{B}^2 \quad \text{and} \quad \mathbf{A}^2 - \mathbf{B}^2 \neq (\mathbf{A} - \mathbf{B})(\mathbf{A} + \mathbf{B}).$$

12. Let $\mathbf{A}(t) = \begin{bmatrix} e^t & e^t \\ \cos t & \sin t \end{bmatrix}$ and $\mathbf{B}(t) = \begin{bmatrix} e^{-t} & e^{-t} \\ t & t^2 \end{bmatrix}$. Determine

(a) $\dfrac{d}{dt}(\mathbf{A}(t) + \mathbf{B}(t))$, (b) $\dfrac{d}{dt}(\mathbf{A}(t)\mathbf{B}(t))$, (c) $\displaystyle\int_0^1 (\mathbf{A}(t) + \mathbf{B}(t))dt$.

13. Let $\mathbf{A}$ and $\mathbf{B}$ commute with each other. Show that

$$\mathbf{A}^r\mathbf{B}^s = \mathbf{B}^s\mathbf{A}^r,$$

where r and s are any two positive integers.

14. *Binomial expansion.* Let $\mathbf{A}$ and $\mathbf{B}$ be two $n \times n$ matrices such that $\mathbf{AB} = \mathbf{BA}$. Show that

$$(\mathbf{A} + \mathbf{B})^k = \sum_{j=0}^{k} \frac{k!}{j!(k-j)!} \mathbf{A}^j\mathbf{B}^{k-j},$$

where k is any positive integer. (By definition, $\mathbf{A}^0 = \mathbf{I}$, where $\mathbf{I}$ is the $n \times n$ identity matrix.)

15. Let $\mathbf{A} = [a_{ij}]$ be an $n \times n$ matrix. The sum of the main diagonal elements a_{ii}, $i = 1, 2, \ldots, n$, is called the *trace* of $\mathbf{A}$. The trace of $\mathbf{A}$ is usually abbreviated as tr $\mathbf{A}$. Thus,

$$\text{tr } \mathbf{A} = a_{11} + a_{22} + \cdots + a_{nn} = \sum_{i=1}^{n} a_{ii}.$$

Let $\mathbf{A}$ and $\mathbf{B}$ be two $n \times n$ matrices.
(a) Show that tr $(\mathbf{A} + \mathbf{B}) = $ tr $(\mathbf{A}) + $ tr $(\mathbf{B})$.
(b) Show that tr $(k\mathbf{A}) = k$ tr $\mathbf{A}$, where k is a scalar.
(c) Show that, in general, tr $\mathbf{A}^2 \neq $ tr $\mathbf{A} \cdot $ tr $\mathbf{A}$.
Let $\mathbf{A}$ be an $m \times n$ matrix and $\mathbf{B}$ an $n \times m$ matrix.
(d) Show that tr $(\mathbf{AB}) = $ tr $(\mathbf{BA})$.

(e) Let $\mathbf{A} = \begin{bmatrix} 1 & 1 & -2 \\ -1 & 2 & 1 \\ 0 & 1 & -1 \end{bmatrix}$. Show that

$$\text{tr } \mathbf{A} + \text{tr } \mathbf{A}^2 + \text{tr } \mathbf{A}^3 = 16.$$

---

[2] More generally, a square matrix $\mathbf{A}$ is said to be a *diagonal matrix* if the elements $a_{ij} = 0$, $i \neq j$. The elements a_{ii} of the main diagonal may or may not be zero.

16. A matrix $\mathbf{A}$ such that $\mathbf{A}^2 = \mathbf{A}$ is called *idempotent*. For an $n \times n$ diagonal matrix $\mathbf{A}$, show that there are 2^n idempotent matrices of order n.

8.7. The Inverse of a Matrix

We shall assume that the reader is familiar with some of the basic results developed in the theory of determinants encountered in a course in college algebra.

Let $\mathbf{A} = [a_{ij}]$ be a square matrix. We shall denote the determinant of $\mathbf{A}$ by det $\mathbf{A}$. A square matrix $\mathbf{A}$ is called *nonsingular* if det $\mathbf{A} \neq 0$. If det $\mathbf{A} = 0$, $\mathbf{A}$ is called a *singular* matrix. If $\mathbf{A}$ and $\mathbf{B}$ are two arbitrary square matrices (of the same order) such that $\mathbf{AB} = \mathbf{BA} = \mathbf{I}$, then $\mathbf{B}$ is called the *inverse* of $\mathbf{A}$ (provided it exists). The inverse of $\mathbf{A}$ is usually denoted by $\mathbf{A}^{-1}$. Thus, $\mathbf{B} = \mathbf{A}^{-1}$ and we have

$$\mathbf{AA}^{-1} = \mathbf{A}^{-1}\mathbf{A} = \mathbf{I}. \tag{8.7.1}$$

Let M_{ij} denote the cofactor (signed minor) of the element a_{ij} in the det $\mathbf{A}$. The following result is known.

$$\sum_{j=1}^{n} a_{ij} M_{kj} = \det \mathbf{A} \quad \text{when } i = k, \tag{8.7.2}$$
$$= 0 \quad \text{when } i \neq k.$$

If the matrix $\mathbf{A}$ is nonsingular, then det $\mathbf{A} \neq 0$. Thus, we may write (8.7.2) as

$$\sum_{j=1}^{n} a_{ij} \frac{M_{kj}}{\det \mathbf{A}} = 1 \quad \text{when } i = k,$$
$$= 0 \quad \text{when } i \neq k. \tag{8.7.3}$$

However, (8.7.3) may be written equivalently in the form

$$\mathbf{A} \frac{[M_{ji}]}{\det \mathbf{A}} = \mathbf{I}. \tag{8.7.4}$$

Hence, in view of (8.7.4), we have

$$\mathbf{A}^{-1} = \frac{[M_{ji}]}{\det \mathbf{A}} = \frac{[M_{ij}]^T}{\det \mathbf{A}}. \tag{8.7.5}$$

The matrix $[M_{ji}] = [M_{ij}]^T$ is called the *adjoint* of the matrix $\mathbf{A}$. Thus, we see that whenever a square matrix $\mathbf{A}$ is nonsingular, $\mathbf{A}^{-1}$ exists.

The inverse of a nonsingular matrix is unique. [*See* Exercise 8.7.1.]

Example 8.7.1. Let us find the inverse of the matrix

$$\mathbf{A} = \begin{bmatrix} 1 & 2 & 3 \\ 0 & 1 & -1 \\ -1 & 0 & -2 \end{bmatrix}. \tag{8.7.6}$$

SOLUTION.

$$\det \mathbf{A} = \begin{vmatrix} 1 & 2 & 3 \\ 0 & 1 & -1 \\ -1 & 0 & -2 \end{vmatrix} = 3. \tag{8.7.7}$$

Hence, the matrix $\mathbf{A}$ is nonsingular. We shall now compute the cofactors M_{ij} of the elements a_{ij} in $\det \mathbf{A}$. Now, the cofactor M_{ij} is defined by

$$M_{ij} = (-1)^{i+j} C_{ij}, \tag{8.7.8}$$

where C_{ij} is the minor of the element a_{ij} in the matrix $\mathbf{A}$, which is the determinant of the matrix of order $n - 1$ obtained by deleting the ith row and jth column of the matrix $\mathbf{A}$. Consequently, we have

$$M_{11} = (-1)^{1+1}\begin{vmatrix} 1 & -1 \\ 0 & -2 \end{vmatrix} = -2, \qquad M_{12} = (-1)^{1+2}\begin{vmatrix} 0 & -1 \\ -1 & -2 \end{vmatrix} = 1,$$

$$M_{21} = (-1)^{2+1}\begin{vmatrix} 2 & 3 \\ 0 & -2 \end{vmatrix} = 4, \qquad M_{22} = (-1)^{2+2}\begin{vmatrix} 1 & 3 \\ -1 & -2 \end{vmatrix} = 1,$$

$$M_{31} = (-1)^{3+1}\begin{vmatrix} 2 & 3 \\ 1 & -1 \end{vmatrix} = -5, \qquad M_{32} = (-1)^{3+2}\begin{vmatrix} 1 & 3 \\ 0 & -1 \end{vmatrix} = 1,$$

$$M_{13} = (-1)^{1+3}\begin{vmatrix} 0 & 1 \\ -1 & 0 \end{vmatrix} = 1,$$

$$M_{23} = (-1)^{2+3}\begin{vmatrix} 1 & 2 \\ -1 & 0 \end{vmatrix} = -2, \tag{8.7.9}$$

$$M_{33} = (-1)^{3+3}\begin{vmatrix} 1 & 2 \\ 0 & 1 \end{vmatrix} = 1.$$

Thus,

$$[M_{ij}] = \begin{bmatrix} M_{11} & M_{12} & M_{13} \\ M_{21} & M_{22} & M_{23} \\ M_{31} & M_{32} & M_{33} \end{bmatrix} = \begin{bmatrix} -2 & 1 & 1 \\ 4 & 1 & -2 \\ -5 & 1 & 1 \end{bmatrix}$$

and

$$[M_{ji}] = [M_{ij}]^T = \begin{bmatrix} -2 & 4 & -5 \\ 1 & 1 & 1 \\ 1 & -2 & 1 \end{bmatrix}. \tag{8.7.10}$$

Hence, in view of (8.7.5), (8.7.7), and (8.7.10), we obtain

$$\mathbf{A}^{-1} = \frac{[M_{ij}]^T}{\det \mathbf{A}} = \begin{bmatrix} -\frac{2}{3} & \frac{4}{3} & -\frac{5}{3} \\ \frac{1}{3} & \frac{1}{3} & \frac{1}{3} \\ \frac{1}{3} & -\frac{2}{3} & \frac{1}{3} \end{bmatrix}.$$

[*Note* 1.]

$$\begin{bmatrix} 1 & 2 & 3 \\ 0 & 1 & -1 \\ -1 & 0 & -2 \end{bmatrix} \begin{bmatrix} -\frac{2}{3} & \frac{4}{3} & -\frac{5}{3} \\ \frac{1}{3} & \frac{1}{3} & \frac{1}{3} \\ \frac{1}{3} & -\frac{2}{3} & \frac{1}{3} \end{bmatrix} = \begin{bmatrix} 1 & 0 & 0 \\ 0 & 1 & 0 \\ 0 & 0 & 1 \end{bmatrix}.$$

Thus, the computed inverse matrix is indeed correct.

[*Note* 2.]

One may obtain the value of det $\mathbf{A}$ by utilizing Formula (8.7.2), namely,

$$\det \mathbf{A} = \sum_{j=1}^{3} a_{ij} M_{ij}. \tag{8.7.11}$$

For instance, when $i = 1$ in (8.7.11), we obtain

$$\det \mathbf{A} = \sum_{j=1}^{3} a_{1j} M_{1j} = a_{11} M_{11} + a_{12} M_{12} + a_{13} M_{13}. \tag{8.7.12}$$

From (8.7.6) and (8.7.9), we see that $a_{11} = 1$, $a_{12} = 2$, $a_{13} = 3$, $M_{11} = -2$, $M_{12} = 1$, and $M_{13} = 1$. Substituting these values into (8.7.12), we obtain

$$\det \mathbf{A} = 1 \cdot (-2) + 2 \cdot 1 + 3 \cdot 1 = 3.$$

Thus, we may utilize Formula (8.7.11) to check the cofactors found in (8.7.9).

We shall now outline another method for finding the inverse of a non-singular square matrix $\mathbf{A}$. This method has definite advantages over the method given above, especially when $\mathbf{A}$ has order greater than three. However, this method may involve some tedious arithmetical manipulations. For a justification of this method, see books on matrix theory. [*See*, for example, [21], pp. 113–123.]

Definition 8.7.1. A nonzero matrix $\mathbf{A}$ is said to have *rank r* if at least one of its r-square minors is different from zero while every $(r + 1)$-square minor, if any, is zero. A zero matrix is said to have rank 0.

For example, the matrix $\mathbf{A} = \begin{bmatrix} 1 & 2 & 3 \\ -2 & 3 & 1 \\ 3 & -1 & 2 \end{bmatrix}$ has rank 2, since

$\begin{vmatrix} 1 & 2 \\ -2 & 3 \end{vmatrix} = 7 \neq 0$, while det $\mathbf{A} = 0$. The matrix $\mathbf{A} = \begin{bmatrix} 1 & 2 & 3 \\ -2 & -4 & -6 \\ 3 & 6 & 9 \end{bmatrix}$

has rank 1, since $|1| = 1 \neq 0$, while every 2-square minor is zero.

Consider the matrices

$$\mathbf{A} = \begin{bmatrix} a_{11} & a_{12} & a_{13} \\ a_{21} & a_{22} & a_{23} \\ a_{31} & a_{32} & a_{33} \end{bmatrix} \quad \text{and} \quad \mathbf{I} = \begin{bmatrix} 1 & 0 & 0 \\ 0 & 1 & 0 \\ 0 & 0 & 1 \end{bmatrix}. \tag{8.7.13}$$

Suppose that we wish to interchange, say, the first and third rows of the matrix $\mathbf{A}$. This may be accomplished as follows. Let $\mathbf{E}_1$ denote the matrix obtained from the identity matrix $\mathbf{I}$ by interchanging the first and third rows,

that is, $\mathbf{E}_1 = \begin{bmatrix} 0 & 0 & 1 \\ 0 & 1 & 0 \\ 1 & 0 & 0 \end{bmatrix}$. Then,

$$\mathbf{E}_1\mathbf{A} = \begin{bmatrix} 0 & 0 & 1 \\ 0 & 1 & 0 \\ 1 & 0 & 0 \end{bmatrix} \begin{bmatrix} a_{11} & a_{12} & a_{13} \\ a_{21} & a_{22} & a_{23} \\ a_{31} & a_{32} & a_{33} \end{bmatrix} = \begin{bmatrix} a_{31} & a_{32} & a_{33} \\ a_{21} & a_{22} & a_{23} \\ a_{11} & a_{12} & a_{13} \end{bmatrix}.$$

Suppose that we wish to multiply the second row of the matrix $\mathbf{A}$ by a nonzero scalar α. This may be accomplished as follows. Let $\mathbf{E}_2$ denote the matrix obtained by multiplying the elements of the second row of the

matrix $\mathbf{I}$ by α, that is, $\mathbf{E}_2 = \begin{bmatrix} 1 & 0 & 0 \\ 1 & \alpha & 0 \\ 0 & 0 & 1 \end{bmatrix}$. Then,

$$\mathbf{E}_2\mathbf{A} = \begin{bmatrix} 1 & 0 & 0 \\ 0 & \alpha & 0 \\ 0 & 0 & 1 \end{bmatrix} \begin{bmatrix} a_{11} & a_{12} & a_{13} \\ a_{21} & a_{22} & a_{23} \\ a_{31} & a_{32} & a_{33} \end{bmatrix} = \begin{bmatrix} a_{11} & a_{12} & a_{13} \\ \alpha a_{21} & \alpha a_{22} & \alpha a_{23} \\ a_{31} & a_{32} & a_{33} \end{bmatrix}.$$

Suppose that we wish to add k times the third row to the first row of the matrix $\mathbf{A}$. This may be accomplished as follows. Let $\mathbf{E}_3$ denote the matrix obtained from the identity matrix $\mathbf{I}$ by adding k times the third row

to the first row, that is, $\mathbf{E}_3 = \begin{bmatrix} 1 & 0 & k \\ 0 & 1 & 0 \\ 0 & 0 & 1 \end{bmatrix}$. Then,

$$\mathbf{E}_3\mathbf{A} = \begin{bmatrix} 1 & 0 & k \\ 0 & 1 & 0 \\ 0 & 0 & 1 \end{bmatrix} \begin{bmatrix} a_{11} & a_{12} & a_{13} \\ a_{21} & a_{22} & a_{23} \\ a_{31} & a_{32} & a_{33} \end{bmatrix}$$

$$= \begin{bmatrix} a_{11} + ka_{31} & a_{12} + ka_{32} & a_{13} + ka_{33} \\ a_{21} & a_{22} & a_{23} \\ a_{31} & a_{32} & a_{33} \end{bmatrix}.$$

Let us note the significance of $\mathbf{E}_3\,\mathbf{E}_2\,\mathbf{E}_1\mathbf{A}$:

$$\mathbf{E}_3\,\mathbf{E}_2\,\mathbf{E}_1\mathbf{A} = \mathbf{E}_3\,\mathbf{E}_2 \begin{bmatrix} a_{31} & a_{32} & a_{33} \\ a_{21} & a_{22} & a_{23} \\ a_{11} & a_{12} & a_{13} \end{bmatrix} = \mathbf{E}_3 \begin{bmatrix} a_{31} & a_{32} & a_{33} \\ \alpha a_{21} & \alpha a_{22} & \alpha a_{23} \\ a_{11} & a_{12} & a_{13} \end{bmatrix}$$

$$= \begin{bmatrix} a_{31} + ka_{11} & a_{32} + ka_{12} & a_{33} + ka_{13} \\ \alpha a_{21} & \alpha a_{22} & \alpha a_{23} \\ a_{11} & a_{12} & a_{13} \end{bmatrix}.$$

Let

$$\mathbf{F}_1 = \begin{bmatrix} 0 & 0 & 1 \\ 0 & 1 & 0 \\ 1 & 0 & 0 \end{bmatrix}, \quad \mathbf{F}_2 = \begin{bmatrix} 1 & 0 & 0 \\ 0 & \alpha & 0 \\ 0 & 0 & 1 \end{bmatrix},$$

$$\mathbf{F}_3 = \begin{bmatrix} 1 & 0 & 0 \\ 0 & 1 & 0 \\ k & 0 & 1 \end{bmatrix}, \quad \text{and} \quad \mathbf{A} = \begin{bmatrix} a_{11} & a_{12} & a_{13} \\ a_{21} & a_{22} & a_{23} \\ a_{31} & a_{32} & a_{33} \\ a_{41} & a_{42} & a_{43} \end{bmatrix}. \quad (8.7.14)$$

What is the significance of $\mathbf{AF}_1$, $\mathbf{AF}_2$, $\mathbf{AF}_3$, and $\mathbf{AF}_1\mathbf{F}_2\mathbf{F}_3$?

Definition 8.7.2. *Elementary Row Operations.* The elementary row operations on a matrix $\mathbf{A}$ are

(i) the interchange of any two rows,
(ii) the multiplication of a row by a nonzero scalar,
(iii) the addition of any multiple of one row to another row.

Definition 8.7.3. Let $\mathbf{A}$ and $\mathbf{B}$ be two $m \times n$ matrices. The matrix $\mathbf{B}$ is said to be *row equivalent* to the matrix $\mathbf{A}$ if $\mathbf{B}$ can be obtained from $\mathbf{A}$ by a finite number of elementary row operations.

Definition 8.7.4. An $n \times n$ matrix obtained from the $n \times n$ identity matrix $\mathbf{I}$ by means of one elementary row operation on $\mathbf{I}$ is called an *elementary matrix*.

If $\mathbf{I} = \begin{bmatrix} 1 & 0 & 0 \\ 0 & 1 & 0 \\ 0 & 0 & 1 \end{bmatrix}$, then some examples of elementary matrices are

$$\begin{bmatrix} 0 & 1 & 0 \\ 1 & 0 & 0 \\ 0 & 0 & 1 \end{bmatrix}, \quad \begin{bmatrix} 1 & 0 & 0 \\ 0 & 0 & 1 \\ 0 & 1 & 0 \end{bmatrix}, \quad \begin{bmatrix} 0 & 0 & 1 \\ 0 & 1 & 0 \\ 1 & 0 & 0 \end{bmatrix}, \quad \begin{bmatrix} \alpha & 0 & 0 \\ 0 & 1 & 0 \\ 0 & 0 & 1 \end{bmatrix},$$

$$\begin{bmatrix} 1 & 0 & 0 \\ 0 & \alpha & 0 \\ 0 & 0 & 1 \end{bmatrix}, \quad \begin{bmatrix} 1 & 0 & 0 \\ 0 & 1 & 0 \\ 0 & 0 & \alpha \end{bmatrix}, \quad \begin{bmatrix} 1 & 0 & k \\ 0 & 1 & 0 \\ 0 & 0 & 1 \end{bmatrix}, \quad \text{and} \quad \begin{bmatrix} 1 & 0 & 0 \\ 0 & 1 & k \\ 0 & 0 & 1 \end{bmatrix}.$$

The following results are known. [*See*, for example, [53], pp. 99–106.]

Theorem 8.7.1. *Let $\mathbf{A}$ be an $m \times n$ matrix. Each elementary row operation on $\mathbf{A}$ can be effected by multiplying $\mathbf{A}$ on the left by an elementary matrix.*

Theorem 8.7.2. *Let $\mathbf{A}$ and $\mathbf{B}$ be two $m \times n$ matrices. If the matrix $\mathbf{B}$ is row equivalent to the matrix $\mathbf{A}$, then $\mathbf{B} = \mathbf{EA}$, where $\mathbf{E}$ is a product of elementary matrices.*

Theorem 8.7.3. *If a square matrix is reduced to the identity matrix by a sequence of elementary row operations, the same sequence of elementary row* *operations performed on the identity matrix produces the inverse of the given matrix.*

Example 8.7.2.   Let us utilize Theorem 8.7.3 to find the inverse of the matrix **A** given in Example 8.7.1.

SOLUTION.   We shall perform a single elementary row operation at a time. We shall exhibit the successive matrices row equivalent to **A** in the left-hand column, and the successive matrices row equivalent to **I** in the right-hand column.

$$\mathbf{A} = \begin{bmatrix} 1 & 2 & 3 \\ 0 & 1 & -1 \\ -1 & 0 & -2 \end{bmatrix}, \qquad \mathbf{I} = \begin{bmatrix} 1 & 0 & 0 \\ 0 & 1 & 0 \\ 0 & 0 & 1 \end{bmatrix}.$$

Add the first row to the third row:

$$\begin{bmatrix} 1 & 2 & 3 \\ 0 & 1 & -1 \\ 0 & 2 & 1 \end{bmatrix}, \qquad \begin{bmatrix} 1 & 0 & 0 \\ 0 & 1 & 0 \\ 1 & 0 & 1 \end{bmatrix}.$$

Add -2 times the second row to the third row:

$$\begin{bmatrix} 1 & 2 & 3 \\ 0 & 1 & -1 \\ 0 & 0 & 3 \end{bmatrix}, \qquad \begin{bmatrix} 1 & 0 & 0 \\ 0 & 1 & 0 \\ 1 & -2 & 1 \end{bmatrix}.$$

Add -2 times the second row to the first row:

$$\begin{bmatrix} 1 & 0 & 5 \\ 0 & 1 & -1 \\ 0 & 0 & 3 \end{bmatrix}, \qquad \begin{bmatrix} 1 & -2 & 0 \\ 0 & 1 & 0 \\ 1 & -2 & 1 \end{bmatrix}.$$

Multiply the third row by $\frac{1}{3}$:

$$\begin{bmatrix} 1 & 0 & 5 \\ 0 & 1 & -1 \\ 0 & 0 & 1 \end{bmatrix}, \qquad \begin{bmatrix} 1 & -2 & 0 \\ 0 & 1 & 0 \\ \frac{1}{3} & -\frac{2}{3} & \frac{1}{3} \end{bmatrix}.$$

Add the third row to the second row:

$$\begin{bmatrix} 1 & 0 & 5 \\ 0 & 1 & 0 \\ 0 & 0 & 1 \end{bmatrix}, \qquad \begin{bmatrix} 1 & -2 & 0 \\ \frac{1}{3} & \frac{1}{3} & \frac{1}{3} \\ \frac{1}{3} & -\frac{2}{3} & \frac{1}{3} \end{bmatrix}.$$

Add -5 times the third row to the first row:

$$\begin{bmatrix} 1 & 0 & 0 \\ 0 & 1 & 0 \\ 0 & 0 & 1 \end{bmatrix} = \mathbf{I}, \qquad \begin{bmatrix} -\frac{2}{3} & \frac{4}{3} & -\frac{5}{3} \\ \frac{1}{3} & \frac{1}{3} & \frac{1}{3} \\ \frac{1}{3} & -\frac{2}{3} & \frac{1}{3} \end{bmatrix} = \mathbf{A}^{-1}.$$

Example 8.7.3. Let us find the inverse of the matrix $\mathbf{A} = \begin{bmatrix} 3 & 5 \\ 2 & 4 \end{bmatrix}$.

SOLUTION.

$$\mathbf{A} = \begin{bmatrix} 3 & 5 \\ 2 & 4 \end{bmatrix}, \qquad \mathbf{I} = \begin{bmatrix} 1 & 0 \\ 0 & 1 \end{bmatrix}.$$

$$\begin{bmatrix} 2 & 4 \\ 3 & 5 \end{bmatrix}, \qquad \begin{bmatrix} 0 & 1 \\ 1 & 0 \end{bmatrix}.$$

$$\begin{bmatrix} 1 & 2 \\ 3 & 5 \end{bmatrix}, \qquad \begin{bmatrix} 0 & \frac{1}{2} \\ 1 & 0 \end{bmatrix}.$$

$$\begin{bmatrix} 1 & 2 \\ 0 & -1 \end{bmatrix}, \qquad \begin{bmatrix} 0 & \frac{1}{2} \\ 1 & -\frac{3}{2} \end{bmatrix}.$$

$$\begin{bmatrix} 1 & 2 \\ 0 & 1 \end{bmatrix}, \qquad \begin{bmatrix} 0 & \frac{1}{2} \\ -1 & \frac{3}{2} \end{bmatrix}.$$

$$\begin{bmatrix} 1 & 0 \\ 0 & 1 \end{bmatrix} = \mathbf{I}, \qquad \begin{bmatrix} 2 & -\frac{5}{2} \\ -1 & \frac{3}{2} \end{bmatrix} = \mathbf{A}^{-1}.$$

Remark 8.7.1. If in Definition 8.7.2, we replace "row" or "rows" by "column" or "columns," we obtain the definition of *elementary column operations*. The above theorems are also true for elementary column operations. In Theorem 8.7.1, we replace "*row*" by "*column*" and "*on the left*" by "*on the right*." In Theorem 8.7.2, we replace "*row*" by "*column*" and "$\mathbf{B} = \mathbf{EA}$, *where* $\mathbf{E}$" by "$\mathbf{B} = \mathbf{AF}$, *where* $\mathbf{F}$." In Theorem 8.7.3, we replace "*row*" by "*column*."

The *elementary transformations* of a matrix are made up of elementary row (or column) operations. An $n \times n$ matrix obtained from the $n \times n$ identity matrix $\mathbf{I}$ by means of one elementary row (or column) operation on $\mathbf{I}$ is called an *elementary matrix*. If an $m \times n$ matrix $\mathbf{A}$ can be obtained from an $m \times n$ matrix $\mathbf{B}$ by a finite number of elementary row and column operations, the matrix $\mathbf{A}$ is said to be *equivalent*[3] to the matrix $\mathbf{B}$. We write $\mathbf{A} \sim \mathbf{B}$.

The following results are known. [*See* [21] or [53].]

Theorem 8.7.4. *When a sequence of elementary transformations is applied to a matrix* $\mathbf{A}$, *the resulting matrix has the same rank and order as* $\mathbf{A}$.

Theorem 8.7.5. *Two* $m \times n$ *matrices are equivalent if and only if they have the same rank.*

---

[3] In textbooks on linear algebra, one usually defines two matrices as being *equivalent* if they represent the same transformation with respect to different bases.

By means of elementary transformations, any matrix $\mathbf{A}$ of rank $r > 0$ can be reduced to one of the forms

$$\begin{bmatrix} \mathbf{I}_r \\ \mathbf{0} \end{bmatrix}, \qquad [\mathbf{I}_r \quad \mathbf{0}], \qquad \begin{bmatrix} \mathbf{I}_r & \mathbf{0} \\ \mathbf{0} & \mathbf{0} \end{bmatrix}, \qquad \text{or} \quad \mathbf{I}_r, \qquad (8.7.15)$$

where $\mathbf{I}_r$ is the identity matrix of order r.   When a matrix has been reduced to a form (8.7.15) by a sequence of elementary transformations, we say that it has been reduced to *normal form*.

Example 8.7.4.   Let us reduce the following matrix to normal form by elementary transformations:

$$\mathbf{A} = \begin{bmatrix} 1 & 2 & -3 & 3 \\ -1 & 3 & 4 & 2 \\ 2 & -1 & -3 & 1 \\ 3 & -2 & -5 & 1 \end{bmatrix}.$$

SOLUTION.   Add the first row to the second row; add -2 times the first row to the third row; add -3 times the first row to the fourth row, and we obtain

$$\begin{bmatrix} 1 & 2 & -3 & 3 \\ 0 & 5 & 1 & 5 \\ 0 & -5 & 3 & -5 \\ 0 & -8 & 4 & -8 \end{bmatrix}.$$

Add -2 times the first column to the second column; add 3 times the first column to the third column; add -3 times the first column to the fourth column, and we obtain

$$\begin{bmatrix} 1 & 0 & 0 & 0 \\ 0 & 5 & 1 & 5 \\ 0 & -5 & 3 & -5 \\ 0 & -8 & 4 & -8 \end{bmatrix}.$$

Add -1 times the second column to the fourth column:

$$\begin{bmatrix} 1 & 0 & 0 & 0 \\ 0 & 5 & 1 & 0 \\ 0 & -5 & 3 & 0 \\ 0 & -8 & 4 & 0 \end{bmatrix}.$$

Interchange the second and third column:

$$\begin{bmatrix} 1 & 0 & 0 & 0 \\ 0 & 1 & 5 & 0 \\ 0 & 3 & -5 & 0 \\ 0 & 4 & -8 & 0 \end{bmatrix}.$$

Add -3 times the second row to the third row; add -4 times the second row to fourth row, and we obtain

$$\begin{bmatrix} 1 & 0 & 0 & 0 \\ 0 & 1 & 5 & 0 \\ 0 & 0 & -20 & 0 \\ 0 & 0 & -28 & 0 \end{bmatrix}.$$

Multiply the third row by $-\frac{1}{20}$:

$$\begin{bmatrix} 1 & 0 & 0 & 0 \\ 0 & 1 & 5 & 0 \\ 0 & 0 & 1 & 0 \\ 0 & 0 & -28 & 0 \end{bmatrix}.$$

Add -5 times the third row to the second row; add 28 times the third row to the fourth row, and we obtain the desired normal form

$$\begin{bmatrix} 1 & 0 & 0 & 0 \\ 0 & 1 & 0 & 0 \\ 0 & 0 & 1 & 0 \\ 0 & 0 & 0 & 0 \end{bmatrix} = \begin{bmatrix} \mathbf{I}_3 & \mathbf{0} \\ \mathbf{0} & \mathbf{0} \end{bmatrix}. \tag{8.7.16}$$

In view of Theorem 8.7.4 and (8.7.16), we see that the rank of the matrix $\mathbf{A}$ is three.   The above discussion may be displayed conveniently as follows:

$$\mathbf{A} = \begin{bmatrix} 1 & 2 & -3 & 3 \\ -1 & 3 & 4 & 2 \\ 2 & -1 & -3 & 1 \\ 3 & -2 & -5 & 1 \end{bmatrix} \sim \begin{bmatrix} 1 & 2 & -3 & 3 \\ 0 & 5 & 1 & 5 \\ 0 & -5 & 3 & -5 \\ 0 & -8 & 4 & -8 \end{bmatrix} \sim \begin{bmatrix} 1 & 0 & 0 & 0 \\ 0 & 5 & 1 & 5 \\ 0 & -5 & 3 & -5 \\ 0 & -8 & 4 & -8 \end{bmatrix}$$

$$\sim \begin{bmatrix} 1 & 0 & 0 & 0 \\ 0 & 5 & 1 & 0 \\ 0 & -5 & 3 & 0 \\ 0 & -8 & 4 & 0 \end{bmatrix} \sim \begin{bmatrix} 1 & 0 & 0 & 0 \\ 0 & 1 & 5 & 0 \\ 0 & 3 & -5 & 0 \\ 0 & 4 & -8 & 0 \end{bmatrix} \sim \begin{bmatrix} 1 & 0 & 0 & 0 \\ 0 & 1 & 5 & 0 \\ 0 & 0 & -20 & 0 \\ 0 & 0 & -28 & 0 \end{bmatrix}$$

$$\sim \begin{bmatrix} 1 & 0 & 0 & 0 \\ 0 & 1 & 5 & 0 \\ 0 & 0 & 1 & 0 \\ 0 & 0 & -28 & 0 \end{bmatrix} \sim \begin{bmatrix} 1 & 0 & 0 & 0 \\ 0 & 1 & 0 & 0 \\ 0 & 0 & 1 & 0 \\ 0 & 0 & 0 & 0 \end{bmatrix} = \begin{bmatrix} \mathbf{I}_3 & \mathbf{0} \\ \mathbf{0} & \mathbf{0} \end{bmatrix}.$$

Exercises 8.7

1. Prove that the inverse of a nonsingular matrix is unique.
2. Let $\mathbf{A}$ and $\mathbf{B}$ be nonsingular matrices of the same order.   Prove:
   (a) $\mathbf{AB}$ is a nonsingular matrix,
   (b) $(\mathbf{AB})^{-1} = \mathbf{B}^{-1}\mathbf{A}^{-1}$.
3. Let $\mathbf{A}$ be a nonsingular matrix.   Prove $(\mathbf{A}^{-1})^{-1} = \mathbf{A}$.

In Exercises 4 through 13, find the inverse of each nonsingular matrix.

4. $\begin{bmatrix} 1 & 2 \\ 1 & 3 \end{bmatrix}$.

5. $\begin{bmatrix} i & 1 \\ 1 & i \end{bmatrix}$, $i^2 = -1$.

6. $\begin{bmatrix} -1 & 3 & 0 \\ 0 & 1 & 1 \\ 2 & -1 & 4 \end{bmatrix}$.

7. $\begin{bmatrix} 1 & 2 & 3 \\ 2 & 3 & 5 \\ 3 & -1 & 8 \end{bmatrix}$.

8. $\begin{bmatrix} 5 & 0 & -3 \\ 3 & 4 & -1 \\ 1 & -2 & 0 \end{bmatrix}$.

9. $\begin{bmatrix} 0 & 0 & 1 \\ 2 & 4 & -6 \\ 3 & 5 & -10 \end{bmatrix}$.

10. $\begin{bmatrix} 1 & -2 & 1 & -1 \\ -1 & 1 & 2 & 1 \\ 0 & 0 & 1 & 1 \\ -2 & 6 & -3 & 6 \end{bmatrix}$.

11. $\begin{bmatrix} 1 & -2 & 0 & 1 \\ 0 & 1 & -1 & 0 \\ -2 & 3 & -4 & -2 \\ -1 & 5 & -3 & -4 \end{bmatrix}$.

12. $\begin{bmatrix} \alpha_1 & 0 & 0 & 0 & 0 \\ 0 & \alpha_2 & 0 & 0 & 0 \\ 0 & 0 & \alpha_3 & 0 & 0 \\ 0 & 0 & 0 & \alpha_4 & 0 \\ 0 & 0 & 0 & 0 & \alpha_5 \end{bmatrix}$, $\alpha_1 \alpha_2 \alpha_3 \alpha_4 \alpha_5 \neq 0$.

13. $\begin{bmatrix} 1 & 1 & 1 & 1 & 1 \\ 0 & 1 & 1 & 1 & 1 \\ 0 & 0 & 1 & 1 & 1 \\ 0 & 0 & 0 & 1 & 1 \\ 0 & 0 & 0 & 0 & 1 \end{bmatrix}$.

In Exercises 14 through 19, find the rank of each matrix.

14. $\begin{bmatrix} 1 & 2 & 5 \\ 2 & 3 & 10 \end{bmatrix}$.

15. $\begin{bmatrix} 2 & 6 \\ -1 & -3 \\ 3 & 9 \end{bmatrix}$.

16. $\begin{bmatrix} -1 & 2 & 1 \\ 0 & 3 & 3 \\ 2 & -1 & 1 \end{bmatrix}$.

17. $\begin{bmatrix} -2 & 4 & 6 \\ 1 & -2 & -3 \\ -3 & 6 & 9 \end{bmatrix}$.

18. $\begin{bmatrix} 2 & 1 & 0 & 3 \\ 1 & 0 & -2 & -1 \\ 3 & -1 & 1 & 3 \\ -1 & 2 & 4 & 5 \end{bmatrix}$.

19. $\begin{bmatrix} 3 & 1 & 4 \\ -1 & 3 & 2 \\ 2 & 4 & 6 \\ 4 & -1 & 3 \end{bmatrix}$.

20. Let $\mathbf{A} = \begin{bmatrix} -1 & 2 & 0 & 4 \\ 2 & 0 & 4 & 4 \\ -2 & 1 & -3 & -1 \\ -4 & 3 & -5 & 1 \end{bmatrix}$ and $\mathbf{B} = \begin{bmatrix} 1 & 3 & 2 & 8 \\ -2 & 1 & 3 & 5 \\ 3 & 0 & -3 & -3 \\ -1 & 2 & 3 & 7 \end{bmatrix}$.

Show that $\mathbf{A} \sim \mathbf{B}$.

21. Let $\mathbf{A}$ and $\mathbf{B}$ be nonsingular matrices of the same rank.
    (a) Show that if $\mathbf{AB} = \mathbf{AC}$, then $\mathbf{B} = \mathbf{C}$.   (Thus, the cancellation law is valid.)
    Show that if $\mathbf{AB} = \mathbf{BA}$, then
    (b) $\mathbf{A}^{-1}\mathbf{B} = \mathbf{BA}^{-1}$,   (c) $\mathbf{AB}^{-1} = \mathbf{B}^{-1}\mathbf{A}$,   (d) $\mathbf{A}^{-1}\mathbf{B}^{-1} = \mathbf{B}^{-1}\mathbf{A}^{-1}$.

22. Let $\mathbf{A}$ and $\mathbf{B}$ be nonsingular symmetric matrices.
    (a) Show that $\mathbf{A}^{-1}$ is symmetric.
    Show that if $\mathbf{AB} = \mathbf{BA}$, then
    (b) $\mathbf{A}^{-1}\mathbf{B}$,   (c) $\mathbf{B}^{-1}\mathbf{A}$,   (d) $\mathbf{A}^{-1}\mathbf{B}^{-1}$ are symmetric.

23. Suppose that a matrix function $\mathbf{A}$ is differentiable on an interval J, and $\mathbf{A}^{-1}(t)$ exists for every t in J.
    (a) Show that

$$\frac{d}{dt}\mathbf{A}^{-1}(t) = -\mathbf{A}^{-1}(t) \cdot \frac{d}{dt}\mathbf{A}(t) \cdot \mathbf{A}^{-1}(t).$$

    (b) Find the derivative of the inverse of $\mathbf{A}$ on the interval I: $-\infty < t < \infty$, where

$$\mathbf{A}(t) = \begin{bmatrix} \cos t & -\sin t \\ \sin t & \cos t \end{bmatrix}.$$

8.8. Some Basic Properties of Vectors

An ordered n-tuple of the form $[a_1, a_2, \ldots, a_n]$, whose elements a_i are real or complex numbers is called an *n-dimensional row vector*, or *row vector*, and we shall denote it by $\mathbf{a}$:

$$\mathbf{a} = [a_1, a_2, \ldots, a_n]. \tag{8.8.1}$$

The element a_i is called the ith *component* of $\mathbf{a}$.   A row vector $\mathbf{a}$ is likewise considered a $1 \times n$ matrix.   On the other hand, an array of the form

$$\mathbf{b} = \begin{bmatrix} b_1 \\ b_2 \\ \vdots \\ b_n \end{bmatrix} \tag{8.8.2}$$

is called an *n-dimensional column vector*, or *column vector*.   It is actually an $n \times 1$ matrix.

It is often useful to view a matrix $\mathbf{A}$ of order $p \times q$ as defining p row vectors or q column vectors.   When the matrix $\mathbf{A}$ is considered as p row vectors, the elements of a row are the components of a q-dimensional vector; while when the matrix $\mathbf{A}$ is considered as q column vectors, the elements of a column are the components of a p-dimensional vector.

These vectors are matrices in their own right, consequently, all the usual rules of matrix algebra can be applied to them.   First, we notice that the transpose of a row vector is a column vector (and vice versa).   Thus,

$$\mathbf{a}^T = [a_1, a_2, \ldots, a_n]^T = \begin{bmatrix} a_1 \\ a_2 \\ \vdots \\ a_n \end{bmatrix}.$$

If $\mathbf{x} = [x_1, x_2, \ldots, x_n]$ and $\mathbf{y} = [y_1, y_2, \ldots, y_n]$, then

$$\mathbf{x} \pm \mathbf{y} = [x_1 \pm y_1, x_2 \pm y_2, \ldots, x_n \pm y_n], \tag{8.8.3}$$

$$\alpha\mathbf{x} = [\alpha x_1, \alpha x_2, \ldots, \alpha x_n], \tag{8.8.4}$$

where α is a scalar.

A vector, each of whose components is zero, is called a *zero vector*, and we denote it by $\mathbf{0}$.   (It can be either a row or a column.)

The *product of two vectors* $\mathbf{x}$ and $\mathbf{y}$ (with the same number of components) is really covered by Definition 8.6.4.   Let $\mathbf{x} = [x_1, x_2, \ldots, x_n]$, $\mathbf{y} = [y_1, y_2, \ldots, y_n]$.   Then,

$$\mathbf{x}\mathbf{y}^T = [x_1, x_2, \ldots, x_n] \begin{bmatrix} y_1 \\ y_2 \\ \vdots \\ y_n \end{bmatrix} = [x_1 y_1 + x_2 y_2 + \cdots + x_n y_n]$$

$$= x_1 y_1 + x_2 y_2 + \cdots + x_n y_n$$

$$= \sum_{i=1}^{n} x_i y_i. \tag{8.8.5}$$

This is the familiar scalar (or dot) product of vector algebra.   For example, if $\mathbf{x} = [1, 0, -1]$ and $\mathbf{y} = [2, 1, 3]$, then $\mathbf{x}\mathbf{y}^T = 1 \cdot 2 + 0 \cdot 1 + (-1) \cdot 3 = -1$. If $\mathbf{x} = [1, i, -2i]$ and $\mathbf{y} = [-i, 1 + i, -3]$, where $i^2 = -1$, then $\mathbf{x}\mathbf{y}^T = 1 \cdot (-i) + i \cdot (1 + i) + (-2i)(-3) = -1 + 6i$.   It is this " scalar product" (8.8.5) that we are mostly concerned with, however, we might mention that a column vector times a row vector yields a square matrix:

$$\mathbf{x}^T\mathbf{y} = \begin{bmatrix} x_1 \\ x_2 \\ \vdots \\ x_n \end{bmatrix} [y_1, y_2, \ldots, y_n] = \begin{bmatrix} x_1 y_1 & x_1 y_2 & \cdots & x_1 y_n \\ x_2 y_1 & x_2 y_2 & \cdots & x_2 y_n \\ \vdots & \vdots & & \vdots \\ x_n y_1 & x_n y_2 & \cdots & x_n y_n \end{bmatrix}.$$

One should verify that both product types are consistent with Definition 8.6.4.

Let $\mathbf{x}$, $\mathbf{y}$, $\mathbf{z}$ be n-dimensional row vectors, and α a scalar.   We have [*see* Exercise 8.8.1]

$$\mathbf{x}\mathbf{y}^T = \mathbf{y}\mathbf{x}^T, \tag{8.8.6}$$

$$\mathbf{x}(\mathbf{y} + \mathbf{z})^T = \mathbf{x}\mathbf{y}^T + \mathbf{x}\mathbf{z}^T, \tag{8.8.7}$$

$$(\alpha\mathbf{x})\mathbf{y}^T = \alpha(\mathbf{x}\mathbf{y}^T), \qquad \mathbf{x}(\alpha\mathbf{y}^T) = \alpha(\mathbf{x}\mathbf{y}^T). \tag{8.8.8}$$

We symbolize the complex conjugate transpose of a vector $\mathbf{a} = [a_1, a_2, \ldots, a_n]$ by $\mathbf{a}^*$. Thus,

$$\mathbf{a}^* = \bar{\mathbf{a}}^T = \begin{bmatrix} \bar{a}_1 \\ \bar{a}_2 \\ \vdots \\ \bar{a}_n \end{bmatrix}, \tag{8.8.9}$$

where, as usual, $\bar{a}_i$ denotes the complex conjugate of the element a_i $(i = 1, 2, \ldots, n)$. Let $\mathbf{x}$ and $\mathbf{y}$ be n-dimensional column vectors:

$$\mathbf{x} = \begin{bmatrix} x_1 \\ x_2 \\ \vdots \\ x_n \end{bmatrix}, \qquad \mathbf{y} = \begin{bmatrix} y_1 \\ y_2 \\ \vdots \\ y_n \end{bmatrix}.$$

The *inner product* of $\mathbf{x}$ and $\mathbf{y}$, *in that order*, is written $(\mathbf{x}, \mathbf{y})$ and is defined by

$$(\mathbf{x}, \mathbf{y}) = \mathbf{x}^*\mathbf{y} = [\bar{x}_1, \bar{x}_2, \ldots, \bar{x}_n] \begin{bmatrix} y_1 \\ y_2 \\ \vdots \\ y_n \end{bmatrix}$$

$$= \bar{x}_1 y_1 + \bar{x}_2 y_2 + \cdots + \bar{x}_n y_n = \sum_{k=1}^{n} \bar{x}_k y_k. \tag{8.8.10}$$

[*Note:* $\mathbf{x}$ and $\mathbf{y}$ need not have been defined as columns, in which case the term $\mathbf{x}^*\mathbf{y}$ in Equation (8.8.10) would assume a different form. Thus, if $\mathbf{x}$ and $\mathbf{y}$ were row vectors, the term $\mathbf{x}^*\mathbf{y}$ would be replaced by $\bar{\mathbf{x}}\mathbf{y}^T$ to give the indicated scalar product. In all cases, however, the summation in (8.8.10) gives the inner product, regardless of how $\mathbf{x}$ and $\mathbf{y}$ are defined.]

When $\mathbf{x}$ and $\mathbf{y}$ are restricted to the real domain, we see that $(\mathbf{x}, \mathbf{y}) = \mathbf{x}^T\mathbf{y}$. Also, when $\mathbf{y} = \mathbf{x}$, that is, $y_i = x_i$, $i = 1, 2, \ldots, n$, (8.8.10) reduces to

$$(\mathbf{x}, \mathbf{x}) = \sum_{k=1}^{n} \bar{x}_k x_k = \sum_{k=1}^{n} |x_k|^2, \tag{8.8.11}$$

which is a nonnegative (real) number. For example, if $\mathbf{x} = \begin{bmatrix} i \\ 1 - i \\ 3 \end{bmatrix}$, then

$$(\mathbf{x}, \mathbf{x}) \sum_{k=1}^{3} \bar{x}_k x_k = \bar{x}_1 x_1 + \bar{x}_2 x_2 + \bar{x}_3 x_3 = i \cdot i + \overline{(1 - i)} \cdot (1 - i) + 3 \cdot 3$$

$$= i^2 + (1 + i) \cdot (1 - i) + 9 = - i^2 + 1 - i^2 + 9 = 12.$$

The inner product is often defined by $(\mathbf{x}, \mathbf{y}) = \mathbf{x}^T\bar{\mathbf{y}} = \sum_{k=1}^{n} x_k \bar{y}_k$, which gives the conjugate of our definition.

If $\mathbf{x}$, $\mathbf{y}$, $\mathbf{z}$ are n-dimensional column vectors and α is a scalar, then we have [*see* Exercise 8.8.2]

$$(\mathbf{x}, \mathbf{y}) = \overline{(\mathbf{y}, \mathbf{x})}, \tag{8.8.12}$$

$$(\mathbf{x}, \mathbf{y} + \mathbf{z}) = (\mathbf{x}, \mathbf{y}) + (\mathbf{x}, \mathbf{z}), \tag{8.8.13}$$

$$(\alpha\mathbf{x}, \mathbf{y}) = \bar{\alpha}(\mathbf{x}, \mathbf{y}), \qquad (\mathbf{x}, \alpha\mathbf{y}) = \alpha(\mathbf{x}, \mathbf{y}). \tag{8.8.14}$$

The concept of linear dependence and linear independence of vectors is similar to that given for functions. [*See* Definition 4.3.5 and 4.3.6.] Let

$$\mathbf{x}^{(1)} = \begin{bmatrix} x_{11} \\ x_{21} \\ \vdots \\ x_{n1} \end{bmatrix}, \qquad \mathbf{x}^{(2)} = \begin{bmatrix} x_{12} \\ x_{22} \\ \vdots \\ x_{n2} \end{bmatrix}, \ \ldots, \qquad \mathbf{x}^{(k)} = \begin{bmatrix} x_{1k} \\ x_{2k} \\ \vdots \\ x_{nk} \end{bmatrix}, \tag{8.8.15}$$

denote k column vectors, each having n components. We shall abbreviate this statement by saying: *k n-vectors* $\mathbf{x}^{(1)}$, $\mathbf{x}^{(2)}$, $\ldots$, $\mathbf{x}^{(k)}$.

Definition 8.8.1. The k n-vectors $\mathbf{x}^{(1)}$, $\mathbf{x}^{(2)}$, $\ldots$, $\mathbf{x}^{(k)}$ are said to be *linearly dependent* if there exist constants c_1, c_2, $\ldots$, c_k, not all zero, such that

$$c_1\mathbf{x}^{(1)} + c_2\mathbf{x}^{(2)} + \cdots + c_k\mathbf{x}^{(k)} = \mathbf{0}. \tag{8.8.16}$$

If, however, $\sum_{j=1}^{k} c_j \mathbf{x}^{(j)} = \mathbf{0}$ only when all the constants c_j are zero, the k n-vectors $\mathbf{x}^{(1)}$, $\mathbf{x}^{(2)}$, $\ldots$, $\mathbf{x}^{(k)}$ are said to be *linearly independent*.

Clearly, the constants c_j $(j = 1, \ldots, k)$ may be real, imaginary, or both.

Example 8.8.1.   Let us show that the vectors

$$\mathbf{x}^{(1)} = \begin{bmatrix} 1 \\ -1 \\ 2 \end{bmatrix}, \qquad \mathbf{x}^{(2)} = \begin{bmatrix} 2 \\ 3 \\ 1 \end{bmatrix}, \qquad \mathbf{x}^{(3)} = \begin{bmatrix} -1 \\ 2 \\ -3 \end{bmatrix},$$

are linearly independent.

SOLUTION.   The linear combination $c_1\mathbf{x}^{(1)} + c_2\mathbf{x}^{(2)} + c_3\mathbf{x}^{(3)} = \mathbf{0}$ can be written as

$$\begin{cases} c_1 + 2c_2 - c_3 = 0, \\ -c_1 + 3c_2 + 2c_3 = 0, \\ 2c_1 + c_2 - 3c_3 = 0. \end{cases} \tag{8.8.17}$$

The determinant of coefficients of the system (8.8.17) is

$$\begin{vmatrix} 1 & 2 & -1 \\ -1 & 3 & 2 \\ 2 & 1 & -3 \end{vmatrix} = -2 \neq 0.$$

Hence, in view of Theorem II of Section 4.4, the only solution of the system (8.8.17) is the trivial solution: $c_1 = 0$, $c_2 = 0$, $c_3 = 0$. Hence, the given vectors are linearly independent.

In view of Theorems II and III of Section 4.4, we have the following results.

Theorem 8.8.1. *A necessary and sufficient condition that n nonzero n-vectors* $\mathbf{x}^{(1)}$, $\mathbf{x}^{(2)}$, ..., $\mathbf{x}^{(n)}$ *be linearly dependent is that*

$$\begin{vmatrix} x_{11} & x_{12} & \cdots & x_{1n} \\ x_{21} & x_{22} & \cdots & x_{2n} \\ \vdots & \vdots & & \vdots \\ x_{n1} & x_{n2} & \cdots & x_{nn} \end{vmatrix} = 0. \tag{8.8.18}$$

Theorem 8.8.2. *The k n-vectors* $\mathbf{x}^{(1)}$, $\mathbf{x}^{(2)}$, ..., $\mathbf{x}^{(k)}$ *are linearly dependent if* $k > n$.

Consider the matrix,

$$\begin{bmatrix} x_{11} & x_{12} & \cdots & x_{1k} \\ x_{21} & x_{22} & \cdots & x_{2k} \\ \vdots & \vdots & & \vdots \\ x_{n1} & x_{n2} & \cdots & x_{nk} \end{bmatrix}, \tag{8.8.19}$$

associated with the k n-vectors $\mathbf{x}^{(1)}$, $\mathbf{x}^{(2)}$, ..., $\mathbf{x}^{(k)}$ displayed in (8.8.15). The following results may be established. [*See*, for example, [21], pp. 153–154.]

Theorem 8.8.3. *A necessary and sufficient condition that k n-vectors* $\mathbf{x}^{(1)}$, $\mathbf{x}^{(2)}$, ..., $\mathbf{x}^{(k)}$ *be linearly dependent is that the matrix (8.8.19) has rank* $r < k$.

Theorem 8.8.4. *A necessary and sufficient condition that k n-vectors* $\mathbf{x}^{(1)}$, $\mathbf{x}^{(2)}$, ..., $\mathbf{x}^{(k)}$ *be linearly independent is that the matrix (8.8.19) has rank* $r = k$.

Remark 8.8.1. When $k \leq n$ in (8.8.19), it can be shown that if the rank of the matrix (8.8.19) is $r < k$, then there are exactly r vectors of the set which are linearly independent. The remaining $k - r$ vectors can be expressed as a linear combination of these r vectors.

We illustrate this remark in the following example.

Example 8.8.2. Let

$$\mathbf{x}^{(1)} = \begin{bmatrix} 1 \\ 2 \\ -1 \\ 3 \end{bmatrix}, \quad \mathbf{x}^{(2)} = \begin{bmatrix} 2 \\ 1 \\ 3 \\ -1 \end{bmatrix}, \quad \mathbf{x}^{(3)} = \begin{bmatrix} 4 \\ 5 \\ 1 \\ 5 \end{bmatrix}, \quad \mathbf{x}^{(4)} = \begin{bmatrix} -1 \\ 4 \\ -9 \\ 11 \end{bmatrix}.$$

$$\tag{8.8.20}$$

Consider the matrix associated with these vectors, namely,

$$\begin{bmatrix} 1 & 2 & 4 & -1 \\ 2 & 1 & 5 & 4 \\ -1 & 3 & 1 & -9 \\ 3 & -1 & 5 & 11 \end{bmatrix}. \qquad (8.8.21)$$

The reader may show that the rank of the matrix (8.8.21) is two. Thus, $r = 2 < 4 = k$, and $k - r = 4 - 2 = 2$. Hence, there are exactly two linearly independent vectors of the set given in (8.8.20). Since the rank of the matrix comprised of the vectors, say $\mathbf{x}^{(1)}$ and $\mathbf{x}^{(2)}$, is two, in view of Theorem 8.8.4 with $k = 2$ and $n = 4$, the vectors $\mathbf{x}^{(1)}$ and $\mathbf{x}^{(2)}$ are linearly independent. We shall now show that $\mathbf{x}^{(3)}$ and $\mathbf{x}^{(4)}$ can be expressed as a linear combination of $\mathbf{x}^{(1)}$ and $\mathbf{x}^{(2)}$. Since the vectors $\mathbf{x}^{(1)}, \mathbf{x}^{(2)}, \mathbf{x}^{(3)}$, and $\mathbf{x}^{(4)}$ are linearly dependent, there exist constants c_1, c_2, c_3, and c_4, not all zero, such that

$$c_1 \mathbf{x}^{(1)} + c_2 \mathbf{x}^{(2)} + c_3 \mathbf{x}^{(3)} + c_4 \mathbf{x}^{(4)} = \mathbf{0}. \qquad (8.8.22)$$

This linear combination can be expressed as

$$\begin{cases} c_1 + 2c_2 + 4c_3 - c_4 = 0, \\ 2c_1 + c_2 + 5c_3 + 4c_4 = 0, \\ -c_1 + 3c_2 + c_3 - 9c_4 = 0, \\ 3c_1 - c_2 + 5c_3 + 11c_4 = 0. \end{cases} \qquad (8.8.23)$$

Since the determinant of the coefficients of c_1 and c_2 in the first two equations in (8.8.23) is different from zero, we may utilize these two equations to solve for c_1 and c_2 in terms of c_3 and c_4. Thus, we may consider the system

$$\begin{cases} c_1 + 2c_2 = -4c_3 + c_4, \\ 2c_1 + c_2 = -5c_3 - 4c_4. \end{cases}$$

Solving this system for c_1 and c_2, we obtain

$$c_1 = -2c_3 - 3c_4, \qquad c_2 = -c_3 + 2c_4. \qquad (8.8.24)$$

In (8.8.24), taking $c_4 = 0$, we have $c_1 = -2c_3$, $c_2 = -c_3$. Substituting these values into (8.8.22) and then dividing by c_3, we obtain

$$\mathbf{x}^{(3)} = 2\mathbf{x}^{(1)} + \mathbf{x}^{(2)}.$$

Similarly, if we take $c_3 = 0$ in (8.8.24), we obtain

$$\mathbf{x}^{(4)} = 3\mathbf{x}^{(1)} - 2\mathbf{x}^{(2)}.$$

Vector functions. Suppose that the components x_i of an n-vector $\mathbf{x}$ are functions of the variable t on the interval J: $\alpha \le t \le \beta$. Then the n-vector $\mathbf{x}$ will depend upon t, and we write

$$\mathbf{x}(t) = \begin{bmatrix} x_1(t) \\ x_2(t) \\ \vdots \\ x_n(t) \end{bmatrix}. \tag{8.8.25}$$

A *vector function* $\mathbf{x}$ is a correspondence which assigns to each value t in J a unique n-vector $\mathbf{x}(t)$.

Remark 8.8.2. Continuity, differentiability, and integrability of vector functions are formulated as for matrix functions.

Definition 8.8.2. The k n-vectors $\mathbf{x}^{(1)}(t)$, $\mathbf{x}^{(2)}(t)$, $\ldots$, $\mathbf{x}^{(k)}(t)$ are said to be *linearly dependent* on the interval J: $\alpha \le t \le \beta$, if there exist constants $c_1, c_2, \ldots, c_k$, not all zero, such that

$$c_1 \mathbf{x}^{(1)}(t) + c_2 \mathbf{x}^{(2)}(t) + \cdots + c_k \mathbf{x}^{(k)}(t) = \mathbf{0}, \tag{8.8.26}$$

for every t in J. If, however, $\sum_{j=1}^{k} c_j \mathbf{x}^{(j)}(t) = \mathbf{0}$ only when all the constants c_j are zero, for every t in J, the k n-vectors $\mathbf{x}^{(1)}(t)$, $\mathbf{x}^{(2)}(t)$, $\ldots$, $\mathbf{x}^{(k)}(t)$, are said to be *linearly independent*.

Example 8.8.3. Let us show that the vector functions $\mathbf{x}^{(1)}$, $\mathbf{x}^{(2)}$ defined by

$$\mathbf{x}^{(1)}(t) = \begin{bmatrix} t \\ 1 \end{bmatrix}, \qquad \mathbf{x}^{(2)}(t) = \begin{bmatrix} t^2 \\ t \end{bmatrix}$$

are linearly independent on the interval J: $-\infty < t < \infty$.

SOLUTION. The linear combination $c_1 \mathbf{x}^{(1)}(t) + c_2 \mathbf{x}^{(2)}(t) = \mathbf{0}$ can be expressed as

$$\begin{cases} tc_1 + t^2 c_2 = 0, \\ c_1 + tc_2 = 0. \end{cases} \tag{8.8.27}$$

The only constants that will satisfy the system (8.8.27) for *every* t in J are $c_1 = 0$ and $c_2 = 0$. Thus, the given vectors are linearly independent on J.

Note, in this example, that for each t in J, we can find different sets of constants c_1, c_2 not both zero, which will satisfy the system (8.8.27). Thus, $\mathbf{x}^{(1)}(t)$ and $\mathbf{x}^{(2)}(t)$ are linearly dependent at each point t in J. However, we cannot find a set of nonzero constants c_1, c_2 which will satisfy the system (8.8.27) for every t in J.

Exercises 8.8

1. Prove (8.8.6), (8.8.7), and (8.8.8).
2. Prove (8.8.12), (8.8.13), and (8.8.14).
3. If $\mathbf{x} = [-i, 1 + i, 2 + i, 2]$, $\quad \mathbf{y} = [i, 1 - i, 2 - i, 2i]$, $\quad i^2 = -1$, find
   (a) $\mathbf{x} + \mathbf{y}$, (b) $\mathbf{x} - \mathbf{y}$, (c) $(1 + i)\mathbf{x}$, (d) $-2i\mathbf{y}$, (e) $\mathbf{x}\mathbf{y}^T$, (f) $\mathbf{y}\mathbf{x}^T$,
   (g) $(\mathbf{x}, \mathbf{y})$, (h) $(\mathbf{y}, \mathbf{x})$, (i) $(\mathbf{x}, \mathbf{x})$, (j) $(\mathbf{y}, \mathbf{y})$, (k) $(2i\mathbf{x}, \mathbf{y})$, (l) $(\mathbf{x}, 2i\mathbf{y})$.

4. Determine whether each set of vectors given below is linearly dependent. If linearly dependent, express the linearly dependent vectors in terms of the linearly independent vectors. In order to save space, we shall display each vector as a row vector. However, for computational purposes, you may consider them as column vectors. Or, if you wish, simply take the transpose of each given vector.

(a) $\mathbf{x}^{(1)} = [1, 2, 3]$, $\mathbf{x}^{(2)} = [-1, -2, 4]$, $\mathbf{x}^{(3)} = [3, 1, 5]$.

(b) $\mathbf{x}^{(1)} = [2, -1, 3]$, $\mathbf{x}^{(2)} = [-1, 4, 2]$, $\mathbf{x}^{(3)} = [3, 2, 8]$.

(c) $\mathbf{x}^{(1)} = [2, 4, 6]$, $\mathbf{x}^{(2)} = [-1, 6, 9]$, $\mathbf{x}^{(3)} = [-1, 2, 3]$.

(d) $\mathbf{x}^{(1)} = [0, 0, -12, 2]$, $\mathbf{x}^{(2)} = [-1, 2, -3, 0]$, $\mathbf{x}^{(3)} = [3, -6, -9, 3]$.

(e) $\mathbf{x}^{(1)} = [-2, 4, -6, 8]$, $\mathbf{x}^{(2)} = [-4, 11, -3, -2]$,
    $\mathbf{x}^{(3)} = [1, -3, 0, 2]$, $\mathbf{x}^{(4)} = [0, -3, -9, 18]$.

(f) $\mathbf{x}^{(1)} = [1, 0, 0, 0, 0]$, $\mathbf{x}^{(2)} = [0, 2, 0, 0, 0]$, $\mathbf{x}^{(3)} = [0, 0, 3, 0, 0]$,
    $\mathbf{x}^{(4)} = [0, 0, 0, 4, 0]$, $\mathbf{x}^{(5)} = [0, 0, 0, 0, 5]$.

(g) $\mathbf{x}^{(1)} = [0, 0, 4, -4]$, $\mathbf{x}^{(2)} = [-2, 2, 4, -4]$, $\mathbf{x}^{(3)} = [1, -1, 18, -18]$,
    $\mathbf{x}^{(4)} = [3, -3, 6, -6]$.

(h) $\mathbf{x}^{(1)} = [1, -i, 2]$, $\mathbf{x}^{(2)} = [i, -1, 3]$, $\mathbf{x}^{(3)} = [1, -2, -i]$, $i^2 = -1$.

(i) $\mathbf{x}^{(1)} = [1 + i, -1 - i, 2]$, $\mathbf{x}^{(2)} = [-1 + 2i, 2 - 3i, -i]$,
    $\mathbf{x}^{(3)} = [i, 1, 5]$.

(j) $\mathbf{x}^{(1)} = [-1 - 2i, 2 - i, 0, 0]$, $\mathbf{x}^{(2)} = [0, 0, 2 + 4i, -4 + 2i]$,
    $\mathbf{x}^{(3)} = [-i, 1, i, -1]$, $\mathbf{x}^{(4)} = [1, i, 2i, -2]$.

5. Determine whether each set of vector functions defined below is linearly dependent on the interval $J: -\infty < t < \infty$. If linearly dependent, express the linear dependent vectors in terms of the linearly independent vectors. As in Exercise 4, we shall display each vector as a row vector.

(a) $\mathbf{x}^{(1)}(t) = [e^t, t]$, $\mathbf{x}^{(2)}(t) = [e^{-t}, 1]$.

(b) $\mathbf{x}^{(1)}(t) = [e^t, e^{-t}, t]$, $\mathbf{x}^{(2)}(t) = [-2e^t, -e^{-t}, 3t]$,
    $\mathbf{x}^{(3)}(t) = [-4e^t, -e^{-t}, 11t]$.

(c) $\mathbf{x}^{(1)}(t) = [\sin^2 t, 2\cos^2 t, t + 1]$, $\mathbf{x}^{(2)}(t) = [2\sin^2 t, 3\cos^2 t, t]$,
    $\mathbf{x}^{(3)}(t) = [\sin^2 t, \cos^2 t, -1]$.

(d) $\mathbf{x}^{(1)}(t) = [1, t, t^2]$, $\mathbf{x}^{(2)}(t) = [-t, -t^2, -t^3]$, $\mathbf{x}^3(t) = [t^2, t^3, t^4]$.

(e) $\mathbf{x}^{(1)}(t) = [\sin 2t, \cos 2t, 2]$, $\mathbf{x}^{(2)}(t) = [4\sin t \cos t, 10\cos^2 t - 5, 0]$,
    $\mathbf{x}^{(3)}(t) = [\sin t \cos t, 4\cos^2 t - 2, -1]$.

(f) $\mathbf{x}^{(1)}(t) = [1, 0, 0, 0]$, $\mathbf{x}^{(2)}(t) = [0, e^t, 0, 0]$, $\mathbf{x}^{(3)}(t) = [0, 0, e^{2t}, 0]$,
    $\mathbf{x}^{(4)}(t) = [0, 0, 0, e^{3t}]$.

(g) $\mathbf{x}^{(1)}(t) = [e^t, \sin^2 t, e^{-t}, t]$, $\mathbf{x}^{(2)}(t) = [2e^t, 2 - 2\cos^2 t, -e^{-t}, 2t]$,
    $\mathbf{x}^{(3)}(t) = [-4e^t, -4\sin^2 t, -e^{-t}, -4t]$.

(h) $\mathbf{x}^{(1)}(t) = [e^t, e^{-t}, e^{2t}, e^{-2t}, 1]$, $\mathbf{x}^{(2)}(t) = [-e^t, e^{-t}, -e^{2t}, e^{-2t}, -1]$.

6. Consider the k n-vectors $\mathbf{x}^{(1)}, \mathbf{x}^{(2)}, \ldots, \mathbf{x}^{(k)}$. Show that if any subcollection of the $\mathbf{x}^{(i)}$s are linearly dependent, then the k n-vectors are also linearly dependent.

7. Show that if the k n-vectors $\mathbf{x}^{(1)}, \mathbf{x}^{(2)}, \ldots, \mathbf{x}^{(k)}$ are linearly independent, then any subcollection of the $\mathbf{x}^{(i)}$s is also linearly independent.

8. Show that k n-vectors $\mathbf{x}^{(1)}, \mathbf{x}^{(2)}, \ldots, \mathbf{x}^{(k)}$ are linearly dependent if and only if at least one of these vectors can be expressed as a linear combination of the rest.

9. Suppose that k n-vectors $\mathbf{x}^{(1)}, \mathbf{x}^{(2)}, \ldots, \mathbf{x}^{(k)}$ are linearly independent. Show that $(k + 1)$ n-vectors $\mathbf{x}^{(1)}, \mathbf{x}^{(2)}, \ldots, \mathbf{x}^{(k)}, \mathbf{x}^{(k+1)}$ are linearly independent if and only if $\mathbf{x}^{(k+1)}$ cannot be expressed as a linear combination of the k n-vectors $\mathbf{x}^{(1)}, \mathbf{x}^{(2)}, \ldots, \mathbf{x}^{(k)}$.

10. *The multiplication of a vector by a matrix is a linear operation.*   Let $\mathbf{A}$ and $\mathbf{B}$ be $m \times n$ matrices, $\mathbf{x}$ and $\mathbf{y}$ be n-dimensional column vectors, and α and β be scalars.   Show that

(a) $\mathbf{A}(\alpha\mathbf{x} + \beta\mathbf{y}) = \alpha\mathbf{A}\mathbf{x} + \beta\mathbf{A}\mathbf{y}$,   (b) $(\alpha\mathbf{A} + \beta\mathbf{B})\mathbf{x} = \alpha\mathbf{A}\mathbf{x} + \beta\mathbf{B}\mathbf{x}$.

8.9. Some Basic Results on Systems of Linear Equations

A system of m linear equations in n unknowns $x_1, x_2, \ldots, x_n$ is given by

$$\begin{cases} a_{11}x_1 + a_{12}x_2 + \cdots + a_{1n}x_n = b_1, \\ a_{21}x_1 + a_{22}x_2 + \cdots + a_{2n}x_n = b_2, \\ \quad\vdots \qquad\quad \vdots \qquad\qquad\quad \vdots \\ a_{m1}x_1 + a_{m2}x_2 + \cdots + a_{mn}x_n = b_m, \end{cases} \tag{8.9.1}$$

where the a_{ij} and b_i are either constants or quantities that are independent of the n unknowns.   Clearly, in (8.9.1), $m < n$ or $m = n$ or $m > n$.   System (8.9.1) can also be written as

$$\sum_{j=1}^{n} a_{ij}x_j = b_i, \qquad i = 1, 2, \ldots, m. \tag{8.9.2}$$

Letting

$$\mathbf{A} = \begin{bmatrix} a_{11} & a_{12} & \cdots & a_{1n} \\ a_{21} & a_{22} & \cdots & a_{2n} \\ \vdots & \vdots & & \vdots \\ a_{m1} & a_{m2} & \cdots & a_{mn} \end{bmatrix}, \qquad \mathbf{x} = \begin{bmatrix} x_1 \\ x_2 \\ \vdots \\ x_n \end{bmatrix}, \qquad \mathbf{b} = \begin{bmatrix} b_1 \\ b_2 \\ \vdots \\ b_m \end{bmatrix}, \tag{8.9.3}$$

we may also write the system (8.9.1) equivalently in matrix form as

$$\mathbf{A}\mathbf{x} = \mathbf{b}. \tag{8.9.4}$$

The matrix $\mathbf{A}$ is called the *coefficient matrix* of the system (8.9.1).   The $m \times (n + 1)$ matrix

$$\begin{bmatrix} a_{11} & a_{12} & \cdots & a_{1n} & b_1 \\ a_{21} & a_{22} & \cdots & a_{2n} & b_2 \\ \vdots & \vdots & & \vdots & \vdots \\ a_{m1} & a_{m2} & \cdots & a_{mn} & b_m \end{bmatrix}, \tag{8.9.5}$$

is called the *augmented matrix* of the system (8.9.1), and we shall denote it by $[\mathbf{A}\ \mathbf{b}]$.

Any set of values $x_1, x_2, \ldots, x_n$ which simultaneously satisfies the system (8.9.1) is called a *solution* of the system.   The system (8.9.1) is said to be *consistent* if it has one or more solutions; otherwise, it is *inconsistent*.   If the vector $\mathbf{b} = \mathbf{0}$, the system (8.9.1) is said to be *homogeneous*; otherwise, it is *nonhomogeneous.*   Thus, a system of m homogeneous linear equations in the n unknowns $x_1, x_2, \ldots, x_n$ has the form

$$\sum_{j=1}^{n} a_{ij} x_j = 0, \qquad i = 1, 2, \ldots, m, \tag{8.9.6}$$

or

$$\mathbf{Ax} = \mathbf{0}. \tag{8.9.7}$$

Consider the system given in Example 8.3.2, namely,

$$\begin{cases} x_1 - x_2 + x_3 = 2, \\ 2x_1 - 3x_2 + 4x_3 = 8, \\ 3x_1 - 4x_2 + 5x_3 = 10. \end{cases} \tag{8.9.8}$$

The solution of the system (8.9.8) was found to be

$$x_1 = x_3 - 2, \qquad x_2 = 2x_3 - 4, \qquad x_3 = x_3, \tag{8.9.9}$$

where x_3 is arbitrary.   Since all solutions of the system (8.9.8) can be obtained from (8.9.9) by varying x_3, we call (8.9.9) a *complete solution.*   The solution obtained by assigning a particular value to x_3 in (8.9.9) is called a *particular solution* of the system (8.9.8).   We may also write (8.9.9) equivalently as

$$x_1 = t - 2, \qquad x_2 = 2t - 4, \qquad x_3 = t, \tag{8.9.10}$$

where t is arbitrary.

The following results are known.   [*See,* for example, [21], pp. 140–167.]

Theorem 8.9.1.   *A necessary and sufficient condition that the system* $\mathbf{Ax} = \mathbf{b}$ *be consistent is that the matrix* $\mathbf{A}$ *and the augmented matrix* $[\mathbf{A}\ \mathbf{b}]$ *have the same rank.*

Theorem 8.9.2.   *A necessary and sufficient condition that the system* $\mathbf{Ax} = \mathbf{b}$ *be consistent is that the vector* $\mathbf{b}$ *is a linear combination of the columns of the matrix* $\mathbf{A}$.

Theorem 8.9.3.   *If the system* $\mathbf{Ax} = \mathbf{b}$ *is consistent, then a complete solution of this system is given by a complete solution of the corresponding homogeneous system* $\mathbf{Ax} = \mathbf{0}$ *plus a particular solution of the system* $\mathbf{Ax} = \mathbf{b}$.

Theorem 8.9.4.   *A necessary and sufficient condition that the homogeneous system* $\mathbf{Ax} = \mathbf{0}$ *have nontrivial solutions is that the rank of the matrix* $\mathbf{A}$ *is less than the number of components of the vector* $\mathbf{x}$, *namely,* n.

Theorem 8.9.5.   *If in the homogeneous system* $\mathbf{Ax} = \mathbf{0}$, $\mathbf{A}$ *is an* $n \times n$ *matrix, then a necessary and sufficient condition that this system have nontrivial solutions is that* $\mathbf{A}$ *is a singular matrix, that is,* $\det \mathbf{A} = 0$.

Theorem 8.9.6.   *If in the homogeneous system* $\mathbf{A}\mathbf{x} = \mathbf{0}$, $\mathbf{A}$ *is an* $m \times n$ *matrix with* $m < n$, *then this system always has nontrivial solutions.*

Remark 8.9.1.   It turns out that, given any consistent system of rank r, the result of solving for r of the variables in terms of the remaining $n - r$ variables is a *complete solution* of the system.   For one can obtain all the solutions of the given system by assigning arbitrary values to the $n - r$ variables.   [*See* Example 8.9.2.]

Example 8.9.1.   Let us solve the system

$$\begin{cases} x_1 + 2x_2 + x_3 = 1, \\ -2x_1 + x_2 - x_3 = 2, \\ 4x_1 - x_2 + 2x_3 = -3. \end{cases} \tag{8.9.11}$$

SOLUTION.   We may write the system (8.9.11) equivalently as

$$\mathbf{A}\mathbf{x} = \mathbf{b}, \tag{8.9.12}$$

where

$$\mathbf{A} = \begin{bmatrix} 1 & 2 & 1 \\ -2 & 1 & -1 \\ 4 & -1 & 2 \end{bmatrix}, \qquad \mathbf{x} = \begin{bmatrix} x_1 \\ x_2 \\ x_3 \end{bmatrix}, \qquad \mathbf{b} = \begin{bmatrix} 1 \\ 2 \\ -3 \end{bmatrix}. \tag{8.9.13}$$

Now,

$$\det \mathbf{A} = \begin{vmatrix} 1 & 2 & 1 \\ -2 & 1 & -1 \\ 4 & -1 & 2 \end{vmatrix} = -1 \neq 0.$$

Thus, the matrix $\mathbf{A}$ is nonsingular and $\mathbf{A}^{-1}$ exists.   Multiplying (8.9.12) on the left by $\mathbf{A}^{-1}$, we obtain

$$\mathbf{x} = \mathbf{A}^{-1}\mathbf{b}. \tag{8.9.14}$$

The reader may show that $\mathbf{A}^{-1} = \begin{bmatrix} -1 & 5 & 3 \\ 0 & 2 & 1 \\ 2 & -9 & -5 \end{bmatrix}$.   In view of (8.9.13) and (8.9.14), we have

$$\begin{bmatrix} x_1 \\ x_2 \\ x_3 \end{bmatrix} = \begin{bmatrix} -1 & 5 & 3 \\ 0 & 2 & 1 \\ 2 & -9 & -5 \end{bmatrix} \begin{bmatrix} 1 \\ 2 \\ -3 \end{bmatrix} = \begin{bmatrix} 0 \\ 1 \\ -1 \end{bmatrix}.$$

Hence, the unique solution of the system (8.9.11) is $x_1 = 0$, $x_2 = 1$, $x_3 = -1$.

Example 8.9.2.   Let us solve the system

$$\begin{cases} x_1 - x_2 + 2x_3 + 3x_4 = 5, \\ x_1 - 2x_2 + x_3 + 5x_4 = 5, \\ 2x_1 + x_2 + 7x_3 = 10. \end{cases} \tag{8.9.15}$$

SOLUTION.   We may write the system (8.9.15) equivalently as

$$\mathbf{Ax} = \mathbf{b}, \qquad (8.9.16)$$

where

$$\mathbf{A} = \begin{bmatrix} 1 & -1 & 2 & 3 \\ 1 & -2 & 1 & 5 \\ 2 & 1 & 7 & 0 \end{bmatrix}, \qquad \mathbf{x} = \begin{bmatrix} x_1 \\ x_2 \\ x_3 \\ x_4 \end{bmatrix}, \qquad \mathbf{b} = \begin{bmatrix} 5 \\ 5 \\ 10 \end{bmatrix}. \qquad (8.9.17)$$

The reader may verify that the matrix $\mathbf{A}$ and the augmented matrix

$$[\mathbf{A}\ \mathbf{b}] = \begin{bmatrix} 1 & -1 & 2 & 3 & 5 \\ 1 & -2 & 1 & 5 & 5 \\ 2 & 1 & 7 & 0 & 10 \end{bmatrix}$$

have the same rank, namely, two.   In view of Theorem 8.9.1, the system
(8.9.15) is consistent.

We shall now determine a complete solution of the system (8.9.15).   First,
we consider the homogeneous system $\mathbf{Ax} = \mathbf{0}$, that is,

$$\begin{cases} x_1 - x_2 + 2x_3 + 3x_4 = 0, \\ x_1 - 2x_2 + x_3 + 5x_4 = 0, \\ 2x_1 + x_2 + 7x_3 = 0. \end{cases} \qquad (8.9.18)$$

In view of Theorem 8.9.6 with $m = 3 < 4 = n$, the system (8.9.18) has nontrivial solutions.   Since the rank of the matrix $\mathbf{A}$ is two and the determinant
of the coefficients of x_1 and x_2 in the first two equations in (8.9.18) is different
from zero, we may utilize these two equations to solve for x_1 and x_2 in terms
of x_3 and x_4.   Thus, we may consider the system

$$\begin{cases} x_1 - x_2 = -2x_3 - 3x_4, \\ x_1 - 2x_2 = -x_3 - 5x_4. \end{cases} \qquad (8.9.19)$$

Solving the system (8.9.19), we obtain

$$x_1 = -3x_3 - x_4, \qquad x_2 = -x_3 + 2x_4, \qquad (8.9.20)$$

where x_3 and x_4 are arbitrary.   We may also write (8.9.20) equivalently as

$$x_1 = -3t_1 - t_2, \qquad x_2 = -t_1 + 2t_2, \qquad x_3 = t_1, \qquad x_4 = t_2,$$

or $\hspace{12cm} (8.9.21)$

$$x_1 = 3t_1 + t_2, \qquad x_2 = t_1 - 2t_2, \qquad x_3 = -t_1, \qquad x_4 = -t_2,$$

where t_1 and t_2 are arbitrary.   Thus, a complete solution of the system
(8.9.18) is given by (8.9.21).

Next, we shall obtain a particular solution of the system (8.9.15). Setting $x_3 = x_4 = 0$ in (8.9.15) and then considering the first two equations, we have

$$\begin{cases} x_1 - \ x_2 = 5, \\ x_1 - 2x_2 = 5. \end{cases} \tag{8.9.22}$$

Solving the system (8.9.22), we obtain $x_1 = 5$ and $x_2 = 0$. Thus, a particular solution of the system (8.9.15) is $x_1 = 5$, $x_2 = 0$, $x_3 = 0$, $x_4 = 0$. In view of (8.9.21) and Theorem 8.9.3, a complete solution of the system (8.9.15) is given by

$$x_1 = 3t_1 + t_2 + 5, \qquad x_2 = t_1 - 2t_2 + 0 = t_1 - 2t_2, \qquad x_3 = -t_1 + 0 = -t_1,$$
$$x_4 = -t_2 + 0 = -t_2. \tag{8.9.23}$$

We leave it for the reader to verify that if he substitutes x_1, x_2, x_3, x_4 of (8.9.23) into (8.9.15), the system (8.9.15) is satisfied identically.

Remark 8.9.2. The reader should note the similarity between finding a complete solution of a system of nonhomogeneous linear equations and finding a general solution of a nonhomogeneous linear differential equation.

Exercises 8.9

1. Prove that a square matrix $\mathbf{A}$ is singular if and only if the columns (or rows) of $\mathbf{A}$ are linearly dependent.

In Exercises 2 through 5, employ the method of Example 8.9.1 to find the unique solution of each system of equations.

2. $\begin{cases} x_1 - 2x_2 = 5, \\ 3x_1 + \ x_2 = 2. \end{cases}$

3. $\begin{cases} x_1 - \ x_2 + \ x_3 = 2, \\ -x_1 + 2x_2 - 2x_3 = 1, \\ 2x_1 - 3x_2 + 4x_3 = -4. \end{cases}$

4. $\begin{cases} 4x_1 - 9x_2 + 4x_3 = -3, \\ 2x_1 - 3x_2 + \ x_3 = 4, \\ -x_1 + 2x_2 - \ x_3 = 1. \end{cases}$

5. $\begin{cases} 2x_1 - 3x_2 + 6x_3 - \ 7x_4 = 2, \\ -2x_1 + 3x_2 - 5x_3 + \ 6x_4 = -1, \\ x_1 - \ x_2 + 2x_3 - \ 3x_4 = 0, \\ 3x_1 - 4x_2 + 6x_3 - 11x_4 = -6. \end{cases}$

In Exercises 6 through 10, determine whether each system of equations is consistent. If consistent, find its complete solution.

6. $\begin{cases} x_1 + \ x_2 + 2x_3 = 1, \\ 2x_1 - 3x_2 - \ x_3 = 2, \\ 3x_1 - 4x_2 - \ x_3 = -3. \end{cases}$

7. $\begin{cases} 2x_1 - 3x_2 + 3x_3 = 1, \\ x_1 + 2x_2 + 5x_3 = 4, \\ 3x_1 - 5x_2 + 4x_3 = 1. \end{cases}$

8. $\begin{cases} 3x_1 + \ x_2 + 4x_3 + 2x_4 = -3, \\ 2x_1 - 3x_2 - \ x_3 + 5x_4 = 9, \\ 4x_1 - 2x_2 + 2x_3 + 6x_4 = 6. \end{cases}$

9. $\begin{cases} x_1 + 2x_2 - 2x_3 - \ x_4 = 0, \\ 2x_1 - 3x_2 + 5x_3 + 2x_4 = 8, \\ 4x_1 + 5x_2 + \ x_3 - \ x_4 = 6. \end{cases}$

10. $\begin{cases} 3x_1 + x_2 - x_3 + 5x_4 = 5, \\ x_1 - x_2 - x_3 + x_4 = -1, \\ 4x_1 + 6x_2 + x_3 + 9x_4 = 9, \\ 6x_1 - 2x_2 - 4x_3 + 8x_4 = 8. \end{cases}$

11. Let $\mathbf{A}$ be an $m \times n$ matrix. Suppose that $\mathbf{Ax} = \mathbf{0}$ for all n-dimensional column vectors $\mathbf{x}$. Show that $\mathbf{A}$ is a zero matrix.

8.10. Theory of Systems of Linear First-Order Differential Equations

Consider the system of n linear first-order differential equations

$$\begin{cases} x_1' = a_{11}(t)x_1 + a_{12}(t)x_2 + \cdots + a_{1n}(t)x_n + g_1(t), \\ x_2' = a_{21}(t)x_1 + a_{22}(t)x_2 + \cdots + a_{2n}(t)x_n + g_2(t), \\ \vdots \\ x_n' = a_{n1}(t)x_1 + a_{n2}(t)x_2 + \cdots + a_{nn}(t)x_n + g_n(t). \end{cases} \tag{8.10.1}$$

Letting

$$\mathbf{A}(t) = \begin{bmatrix} a_{11}(t) & a_{12}(t) & \cdots & a_{1n}(t) \\ a_{21}(t) & a_{22}(t) & \cdots & a_{2n}(t) \\ \vdots & \vdots & & \vdots \\ a_{n1}(t) & a_{n2}(t) & \cdots & a_{nn}(t) \end{bmatrix},$$

$$\mathbf{x} = \begin{bmatrix} x_1 \\ x_2 \\ \vdots \\ x_n \end{bmatrix} = \begin{bmatrix} x_1(t) \\ x_2(t) \\ \vdots \\ x_n(t) \end{bmatrix} = \mathbf{x}(t), \qquad \mathbf{g}(t) = \begin{bmatrix} g_1(t) \\ g_2(t) \\ \vdots \\ g_n(t) \end{bmatrix}, \tag{8.10.2}$$

we may write the system (8.10.1) equivalently in matrix form as

$$\mathbf{x}' = \mathbf{A}(t)\mathbf{x} + \mathbf{g}(t). \tag{8.10.3}$$

In the ensuing discussion, unless stated to the contrary, we shall assume that the matrix function $\mathbf{A}$ and the vector function $\mathbf{g}$ are continuous functions of t defined on a common interval J: $a \leq t \leq b$. Also, $\mathbf{A}(t) \neq \mathbf{0}$ for every t in J.

If $\mathbf{g}(t) = \mathbf{0}$ for every t in J, then the system (8.10.3) is called *homogeneous*. If at least one of the components of the vector $\mathbf{g}(t)$ is not identically zero in J, then the system (8.10.3) is called *nonhomogeneous*.

Definition 8.10.1. A vector function $\mathbf{x} = \boldsymbol{\phi}(t)$ defined on an interval J: $a \leq t \leq b$, is said to be a *solution* of Equation (8.10.3), if its components satisfy the system (8.10.1) for each t in J.

For convenience, we shall now state Theorem 8.2.2 in the following equivalent matrix form.

Theorem 8.10.1. *Let $\mathbf{A}(t)$ and $\mathbf{g}(t)$ be continuous on an interval J: $a \leq t \leq b$, containing the point t_0. Then there exists a unique solution $\mathbf{x} = \boldsymbol{\phi}(t)$ satisfying Equation (8.10.3) for each t in J and the initial condition $\boldsymbol{\phi}(t_0) = \boldsymbol{\eta}$.*

Note that the initial condition $\boldsymbol{\phi}(t_0) = \boldsymbol{\eta}$ asserts that

$$\phi_1(t_0) = \eta_1, \qquad \phi_2(t_0) = \eta_2, \quad \ldots, \quad \phi_n(t_0) = \eta_n, \tag{8.10.4}$$

where $\eta_1, \eta_2, \ldots, \eta_n$ are given constants.

Let us consider the homogeneous equation

$$\mathbf{x}' = \mathbf{A}(t)\mathbf{x}. \tag{8.10.5}$$

We shall now state some basic results concerning Equation (8.10.5).   These results bear close resemblance to those given in Sections 4.2, 4.4, 6.2, and 6.4, and their proofs may be modeled, with some modifications, after the proofs given in those sections.   Thus, we shall give proofs for only a few of these results.   The remaining proofs will be left as exercises for the reader.   [*See* Exercises 8.10.1, 8.10.2, and 8.10.3.]

We shall also denote the solution of (8.10.5) by the vector function $\mathbf{x} = \mathbf{x}(t)$. Let us denote n solutions of (8.10.5) by

$$\mathbf{x}^{(1)}(t) = \begin{bmatrix} x_{11}(t) \\ x_{21}(t) \\ \vdots \\ x_{n1}(t) \end{bmatrix}, \qquad \mathbf{x}^{(2)}(t) = \begin{bmatrix} x_{12}(t) \\ x_{22}(t) \\ \vdots \\ x_{n2}(t) \end{bmatrix}, \quad \ldots, \quad \mathbf{x}^{(n)}(t) = \begin{bmatrix} x_{1n}(t) \\ x_{2n}(t) \\ \vdots \\ x_{nn}(t) \end{bmatrix}.$$

$$\tag{8.10.6}$$

Theorem 8.10.2. *Let $\mathbf{A}(t)$ be continuous on an interval J: $a \leq t \leq b$, and let t_0 be any fixed point in J. Let $\mathbf{x} = \mathbf{x}(t)$ be a solution of the system (8.10.5) such that $\mathbf{x}(t_0) = \mathbf{0}$. Then $\mathbf{x}(t) = \mathbf{0}$ for every t in J.*

Theorem 8.10.3. *Let $\mathbf{x}^{(1)}(t)$, $\mathbf{x}^{(2)}(t)$, $\ldots$, $\mathbf{x}^{(n)}(t)$ be n solutions of the system (8.10.5). Then the linear combination of these n solutions*

$$c_1\mathbf{x}^{(1)}(t) + c_2\mathbf{x}^{(2)}(t) + \cdots + c_n\mathbf{x}^{(n)}(t),$$

where $c_1, c_2, \ldots, c_n$ are arbitrary constants, is also a solution of the system (8.10.5).

PROOF.   Utilizing Remark 8.8.2, (8.6.23), (8.6.22), (8.10.5), and Exercise 8.8.10, we obtain

$$\frac{d}{dt}\left(\sum_{i=1}^{n} c_i \mathbf{x}^{(i)}(t) \right) = \sum_{i=1}^{n} \frac{d}{dt}\left(c_i \mathbf{x}^{(i)}(t) \right) = \sum_{i=1}^{n} c_i \frac{d\mathbf{x}^{(i)}(t)}{dt}$$

$$= \sum_{i=1}^{n} c_i \mathbf{A}(t)\mathbf{x}^{(i)}(t) = \mathbf{A}(t)\left(\sum_{i=1}^{n} c_i \mathbf{x}^{(i)}(t) \right).$$

Thus, $\sum_{i=1}^{n} c_i \mathbf{x}^{(i)}(t)$ is also a solution of the system (8.10.5).

Example 8.10.1. Consider the homogeneous system

$$\mathbf{x}' = \begin{bmatrix} \frac{1}{2} & \frac{1}{2} \\ -\frac{3}{2} & \frac{5}{2} \end{bmatrix} \mathbf{x}. \tag{8.10.7}$$

The vectors $\mathbf{x}^{(1)}(t) = \begin{bmatrix} e^t \\ e^t \end{bmatrix}$ and $\mathbf{x}^{(2)}(t) = \begin{bmatrix} e^{2t} \\ 3e^{2t} \end{bmatrix}$ are solutions of Equation (8.10.7) on the interval J: $-\infty < t < \infty$, for, substituting these vectors into (8.10.7), we see that

$$\begin{bmatrix} e^t \\ e^t \end{bmatrix} = \begin{bmatrix} \frac{1}{2} & \frac{1}{2} \\ -\frac{3}{2} & \frac{5}{2} \end{bmatrix} \begin{bmatrix} e^t \\ e^t \end{bmatrix} = \begin{bmatrix} e^t \\ e^t \end{bmatrix}$$

and

$$\begin{bmatrix} 2e^{2t} \\ 6e^{2t} \end{bmatrix} = \begin{bmatrix} \frac{1}{2} & \frac{1}{2} \\ -\frac{3}{2} & \frac{5}{2} \end{bmatrix} \begin{bmatrix} e^{2t} \\ 3e^{2t} \end{bmatrix} = \begin{bmatrix} 2e^{2t} \\ 6e^{2t} \end{bmatrix}.$$

Thus, $\mathbf{x} = c_1 \mathbf{x}^{(1)}(t) + c_2 \mathbf{x}^{(2)}(t)$, where c_1 and c_2 are arbitrary constants, is also a solution of Equation (8.10.7) for every t in J.

Theorem 8.10.4. *Let* $\mathbf{A}(t)$ *be continuous on an interval* J: $a \leq t \leq b$.

(a) *The system* (8.10.5) *always possesses n linearly independent solutions on J.*
(b) *If* $\mathbf{x}^{(1)}(t)$, $\mathbf{x}^{(2)}(t)$, $\ldots$, $\mathbf{x}^{(n)}(t)$ *are n linearly independent solutions of the system* (8.10.5) *on J, then every solution of this system can be expressed as a linear combination*

$$c_1 \mathbf{x}^{(1)}(t) + c_2 \mathbf{x}^{(2)}(t) + \cdots + c_n \mathbf{x}^{(n)}(t),$$

of these n linearly independent solutions, for suitably chosen constants $c_1, c_2, \ldots, c_n$.

Definition 8.10.2. Let $\mathbf{x}^{(1)}(t)$, $\mathbf{x}^{(2)}(t)$, $\ldots$, $\mathbf{x}^{(n)}(t)$ be n linearly independent solutions of the system (8.10.5) on an interval J: $a \leq t \leq b$. The *general solution* $\mathbf{x} = \mathbf{x}(t)$ of the system (8.10.5) is defined by

$$\mathbf{x} = c_1 \mathbf{x}^{(1)}(t) + c_2 \mathbf{x}^{(2)}(t) + \cdots + c_n \mathbf{x}^{(n)}(t), \tag{8.10.8}$$

where $c_1, c_2, \ldots, c_n$ are arbitrary constants.

Definition 8.10.3. Any set of n linearly independent solutions of the system (8.10.5) on an interval J: $a \leq t \leq b$, is said to form a *fundamental set* of this system on J.

Definition 8.10.4. Let t_0 be any fixed point in an interval J: $a \leq t \leq b$. The set of n linearly independent solutions $\mathbf{x}^{(1)}(t)$, $\mathbf{x}^{(2)}(t)$, $\ldots$, $\mathbf{x}^{(n)}(t)$ of the system (8.10.5) on J which satisfies the set of initial conditions

$$\mathbf{x}^{(1)}(t_0) = \mathbf{e}^{(1)} = \begin{bmatrix} 1 \\ 0 \\ \vdots \\ 0 \end{bmatrix}, \qquad \mathbf{x}^{(2)}(t_0) = \mathbf{e}^{(2)} = \begin{bmatrix} 0 \\ 1 \\ \vdots \\ 0 \end{bmatrix}, \quad \ldots, \quad \mathbf{x}^{(n)}(t_0) = \mathbf{e}^{(n)} = \begin{bmatrix} 0 \\ 0 \\ \vdots \\ 1 \end{bmatrix},$$

$$(8.10.9)$$

is called the *basic set of solutions* at t_0 of the system (8.10.5).

Definition 8.10.5. Let the n n-vector functions $\mathbf{x}^{(1)}$, $\mathbf{x}^{(2)}$, $\ldots$, $\mathbf{x}^{(n)}$ be defined as in (8.10.6) on an interval J: $a \le t \le b$. The determinant

$$W(\mathbf{x}^{(1)}, \mathbf{x}^{(2)}, \ldots, \mathbf{x}^{(n)}) = \begin{vmatrix} x_{11} & x_{12} & \cdots & x_{1n} \\ x_{21} & x_{22} & \cdots & x_{2n} \\ \vdots & \vdots & & \vdots \\ x_{n1} & x_{n2} & \cdots & x_{nn} \end{vmatrix} \qquad (8.10.10)$$

is called the *Wronskian* of the n n-vector functions $\mathbf{x}^{(1)}$, $\mathbf{x}^{(2)}$, $\ldots$, $\mathbf{x}^{(n)}$. Its value at any point t in J will be denoted by $W(\mathbf{x}^{(1)}, \mathbf{x}^{(2)}, \ldots, \mathbf{x}^{(n)}; t)$. Thus

$$W(\mathbf{x}^{(1)}, \mathbf{x}^{(2)}, \ldots, \mathbf{x}^{(n)}; t) = \begin{vmatrix} x_{11}(t) & x_{12}(t) & \cdots & x_{1n}(t) \\ x_{21}(t) & x_{22}(t) & \cdots & x_{2n}(t) \\ \vdots & \vdots & & \vdots \\ x_{n1}(t) & x_{n2}(t) & \cdots & x_{nn}(t) \end{vmatrix}. \qquad (8.10.11)$$

Example 8.10.2. If $\mathbf{x}^{(1)}(t) = \begin{bmatrix} 1 \\ t \end{bmatrix}$, $\mathbf{x}^{(2)}(t) = \begin{bmatrix} e^t \\ e^{2t} \end{bmatrix}$ in J: $-\infty < t < \infty$,

then

$$W(\mathbf{x}^{(1)}, \mathbf{x}^{(2)}; t) = \begin{vmatrix} 1 & e^t \\ t & e^{2t} \end{vmatrix} = e^{2t} - te^t,$$

and

$$W(\mathbf{x}^{(1)}, \mathbf{x}^{(2)}; 0) = \begin{vmatrix} 1 & 1 \\ 0 & 1 \end{vmatrix} = 1.$$

Example 8.10.3. If $\mathbf{x}^{(1)}(t) = \begin{bmatrix} 1 \\ t \end{bmatrix}$, $\mathbf{x}^{(2)}(t) = \begin{bmatrix} t \\ t^2 \end{bmatrix}$ in J: $-\infty < t < \infty$,

then

$$W(\mathbf{x}^{(1)}, \mathbf{x}^{(2)}; t) = \begin{vmatrix} 1 & t \\ t & t^2 \end{vmatrix} = 0.$$

Note that, even though the Wronskian of these vector functions is identically zero in J, they are linearly independent on J. [*See* Example 8.8.3.]

Let

$$\mathbf{X}(t) = \begin{bmatrix} x_{11}(t) & x_{12}(t) & \cdots & x_{1n}(t) \\ x_{21}(t) & x_{22}(t) & \cdots & x_{2n}(t) \\ \vdots & \vdots & & \vdots \\ x_{n1}(t) & x_{n2}(t) & \cdots & x_{nn}(t) \end{bmatrix}. \qquad (8.10.12)$$

Thus,

$$\det \mathbf{X}(t) = W(\mathbf{x}^{(1)}, \mathbf{x}^{(2)}, \ldots, \mathbf{x}^{(n)}; t). \tag{8.10.13}$$

Theorem 8.10.5. *Let* $\mathbf{x}^{(1)} = \mathbf{x}^{(1)}(t)$, $\mathbf{x}^{(2)} = \mathbf{x}^{(2)}(t)$, $\ldots$, $\mathbf{x}^{(n)} = \mathbf{x}^{(n)}(t)$ *be n solutions of the system* (8.10.5) *on an interval J: $a \leq t \leq b$. A necessary and sufficient condition that* $\mathbf{x}^{(1)}$, $\mathbf{x}^{(2)}$, $\ldots$, $\mathbf{x}^{(n)}$ *be linearly independent on J is that* $W(\mathbf{x}^{(1)}, \mathbf{x}^{(2)}, \ldots, \mathbf{x}^{(n)}; t) \neq 0$ *for every t in J.*

PROOF. *Sufficiency.* This condition asserts that if $W(\mathbf{x}^{(1)}, \mathbf{x}^{(2)}, \ldots, \mathbf{x}^{(n)}; t) \neq 0$ for every t in J, then $\mathbf{x}^{(1)}$, $\mathbf{x}^{(2)}$, $\ldots$, $\mathbf{x}^{(n)}$ are linearly independent on J. To show that $\mathbf{x}^{(1)}$, $\mathbf{x}^{(2)}$, $\ldots$, $\mathbf{x}^{(n)}$ are linearly independent on J, we must show that the only constants $c_1, c_2, \ldots, c_n$ such that

$$c_1 \mathbf{x}^{(1)}(t) + c_2 \mathbf{x}^{(2)}(t) + \cdots + c_n \mathbf{x}^{(n)}(t) = \mathbf{0} \tag{8.10.14}$$

for every t in J are the constants $c_1 = c_2 = \cdots = c_n = 0$. [*See* Definition 8.8.2.]

We may write (8.10.14) equivalently as

$$\mathbf{X}(t)\mathbf{c} = \mathbf{0}, \tag{8.10.15}$$

where $\mathbf{c}^T = [c_1, c_2, \ldots, c_n]$. By hypothesis, $W(\mathbf{x}^{(1)}, \mathbf{x}^{(2)}, \ldots, \mathbf{x}^{(n)}; t) \neq 0$ for every t in J. Thus, in view of (8.10.13), $\det \mathbf{X}(t) \neq 0$ for every t in J, and $\det \mathbf{X}(t_0) \neq 0$ for any fixed point t_0 in J. Consequently, $\mathbf{X}(t_0)$ is a nonsingular matrix. [*See* second paragraph of Section 8.7.] In view of Theorem 8.9.5, the homogeneous system

$$\mathbf{X}(t_0)\mathbf{c} = \mathbf{0}$$

has only the trivial solution $\mathbf{c} = \mathbf{0}$, that is, $c_1 = c_2 = \cdots = c_n = 0$. Thus, the vectors $\mathbf{x}^{(1)}$, $\mathbf{x}^{(2)}$, $\ldots$, $\mathbf{x}^{(n)}$ are linearly independent on J.

Necessity. This condition asserts that if $\mathbf{x}^{(1)}$, $\mathbf{x}^{(2)}$, $\ldots$, $\mathbf{x}^{(n)}$ are n linearly independent solutions of the system (8.10.5) on J, then $W(\mathbf{x}^{(1)}, \mathbf{x}^{(2)}, \ldots, \mathbf{x}^{(n)}; t) \neq 0$ for every t in J. We shall establish this result by contradiction. Suppose that $\mathbf{x}^{(1)}$, $\mathbf{x}^{(2)}$, $\ldots$, $\mathbf{x}^{(n)}$ are linearly independent on J and there exists a point t_0 in J such that $W(\mathbf{x}^{(1)}, \mathbf{x}^{(2)}, \ldots, \mathbf{x}^{(n)}; t_0) = 0$, that is, $\det \mathbf{X}(t_0) = 0$. Consider the system of n homogeneous linear equations in the unknowns $\alpha_1, \alpha_2, \ldots, \alpha_n$:

$$\mathbf{X}(t_0)\boldsymbol{\alpha} = \mathbf{0}, \tag{8.10.16}$$

where $\boldsymbol{\alpha}^T = [\alpha_1, \alpha_2, \ldots, \alpha_n]$. Since $\det \mathbf{X}(t_0) = 0$, it follows from Theorem 8.9.5 that the system (8.10.16) has a nontrivial solution, say $\alpha_1 = c_1$, $\alpha_2 = c_2, \ldots, \alpha_n = c_n$.

Let us define a vector function $\mathbf{g} = \begin{bmatrix} g_{11} \\ g_{21} \\ \vdots \\ g_{n1} \end{bmatrix}$ by

$$\mathbf{g}(t) = \sum_{i=1}^{n} \mathbf{x}^{(i)}(t)c_i. \tag{8.10.17}$$

In view of Theorem 8.10.3, the vector $\mathbf{g}(t)$ is also a solution of the system (8.10.5). Moreover, $\mathbf{g}(t_0) = \mathbf{0}$. (Why?) Thus, from Theorem 8.10.2, $\mathbf{g}(t) = \mathbf{0}$ for every t in J. Consequently, we have for every t in J,

$$c_1\mathbf{x}^{(1)}(t) + c_2\,\mathbf{x}^{(2)}(t) + \cdots + c_n\,\mathbf{x}^{(n)}(t) = \mathbf{0}. \tag{8.10.18}$$

Since $c_1.\ c_2,\ \ldots,\ c_n$ are not all zero, the linear combination (8.10.8) asserts that $\mathbf{x}^{(1)}$, $\mathbf{x}^{(2)}$, $\ldots$, $\mathbf{x}^{(n)}$ are linearly dependent on J. This, however, contradicts the fact that $\mathbf{x}^{(1)}$, $\mathbf{x}^{(2)}$, $\ldots$, $\mathbf{x}^{(n)}$ are linearly independent on J. Thus, the assumption that there exists a point t_0 in J such that $W(\mathbf{x}^{(1)}, \mathbf{x}^{(2)}, \ldots, \mathbf{x}^{(n)};\ t_0) = 0$ is false. Consequently, $W(\mathbf{x}^{(1)}, \mathbf{x}^{(2)}, \ldots, \mathbf{x}^{(n)};\ t) \neq 0$ for every t in J, and the theorem is established.

Theorem 8.10.6. *Let* $\mathbf{x}^{(1)} = \mathbf{x}^{(1)}(t)$, $\mathbf{x}^{(2)} = \mathbf{x}^{(2)}(t)$, $\ldots$, $\mathbf{x}^{(n)} = \mathbf{x}^{(n)}(t)$ *be n solutions of the system (8.10.5) on an interval* J: $a \leq t \leq b$, *and let t_0 be any fixed point in J. A necessary and sufficient condition that* $\mathbf{x}^{(1)}$, $\mathbf{x}^{(2)}$, $\ldots$, $\mathbf{x}^{(n)}$ *be linearly independent on J is that* $W(\mathbf{x}^{(1)}, \mathbf{x}^{(2)}, \ldots, \mathbf{x}^{(n)};\ t_0) \neq 0$.

Consider the homogeneous system of Example 8.10.1:

$$\mathbf{x}' = \begin{bmatrix} \frac{1}{2} & \frac{1}{2} \\ -\frac{3}{2} & \frac{5}{2} \end{bmatrix} \mathbf{x}. \tag{8.10.7}$$

We already verified that the vectors $\mathbf{x}^{(1)}(t) = \begin{bmatrix} e^t \\ e^t \end{bmatrix}$, $\mathbf{x}^{(2)}(t) = \begin{bmatrix} e^{2t} \\ 3e^{2t} \end{bmatrix}$ were solutions of Equation (8.10.7) on the interval J: $-\infty < t < \infty$. To show thatthese vectors are linearly independent on J, all we need to demonstrate is that the Wronskian of these vectors is different from zero *at a point t_0 of* J. Let us take $t_0 = 0$. Thus,

$$W(\mathbf{x}^{(1)}, \mathbf{x}^{(2)};\ 0) = \begin{vmatrix} 1 & 1 \\ 1 & 3 \end{vmatrix} = 2 \neq 0.$$

Hence, the general solution $\mathbf{x} = \mathbf{x}(t)$ of the system (8.10.7) on the interval J is given by [*see* Definition 8.10.2]

$$\mathbf{x} = c_1\mathbf{x}^{(1)}(t) + c_2\,\mathbf{x}^{(2)}(t) = c_1\begin{bmatrix} e^t \\ e^t \end{bmatrix} + c_2\begin{bmatrix} e^{2t} \\ 3e^{2t} \end{bmatrix} = c_1\begin{bmatrix} 1 \\ 1 \end{bmatrix} e^t + c_2\begin{bmatrix} 1 \\ 3 \end{bmatrix} e^{2t},$$

where c_1 and c_2 are arbitrary constants.

The following theorem follows from Theorems 8.10.5 and 8.10.6. However, we shall give an alternative proof which will lead us to develop a formula for $W(\mathbf{x}^{(1)}, \mathbf{x}^{(2)}, \ldots, \mathbf{x}^{(n)};\ t)$.

Theorem 8.10.7. *Let* $\mathbf{x}^{(1)} = \mathbf{x}^{(1)}(t)$, $\mathbf{x}^{(2)} = \mathbf{x}^{(2)}(t)$, $\ldots$, $\mathbf{x}^{(n)} = \mathbf{x}^{(n)}(t)$ *be n solutions of the system (8.10.5) on an interval* J: $a \leq t \leq b$. *Then either*

$$W(\mathbf{x}^{(1)}, \mathbf{x}^{(2)}, \ldots, \mathbf{x}^{(n)};\ t) = 0 \tag{8.10.19}$$

for every t in J, or

$$W(\mathbf{x}^{(1)}, \mathbf{x}^{(2)}, \ldots, \mathbf{x}^{(n)}; t) \neq 0 \qquad (8.10.20)$$

for every t in J.

PROOF. From (8.10.11), we have

$$W(\mathbf{x}^{(1)}, \mathbf{x}^{(2)}, \ldots, \mathbf{x}^{(n)}; t) = \begin{vmatrix} x_{11}(t) & x_{12}(t) & \cdots & x_{1n}(t) \\ x_{21}(t) & x_{22}(t) & \cdots & x_{2n}(t) \\ \vdots & \vdots & & \vdots \\ x_{n1}(t) & x_{n2}(t) & \cdots & x_{nn}(t) \end{vmatrix}. \qquad (8.10.21)$$

In view of Remark 6.4.1, the derivative of the above determinant with respect to t is

$$W' = \begin{vmatrix} x'_{11}(t) & x'_{12}(t) & \cdots & x'_{1n}(t) \\ x_{21}(t) & x_{22}(t) & \cdots & x_{2n}(t) \\ \vdots & \vdots & & \vdots \\ x_{n1}(t) & x_{n2}(t) & \cdots & x_{nn}(t) \end{vmatrix} + \begin{vmatrix} x_{11}(t) & x_{12}(t) & \cdots & x_{1n}(t) \\ x'_{21}(t) & x'_{22}(t) & \cdots & x'_{2n}(t) \\ \vdots & \vdots & & \vdots \\ x_{n1}(t) & x_{n2}(t) & \cdots & x_{nn}(t) \end{vmatrix}$$

$$+ \cdots + \begin{vmatrix} x_{11}(t) & x_{12}(t) & \cdots & x_{1n}(t) \\ x_{21}(t) & x_{22}(t) & \cdots & x_{2n}(t) \\ \vdots & \vdots & & \vdots \\ x'_{n1}(t) & x'_{n2}(t) & \cdots & x'_{nn}(t) \end{vmatrix}. \qquad (8.10.22)$$

Since $\mathbf{x}^{(1)}, \mathbf{x}^{(2)}, \ldots, \mathbf{x}^{(n)}$ are solutions of the system (8.10.5), we may write (8.10.22) equivalently as [*see* Exercise 8.10.4]

$$W' = \begin{vmatrix} \sum_{i=1}^{n} a_{1i}(t)x_{i1}(t) & \sum_{i=1}^{n} a_{1i}(t)x_{i2}(t) & \cdots & \sum_{i=1}^{n} a_{1i}(t)x_{in}(t) \\ x_{21}(t) & x_{22}(t) & \cdots & x_{2n}(t) \\ \vdots & \vdots & & \vdots \\ x_{n1}(t) & x_{n2}(t) & \cdots & x_{nn}(t) \end{vmatrix}$$

$$+ \begin{vmatrix} x_{11}(t) & x_{12}(t) & \cdots & x_{1n}(t) \\ \sum_{i=1}^{n} a_{2i}(t)x_{i1}(t) & \sum_{i=1}^{n} a_{2i}(t)x_{i2}(t) & \cdots & \sum_{i=1}^{n} a_{2i}(t)x_{in}(t) \\ \vdots & \vdots & & \vdots \\ x_{n1}(t) & x_{n2}(t) & \cdots & x_{nn}(t) \end{vmatrix}$$

$$+ \cdots + \begin{vmatrix} x_{11}(t) & x_{12}(t) & \cdots & x_{1n}(t) \\ x_{21}(t) & x_{22}(t) & \cdots & x_{2n}(t) \\ \vdots & \vdots & & \vdots \\ \sum_{i=1}^{n} a_{ni}(t)x_{i1}(t) & \sum_{i=1}^{n} a_{ni}(t)x_{i2}(t) & \cdots & \sum_{i=1}^{n} a_{ni}(t)x_{in}(t) \end{vmatrix}.$$

$$(8.10.23)$$

Concerning the first determinant in the right-hand side of (8.10.23): Add $-a_{12}(t)$ times the second row to the first row; add $-a_{13}(t)$ times the third row to the first row; ...; add $-a_{1n}(t)$ times the last row to the first row, and we obtain the equivalent determinant

$$\begin{vmatrix} a_{11}(t)x_{11}(t) & a_{11}(t)x_{12}(t) & \cdots & a_{11}(t)x_{1n}(t) \\ x_{21}(t) & x_{22}(t) & \cdots & x_{2n}(t) \\ \vdots & \vdots & & \vdots \\ x_{n1}(t) & x_{n2}(t) & \cdots & x_{nn}(t) \end{vmatrix}$$

$$= a_{11}(t)W(\mathbf{x}^{(1)}, \mathbf{x}^{(2)}, \ldots, \mathbf{x}^{(n)}; t). \qquad (8.10.24)$$

In a similar fashion, we leave it for the reader to show that [*see* Exercise 8.10.5] the second determinant in (8.10.23) is equivalent to

$$\begin{vmatrix} x_{11}(t) & x_{12}(t) & \cdots & x_{1n}(t) \\ a_{22}(t)x_{21}(t) & a_{22}(t)x_{22}(t) & \cdots & a_{22}(t)x_{2n}(t) \\ \vdots & \vdots & & \vdots \\ x_{n1}(t) & x_{n2}(t) & \cdots & x_{nn}(t) \end{vmatrix}$$

$$= a_{22}(t)W(\mathbf{x}^{(1)}, \mathbf{x}^{(2)}, \ldots, \mathbf{x}^{(n)}; t),$$

and the nth determinant is equivalent to

$$\begin{vmatrix} x_{11}(t) & x_{12}(t) & \cdots & x_{1n}(t) \\ x_{21}(t) & x_{22}(t) & \cdots & x_{2n}(t) \\ \vdots & \vdots & & \vdots \\ a_{nn}(t)x_{n1}(t) & a_{nn}(t)x_{n2}(t) & \cdots & a_{nn}(t)x_{nn}(t) \end{vmatrix}$$

$$= a_{nn}(t)W(\mathbf{x}^{(1)}, \mathbf{x}^{(2)}, \ldots, \mathbf{x}^{(n)}; t). \qquad (8.10.25)$$

In view of (8.10.24) and (8.10.25), (8.10.23) becomes

$$W' = (a_{11}(t) + a_{22}(t) + \cdots + a_{nn}(t))W = \left(\sum_{i=1}^{n} a_{ii}(t) \right) W. \qquad (8.10.26)$$

The general solution of the first-order differential equation (8.10.26) is

$$W(\mathbf{x}^{(1)}, \mathbf{x}^{(2)}, \ldots, \mathbf{x}^{(n)}; t) = c \exp \left[\int \sum_{i=1}^{n} a_{ii}(t)dt \right], \qquad (8.10.27)$$

where c is an arbitrary constant. Since $\exp \left[\int \sum_{i=1}^{n} a_{ii}(t)dt \right] \neq 0$ for every t in J, it follows from (8.10.27) that if $c = 0$, then $W(\mathbf{x}^{(1)}, \mathbf{x}^{(2)}, \ldots, \mathbf{x}^{(n)}; t) = 0$ for every t in J; and if $c \neq 0$, then $W(\mathbf{x}^{(1)}, \mathbf{x}^{(2)}, \ldots, \mathbf{x}^{(n)}; t) \neq 0$ for every t in J. Thus, the theorem is established.

Remark 8.10.1. We may also write (8.10.27) as

$$W(\mathbf{x}^{(1)}, \mathbf{x}^{(2)}, \ldots, \mathbf{x}^{(n)}; t) = C \exp \left[\int_{t_0}^{t} \sum_{i=1}^{n} a_{ii}(\xi)d\xi \right], \qquad (8.10.28)$$

where t_0 is a fixed point in J. Thus, $W(\mathbf{x}^{(1)}, \mathbf{x}^{(2)}, \ldots, \mathbf{x}^{(n)}; t_0) = Ce^0 = C$, and (8.10.28) becomes

$$W(\mathbf{x}^{(1)}, \mathbf{x}^{(2)}, \ldots, \mathbf{x}^{(n)}; t) = W(\mathbf{x}^{(1)}, \mathbf{x}^{(2)}, \ldots, \mathbf{x}^{(n)}; t_0) \exp\left[\int_{t_0}^{t} \sum_{i=1}^{n} a_{ii}(\xi)d\xi\right].$$

$$(8.10.29)$$

This result is known as *Abel's identity* or *Liouville's formula*, and is very useful.

Since the sum of the diagonal elements of a square matrix $\mathbf{A}$ is the trace of $\mathbf{A}$ [*see* Exercise 8.6.15], we may also write (8.10.29) as

$$W(\mathbf{x}^{(1)}, \mathbf{x}^{(2)}, \ldots, \mathbf{x}^{(n)}; t) = W(\mathbf{x}^{(1)}, \mathbf{x}^{(2)}, \ldots, \mathbf{x}^{(n)}; t_0) \exp\left[\int_{t_0}^{t} \operatorname{tr} \mathbf{A}(\xi)d\xi\right].$$

$$(8.10.30)$$

Exercises 8.10

1. Prove Theorem 8.10.2.
2. Prove Theorem 8.10.4.
3. Prove Theorem 8.10.6.
4. Establish (8.10.23).
5. Establish (8.10.25).
6. Let the n n-vector functions $\mathbf{x}^{(1)}, \mathbf{x}^{(2)}, \ldots, \mathbf{x}^{(n)}$ be defined on an interval J: $a \leq t \leq b$.
   (a) Let t_0 be any fixed point in J. If $W(\mathbf{x}^{(1)}, \mathbf{x}^{(2)}, \ldots, \mathbf{x}^{(n)}; t_0) \neq 0$, show that $\mathbf{x}^{(1)}, \mathbf{x}^{(2)}, \ldots, \mathbf{x}^{(n)}$ are linearly independent on J.
   (b) If $\mathbf{x}^{(1)}, \mathbf{x}^{(2)}, \ldots, \mathbf{x}^{(n)}$ are linearly dependent on J, show that $W(\mathbf{x}^{(1)}, \mathbf{x}^{(2)}, \ldots, \mathbf{x}^{(n)}; t) = 0$ for every t in J.
7. Determine whether the vector functions $\mathbf{x}^{(1)}, \mathbf{x}^{(2)}, \mathbf{x}^{(3)}, \mathbf{x}^{(4)}$ defined below are linearly independent on the interval J: $-\infty < t < \infty$.

(a)
$$\mathbf{x}^{(1)}(t) = \begin{bmatrix} 1 \\ t \\ t^2 \\ t^3 \end{bmatrix}, \qquad \mathbf{x}^{(2)}(t) = \begin{bmatrix} \sin t \\ \cos t \\ e^t - 1 \\ t \end{bmatrix},$$

$$\mathbf{x}^{(3)}(t) = \begin{bmatrix} \sin 2t \\ 0 \\ \cos 2t \\ e^{2t} - 1 \end{bmatrix}, \qquad \mathbf{x}^{(4)}(t) = \begin{bmatrix} \sin t \\ t \cos t \\ t^2 \cos 2t \\ e^t \end{bmatrix}.$$

(b)
$$\mathbf{x}^{(1)}(t) = \begin{bmatrix} 1 \\ t \\ t^2 \end{bmatrix}, \qquad \mathbf{x}^{(2)}(t) = \begin{bmatrix} t^3 \\ t^4 \\ t^5 \end{bmatrix}, \qquad \mathbf{x}^{(3)}(t) = \begin{bmatrix} t^6 \\ t^7 \\ t^8 \end{bmatrix}.$$

8. (a) Verify that the general solution $\mathbf{x} = \mathbf{x}(t)$ of the homogeneous system

$$\mathbf{x}' = \begin{bmatrix} 1 & -1 & 0 \\ 1 & 2 & 1 \\ -2 & 1 & -1 \end{bmatrix} \mathbf{x}$$

on the interval J: $-\infty < t < \infty$, is given by

$$\mathbf{x} = c_1 \mathbf{x}^{(1)}(t) + c_2 \mathbf{x}^{(2)}(t) + c_3 \mathbf{x}^{(3)}(t),$$

where c_1, c_2, c_3 are arbitrary constants and

$$\mathbf{x}^{(1)}(t) = \begin{bmatrix} -1 \\ 1 \\ 1 \end{bmatrix} e^{2t}, \qquad \mathbf{x}^{(2)}(t) = \begin{bmatrix} \frac{1}{2} \\ 1 \\ -\frac{7}{2} \end{bmatrix} e^{-t}, \qquad \mathbf{x}^{(3)}(t) = \begin{bmatrix} 1 \\ 0 \\ -1 \end{bmatrix} e^{t}.$$

(b) Using Abel's identity, show that

$$W(\mathbf{x}^{(1)}, \mathbf{x}^{(2)}, \mathbf{x}^{(3)}; t) = -3e^{2t}.$$

9. Let $\mathbf{u}^{(1)}(t), \mathbf{u}^{(2)}(t), \ldots, \mathbf{u}^{(n)}(t)$ and $\mathbf{v}^{(1)}(t), \mathbf{v}^{(2)}(t), \ldots, \mathbf{v}^{(n)}(t)$ be two fundamental sets of solutions of Equation (8.10.5) on the interval J: $a \leq t \leq b$. Show that

$$W(\mathbf{u}^{(1)}, \mathbf{u}^{(2)}, \ldots, \mathbf{u}^{(n)}; t) = k W(\mathbf{v}^{(1)}, \mathbf{v}^{(2)}, \ldots, \mathbf{v}^{(n)}; t),$$

where k is a constant different from zero.

8.11. Linear Homogeneous Systems with Constant Coefficients

Consider the homogeneous system

$$\begin{cases} x_1' = a_{11} x_1 + a_{12} x_2 + \cdots + a_{1n} x_n, \\ x_2' = a_{21} x_1 + a_{22} x_2 + \cdots + a_{2n} x_n, \\ \quad \vdots \\ x_n' = a_{n1} x_1 + a_{n2} x_2 + \cdots + a_{nn} x_n, \end{cases} \tag{8.11.1}$$

or equivalently

$$\mathbf{x}' = \mathbf{A}\mathbf{x}, \tag{8.11.2}$$

where the elements a_{ij} of the matrix $\mathbf{A}$ are constants. Reasoning as in the discussion given in Section 4.6 or 6.6, it appears plausible to seek a solution of the system (8.11.2) of the form

$$\mathbf{x} = \boldsymbol{\xi} e^{\lambda t}, \tag{8.11.3}$$

where λ and the constant vector $\boldsymbol{\xi}^T = [\xi_1, \xi_2, \ldots, \xi_n]$ are to be determined. Substituting (8.11.3) into (8.11.2), we have

$$\boldsymbol{\xi} \lambda e^{\lambda t} = \mathbf{A} \boldsymbol{\xi} e^{\lambda t}. \tag{8.11.4}$$

Since the scalar function $e^{\lambda t} \neq 0$ for all t, we may write (8.11.4) as $\mathbf{A}\boldsymbol{\xi} = \lambda\boldsymbol{\xi}$ or

$$(\mathbf{A} - \lambda\mathbf{I})\boldsymbol{\xi} = \mathbf{0}, \tag{8.11.5}$$

where $\mathbf{I}$ is an $n \times n$ identity matrix. [*See* (8.6.14).] In view of Theorem 8.9.5, a necessary and sufficient condition that the system (8.11.5) have a nontrivial solution ($\boldsymbol{\xi} \neq \mathbf{0}$) is that

$$\det [\mathbf{A} - \lambda\mathbf{I}] = 0. \tag{8.11.6}$$

Equation (8.11.6) may be written as

$$\begin{vmatrix} a_{11} - \lambda & a_{12} & \cdots & a_{1n} \\ a_{21} & a_{22} - \lambda & \cdots & a_{2n} \\ \vdots & \vdots & & \vdots \\ a_{n1} & a_{n2} & \cdots & a_{nn} - \lambda \end{vmatrix} = 0. \tag{8.11.7}$$

Equation (8.11.6) [or (8.11.7)] is usually called the *characteristic equation* or the *auxiliary equation* of the system (8.11.2) and is a polynomial equation in λ of degree n (obtained by expanding the above determinant). The polynomial p defined by

$$p(\lambda) = \det [\mathbf{A} - \lambda\mathbf{I}] \tag{8.11.8}$$

is usually called the *characteristic polynomial* of the system (8.11.2). Thus, if λ is a root of the equation $p(\lambda) = 0$, then $\mathbf{x} = \boldsymbol{\xi}e^{\lambda t}$ is a solution of the system (8.11.2). The roots $\lambda_1, \lambda_2, \ldots, \lambda_n$ of the characteristic equation $p(\lambda) = 0$ are called the *characteristic roots* or *eigenvalues* of the matrix $\mathbf{A}$. The vectors $\boldsymbol{\xi}^{(1)}$, $\boldsymbol{\xi}^{(2)}$, $\ldots$, $\boldsymbol{\xi}^{(n)}$ associated with the eigenvalues $\lambda_1, \lambda_2, \ldots, \lambda_n$ are called *characteristic vectors* or *eigenvectors* of the matrix $\mathbf{A}$, and they satisfy the system (8.11.5):

$$\mathbf{A}\boldsymbol{\xi}^{(i)} = \lambda_i \boldsymbol{\xi}^{(i)}, \qquad i = 1, 2, \ldots, n. \tag{8.11.9}$$

From (8.11.9), we see that any scalar multiple of an eigenvector is again an eigenvector of the system (8.11.2). For these special vectors $\boldsymbol{\xi}$, one may view (8.11.9) as a linear transformation mapping a vector $\boldsymbol{\xi}$ into a scalar multiple of itself, namely, $\lambda\boldsymbol{\xi}$.

The form of the solutions of the system (8.11.2) will depend on the nature of the roots of the characteristic equation $p(\lambda) = 0$. When the degree of the equation $p(\lambda) = 0$ is greater than or equal to three, the task of finding its roots is usually very difficult. For $n \geq 3$, the systems (8.11.2) will be chosen with care and thus, may not be typical of application problems.

Unless stated to the contrary, we shall take the elements a_{ij} of the matrix $\mathbf{A}$ to be real constants.

Suppose that all the roots of the characteristic equation $p(\lambda) = 0$ are real and simple. Then

$$\mathbf{x}^{(1)}(t) = \boldsymbol{\xi}^{(1)}e^{\lambda_1 t}, \; \mathbf{x}^{(2)}(t) = \boldsymbol{\xi}^{(2)}e^{\lambda_2 t}, \ldots, \mathbf{x}^{(n)}(t) = \boldsymbol{\xi}^{(n)}e^{\lambda_n t} \tag{8.11.10}$$

are solutions of the system (8.11.2).   Consider the Wronskian of these solutions on the interval $J: -\infty < t < \infty$, namely,

$$W(\mathbf{x}^{(1)}, \mathbf{x}^{(2)}, \ldots, \mathbf{x}^{(n)}; t) = \begin{vmatrix} \xi_1^{(1)}e^{\lambda_1 t} & \xi_1^{(2)}e^{\lambda_2 t} & \cdots & \xi_1^{(n)}e^{\lambda_n t} \\ \xi_2^{(1)}e^{\lambda_1 t} & \xi_2^{(2)}e^{\lambda_2 t} & \cdots & \xi_2^{(n)}e^{\lambda_n t} \\ \vdots & \vdots & & \vdots \\ \xi_n^{(1)}e^{\lambda_1 t} & \xi_n^{(2)}e^{\lambda_2 t} & \cdots & \xi_n^{(n)}e^{\lambda_n t} \end{vmatrix}$$

$$= e^{(\lambda_1 + \lambda_2 + \cdots + \lambda_n)t} \begin{vmatrix} \xi_1^{(1)} & \xi_1^{(2)} & \cdots & \xi_1^{(n)} \\ \xi_2^{(1)} & \xi_2^{(2)} & \cdots & \xi_2^{(n)} \\ \vdots & \vdots & & \vdots \\ \xi_n^{(1)} & \xi_n^{(2)} & \cdots & \xi_n^{(n)} \end{vmatrix}. \tag{8.11.11}$$

If we can show that $W(\mathbf{x}^{(1)}, \mathbf{x}^{(2)}, \ldots, \mathbf{x}^{(n)}; t_0) \neq 0$ for any fixed point t_0 in J, then in view of Theorem 8.10.6, the solutions $\mathbf{x}^{(1)}(t), \mathbf{x}^{(2)}(t), \ldots, \mathbf{x}^{(n)}(t)$ of the system (8.11.2) will be linearly independent on J.   Thus, these solutions will then form a fundamental set of the system (8.11.2).   Consequently, we will then have that the general solution $\mathbf{x} = \mathbf{x}(t)$ of the system (8.11.2) is given by

$$\mathbf{x} = c_1 \xi^{(1)}e^{\lambda_1 t} + c_2 \xi^{(2)}e^{\lambda_2 t} + \cdots + c_n \xi^{(n)}e^{\lambda_n t}, \tag{8.11.12}$$

where $c_1, c_2, \ldots, c_n$ are arbitrary constants.

In (8.11.11), $e^{(\lambda_1 + \lambda_2 + \cdots + \lambda_n)t} \neq 0$ for all t in J.   Thus, if we can show that

$$\begin{vmatrix} \xi_1^{(1)} & \xi_1^{(2)} & \cdots & \xi_1^{(n)} \\ \xi_2^{(1)} & \xi_2^{(2)} & \cdots & \xi_2^{(n)} \\ \vdots & \vdots & & \vdots \\ \xi_n^{(1)} & \xi_n^{(2)} & \cdots & \xi_n^{(n)} \end{vmatrix} \neq 0, \tag{8.11.13}$$

we will then have $W(\mathbf{x}^{(1)}, \mathbf{x}^{(2)}, \ldots, \mathbf{x}^{(n)}; t) \neq 0$ for all t in J.   To establish the result in (8.11.13), we must show that the eigenvectors $\xi^{(1)}, \xi^{(2)}, \ldots, \xi^{(n)}$ are linearly independent.   [*See* Theorem 8.8.1.]   For a method of establishing this result, see the hint to Exercise 8.11.1.

Summarizing, we have the following result.

Theorem 8.11.1.   *Consider the system* (8.11.2). *Suppose that the eigenvalues $\lambda_1, \lambda_2, \ldots, \lambda_n$ of the characteristic Equation (8.11.6) are real and simple.   Then the general solution $x = x(t)$ of the system* (8.11.2) *on the interval J: $-\infty < t < \infty$ isgiven by* (8.11.12).

Example 8.11.1.   Let us find the general solution of the system

$$\begin{cases} x_1' = \quad x_1 + \ x_2 - 2x_3, \\ x_2' = -x_1 + 2x_2 + \ x_3, \\ x_3' = \qquad \quad x_2 - \ x_3, \end{cases} \tag{8.11.14}$$

or

$$\mathbf{x}' = \begin{bmatrix} 1 & 1 & -2 \\ -1 & 2 & 1 \\ 0 & 1 & -1 \end{bmatrix} \mathbf{x}. \tag{8.11.15}$$

SOLUTION.   Observe that

$$\mathbf{A} - \lambda\mathbf{I} = \begin{bmatrix} 1 & 1 & -2 \\ -1 & 2 & 1 \\ 0 & 1 & -1 \end{bmatrix} - \lambda \begin{bmatrix} 1 & 0 & 0 \\ 0 & 1 & 0 \\ 0 & 0 & 1 \end{bmatrix}$$

$$= \begin{bmatrix} 1 & 1 & -2 \\ -1 & 2 & 1 \\ 0 & 1 & -1 \end{bmatrix} - \begin{bmatrix} \lambda & 0 & 0 \\ 0 & \lambda & 0 \\ 0 & 0 & \lambda \end{bmatrix}$$

$$= \begin{bmatrix} 1 - \lambda & 1 & -2 \\ -1 & 2 - \lambda & 1 \\ 0 & 1 & -1 - \lambda \end{bmatrix}.$$

Thus, with $\boldsymbol{\xi}^T = [\xi_1, \xi_2, \xi_3]$, Equation (8.11.5) becomes

$$\begin{bmatrix} 1 - \lambda & 1 & -2 \\ -1 & 2 - \lambda & 1 \\ 0 & 1 & -1 - \lambda \end{bmatrix} \begin{bmatrix} \xi_1 \\ \xi_2 \\ \xi_3 \end{bmatrix} = \begin{bmatrix} 0 \\ 0 \\ 0 \end{bmatrix}, \tag{8.11.16}$$

which can be written in scalar form as

$$\begin{cases} (1 - \lambda)\xi_1 + \xi_2 - 2\xi_3 = 0, \\ -\xi_1 + (2 - \lambda)\xi_2 + \xi_3 = 0, \\ \xi_2 - (1 + \lambda)\xi_3 = 0. \end{cases} \tag{8.11.17}$$

The system (8.11.15) has a nontrivial solution $[(\xi_1, \xi_2, \xi_3) \neq (0, 0, 0)]$ if and only if

$$\begin{vmatrix} 1 - \lambda & 1 & -2 \\ -1 & 2 - \lambda & 1 \\ 0 & 1 & -(1 + \lambda) \end{vmatrix} = 0. \tag{8.11.18}$$

Note that (8.11.18) may be obtained directly from (8.11.7).   From (8.11.18), we obtain, upon expanding the determinant, $-\lambda^3 + 2\lambda^2 + \lambda - 2 = 0$ or

$$\lambda^3 - 2\lambda^2 - \lambda + 2 = 0. \tag{8.11.19}$$

Utilizing synthetic division, the characteristic roots (eigenvalues) of the characteristic Equation (8.11.19) are found to be $\lambda_1 = -1$, $\lambda_2 = 1$, and $\lambda_3 = 2$.

Substituting $\lambda = \lambda_1 = -1$ into (8.11.17), we have

$$\begin{cases} 2\xi_1 + \xi_2 - 2\xi_3 = 0, \\ -\xi_1 + 3\xi_2 + \xi_3 = 0, \\ \xi_2 = 0. \end{cases} \tag{8.11.20}$$

Solving the system (8.11.20), we obtain $\xi_2 = 0$ and $\xi_3 = \xi_1$, where ξ_1 is arbitrary. Thus, $\xi^T = [\xi_1, 0, \xi_1] = \xi_1[1, 0, 1]$. A corresponding solution of the system (8.11.15) is given by

$$\mathbf{x}^{(1)}(t) = \xi e^{\lambda_1 t} = \begin{bmatrix} 1 \\ 0 \\ 1 \end{bmatrix} e^{-t}. \tag{8.11.21}$$

Substituting $\lambda = \lambda_2 = 1$ into (8.11.17), we have

$$\begin{cases} \xi_2 - 2\xi_3 = 0, \\ -\xi_1 + \xi_2 + \xi_3 = 0, \\ \xi_2 - 2\xi_3 = 0. \end{cases} \tag{8.11.22}$$

Solving the system (8.11.22), we obtain $\xi_1 = 3\xi_3$ and $\xi_2 = 2\xi_3$, where ξ_3 is arbitrary. Thus, $\xi^T = [3\xi_3, 2\xi_3, \xi_3] = \xi_3[3, 2, 1]$. A second solution of the system (8.11.15) is given by

$$\mathbf{x}^{(2)}(t) = \xi e^{\lambda_2 t} = \begin{bmatrix} 3 \\ 2 \\ 1 \end{bmatrix} e^{t}. \tag{8.11.23}$$

Similarly, substituting $\lambda = \lambda_3 = 2$ into (8.11.17), we find that $\xi^T = \xi_3[1, 3, 1]$ where ξ_3 is arbitrary. A third solution of the system (8.11.15) is given by

$$\mathbf{x}^{(3)}(t) = \xi e^{\lambda_3 t} = \begin{bmatrix} 1 \\ 3 \\ 1 \end{bmatrix} e^{2t}. \tag{8.11.24}$$

In view of (8.11.21), (8.11.23), and (8.11.24), the general solution of the system (8.11.15) on the interval $J: -\infty < t < \infty$ is given by

$$\mathbf{x} = c_1 \begin{bmatrix} 1 \\ 0 \\ 1 \end{bmatrix} e^{-t} + c_2 \begin{bmatrix} 3 \\ 2 \\ 1 \end{bmatrix} e^{t} + c_3 \begin{bmatrix} 1 \\ 3 \\ 1 \end{bmatrix} e^{2t}, \tag{8.11.25}$$

where c_1, c_2, and c_3 are arbitrary constants.

Remark 8.11.1. If all the roots of the characteristic equation $p(\lambda) = 0$ are simple, but some are imaginary, we follow a similar procedure as given above. Since the coefficients of the equation $p(\lambda) = 0$ are real, if $\lambda_j = \alpha_j + i\beta_j$ are roots of $p(\lambda) = 0$, so are $\lambda_j = \alpha_j - i\beta_j$. To obtain the real solutions when the eigenvalues are imaginary, we may utilize Euler's formula (4.6.40). Another method will be given in Section 8.13.

Example 8.11.2.   Let us find the general solution of the system

$$\mathbf{x}' = \begin{bmatrix} 1 & 2 \\ -2 & 1 \end{bmatrix} \mathbf{x}. \tag{8.11.26}$$

SOLUTION. $\mathbf{A} - \lambda \mathbf{I} = \begin{bmatrix} 1 & 2 \\ -2 & 1 \end{bmatrix} - \begin{bmatrix} \lambda & 0 \\ 0 & \lambda \end{bmatrix} = \begin{bmatrix} 1 - \lambda & 2 \\ -2 & 1 - \lambda \end{bmatrix}.$

With $\boldsymbol{\xi}^T = [\xi_1, \xi_2]$, $(\mathbf{A} - \lambda \mathbf{I})\, \boldsymbol{\xi} = \mathbf{0}$ can be expressed as

$$\begin{cases} (1 - \lambda)\xi_1 + \quad\quad 2\xi_2 = 0, \\ \quad\quad -2\xi_1 + (1 - \lambda)\xi_2 = 0. \end{cases} \tag{8.11.27}$$

The system (8.11.27) has a nontrivial solution if and only if

$$\begin{vmatrix} 1 - \lambda & 2 \\ -2 & 1 - \lambda \end{vmatrix} = \lambda^2 - 2\lambda + 5 = 0.$$

Thus, the characteristic roots are found to be $\lambda_1 = 1 + 2i$, $\lambda_2 = 1 - 2i$. Substituting $\lambda = \lambda_1 = 1 + 2i$ into (8.11.27), we have

$$\begin{cases} -2i\xi_1 + \quad 2\xi_2 = 0, \\ -2\xi_1 - 2i\xi_2 = 0. \end{cases} \tag{8.11.28}$$

Solving the system (8.11.28), we obtain $\xi_2 = i\xi_1$, where ξ_1 is arbitrary. Thus, $\boldsymbol{\xi}^T = [\xi_1, i\xi_1] = \xi_1[1, i]$. A solution of the system (8.11.26) is given by

$$\mathbf{x}^{(1)}(t) = \boldsymbol{\xi} e^{\lambda_1 t} = \begin{bmatrix} 1 \\ i \end{bmatrix} e^{(1 + 2i)t}. \tag{8.11.29}$$

Substituting $\lambda = \lambda_2 = 1 - 2i$ into (8.11.27), we find that $\boldsymbol{\xi}^T = \xi_1[1, -i]$, where ξ_1 is arbitrary. Note that the eigenvector associated with the eigenvalue λ_2 is the complex conjugate of the eigenvector associated with the eigenvalue λ_1. More generally, if $\mathbf{A}$ is a real matrix and λ_1, λ_2 are a conjugate pair of eigenvalues of $\mathbf{A}$, then their corresponding eigenvectors are likewise conjugate up to a constant multiple. [*See* Exercise 8.11.2.]   A second solution of the system (8.11.26) is given by

$$\mathbf{x}^{(2)}(t) = \boldsymbol{\xi} e^{\lambda_2 t} = \begin{bmatrix} 1 \\ -i \end{bmatrix} e^{(1 - 2i)t}. \tag{8.11.30}$$

Thus, the general solution of the system (8.11.26) is

$$\mathbf{x} = C_1 \begin{bmatrix} 1 \\ i \end{bmatrix} e^{(1 + 2i)t} + C_2 \begin{bmatrix} 1 \\ -i \end{bmatrix} e^{(1 - 2i)t}, \tag{8.11.31}$$

where C_1 and C_2 are arbitrary constants.   The solution in (8.11.31) is complex-valued.   We shall render it real-valued by utilizing Euler's formula.   We may write (8.11.31) as

$$x = e^t C_1 \begin{bmatrix} 1 \\ i \end{bmatrix} (\cos 2t + i \sin 2t) + e^t C_2 \begin{bmatrix} 1 \\ -i \end{bmatrix} (\cos 2t - i \sin 2t)$$

$$= e^t \begin{bmatrix} C_1 \cos 2t + iC_1 \sin 2t \\ -C_1 \sin 2t + iC_1 \cos 2t \end{bmatrix} + e^t \begin{bmatrix} C_2 \cos 2t - iC_2 \sin 2t \\ -C_2 \sin 2t - iC_2 \cos 2t \end{bmatrix}$$

$$= e^t \begin{bmatrix} (C_1 + C_2) \cos 2t \\ -(C_1 + C_2) \sin 2t \end{bmatrix} + e^t \begin{bmatrix} i(C_1 - C_2) \sin 2t \\ i(C_1 - C_2) \cos 2t \end{bmatrix}.$$

Letting $c_1 = C_1 + C_2$ and $c_2 = i(C_1 - C_2)$ be two new arbitrary constants, the above expression becomes

$$\mathbf{x} = e^t \begin{bmatrix} c_1 \cos 2t \\ -c_1 \sin 2t \end{bmatrix} + e^t \begin{bmatrix} c_2 \sin 2t \\ c_2 \cos 2t \end{bmatrix},$$

or

$$\mathbf{x} = c_1 \begin{bmatrix} \cos 2t \\ -\sin 2t \end{bmatrix} e^t + c_2 \begin{bmatrix} \sin 2t \\ \cos 2t \end{bmatrix} e^t, \qquad (8.11.32)$$

which is the desired real-valued solution of the system (8.11.26).

Remark 8.11.2.   The case when the characteristic equation $p(\lambda) = 0$ has repeated roots will be discussed in the next section.

Remark 8.11.3.   The determination of the characteristic equation of an $n \times n$ matrix $\mathbf{A}$ for $n \geq 3$ by expanding the determinant in (8.11.7) may become cumbersome.   We shall give below a useful recursive formula for determining the coefficients of the characteristic equation.   [*See* (8.11.45).] Computer programs are available that use such recursive formulas for finding the characteristic equation.   For programming purposes it is more expedient to consider the characteristic polynomial $P(\lambda) = \det [\lambda \mathbf{I} - \mathbf{A}]$ rather than $p(\lambda) = \det [\mathbf{A} - \lambda \mathbf{I}]$.   However, since $p(\lambda) = (-1)^n P(\lambda)$, the characteristic equations $p(\lambda) = 0$ and $P(\lambda) = 0$ have the same characteristic roots.

Since $P(\lambda)$ is a polynomial of degree n in λ with coefficient of λ^n equal 1, we may express it as

$$P(\lambda) = \det [\lambda \mathbf{I} - \mathbf{A}] = \lambda^n + a_1 \lambda^{n-1} + a_2 \lambda^{n-2} + \cdots + a_{n-1} \lambda + a_n. \qquad (8.11.33)$$

As we shall presently see, the coefficients a_1 and a_n have special significance. Let $\lambda_1, \lambda_2, \ldots, \lambda_n$ be the characteristic roots of $P(\lambda) = 0$.   We may then write (8.11.33) as

$$P(\lambda) = (\lambda - \lambda_1)(\lambda - \lambda_2) \cdots (\lambda - \lambda_n). \qquad (8.11.34)$$

Substituting $\lambda = 0$ into (8.11.33), we obtain

$$P(0) = \det(-\mathbf{A}) = a_n. \tag{8.11.35}$$

But, $\det(-\mathbf{A}) = (-1)^n \det \mathbf{A}$.   Thus

$$a_n = (-1)^n \det \mathbf{A}. \tag{8.11.36}$$

Substituting $\lambda = 0$ into (8.11.34), we find

$$P(0) = (-1)^n \lambda_1 \lambda_2 \cdots \lambda_n. \tag{8.11.37}$$

In view of (8.11.35) through (8.11.37), we see that

$$\det \mathbf{A} = \lambda_1 \lambda_2 \cdots \lambda_n. \tag{8.11.38}$$

Thus, $\det \mathbf{A}$ is equal to the product of the characteristic roots of the matrix $\mathbf{A}$.
Expanding $P(\lambda)$ of (8.11.34), we obtain

$$P(\lambda) = \lambda^n - (\lambda_1 + \lambda_2 + \cdots + \lambda_n)\lambda^{n-1} + \cdots + (-1)^n \lambda_1 \lambda_2 \cdots \lambda_n. \tag{8.11.39}$$

From (8.11.33) and (8.11.39), the coefficient of λ^{n-1} is given by

$$a_1 = -(\lambda_1 + \lambda_2 + \cdots + \lambda_n). \tag{8.11.40}$$

Now, expanding $P(\lambda) = \det[\lambda \mathbf{I} - \mathbf{A}]$ so as to obtain the coefficient of λ^{n-1},
we get

$$P(\lambda) = \lambda^n - (a_{11} + a_{22} + \cdots + a_{nn})\lambda^{n-1} + \cdots. \tag{8.11.41}$$

From (8.11.33) and (8.11.41), the coefficient of λ^{n-1} is also given by

$$a_1 = -(a_{11} + a_{22} + \cdots + a_{nn}). \tag{8.11.42}$$

But $a_{11} + a_{22} + \cdots + a_{nn} = \operatorname{tr} \mathbf{A}$.   [*See* Exercise 8.6.15.]   Consequently, in
view of (8.11.40) and (8.11.42), we have

$$a_1 = -\operatorname{tr} \mathbf{A}, \tag{8.11.43}$$

and

$$\lambda_1 + \lambda_2 + \cdots + \lambda_n = \operatorname{tr} \mathbf{A} = a_{11} + a_{22} + \cdots + a_{nn}. \tag{8.11.44}$$

If we denote the traces of $\mathbf{A}, \mathbf{A}^2, \ldots, \mathbf{A}^n$ by $T_1, T_2, \ldots, T_n$, respectively, then
a *recursive formula* for expressing the coefficients of the characteristic equation
$P(\lambda) = 0$ in terms of the various T_n is given by

$$\begin{cases} a_1 = -T_1, \\ a_2 = -\frac{1}{2}(a_1 T_1 + T_2), \\ a_3 = -\frac{1}{3}(a_2 T_1 + a_1 T_2 + T_3), \\ \quad \vdots \\ a_n = -\frac{1}{n}(a_{n-1} T_1 + a_{n-2} T_2 + \cdots + a_1 T_{n-1} + T_n). \end{cases} \tag{8.11.45}$$

[*See* [56], pp. 303–305, for a proof of this result.]

Example 8.11.3. Let us find the characteristic equation of the matrix $\mathbf{A}$ given in (8.11.15) by utilizing the recursive Formula (8.11.45).

SOLUTION. Since $\mathbf{A} = \begin{bmatrix} 1 & 1 & -2 \\ -1 & 2 & 1 \\ 0 & 1 & -1 \end{bmatrix}$, the reader may verify that $\mathbf{A}^2 = \begin{bmatrix} 0 & 1 & 1 \\ -3 & 4 & 3 \\ -1 & 1 & 2 \end{bmatrix}$ and $\mathbf{A}^3 = \begin{bmatrix} -1 & 3 & 0 \\ -7 & 8 & 7 \\ -2 & 3 & 1 \end{bmatrix}$. Thus, $T_1 = \mathrm{Tr}\,\mathbf{A} = 1 + 2 - 1 = 2$, $T_2 = \mathrm{Tr}\,\mathbf{A}^2 = 0 + 4 + 2 = 6$, and $T_3 = \mathrm{Tr}\,\mathbf{A}^3 = -1 + 8 + 1 = 8$. Utilizing (8.11.45), we obtain

$$\begin{cases} a_1 = -T_1 = -2, \\ a_2 = -\tfrac{1}{2}(a_1 T_1 + T_2) = -\tfrac{1}{2}(-2 \cdot 2 + 6) = -1, \\ a_3 = -\tfrac{1}{3}(a_2 T_1 + a_1 T_2 + T_3) = -\tfrac{1}{3}(-1 \cdot 2 - 2 \cdot 6 + 8) = 2. \end{cases}$$

Substituting these values into the characteristic equation $\lambda^3 + a_1 \lambda^2 + a_2 \lambda + a_3 = 0$, we obtain

$$\lambda^3 - 2\lambda^2 - \lambda + 2 = 0,$$

which agrees with the characteristic equation found in (8.11.19).

Since $p(\lambda) = (-1)^n P(\lambda)$, utilizing (8.11.45) we leave it as an exercise for the reader [*see* Exercise 8.11.25] to show that, if we take for the characteristic polynomial

$$p(\lambda) = \det\,[\mathbf{A} - \lambda\mathbf{I}] = a_0 \lambda^n + a_1 \lambda^{n-1} + a_2 \lambda^{n-2} + \cdots + a_{n-1}\lambda + a_n, \quad (8.11.46)$$

instead of $P(\lambda)$, a recursive formula for expressing the coefficients of the characteristic equation $p(\lambda) = 0$, in terms of the various T_n, is given by

$$\begin{cases} a_0 = (-1)^n, \\ a_1 = -(-1)^n T_1, \\ a_2 = -\tfrac{1}{2}[a_1 T_1 + (-1)^n T_2], \\ a_3 = -\tfrac{1}{3}[a_2 T_1 + a_1 T_2 + (-1)^n T_3], \\ \quad \vdots \\ a_n = -\dfrac{1}{n}[a_{n-1}T_1 + a_{n-2}T_2 + \cdots + a_1 T_{n-1} + (-1)^n T_n]. \end{cases} \quad (8.11.47)$$

Due to the presence of the term $(-1)^n$ in (8.11.47), the programming of the recursive Formula (8.11.45) is preferred.

Exercises 8.11

1. Show that the eigenvectors $\xi^{(1)}$, $\xi^{(2)}, \ldots, \xi^{(n)}$ associated with the simple eigenvalues λ_1, $\lambda_2, \ldots, \lambda_n$ of the system (8.11.2) are linearly independent.
2. (a) If the matrices A and B are conformable for multiplication, show that

$$\overline{AB} = \overline{A}\,\overline{B}.$$

(b) If A is a real matrix and λ_1, λ_2 are a complex conjugate pair of eigenvalues of A, show that the corresponding eigenvectors $\xi^{(1)}$, $\xi^{(2)}$ satisfy

$$\xi^{(1)} = k\xi^{(2)},$$

where k is a scalar.

In Exercises 3 through 12, find the general solution of each system.

3. $\mathbf{x}' = \begin{bmatrix} -2 & -8 & -12 \\ 1 & 4 & 4 \\ 0 & 0 & 1 \end{bmatrix} \mathbf{x}.$

4. $\mathbf{x}' = \begin{bmatrix} 1 & 1 & 2 \\ 0 & 2 & 2 \\ -1 & 1 & 3 \end{bmatrix} \mathbf{x}$

5. $\mathbf{x}' = \begin{bmatrix} 2 & 1 & -2 \\ 0 & 3 & -2 \\ 3 & 1 & -3 \end{bmatrix} \mathbf{x}.$

6. $\mathbf{x}' = \begin{bmatrix} 3 & -1 & 1 \\ -1 & 5 & -1 \\ 1 & -1 & 3 \end{bmatrix} \mathbf{x}.$

7. $\mathbf{x}' = \begin{bmatrix} 1 & 1 & -1 \\ 1 & -1 & 1 \\ -1 & 1 & 1 \end{bmatrix} \mathbf{x}.$

8. $\mathbf{x}' = \begin{bmatrix} 1 & 1 & 8 \\ 0 & 3 & 0 \\ 2 & 1 & 1 \end{bmatrix} \mathbf{x}.$

9. $\mathbf{x}' = \begin{bmatrix} 1 & 1 & 1 \\ -1 & -1 & 0 \\ -1 & 0 & -1 \end{bmatrix} \mathbf{x}.$

10. $\mathbf{x}' = \begin{bmatrix} -5 & 5 & 2 \\ -10 & 5 & 4 \\ -20 & 10 & 9 \end{bmatrix} \mathbf{x}.$

11. $\mathbf{x}' = \begin{bmatrix} 1 & -1 & 1 & -1 \\ 0 & -1 & 2 & -2 \\ 0 & 0 & 2 & -3 \\ 0 & 0 & 0 & -2 \end{bmatrix} \mathbf{x}.$

12. $\mathbf{x}' = \begin{bmatrix} \alpha_1 & 0 & 0 & \cdots & 0 \\ 0 & \alpha_2 & 0 & \cdots & 0 \\ 0 & 0 & \alpha_3 & \cdots & 0 \\ \vdots & \vdots & \vdots & & \vdots \\ 0 & 0 & 0 & \cdots & \alpha_n \end{bmatrix} \mathbf{x},$ where α_i, $i = 1, \ldots, n$, are distinct numbers.

[*Note:* The elements α_i along the diagonal are the eigenvalues of the given diagonal matrix.]

In Exercises 13 through 16, solve each initial-value problem.

13. $\mathbf{x}' = \begin{bmatrix} 1 & 2 \\ 4 & 3 \end{bmatrix} \mathbf{x}, \quad \mathbf{x}(0) = \begin{bmatrix} 1 \\ 5 \end{bmatrix}.$ 14. $\mathbf{x}' = \begin{bmatrix} 2 & 1 \\ -1 & 2 \end{bmatrix} \mathbf{x}, \quad \mathbf{x}(0) = \begin{bmatrix} 2 \\ -3 \end{bmatrix}.$

15. $\mathbf{x}' = \begin{bmatrix} -1 & 1 & 1 \\ 1 & -1 & 1 \\ 1 & 1 & 1 \end{bmatrix} \mathbf{x}, \quad \mathbf{x}(0) = \begin{bmatrix} 1 \\ 5 \\ 3 \end{bmatrix}.$

16. $\mathbf{x}' = \begin{bmatrix} 2 & -1 & 1 \\ 1 & 0 & 1 \\ -2 & 0 & -1 \end{bmatrix} \mathbf{x}, \quad \mathbf{x}(0) = \begin{bmatrix} -1 \\ 2 \\ 7 \end{bmatrix}.$

17. Show that the eigenvalues of a triangular matrix:

$$\begin{bmatrix} a_{11} & a_{12} & a_{13} & \cdots & a_{1n} \\ 0 & a_{22} & a_{23} & \cdots & a_{2n} \\ 0 & 0 & a_{33} & \cdots & a_{3n} \\ \vdots & \vdots & \vdots & & \vdots \\ 0 & 0 & 0 & \cdots & a_{nn} \end{bmatrix}$$

are simply the diagonal elements of the matrix.

18. Suppose that $\lambda_1, \lambda_2, \ldots, \lambda_n$ are eigenvalues of an n-square matrix $\mathbf{A}$ and k ($\neq 0$) is a scalar.
    (a) Show that $k\lambda_1, k\lambda_2, \ldots, k\lambda_n$ are eigenvalues of the matrix $k\mathbf{A}$.
    (b) Show that $\lambda_1 - k, \lambda_2 - k, \ldots, \lambda_n - k$ are eigenvalues of the matrix $\mathbf{A} - k\mathbf{I}$.

19. Suppose that $\mathbf{A}$ and $\mathbf{B}$ are $n \times n$ matrices and $\mathbf{A}$ is nonsingular.
    (a) Show that $\mathbf{A}^{-1}\mathbf{B}$ and $\mathbf{B}\mathbf{A}^{-1}$ have the same eigenvalues.
    (b) Show that $\mathbf{B}$ and $\mathbf{A}^{-1}\mathbf{B}\mathbf{A}$ have the same eigenvalues.

20. Show that $\lambda = 0$ is an eigenvalue of an n-square matrix $\mathbf{A}$ if and only if $\mathbf{A}$ is singular.

21. A vector $\mathbf{x}$ such that the inner product $(\mathbf{x}, \mathbf{x}) = \mathbf{x}^*\mathbf{x} = 1$ is called a *unit vector*. If $\mathbf{x}$ is a unit vector such that $\mathbf{A}\mathbf{x} = \lambda\mathbf{x}$, show that $\mathbf{x}^*\mathbf{A}\mathbf{x} = \lambda$.

22. Let $\mathbf{A}$ be an $n \times n$ Hermitian matrix, that is, $\mathbf{A}^* = \bar{\mathbf{A}}^T = \mathbf{A}$, and let $\mathbf{x}$ and $\mathbf{y}$ be n-dimensional column vectors.
    (a) Show that det $\mathbf{A}$ is real.
    (b) Show that $(\mathbf{A}\mathbf{x}, \mathbf{y}) = (\mathbf{x}, \mathbf{A}\mathbf{y})$.
       Let $\boldsymbol{\xi}$ be an eigenvector associated with the eigenvalue λ of the matrix $\mathbf{A}$.
    (c) Show that $\bar{\lambda} = \lambda$.
       Thus, the *eigenvalues of a Hermitian matrix are real*. Since a real symmetric matrix is a special case of a Hermitian matrix, it follows that the *eigenvalues of a real symmetric matrix are also real*.
       Two vectors $\mathbf{x}^{(1)}, \mathbf{x}^{(2)}$ are said to be *orthogonal* if their inner product is zero, that is, $(\mathbf{x}^{(1)}, \mathbf{x}^{(2)}) = 0$.
    (d) Show that the eigenvectors $\boldsymbol{\xi}^{(1)}$ and $\boldsymbol{\xi}^{(2)}$ associated with the simple eigenvalues of the matrix $\mathbf{A}$ are orthogonal.
    (e) Show that $\mathbf{x}\mathbf{y}^* + \mathbf{y}\mathbf{x}^*$ is Hermitian.
    (f) Is the sum of two conformable Hermitian matrices a Hermitian matrix?

23. Find the general solution of the system

$$\mathbf{x}' = \begin{bmatrix} 1 & 0 & 0 \\ 0 & 0 & 3 - 4i \\ 0 & 3 + 4i & 0 \end{bmatrix} \mathbf{x}, \qquad i^2 = -1.$$

24. Find the general solution of the system

$$\mathbf{x}' = \begin{bmatrix} 1 & 1 & -i \\ 1 & 0 & -1 \\ i & -1 & 2 \end{bmatrix} \mathbf{x}, \qquad i^2 = -1.$$

25. Establish the recursive Formula (8.11.47).
26. Let $\mathbf{A}$ and $\mathbf{B}$ be two $n \times n$ matrices.
    (a) Show that $\mathbf{A}$ and $\mathbf{A}^T$ have the same characteristic polynomial.
    (b) Show that $\mathbf{AB}$ and $\mathbf{BA}$ have the same characteristic polynomial.
27. Two $n \times n$ matrices $\mathbf{A}$ and $\mathbf{B}$ are *similar* if there exists a nonsingular matrix $\mathbf{P}$ such that $\mathbf{A} = \mathbf{P}^{-1}\mathbf{BP}$. Show that similar matrices have the same characteristic polynomial.

8.12. Repeated Roots

When the characteristic equation $p(\lambda) = 0$ of the system (8.11.2) has repeated roots (eigenvalues), the theory for finding the linearly independent solutions requires advanced knowledge of linear algebra. Thus, the reader may either solve the system (8.11.2) by the method given in Section 8.4, which works regardless of whether or not the roots of $p(\lambda) = 0$ are repeated, or else accept the procedure given below. The reader is urged to review Remark 6.6.2.

In view of (6.6.12), if $\lambda = r$ is a root of multiplicity s of the characteristic equation $p(\lambda) = 0$ of the system (8.11.2), then $p(r) = 0$, $p'(r) = 0$, $\ldots, p^{(s-1)}(r) = 0$ and $p^{(s)}(r) \neq 0$. Now, if the eigenvalue r gives rise to s linearly independent eigenvectors, $\xi^{(1)}$, $\xi^{(2)}$, $\ldots$, $\xi^{(s)}$, then $\mathbf{x}^{(1)}(t) = \xi^{(1)}e^{rt}$, $\mathbf{x}^{(2)}(t) = \xi^{(2)}e^{rt}$, $\ldots$, $\mathbf{x}^{(s)}(t) = \xi^{(s)}e^{rt}$ are s linearly independent solutions of the system (8.11.2). For this case, we then have the desired number of linearly independent solutions of (8.11.2) of the form ξe^{rt}. However, if the eigenvalue r does not give rise to s linearly independent eigenvectors, then there will be less than s linearly independent solutions of the form ξe^{rt}. The situation now is somewhat similar to the results given in Theorems 6.6.2 and 6.6.3 for the repeated roots.

It is known that if $\lambda = r$ is a root of multiplicity s of the characteristic equation $p(\lambda) = 0$ of the system (8.11.2), then the rank of the matrix $\mathbf{A} - r\mathbf{I}$ is not less than $n - s$. [*See* Exercise 8.12.11 for a special case.] Also, the number q of linearly independent eigenvectors associated with the eigenvalue r is not greater than s. That is,

$$1 \leqq q \leqq s. \tag{8.12.1}$$

Moreover, if only one eigenvector is associated with the eigenvalue r, the sought s linearly independent solutions $\mathbf{x}^{(1)}(t)$, $\mathbf{x}^{(2)}(t)$, ..., $\mathbf{x}^{(s)}(t)$, are of the form

$$\mathbf{x}^{(1)}(t) = \boldsymbol{\xi} e^{rt}, \tag{8.12.2}$$

where $\boldsymbol{\xi}$ is determined by the equation

$$(\mathbf{A} - r\mathbf{I})\boldsymbol{\xi} = \mathbf{0}. \tag{8.12.3}$$

$$\mathbf{x}^{(2)}(t) = \boldsymbol{\eta}^{(1)} t e^{rt} + \boldsymbol{\eta}^{(2)} e^{rt}, \tag{8.12.4}$$

where $\boldsymbol{\eta}^{(1)}$ and $\boldsymbol{\eta}^{(2)}$ are determined by the equations

$$\begin{cases} (\mathbf{A} - r\mathbf{I})\boldsymbol{\eta}^{(1)} = \mathbf{0}, \\ (\mathbf{A} - r\mathbf{I})\boldsymbol{\eta}^{(2)} = \boldsymbol{\eta}^{(1)}. \end{cases} \tag{8.12.5}$$

$$\vdots$$

$$\mathbf{x}^{(s)}(t) = \boldsymbol{\zeta}^{(1)} \frac{t^{s-1}}{(s-1)!} e^{rt} + \boldsymbol{\zeta}^{(2)} \frac{t^{s-2}}{(s-2)!} e^{rt} + \cdots + \boldsymbol{\zeta}^{(s-1)} t e^{rt} + \boldsymbol{\zeta}^{(s)} e^{rt}, \tag{8.12.6}$$

where $\boldsymbol{\zeta}^{(k)}$, $k = 1, \ldots, s$, are determined by the equations

$$\begin{cases} (\mathbf{A} - r\mathbf{I})\boldsymbol{\zeta}^{(1)} = \mathbf{0}, \\ (\mathbf{A} - r\mathbf{I})\boldsymbol{\zeta}^{(2)} = \boldsymbol{\zeta}^{(1)}, \\ (\mathbf{A} - r\mathbf{I})\boldsymbol{\zeta}^{(3)} = \boldsymbol{\zeta}^{(2)}, \\ \qquad\qquad \vdots \\ (\mathbf{A} - r\mathbf{I})\boldsymbol{\zeta}^{(s)} = \boldsymbol{\zeta}^{(s-1)}. \end{cases} \tag{8.12.7}$$

Notice in the first system in (8.12.7) that $\det (\mathbf{A} - r\mathbf{I}) = 0$, since we have at least one nontrivial solution of the form (8.12.2). Therefore, the remaining nonhomogeneous systems in (8.12.7) cannot be solved by Cramer's rule. Evidently, consistency of these systems requires that [*see* Theorem 8.9.1]

$$\text{rank } (\mathbf{A} - r\mathbf{I}) = \text{rank } (\mathbf{A} - r\mathbf{I} \quad \boldsymbol{\zeta}^{(k)}), \qquad k = 1, 2, \ldots, s - 1. \tag{8.12.10}$$

We might point out once again that a systematic and rigorous treatment of multiple roots is farther than we care to go in a brief discussion. Fortunately, as it turns out, one can, upon solving $(\mathbf{A} - r\mathbf{I})\boldsymbol{\xi} = \mathbf{0}$, determine the linearly independent forms by the number of arbitrary constants in the vector $\boldsymbol{\xi}$. We can best illustrate this by a few examples.

Remark 8.12.1. Before proceeding with the examples, observe that the first system in (8.12.5) is equivalent to the system (8.12.3). Thus, we may take $\boldsymbol{\eta}^{(1)} = \boldsymbol{\xi}$, and the second system in (8.12.5) now becomes

$$(\mathbf{A} - r\mathbf{I})\boldsymbol{\eta}^{(2)} = \boldsymbol{\xi}. \tag{8.12.8}$$

Since $\boldsymbol{\xi}$ is known from (8.12.3), $\boldsymbol{\eta}^{(2)}$ may now be determined by (8.12.8). Also, since the first system in (8.12.7) is equivalent to the system (8.12.3), with $\boldsymbol{\zeta}^{(1)} = \boldsymbol{\xi}$, the second system in (8.12.7) becomes

$$(\mathbf{A} - r\mathbf{I})\boldsymbol{\zeta}^{(2)} = \boldsymbol{\xi}. \tag{8.12.9}$$

In view of (8.12.8) and (8.12.9), we may take $\boldsymbol{\zeta}^{(2)} = \boldsymbol{\eta}^{(2)}$, and $\boldsymbol{\zeta}^{(3)}$ may now be determined from the third system in (8.12.7). Continuing in this fashion, we may determine $\boldsymbol{\zeta}^{(4)}, \ldots, \boldsymbol{\zeta}^{(s)}$ in (8.12.7).

Example 8.12.1. Let us find the general solution of the system

$$\mathbf{x}' = \begin{bmatrix} 1 & 0 & 0 \\ 0 & 1 & 0 \\ 0 & 0 & 1 \end{bmatrix} \mathbf{x}. \tag{8.12.11}$$

SOLUTION. $\mathbf{A} - \lambda\mathbf{I} = \begin{bmatrix} 1-\lambda & 0 & 0 \\ 0 & 1-\lambda & 0 \\ 0 & 0 & 1-\lambda \end{bmatrix}$.   With $\boldsymbol{\xi}^T = [\xi_1, \, \xi_2, \, \xi_3]$,

$(\mathbf{A} - \lambda\mathbf{I})\boldsymbol{\xi} = \mathbf{0}$ can be expressed as

$$\begin{cases} (1-\lambda)\xi_1 & = 0, \\ (1-\lambda)\xi_2 & = 0, \\ (1-\lambda)\xi_3 = 0. \end{cases} \tag{8.12.12}$$

The system (8.12.12) has a nontrivial solution if and only if

$$\begin{vmatrix} 1-\lambda & 0 & 0 \\ 0 & 1-\lambda & 0 \\ 0 & 0 & 1-\lambda \end{vmatrix} = (1-\lambda)^3 = 0.$$

Thus, $\lambda = 1$ is a root of multiplicity three of the system (8.12.11). Substituting $\lambda = 1$ into (8.12.12), we have

$$\begin{cases} 0 \cdot \xi_1 = 0, \\ 0 \cdot \xi_2 = 0, \\ 0 \cdot \xi_3 = 0. \end{cases} \tag{8.12.13}$$

Hence, ξ_1, ξ_2, and ξ_3 are all arbitrary. We therefore have three linearly independent solutions of the form $\mathbf{x} = \boldsymbol{\xi}e^t$. All we need to do is to choose *any* three column rectors $\boldsymbol{\xi}^{(1)}$, $\boldsymbol{\xi}^{(2)}$, and $\boldsymbol{\xi}^{(3)}$ with determinant $|\boldsymbol{\xi}^{(1)} \, \boldsymbol{\xi}^{(2)} \, \boldsymbol{\xi}^{(3)}|$

different from zero.   Thus, since $\begin{vmatrix} 1 & 0 & 0 \\ 0 & 1 & 0 \\ 0 & 0 & 1 \end{vmatrix} \neq 0$, we have

$$\mathbf{x}^{(1)}(t) = \begin{bmatrix} 1 \\ 0 \\ 0 \end{bmatrix} e^t, \qquad \mathbf{x}^{(2)}(t) = \begin{bmatrix} 0 \\ 1 \\ 0 \end{bmatrix} e^t, \qquad \mathbf{x}^{(3)}(t) = \begin{bmatrix} 0 \\ 0 \\ 1 \end{bmatrix} e^t. \tag{8.12.14}$$

The general solution of the system (8.12.11) on the interval J: $-\infty < t < \infty$ is given by

$$\mathbf{x} = c_1 \mathbf{x}^{(1)}(t) + c_2 \mathbf{x}^{(2)}(t) + c_3 \mathbf{x}^{(3)}(t), \qquad (8.12.15)$$

where $\mathbf{x}^{(1)}(t)$, $\mathbf{x}^{(2)}(t)$, $\mathbf{x}^{(3)}(t)$ are given in (8.12.14), and c_1, c_2, c_3 are arbitrary constants.

Example 8.12.2.   Let us find the general solution of the system

$$\mathbf{x}' = \begin{bmatrix} 3 & 0 & 0 \\ 0 & 7 & 5 \\ 0 & -4 & -2 \end{bmatrix} \mathbf{x}. \qquad (8.12.16)$$

SOLUTION. $\mathbf{A} - \lambda \mathbf{I} = \begin{bmatrix} 3 - \lambda & 0 & 0 \\ 0 & 7 - \lambda & 5 \\ 0 & -4 & -2 - \lambda \end{bmatrix}.$

With $\boldsymbol{\xi}^T = [\xi_1, \xi_2, \xi_3]$, $(\mathbf{A} - \lambda \mathbf{I})\boldsymbol{\xi} = \mathbf{0}$ becomes

$$\begin{cases} (3 - \lambda)\xi_1 & = 0, \\ (7 - \lambda)\xi_2 + \quad 5\xi_3 = 0, \\ -4\xi_2 - (2 + \lambda)\xi_3 = 0. \end{cases} \qquad (8.12.17)$$

The system (8.12.17) has a nontrivial solution if and only if

$$\begin{vmatrix} 3 - \lambda & 0 & 0 \\ 0 & 7 - \lambda & 5 \\ 0 & -4 & -(2 + \lambda) \end{vmatrix} = (3 - \lambda)[(\lambda - 3)(\lambda - 2)] = 0.$$

Thus, $\lambda = 3$ is a root of multiplicity two, while $\lambda = 2$ is a simple root.
  Substituting $\lambda = 3$ into (8.12.17), we have

$$\begin{cases} 0 \cdot \xi_1 & = 0, \\ 4\xi_2 + 5\xi_3 = 0, \\ -4\xi_2 - 5\xi_3 = 0. \end{cases} \qquad (8.12.18)$$

Solving the system (8.12.18), we see that ξ_1 is arbitrary and $\xi_3 = -\frac{4}{5}\xi_2$, where ξ_2 is also arbitrary.   Thus, $\boldsymbol{\xi}^T = [\xi_1, \xi_2, \xi_3] = [\xi_1, \xi_2, -\frac{4}{5}\xi_2]$.   We can obviously get two linearly independent solutions of the form $\mathbf{x} = \boldsymbol{\xi}e^{3t}$, but we must make sure that $\boldsymbol{\xi}^{(1)}$ and $\boldsymbol{\xi}^{(2)}$ are not linearly dependent.   Thus, with $\xi_1 = 1$, $\xi_2 = 5$, we have $\boldsymbol{\xi}^{(1)} = \begin{bmatrix} 1 \\ 5 \\ -4 \end{bmatrix}$, and with $\xi_1 = 1$, $\xi_2 = 0$, we have

$\boldsymbol{\xi}^{(2)} = \begin{bmatrix} 1 \\ 0 \\ 0 \end{bmatrix}$.   Therefore, the eigenvalue of multiplicity two has given rise to

two linearly independent solutions of the system (8.12.16), namely,

$$\mathbf{x}^{(1)}(t) = \begin{bmatrix} 1 \\ 5 \\ -4 \end{bmatrix} e^{3t}, \qquad \mathbf{x}^{(2)}(t) = \begin{bmatrix} 1 \\ 0 \\ 0 \end{bmatrix} e^{3t}. \tag{8.12.19}$$

Substituting $\lambda = 2$ into (8.12.17), we have

$$\begin{cases} \xi_1 & = 0, \\ 5\xi_2 + 5\xi_3 = 0, \\ -4\xi_2 - 4\xi_3 = 0. \end{cases} \tag{8.12.20}$$

Solving the system (8.12.20), we get $\xi_1 = 0$, $\xi_3 = -\xi_2$ where ξ_2 is arbitrary. Thus,

$$\boldsymbol{\xi}^{(3)} = \begin{bmatrix} \xi_1 \\ \xi_2 \\ \xi_3 \end{bmatrix} = \begin{bmatrix} 0 \\ \xi_2 \\ -\xi_2 \end{bmatrix} = \xi_2 \begin{bmatrix} 0 \\ 1 \\ -1 \end{bmatrix}.$$

A third linearly independent solution of the system (8.12.16) is given by

$$\mathbf{x}^{(3)}(t) = \begin{bmatrix} 0 \\ 1 \\ -1 \end{bmatrix} e^{2t}. \tag{8.12.21}$$

The general solution of the system (8.12.16) on the interval J: $-\infty < t < \infty$ is given by

$$\mathbf{x} = c_1 \mathbf{x}^{(1)}(t) + c_2 \mathbf{x}^{(2)}(t) + c_3 \mathbf{x}^{(3)}(t), \tag{8.12.22}$$

where $\mathbf{x}^{(1)}(t)$, $\mathbf{x}^{(2)}(t)$, $\mathbf{x}^{(3)}(t)$ are given, respectively, by (8.12.19) and (8.12.21) and c_1, c_2, c_3 are arbitrary constants.

Example 8.12.3.   Let us find the general solution of the system

$$\mathbf{x}' = \begin{bmatrix} 0 & 0 & 1 \\ 1 & 0 & -3 \\ 0 & 1 & 3 \end{bmatrix} \mathbf{x}. \tag{8.12.23}$$

SOLUTION.   With $\boldsymbol{\xi}^T = [\xi_1, \xi_2, \xi_3]$, we may express $(\mathbf{A} - \lambda\mathbf{I})\,\boldsymbol{\xi} = \mathbf{0}$ as

$$\begin{cases} -\lambda\xi_1 & + & \xi_3 = 0, \\ \xi_1 - \lambda\xi_2 - & & 3\xi_3 = 0, \\ & \xi_2 + (3 - \lambda)\xi_3 = 0. \end{cases} \tag{8.12.24}$$

The system (8.12.24) has a nontrivial solution if and only if

$$\begin{vmatrix} -\lambda & 0 & 1 \\ 1 & -\lambda & -3 \\ 0 & 1 & 3 - \lambda \end{vmatrix} = (\lambda - 1)^3 = 0.$$

Thus, $\lambda = r = 1$ is a root of multiplicity three of the system (8.12.23).

Substituting $\lambda = 1$ into (8.12.24), we have

$$\begin{cases} -\xi_1 \qquad\quad + \xi_3 = 0, \\ \quad\ \xi_1 - \xi_2 - 3\xi_3 = 0, \\ \qquad\quad\ \xi_2 + 2\xi_3 = 0. \end{cases} \tag{8.12.25}$$

Solving the System (8.12.25), we obtain $\xi_1 = \xi_3$, $\xi_2 = -2\xi_3$, where ξ_3 is arbitrary. Thus, $\xi^T = [\xi_1, \xi_2, \xi_3] = [\xi_3, -2\xi_3, \xi_3] = \xi_3[1, -2, 1]$. Since ξ involves one arbitrary constant, namely ξ_3, only one solution of the form ξe^t is available. Consequently, one solution of the system (8.12.23) is given by

$$\mathbf{x}^{(1)}(t) = \xi e^{rt} = \begin{bmatrix} 1 \\ -2 \\ 1 \end{bmatrix} e^t. \tag{8.12.26}$$

To find a second solution of the system (8.12.23), we utilize (8.12.4) and (8.12.5) with $r = 1$. We have

$$\mathbf{x}^{(2)}(t) = \boldsymbol{\eta}^{(1)} t e^t + \boldsymbol{\eta}^{(2)} e^t, \tag{8.12.27}$$

where $\boldsymbol{\eta}^{(1)}$, $\boldsymbol{\eta}^{(2)}$ are determined by the equations

$$\begin{cases} (\mathbf{A} - \mathbf{I})\,\boldsymbol{\eta}^{(1)} = \mathbf{0}, \\ (\mathbf{A} - \mathbf{I})\,\boldsymbol{\eta}^{(2)} = \boldsymbol{\eta}^{(1)}. \end{cases} \tag{8.12.28}$$

Note that the first system in (8.12.28) is equivalent to the system (8.12.25).

Thus, we may take $\boldsymbol{\eta}^{(1)} = \xi = \begin{bmatrix} 1 \\ -2 \\ 1 \end{bmatrix}$. Letting $\boldsymbol{\eta}^{(2)} = \begin{bmatrix} \eta_1 \\ \eta_2 \\ \eta_3 \end{bmatrix}$, we may now write the second system in (8.12.28) as

$$\begin{bmatrix} -1 & 0 & 1 \\ 1 & -1 & -3 \\ 0 & 1 & 2 \end{bmatrix} \begin{bmatrix} \eta_1 \\ \eta_2 \\ \eta_3 \end{bmatrix} = \begin{bmatrix} 1 \\ -2 \\ 1 \end{bmatrix},$$

which can be expressed as

$$\begin{cases} -\eta_1 \qquad\quad + \eta_3 = 1, \\ \quad\ \eta_1 - \eta_2 - 3\eta_3 = -2, \\ \qquad\quad\ \eta_2 + 2\eta_3 = 1. \end{cases} \tag{8.12.29}$$

In view of Theorem 8.9.1, we see that the system (8.12.29) is indeed consistent, and its solution is given by $\eta_1 = -1 + \eta_3$, $\eta_2 = 1 - 2\eta_3$, where η_3 is arbitrary. Thus, taking $\eta_3 = 0$, we have

$$\boldsymbol{\eta}^{(2)} = \begin{bmatrix} \eta_1 \\ \eta_2 \\ \eta_3 \end{bmatrix} = \begin{bmatrix} -1 \\ 1 \\ 0 \end{bmatrix}. \tag{8.12.30}$$

Substituting $\boldsymbol{\eta}^{(1)}$ and $\boldsymbol{\eta}^{(2)}$ into (8.12.27), a second solution of the system (8.12.23) is given by

$$\mathbf{x}^{(2)}(t) = \begin{bmatrix} 1 \\ -2 \\ 1 \end{bmatrix} te^t + \begin{bmatrix} -1 \\ 1 \\ 0 \end{bmatrix} e^t. \tag{8.12.31}$$

To find a third solution of the system (8.12.23), we utilize (8.12.6) and (8.12.7) with $s = 3$ and $r = 1$. We have

$$\mathbf{x}^{(3)}(t) = \boldsymbol{\zeta}^{(1)} \frac{t^2}{2} e^t + \boldsymbol{\zeta}^{(2)} te^t + \boldsymbol{\zeta}^{(3)} e^t, \tag{8.12.32}$$

where $\boldsymbol{\zeta}^{(1)}$, $\boldsymbol{\zeta}^{(2)}$, $\boldsymbol{\zeta}^{(3)}$ are determined by the equations

$$\begin{cases} (\mathbf{A} - \mathbf{I})\,\boldsymbol{\zeta}^{(1)} = \mathbf{0}, \\ (\mathbf{A} - \mathbf{I})\,\boldsymbol{\zeta}^{(2)} = \boldsymbol{\zeta}^{(1)}, \\ (\mathbf{A} - \mathbf{I})\,\boldsymbol{\zeta}^{(3)} = \boldsymbol{\zeta}^{(2)}. \end{cases} \tag{8.12.33}$$

The first system in (8.12.33) is equivalent to the system (8.12.25). Hence, we may take $\boldsymbol{\zeta}^{(1)T} = \boldsymbol{\xi}^T = [1, -2, 1]$. With $\boldsymbol{\zeta}^{(1)} = \boldsymbol{\xi}$, the second system in (8.12.33) is equivalent to the system (8.12.29). Thus, we may take $\boldsymbol{\zeta}^{(2)} = \boldsymbol{\eta}^{(2)}$, where $\boldsymbol{\eta}^{(2)}$ is given in (8.12.30). Consequently, we may write the third system in (8.12.33) as $(\mathbf{A} - \mathbf{I})\,\boldsymbol{\zeta}^{(3)} = \boldsymbol{\eta}^{(2)}$. Letting $\boldsymbol{\zeta}^{(3)T} = [\zeta_1, \zeta_2, \zeta_3]$, we then have

$$\begin{bmatrix} -1 & 0 & 1 \\ 1 & -1 & -3 \\ 0 & 1 & 2 \end{bmatrix} \begin{bmatrix} \zeta_1 \\ \zeta_2 \\ \zeta_3 \end{bmatrix} = \begin{bmatrix} -1 \\ 1 \\ 0 \end{bmatrix},$$

which can be written as

$$\begin{cases} -\zeta_1 \qquad\quad + \zeta_3 = -1, \\ \zeta_1 - \zeta_2 - 3\zeta_3 = 1, \\ \zeta_2 + 2\zeta_3 = 0. \end{cases} \tag{8.12.34}$$

In view of Theorem 8.9.1, we see that the system (8.12.34) is consistent, and its solution is given by $\zeta_1 = 1 + \zeta_3$, $\zeta_2 = -2\zeta_3$, where ζ_3 is arbitrary. Taking $\zeta_3 = 0$, we have $\boldsymbol{\zeta}^{(3)T} = [\zeta_1, \zeta_2, \zeta_3] = [1, 0, 0]$. Substituting $\boldsymbol{\zeta}^{(1)} = \boldsymbol{\xi}$, $\boldsymbol{\zeta}^{(2)} = \boldsymbol{\eta}^{(2)}$ and $\boldsymbol{\zeta}^{(3)}$ just found into (8.12.32), a third solution of the system (8.12.23) is given by

$$\mathbf{x}^{(3)}(t) = \frac{1}{2} \begin{bmatrix} 1 \\ -2 \\ 1 \end{bmatrix} t^2 e^t + \begin{bmatrix} -1 \\ 1 \\ 0 \end{bmatrix} te^t + \begin{bmatrix} 1 \\ 0 \\ 0 \end{bmatrix} e^t. \tag{8.12.35}$$

Clearly, the solutions $\mathbf{x}^{(1)}(t)$, $\mathbf{x}^{(2)}(t)$, $\mathbf{x}^{(3)}(t)$ are linearly independent on the interval J: $-\infty < t < \infty$. Thus, the general solution of the system (8.12.23) on the interval J is given by

$$\mathbf{x} = c_1\mathbf{x}^{(1)}(t) + c_2\,\mathbf{x}^{(2)}(t) + c_3\,\mathbf{x}^{(3)}(t), \tag{8.12.36}$$

where $\mathbf{x}^{(1)}(t)$, $\mathbf{x}^{(2)}(t)$, $\mathbf{x}^{(3)}(t)$ are given, respectively, by (8.12.26), (8.12.31), (8.12.35) and c_1, c_2, c_3 are arbitrary constants.

Exercises 8.12

In Exercises 1 through 7, find the general solution of each system.

1. $\mathbf{x}' = \begin{bmatrix} 2 & 1 \\ -1 & 4 \end{bmatrix}\mathbf{x}.$

2. $\mathbf{x}' = \begin{bmatrix} 1 & -1 \\ 1 & 3 \end{bmatrix}\mathbf{x}.$

3. $\mathbf{x}' = \begin{bmatrix} 0 & 0 & -1 \\ 0 & 1 & 0 \\ 1 & 0 & 2 \end{bmatrix}\mathbf{x}.$

4. $\mathbf{x}' = \begin{bmatrix} 1 & 0 & -1 \\ 2 & 2 & 2 \\ 2 & 1 & 2 \end{bmatrix}\mathbf{x}.$

5. $\mathbf{x}' = \begin{bmatrix} 2 & 1 & 1 \\ 2 & 3 & 2 \\ 1 & 1 & 2 \end{bmatrix}\mathbf{x}.$

6. $\mathbf{x}' = \begin{bmatrix} 3 & 2 & 1 \\ -1 & 0 & -1 \\ 1 & 1 & 2 \end{bmatrix}\mathbf{x}.$

7. $\mathbf{x}' = \begin{bmatrix} 3 & 2 & 1 & 2 \\ 2 & 3 & 1 & 2 \\ 2 & 2 & 2 & 2 \\ -4 & -1 & -1 & -1 \end{bmatrix}\mathbf{x}.$

In Exercises 8 through 10, solve each initial-value problem.

8. $\mathbf{x}' = \begin{bmatrix} 4 & 1 \\ -1 & 2 \end{bmatrix}\mathbf{x}, \quad \mathbf{x}(0) = \begin{bmatrix} 2 \\ -1 \end{bmatrix}.$

9. $\mathbf{x}' = \begin{bmatrix} 0 & 0 & -2 \\ 0 & 1 & 0 \\ 1 & 0 & 3 \end{bmatrix}\mathbf{x}, \quad \mathbf{x}(0) = \begin{bmatrix} 3 \\ 1 \\ -1 \end{bmatrix}.$

10. $\mathbf{x}' = \begin{bmatrix} 1 & 2 & 1 \\ 0 & 2 & 0 \\ -1 & 0 & 3 \end{bmatrix}\mathbf{x}, \quad \mathbf{x}(0) = \begin{bmatrix} 0 \\ -1 \\ 1 \end{bmatrix}.$

11. Consider the characteristic polynomial

$$p(\lambda) = \det\,[\mathbf{A} - \lambda\mathbf{I}] = \begin{vmatrix} a_{11}-\lambda & a_{12} & a_{13} \\ a_{21} & a_{22}-\lambda & a_{23} \\ a_{31} & a_{32} & a_{33}-\lambda \end{vmatrix}$$

associated with Equation (8.11.2) when $\mathbf{A}$ is a 3×3 matrix. Show that

(a) $p'(\lambda) = -\left\{ \begin{vmatrix} a_{22}-\lambda & a_{23} \\ a_{32} & a_{33}-\lambda \end{vmatrix} + \begin{vmatrix} a_{11}-\lambda & a_{13} \\ a_{31} & a_{33}-\lambda \end{vmatrix} + \begin{vmatrix} a_{11}-\lambda & a_{12} \\ a_{21} & a_{22}-\lambda \end{vmatrix} \right\}$

$\qquad\qquad \equiv -(\text{sum of the principal minors of } \mathbf{A} - \lambda\mathbf{I} \text{ of order two});$

(b) $p''(\lambda) = 2!\{(a_{11} - \lambda) + (a_{22} - \lambda) + (a_{33} - \lambda)\}$
$\qquad\qquad \equiv 2! \text{ times the sum of the principal minors of } \mathbf{A} - \lambda\mathbf{I} \text{ of order one;}$

(c) $p'''(\lambda) = -3!, \qquad p^{(k)}(\lambda) = 0, \qquad k = 4, 5, \dots;$

(d) if $\lambda = r$ is a root of multiplicity $s = 2$ of $p(\lambda) = 0$, then the rank of $\mathbf{A} - r\mathbf{I}$ is not less than $n - s = 3 - 2 = 1$.

(These results can be generalized.)

8.13. Fundamental Matrices

Let us consider the system given in (8.10.5), namely,

$$\mathbf{x}' = \mathbf{A}(t)\mathbf{x}. \tag{8.13.1}$$

Suppose that

$$\mathbf{x}^{(1)}(t) = \begin{bmatrix} x_1^{(1)}(t) \\ x_2^{(1)}(t) \\ \vdots \\ x_n^{(1)}(t) \end{bmatrix}, \quad \mathbf{x}^{(2)}(t) = \begin{bmatrix} x_1^{(2)}(t) \\ x_2^{(2)}(t) \\ \vdots \\ x_n^{(2)}(t) \end{bmatrix}, \quad \dots, \quad \mathbf{x}^{(n)}(t) = \begin{bmatrix} x_1^{(n)}(t) \\ x_2^{(n)}(t) \\ \vdots \\ x_n^{(n)}(t) \end{bmatrix} \tag{8.13.2}$$

are any n linearly independent solutions of the system (8.13.1) on an interval $J: \ a \leq t \leq b$. The matrix $\boldsymbol{\Phi}$ defined by

$$\boldsymbol{\Phi}(t) = \begin{bmatrix} x_1^{(1)}(t) & x_1^{(2)}(t) & \cdots & x_1^{(n)}(t) \\ x_2^{(1)}(t) & x_2^{(2)}(t) & \cdots & x_2^{(n)}(t) \\ \vdots & \vdots & & \vdots \\ x_n^{(1)}(t) & x_n^{(2)}(t) & \cdots & x_n^{(n)}(t) \end{bmatrix} \tag{8.13.3}$$

is called a *fundamental matrix* of the system (8.13.1). Since the matrix $\boldsymbol{\Phi}$ has for its columns the n linearly independent vectors $\mathbf{x}^{(1)}(t)$, $\mathbf{x}^{(2)}(t)$, $\dots$, $\mathbf{x}^{(n)}(t)$, it follows from Theorem 8.10.5 that $\boldsymbol{\Phi}(t)$ is a nonsingular matrix for every t in J. Thus, $\boldsymbol{\Phi}^{-1}(t)$ exists for every t in J.

In view of Definition 8.10.2, *the general solution* $\mathbf{x} = \mathbf{x}(t)$ *of the system* (8.13.1) *on the interval J is given by*

$$\mathbf{x} = c_1\mathbf{x}^{(1)}(t) + c_2\mathbf{x}^{(2)}(t) + \cdots + c_n\mathbf{x}^{(n)}(t), \tag{8.13.4}$$

where $c_1, c_2, \dots, c_n$ *are arbitrary constants.* Utilizing (8.13.3), we may now write (8.13.4) equivalently as

$$\mathbf{x} = \boldsymbol{\Phi}(t)\mathbf{c}, \tag{8.13.5}$$

where $\mathbf{c}^T = [c_1, c_2, \ldots, c_n]$ is a constant vector with arbitrary components $c_1, c_2, \ldots, c_n$.

Suppose now that we wish to obtain the solution of the system (8.13.1) subject to the initial condition

$$\mathbf{x}(t_0) = \mathbf{x}^0, \tag{8.13.6}$$

where t_0 is a point in J and $\mathbf{x}^0$ is a given initial vector. In view of (8.13.6), with $t = t_0$ in (8.13.5), we obtain

$$\mathbf{\Phi}(t_0)\mathbf{c} = \mathbf{x}^0. \tag{8.13.7}$$

Multiplying (8.13.7) on the left by $\mathbf{\Phi}^{-1}(t_0)$, we obtain $\mathbf{c} = \mathbf{\Phi}^{-1}(t_0)\mathbf{x}^0$. Substituting this value of $\mathbf{c}$ into (8.13.5), we have

$$\mathbf{x} = \mathbf{\Phi}(t)\mathbf{\Phi}^{-1}(t_0)\mathbf{x}^0. \tag{8.13.8}$$

Thus, *the solution of the system* (8.13.1) *subject to the initial condition* (8.13.6) *is given on J by* (8.13.8).

Example 8.13.1.   Let us solve the initial-value problem

$$\mathbf{x}' = \begin{bmatrix} 2 & 2 \\ 1 & 3 \end{bmatrix} \mathbf{x}, \qquad \mathbf{x}(0) = \begin{bmatrix} 1 \\ 4 \end{bmatrix}. \tag{8.13.9}$$

SOLUTION.   The reader may show that two linearly independent solutions of the system in (8.13.9) on J: $\;-\infty < t < \infty$, are given by

$$\mathbf{x}^{(1)}(t) = \begin{bmatrix} -2 \\ 1 \end{bmatrix} e^t, \qquad \mathbf{x}^{(2)}(t) = \begin{bmatrix} 1 \\ 1 \end{bmatrix} e^{4t}.$$

Thus,

$$\mathbf{\Phi}(t) = \begin{bmatrix} -2e^t & e^{4t} \\ e^t & e^{4t} \end{bmatrix}, \qquad \mathbf{\Phi}(0) = \begin{bmatrix} -2 & 1 \\ 1 & 1 \end{bmatrix}$$

and

$$\mathbf{\Phi}^{-1}(0) = \frac{1}{3} \begin{bmatrix} -1 & 1 \\ 1 & 2 \end{bmatrix}.$$

Substituting $\mathbf{\Phi}(t)$, $\mathbf{\Phi}^{-1}(0)$ and the initial condition given in (8.13.9) into (8.13.8), we have

$$\mathbf{x} = \frac{1}{3} \begin{bmatrix} -2e^t & e^{4t} \\ e^t & e^{4t} \end{bmatrix} \begin{bmatrix} -1 & 1 \\ 1 & 2 \end{bmatrix} \begin{bmatrix} 1 \\ 4 \end{bmatrix}. \tag{8.13.10}$$

Performing the indicated multiplications in (8.13.10), we obtain the desired solution of our initial-value problem on J, namely,

$$\mathbf{x} = \begin{bmatrix} -2e^t + 3e^{4t} \\ e^t + 3e^{4t} \end{bmatrix} = \begin{bmatrix} -2 \\ 1 \end{bmatrix} e^t + 3 \begin{bmatrix} 1 \\ 1 \end{bmatrix} e^{4t}.$$

For the system (8.11.26) of Example 8.11.2, we found two linearly independent solutions of the form [*see* (8.11.29), (8.11.30)]

$$\mathbf{x}^{(1)}(t) = \begin{bmatrix} 1 \\ i \end{bmatrix} e^{(1+2i)t}, \qquad \mathbf{x}^{(2)}(t) = \begin{bmatrix} 1 \\ -i \end{bmatrix} e^{(1-2i)t}. \qquad (8.13.11)$$

We shall now give an equivalent method to obtain the solution found in (8.11.32).

Since $\mathbf{x}^{(1)}(t) = \begin{bmatrix} x_1^{(1)}(t) \\ x_2^{(1)}(t) \end{bmatrix}$, $\mathbf{x}^{(2)}(t) = \begin{bmatrix} x_1^{(2)}(t) \\ x_2^{(2)}(t) \end{bmatrix}$, in view of (8.13.11) we have

$$x_1^{(1)}(t) = e^t e^{2it}, \quad x_2^{(1)}(t) = ie^t e^{2it}, \quad x_1^{(2)}(t) = e^t e^{-2it}, \quad x_2^{(2)}(t) = -ie^t e^{-2it}.$$
$$(8.13.12)$$

Consider a new fundamental system of solutions $\tilde{\mathbf{x}}^{(1)}(t)$, $\tilde{\mathbf{x}}^{(2)}(t)$, where the components of $\tilde{\mathbf{x}}^{(1)}(t)$ and $\tilde{\mathbf{x}}^{(2)}(t)$ are given by

$$\tilde{x}_1^{(1)}(t) = \frac{x_1^{(1)}(t) + x_1^{(2)}(t)}{2}, \qquad \tilde{x}_1^{(2)}(t) = \frac{x_1^{(1)}(t) - x_1^{(2)}(t)}{2i},$$

$$(8.13.13)$$

$$\tilde{x}_2^{(1)}(t) = \frac{x_2^{(1)}(t) + x_2^{(2)}(t)}{2}, \qquad \tilde{x}_2^{(2)}(t) = \frac{x_2^{(1)}(t) - x_2^{(2)}(t)}{2i}.$$

Utilizing Euler's formula

$$e^{\pm i\beta t} = \cos \beta t \pm i \sin \beta t, \qquad (8.13.14)$$

where β is a real number, we obtain from (8.13.12) and (8.13.13):

$$\tilde{x}_1^{(1)}(t) = e^t \frac{(e^{2it} + e^{-2it})}{2} = e^t \cos 2t,$$

$$\tilde{x}_2^{(1)}(t) = ie^t \frac{(e^{2it} - e^{-2it})}{2} = -e^t \sin 2t,$$

$$(8.13.15)$$

$$\tilde{x}_1^{(2)}(t) = e^t \frac{(e^{2it} - e^{-2it})}{2i} = e^t \sin 2t,$$

$$\tilde{x}_2^{(2)}(t) = ie^t \frac{(e^{2it} + e^{-2it})}{2i} = e^t \cos 2t.$$

Thus, the general solution of the system (8.11.26) is given by

$$\mathbf{x} = c_1 \tilde{\mathbf{x}}^{(1)}(t) + c_2 \tilde{\mathbf{x}}^{(2)}(t) = c_1 \begin{bmatrix} \tilde{x}_1^{(1)}(t) \\ \tilde{x}_2^{(1)}(t) \end{bmatrix} + c_2 \begin{bmatrix} \tilde{x}_1^{(2)}(t) \\ \tilde{x}_2^{(2)}(t) \end{bmatrix}$$

$$= c_1 \begin{bmatrix} \cos 2t \\ -\sin 2t \end{bmatrix} e^t + c_2 \begin{bmatrix} \sin 2t \\ \cos 2t \end{bmatrix} e^t, \qquad (8.13.16)$$

which is the same as the solution given in (8.11.32).

Now, we may express (8.13.13) as

$$
\begin{bmatrix} \tilde{x}_1^{(1)}(t) & \tilde{x}_1^{(2)}(t) \\ \tilde{x}_2^{(1)}(t) & \tilde{x}_2^{(2)}(t) \end{bmatrix} = \begin{bmatrix} x_1^{(1)}(t) & x_1^{(2)}(t) \\ x_2^{(1)}(t) & x_2^{(2)}(t) \end{bmatrix} \begin{bmatrix} \dfrac{1}{2} & \dfrac{1}{2i} \\ \dfrac{1}{2} & -\dfrac{1}{2i} \end{bmatrix}. \tag{8.13.17}
$$

A fundamental matrix associated with the linearly independent solutions given in (8.13.11) is

$$
\boldsymbol{\Phi}(t) = \begin{bmatrix} e^t e^{2it} & e^t e^{-2it} \\ ie^t e^{2it} & -ie^t e^{-2it} \end{bmatrix}. \tag{8.13.18}
$$

Clearly, $\boldsymbol{\Phi}(t)$ is complex-valued. In view of (8.13.17), we may render $\boldsymbol{\Phi}(t)$ real-valued by considering the matrix $\tilde{\boldsymbol{\Phi}}$, where

$$
\tilde{\boldsymbol{\Phi}}(t) = \begin{bmatrix} \tilde{x}_1^{(1)}(t) & \tilde{x}_1^{(2)}(t) \\ \tilde{x}_2^{(1)}(t) & \tilde{x}_2^{(2)}(t) \end{bmatrix}. \tag{8.13.19}
$$

Thus,

$$
\tilde{\boldsymbol{\Phi}}(t) = \boldsymbol{\Phi}(t)\mathbf{K}, \tag{8.13.20}
$$

where

$$
\mathbf{K} = \begin{bmatrix} \dfrac{1}{2} & \dfrac{1}{2i} \\ \dfrac{1}{2} & -\dfrac{1}{2i} \end{bmatrix}. \tag{8.13.21}
$$

Since $\boldsymbol{\Phi}(t)$ and $\mathbf{K}$ are nonsingular, so is $\tilde{\boldsymbol{\Phi}}(t)$ nonsingular. Hence, $\tilde{\boldsymbol{\Phi}}(t)$ is also a fundamental matrix of the system (8.11.26).

More generally, if $\lambda_1 = \alpha + i\beta$, $\lambda_2 = \alpha - i\beta$, $\lambda_3 = \gamma + i\delta$, $\lambda_4 = \gamma - i\delta$ are four imaginary roots of the characteristic equation $p(\lambda) = 0$ associated with the system (8.13.1), and if the system (8.13.1) has a complex-valued fundamental matrix of the form

$$
\boldsymbol{\Phi}(t) = \begin{bmatrix} \xi_1^{(1)} e^{\lambda_1 t} & \bar{\xi}_1^{(1)} e^{\lambda_2 t} & \xi_1^{(2)} e^{\lambda_3 t} & \bar{\xi}_1^{(2)} e^{\lambda_4 t} & x_1^{(5)}(t) & x_1^{(6)}(t) & \cdots & x_1^{(n)}(t) \\ \xi_2^{(1)} e^{\lambda_1 t} & \bar{\xi}_2^{(1)} e^{\lambda_2 t} & \xi_2^{(2)} e^{\lambda_3 t} & \bar{\xi}_2^{(2)} e^{\lambda_4 t} & x_2^{(5)}(t) & x_2^{(6)}(t) & \cdots & x_2^{(n)}(t) \\ \vdots & \vdots & \vdots & \vdots & \vdots & \vdots & & \vdots \\ \xi_n^{(1)} e^{\lambda_1 t} & \bar{\xi}_n^{(1)} e^{\lambda_2 t} & \xi_n^{(2)} e^{\lambda_3 t} & \bar{\xi}_n^{(2)} e^{\lambda_4 t} & x_n^{(5)}(t) & x_n^{(6)}(t) & \cdots & x_n^{(n)}(t) \end{bmatrix} \tag{8.13.22}
$$

where $\mathbf{x}^{(5)}(t)$, $\mathbf{x}^{(6)}(t)$, $\ldots$, $\mathbf{x}^{(n)}(t)$ are real-valued, then $\boldsymbol{\Phi}(t)$ may be rendered real-valued by considering an (equivalent) matrix $\tilde{\boldsymbol{\Phi}}$ given by

$$
\tilde{\boldsymbol{\Phi}}(t) = \boldsymbol{\Phi}(t)\mathbf{K}, \tag{8.13.23}
$$

where

$$\mathbf{K} = \begin{bmatrix} \dfrac{1}{2} & \dfrac{1}{2i} & 0 & 0 & 0 & 0 & \cdots & 0 \\[2mm] \dfrac{1}{2} & -\dfrac{1}{2i} & 0 & 0 & 0 & 0 & \cdots & 0 \\[2mm] 0 & 0 & \dfrac{1}{2} & \dfrac{1}{2i} & 0 & 0 & \cdots & 0 \\[2mm] 0 & 0 & \dfrac{1}{2} & -\dfrac{1}{2i} & 0 & 0 & \cdots & 0 \\[2mm] 0 & 0 & 0 & 0 & 1 & 0 & \cdots & 0 \\ 0 & 0 & 0 & 0 & 0 & 1 & \cdots & 0 \\ \vdots & \vdots & \vdots & \vdots & \vdots & \vdots & & \vdots \\ 0 & 0 & 0 & 0 & 0 & 0 & \cdots & 1 \end{bmatrix}. \qquad (8.13.24)$$

Example 8.13.2. Let us find the general solution of the system

$$\mathbf{x}' = \begin{bmatrix} 1 & -1 & -1 \\ 1 & -1 & 0 \\ 1 & 0 & -1 \end{bmatrix} \mathbf{x}. \qquad (8.13.25)$$

SOLUTION.   Let $\boldsymbol{\xi}^T = [\xi_1, \xi_2, \xi_3]$.   We may express $(\mathbf{A} - \lambda\mathbf{I})\,\boldsymbol{\xi} = \mathbf{0}$ as

$$\begin{cases} (1 - \lambda)\,\xi_1 - & \xi_2 - & \xi_3 = 0, \\ \xi_1 - (1 + \lambda)\,\xi_2 & & = 0, \\ \xi_1 & - (1 + \lambda)\,\xi_3 = 0. \end{cases} \qquad (8.13.26)$$

The characteristic equation

$$\begin{vmatrix} 1 - \lambda & -1 & -1 \\ 1 & -(1 + \lambda) & 0 \\ 1 & 0 & -(1 + \lambda) \end{vmatrix} = 0$$

has roots $\lambda_1 = i$, $\lambda_2 = -i$, $\lambda_3 = -1$.   Substituting $\lambda = \lambda_1 = i$ into $(8.13.26)$, we find $\boldsymbol{\xi}^T = \xi_1\left[1, \dfrac{1}{1 + i}, \dfrac{1}{1 + i}\right]$, where ξ_1 is arbitrary.   Taking $\xi_1 = 1 + i$, one solution of the system $(8.13.25)$ is given by $\mathbf{x}^{(1)T}(t) = [1 + i, 1, 1]e^{it}$.   For $\lambda = \lambda_2 = -i$, a second solution of the system $(8.13.25)$ is given by $\mathbf{x}^{(2)T}(t) = [1 - i, 1, 1]e^{-it}$.   For $\lambda = \lambda_3 = -1$, a third solution of the system $(8.13.25)$ is of the form $\mathbf{x}^{(3)T}(t) = [0, 1, -1]e^{-t}$.   Clearly, the solutions $\mathbf{x}^{(1)}(t)$, $\mathbf{x}^{(2)}(t)$, $\mathbf{x}^{(3)}(t)$ are linearly independent on the interval $J:\;\; -\infty < t < \infty$.   Thus, a fundamental matrix for the system $(8.13.25)$ is

$$\boldsymbol{\Phi}(t) = \begin{bmatrix} (1 + i)e^{it} & (1 - i)e^{-it} & 0 \\ e^{it} & e^{-it} & e^{-t} \\ e^{it} & e^{-it} & -e^{-t} \end{bmatrix},$$

which is complex-valued.   To render $\boldsymbol{\Phi}(t)$ real-valued, we consider a fundamental matrix $\tilde{\boldsymbol{\Phi}}$, where

$$\boldsymbol{\Phi}(t) = \begin{bmatrix} (1+i)e^{it} & (1-i)e^{-it} & 0 \\ e^{it} & e^{-it} & e^{-t} \\ e^{it} & e^{-it} & -e^{-t} \end{bmatrix} \begin{bmatrix} \dfrac{1}{2} & \dfrac{1}{2i} & 0 \\ \dfrac{1}{2} & -\dfrac{1}{2i} & 0 \\ 0 & 0 & 1 \end{bmatrix}. \qquad (8.13.27)$$

Performing the indicated multiplication in (8.13.27), and utilizing the formulas $\cos t = (e^{it} + e^{-it})/2$, $\sin t = (e^{it} - e^{-it})/2i$, we obtain

$$\tilde{\boldsymbol{\Phi}}(t) = \begin{bmatrix} \cos t - \sin t & \sin t + \cos t & 0 \\ \cos t & \sin t & e^{-t} \\ \cos t & \sin t & -e^{-t} \end{bmatrix}. \qquad (8.13.28)$$

Thus, the general solution of the system (8.13.25) on the interval J is given by $\mathbf{x} = \tilde{\boldsymbol{\Phi}}(t)\mathbf{c}$, where $\mathbf{c}^T = [c_1, c_2, c_3]$ and c_1, c_2, c_3 are arbitrary constants.   In view of (8.13.28), we may also write the general solution as

$$\mathbf{x} = c_1 \begin{bmatrix} \cos t - \sin t \\ \cos t \\ \cos t \end{bmatrix} + c_2 \begin{bmatrix} \sin t + \cos t \\ \sin t \\ \sin t \end{bmatrix} + c_3 \begin{bmatrix} 0 \\ 1 \\ -1 \end{bmatrix} e^{-t}.$$

Exercises 8.13

In Exercises 1 through 3, find the general solution of each system.

1. $\mathbf{x}' = \begin{bmatrix} 5 & 1 \\ -5 & 3 \end{bmatrix} \mathbf{x}.$

2. $\mathbf{x}' = \begin{bmatrix} 4 & -4 & 0 \\ 1 & 2 & 1 \\ 0 & 2 & 4 \end{bmatrix} \mathbf{x}.$

3. $\mathbf{x}' = \begin{bmatrix} 0 & -1 & 0 & 0 \\ 1 & 0 & 0 & 0 \\ 0 & 0 & 0 & -1 \\ 0 & 0 & 1 & 0 \end{bmatrix} \mathbf{x}.$

In Exercises 4 through 10, solve each initial-value problem.

4. $\mathbf{x}' = \begin{bmatrix} 1 & -1 \\ 4 & 5 \end{bmatrix} \mathbf{x}, \qquad \mathbf{x}(0) = \begin{bmatrix} 1 \\ -2 \end{bmatrix}.$

5. $\mathbf{x}' = \begin{bmatrix} 2 & 0 & -2 \\ 0 & 2 & -1 \\ 3 & -2 & 2 \end{bmatrix} \mathbf{x}, \qquad \mathbf{x}(0) = \begin{bmatrix} 4 \\ 2 \\ -2 \end{bmatrix}.$

6. $\mathbf{x}' = \begin{bmatrix} 1 & 2 & -1 & -1 \\ -4 & 0 & 1 & 4 \\ -1 & 5 & -2 & -1 \\ -4 & -4 & 3 & 6 \end{bmatrix} \mathbf{x}, \qquad \mathbf{x}(0) = \begin{bmatrix} 1 \\ 3 \\ -2 \\ 3 \end{bmatrix}.$

7. $\mathbf{x}' = \begin{bmatrix} 1 & -2 & 0 & 0 \\ 1 & -1 & 0 & 0 \\ 0 & 0 & 2 & -4 \\ 0 & 0 & 2 & -2 \end{bmatrix} \mathbf{x}, \qquad \mathbf{x}(0) = \begin{bmatrix} 2 \\ 1 \\ -2 \\ -1 \end{bmatrix}.$

8. $\mathbf{x}' = \begin{bmatrix} 0 & 4 & 0 & 0 \\ 0 & 0 & 0 & 1 \\ -1 & 0 & 0 & -5 \\ 0 & 0 & 1 & 0 \end{bmatrix} \mathbf{x}, \qquad \mathbf{x}(0) = \begin{bmatrix} 2 \\ -1 \\ 1 \\ -2 \end{bmatrix}.$

9. $\mathbf{x}' = \begin{bmatrix} -1 & -1 & -1 & -3 & -1 \\ -2 & 0 & 1 & -4 & 2 \\ 0 & 0 & 1 & 0 & 0 \\ 3 & 1 & -1 & 5 & -1 \\ 2 & 0 & 1 & 2 & 2 \end{bmatrix} \mathbf{x}, \qquad \mathbf{x}(0) = \begin{bmatrix} 2 \\ 2 \\ -1 \\ -1 \\ 1 \end{bmatrix}.$

(Two roots of the characteristic equation are 1 and $1 + i$.)

10. $\mathbf{x}' = \begin{bmatrix} 1 & 0 & 0 & 0 & 0 \\ 0 & 6 & -58/3 & -9 & 30 \\ 0 & 0 & 0 & 0 & 6 \\ 0 & 2 & -6 & 0 & 0 \\ 0 & 0 & 0 & 1 & 0 \end{bmatrix} \mathbf{x}, \qquad \mathbf{x}(0) = \begin{bmatrix} 2 \\ 2 \\ -1 \\ -1 \\ 1 \end{bmatrix}.$

11. Let

$$A = \begin{bmatrix} -1 & -1 & -1 & -3 & -1 \\ -2 & 0 & 1 & -4 & 2 \\ 0 & 0 & 1 & 0 & 0 \\ 3 & 1 & -1 & 5 & -1 \\ 2 & 0 & 1 & 2 & 2 \end{bmatrix}$$

be the coefficient matrix given in Exercise 9, and let

$$B = \begin{bmatrix} 1 & 0 & 0 & 0 & 0 \\ 0 & 1 & -1 & 0 & 0 \\ 0 & 1 & 1 & 0 & 0 \\ 0 & 0 & 0 & 2 & 2 \\ 0 & 0 & 0 & -2 & 2 \end{bmatrix}, \qquad P = \begin{bmatrix} 0 & 0 & 1 & 0 & 0 \\ 1 & 0 & 0 & 0 & 1 \\ 0 & 1 & 0 & 1 & 0 \\ 0 & 0 & 1 & 0 & 1 \\ 1 & 0 & 0 & 1 & 0 \end{bmatrix}.$$

(a) Show that the matrix P is nonsingular and

$$P^{-1} = \begin{bmatrix} 1 & 1 & 0 & -1 & 0 \\ 1 & 1 & 1 & -1 & -1 \\ 1 & 0 & 0 & 0 & 0 \\ -1 & -1 & 0 & 1 & 1 \\ -1 & 0 & 0 & 1 & 0 \end{bmatrix}.$$

(b) Verify that

$$A = P^{-1}BP.$$

Thus [*see* Exercise 8.11.27] the matrices **A** and **B** are similar, and have the same characteristic polynomial and eigenvalues, namely,

$$\lambda_1 = 1, \qquad \lambda_2 = 1 + i, \qquad \lambda_3 = 1 - i, \qquad \lambda_4 = 2 + 2i, \qquad \lambda_5 = 2 - 2i.$$

(c) Let $\boldsymbol{\xi}^{(i)}$ and $\boldsymbol{\eta}^{(i)}$ $(i = 1, \ldots, 5)$ denote the eigenvectors associated with the eigenvalues λ_i $(i = 1, \ldots, 5)$ of the matrices **A** and **B**, respectively. Show that $\boldsymbol{\xi}^{(1)T} = [1, 1, 1, -1, -1]$, $\boldsymbol{\xi}^{(2)T} = [-1, -1 + i, 0, 1, 0]$, $\boldsymbol{\xi}^{(3)T} = [-1, -1 - i, 0, 1, 0]$, $\boldsymbol{\xi}^{(4)T} = [-1, -1 - i, 0, 1 + i, 1]$, $\boldsymbol{\xi}^{(5)T} = [-1, -1 + i, 0, 1 - i, 1]$, and $\boldsymbol{\eta}^{(1)T} = [1, 0, 0, 0, 0]$, $\boldsymbol{\eta}^{(2)T} = [0, -1, i, 0, 0]$, $\boldsymbol{\eta}^{(3)T} = [0, -1, -i, 0, 0]$, $\boldsymbol{\eta}^{(4)T} = [0, 0, 0, 1, i]$, $\boldsymbol{\eta}^{(5)T} = [0, 0, 0, 1, -i]$.

(d) Verify directly that $\boldsymbol{\xi}^{(i)} = P^{-1}\boldsymbol{\eta}^{(i)}$ $(i = 1, \ldots, 5)$.

(e) Let $\boldsymbol{\Phi}_A(t)$ and $\boldsymbol{\Phi}_B(t)$ denote fundamental matrices of the systems $\mathbf{x}' = \mathbf{A}\mathbf{x}$ and $\mathbf{y}' = \mathbf{B}\mathbf{y}$, respectively. Verify directly that $\boldsymbol{\Phi}_A(t) = P^{-1}\boldsymbol{\Phi}_B(t)$.

(f) Verify directly that the solution of the initial-value problem of Exercise 9, namely,

$$\mathbf{x}' = \mathbf{A}\mathbf{x}, \qquad \mathbf{x}^T(0) = [2, 2, -1, -1, 1],$$

can also be obtained by utilizing the simpler formula

$$\mathbf{x} = P^{-1}\boldsymbol{\Phi}_B(t)\mathbf{C},$$

where

$$\mathbf{C} = \boldsymbol{\Phi}_B^{-1}(0)P\mathbf{x}(0).$$

[*Note:* In finding **x**, first find $P^{-1}\boldsymbol{\Phi}_B(t)$, second find **C**, and finally perform the product $P^{-1}\boldsymbol{\Phi}_B(t) \cdot \mathbf{C}$. The main difficulty in determining whether two matrices **A** and **B** are similar is the determination of the nonsingular matrix **P** so that $\mathbf{A} = P^{-1}BP$.]

(g) Show that the results stated in (d), (e), and (f) are true for any two $n \times n$ matrices **A** and **B** which are similar. In (d) and (e), we actually have $\boldsymbol{\xi}^{(i)} = kP^{-1}\boldsymbol{\eta}^{(i)}$ and $\boldsymbol{\Phi}_A(t) = kP^{-1}\boldsymbol{\Phi}_B(t)$, where k is a nonzero scalar.

12. Consider the linear homogeneous system

$$(1) \qquad\qquad \mathbf{x}' = \mathbf{A}\mathbf{x},$$

where **A** is an $n \times n$ constant matrix. Let $\boldsymbol{\xi}^{(1)}, \boldsymbol{\xi}^{(2)}, \ldots, \boldsymbol{\xi}^{(n)}$ be the eigenvectors associated with the simple eigenvalues $\lambda_1, \lambda_2, \ldots, \lambda_n$ of the matrix **A**. Thus,

$$(2) \qquad\qquad \mathbf{A}\boldsymbol{\xi}^{(i)} = \lambda_i \boldsymbol{\xi}^{(i)}, \qquad i = 1, 2, \ldots, n.$$

Consider a matrix **P** whose columns are the eigenvectors $\boldsymbol{\xi}^{(1)}, \boldsymbol{\xi}^{(2)}, \ldots, \boldsymbol{\xi}^{(n)}$, that is,

$$(3) \qquad\qquad P = \begin{bmatrix} \xi_1^{(1)} & \xi_1^{(2)} & \cdots & \xi_1^{(n)} \\ \xi_1^{(1)} & \xi_2^{(2)} & \cdots & \xi_2^{(n)} \\ \vdots & \vdots & & \vdots \\ \xi_n^{(1)} & \xi_n^{(2)} & \cdots & \xi_n^{(n)} \end{bmatrix}.$$

Clearly, $\det \mathbf{P} \neq 0$, and $\mathbf{P}^{-1}$ exists.   Let

(4)
$$\mathbf{x} = \mathbf{Py}.$$

(a) Utilizing (1), (2), (3), (4), show that

(5)
$$\mathbf{y}' = \mathbf{Dy},$$

where

(6)
$$\mathbf{D} = \mathbf{P}^{-1}\mathbf{AP}, \qquad \mathbf{D} = \begin{bmatrix} \lambda_1 & 0 & \cdots & 0 \\ 0 & \lambda_2 & \cdots & 0 \\ \vdots & \vdots & & \vdots \\ 0 & 0 & \cdots & \lambda_n \end{bmatrix}.$$

Hence, $\mathbf{D}$ and $\mathbf{A}$ are similar matrices, and the transformation given in (6) is called a *similar transformation*.   In view of Exercise 8.11.12, one may now readily obtain the general solution of the system (5), which in turn will enable one to obtain the general solution of (1) by using (4).   Thus, we have another point of view for solving the system (1).   However, it should be observed that solving the system (5) required knowing the eigenvalues and eigenvectors of the matrix $\mathbf{A}$.

(b) Utilize the above method to solve Exercises 8.11.3 and 8.11.11.

8.14. Nonhomogeneous Linear Systems

Let us consider the nonhomogeneous linear system

$$\mathbf{x}' = \mathbf{A}(t)\mathbf{x} + \mathbf{g}(t), \tag{8.14.1}$$

where the matrix function $\mathbf{A}$ and the vector function $\mathbf{g}$ are assumed to be continuous on a common interval J: $a \leq t \leq b$, and $\mathbf{A}(t) \neq \mathbf{0}$ for every t in J.   The theorems in Sections 4.7 or 6.7 extend to the system (8.14.1).   [*See* Exercises 8.14.1, 8.14.2, and 8.14.3.]   Thus, as in Section 4.7 or 6.7, let $\mathbf{x}^{(1)}(t)$, $\mathbf{x}^{(2)}(t), \ldots, \mathbf{x}^{(n)}(t)$ be n linearly independent solutions of the homogeneous system

$$\mathbf{x}' = \mathbf{A}(t)\mathbf{x} \tag{8.14.2}$$

on the interval J, and let $\mathbf{x}_p$ be a solution of the system (8.14.1) on J.   Then, the *general solution* $\mathbf{x} = \mathbf{x}(t)$ of the system (8.14.1) is defined by

$$\mathbf{x} = \mathbf{x}_h(t) + \mathbf{x}_p(t), \tag{8.14.3}$$

where

$$\mathbf{x}_h(t) = c_1 \mathbf{x}^{(1)}(t) + c_2 \mathbf{x}^{(2)}(t) + \cdots + c_n \mathbf{x}^{(n)}(t), \tag{8.14.4}$$

and $c_1, c_2, \ldots, c_n$ are arbitrary constants.   We may also write (8.14.4) as

$$\mathbf{x}_h(t) = \mathbf{\Phi}(t)\mathbf{c}, \tag{8.14.5}$$

where the fundamental matrix $\boldsymbol{\Phi}$ is given in (8.13.3) and $\mathbf{c}^T = [c_1, c_2, \ldots, c_n]$. The functions $\mathbf{x}_h$ and $\mathbf{x}_p$ are called, respectively, the *complementary function* and a *particular solution (or integral)* of the system (8.14.1).

If $\boldsymbol{\Phi}$ is a fundamental matrix of a system (8.14.2), then in view of (8.14.5), we may express (8.14.2) as $(\boldsymbol{\Phi}'(t) - \mathbf{A}(t)\boldsymbol{\Phi}(t))\mathbf{c} = 0$. Since $\mathbf{c}$ is an arbitrary constant vector, it follows from Exercise 8.9.11 that $\boldsymbol{\Phi}'(t) - \mathbf{A}(t)\boldsymbol{\Phi}(t) = \mathbf{0}$. Thus, $\boldsymbol{\Phi}$ satisfies the equation

$$\boldsymbol{\Phi}'(t) = \mathbf{A}(t)\boldsymbol{\Phi}(t). \tag{8.14.6}$$

Methods of finding a particular solution $\mathbf{x}_p$ of the system (8.14.1) are essentially the same as those given in Sections 4.8, 4.9, 6.8, and 6.9. Once we have determined the complementary function $\mathbf{x}_h$ of (8.14.1), we may utilize the method of variation of parameters to find $\mathbf{x}_p$. If $\mathbf{A}$ is a constant matrix and the components of the vector $\mathbf{g}$ are functions which have a finite number of linearly independent derivatives, we may also utilize the method of undetermined coefficients to find a particular solution $\mathbf{x}_p$ of (8.14.1). We shall now explain *the method of variation of parameters (or constants) for the system* (8.14.1). In Example 8.14.1, Method 2, we illustrate the method of undetermined coefficients.

Let $\boldsymbol{\Phi}$ be a fundamental matrix of (8.14.2). The general solution of (8.14.2) is given by

$$\mathbf{x}_h(t) = \boldsymbol{\Phi}(t)\mathbf{c}. \tag{8.14.7}$$

Let us assume a particular solution $\mathbf{x}_p(t)$ of the system (8.14.1) on J has the form

$$\mathbf{x}_p(t) = \boldsymbol{\Phi}(t)\mathbf{v}(t). \tag{8.14.8}$$

In view of Remark 8.8.2 and Formula (8.6.24), upon differentiating (8.14.8) we obtain

$$\mathbf{x}_p'(t) = \boldsymbol{\Phi}(t)\mathbf{v}'(t) + \boldsymbol{\Phi}'(t)\mathbf{v}(t). \tag{8.14.9}$$

Substituting $\mathbf{x}_p$ and $\mathbf{x}_p'$ of (8.14.8) and (8.14.9) into (8.14.1), we have

$$\boldsymbol{\Phi}(t)\mathbf{v}'(t) + \boldsymbol{\Phi}'(t)\mathbf{v}(t) = \mathbf{A}(t)\boldsymbol{\Phi}(t)\mathbf{v}(t) + \mathbf{g}(t). \tag{8.14.10}$$

Utilizing (8.14.6), Equation (8.14.10) becomes

$$\boldsymbol{\Phi}(t)\mathbf{v}'(t) = \mathbf{g}(t). \tag{8.14.11}$$

Since $\boldsymbol{\Phi}(t)$ is nonsingular on J, $\boldsymbol{\Phi}^{-1}(t)$ exists for every t in J. Thus, multiplying (8.14.11) on the left by $\boldsymbol{\Phi}^{-1}(t)$, we obtain

$$\mathbf{v}'(t) = \boldsymbol{\Phi}^{-1}(t)\mathbf{g}(t). \tag{8.14.12}$$

Integrating (8.14.12) and suppressing the constant vector of integration $\mathbf{k}$, since $\boldsymbol{\Phi}(t)\mathbf{k}$ can ultimately be absorbed in the complementary function $\mathbf{x}_h$, we obtain

$$\mathbf{v}(t) = \int \boldsymbol{\Phi}^{-1}(t)\mathbf{g}(t)dt. \tag{8.14.13}$$

Substituting $\mathbf{v}(t)$ of (8.14.13) into (8.14.8), we have

$$\mathbf{x}_p(t) = \boldsymbol{\Phi}(t) \int \boldsymbol{\Phi}^{-1}(t)\mathbf{g}(t)dt. \tag{8.14.14}$$

Thus, if the system (8.14.1) has a solution of the form given in (8.14.8), then $\mathbf{x}_p$ is given by (8.14.14). We leave it as an exercise for the reader [*see* Exercise 8.14.4] to show that if $\mathbf{x}_p$ is given by (8.14.14) and $\boldsymbol{\Phi}$ is a fundamental matrix of (8.14.2) on the interval J, then $\mathbf{x}_p$ satisfies the system (8.14.1).

In view of (8.14.3), (8.14.5), and (8.14.14), *the general solution of the system* (8.14.1) *on J is given by*

$$\mathbf{x} = \boldsymbol{\Phi}(t)\mathbf{c} + \boldsymbol{\Phi}(t) \int \boldsymbol{\Phi}^{-1}(t)\mathbf{g}(t)dt. \tag{8.14.15}$$

Since $\mathbf{c}$ is an arbitrary constant vector, we may also write (8.14.15) as

$$\mathbf{x} = \boldsymbol{\Phi}(t)\mathbf{c} + \boldsymbol{\Phi}(t) \int_{t_0}^{t} \boldsymbol{\Phi}^{-1}(s)\mathbf{g}(s)ds, \tag{8.14.16}$$

where t and t_0 belong to J. (Why?)

Suppose now that we wish to solve the system (8.14.1) subject to the initial condition

$$\mathbf{x}(t_0) = \mathbf{x}^0. \tag{8.14.17}$$

In view of (8.14.17), with $t = t_0$ in (8.14.16), we obtain $\mathbf{x}^0 = \boldsymbol{\Phi}(t_0)\mathbf{c}$. Thus, $\mathbf{c} = \boldsymbol{\Phi}^{-1}(t_0)\mathbf{x}^0$. Substituting this value of $\mathbf{c}$ into (8.14.16), *the solution of the given initial-value problem on J is given by*

$$\mathbf{x} = \boldsymbol{\Phi}(t)\boldsymbol{\Phi}^{-1}(t_0)\mathbf{x}^0 + \boldsymbol{\Phi}(t) \int_{t_0}^{t} \boldsymbol{\Phi}^{-1}(s)\mathbf{g}(s)ds. \tag{8.14.18}$$

Example 8.14.1. Let us find the general solution of the system

$$\mathbf{x}' = \begin{bmatrix} 1 & 4 \\ 1 & 1 \end{bmatrix} \mathbf{x} + \begin{bmatrix} -10 \\ 1 \end{bmatrix} e^{3t} \tag{8.14.19}$$

on the interval J: $-\infty < t < \infty$.

SOLUTION. *Method* 1. [Utilizing Formula (8.14.15).] Consider the homogeneous system

$$\mathbf{x}' = \begin{bmatrix} 1 & 4 \\ 1 & 1 \end{bmatrix} \mathbf{x}. \tag{8.14.20}$$

Let $\xi^T = [\xi_1, \xi_2]$. The expression $(\mathbf{A} - \lambda\mathbf{I})\xi = \mathbf{0}$ can be written as

$$\begin{cases} (1 - \lambda)\xi_1 + \qquad 4\xi_2 = 0, \\ \qquad \xi_1 + (1 - \lambda)\xi_2 = 0. \end{cases} \tag{8.14.21}$$

The characteristic equation

$$p(\lambda) = \begin{vmatrix} 1 - \lambda & 4 \\ 1 & 1 - \lambda \end{vmatrix} = 0, \tag{8.14.22}$$

has roots $\lambda = -1$ and $\lambda = 3$. Substituting, respectively, these values of λ into (8.14.21), we obtain $\xi^{(1)T} = [2, -1]$ and $\xi^{(2)T} = [2, 1]$. Thus, two linearly independent solutions of the homogeneous system (8.14.20) are given by

$$\mathbf{x}^{(1)}(t) = \begin{bmatrix} 2 \\ -1 \end{bmatrix} e^{-t}, \qquad \mathbf{x}^{(2)}(t) = \begin{bmatrix} 2 \\ 1 \end{bmatrix} e^{3t}. \tag{8.14.23}$$

Consequently, a fundamental matrix $\mathbf{\Phi}(t)$ of the system (8.14.20) is given by

$$\mathbf{\Phi}(t) = \begin{bmatrix} 2e^{-t} & 2e^{3t} \\ -e^{-t} & e^{3t} \end{bmatrix}. \tag{8.14.24}$$

Utilizing the methods given in Section 8.7, we find

$$\mathbf{\Phi}^{-1}(t) = \frac{1}{4} \begin{bmatrix} e^{t} & -2e^{t} \\ e^{-3t} & 2e^{-3t} \end{bmatrix}. \tag{8.14.25}$$

We are given that $\mathbf{g}(t) = \begin{bmatrix} -10 \\ 1 \end{bmatrix} e^{3t}$. Substituting $\mathbf{\Phi}$, $\mathbf{\Phi}^{-1}$, and $\mathbf{g}$ into (8.14.14), we have

$$\mathbf{x}_p(t) = \begin{bmatrix} 2e^{-t} & 2e^{3t} \\ -e^{-t} & e^{3t} \end{bmatrix} \int \frac{1}{4} \begin{bmatrix} e^{t} & -2e^{t} \\ e^{-3t} & 2e^{-3t} \end{bmatrix} \begin{bmatrix} -10e^{3t} \\ e^{3t} \end{bmatrix} dt. \tag{8.14.26}$$

In (8.14.26), performing the indicated operations and utilizing Formula (8.6.21), we obtain

$$\mathbf{x}_p(t) = -\frac{1}{4} \begin{bmatrix} 16t + 6 \\ 8t - 3 \end{bmatrix} e^{3t}. \tag{8.14.27}$$

Substituting $\mathbf{\Phi}$ and $\mathbf{x}_p$ of (8.14.24) and (8.14.27) into (8.14.15), the general solution of the given system on J is

$$\mathbf{x} = \begin{bmatrix} 2e^{-t} & 2e^{3t} \\ -e^{-t} & e^{3t} \end{bmatrix} \mathbf{c} - \frac{1}{4} \begin{bmatrix} 16t + 6 \\ 8t - 3 \end{bmatrix} e^{3t},$$

or

$$\mathbf{x} = c_1 \begin{bmatrix} 2 \\ -1 \end{bmatrix} e^{-t} + c_2 \begin{bmatrix} 2 \\ 1 \end{bmatrix} e^{3t} - \frac{1}{4} \begin{bmatrix} 16t + 6 \\ 8t - 3 \end{bmatrix} e^{3t}. \tag{8.14.28}$$

Method 2.   (*Undetermined coefficients.*)   In view of (8.14.22), the characteristic polynomial of the system (8.14.19) is

$$p(\lambda) = \lambda^2 - 2\lambda - 3. \tag{8.14.29}$$

The vector $\mathbf{g}(t)$ in (8.14.19) is of the form $\mathbf{g}(t) = \mathbf{k}e^{\alpha t}$, where $\mathbf{k}^T = [-10, 1]$ and $\alpha = 3$.

Note that $p(\alpha) = p(3) = 9 - 6 - 3 = 0$ and $p'(\alpha) = p'(3) = 6 - 2 \neq 0$.   Thus, in view of Formula (4.8.16), a particular solution $\mathbf{x}_p$ of the system (8.14.19) may be taken of the form

$$\mathbf{x}_p(t) = \mathbf{a}e^{3t} + \mathbf{b}te^{3t}. \tag{8.14.30}$$

Note, we have slightly modified Formula (4.8.16) to incorporate the term $\mathbf{a}e^{3t}$.   For scalar differential equations, the term ae^{3t} was not retained, since it could ultimately be absorbed in the complementary function.   However, for systems, the term $\mathbf{a}e^{3t}$ is necessary, since it generally cannot be absorbed in the complementary function of the given system, unless it is a multiple of that part of the complementary function involving e^{3t}.

Let

$$\mathbf{a} = \begin{bmatrix} a_1 \\ a_2 \end{bmatrix} \quad \text{and} \quad \mathbf{b} = \begin{bmatrix} b_1 \\ b_2 \end{bmatrix}. \tag{8.14.31}$$

Differentiating (8.14.30), we obtain

$$\mathbf{x}'_p(t) = 3\mathbf{a}e^{3t} + (3\mathbf{b}t + \mathbf{b})e^{3t}. \tag{8.14.32}$$

Substituting $\mathbf{x}_p$ and $\mathbf{x}'_p$ into (8.14.19) and using (8.14.31), we have

$$\begin{bmatrix} 3b_1 t + 3a_1 + b_1 \\ 3b_2 t + 3a_2 + b_2 \end{bmatrix} e^{3t} = \begin{bmatrix} 1 & 4 \\ 1 & 1 \end{bmatrix} \begin{bmatrix} b_1 t + a_1 \\ b_2 t + a_2 \end{bmatrix} e^{3t} + \begin{bmatrix} -10 \\ 1 \end{bmatrix} e^{3t}. \tag{8.14.33}$$

In (8.14.33), dividing by e^{3t}, performing the indicated operations, and using the definition of equality of matrices, we obtain

$$\begin{cases} 3b_1 t + 3a_1 + b_1 = (4b_2 + b_1)t + 4a_2 + a_1 - 10, \\ 3b_2 t + 3a_2 + b_2 = (b_2 + b_1)t + a_2 + a_1 + 1. \end{cases} \tag{8.14.34}$$

Equation (8.14.34) will be an identity in t provided that

$$3b_1 = 4b_2 + b_1, \tag{8.14.35}$$

$$3a_1 + b_1 = 4a_2 + a_1 - 10, \tag{8.14.36}$$

$$3b_2 = b_2 + b_1, \tag{8.14.37}$$

$$3a_2 + b_2 = a_2 + a_1 + 1. \tag{8.14.38}$$

From (8.14.35) or (8.14.37), we see that $b_1 = 2b_2$.   Substituting this value of b_1 into (8.14.36), we obtain $2a_1 - 4a_2 = -2b_2 - 10$, or

$$a_1 - 2a_2 = -b_2 - 5. \tag{8.14.39}$$

From (8.14.38), we have

$$a_1 - 2a_2 = b_2 - 1. \tag{8.14.40}$$

In view of (8.14.39) and (8.14.40), it follows that $-b_2 - 5 = b_2 - 1$. Thus, $b_2 = -2$, and $b_1 = 2b_2 = -4$. Hence,

$$\mathbf{b} = \begin{bmatrix} b_1 \\ b_2 \end{bmatrix} = \begin{bmatrix} -4 \\ -2 \end{bmatrix} = -\begin{bmatrix} 4 \\ 2 \end{bmatrix} = -\frac{1}{4}\begin{bmatrix} 16 \\ 8 \end{bmatrix}. \tag{8.14.41}$$

Substituting $b_2 = -2$ into Equation (8.14.39) or (8.14.40) gives rise to the same equation, namely, $a_1 - 2a_2 = -3$ or

$$a_1 = 2a_2 - 3, \tag{8.14.42}$$

where a_2 may be considered as being arbitrary. Now, if we choose $a_2 = \frac{3}{4}$, then $a_1 = -\frac{6}{4}$, and, consequently, we have

$$\mathbf{a} = \begin{bmatrix} a_1 \\ a_2 \end{bmatrix} = \begin{bmatrix} -\frac{6}{4} \\ \frac{3}{4} \end{bmatrix} = -\frac{1}{4}\begin{bmatrix} 6 \\ -3 \end{bmatrix}. \tag{8.14.43}$$

Substituting the values of $\mathbf{b}$ and $\mathbf{a}$ of (8.14.41) and (8.14.43) into (8.14.30), we have

$$\mathbf{x}_p(t) = -\frac{1}{4}\begin{bmatrix} 6 \\ -3 \end{bmatrix} e^{3t} - \frac{1}{4}\begin{bmatrix} 16t \\ 8t \end{bmatrix} e^{3t} = -\frac{1}{4}\begin{bmatrix} 16t + 6 \\ 8t - 3 \end{bmatrix} e^{3t}, \tag{8.14.44}$$

which is the same as the particular solution found in (8.14.27). Thus, the general solution of the system (8.14.19) is given by (8.14.28).

Since a_2 is arbitrary in (8.14.42), there exist infinitely many particular solutions of the system (8.14.19). The choice of $a_2 = \frac{3}{4}$ was made simply to have $\mathbf{x}_p$ agree with the one already found. [*See* (8.14.27).] However, any other choice of a_2 would have given us a particular solution of the system (8.14.19): For instance, if we take $a_2 = 2$, then $a_1 = 1$, and another particular solution of (8.14.19) is given by

$$\mathbf{x}_p(t) = \begin{bmatrix} 1 \\ 2 \end{bmatrix} e^{3t} - \begin{bmatrix} 4t \\ 2t \end{bmatrix} e^{3t} = \begin{bmatrix} 1 - 4t \\ 2 - 2t \end{bmatrix} e^{3t}. \tag{8.14.45}$$

The reader is urged to verify that the particular solution $\mathbf{x}_p$ of (8.14.45) does indeed satisfy the system (8.14.19).

Exercises 8.14

1. Let $\mathbf{u}(t)$ and $\mathbf{v}(t)$ be two solutions of the system (8.14.1) on the interval J: $a \leq t \leq b$. Show that $\mathbf{u}(t) - \mathbf{v}(t)$ is a solution of the system (8.14.2) on J.
2. Let $\mathbf{x}^{(1)}(t), \mathbf{x}^{(2)}(t), \ldots, \mathbf{x}^{(n)}(t)$, be n linearly independent solutions of the system (8.14.2) on the interval J: $a \leq t \leq b$, and let $\mathbf{v}(t)$ be a solution of the system

(8.14.1) on J.   Show that every solution of the system (8.14.1) on J can be expressed in the form: $c_1 \mathbf{x}^{(1)}(t) + c_2 \mathbf{x}^{(2)}(t) + \cdots + c_n \mathbf{x}^{(n)}(t) + \mathbf{v}(t)$, for suitably chosen constants $c_1, c_2, \ldots, c_n$.

3. *Principle of superposition for particular solutions of the system* (8.14.1) *on the interval* J: $a \le t \le b$.   Let $\mathbf{x}_{p_i}(t)$, $i = 1, \ldots, n$, be particular solutions of the systems

$$\begin{cases} \mathbf{x}' = \mathbf{A}(t)\mathbf{x} + \mathbf{g}_1(t), \\ \mathbf{x}' = \mathbf{A}(t)\mathbf{x} + \mathbf{g}_2(t), \\ \vdots \\ \mathbf{x}' = \mathbf{A}(t)\mathbf{x} + \mathbf{g}_n(t), \end{cases}$$

on the interval J.   Show that $\mathbf{x}_p(t) = \mathbf{x}_{p_1}(t) + \cdots + \mathbf{x}_{p_n}(t)$ is a particular solution of the system $\mathbf{x}' = \mathbf{A}(t)\mathbf{x} + \mathbf{g}(t)$ on the interval J, where $\mathbf{g}(t) = \mathbf{g}_1(t) + \cdots + \mathbf{g}_n(t)$.

4. Show that if $\mathbf{x}_p$ is given by (8.14.14) and $\boldsymbol{\Phi}$ is a fundamental matrix of (8.14.2) on the interval J: $a \le t \le b$, then $\mathbf{x}_p$ satisfies the system (8.14.1).

In Exercises 5 through 9, utilize the method of variation of parameters to solve each system.

5. $\mathbf{x}' = \begin{bmatrix} -5 & -2 \\ 4 & 1 \end{bmatrix} \mathbf{x} + \begin{bmatrix} 2e^t \\ 12e^{3t} \end{bmatrix}.$ 6. $\mathbf{x}' = \begin{bmatrix} 1 & -1 \\ 5 & -3 \end{bmatrix} \mathbf{x} + \begin{bmatrix} e^{-t} \sin t \\ -e^{-t} \cos t \end{bmatrix}.$

7. $\mathbf{x}' = \begin{bmatrix} -1 & 1 \\ -4 & 3 \end{bmatrix} \mathbf{x} + \begin{bmatrix} e^t/t \\ 0 \end{bmatrix},$ $t > 0.$

8. $\mathbf{x}' = \begin{bmatrix} 2 & 1 \\ -5 & -2 \end{bmatrix} \mathbf{x} + \begin{bmatrix} \tan t \\ 1 \end{bmatrix},$ $0 < t < \dfrac{\pi}{2}.$

9. $\mathbf{x}' = \begin{bmatrix} 1 & 0 & -4 \\ 0 & 1 & 2 \\ 1 & 1 & 4 \end{bmatrix} \mathbf{x} + \begin{bmatrix} -e^{2t} \\ -e^{3t} \\ 4te^{4t} \end{bmatrix}.$

In Exercises 10 through 14, utilize the method of undetermined coefficients to solve each system.

10. $\mathbf{x}' = \begin{bmatrix} -1 & 1 \\ 4 & 2 \end{bmatrix} \mathbf{x} + \begin{bmatrix} 6e^t \\ -4e^{-t} \end{bmatrix}.$

[*Note:* To find a particular solution of the given system, find a particular solution

of the systems $\mathbf{x}' = \begin{bmatrix} -1 & 1 \\ 4 & 2 \end{bmatrix} \mathbf{x} + \begin{bmatrix} 6 \\ 0 \end{bmatrix} e^t$, $\mathbf{x}' = \begin{bmatrix} -1 & 1 \\ 4 & 2 \end{bmatrix} \mathbf{x} + \begin{bmatrix} 0 \\ -4 \end{bmatrix} e^{-t}$, and

then use the principle of superposition.]

11. $\mathbf{x}' = \begin{bmatrix} -1 & -8 \\ 1 & 3 \end{bmatrix} \mathbf{x} + \begin{bmatrix} e^t \sin t \\ e^t \cos t \end{bmatrix}.$ 12. $\mathbf{x}' = \begin{bmatrix} 5 & 3 \\ -3 & -1 \end{bmatrix} \mathbf{x} + \begin{bmatrix} 4t^2 - \frac{1}{2} \\ -6t \end{bmatrix}.$

13. $\mathbf{x}' = \begin{bmatrix} 3 & 2 \\ -2 & -1 \end{bmatrix} \mathbf{x} + \begin{bmatrix} e^t \\ \sin t - \cos t \end{bmatrix}.$

14. $\mathbf{x}' = \begin{bmatrix} -1 & 1 & 2 \\ 4 & -1 & 2 \\ 0 & 0 & -1 \end{bmatrix} \mathbf{x} + \begin{bmatrix} 2e^t \\ -9t \\ 2\cos t \end{bmatrix}$.

In Exercises 15 and 16, solve each initial-value problem.

15. $\mathbf{x}' = \begin{bmatrix} -2 & 5 \\ -5 & 8 \end{bmatrix} \mathbf{x} + \frac{1}{5} \begin{bmatrix} e^{4t} \\ e^{3t} \end{bmatrix}$, $\mathbf{x}(0) = \begin{bmatrix} 1 \\ 2 \end{bmatrix}$.

16. $\mathbf{x}' = \begin{bmatrix} 1 & -1 & -1 \\ 1 & -1 & 0 \\ 1 & 0 & -1 \end{bmatrix} \mathbf{x} + \begin{bmatrix} 1 \\ \sin t - \cos t \\ \sin t + \cos t \end{bmatrix}$, $\mathbf{x}(0) = \begin{bmatrix} 1 \\ 2 \\ -1 \end{bmatrix}$.

[*Note:* A fundamental matrix of the homogeneous system is given by (8.13.28).]

17. As in Exercise 1.3.9, the system

(1) $$t\mathbf{x}' = \mathbf{A}\mathbf{x}, \qquad t > 0,$$

where $\mathbf{A}$ is an $n \times n$ constant matrix, can be solved by assuming a solution of the form

(2) $$\mathbf{x} = \boldsymbol{\xi} t^\lambda$$

where λ and the constant vector $\boldsymbol{\xi}$ are to be determined. Substituting (2) into (1), show that

(3) $$(\mathbf{A} - \lambda \mathbf{I})\boldsymbol{\xi} = \mathbf{0}.$$

The system (3) has a nontrivial solution ($\boldsymbol{\xi} \neq \mathbf{0}$) if and only if

(4) $$\det [\mathbf{A} - \lambda \mathbf{I}] = 0.$$

Thus, if λ is a root of the characteristic Equation (4), then $\mathbf{x} = \boldsymbol{\xi} t^\lambda$ is a solution of the system (1).

Suppose that the roots of the characteristic Equation (4) are all simple. If $\boldsymbol{\Phi}(t)$ is a fundamental matrix of the system (1), utilize the method of variation of parameters to show that *a particular solution* $\mathbf{x}_p$ *of the system*

(5) $$t\mathbf{x}' = \mathbf{A}\mathbf{x} + \mathbf{g}(t), \qquad t > 0,$$

is given by

(6) $$\mathbf{x}_p(t) = \boldsymbol{\Phi}(t) \int \boldsymbol{\Phi}^{-1}(t) \left(\frac{\mathbf{g}(t)}{t} \right) dt.$$

18. Solve the initial-value problem

$$t\mathbf{x}' = \begin{bmatrix} -1 & 3 \\ -2 & 4 \end{bmatrix} \mathbf{x} + \begin{bmatrix} 2t^3 \\ 6t^4 \end{bmatrix}, \qquad \mathbf{x}(1) = \begin{bmatrix} 1 \\ -1 \end{bmatrix}, \qquad t > 0.$$

19. Consider the system

(1) $$t\mathbf{x}' = \mathbf{A}\mathbf{x}, \qquad t > 0,$$

where $\mathbf{A} = \begin{bmatrix} a_{11} & a_{12} \\ a_{21} & a_{22} \end{bmatrix}$. Suppose that the characteristic equation

(2)
$$\det\,[\mathbf{A} - \lambda\mathbf{I}] = 0$$

of the system (1) has repeated roots say, $\lambda = \alpha, \alpha$.   Then, one solution of (1) is given by

(3)
$$\mathbf{x}^{(1)}(t) = \boldsymbol{\xi}t^{\alpha},$$

where $\boldsymbol{\xi}^{T} = [\xi_1, \xi_2]$ is the eigenvector associated with the eigenvalue $\lambda = \alpha$.   To obtain a second linearly independent solution of the system (1), we let

(4)
$$\mathbf{x}^{(2)}(t) = \begin{bmatrix} 1 & \xi_1 t^{\alpha} \\ 0 & \xi_2 t^{\alpha} \end{bmatrix}\mathbf{v}(t),$$

where the vector function $\mathbf{v}$ is to be determined.   The matrix in (4) is obtained from the identity matrix $\begin{bmatrix} 1 & 0 \\ 0 & 1 \end{bmatrix}$ by replacing the second column by the solution $\mathbf{x}^{(1)}(t)$ given in (3).   This is known as *the method of reduction of order* concerning a system of equations.

Substituting $\mathbf{x} = \mathbf{x}^{(2)}(t)$ of (4) into (1), show that

(5)
$$\begin{bmatrix} t & \xi_1 t^{\alpha+1} \\ 0 & \xi_2 t^{\alpha+1} \end{bmatrix}\mathbf{v}' = \begin{bmatrix} a_{11} & 0 \\ a_{21} & 0 \end{bmatrix}\mathbf{v}.$$

Letting $\mathbf{v} = \begin{bmatrix} v_1(t) \\ v_2(t) \end{bmatrix}$, show that (5) can be expressed as

(6)
$$tv_1' + \xi_1 t^{\alpha+1}v_2' = a_{11}v_1,$$
(7)
$$\xi_2 t^{\alpha+1}v_2' = a_{21}v_1.$$

Case 1.   When $\xi_2 \neq 0$ in (7).   Solving for v_2' and then substituting into (6), show that

(8)
$$\frac{v_1'}{v_1} = \frac{k}{t},$$

where

(9)
$$k = a_{11} - \frac{\xi_1}{\xi_2}a_{21}.$$

Integrating (8) and suppressing the constant of integration, we obtain

(10)
$$v_1(t) = t^{k}.$$

Substituting v_1 of (10) into (7), then integrating and suppressing the constant of integration, show that

(11)
$$v_2(t) = \begin{cases} \dfrac{a_{21}}{\xi_2(k - \alpha)}\,t^{k-\alpha}, & \text{if } k \neq \alpha, \\[2ex] \dfrac{a_{21}}{\xi_2}\,\ln t, & \text{if } k = \alpha. \end{cases}$$

Consequently, $\mathbf{v}(t)$ is now determined, and

$$(12) \qquad \mathbf{x}^{(2)}(t) = \begin{bmatrix} 1 & \xi_1 t^{\alpha} \\ 0 & \xi_2 t^{\alpha} \end{bmatrix} \begin{bmatrix} v_1(t) \\ v_2(t) \end{bmatrix},$$

where v_1 and v_2 are given, respectively, in (10) and (11).

Case 2. When $\xi_2 = 0$ in (7). Show that the method of reduction of order does not give rise to a second linearly independent solution of the system (1). In this case, a second linearly independent solution of (1) may be determined as follows. Since one solution of the system (1) is of the form $\mathbf{x}^{(1)}(t) = \xi t^{\lambda}$, where $\lambda = \alpha$, we have

$$(13) \quad \begin{cases} (a_{11} - \lambda)\xi_1 + a_{12}\xi_2 = 0, \\ a_{21}\xi_1 + (a_{22} - \lambda)\xi_2 = 0, \end{cases} \quad \begin{vmatrix} a_{11} - \lambda & a_{12} \\ a_{21} & a_{22} - \lambda \end{vmatrix}$$

$$= (a_{11} - \lambda)(a_{22} - \lambda) - a_{12}a_{21} = 0.$$

Since $\xi = \begin{bmatrix} \xi_1 \\ \xi_2 \end{bmatrix}$ and $\xi_2 = 0$, we must have $\xi_1 \neq 0$. (Why?) In view of (13), show that $a_{21} = 0$, $a_{11} = \alpha$, and $a_{22} = \alpha$. Let

$$(14) \qquad \mathbf{x}^{(2)}(t) = \begin{bmatrix} x_{12}(t) \\ x_{22}(t) \end{bmatrix}.$$

Substituting (14) into (1), show that

$$(15) \qquad \begin{cases} tx_{12}' = \alpha x_{12} + a_{12}x_{22}, \end{cases}$$
$$(16) \qquad \begin{cases} tx_{22}' = \alpha x_{22}. \end{cases}$$

Integrating (16) and suppressing the constant of integration, we obtain

$$(17) \qquad x_{22}(t) = t^{\alpha}.$$

Substituting x_{22} of (17) into (15), then integrating and suppressing the constant of integration, show that

$$(18) \qquad x_{12}(t) = a_{12} t^{\alpha} \ln t.$$

Thus,

$$(19) \qquad \mathbf{x}^{(2)}(t) = \begin{bmatrix} x_{12}(t) \\ x_{22}(t) \end{bmatrix} = \begin{bmatrix} a_{12} \ln t \\ 1 \end{bmatrix} t^{\alpha}.$$

Clearly, $\mathbf{x}^{(1)}(t)$ and $\mathbf{x}^{(2)}(t)$ are linearly independent on the interval J: $t > 0$.

20. Solve the initial-value problem

$$t\mathbf{x}' = \begin{bmatrix} -1 & 2 \\ -2 & 3 \end{bmatrix} \mathbf{x} + \begin{bmatrix} t^2 \\ 2t^3 \end{bmatrix}, \qquad \mathbf{x}(1) = \begin{bmatrix} 1 \\ 2 \end{bmatrix}, \qquad t > 0.$$

21. Solve the initial-value problem

$$t\mathbf{x}' = \begin{bmatrix} 1 & 1 \\ 0 & 1 \end{bmatrix} \mathbf{x} + \begin{bmatrix} \ln t \\ 1 \end{bmatrix} t, \qquad \mathbf{x}(1) = \begin{bmatrix} 2 \\ 3 \end{bmatrix}, \qquad t > 0.$$

8.15. Solutions of Linear Systems Utilizing the Exponential Function e^{tA}

In this section we shall develop the solution of the linear homogeneous system

$$\mathbf{x}' = \mathbf{A}\mathbf{x} \tag{8.15.1}$$

from the point of view of involving the exponential function e^{tA}, where $\mathbf{A}$ is an $n \times n$ constant matrix. Also, we shall discuss the solution of the non-homogeneous linear system.

Define $\mathbf{A}^0 = \mathbf{I}$, where $\mathbf{I}$ is an $n \times n$ identity matrix. In view of the associative law (8.6.11), we have for any two positive integers r and s,

$$\mathbf{A}^r\mathbf{A}^s = \mathbf{A}^s\mathbf{A}^r = \mathbf{A}^{r+s}. \tag{8.15.2}$$

An expression of the form

$$P_m(\mathbf{A}) = a_0\mathbf{I} + a_1\mathbf{A} + a_2\mathbf{A}^2 + \cdots + a_m\mathbf{A}^m = \sum_{k=0}^{m} a_k\mathbf{A}^k, \tag{8.15.3}$$

where a_k $(k = 0, 1, \ldots, m)$ are scalars, is called a *polynomial in* $\mathbf{A}$ *of degree m.* Clearly, $P_m(\mathbf{A})$ is an $n \times n$ matrix whenever $\mathbf{A}$ is an $n \times n$ matrix.

If $\mathbf{D}$ is an $n \times n$ diagonal matrix:

$$\mathbf{D} = \begin{bmatrix} \lambda_1 & 0 & \cdots & 0 \\ 0 & \lambda_2 & \cdots & 0 \\ \vdots & \vdots & & \vdots \\ 0 & 0 & \cdots & \lambda_n \end{bmatrix}, \tag{8.15.4}$$

then for any positive integer r, we have [*see* Exercise 8.6.10(b)]

$$\mathbf{D}^r = \begin{bmatrix} \lambda_1^{\,r} & 0 & \cdots & 0 \\ 0 & \lambda_2^{\,r} & \cdots & 0 \\ \vdots & \vdots & & \vdots \\ 0 & 0 & \cdots & \lambda_n^{\,r} \end{bmatrix}. \tag{8.15.5}$$

Utilizing (8.15.3), (8.15.5), (8.6.5), and (8.6.4), we obtain [*see* Exercise 8.15.1]

$$P_m(\mathbf{D}) = \begin{bmatrix} P_m(\lambda_1) & 0 & \cdots & 0 \\ 0 & P_m(\lambda_2) & \cdots & 0 \\ \vdots & \vdots & & \vdots \\ 0 & 0 & \cdots & P_m(\lambda_n) \end{bmatrix}. \tag{8.15.6}$$

Example 8.15.1. If $P_2(\mathbf{A}) = 3\mathbf{I} + 2\mathbf{A} - \mathbf{A}^2$ and $\mathbf{D} = \begin{bmatrix} 1 & 0 \\ 0 & 2 \end{bmatrix}$, then

$$P_2(\mathbf{D}) = \begin{bmatrix} P_2(1) & 0 \\ 0 & P_2(2) \end{bmatrix} = \begin{bmatrix} 4 & 0 \\ 0 & 3 \end{bmatrix}.$$

From calculus, $e^x = \sum_{n=0}^{\infty} x^n/n!$, $-\infty < x < \infty$. Thus, it is somewhat plausible to define $e^{\mathbf{B}}$ by

$$e^{\mathbf{B}} = \mathbf{I} + \frac{1}{1!}\mathbf{B} + \frac{1}{2!}\mathbf{B}^2 + \cdots + \frac{1}{k!}\mathbf{B}^k + \cdots, \qquad (8.15.7)$$

where $\mathbf{B}$ denotes an $n \times n$ constant matrix. The sum indicated in (8.15.7), which involves the addition of an infinite number of matrices has, in itself, no meaning. In order to assign a meaning to the sum of such infinite series of matrices, we consider the associated sequence of partial sums $\mathbf{S}_0$, $\mathbf{S}_1$, $\mathbf{S}_2$, ..., $\mathbf{S}_m$, ..., where

$$
\begin{cases}
\mathbf{S}_0 = \mathbf{I}, \\[2mm]
\mathbf{S}_1 = \mathbf{I} + \dfrac{1}{1!}\mathbf{B}, \\[2mm]
\mathbf{S}_2 = \mathbf{I} + \dfrac{1}{1!}\mathbf{B} + \dfrac{1}{2!}\mathbf{B}^2, \\[2mm]
\quad\vdots \\[2mm]
\mathbf{S}_m = \mathbf{I} + \dfrac{1}{1!}\mathbf{B} + \dfrac{1}{2!}\mathbf{B}^2 + \cdots + \dfrac{1}{m!}\mathbf{B}^m, \\[2mm]
\quad\vdots
\end{cases}
$$

Note that the elements of the sequence of partial sums are polynomials in $\mathbf{B}$. Now, if the $\lim_{m\to\infty} \mathbf{S}_m$ converges in the sense that each element of $\mathbf{S}_m$ converges, then we define

$$e^{\mathbf{B}} = \lim_{m\to\infty} \mathbf{S}_m = \lim_{m\to\infty} \sum_{k=0}^{m} \frac{\mathbf{B}^k}{k!}. \qquad (8.15.8)$$

We shall now show that if $\mathbf{B}$ is an $n \times n$ constant matrix, whose elements b_{ij} are uniformly bounded, that is, $|b_{ij}| < M$, where M is a constant for $i, j = 1, 2, \ldots, n$, then $\lim_{m\to\infty} \mathbf{S}_m$ exists. (Here, the symbol "$|\ \ |$" denotes the absolute value.)

Since $\mathbf{B}^2 = \mathbf{BB}$, in view of Definition 8.6.4, we see that an element of $\mathbf{B}^2$, say B_{ij}, is given by

$$B_{ij} = \sum_{k=1}^{n} b_{ik} b_{kj}, \qquad i, j = 1, 2, \ldots, n.$$

Thus,

$$|B_{ij}| \leq \sum_{k=1}^{n} |b_{ik}||b_{kj}| < \sum_{k=1}^{n} M^2 = nM^2.$$

Consequently, each element in $\mathbf{B}^2$ is bounded by nM^2. Similarly, since $\mathbf{B}^3 = \mathbf{B}^2 \cdot \mathbf{B}$, we see that each element of $\mathbf{B}^3$ is bounded by $n^2 M^3$. By induction, one may claim that each element of $\mathbf{B}^k$ is bounded by $n^{k-1}M^k$. Hence, every element of $\mathbf{S}_m - \mathbf{I}$ is bounded by the series

$$\sum_{k=1}^{m} \frac{n^{k-1}M^k}{k!} = \frac{1}{n} \sum_{k=1}^{m} \frac{(nM)^k}{k!}.$$

Thus, each term of $\mathbf{S}_m$ converges, since each term of $\mathbf{S}_m - \mathbf{I}$ is a series bounded in absolute value term by term by the elements of the series $\dfrac{1}{n} \displaystyle\sum_{k=1}^{m} \dfrac{(nM)^k}{k!}$ which converges to $\dfrac{1}{n}(e^{nM} - 1)$ as $m \to \infty$.

In view of (8.15.8), we see that $e^{\mathbf{o}} = \mathbf{I}$.   If $\mathbf{A}$ and $\mathbf{B}$ are two $n \times n$ matrices such that $\mathbf{AB} = \mathbf{BA}$, then utilizing Theorem 7.2.8 and Exercise 8.6.14, we leave it as an exercise for the reader [*see* Exercise 8.15.2] to show that the law of exponents holds, namely,

$$e^{\mathbf{A}} \cdot e^{\mathbf{B}} = e^{\mathbf{A}+\mathbf{B}}. \tag{8.15.9}$$

Note that, in general, the result given in (8.15.9) is not true unless $\mathbf{A}$ and $\mathbf{B}$ commute with each other.

If $\mathbf{B} = -\mathbf{A}$ in (8.15.9), then $e^{\mathbf{A}} \cdot e^{-\mathbf{A}} = e^{\mathbf{A}-\mathbf{A}} = e^{\mathbf{o}} = \mathbf{I}$.   Thus,

$$e^{-\mathbf{A}} = (e^{\mathbf{A}})^{-1}. \tag{8.15.10}$$

Let $\mathbf{A}$ be an $n \times n$ constant matrix and let t be any real number.   We define $e^{t\mathbf{A}}$ by

$$e^{t\mathbf{A}} = \lim_{m \to \infty} \sum_{k=0}^{m} \frac{t^k \mathbf{A}^k}{k!}. \tag{8.15.11}$$

We leave it as an exercise for the reader [*see* Exercise 8.15.3] to show that

$$\frac{d}{dt}(e^{t\mathbf{A}}) = \mathbf{A}e^{t\mathbf{A}} = e^{t\mathbf{A}}\mathbf{A}. \tag{8.15.12}$$

From the above results we see that the exponential function $e^{t\mathbf{A}}$ retains the same basic properties of the exponential function $e^{\alpha x}$, except for the law of exponents involving matrices, we require that $\mathbf{AB} = \mathbf{BA}$.

Let $\mathbf{A}$ be an $n \times n$ constant matrix.   The characteristic polynomial of the matrix $\mathbf{A}$ is defined by [*see* (8.11.8)]

$$p(\lambda) \equiv \det[\mathbf{A} - \lambda\mathbf{I}] \equiv a_0 + a_1\lambda + a_2\lambda^2 + \cdots + a_n\lambda^n, \tag{8.15.13}$$

where $a_n = (-1)^n$.   [*Note:* $a_n = (-1)^n$ refers only to the last coefficient.]

We make the following observations in preparation for the forthcoming important theorem.   Let $\mathbf{A} = \begin{bmatrix} 1 & 0 & 1 \\ 0 & 1 & -1 \\ -1 & 2 & 2 \end{bmatrix}$, then

$$p(\lambda) = \begin{vmatrix} 1-\lambda & 0 & 1 \\ 0 & 1-\lambda & -1 \\ -1 & 2 & 2-\lambda \end{vmatrix} = 5 - 8\lambda + 4\lambda^2 - \lambda^3.$$

The reader may verify that

$$\mathbf{A}^2 = \begin{bmatrix} 0 & 2 & 3 \\ 1 & -1 & -3 \\ -3 & 6 & 1 \end{bmatrix}, \qquad \mathbf{A}^3 = \begin{bmatrix} -3 & 8 & 4 \\ 4 & -7 & -4 \\ -4 & 8 & -7 \end{bmatrix}.$$

Thus,

$$p(\mathbf{A}) = 5\mathbf{I} - 8\mathbf{A} + 4\mathbf{A}^2 - \mathbf{A}^3$$

$$= 5\begin{bmatrix} 1 & 0 & 0 \\ 0 & 1 & 0 \\ 0 & 0 & 1 \end{bmatrix} - 8\begin{bmatrix} 1 & 0 & 1 \\ 0 & 1 & -1 \\ -1 & 2 & 2 \end{bmatrix} + 4\begin{bmatrix} 0 & 2 & 3 \\ 1 & -1 & -3 \\ -3 & 6 & 1 \end{bmatrix}$$

$$- \begin{bmatrix} -3 & 8 & 4 \\ 4 & -7 & -4 \\ -4 & 8 & -7 \end{bmatrix} = \begin{bmatrix} 0 & 0 & 0 \\ 0 & 0 & 0 \\ 0 & 0 & 0 \end{bmatrix} = \mathbf{0}.$$

Hence, the matrix $\mathbf{A}$ satisfies its own characteristic equation $p(\mathbf{A}) = \mathbf{0}$.
Let $\mathbf{B}$ denote the adjoint of the characteristic matrix:

$$\mathbf{A} - \lambda\mathbf{I} = \begin{bmatrix} 1-\lambda & 0 & 1 \\ 0 & 1-\lambda & -1 \\ -1 & 2 & 2-\lambda \end{bmatrix}.$$

Utilizing the method given in Section 8.7, we find

$$\mathbf{B} = \begin{bmatrix} 4-3\lambda+\lambda^2 & 2 & -1+\lambda \\ -1 & 3-3\lambda+\lambda^2 & 1-\lambda \\ 1-\lambda & -2+2\lambda & 1-2\lambda+\lambda^2 \end{bmatrix}$$

$$= \begin{bmatrix} 4 & 2 & -1 \\ -1 & 3 & 1 \\ 1 & -2 & 1 \end{bmatrix} + \begin{bmatrix} -3 & 0 & 1 \\ 0 & -3 & -1 \\ -1 & 2 & -2 \end{bmatrix}\lambda + \begin{bmatrix} 1 & 0 & 0 \\ 0 & 1 & 0 \\ 0 & 0 & 1 \end{bmatrix}\lambda^2$$

$$\equiv B_0 + B_1\lambda + B_2\lambda^2.$$

Thus, $\mathbf{B}$, the adjoint of the characteristic matrix $\mathbf{A} - \lambda\mathbf{I}$, can be expressed as a
matrix polynomial of degree at most $n - 1 = 3 - 1 = 2$ in λ. This result is
true for the general case when the characteristic matrix $\mathbf{A} - \lambda\mathbf{I}$ is a polynomial of degree n.

The following theorem is important and useful.

Theorem 8.15.1. *Cayley-Hamilton. Every $n \times n$ matrix $\mathbf{A}$ satisfies its own
characteristic equation. That is, $p(\mathbf{A}) = \mathbf{0}$.*

PROOF.   Let $\mathbf{B}$ denote the adjoint of the characteristic matrix $\mathbf{A} - \lambda\mathbf{I}$.   Then $\mathbf{B}$ can be expressed as a matrix polynomial of degree not exceeding $n - 1$ in λ.   Let

$$\mathbf{B} = \mathbf{B}_0 + \mathbf{B}_1\lambda + \mathbf{B}_2\lambda^2 + \cdots + \mathbf{B}_{n-1}\lambda^{n-1} = \sum_{k=0}^{n-1} \mathbf{B}_k\lambda^k, \qquad (8.15.14)$$

where $\mathbf{B}_j$ is the matrix whose elements are the coefficients of λ^j in the corresponding elements of $\mathbf{B}$.   In view of (8.7.5) with $[M_{ij}]^T$ replaced by $\mathbf{B}$ and $\mathbf{A}$ replaced by $\mathbf{A} - \lambda\mathbf{I}$, we obtain $(\mathbf{A} - \lambda\mathbf{I})\mathbf{B} = \det[\mathbf{A} - \lambda\mathbf{I}]\,\mathbf{I}$ or

$$(\mathbf{A} - \lambda\mathbf{I})\mathbf{B} = p(\lambda)\mathbf{I}. \qquad (8.15.15)$$

In view of (8.15.13) and (8.15.14), we may express (8.15.15) as

$$\sum_{k=0}^{n-1} \mathbf{A}\mathbf{B}_k\lambda^k - \sum_{k=0}^{n-1} \mathbf{B}_k\lambda^{k+1} = \sum_{k=0}^{n} (a_k\mathbf{I})\lambda^k,$$

which we may consider as being an identity in λ.   Thus, upon equating corresponding coefficients, we obtain

$$\begin{cases} \mathbf{A}\mathbf{B}_0 & = a_0\mathbf{I}, \\ \mathbf{A}\mathbf{B}_1 - \mathbf{B}_0 & = a_1\mathbf{I}, \\ \mathbf{A}\mathbf{B}_2 - \mathbf{B}_1 & = a_2\mathbf{I}, \\ & \;\;\vdots \\ \mathbf{A}\mathbf{B}_{n-1} - \mathbf{B}_{n-2} & = a_{n-1}\mathbf{I}, \\ -\mathbf{B}_{n-1} & = a_n\mathbf{I}. \end{cases} \qquad (8.15.16)$$

Multiplying each equation in (8.15.16) on the left, respectively, by $\mathbf{I}, \mathbf{A}, \mathbf{A}^2, \ldots, \mathbf{A}^{n-1}, \mathbf{A}^n$ and then adding, the terms on the left cancel in pairs, and we obtain

$$\mathbf{0} = a_0\mathbf{I} + a_1\mathbf{A} + a_2\mathbf{A}^2 + \cdots + a_{n-1}\mathbf{A}^{n-1} + a_n\mathbf{A}^n. \qquad (8.15.17)$$

That is, $p(\mathbf{A}) = \mathbf{0}$, which was to be proved.

Remark 8.15.1.   An important application of the Cayley-Hamilton theorem is that it enables one to compute higher powers of a matrix $\mathbf{A}$.   For example, if we have already computed $\mathbf{A}^2, \mathbf{A}^3, \ldots, \mathbf{A}^{n-1}$, we may compute $\mathbf{A}^n$ by utilizing Formula (8.15.17).   For, since $a_n = (-1)^n \neq 0$, we have

$$\mathbf{A}^n = -\frac{a_0}{a_n}\mathbf{I} - \frac{a_1}{a_n}\mathbf{A} - \cdots - \frac{a_{n-1}}{a_n}\mathbf{A}^{n-1}. \qquad (8.15.18)$$

We may now compute $\mathbf{A}^{n+1}$ as follows.   Multiplying (8.15.18) by $\mathbf{A}$, we obtain

$$\mathbf{A}^{n+1} = -\frac{a_0}{a_n}\mathbf{A} - \frac{a_1}{a_n}\mathbf{A}^2 - \cdots - \frac{a_{n-1}}{a_n}\mathbf{A}^n. \qquad (8.15.19)$$

Substituting $\mathbf{A}^n$ of (8.15.18) into (8.15.19), and then combining similar powers of $\mathbf{A}$, we find

$$\mathbf{A}^{n+1} = \frac{a_{n-1}a_0}{a_n^2}\mathbf{I} + \left(\frac{a_{n-1}a_1}{a_n^2} - \frac{a_0}{a_n}\right)\mathbf{A} + \cdots + \left(\frac{a_{n-1}^2}{a_n^2} - \frac{a_{n-2}}{a_n}\right)\mathbf{A}^{n-1}. \quad (8.15.20)$$

Continuing in this fashion, we see that any positive integral power of $\mathbf{A}$ can be expressed as a linear combination of $\mathbf{I}$, $\mathbf{A}$, $\mathbf{A}^2$, ..., $\mathbf{A}^{n-1}$.

Also, if $\mathbf{A}^{-1}$ exists, then multiplying (8.15.17) by $\mathbf{A}^{-1}$, we obtain

$$\mathbf{A}^{-1} = -\frac{a_1}{a_0}\mathbf{I} - \frac{a_2}{a_0}\mathbf{A} - \cdots - \frac{a_n}{a_0}\mathbf{A}^{n-1}, \qquad a_0 \neq 0. \quad (8.15.21)$$

Multiplying (8.15.21) by $\mathbf{A}^{-1}$, we have

$$\mathbf{A}^{-2} = -\frac{a_1}{a_0}\mathbf{A}^{-1} - \frac{a_2}{a_0}\mathbf{I} - \cdots - \frac{a_n}{a_0}\mathbf{A}^{n-2}. \quad (8.15.22)$$

Substituting $\mathbf{A}^{-1}$ of (8.15.21) into (8.15.22), and then combining similar powers of $\mathbf{A}$, we obtain

$$\mathbf{A}^{-2} = \left(\frac{a_1^2}{a_0^2} - \frac{a_2}{a_0}\right)\mathbf{I} + \left(\frac{a_1a_2}{a_0^2} - \frac{a_3}{a_0}\right)\mathbf{A} + \cdots + \left(\frac{a_1a_{n-1}}{a_0^2} - \frac{a_n}{a_0}\right)\mathbf{A}^{n-2}$$

$$+ \frac{a_1a_n}{a_0^2}\mathbf{A}^{n-1}. \quad (8.15.23)$$

Continuing in this fashion, we see that if $\mathbf{A}^{-1}$ exists, then any negative integral power of $\mathbf{A}$ can be expressed as a linear combination of $\mathbf{I}$, $\mathbf{A}$, $\mathbf{A}^2$, ..., $\mathbf{A}^{n-1}$.

The above procedures may be utilized in computer science to program $\mathbf{A}^p$ and $\mathbf{A}^{-p}$, for large integral values of p.

Theorem 8.15.2. *Let $\mathbf{A}$ be an $n \times n$ matrix with characteristic polynomial $p(\lambda)$. If $f(\lambda)$ is a polynomial of degree $m \geq n$, then $f(\mathbf{A}) = r(\mathbf{A})$, where $r(\lambda)$ is the remainder when $f(\lambda)$ is divided by $p(\lambda)$.*

PROOF.   From algebra we have

$$\frac{f(\lambda)}{p(\lambda)} = q(\lambda) + \frac{r(\lambda)}{p(\lambda)},$$

or

$$f(\lambda) = p(\lambda)q(\lambda) + r(\lambda), \quad (8.15.24)$$

where $r(\lambda)$ is a polynomial of degree at most $m - 1$ in λ. Since (8.15.24) is a polynomial identity in λ, it will also be valid if we replace λ by the matrix $\mathbf{A}$. Thus,

$$f(\mathbf{A}) = p(\mathbf{A})q(\mathbf{A}) + r(\mathbf{A}).$$

By Theorem 8.15.1, $p(\mathbf{A}) = \mathbf{0}$.   Consequently, $f(\mathbf{A}) = \mathbf{0} \cdot q(\mathbf{A}) + r(\mathbf{A}) = r(\mathbf{A})$, and the theorem is established.

The above theorem is applicable only when we are dealing with polynomial functions of $\mathbf{A}$.   If we wish to determine $f(\mathbf{A})$ when f is an analytic function of λ, some basic concepts of complex variables are needed.   Thus, we shall simply state the following important theorem.   [*See* [15], pp. 120–121 for a proof.]

Theorem 8.15.3.   *Suppose that* $\mathbf{A}$ *is an* $n \times n$ *matrix with simple eigenvalues* $\lambda_1, \lambda_2, \ldots, \lambda_n$ *such that* $|\lambda_1| \leq |\lambda_2| \leq \cdots \leq |\lambda_n|$. *Let* C: $|z| = \rho$ *be a circle (in the complex plane) with center at the origin and radius* $\rho > |\lambda_n|$. *If* f *is an analytic function of* λ *in the interior of* C, *then* $f(\mathbf{A}) = r(\mathbf{A})$, *where* $r(\mathbf{A})$ *is a polynomial of degree* $n - 1$ *for which*

$$f(\lambda_k) = r(\lambda_k), \qquad k = 1, 2, \ldots, n. \tag{8.15.25}$$

If $\lambda_1 = \lambda_2 = \cdots = \lambda_v$, *so that* λ_v *is an eigenvalue of multiplicity* v *for the matrix* $\mathbf{A}$, *then* $r(\mathbf{A})$ *is that polynomial of degree* $n - 1$ *for which* $f(\lambda_k) = r(\lambda_k)$, $k = v + 1, v + 2, \ldots, n$, *and also* $f^{(j)}(\lambda) = r^{(j)}(\lambda)$, $j = 0, 1, 2, \ldots, v - 1$.

Note that we may express $r(\mathbf{A})$ as

$$r(\mathbf{A}) = \alpha_0 \mathbf{I} + \alpha_1 \mathbf{A} + \alpha_2 \mathbf{A}^2 + \cdots + \alpha_{n-1} \mathbf{A}^{n-1}. \tag{8.15.26}$$

Example 8.15.2.   Let us evaluate $e^{\mathbf{A}}$, when

$$\mathbf{A} = \begin{bmatrix} -1 & 2 \\ -3 & 4 \end{bmatrix}. \tag{8.15.27}$$

SOLUTION.   The reader may show that the eigenvalues of the matrix $\mathbf{A}$ are $\lambda_1 = 1$, $\lambda_2 = 2$.   Take $f(\lambda) = e^{\lambda}$.   Thus, $f(\mathbf{A}) = e^{\mathbf{A}}$ and $r(\mathbf{A}) = \alpha_0 \mathbf{I} + \alpha_1 \mathbf{A}$.   Consequently, we have

$$e^{\mathbf{A}} = \alpha_0 \mathbf{I} + \alpha_1 \mathbf{A}, \tag{8.15.28}$$

where α_1 and α_2 are determined from $f(\lambda_k) = r(\lambda_k)$, that is, from

$$e^{\lambda_k} = \alpha_0 + \alpha_1 \lambda_k, \qquad k = 1, 2. \tag{8.15.29}$$

Substituting $\lambda_1 = 1$ and $\lambda_2 = 2$, respectively, into (8.15.29), we obtain

$$\begin{cases} e = \alpha_0 + \alpha_1, \\ e^2 = \alpha_0 + 2\alpha_1. \end{cases} \tag{8.15.30}$$

Solving the System (8.15.30) for α_1 and α_2, we find

$$\alpha_0 = 2e - e^2, \qquad \alpha_1 = e^2 - e.$$

Substituting these values of α_0 and α_1 into (8.15.28), we obtain the desired evaluation of $e^{\mathbf{A}}$, namely,

$$e^{\mathbf{A}} = (2e - e^2)\begin{bmatrix} 1 & 0 \\ 0 & 1 \end{bmatrix} + (e^2 - e)\begin{bmatrix} -1 & 2 \\ -3 & 4 \end{bmatrix} = \begin{bmatrix} 3e - 2e^2 & -2e + 2e^2 \\ 3e - 3e^2 & -2e + 3e^2 \end{bmatrix}.$$

$$(8.15.31)$$

Example 8.15.3.   Let us evaluate $e^{t\mathbf{A}}$, where the matrix $\mathbf{A}$ is given in (8.15.27).

SOLUTION.   The eigenvalues of $t\mathbf{A}$ are given by $\lambda_1 = t$, $\lambda_2 = 2t$.   Thus, the evaluation of $e^{t\mathbf{A}}$ may be obtained from (8.15.31) by replacing $\mathbf{A}$ by $t\mathbf{A}$ and the exponents 1 and 2 of e by t and $2t$, that is,

$$e^{t\mathbf{A}} = \begin{bmatrix} 3e^t - 2e^{2t} & -2e^t + 2e^{2t} \\ 3e^t - 3e^{2t} & -2e^t + 3e^{2t} \end{bmatrix}. \tag{8.15.32}$$

To see this more clearly, write (8.15.28) and (8.15.29) as

$$e^{t\mathbf{A}} = \alpha\mathbf{I} + \beta\mathbf{A}, \qquad \alpha = \alpha_0, \quad \beta = t\alpha_1, \tag{8.15.33}$$

and

$$e^{t\lambda_k} = \alpha + \beta\lambda_k, \qquad k = 1, 2. \tag{8.15.34}$$

Substituting $\lambda = 1$ and $\lambda = 2$, respectively, into (8.15.34) and then solving the resulting system simultaneously for α and β, we obtain

$$\alpha = 2e^t - e^{2t}, \qquad \beta = e^{2t} - e^t.$$

Substituting these values of α and β into (8.15.33), we obtain (8.15.32).

Example 8.15.4.   Let us evaluate $e^{t\mathbf{A}}$, when

$$\mathbf{A} = \begin{bmatrix} 3 & 0 & 0 \\ 0 & 7 & 5 \\ 0 & -4 & -2 \end{bmatrix}. \tag{8.15.35}$$

SOLUTION.   From Example 8.12.2, the eigenvalues of $\mathbf{A}$ are $\lambda_1 = 3$, $\lambda_2 = 3$, and $\lambda_3 = 2$.   Hence, $\lambda = 3$ is an eigenvalue of multiplicity two.   Take $f(\lambda) = e^{t\lambda}$.   Thus,

$$e^{t\mathbf{A}} = \alpha\mathbf{I} + \beta\mathbf{A} + \gamma\mathbf{A}^2. \tag{8.15.36}$$

In view of Theorem 8.15.3, α, β, and γ are determined from

$$f(\lambda_k) = r(\lambda_k) \quad \text{or} \quad e^{t\lambda_k} = \alpha + \beta\lambda_k + \gamma\lambda_k^2, \qquad k = 3, \tag{8.15.37}$$

and

$$f^{(j)}(\lambda) = r^{(j)}(\lambda) \tag{8.15.38}$$

for $j = 0, 1$.   When $j = 0$, $f^{(0)}(\lambda) = r^{(0)}(\lambda)$ which means that $f(\lambda) = r(\lambda)$.   Thus,

$$e^{t\lambda} = \alpha + \beta\lambda + \gamma\lambda^2. \tag{8.15.39}$$

When $j = 1$, we have $f'(\lambda) = r'(\lambda)$.   Differentiating (8.15.39) with respect to λ, we obtain

$$te^{t\lambda} = \beta + 2\gamma\lambda. \tag{8.15.40}$$

Substituting $\lambda_3 = 2$ into (8.15.37) and $\lambda = 3$ into (8.15.39) and (8.15.40), we obtain

$$\begin{cases} \alpha + 2\beta + 4\gamma = e^{2t}, \\ \alpha + 3\beta + 9\gamma = e^{3t}, \\ \quad\ \ \beta + 6\gamma = te^{3t}. \end{cases} \tag{8.15.41}$$

Solving the system (8.15.41) for α, β, and γ, we find

$$\alpha = 9e^{2t} - 8e^{3t} + 6te^{3t}, \qquad \beta = -6e^{2t} + 6e^{3t} - 5te^{3t}, \tag{8.15.42}$$
$$\gamma = e^{2t} - e^{3t} + te^{3t}.$$

The reader may verify that

$$\mathbf{A}^2 = \begin{bmatrix} 9 & 0 & 0 \\ 0 & 29 & 25 \\ 0 & -20 & -16 \end{bmatrix}. \tag{8.15.43}$$

In view of (8.15.35), (8.15.42), and (8.15.43), we may write (8.15.36) as

$$e^{t\mathbf{A}} = \alpha \begin{bmatrix} 1 & 0 & 0 \\ 0 & 1 & 0 \\ 0 & 0 & 1 \end{bmatrix} + \beta \begin{bmatrix} 3 & 0 & 0 \\ 0 & 7 & 5 \\ 0 & -4 & -2 \end{bmatrix} + \gamma \begin{bmatrix} 9 & 0 & 0 \\ 0 & 29 & 25 \\ 0 & -20 & -16 \end{bmatrix}$$

$$= \begin{bmatrix} \alpha + 3\beta + 9\gamma & 0 & 0 \\ 0 & \alpha + 7\beta + 29\gamma & 5\beta + 25\gamma \\ 0 & -4\beta - 20\gamma & \alpha - 2\beta - 16\gamma \end{bmatrix}$$

$$= \begin{bmatrix} e^{3t} & 0 & 0 \\ 0 & 5e^{3t} - 4e^{2t} & 5e^{3t} - 5e^{2t} \\ 0 & -4e^{3t} + 4e^{2t} & -4e^3 + 5e^{2t} \end{bmatrix}. \tag{8.15.44}$$

We are now in a position to solve the system (8.15.1) by utilizing the exponential function $e^{t\mathbf{A}}$.   Let $\mathbf{x}(t)$ be any solution of the system

$$\mathbf{x}' = \mathbf{A}\mathbf{x} \tag{8.15.45}$$

where $\mathbf{A}$ is an $n \times n$ constant matrix.   Consider the vector function $\mathbf{v}$ defined by

$$\mathbf{v}(t) = e^{-t\mathbf{A}}\mathbf{x}. \tag{8.15.46}$$

Differentiating (8.15.46) and using (8.15.45), we find

$$\mathbf{v}'(t) = e^{-t\mathbf{A}}\mathbf{x}' - e^{-t\mathbf{A}}\mathbf{A}\mathbf{x} = e^{-t\mathbf{A}}(\mathbf{x}' - \mathbf{A}\mathbf{x}) = e^{-t\mathbf{A}} \cdot \mathbf{0} = \mathbf{0}.$$

Thus $\mathbf{v}(t) = \mathbf{c}$, where $\mathbf{c}$ is an arbitrary constant vector. Multiplying (8.15.46) on the left by $e^{t\mathbf{A}}$ and observing that $e^{t\mathbf{A}} \cdot e^{-t\mathbf{A}} = e^{0} = \mathbf{I}$, we obtain $\mathbf{x} = e^{t\mathbf{A}}\mathbf{v}(t)$ or

$$\mathbf{x} = e^{t\mathbf{A}}\mathbf{c}. \tag{8.15.47}$$

Thus, *the general solution* $\mathbf{x} = \mathbf{x}(t)$ *of the system* (8.13.45) *on the interval J*: $-\infty < t < \infty$, *is given by* (8.15.47).

Suppose that we wish to solve the system (8.15.45) subject to the initial condition

$$\mathbf{x}(t_0) = \mathbf{x}^0, \tag{8.15.48}$$

where t_0 is a point in J and $\mathbf{x}^0$ is a given initial vector. In view of (8.15.48), with $t = t_0$ in (8.15.47), we obtain $\mathbf{x}^0 = e^{t_0\mathbf{A}}\mathbf{c}$. Thus, $\mathbf{c} = e^{-t_0\mathbf{A}}\mathbf{x}^0$. Consequently, (8.15.47) becomes

$$\mathbf{x} = e^{(t-t_0)\mathbf{A}}\mathbf{x}^0. \tag{8.15.49}$$

Thus, *the unique solution* $\mathbf{x} = \mathbf{x}(t)$ *of the given initial-value problem on the interval J is given by* (8.15.49).

Let us consider now the nonhomogeneous linear system

$$\mathbf{x}' = \mathbf{A}\mathbf{x} + \mathbf{g}(t) \tag{8.15.50}$$

where $\mathbf{A}$ is an $n \times n$ constant matrix and the vector function $\mathbf{g}$ is assumed continuous on the interval J: $a \leq x \leq b$. Then, following the method of variation of parameters given in Section 8.14 with $\mathbf{x}_p(t) = e^{\mathbf{A}t}\mathbf{v}(t)$, we leave it as an exercise for the reader [*see* Exercise 8.15.4] to show that *the general solution* $\mathbf{x} = \mathbf{x}(t)$ *of the system* (8.15.50) *on the interval J is given by*

$$\mathbf{x} = \mathbf{x}_h(t) + \mathbf{x}_p(t), \tag{8.15.51}$$

where

$$\mathbf{x}_h(t) = e^{t\mathbf{A}}\mathbf{c}, \quad \mathbf{x}_p(t) = e^{t\mathbf{A}} \int^t e^{-s\mathbf{A}}\mathbf{g}(s)ds = \int^t e^{(t-s)\mathbf{A}}\mathbf{g}(s)ds. \tag{8.15.52}$$

Suppose that we wish to solve the system (8.15.50) subject to the initial condition given in (8.15.48). Then [*see* Exercise 8.15.5], *the unique solution* $\mathbf{x} = \mathbf{x}(t)$ *of the given initial-value problem on the interval J is given by*

$$\mathbf{x} = e^{(t-t_0)\mathbf{A}}\mathbf{x}^0 + e^{t\mathbf{A}} \int_{t_0}^t e^{-s\mathbf{A}}\mathbf{g}(s)ds$$

or $\tag{8.15.53}$

$$\mathbf{x} = e^{(t-t_0)\mathbf{A}}\mathbf{x}^0 + \int_{t_0}^t e^{(t-s)\mathbf{A}}\mathbf{g}(s)ds,$$

where t and t_0 belong to J.

Example 8.15.5. Let us solve the initial-value problem

$$\mathbf{x}' = \begin{bmatrix} -2 & 5 \\ -5 & 8 \end{bmatrix}\mathbf{x} + \begin{bmatrix} e^{4t} \\ e^{3t} \end{bmatrix}, \quad \mathbf{x}(0) = \begin{bmatrix} 1 \\ 2 \end{bmatrix}, \tag{8.15.54}$$

on the interval J: $-\infty < t < \infty$.

SOLUTION.   The eigenvalues of the matrix $\mathbf{A} = \begin{bmatrix} -2 & 5 \\ -5 & 8 \end{bmatrix}$ are found to be $\lambda_1 = 3$, $\lambda_2 = 3$. Thus, $\lambda = 3$ is an eigenvalue of multiplicity two of $\mathbf{A}$. Utilizing Theorem 8.15.3 [see also Example 8.15.3], and setting $f(\lambda) = e^{t\lambda}$, we have

$$e^{t\mathbf{A}} = \alpha\mathbf{I} + \beta\mathbf{A}, \tag{8.15.55}$$

where α and β are determined from

$$\begin{cases} e^{t\lambda} = \alpha + \beta\lambda, \\ te^{t\lambda} = \beta. \end{cases} \tag{8.15.56}$$

Substituting $\lambda = 3$ into (8.15.56), we find

$$\alpha = e^{3t} - 3te^{3t}, \quad \beta = te^{3t}. \tag{8.15.57}$$

Substituting the matrix $\mathbf{A}$ and α and β found in (8.15.57) into (8.15.55), we obtain

$$e^{t\mathbf{A}} = \begin{bmatrix} e^{3t} - 5te^{3t} & 5te^{3t} \\ -5te^{3t} & e^{3t} + 5te^{3t} \end{bmatrix}. \tag{8.15.58}$$

Replacing t by $-s$ in (8.15.58), we obtain

$$e^{-s\mathbf{A}} = \begin{bmatrix} e^{-3s} + 5se^{-3s} & -5se^{-3s} \\ 5se^{-3s} & e^{-3s} - 5se^{-3s} \end{bmatrix}.$$

Using (8.15.53) to find the solution $\mathbf{x} = \mathbf{x}(t)$ of the given initial-value problem on J, we obtain

$$\mathbf{x} = e^{t\mathbf{A}}\begin{bmatrix} 1 \\ 2 \end{bmatrix} + e^{t\mathbf{A}}\int_0^t \begin{bmatrix} e^{-3s} + 5se^{-3s} & -5se^{-3s} \\ 5se^{-3s} & e^{-3s} - 5se^{-3s} \end{bmatrix}\begin{bmatrix} e^{4s} \\ e^{3s} \end{bmatrix} ds$$

$$= e^{t\mathbf{A}}\begin{bmatrix} 1 \\ 2 \end{bmatrix} + e^{t\mathbf{A}}\int_0^t \begin{bmatrix} (1 + 5s)e^s - 5s \\ 5se^s - 5s + 1 \end{bmatrix} ds$$

$$= e^{t\mathbf{A}}\begin{bmatrix} 1 \\ 2 \end{bmatrix} + e^{t\mathbf{A}}\begin{bmatrix} 5te^t - 4e^t - \frac{5}{2}t^2 + 4 \\ 5te^t - 5e^t - \frac{5}{2}t^2 + t + 5 \end{bmatrix}$$

$$= \begin{bmatrix} e^{3t} - 5te^{3t} & 5te^{3t} \\ -5te^{3t} & e^{3t} + 5te^{3t} \end{bmatrix}\begin{bmatrix} 5te^t - 4e^t - \frac{5}{2}t^2 + 5 \\ 5te^t - 5e^t - \frac{5}{2}t^2 + t + 7 \end{bmatrix}$$

$$= \begin{bmatrix} -4e^t + \frac{5}{2}t^2 + 10t + 5 \\ -5e^t + \frac{5}{2}t^2 + 11t + 7 \end{bmatrix}e^{3t}.$$

Exercises 8.15

1. Establish (8.15.6).
2. Establish (8.15.9).
3. Establish (8.15.12).
4. Establish (8.15.52).
5. Establish (8.15.53).
6. Let $\mathbf{A}$ be an $n \times n$ constant matrix and $f(\lambda)$ be a polynomial in λ. Let $\lambda_1, \lambda_2, \ldots,$ λ_n denote the eigenvalues of the matrix $\mathbf{A}$. Show that the eigenvalues of the matrix $f(\mathbf{A})$ are given by $f(\lambda_1), f(\lambda_2), \ldots, f(\lambda_n)$.

In Exercises 7 through 9, use the technique of this section to find the general solution of each system.

7. $\mathbf{x}' = \begin{bmatrix} 1 & -1 \\ 2 & 4 \end{bmatrix} \mathbf{x}.$

8. $\mathbf{x}' = \begin{bmatrix} 1 & 0 & -1 \\ 2 & 2 & 2 \\ 2 & 1 & 2 \end{bmatrix} \mathbf{x}.$

9. $\mathbf{x}' = \begin{bmatrix} 1 & 0 & 0 & 0 \\ 0 & 1 & 0 & 0 \\ 0 & 0 & 1 & -2 \\ 0 & 0 & 1 & -1 \end{bmatrix} \mathbf{x}.$

In Exercises 10 through 16, use the technique of this section to solve each initial-value problem.

10. $\mathbf{x}' = \begin{bmatrix} -2 & 1 \\ -4 & 3 \end{bmatrix} \mathbf{x} + 9 \begin{bmatrix} e^{3t} \\ 2 \end{bmatrix}, \qquad \mathbf{x}(0) = \begin{bmatrix} 3 \\ -3 \end{bmatrix}.$

11. $\mathbf{x}' = \begin{bmatrix} 2 & 1 \\ -1 & 2 \end{bmatrix} \mathbf{x} + \begin{bmatrix} e^{2t} \cos t \\ -e^{2t} \sin t \end{bmatrix}, \qquad \mathbf{x}(0) = \begin{bmatrix} 1 \\ -1 \end{bmatrix}.$

12. $\mathbf{x}' = \begin{bmatrix} -1 & 0 & 0 \\ 0 & -1 & 0 \\ 0 & 0 & -1 \end{bmatrix} \mathbf{x} + \begin{bmatrix} 2te^{-t} \\ 3t^2e^{-t} \\ 4t^3e^{-t} \end{bmatrix}, \qquad \mathbf{x}(0) = \begin{bmatrix} 1 \\ 2 \\ 3 \end{bmatrix}.$

13. $\mathbf{x}' = \begin{bmatrix} 0 & 0 & 1 \\ 0 & 1 & 0 \\ 1 & 0 & 0 \end{bmatrix} \mathbf{x} + \begin{bmatrix} e^{-t} \\ te^t \\ e^t \end{bmatrix}, \qquad \mathbf{x}(0) = \tfrac{1}{4} \begin{bmatrix} -1 \\ 0 \\ 1 \end{bmatrix}.$

14. $\mathbf{x}' = \begin{bmatrix} -3 & 0 & -1 \\ 0 & -2 & 0 \\ 1 & 1 & -1 \end{bmatrix} \mathbf{x} + \begin{bmatrix} e^{-2t} \\ \dfrac{4}{t} e^{-2t} \\ e^{-2t} \end{bmatrix}, \qquad \mathbf{x}(1) = \begin{bmatrix} e^{-2} \\ 0 \\ -e^{-2} \end{bmatrix}, \quad t > 0.$

15. $\mathbf{x}' = \begin{bmatrix} 2 & 1 & 1 \\ 2 & 3 & 2 \\ 1 & 1 & 2 \end{bmatrix} \mathbf{x} + 16 \begin{bmatrix} e^{5t} \\ e^t \\ 0 \end{bmatrix}, \qquad \mathbf{x}(0) = \begin{bmatrix} 2 \\ -4 \\ -2 \end{bmatrix}.$

$$16.\ \mathbf{x}' = \begin{bmatrix} -1 & 1 & 1 \\ -1 & 1 & 0 \\ -1 & 0 & 1 \end{bmatrix} \mathbf{x} + 2 \begin{bmatrix} e^t \\ e^t \cos t \\ -e^t \sin t \end{bmatrix}, \qquad \mathbf{x}(0) = \tfrac{1}{5} \begin{bmatrix} 6 \\ -3 \\ 7 \end{bmatrix}.$$

17. Let $\mathbf{A} = [a_{ij}]$ be an $n \times n$ constant matrix whose elements a_{ij} are uniformly bounded. Following a similar analysis in defining the exponential function $e^{\mathbf{A}}$, namely,

$$(1) \qquad\qquad e^{\mathbf{A}} = \lim_{m \to \infty} \sum_{k=0}^{m} \frac{\mathbf{A}^k}{k!},$$

we may define

$$(2) \qquad \sin \mathbf{A} = \lim_{m \to \infty} \sum_{k=0}^{m} \frac{(-1)^k \mathbf{A}^{2k+1}}{(2k+1)!}, \qquad \cos \mathbf{A} = \lim_{m \to \infty} \sum_{k=0}^{m} \frac{(-1)^k \mathbf{A}^{2k}}{(2k)!},$$

$$(3) \qquad \sinh \mathbf{A} = \lim_{m \to \infty} \sum_{k=0}^{m} \frac{\mathbf{A}^{2k+1}}{(2k+1)!}, \qquad \cosh \mathbf{A} = \lim_{m \to \infty} \sum_{k=0}^{m} \frac{\mathbf{A}^{2k}}{(2k)!}.$$

Also, Euler's formula becomes

$$(4) \qquad\qquad e^{i\mathbf{A}} = \cos \mathbf{A} + i \sin \mathbf{A}, \qquad i^2 = -1.$$

Verify the following formulas.

$$(5) \qquad\qquad e^{-i\mathbf{A}} = \cos \mathbf{A} - i \sin \mathbf{A},$$

$$(6) \qquad \sin \mathbf{A} = \frac{e^{i\mathbf{A}} - e^{-i\mathbf{A}}}{2i}, \qquad \cos \mathbf{A} = \frac{e^{i\mathbf{A}} + e^{-i\mathbf{A}}}{2},$$

$$(7) \qquad \sinh \mathbf{A} = \frac{e^{\mathbf{A}} - e^{-\mathbf{A}}}{2}, \qquad \cosh \mathbf{A} = \frac{e^{\mathbf{A}} + e^{-\mathbf{A}}}{2}.$$

The scalar trigonometric identities extend to the corresponding matrix trigonometric identities. For example, verify the following identities:

$$(8) \qquad \sin^2 \mathbf{A} + \cos^2 \mathbf{A} = \mathbf{I}, \qquad \cosh^2 \mathbf{A} - \sinh^2 \mathbf{A} = \mathbf{I}.$$

18. Let $\mathbf{A} = \begin{bmatrix} \dfrac{\pi}{4} & 1 \\ 0 & \dfrac{5}{4}\pi \end{bmatrix}$. Show that

$$(a)\ \sin \mathbf{A} = \frac{1}{\sqrt{2}} \begin{bmatrix} 1 & -\dfrac{2}{\pi} \\ 0 & -1 \end{bmatrix}, \qquad (b)\ \cos \mathbf{A} = \frac{1}{\sqrt{2}} \begin{bmatrix} 1 & -\dfrac{2}{\pi} \\ 0 & -1 \end{bmatrix}.$$

(c) Utilizing the results of (a) and (b), verify that $\sin^2 \mathbf{A} + \cos^2 \mathbf{A} = \mathbf{I}$.

Suggested Readings

Boyce and DiPrima [9]
Brauer and Nohel [10]
DeRusso, Roy, and Close [14]
Friedman [15]
Hohn [21]

Ince [23]
Tenenbaum and Pollard [46]
Weiss [53]
Zadeh and Desoer [56]

appendix 1 SOME BASIC DEFINITIONS AND CONCEPTS OF POINT SETS

The terms *aggregate*, *class*, *collection*, and *family* are used synonymously with the term *set*. Let R^2 *denote the set of all points in the xy-plane.* We shall be concerned primarily with sets that are "contained" in R^2.[1] The points of these sets will also be referred to as the elements of the sets. However, for a general set, an element of a set may represent anything; for instance, it may represent a person, a cow, a color, an automobile, and so on. An ordered pair of real numbers will be denoted by (x, y). When we say the number pairs are ordered, we mean (x, y) and (y, x) are to be considered different unless $x = y$. We shall now define what is meant by a real-valued function of a real variable x.

Definition 1. A *function* is a set of ordered pairs of real numbers (x, y) such that to each value of the first variable (x) there corresponds (by some rule) a unique value of the second variable (y) (that is, no two distinct ordered pairs have the same first element).

The variables x and y in the ordered pair (x, y) are usually called the *independent* and *dependent* variables, respectively.

The set of all values taken on by the independent and dependent variables are called the *domain* and *range* of the function, respectively.

One usually denotes functions by letters. If f is a function, then $f(c)$, read "f at c," denotes the second element of the ordered pair in f whose first element is c; $f(c)$ is called the *value* of the function f at c. If the ordered pair (a, b) belongs to the function f, that is, $f(a) = b$, we also say that b is the *image* of a under f. Thus, we may say that the range of the function is the collection of images of its domain; and the function *maps* its domain onto its range.

Example 1. The following finite sets [denoted by { }] of ordered pairs of real numbers are functions:

---

[1] Much of the content in this appendix applies also to sets in general.

$$f = \{(1, 3), (2, 5), (3, 7), (4, 9)\}. \tag{1}$$

$$g = \{(1, 3), (2, 3), (3, 3), (4, 3), (5, 3)\}. \tag{2}$$

$$h = \{(1, 1), (3, 3), (5, 5)\}. \tag{3}$$

The domain of f is the set $\{1, 2, 3, 4\}$ while the range of f is the set $\{3, 5, 7, 9\}$. Here we have $f(1) = 3, f(2) = 5, f(3) = 7, f(4) = 9$. Figure A1.1 illustrates this correspondence. The domain of g is the set $\{1, 2, 3, 4, 5\}$ while the range of g is the set $\{3\}$. Here we have $3 = g(1) = g(2) = g(3) = g(4) = g(5)$. Figure A1.2 illustrates this correspondence. Finally, the domain of h is the set $\{1, 3, 5\}$ and the range of h is the same set as that of the domain of h. Also, $h(1) = 1, h(3) = 3,$ and $h(5) = 5$. Figure A1.3 illustrates this correspondence.

Definition 2.   A function that maps its domain onto a set consisting of a single element is called a *constant function*.

The range of the function consists of one element. [*See* Figure A1.2.]

Definition 3.   A function that maps each element of its domain into itself is called the *identity function*.

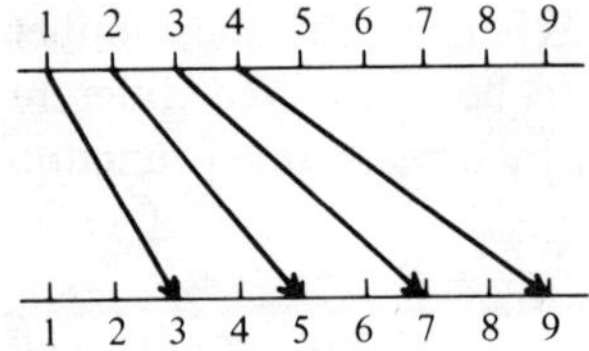

Figure A1.1.

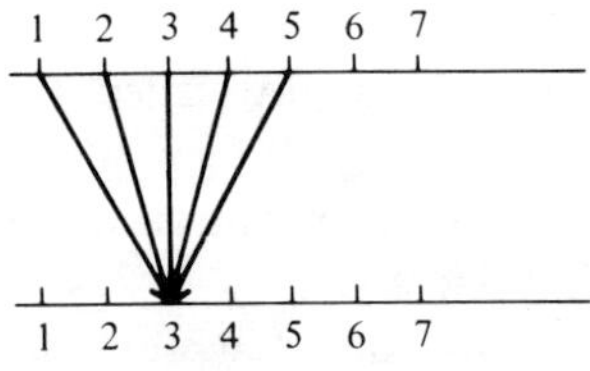

Figure A1.2.

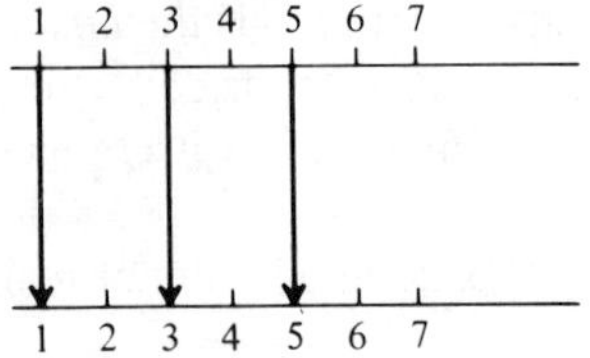

Figure A1.3.

Here the domain and range of the function consist of the same set of elements.   [*See* Figure A.1.3.]

Example 2.   The finite set of ordered pairs of real numbers

$$\{(1, 2), (2, 2), (2, 4), (3, 6), (4, 8)\}, \tag{4}$$

is not a function f since $f(2)$ does not have a unique value.   That is, $f(2) = 2$ and $f(2) = 4$.   Also, we have here $f(1) = 2$, $f(3) = 6$, and $f(4) = 8$.   Figure A1.4 illustrates this correspondence.   Such a set of ordered pairs is known as a relation.[2]

We may also describe functions symbolically.   For instance, the function f in Example 1 may be described as

$$f = \{(x, 2x + 1)\,|\,x \in \{1, 2, 3, 4\}\}. \tag{5}$$

The domain of f is the set $\{1, 2, 3, 4\}$ and the rule of correspondence is

$$f(x) = 2x + 1, \tag{6}$$

where x denotes any of the numbers 1, 2, 3, or 4. In (5), the symbol "$\in$" means "is an element of."

Example 3.   Let R denote the set of all real numbers.   Let f be a function whose domain is R and with rule of correspondence: if $x \in R$, then $f(x) = x^2 - 3x + 1$.   This gives us a complete description of the function f, namely,

$$f = \{(x, x^2 - 3x + 1)\,|\,x \in R\}. \tag{7}$$

This function is an infinite set of ordered pairs.

Since a real-valued function of a real variable x is a set of ordered pairs of real numbers, $(x, f(x))$, we may consider f as being a set of points in R^2. Observe that such functions are special sets in R^2, namely, they enjoy the property of having at most one intersection with each vertical line.

Definition 4.   The *graph* of a function f is the set of points in R^2 whose coordinates form an ordered pair (x, y) that belongs to f.

The graphs of the functions given in Examples 1 and 3 are shown in

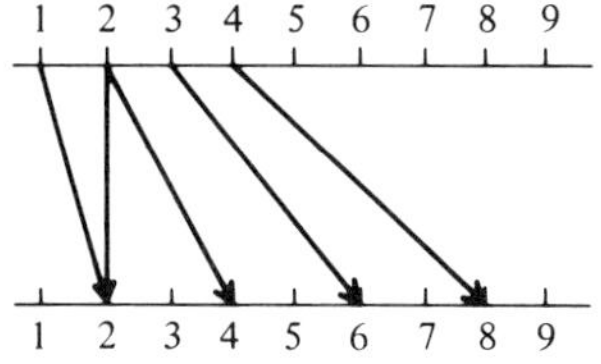

Figure A1.4.

---

[2] A *relation* is any set of ordered pairs of real numbers. The *domain* of the relation consists of all numbers x such that (x, y) is one of the pairs of the relation for some y.   The *range* of the relation consists of all numbers y occurring in any of the pairs of the relation.

Figures A1.5 to A1.8.   Some of the ordered pairs in f of the graph displayed in Figure A1.8 are $(-1, 5)$, $(0, 1)$, $(1, -1)$, $(3/2, -5/4)$, $(2, -1)$, $(3, 1)$, and $(4,5)$.

Definition 5.   A real number x is said to be nonnegative if $x \geqq 0$.

Let the symbols "$\sqrt{\ \ }$" and "$|\ \ |$" denote the nonnegative square root and absolute value, respectively.

Definition 6.   By the *square root function* we mean a function with domain consisting of the set of all nonnegative real numbers and with rule of correspondence: $\sqrt{x}$ is the nonnegative number whose square is x.

For example, $\sqrt{9} \neq \pm 3$, but $\sqrt{9} = 3$.   Also, $\sqrt{9} \neq -3$.

Definition 7.   By the *absolute value* of a function we mean a function with domain consisting of the set of all real numbers and with rule of correspondence:

$$|x| = \begin{cases} x, & \text{if } x \geqq 0 \\ -x, & \text{if } x < 0. \end{cases} \tag{8}$$

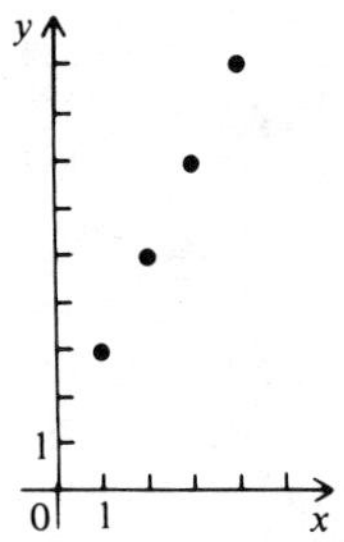

Figure A1.5.
The graph of f

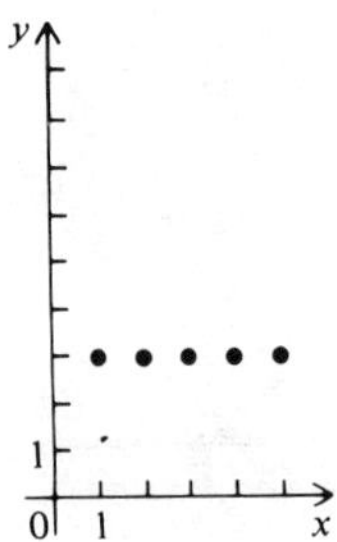

Figure A1.6.
The graph of g

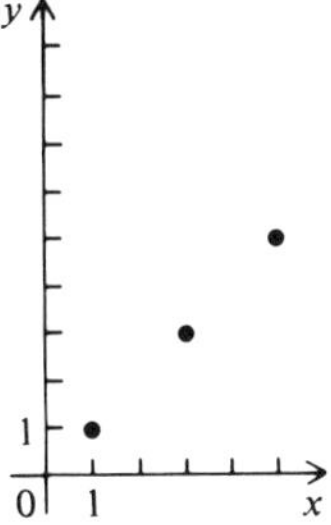

Figure A1.7.
The graph of h

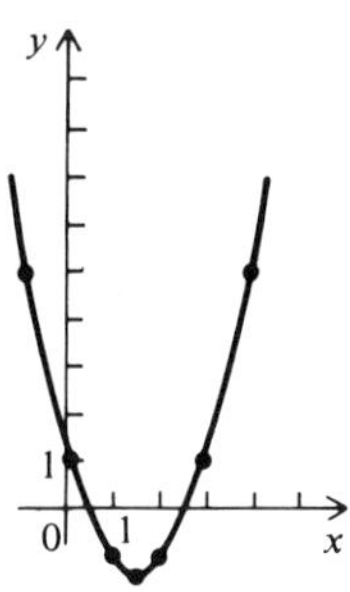

Figure A1.8.
The graph of $f(x) = x^2 - 3x + 1, \; x \in R$

For example, $|3| = 3.$ $|-3| = -(-3) = 3.$ $|0| = 0.$ Thus we, see that

$$\sqrt{x^2} = |x|. \tag{9}$$

Example 4.   Let I denote the interval consisting of all x such that $-c \leqq x \leqq c.$ The graph of the function

$$f_1 = \{(x, \sqrt{c^2 - x^2}) \mid x \in I\} \tag{10}$$

is a semicircle in the upper half plane as shown in Figure A1.9.   The graph of the function

$$f_2 = \{(x, -\sqrt{c^2 - x^2}) \mid x \in I\} \tag{11}$$

is a semicircle in the lower half plane as shown in Figure A1.10.

Definition 8.   A set S is a *subset* of a set T, written $S \subseteqq T$, if every element of S is an element of T.

The expression $S \subseteqq T$ is also read S is *contained* or *included* in T.

By a *nonempty* set we mean a set that contains some elements.

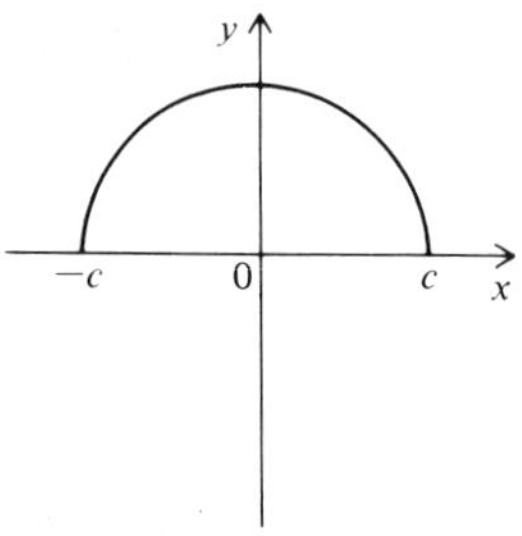

Figure A1.9.
The graph of f_1

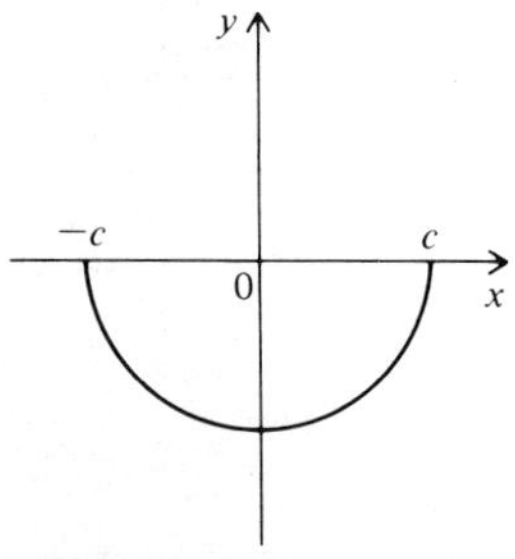

Figure A1.10.
The graph of f_2

Definition 9.   A nonempty set S is a *proper* subset of a set T, written $S \subset T$, if S is contained in T and there exists at least one element of T that is not an element of S.

The expression $S \subset T$ is also read S is *properly contained* or *properly included* in T.

Definition 10.   Two sets, S and T, are said to be *equal*, and we write $S = T$, if and only if[3] every element of S is an element of T and every element of T is an element of S.

Thus, $S = T$ if and only if $S \subseteq T$ and $T \subseteq S$.

Definition 11.   The *union* of two sets, S and T, written $S \cup T$, is the set of all elements that belong to S or to T (or to both S and T).

Definition 12.   The *intersection* of two sets, S and T, written $S \cap T$, is the set of all elements that belong to both S and T.

---

[3] Let B and C be propositions.   *B if and only if C* means that C implies B (if), and B implies C (only if).

Example 5. Let S be the set of real x such that $0 \leq x \leq 4$, that is, $S = \{x \mid 0 \leq x \leq 4\}$, and let $T = \{x \mid -\frac{1}{2} \leq x < 2\}$. Then $S \cup T = \{x \mid -\frac{1}{2} \leq x \leq 4\}$, and $S \cap T = \{x \mid 0 \leq x < 2\}$.

Example 6. Let S and T be the sets as indicated in Figure A1.11. Then $S \cup T$ is the shaded set in Figure A1.12, while $S \cap T$ is the shaded set in Figure A1.13.

Since there might not be any elements belonging to both S and T, it is convenient to make the following definitions.

Definition 13. The *null* or *void* set is the set consisting of no elements.

Definition 14. Two sets S and T are said to be *disjoint* if their intersection is the null set.

Example 7. Let S be the set of all real x such that $|x| \leq 1$, and let T be the set of all real x such that $|x| > 1$. Since $S \cap T$ is the null set, the sets S and T are disjoint.

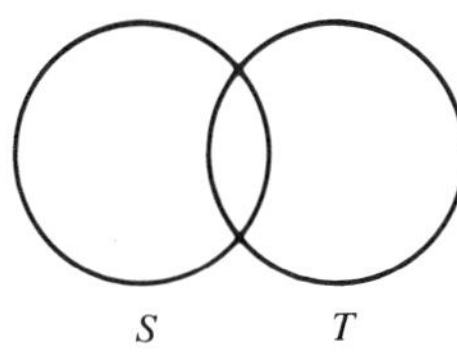

Figure A1.11.

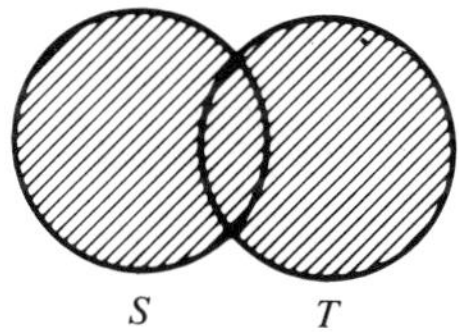

Figure A1.12.
$S \cup T$

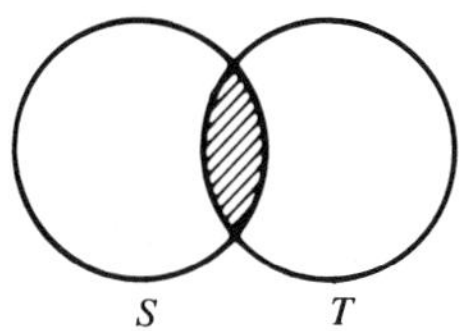

Figure A1.13.
$S \cap T$

Example 8. Let S and T be the sets as indicated in Figure A1.14.   Here $S \cap T$ is the null set, and the sets S and T are disjoint.

Remark 1.   In the ensuing discussion, let S be a set of points contained in R^2.

Definition 15.   In R^2, a *circular neighborhood* of a point P_0 with coordinates (x_0, y_0) is the set of all points (x, y) such that

$$(x - x_0)^2 + (y - y_0)^2 < r^2, \tag{12}$$

where r is any positive (real) number.

We shall denote this circular neighborhood by $N(P_0, r)$.   We observe that this definition asserts that a circular neighborhood of a point P_0 is given by the set of all points (x, y) whose distance from (x_0, y_0) is less than r, that is, the set of all points (x, y) interior to the circle with center at (x_0, y_0) and radius r; it includes the point (x_0, y_0) and excludes the points on the circumference.

Definition 16.   A point P_0 is said to be an *interior point* of a set S if there exists a circular neighborhood $N(P_0, r)$ such that $N(P_0, r) \subset S$.

Definition 17.   The *interior* of a set S is the set consisting of the interior points of S.

Definition 18.   A set S is said to be *open* if each point of S is an interior point of S.

Example 9.   Each of the sets of points (x, y) whose coordinates satisfy
(a) $x^2 + y^2 < 4$,   (b) $x^2 + y^2 > 4$,   (c) $4x^2 + 6y^2 > 8$,
(d) $|x| < \pi$, $y > 0$,   (e) $x < 0$, $|y| < 2\pi$,
is an open set.

Remark 2. R^2 and the null set are open sets.

Definition 19.   The set of all points of R^2 that are not in the set S is called the *complement* of S relative to R^2, and is denoted by $R^2 - S$ or $C(S)$.[4]

Example 10.   The complement of each set in Example 9 is given, respectively, by the set of points (x, y) whose coordinates satisfy
(a) $x^2 + y^2 \geq 4$,   (b) $x^2 + y^2 \leq 4$,   (c) $4x^2 + 6y^2 \leq 8$,
(d) $|x| \geq \pi$ or $y \leq 0$,   (e) $x \geq 0$ or $|y| \geq 2\pi$.

The reader is urged to sketch the sets given in (d) and (e) in Example 9 so that he may readily find the complement of these sets.

Definition 20.   A set of points S is said to be *closed* if the complementary set $C(S)$ is open.

Each set given in Example 10 is a closed set.

---

[4] If T and S are sets in R^2, the *difference* $T - S$ (also called the *complement of S relative to T*), is defined to be the set of all elements of T which are not in S.

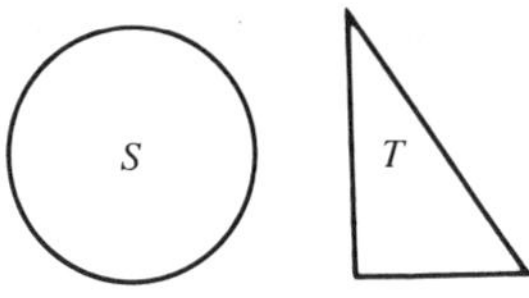

Figure A1.14.

___

Remark 3. R^2 and the null set are closed sets.

Remark 4.   There are sets which are neither open nor closed.   For example, if S is the set of points (x, y) whose coordinates satisfy either $x^2 + y^2 < 4$ or $x = 2$, then S is neither open nor closed.

Definition 21.   A point P_0 is said to be a *boundary point* of a set S if every circular neighborhood $N(P_0, r)$ contains at least one point of S and at least one point of $C(S)$.

Definition 22.   The union of all the boundary points of a set S is called the *boundary* of S, and is denoted by $B(S)$ (or ∂S).

Example 11.   If S is the set of points (x, y) whose coordinates satisfy $1 < x^2 + y^2 \leqq 4$, then some of the boundary points of S are $(2, 0)$, $(0, 2)$, $(-2, 0)$, $(0, -2)$, $(1, 0)$, $(0, 1)$, $(-1, 0)$, $(0, -1)$, $(1, \sqrt{3})$, $(-1, \sqrt{3})$, $((1/\sqrt{2}), (1/\sqrt{2}))$, and $((1/2), (\sqrt{3}/2))$.   Note that the boundary points of S are the points that lie on the circumference of the circle $x^2 + y^2 = 4$ and of the circle $x^2 + y^2 = 1$.   That is,

$$B(S) = \{(x, y) \mid x^2 + y^2 = 4\} \cup \{(x, y) \mid x^2 + y^2 = 1\}.$$

Remark 5.   Observe that no boundary point of an open set can belong to the set; however, every boundary point of a closed set belongs to the set.

Definition 23.   A set of points S is said to be *bounded* if there exists a fixed point P_0 and a number $r > 0$ such that $S \subset N(P_0, r)$.

Example 12.   Each of the sets of points (x, y) whose coordinates satisfy
(a) $x^2 + y^2 < 1$,   (b) $6x^2 + 4y^2 \leqq 1000$,
(c) $|x| \leqq \pi$, $-2\pi < y \leqq 2\pi$,   (d) $y = x^2$, $0 \leqq x \leqq 1$,
(e) $|x| + |y| \leqq 100$,
is a bounded set.

In Example 9, the sets given in (b), (c), (d), and (e) are not bounded.   If in Example 12(e), we replace $|x|$ by x, is the new set also bounded?

Definition 24.   A set of points S which is bounded and closed is called a *compact* set.

In Example 12, the sets given in (b), (d), and (e) are compact sets.   In Example 12, why are the sets given in (a) and (c) not compact sets?

Definition 25.   A point P_0 is said to be a *limit point* of a set S if every circular neighborhood of P_0 contains infinitely many points of S.

Remark 6.   In the above definition of a limit point, the point P_0 may or may not belong to the set.   For example, each point P of the circle $x^2 + y^2 = 1$ is a limit point of the set of points S whose coordinates (x, y) satisfy $x^2 + y^2 < 1$, but P does not belong to S.   On the other hand, each point interior to the circle $x^2 + y^2 = 1$ is a limit point of the set of interior points, and belongs to the set of interior points.   Suppose that S consists of the set of points $\{(1, 0), (\frac{1}{2}, 0), (\frac{1}{3}, 0), \ldots, ((1/n), 0), \ldots\}$, then $(0, 0)$ is a limit point of S but does not belong to S.

Remark 7.   Observe that in Definition 25, all that is really required is that *every* circular neighborhood of P_0 contains at least one point of S which is distinct from P_0.   For, if this condition is satisfied, then the circular neighborhood $N(P_0, r)$: $(x - x_0)^2 + (y - y_0)^2 < r^2$ contains a point P_1: (x_1, y_1), $P_1 \neq P_0$ belonging to S.   Let $(x_1 - x_0)^2 + (y_1 - y_0)^2 = r_1{}^2$.   Similarly, the circular neighborhood $N(P_0, r_1)$: $(x - x_0)^2 + (y - y_0)^2 < r_1{}^2$ contains a point P_2: (x_2, y_2) belonging to S which is different from P_0 and P_1.   This process may be carried out indefinitely.   It follows that the circular neighborhood $N(P_0, r)$ contains an infinite number of points of S.

Definition 26.   Suppose that the function f maps the closed interval I: $a \leqq x \leqq b$ into R^2.   Let $a = x_0 < x_1 < x_2 < \cdots < x_n = b$ be a subdivision of I, and let I_k denote the closed intervals $x_{k-1} \leqq x \leqq x_k$, $k = 1, 2, \ldots, n$. Suppose further that f maps each interval I_k (linearly) onto a line segment. Then the curve represented by the mapping f (that is, the image of f) is said to be a *polygonal curve or polygonal line*.

Figure A1.15 illustrates a polygonal curve represented by the mapping f for the case when $n = 4$.

Definition 27.   Let $I = [a, b]$ be a closed (and bounded) interval on the real line, that is, I is the set of all points t such that $a \leqq t \leqq b$ with a and b finite and $a < b$.   Let $x = f(t)$ and $y = g(t)$ be two continuous real-valued functions of the variable t defined on I.   Suppose that for $t \in I$, at least one of the

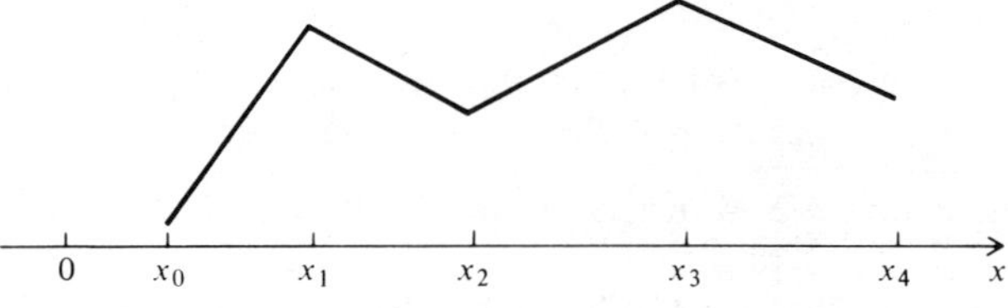

Figure A1.15.

functions f or g is not equal to a constant.   Then, the set of all points (x, y) in R^2 whose coordinates are determined by the relationship

$$x = f(t), \qquad y = g(t), \qquad a \leq t \leq b, \tag{13}$$

is said to define a *continuous (bounded plane) curve or continuous arc*, and the functions given in (13) are called *parametric equations* of the curve.[5]

If the functions f and g in Equation (13) have continuous derivatives f' and g' in I which do not vanish simultaneously for any $t \in I$, then the curve has a continuously turning tangent.   Such a curve is called a *smooth curve* or *smooth arc*. A continuous curve which is composed of a finite number of smooth arcs is called *piecewise* or *sectionally smooth* curve.   Such a curve is also called a *contour*.

In Definitions 28 through 31, let Γ *denote a continuous (bounded plane) curve with parametric equations given by* (13).

Definition 28.   The points $(f(a), g(a))$ and $(f(b), g(b))$ are called, respectively, the *initial* and *terminal* points of the curve Γ.

Definition 29.   A curve Γ is said to be *closed* if its initial and terminal points coincide.

Thus, Γ is closed when $f(a) = f(b)$ and $g(a) = g(b)$, that is, $(f(a), g(a)) = (f(b), g(b))$.

Definition 30.   A curve Γ is said to be *simple* if for any two distinct points t_1, t_2 such that $a \leq t_1 < t_2 \leq b, \quad (t_1, t_2) \neq (a, b)$, we have

$$(f(t_1), g(t_1)) \neq (f(t_2), g(t_2)).$$

Definition 31.   A curve Γ is said to be *simple closed* if it is both simple and closed.

Example 13.   The set of all points (x, y) whose coordinates are determined by the relationship:
1. $x = \cos t, \quad y = \sin t, \quad 0 \leq t \leq 2\pi$, is a simple closed curve (unit circle).
2. $x = \cos t, \quad y = \sin t, \quad 0 \leq t \leq \pi$, is a simple curve (semicircle).
3. $x = \cos t, \quad y = \sin t, \quad 0 \leq t \leq 4\pi$, is a closed curve, but it is not simple closed.
4. $x = 4 \cos t, \quad y = 5 \sin t, \quad 0 \leq t \leq 2\pi$, is a simple closed curve (ellipse).
5. $x = 4 \cos t, \quad y = 5 \sin t, \quad 0 \leq t \leq \dfrac{\pi}{2}$, is a simple curve (elliptic arc).

---

[5] We shall also write Equation (13) as

$$x = x(t), \qquad y = y(t), \qquad a \leq t \leq b.$$

6. $x = f(t) = \begin{cases} t, & 0 \leq t \leq 2 \\ 2, & 2 \leq t \leq 4 \\ 6 - t, & 4 \leq t \leq 6 \\ 0, & 6 \leq t \leq 8, \end{cases}$ $y = g(t) = \begin{cases} 0, & 0 \leq t \leq 2 \\ t - 2, & 2 \leq t \leq 4 \\ 2, & 4 \leq t \leq 6 \\ 8 - t, & 6 \leq t \leq 8, \end{cases}$

is a simple closed curve [square, *see* Figure A1.16].   Also, it is a contour.

7. $x = f(t) = \begin{cases} t, & 0 \leq t \leq 2 \\ 2, & 2 \leq t \leq 4 \\ 6 - t, & 4 \leq t \leq 6, \end{cases}$ $y = g(t) = \begin{cases} 0, & 0 \leq t \leq 2 \\ t - 2, & 2 \leq t \leq 4 \\ 2, & 4 \leq t \leq 6, \end{cases}$

is a simple curve.   Also, it is a contour. (Sketch the curve.)

8. $x = t$, $y = t^2$, $0 \leq t \leq 1$, is a simple curve (parabolic arc).

9. $x = t$, $y = t(1 - t)(2 - t)$, $0 \leq t \leq 2$, is a simple curve.

10. $x = (\frac{1}{2} + \sin t) \sin t$, $y = (\frac{1}{2} + \sin t) \cos t$, $0 \leq t \leq 2\pi$, is a closed curve, but it is not simple closed [limaçon, *see* Figure A1.17].

Definition 32.   Let S be a nonempty open set in R^2.   The set S is said to be *connected* if for any two points $P_1, P_2 \in S$, there exists a polygonal curve γ with initial and terminal points at P_1 and P_2, respectively, and $\gamma \subset S$.

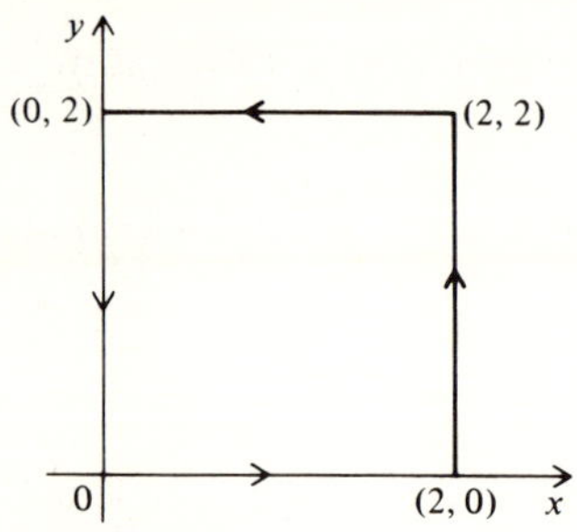

Figure A1.16.
Square

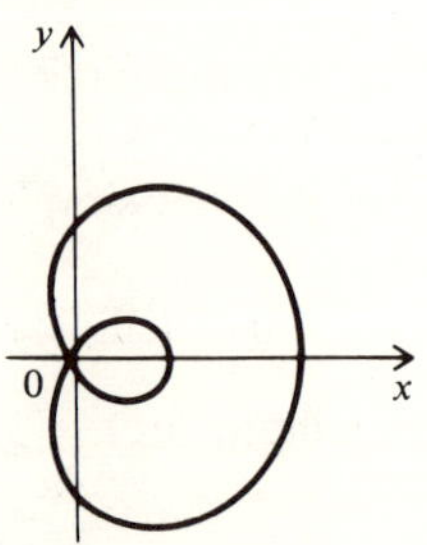

Figure A1.17.
Limaçon

Example 14.   The open set S of all points (x, y) whose coordinates satisfy:
(1) $x^2 + 4y^2 < 16$,   is connected,   (2) $4 < x^2 + y^2 < 25$,   is connected,
(3) $x > 0$, $y > 0$,   is connected.

We shall now define the term "connected set" for sets which are not necessarily open. This definition does not involve joining points by a polygonal curve; however, for open sets the definitions are equivalent.

Definition 33.   A set S in R^2 is said to be *separated* into two sets S_1 and S_2 if (1) S_1 and S_2 are nonempty, (2) S is the union of S_1 and S_2, (3) S_1 and S_2 have no points in common, (4) S_1 contains no limit point of S_2, and (5) S_2 contains no limit point of S_1.

Definition 34.   A set S in R^2 is said to be *connected* if it cannot be written as the union of two separated sets.

Definition 35.   An open connected set S in R^2 *together with* some, none, or all of its boundary points will be called a *region*. If none of the boundary points are included, the region is called an *open region* or a *domain*. If all of the boundary points are included, it will be called a *closed region*.

Remark 8.   Note that a closed region forms a closed set; a closed set does not necessarily form a closed region. For instance, if the set S consists of all points (x, y) whose coordinates satisfy either $1 \leq x^2 + y^2 \leq 2$ or $3 \leq x^2 + y^2 \leq 4$, then S is closed [*see* Exercise 18 (b)]; however, S does not form a closed region, since S is not connected.

Example 15.   Domains and Regions.
1. The interior of a circle, rectangle, or a triangle is a domain.
2. The interior of a circle plus the points on its circumference, the interior of a rectangle plus the points on its perimeter, or the interior of a triangle plus the points on its perimeter is a closed region.
3. The set S of all points (x, y) whose coordinates satisfy:
   (a) $2 < x^2 + y^2 < 4$,   is a domain,   (b) $2 \leq x^2 + y^2 < 4$,   is a region,
   (c) $2 \leq x^2 + y^2 \leq 4$,   is a closed region.
4. The set S of all points (x, y) whose coordinates satisfy:
   (a) $x > 0$, $y > 0$,   is a domain,   (b) $x \geq 0$, $y > 0$,   is a region,
   (c) $x \geq 0$, $y \geq 0$,   is a closed region.

Another important concept here is one that involves regions which are known as "simply connected." Intuitively, these are regions without "holes." Before making this concept more precise, we shall first state, without proof, the following important theorem.

Theorem 1.   *Jordan Curve Theorem.   Let Γ be a simple closed curve.   Then Γ divides the plane into two disjoint domains D_1 and D_2 (one of which is bounded) with Γ the boundary of D_1 as well as the boundary of D_2.*

The bounded domain is called the *interior* of Γ and the unbounded domain is called the *exterior* of Γ.   In Figure A1.18, D_1 is the interior of Γ while D_2 is the exterior of Γ.

Definition 36.   In R^2, a region R is said to be *simply connected* if for every simple closed curve Γ contained in R, the interior of Γ is also contained in R.

A region that is not simply connected is called *multiply connected.*

In Example 15, the regions given in 1, 2, and 4 are all simply connected, while those given in 3 are not simply connected.   Figure A1.19 illustrates a multiply connected region R.   Here the hatched-in areas do not belong to R. Observe, the moment we draw a simple closed curve Γ about any one of the hatched-in areas, the interior of Γ is no longer contained in R.   The region consisting of R^2 with the exception of one or any finite number of points is multiply connected.

Remark 9.   Multiply connected regions can often be made simply connected by introducing as additional boundary lines certain lines joining the inner curves with the outer curve.   This is illustrated in Figure A1.20.   The boundary of R now consists of the curves $\Gamma_1, \Gamma_2, \Gamma_3, \Gamma_4$, and the broken lines (or cuts).   Note that we cannot draw any simple closed curve Γ about the hatched-in areas, since Γ cannot cross any of the broken lines.   These

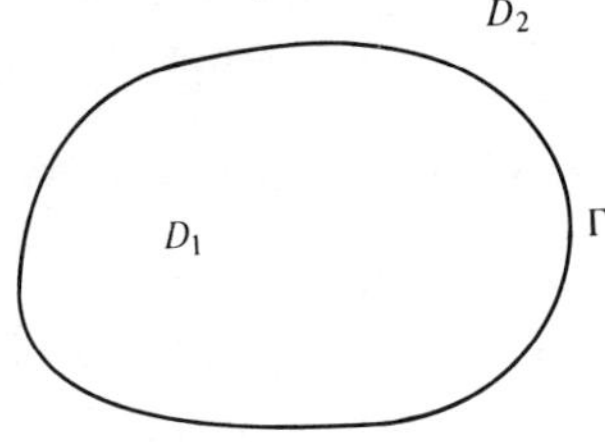

Figure A1.18.
Jordan curve Γ

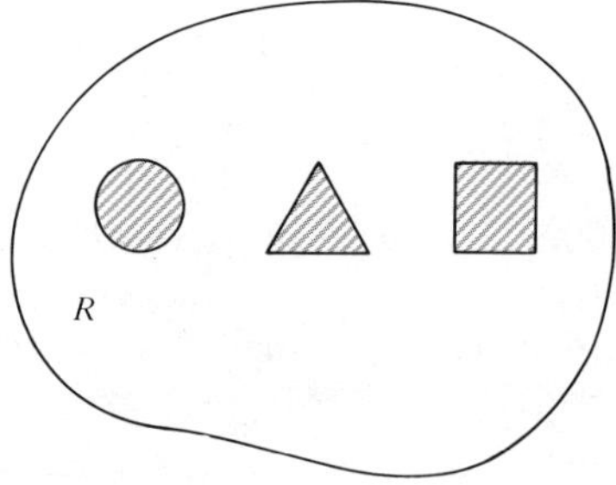

Figure A1.19.
Multiply connected region R

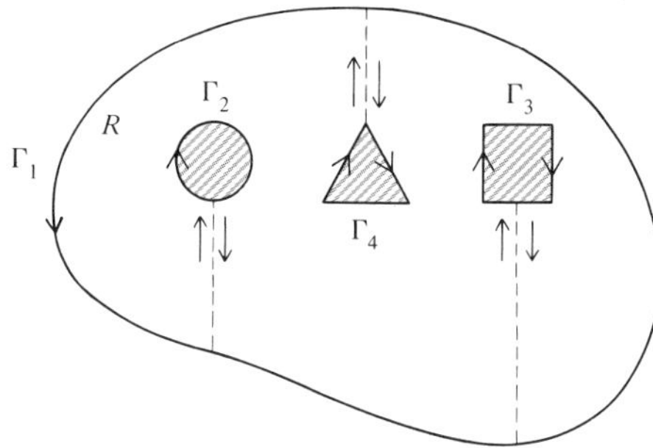

Figure A1.20.

broken lines (or barriers) do not belong to R. They form part of the boundary of R.

Definition 37.   Let S be a point set in R^2. Let P_1: (x_1, y_1) and P_2: (x_2, y_2) be any two points in S. The *diameter* δ of a set S is defined to be

$$\delta(S) = \operatorname*{lub}_{P_1 \in S,\, P_2 \in S} \sqrt{(x_2 - x_1)^2 + (y_2 - y_1)^2}.$$

Here, lub stands for the "least upper bound."

Example 16.   The set S of all points (x, y) whose coordinates satisfy:
1. $x^2 + y^2 \leqq 25$,   has $\delta(S) = 10$,
2. $-1 < x < 1$, $-2 < y < 2$,   has $\delta(S) = 2\sqrt{5}$,

3. $\dfrac{x^2}{4} + \dfrac{y^2}{16} \leqq 1$,   has $\delta(S) = 8$,

4. $x^2 < 2$, $y = 0$;   has $\delta(S) = 2\sqrt{2}$,
5. $x^2 < 2$, $y \geqq 0$,   has infinite diameter, and we write, $\delta(S) = \infty$.

Definition 38.   Let S_1 and S_2 be two point sets in R^2. Let P_1: (x_1, y_1) be any point in S_1 and P_2: (x_2, y_2) be any point in S_2. The *distance between the sets S_1 and S_2* is defined to be

$$d(S_1, S_2) = \operatorname*{glb}_{P_1 \in S_1,\, P_2 \in S_2} \sqrt{(x_2 - x_1)^2 + (y_2 - y_1)^2}.$$

Here, glb stands for the "greatest lower bound."

Example 17.   The sets S_1 and S_2 of all points (x, y) whose coordinates satisfy:
1. S_1: $x^2 + y^2 \leqq 25$, S_2: $y \geqq 10$,   has $d(S_1, S_2) = 5$,
2. S_1: $x^2 + y^2 \leqq 25$, S_2: $y = x^2 + 10$,   has $d(S_1, S_2) = 5$,
3. S_1: $x^2 + y^2 \leqq 25$, S_2: $x = 10, y = 0$,   has $d(S_1, S_2) = 5$,
4. S_1: $x^2 + y^2 \leqq 25$, S_2: $x = y^2$,   has $d(S_1, S_2) = 0$,
5. S_1: $x^2 + y^2 = 1$, S_2: $x^2 + y^2 = 4$,   has $d(S_1, S_2) = 1$.

Exercises

1. Determine whether or not each of the following sets of ordered pairs is a function. For those that are functions, give the domain and range and illustrate by a diagram.
   (a) $\{(1, 1), (2, 2), (3, 4), (4, 7), (5, 11)\}$.
   (b) $\{(0, 0), (-1, -2), (-2, -3), (-3, -5), (-4, -7)\}$.
   (c) $\{(-1, 1), (0, -2), (2, 2), (3, -2)\}$.
   (d) $\{(-3, 1), (-2, 1), (-1, -1), (4, 1)\}$.
   (e) $\{(-4, \frac{1}{2}), (-2, 1), (-1, \frac{3}{2}), (0, 0)\}$.
   (f) $\{(-2, -2), (-1, 2), (0, -2), (1, 2), (0, 0)\}$.

2. Determine the domain and a formula for the rule of correspondence for each of the following functions.

$$f = \{(0, 0), (1, 1), (2, 8), (3, 27), (4, 64), (5, 125), (6, 216)\}.$$
$$g = \{(0, 0), (1, 1), (4, 2), (9, 3), (16, 4), (25, 5)\}.$$
$$h = \{(-3, 3), (-2, 2), (-1, 1), (0, 0), (1, 1), (2, 2), (3, 3)\}.$$
$$F = \{(-2, 0), (-1, -3), (0, -4), (1, -3), (2, 0)\}.$$

3. What is the range of each function given in Exercise 2?

4. Consider a collection of ordered pairs (A, z), where the elements of A are themselves ordered pairs of real numbers and the elements z are real numbers. If no two members of this collection have the same item A as a first element [that is, if it can never happen that there are two members (A_1, z_1) and (A_1, z_2) with $z_1 \neq z_2$], then we call this collection a *(real-valued) function of two (real) variables*. The totality of possible ordered pairs A is called the *domain* of the function. The totality of possible values of z is called the *range* of the function.

   We may view a (real-valued) function of two (real) variables as a mapping f having its domain in R^2 and its range in R. If f is a function of two variables, the value at a point (x, y) in its domain is denoted by $f(x, y)$.

   Let f be a function with domain R^2 and rule of correspondence

$$f(x, y) = x^2 + y^2, \quad \text{that is,} \quad f = \{((x, y), x^2 + y^2) \mid (x, y) \in R^2\}.$$

Find the value of f at

(a) $(0, 0)$, (b) $(1, 1)$, (c) $(2, 3)$,

(d) $(-2, 3)$, (e) $\left(\dfrac{1}{\sqrt{2}}, \dfrac{1}{\sqrt{2}}\right)$, (f) $(\sqrt{3}, \sqrt{2})$.

5. Determine whether or not each of the following sets of ordered pairs is a function.

(a) $\{(2x + 4, x) \mid x \in R\}$. (b) $\{(x^4 + 1, x) \mid x \in R\}$.

(c) $\{(x^3, x) \mid x \in R\}$. (d) $\left\{\left(\dfrac{x^2}{x^2 + 1}, x\right) \mid x \in R\right\}$.

(e) $\{((2x + 4, 3y - 2), (x, y)) \mid (x, y) \in R^2\}$.
(f) $\{((x^4, y + 3), (x, y)) \mid (x, y) \in R^2\}$.
(g) $\{((x + 3, y^4), (x, y)) \mid (x, y) \in R^2\}$.

6. Sketch the graph of each of the following functions.
  (a) $\{(0, 8), (1, 6), (2, 4), (3, 2), (4, 0)\}$.
  (b) $\{(-3, -6), (-2, -3), (-1, 0), (0, 3), (1, 6), (2, 9)\}$.
  (c) $\{(-2, -5), (-1, -2), (0, 1), (1, 4), (2, 7)\}$.
  (d) $\{(-\frac{1}{2}, 0), (0, \frac{1}{2}), (\frac{1}{2}, 1), (1, \frac{3}{2}), (\frac{3}{2}, 2), (2, \frac{5}{2}), (\frac{5}{2}, 3)\}$.

7. Tabulate representative values of f and draw the graph of f, where f is the function with rule of correspondence and domain D_f given below.

  (a) $f(x) = x^2 \sin x, \quad D_f = [0, \pi]$.

  (b) $f(x) = \dfrac{\sin x}{|x|}, \quad D_f = [-\pi, 0)$.

  (c) $f(x) = \begin{cases} 2 - x, & \text{if } x \in [-2, 0] \\ x, & \text{if } x \in (0, 2] \end{cases}, \quad D_f = [-2, 2]$.

  (d) $f(x) = \begin{cases} x^3 + 1, & \text{if } x \in [-1, 1] \\ x - 4, & \text{if } x \in [4, 8] \end{cases}, \quad D_f = [-1, 1] \cup [4, 8]$.

  (e) $f(x) = \begin{cases} x, & \text{if } x \in [-1, 1] \\ x^2, & \text{if } x \in [2, 3] \\ (x - 4)^2, & \text{if } x \in [4, 6] \end{cases}, \quad D_f = [-1, 1] \cup [2, 3] \cup [4, 6]$.

  (f) $f(x) = \begin{cases} x, & \text{if } x \in [0, 1] \\ 1, & \text{if } x \in [1, 3] \\ \sqrt{4 - x}, & \text{if } x \in [3, 4] \end{cases}, \quad D_f = [0, 4]$.

  (g) $f(x) = \begin{cases} x - |x|, & \text{if } x \in [-4, -2] \\ x|x|, & \text{if } x \in [-2, 2] \end{cases}, \quad D_f = [-4, 2]$.

8. Does the following set of ordered pairs determine a function?

$$\left\{\left(\frac{\sin x}{|x|}, x\right) \mid x \in R, \ x \neq 0\right\}.$$

In Exercises 9–13 below, a set S is given.   For each exercise answer the following.
(a) Describe the set S in geometric language with the aid of a figure.
(b) Determine the boundary of the set S.
(c) Determine whether the set S is closed, open, or neither.
(d) Determine whether the set S forms a region.

9. $S = \{(x, y) \mid x^2 + y^2 < 4, x \neq 0\} \cup \{(x, y) \mid -2 < y < 0, x = 0\}$.
10. $S = \{(x, y) \mid x \leq y^2\} \cap \{(x, y) \mid x \leq 1\}$.
11. $S = \{(x, y) \mid 0 < xy \leq 4\} \cap \{(x, y) \mid x > 0\}$.
12. $S = \{(x, y) \mid y = 2 \cos 1/x, x \neq 0\} \cap \{(x, y) \mid x > 0\}$.
13. $S = \{(x, y) \mid x^2 + y^2 < 9, y < 0\} - \{(x, y) \mid x = 1/n, \ n = 1, 2, \ldots, \ -1 \leq y < 0\}$.
  [*Note:* By definition, $A - B = \{(x, y) \mid (x, y) \in A \text{ and } (x, y) \notin B\}$.   The symbol, $\notin$, is read "does not belong to."]

14. (a) $S = \{(x, y) \mid x^2 + y^2 = 4\} \cup \{(x, y) \mid 0 \le x \le 2, y = 0\}$.

      Is S closed? Does S have any interior points?

     (b) Does the set S given in Exercise 12 have any interior points?

15. Let R, S, and T be three sets. With the aid of diagrams [*see*, for instance, Example 6] verify the following.

     (a) COMMUTATIVE LAWS $R \cup S = S \cup R$ and $R \cap S = S \cap R$,

     (b) ASSOCIATIVE LAWS $R \cup (S \cup T) = (R \cup S) \cup T$ and

                                         $R \cap (S \cap T) = (R \cap S) \cap T$,

     (c) DISTRIBUTIVE LAWS $R \cap (S \cup T) = (R \cap S) \cup (R \cap T)$ and

                                          $R \cup (S \cap T) = (R \cup S) \cap (R \cup T)$.

16. Let R, S, and T be three sets. Show that

     (a) $R \cap S \subseteq R$ and $R \cap S \subseteq S$,

     (b) $R \subseteq R \cup S$ and $S \subseteq R \cup S$,

     (c) if $R \subset S$ and $S \subset T$, then $R \subset T$.

17. Let S be a nonempty set in R^2. Prove that a necessary and sufficient condition[6] that S be closed is that S contains all of its limit points.

18. Let S and T be two closed sets contained in R^2. Prove that

     (a) $U = S \cap T$ is closed,

     (b) $U = S \cup T$ is closed.

19. Let S and T be two open sets contained in R^2. Prove that

     (a) $U = S \cap T$ is open,

     (b) $U = S \cup T$ is open.

20. Let S and T be two bounded sets in R^2. Prove that

     (a) $U = S \cap T$ is bounded,

     (b) $U = S \cup T$ is bounded.

21. Let S and T be two connected sets in R^2 having a point P_0 in common. Prove that $U = S \cup T$ is connected.

22. Let S_1 and S_2 be two connected sets in R^2. Show that $S = S_1 \cup S_2$ is not necessarily a connected set.

23. An infinite set is called *countable* or *denumerable* if there exists a one-to-one correspondence between it and the set of all natural numbers. Let G be an arbitrary collection of sets. The *union of all sets in G* is defined to be the set of all those elements which belong to at least one of the sets in G, and is denoted by $\bigcup_{S \in G} S$. The *intersection of all sets in G* is defined to be the set of those elements which belong to every one of the sets in G, and is denoted by $\bigcap_{S \in G} S$. If G is a finite collection, $G = \{S_1, S_2, \ldots, S_n\}$, we write

$$\bigcup_{S \in G} S = \bigcup_{k=1}^{n} S_k = S_1 \cup S_2 \cup \cdots \cup S_n \quad \text{and} \quad \bigcap_{S \in G} S = \bigcap_{k=1}^{n} S_k = S_1 \cap S_2 \cap \cdots \cap S_n.$$

If G is a countable collection, $G = \{S_1, S_2, \ldots\}$, we write

$$\bigcup_{S \in G} S = \bigcup_{k=1}^{\infty} S_k = S_1 \cup S_2 \cup \cdots \quad \text{and} \quad \bigcap_{S \in G} S = \bigcap_{k=1}^{\infty} S_k = S_1 \cap S_2 \cap \cdots.$$

     (a) Prove that the union of any collection of open sets in R^2 is an open set.

---

[6] Let B and C be propositions. A *necessary and sufficient condition for B is C* means that B implies C (necessity), and C implies B (sufficiency).

(b) Prove that the intersection of any collection of closed sets in R^2 is a closed set.

(c) Prove that the union of any collection of connected sets in R^2 having a point P_0 in common is a connected set.

(d) Prove that the intersection of any collection of bounded sets in R^2 is a bounded set.

(e) Let $G = \{S_1, S_2, \ldots\}$ be the collection of open intervals $S_k = \{x \mid -1/k < x < 1 + 1/k\}$, $k = 1, 2, \ldots$. Show that the intersection $T = \bigcap_{k=1}^{\infty} S_k$ is the closed interval $[0, 1]$. (Observe that T, the intersection of the open sets S_k, is not an open set.)

(f) Let $G = \{S_3, S_4, \ldots\}$ be the collection of closed intervals

$$S_k = \left\{ x \mid \frac{1}{k} \leqq x \leqq 1 - \frac{1}{k} \right\}, \quad k = 3, 4, \ldots .$$

Show that the union $T = \bigcup_{k=3}^{\infty} S_k$ is the open interval $(0, 1)$. (Observe that T, the union of the closed sets S_k, is not a closed set.)

(g) Show that the union of an (arbitrary) collection of bounded sets in R^2 is not necessarily a bounded set.

24. Let $P: (x, y)$ be any point in R^2. Definition 19 may be stated more concisely as

$$R^2 - S \equiv C(S) \equiv \{(x, y) \mid P \in R^2, P \notin S\}.$$

Let S and T be two sets in R^2. Prove that

(a) $C(S \cup T) = C(S) \cap C(T)$,

(b) $C(S \cap T) = C(S) \cup C(T)$.

25. Let G be a collection of sets in R^2. Prove that

(a) $C\left(\bigcup_{S \in G} S \right) = \bigcap_{S \in G} C(S)$,

(b) $C\left(\bigcap_{S \in G} S \right) = \bigcup_{S \in G} C(S)$.

The results given in the above two exercises are known as *De Morgan's Laws*.

Suggested Readings

Apostol [3]   Taylor [45]

appendix 2 EXISTENCE THEOREMS

1. Uniform Convergence

Before establishing the forthcoming existence theorems, we shall state a few basic definitions and theorems (without proofs, but including examples) which are usually encountered in a course on advanced calculus. These will be utilized in the proofs of the existence theorems.

Unless explicitly stated to the contrary, the functions and intervals under consideration are assumed to be real-valued and real, respectively.

Definition 1. A *sequence of functions* $\{f_n\}$ is a single-valued correspondence between the positive integers and a collection of functions; that is, corresponding to each positive integer n, a function f_n is given.

Definition 2. Let $\{f_n\}$ be a sequence of functions with each f_n defined for all x in an interval I: $a \leq x \leq b$. The sequence $\{f_n\}$ *converges to function f* on I if for each point $x_0 \in I$, the sequence of numbers $\{f_n(x_0)\}$ converges to the number $f(x_0)$. Function f is the *limit of sequence* $\{f_n\}$ and we write

$$\lim_{n \to \infty} f_n(x) = f(x) \tag{1}$$

for $x \in I$.

Such convergence is also known as pointwise convergence.

Example 1. Consider the sequence of functions $\{f_n\}$ defined for all x in I: $0 \leq x \leq 1$ by

$$f_n(x) = \frac{n^2 x}{n^2 x + 2}, \qquad n = 1, 2, \ldots . \tag{2}$$

Let us show that $\{f_n\}$ converges to function f on I, where f is defined by

$$\begin{cases} f(0) = 0, \\ f(x) = 1, \quad 0 < x \leq 1. \end{cases} \tag{3}$$

SOLUTION.   Observe that $f_n(0) = 0$ for all n.   Thus, $\lim_{n \to \infty} f_n(0) = 0$.   Now, let x_0 be any fixed point in I such that $0 < x_0 \leq 1$.   Then

$$\lim_{n \to \infty} f_n(x_0) = \lim_{n \to \infty} \frac{n^2 x_0}{n^2 x_0 + 2} = \lim_{n \to \infty} \frac{x_0}{x_0 + \dfrac{2}{n^2}} = \frac{x_0}{x_0} = 1.$$

Thus, $f(x) = 1$, $0 < x \leq 1$.   Observe that for each n, function f_n is continuous on I.   But the limit function f is not continuous on I, since it is not continuous at the point $x = 0$.   Figure A2.1 depicts the graphs of the functions f_1, f_4, f_8 defined, respectively, by

$$f_1(x) = \frac{x}{x + 2}, \qquad f_4(x) = \frac{8x}{8x + 1}, \qquad f_8(x) = \frac{32x}{32x + 1}.$$

Also, the graph of the limit function f appears in Figure A2.1.

Definition 3.   Let $\{f_n\}$ be a sequence of functions with each f_n defined for all x in an interval I: $a \leq x \leq b$.   The sequence $\{f_n\}$ *converges uniformly* to function f on I, if for every $\varepsilon > 0$, there exists a positive integer N such that if $n > N$,

$$|f_n(x) - f(x)| < \varepsilon \tag{4}$$

for every $x \in I$.

Remark 1.   There is a very basic difference between convergence (pointwise convergence) and uniform convergence.   In Definition 2, we require that given any $\varepsilon > 0$ and $x \in I$, there be a corresponding N such that $|f_n(x) - f(x)| < \varepsilon$ for all $n > N$.   Here, N depends not only upon ε but also on the particular choice of $x \in I$.   However, in uniform convergence, given any $\varepsilon > 0$, we require a corresponding N such that $|f_n(x) - f(x)| < \varepsilon$ for all $n > N$

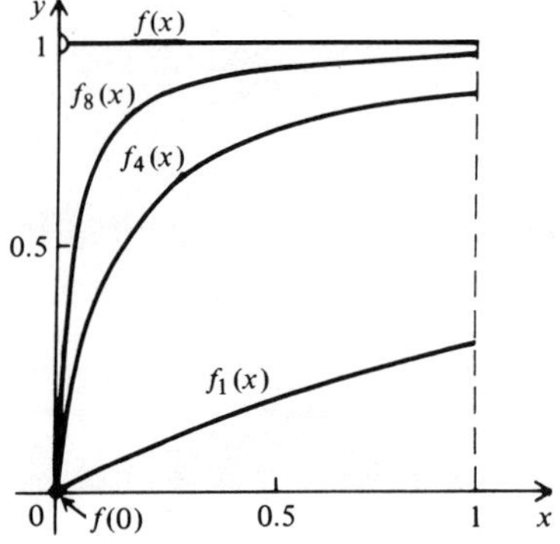

Figure A2.1.

and for all $x \in I$. Thus, in uniform convergence, N depends only upon ε; the given ε works for all $n > N$ and for all $x \in I$.

The inequality (4) may be written as

$$f(x) - \varepsilon < f_n(x) < f(x) + \varepsilon. \tag{5}$$

Geometrically, this asserts that given any $\varepsilon > 0$, the graphs of $f_{N+1}, f_{N+2}, \ldots$ all lie between $f(x) - \varepsilon$ and $f(x) + \varepsilon$ for $n > N$ and for every $x \in I$. [*See* Figure A2.2.]

Example 2.   Consider the sequence of functions $\{f_n\}$ defined for all x in the interval I: $0 \leq x \leq 1$ by

$$f_n(x) = x^3 + \frac{x}{n^2}, \qquad n = 1, 2, \ldots. \tag{6}$$

Let us show that $\{f_n\}$ converges uniformly to function f on I where

$$f(x) = x^3. \tag{7}$$

SOLUTION.   Observe that for every $x \in I$, we have

$$\lim_{n \to \infty} f_n(x) = x^3.$$

Thus, the sequence of functions $\{f_n\}$ converges to f on I.   We shall now show that $\{f_n\}$ converges uniformly to f on I.   To establish this, we proceed as follows.   For all $x \in I$, we have

$$|f_n(x) - f(x)| = \left|\left(x^3 + \frac{x}{n^2}\right) - x^3\right| = \left|\frac{x}{n^2}\right| = \frac{x}{n^2}. \tag{8}$$

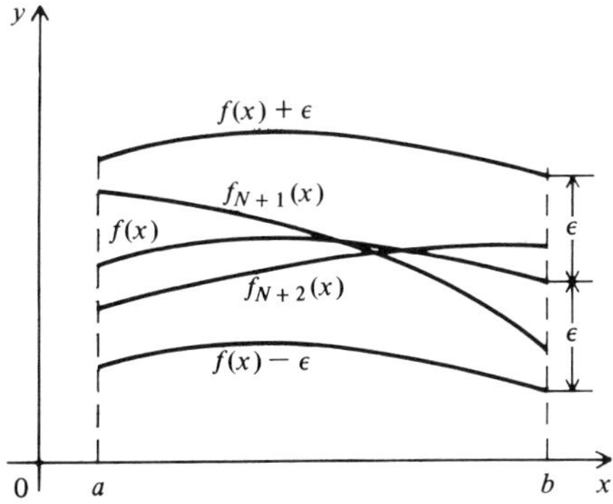

Figure A2.2.
Uniform convergence of a sequence of functions

Since $0 \le x \le 1$ for all $x \in I$, using (8) we have

$$|f_n(x) - f(x)| \le \frac{1}{n^2}. \tag{9}$$

Now, given any $\varepsilon > 0$, let us choose $N > 1/\sqrt{\varepsilon}$. Then, $n > N$ implies $n > 1/\sqrt{\varepsilon}$, or $1/n < \sqrt{\varepsilon}$, and (9) becomes

$$|f_n(x) - f(x)| < \varepsilon \tag{10}$$

for all $n > N$ and all $x \in I$. Therefore, the sequence of functions $\{f_n\}$ converges uniformly to function f on I.

The graphs of the functions f_1, f_2, f_3 defined, respectively, by

$$f_1(x) = x^3 + x, \qquad f_2(x) = x^3 + \frac{x}{4}, \qquad f_3(x) = x^3 + \frac{x}{9},$$

including the graph of the limit function f are shown in Figure A2.3.

In Example 2, if we take $\varepsilon = 0.02$, then a determination of a positive integer N can be accomplished by taking $N = [1/\sqrt{0.02}]$, where $[p]$ denotes the greatest integer less than or equal to the real number p. [*See* Exercise 7.2.4.] Thus,

$$N = \left[\frac{1}{\sqrt{0.02}}\right] = \left[\frac{1}{0.1414}\right] = [7.07] = 7.$$

Hence, whenever $n > N$, that is, for $n = 8, 9, \ldots$, we are assured that $n^2 > 1/0.02$ and that

$$|f_n(x) - f(x)| = \frac{x}{n^2} < 0.02$$

for all $x \in I$.

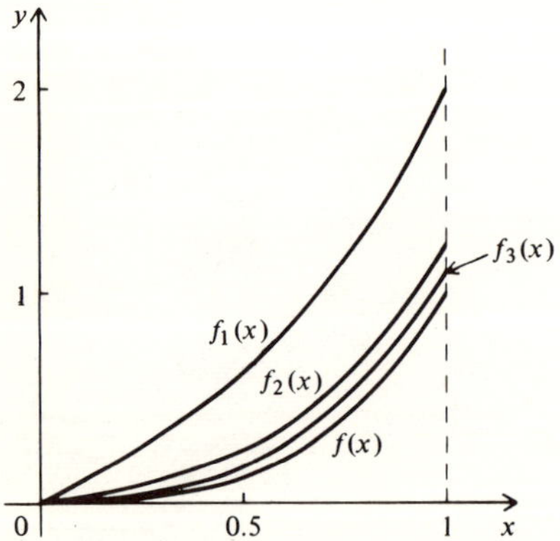

Figure A2.3.

Theorem 1. *If $\{f_n\}$ is a sequence of continuous functions on an interval I: $\ a \leq x \leq b$, and if $\{f_n\}$ converges uniformly to f on I, then f is continuous on I.*

Example 3. In Example 2, each function f_n in the sequence of functions $\{f_n\}$ given in (6) is certainly continuous on I: $\ 0 \leq x \leq 1$. Moreover, we showed that $\{f_n\}$ converges uniformly to f on I. It follows from Theorem 1 that f is continuous on I. For Example 2, we may verify directly that the function f given in (7) is indeed continuous on I.

Theorem 2. *If $\{f_n\}$ is a sequence of continuous functions on I: $\ a \leq x \leq b$, and if $\{f_n\}$ converges uniformly to f on I, then*

$$\lim_{n \to \infty} \int_\alpha^\beta f_n(x)dx = \int_\alpha^\beta \lim_{n \to \infty} f_n(x)dx = \int_\alpha^\beta f(x)dx, \tag{11}$$

where the interval $[\alpha, \beta] \subseteq I$.

[*Note:* The hypotheses of Theorems 1 and 2 are the same.]

Example 4. In Example 3, we showed that the sequence of functions $\{f_n\}$ given in (6) satisfies the hypothesis of Theorem 1 and hence it satisfies the hypothesis of Theorem 2. Thus, for Example 2, we have in view of the conclusion of Theorem 2,

$$\lim_{n \to \infty} \int_0^1 \left(x^3 + \frac{x}{n^2}\right)dx = \int_0^1 \lim_{n \to \infty} \left(x^3 + \frac{x}{n^2}\right)dx = \int_0^1 x^3\, dx. \tag{12}$$

Let us verify directly the result given in (12).

$$\lim_{n \to \infty} \int_0^1 \left(x^3 + \frac{x}{n^2}\right)dx = \lim_{n \to \infty} \left[\frac{x^4}{4} + \frac{x^2}{2n^2}\right]\Bigg|_0^1 = \lim_{n \to \infty} \left[\frac{1}{4} + \frac{1}{2n^2}\right] = \frac{1}{4}.$$

Also,

$$\int_0^1 x^3\, dx = \frac{x^4}{4}\Bigg|_0^1 = \frac{1}{4}.$$

Thus, the conclusion of Theorem 2 has been verified.

Definition 4. Consider the infinite series $\sum_{n=1}^\infty u_n$ where each u_n $(n = 1, 2, \ldots)$ is a function defined on the interval I: $\ a \leq x \leq b$. The series $\sum_{n=1}^\infty u_n$ is said to *converge uniformly* to a function f on I if the sequence of partial sums

$$s_n = \sum_{i=1}^n u_i \tag{13}$$

converges uniformly to f on I.

The following theorem gives a very useful criterion for the uniform convergence of an infinite series.

Theorem 3.   (*Weierstrass M-Test.*)   *Suppose $\sum_{n=1}^{\infty} M_n$ is a convergent series of positive terms and suppose that the series $\sum_{n=1}^{\infty} u_n$ consists of terms u_n, each of which is a function defined on an interval I: $a \leq x \leq b$.   If for all $x \in I$,*

$$|u_n(x)| \leq M_n, \qquad n = 1, 2, \ldots, \tag{14}$$

then the series $\sum_{n=1}^{\infty} u_n$ converges uniformly on I.

Example 5.   Consider the series

$$\sum_{n=1}^{\infty} \frac{x^n \cos nx}{n^2}$$

on the interval I: $0 \leq x \leq 1$.   Let us take $M_n = 1/n^2$.   Now, the series $\sum_{n=1}^{\infty} 1/n^2$ is a convergent series of positive terms.   Note that for all $x \in I$ we have

$$|u_n(x)| = \left| \frac{x^n \cos nx}{n^2} \right| = \frac{|x|^n |\cos nx|}{n^2} \leq \frac{1 \cdot 1}{n^2} = \frac{1}{n^2}, \qquad n = 1, 2, \ldots .$$

Hence, in view of Theorem 3, the series

$$\sum_{n=1}^{\infty} \frac{x^n \cos nx}{n^2}$$

converges uniformly on I.

2. Lipschitz Condition

The reader is urged to review the concepts of a region and a domain given in Definition 35 of Appendix 1.

Definition 5.   Let f be defined in a domain $D \subset R^2$, where R^2 denotes the real xy-plane, and let (x_0, y_0) be a point in D.   The function f is said to be *continuous* at (x_0, y_0) if, given any $\varepsilon > 0$, there exists a $\delta > 0$ such that

$$|f(x, y) - f(x_0, y_0)| < \varepsilon \tag{15}$$

for all points (x, y) in D satisfying $|x - x_0| < \delta$ and $|y - y_0| < \delta$.

Definition 6.   The function f defined on a set $S \subset R^2$ is said to be *bounded* on S if there exists a constant $M > 0$ such that $|f(x, y)| \leq M$ for all $(x, y) \in S$.

   [*Note:* Besides R^2, we shall also be working, more generally, in R^n, or Euclidean n-space, n a positive integer.   The points of R^n are ordered n-tuples of real numbers $(x_1, x_2, \ldots, x_n)$.   The definitions of domain, continuity, boundedness, closure, compactness, and so on, in R^n $(n \neq 2)$ are similar to the corresponding definitions in R^2.]

Theorem 4. *If f is defined and continuous on a compact (bounded and closed) set $S \subset R^n$, then f is bounded on S.*

Example 6. (For $n = 2$.) Consider the function f defined by $f(x, y) = x \sin x + x^2 y^3$ on the rectangle R: $-2 \leqq x \leqq 4$, $1 \leqq y \leqq 3$. Clearly, R is a compact set $\subset R^2$. Consequently, by Theorem 4, the function f is bounded on R. By direct verification we see that $|f(x, y)| \leqq |x| \, |\sin x| + |x|^2 |y|^3 \leqq 4 \cdot 1 + 16 \cdot 27 = 436$ for all $(x, y) \in R$.

Remark 2. In Example 1.4.4 of Chapter 1, we verified that the initial-value problem

$$\begin{cases} \dfrac{dy}{dx} = y^{2/3}, \\[2mm] y(0) = 0, \end{cases} \tag{16}$$

had more than one solution, namely, $y = 0$ and $y = x^3/27$. Thus, even though the function f given by $f(x, y) = y^{2/3}$ is continuous for all x and y, the solution of the initial-value problem (16) is not unique. Now, uniqueness is a very important practical and theoretical property. In order to insure uniqueness of a given initial-value problem, a further condition is required. One such *sufficient* condition is known as a *Lipschitz condition*, which we shall now define.

Definition 7. The function f defined in a region $S \subset R^2$ is said to satisfy a *Lipschitz condition* (with respect to y) in S, if there exists a constant $K > 0$ such that for every two points (x, y_1), (x, y_2) in S,

$$|f(x, y_1) - f(x, y_2)| \leqq K|y_1 - y_2|. \tag{17}$$

The constant K is usually known as a *Lipschitz constant*.

Remark 3. The Lipschitz condition is a strong uniformity condition. Let us write (17) as

$$\left| \frac{f(x, y_1) - f(x, y_2)}{y_1 - y_2} \right| \leqq K, \qquad y_1 \neq y_2. \tag{18}$$

Thus, with fixed x, the resulting function of y has bounded difference quotients, and the bound K is independent of x.

Remark 4. Consider the function f defined by

$$f(x, y) = y^{2/3} \tag{19}$$

in some region $S \subset R^2$ containing the origin. Clearly, the function f is continuous for all points (x, y) in S. Setting $y_1 = 0$ and $y_2 = y$, inequality (17) becomes

$$|f(x, 0) - f(x, y)| = |0 - y^{2/3}| \leqq K|0 - y|,$$

or

$$|y^{2/3}| \leqq K|y|. \tag{20}$$

Thus, a Lipschitz condition is not satisfied for all $(x, y) \in S$. For, suppose that such a K existed. Taking $y = 1/(4K)^3$, the inequality (20) gives

$$\frac{1}{16K^2} \leqq \frac{K}{64K^3} \quad \text{or} \quad 1 \leqq \frac{1}{4},$$

which is absurd. Thus, *the continuity of a function f in x and y does not imply a Lipschitz condition.*

Remark 5. Let the function f satisfy a Lipschitz condition in a region $S \subset R^2$. Then the function f is continuous in y alone in S. To see this, we proceed as follows. The function f is continuous in y alone, if given any $(x, y_1) \in S$ and any $\varepsilon > 0$, there exists a $\delta > 0$ such that

$$|f(x, y_1) - f(x, y_2)| < \varepsilon$$

whenever $(x, y_2) \in S$ and $|y_1 - y_2| < \delta$. If the function f satisfies a Lipschitz condition in S, then

$$|f(x, y_1) - f(x, y_2)| \leqq K|y_1 - y_2|$$

for every two points (x, y_1), $(x, y_2) \in S$. Now, given any $\varepsilon > 0$, let us take $\delta = \varepsilon/K$. Thus,

$$|f(x, y_1) - f(x, y_2)| \leqq K|y_1 - y_2| < \varepsilon \tag{21}$$

whenever $|y_1 - y_2| < \delta = \dfrac{\varepsilon}{K}$.

The above proof actually reveals that the function f is uniformly continuous in y alone in S.

Remark 6. Suppose that the function f is continuous in a convex region $S \subset R^2$.[1] Suppose further that $\partial f/\partial y$ exists and is bounded on S. Then the function f satisfies a Lipschitz condition in S. For, since $\partial f/\partial y$ is bounded on S, in view of Definition 6, we have for all (x, y) in S

$$\left|\frac{\partial f}{\partial y}\right| \leqq K, \qquad \text{for some } K > 0. \tag{22}$$

Let (x, y_1) and (x, y_2) be any two points in S. Then by the Law of the Mean we have

$$f(x, y_1) - f(x, y_2) = \frac{\partial}{\partial y} f(x, Y) \cdot (y_1 - y_2), \tag{23}$$

---

[1] A region $S \subset R^2$ is said to be *convex* if given any two points P_1: (x_1, y_1) and P_2: (x_2, y_2) in S, the line segment joining P_1 and P_2 is also contained in S. The interiors of circles and rectangles are examples of convex regions.

for some Y between y_1 and y_2.   Utilizing (22) and (23), we obtain

$$|f(x, y_1) - f(x, y_2)| = \left| \frac{\partial}{\partial y} f(x, Y) \right| \cdot |y_1 - y_2| \leq K|y_1 - y_2|,$$

which is a Lipschitz condition in S for the function f.   Thus, the existence and boundedness of $\partial f/\partial y$ on a convex region $S \subset R^2$ is a sufficient condition to guarantee us a Lipschitz condition in S for the function f.   Consequently [*see* Theorem 6], we are assured that when it holds for a given initial-value problem, the problem has a unique solution.   However, the existence and boundedness of $\partial f/\partial y$ in S is not a *necessary condition* for a Lipschitz condition in S for the function f.   Thus (in the convex region S), a Lipschitz condition is a *weaker* (that is, *more general*) *condition*.

In most cases that the reader will encounter, it is unlikely that a Lipschitz condition will be satisfied when the partial derivative condition is not satisfied. When this latter condition is not satisfied, one usually begins by trying to prove that a solution is not unique.

The following example shows that a function f may satisfy a Lipschitz condition in a region $S \subset R^2$ and yet $\partial f/\partial y$ does not exist for all (x, y) in S.

Example 7.   Let the region $S \subset R^2$ be the square: $|x| \leq 2$, $|y| \leq 2$.   Let the function f be defined on S by $f(x, y) = x^3|y|$.   Then for every two points (x, y_1) and (x, y_2) in S, we have

$$|f(x, y_1) - f(x, y_2)| = \left| x^3|y_1| - x^3|y_2| \right| \leq |x|^3|y_1 - y_2| \leq 8|y_1 - y_2|,$$

which is a Lipschitz condition in S for the function f.   However, on the intervals $-2 < x < 0$, $y = 0$,   and $0 < x < 2$, $y = 0$, which are contained in S, $\partial f/\partial y$ does not exist since $\partial f/\partial y = x^3$ for $y = 0^+$, and $\partial f/\partial y = -x^3$ for $y = 0^-$.

3. Equivalence of Solutions of an Initial-Value Problem and an Integral Equation. The Picard Method

Let f be a function defined and continuous in a domain $D \subset R^2$, and let (x_0, y_0) be a point of D.   Consider the initial-value problem

$$\begin{cases} \dfrac{dy}{dx} = f(x, y), \\[2mm] y(x_0) = y_0. \end{cases} \tag{24}$$

Let I be an interval containing x_0 in its interior.   Let ϕ be a function defined on I such that $[x, \phi(x)] \in D$ for all $x \in I$.   Thus, $f[x, \phi(x)]$ is defined for all

$x \in I$.   The function ϕ is said to be a *solution of the initial-value problem* (24) *on I*, if ϕ is differentiable for all $x \in I$ and

$$
\begin{cases}
\dfrac{d\phi(x)}{dx} = f[x, \phi(x)], \\[2mm]
\phi(x_0) = y_0,
\end{cases}
\tag{25}
$$

for all $x \in I$.

   We shall now establish the following useful result.

Theorem 5.   *A necessary and sufficient condition that a function ϕ be a solution of the initial-value problem* (24) *on an interval I is that ϕ be a solution of the integral equation*

$$
\phi(x) = y_0 + \int_{x_0}^{x} f[t, \phi(t)]dt
\tag{26}
$$

on I.

PROOF.   *Necessity.*   This condition asserts that if ϕ is a solution of the initial-value problem (24) on an interval I, then ϕ is a solution of the integral equation (26) on I.   Let ϕ be a solution of (24).   In view of (25), we have

$$
\frac{d\phi(x)}{dx} = f[x, \phi(x)]
\tag{27}
$$

for all $x \in I$.   Since ϕ is continuous on I and f is continuous in D, it follows that $f[x, \phi(x)]$ is continuous on I.   Hence, integrating (27) from x_0 to x, we obtain

$$
\phi(x) = \phi(x_0) + \int_{x_0}^{x} f[t, \phi(t)]dt
\tag{28}
$$

for all $x \in I$.   Putting the initial condition in (25) into (28), we see that ϕ is a solution of the integral equation (26) on I.

Sufficiency.   This condition asserts that if ϕ is a solution of the integral equation (26) on I, then ϕ is a solution of the initial-value problem (24) for all $x \in I$.   Since $f[x, \phi(x)]$ is continuous for all $x \in I$, the derivative of the integral on the right-hand side of (26) exists and equals the integrand.   [*See* the Fundamental Theorem of Integral Calculus.]   Consequently, differentiating (26), we obtain

$$
\frac{d\phi(x)}{dx} = f[x, \phi(x)]
$$

for all $x \in I$.   Also, from (26), we see that $\phi(x_0) = y_0$.   Thus, ϕ is a solution of the initial-value problem (24) for all $x \in I$.

Remark 7.   From the above theorem, we see that the study of the initial-value problem (24) has been transformed to the study of the integral equation (26).   This equation will be utilized in establishing the existence theorems.

Example 8. Consider the initial-value problem

$$\begin{cases} \dfrac{dy}{dx} = 2xy, \\[2mm] y(0) = 1, \end{cases} \tag{29}$$

on the interval I: $a \le x \le b$, where $a > -\infty$ and $b < \infty$. Utilizing the method of separation of variables, the reader may readily show that the function ϕ given by $\phi(x) = e^{x^2}$ is a solution of the initial-value problem (29) on I. Since $f(x, y) = 2xy$, we see that $f[x, \phi(x)] = 2x\phi(x) = 2xe^{x^2}$ is continuous for all $x \in I$. Now, using (26) with $\phi(x_0) = \phi(0) = 1$, we also obtain

$$\phi(x) = \phi(0) + \int_0^x f[t, \phi(t)]dt = 1 + \int_0^x 2te^{t^2}\, dt = 1 + e^{t^2}\Big|_0^x = 1 + e^{x^2} - 1 = e^{x^2}.$$

Various methods are known for proving the existence and uniqueness of the initial-value problem (24). We shall utilize the method due to Picard which is known as the *method of successive approximations*. For the present, we shall simply illustrate this method by means of an example. Basically, this method consists of first selecting an arbitrary function ϕ_0 continuous on an interval I containing the point x_0 in its interior. Usually one selects $\phi_0(x) = y_0$, where y_0 is given in (24). We substitute $\phi_0(t)$ for $\phi(t)$ into the integrand of (26), and denote the result of the operation by $\phi_1(x)$. Thus,

$$\phi_1(x) = y_0 + \int_{x_0}^x f[t, \phi_0(t)]dt. \tag{30}$$

The next approximating function is obtained by substituting $\phi_1(t)$ for $\phi_0(t)$ into the integrand of (30). Denoting the result of the operation by $\phi_2(x)$, we have

$$\phi_2(x) = y_0 + \int_{x_0}^x f[t, \phi_1(t)]dt.$$

Continuing in this fashion, we generate a sequence of approximating functions $\{\phi_n\}$ which satisfy the integral equations

$$\begin{cases} \phi_1(x) = y_0 + \displaystyle\int_{x_0}^x f[t, \phi_0(t)]dt, \\[4mm] \phi_2(x) = y_0 + \displaystyle\int_{x_0}^x f[t, \phi_1(t)]dt, \\[4mm] \phi_3(x) = y_0 + \displaystyle\int_{x_0}^x f[t, \phi_2(t)]dt, \\[2mm] \qquad \vdots \\[2mm] \phi_n(x) = y_0 + \displaystyle\int_{x_0}^x f[t, \phi_{n-1}(t)]dt, \\[2mm] \qquad \vdots \end{cases} \tag{31}$$

The functions ϕ_n are called the *Picard iterants*.

It will be shown in Section 4 that if $\{\phi_n\}$ is a sequence of continuous functions on some interval I containing x_0 in its interior, and if $\{\phi_n\}$ converges uniformly on I to function ϕ on I, then

$$\phi(x) = \lim_{n \to \infty} \phi_n(x) = y_0 + \lim_{n \to \infty} \int_{x_0}^{x} f[t, \phi_{n-1}(t)]dt = y_0 + \int_{x_0}^{x} f[t, \phi(t)]dt, \quad (32)$$

on I and $\phi(x_0) = y_0$. Thus, in view of Theorem 5, the function ϕ will then be a solution of the initial-value problem (24) for every $x \in I$. Also, as will be seen in Section 4, to insure that the function ϕ is the unique solution of the initial-value problem (24) for every $x \in I$, we shall require that the function f, besides being continuous in a domain $D \subset R^2$, also satisfy a Lipschitz condition in D.

Example 9.   Let us utilize Picard's method of successive approximations to solve the initial-value problem

$$\begin{cases} \dfrac{dy}{dx} = 2xy, \\[2mm] y(0) = 1, \end{cases} \quad (33)$$

on some interval I containing the point $x = 0$ in its interior.

SOLUTION.   Here $f(x, y) = 2xy$, $x_0 = 0$,   and $y(0) = y_0 = 1$.   Let us take

$$\phi_0(x) = y_0 = 1. \quad (34)$$

Thus, $f[t, \phi_0(t)] = 2t\phi_0(t) = 2t$.   Using the first equation in (31), we obtain

$$\phi_1(x) = 1 + \int_0^x 2t\,dt = 1 + x^2. \quad (35)$$

Observe that $f[t, \phi_1(t)] = 2t\phi_1(t) = 2t(1 + t^2) = 2t + 2t^3$.   Using the second equation in (31), we find that

$$\phi_2(x) = 1 + \int_0^x (2t + 2t^3)dt = 1 + \frac{x^2}{1!} + \frac{x^4}{2!}. \quad (36)$$

Utilizing (36), we see that

$$f[t, \phi_2(t)] = 2t\phi_2(t) = 2t\left(1 + \frac{t^2}{1!} + \frac{t^4}{2!}\right).$$

Utilizing the third equation in (31), we obtain

$$\phi_3(x) = 1 + \int_0^x 2t\left(1 + \frac{t^2}{1!} + \frac{t^4}{2!}\right)dt = 1 + \frac{x^2}{1!} + \frac{x^4}{2!} + \frac{x^6}{3!}. \quad (37)$$

In view of (35), (36), and (37) it appears that the nth approximating function ϕ_n is given by

$$\phi_n(x) = 1 + \frac{x^2}{1!} + \frac{x^4}{2!} + \frac{x^6}{3!} + \cdots + \frac{x^{2n}}{n!} = \sum_{j=0}^{n} \frac{x^{2j}}{j!}. \tag{38}$$

We shall now establish (38) by mathematical induction. By (35), we know that (38) is true for $n = 1$. Let us now assume that (38) is true for $n = k$, that is,

$$\phi_k(x) = \sum_{j=0}^{k} \frac{x^{2j}}{j!}.$$

If we can show that (38) is also true for $n = k + 1$, then by mathematical induction (38) will be true for all positive integral values of n. Observe that

$$f[t, \phi_k(t)] = 2t\phi_k(t) = \sum_{j=0}^{k} 2\frac{t^{2j+1}}{j!}.$$

Thus,

$$\phi_{k+1}(x) = 1 + \sum_{j=0}^{k} \int_0^x 2\frac{t^{2j+1}}{j!}\, dt = 1 + \sum_{j=0}^{k} \frac{x^{2j+2}}{(j+1)!} = \sum_{j=0}^{k+1} \frac{x^{2j}}{j!},$$

and (38) is established.

Observe that the sequence of functions $\{\phi_n\}$, with $\phi_0(x) = 1$, is a sequence of partial sums of the series

$$\sum_{j=0}^{\infty} \frac{x^{2j}}{j!}. \tag{39}$$

By the ratio test, series (39) converges for all (finite) values of x. Thus,

$$\phi(x) = \lim_{n\to\infty} \phi_n(x) = \lim_{n\to\infty} \sum_{j=0}^{n} \frac{x^{2j}}{j!} = \sum_{j=0}^{\infty} \frac{x^{2j}}{j!}. \tag{40}$$

In its interval of convergence, namely, the interval I: $|x| < \infty$, the series (39) may be differentiated term by term. [*See* Theorem 7.2.6.] Consequently, for every $x \in I$, differentiating (40) we obtain

$$\phi'(x) = \sum_{j=1}^{\infty} 2\frac{x^{2j-1}}{(j-1)!}. \tag{41}$$

We shall now verify directly that the function ϕ is a solution of the initial-value problem (33) for all $x \in I$, and is such that $\phi(x_0) = \phi(0) = 1$. Substituting $\phi(x)$ for y and $\phi'(x)$ for y' into the differential equation (33), we see that it is indeed satisfied.

$$\phi'(x) = \sum_{j=1}^{\infty} 2\frac{x^{2j-1}}{(j-1)!} = 2x\sum_{j=0}^{\infty} \frac{x^{2j}}{j!} = 2x\phi(x), \qquad x \in I.$$

In view of (40), it is evident that $\phi(0) = 1$.

For this example, the series (39) is the Taylor series expansion of e^{x^2} about the origin. Consequently, the function ϕ is also given by

$$\phi(x) = e^{x^2} \tag{42}$$

for all $x \in I$.

Remark 8. It should be mentioned that the initial-value problem in Example 9 was chosen with care. The main difficulty of applying Picard's method of successive approximations to solve a given initial-value problem is that, in general, it is extremely difficult or even impossible to obtain an explicit formula for the nth approximating function ϕ_n. However, in theory, the method works very nicely.

4. The Basic Existence and Uniqueness Theorem for the First-Order Differential Equation $y' = f(x, y)$

Theorem 1.4.1 of Chapter 1 will be established as a corollary of the following theorem.

Theorem 6. *Let the function f be defined and continuous in a domain $D \subset R^2$, and let f satisfy a Lipschitz condition (with respect to y) in D, namely,*

$$|f(x, y_1) - f(x, y_2)| \leqq K|y_1 - y_2|, \qquad K > 0, \tag{43}$$

for every two points (x, y_1) and (x, y_2) in D. Let (x_0, y_0) be a point in D. Then an interval

$$I: \quad |x - x_0| \leqq h, \qquad h > 0, \tag{44}$$

exists on which there is a unique solution ϕ satisfying the initial-value problem

$$\left\{ \begin{aligned} &\frac{dy}{dx} = f(x, y), \tag{45} \\[2mm] &y(x_0) = y_0. \tag{46} \end{aligned} \right.$$

PROOF OF THEOREM 6. Because the proof of this theorem is long, we shall divide it into five parts.

(i) *Determination of the interval I.* Let $R \subset D$ be a rectangle defined by

$$R: \quad |x - x_0| \leqq a, \qquad |y - y_0| \leqq b, \qquad a > 0, b > 0. \tag{47}$$

Such a rectangle R exists, since D is an *open* connected set $\subset R^2$. [*Note:* The center of the rectangle R is at $(x_0, y_0) \in D$.] Since the function f is continuous in D, it is also continuous on R, which is a compact set.   Thus, by Theorem 4, f is bounded on R, and [*see* Definition 6] there exists a constant $M > 0$ such that

$$|f(x, y)| \leq M \tag{48}$$

for all $(x, y) \in R$.   Clearly, the function f also satisfies inequality (43) on R. Let us define h by

$$h = \min\left(a, \frac{b}{M}\right). \tag{49}$$

Clearly, $h > 0$.   Observe that if $a \leq b/M$, then $h = a$; however, if $a \geq b/M$, then $h = b/M$.   For the interval I under consideration, we take

$$I: \quad |x - x_0| \leq h, \tag{50}$$

where h is defined by (49).

Consider now the sequence of successive approximations ϕ_0, ϕ_1, ϕ_2, $\ldots$, ϕ_n, $\ldots$ defined by (31) with $\phi_0(x) \equiv y_0$.   That is,

$$\phi_0(x) = y_0,$$
$$\phi_n(x) = y_0 + \int_{x_0}^{x} f[t, \phi_{n-1}(t)]dt, \qquad n = 1, 2, \ldots . \tag{51}$$

(ii) *Each function ϕ_n defined by (51) exists, is continuous, and satisfies the inequality*

$$|\phi_n(x) - y_0| \leq M|x - x_0| \tag{52}$$

for all x in

$$I: \quad |x - x_0| \leq h. \tag{53}$$

Before establishing this result, observe that in view of (49), $h \leq b/M$. Hence, inequality (52) implies that

$$|\phi_n(x) - y_0| \leq Mh \leq b, \tag{54}$$

for all $x \in I$.   Also, since we have $h \leq a$, the points $[x, \phi_n(x)]$ will be in R whenever $x \in I$.   One may give a geometric interpretation of inequality (52).   It asserts that each function ϕ_n lies in a "bow tie" region $S \subset R$, where S is the set of points bounded by the lines

$$L_1: \quad y - y_0 = M(x - x_0), \qquad L_2: \quad y - y_0 = -M(x - x_0) \tag{55}$$

and the vertical lines

$$x - x_0 = h, \qquad x - x_0 = -h. \tag{56}$$

[*See* Figures A2.4 and A2.5.]

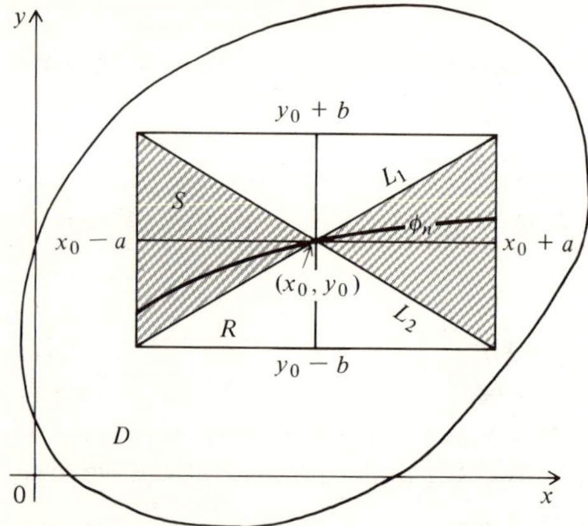

Figure A2.4.
Region S: $h = a$

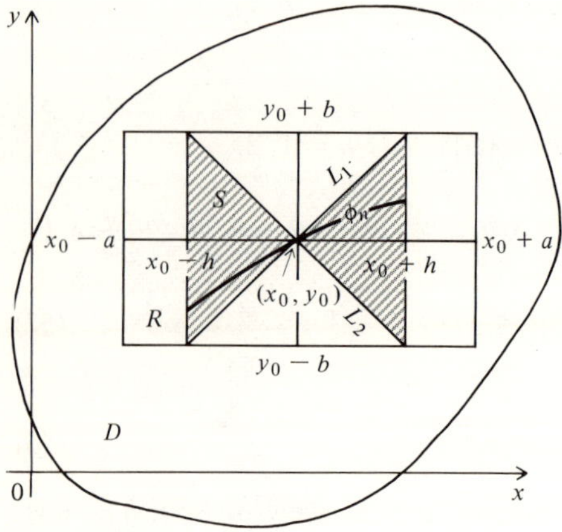

Figure A2.5.
Region S: $h = b/M$

PROOF OF (ii).   Since $\phi_0(x) = y_0$, clearly $\phi_0(x)$ exists, is continuous, and satisfies (52) with $n = 0$ for all $x \in I$.   With $n = 1$ in (51), we have

$$\phi_1(x) = y_0 + \int_{x_0}^{x} f[t, y_0]dt.$$

Since $(x, y_0) \in R$ for all $x \in I$, we have $|f(x, y_0)| \leqq M$.   Thus, for all $x \in I$, we see that

$$|\phi_1(x) - y_0| \leqq \int_{x_0}^{x} |[f(t, y_0)]|\, dt \leqq \int_{x_0}^{x} M\, dt = M|x - x_0|,$$

and $\phi_1(x)$ satisfies inequality (52).   Since the function f is continuous on R, the function g defined by $g(x) = f(x, y_0)$ is continuous on I.   Thus, the above function ϕ_1 exists as a continuous function on I.

We shall now establish result (ii) by mathematical induction.   Suppose that for $n = k$ the function ϕ_k given by (51) exists and is continuous on I; and that inequality (52) holds, that is,

$$|\phi_k(x) - y_0| \leq M|x - x_0|, \qquad x \in I.$$

Thus, the points $[x, \phi_k(x)] \in R$ for $x \in I$.   Consequently, the function g_k defined by $g_k(x) = f[x, \phi_k(x)]$ exists for $x \in I$.   Moreover, $g_k(x)$ is continuous on I, since f is continuous on R and $\phi_k(x)$ is continuous on I.   With $n = k + 1$ in (51), we have

$$\phi_{k+1}(x) = y_0 + \int_{x_0}^{x} f[t, \phi_k(t)]dt.$$

Clearly, $\phi_{k+1}(x)$ exists as a continuous function on I.   Also,

$$|\phi_{k+1}(x) - y_0| \leq \int_{x_0}^{x} |f[t, \phi_k(t)]| \, dt \leq M \int_{x_0}^{x} dt = M|x - x_0|.$$

Hence, $\phi_{k+1}(x)$ satisfies inequality (52) and the inductive proof is complete.

Summarizing, we have that each of the successive approximations $\{\phi_n\}$ defined by (51) exists, is continuous, and satisfies the inequality

$$|\phi_n(x) - y_0| \leq M|x - x_0| \leq b, \tag{57}$$

for all x in the interval I: $|x - x_0| \leq h$.   Consequently, $[x, \phi_n(x)] \in R$ for $x \in I$, and thus $f[x, \phi_n(x)]$ is defined and continuous for all $x \in I$.

In the remaining parts of the proof of the theorem, we shall concern ourselves with the interval $J \subset I$, where

$$J: \quad x_0 \leq x \leq x_0 + h. \tag{58}$$

Clearly, for $x \in J$, $0 \leq x - x_0 \leq h$.   Arguments similar to those given below may be applied to the interval $x_0 - h \leq x \leq x_0$.

(iii) *The sequence of successive approximations $\{\phi_n\}$ defined by (51) converges uniformly on J to a continuous limit function ϕ on J.*

PROOF OF (iii).   We first establish the following inequality by mathematical induction.

$$|\phi_n(x) - \phi_{n-1}(x)| \leq \frac{MK^{n-1}}{n!}(x - x_0)^n \leq \frac{M}{K}\frac{(Kh)^n}{n!}, \tag{59}$$

for all $x \in J$ $(n = 1, 2, \ldots)$.   In view of (51), we have

$$\phi_1(x) - \phi_0(x) = y_0 + \int_{x_0}^{x} f[t, y_0]dt - y_0 = \int_{x_0}^{x} f[t, y_0]dt.$$

Thus,

$$|\phi_1(x) - \phi_0(x)| \leq \int_{x_0}^{x} |f[t, y_0]| \, dt \leq M \int_{x_0}^{x} dt = M(x - x_0) \leq Mh,$$

for all $x \in J$. Hence, inequality (59) is true for $n = 1$. Assume that inequality (59) is true for $n = k$, that is,

$$|\phi_k(x) - \phi_{k-1}(x)| \leq \frac{MK^{k-1}}{k!} (x - x_0)^k \leq \frac{M}{K} \frac{(Kh)^k}{k!}, \tag{60}$$

for all $x \in J$. From (51) we have

$$\phi_{k+1}(x) - \phi_k(x) = y_0 + \int_{x_0}^{x} f[t, \phi_k(t)] dt - y_0 - \int_{x_0}^{x} f[t, \phi_{k-1}(t)] dt$$

$$= \int_{x_0}^{x} \{f[t, \phi_k(t)] - f[t, \phi_{k-1}(t)]\} dt.$$

Thus,

$$|\phi_{k+1}(x) - \phi_k(x)| \leq \int_{x_0}^{x} |f[t, \phi_k(t)] - f[t, \phi_{k-1}(t)]| \, dt. \tag{61}$$

In view of the result of part (ii), $[x, \phi_k(x)]$ and $[x, \phi_{k-1}(x)] \in R$ for $x \in J$. By hypothesis, the function f satisfies a Lipschitz condition in D and hence also in $R \subset D$. Utilizing inequality (43) with $y_1 = \phi_k(x)$ and $y_2 = \phi_{k-1}(x)$, inequality (61) becomes

$$|\phi_{k+1}(x) - \phi_k(x)| \leq \int_{x_0}^{x} K|\phi_k(t) - \phi_{k-1}(t)| \, dt. \tag{62}$$

Substituting the first part of the inequality in (60) into (62), we get

$$|\phi_{k+1}(x) - \phi_k(x)| \leq K \frac{M}{k!} K^{k-1} \int_{x_0}^{x} (t - x_0)^k \, dt \leq \frac{MK^k}{k!} \frac{1}{k+1} (t - x_0)^{k+1} \Big|_{x_0}^{x}$$

$$= \frac{MK^k}{(k+1)!} (x - x_0)^{k+1} \leq \frac{M}{K} \frac{(Kh)^{k+1}}{(k+1)!},$$

for all $x \in J$, and the inductive proof is complete.

A key point in the proof of the theorem is that $\phi_n(x)$ can be written as

$$\phi_n(x) = y_0 + [\phi_1(x) - \phi_0(x)] + [\phi_2(x) - \phi_1(x)] + \cdots$$

$$+ [\phi_n(x) - \phi_{n-1}(x)] = y_0 + \sum_{i=1}^{n} [\phi_i(x) - \phi_{i-1}(x)], \tag{63}$$

with $\phi_0(x) = y_0$. Thus, $\phi_n(x)$ is the nth partial sum of the series

$$y_0 + \sum_{n=1}^{\infty} [\phi_n(x) - \phi_{n-1}(x)], \tag{64}$$

for all $x \in J$. In view of (59), we see that the constant term y_0 and each term $[\phi_n(x) - \phi_{n-1}(x)]$ of the series (64) is dominated by a series of positive terms, namely,

$$|y_0| + \frac{M}{K} \cdot \frac{Kh}{1!} + \frac{M}{K} \cdot \frac{K^2 h^2}{2!} + \frac{M}{K} \cdot \frac{K^3 h^3}{3!} + \cdots = |y_0| + \frac{M}{K} \sum_{n=1}^{\infty} \frac{(Kh)^n}{n!} \tag{65}$$

for all $x \in J$. From Example 7.2.3(1), we know that the series

$$\sum_{n=0}^{\infty} \frac{x^n}{n!} = e^x$$

converges for all finite values of x. Thus, the series (65) converges to

$$|y_0| + \frac{M}{K} (e^{Kh} - 1), \tag{66}$$

for all finite values of Kh. Therefore, by the Weierstrass M-Test [*see* Theorem 3], series (64) converges uniformly for all $x \in J$. Consequently [*see* Definition 4], the sequence of partial sums $\{s_n\} = \{\phi_n\}$ given in (63) converges uniformly to a limit function ϕ for all $x \in J$. Thus, in view of Theorem 1, the limit function ϕ is continuous for all $x \in J$. Part (iii) is now established.

(iv) *The limit function ϕ is a solution of the differential equation (45) for all $x \in J$ and satisfies the initial condition (46).*

PROOF OF (iv). From part (ii), we know that each function ϕ_n satisfies the inequality $|\phi_n(x) - y_0| \leq b$ for all $x \in I$ and hence for all $x \in J \subset I$. Also, in view of part (iii), $|\phi(x) - y_0| \leq b$ for all $x \in J$. Consequently, for all $x \in J$, $[x, \phi_n(x)]$ and $[x, \phi(x)]$ are points $\in R$. Utilizing inequality (43) with $y_1 = \phi(x)$ and $y_2 = \phi_n(x)$, we obtain

$$|f[x, \phi(x)] - f[x, \phi_n(x)]| \leq K|\phi(x) - \phi_n(x)|, \qquad K > 0. \tag{67}$$

Since the sequence $\{\phi_n\}$ converges uniformly to limit function ϕ on J, we have [*see* Definition 3] that given any $\varepsilon > 0$, there exists a positive integer N such that if $n > N$

$$|\phi(x) - \phi_n(x)| < \frac{\varepsilon}{K} \tag{68}$$

for all $x \in J$. Substituting (68) into (67), we obtain if $n > N$

$$|f[x, \phi(x)] - f[x, \phi_n(x)]| < \varepsilon$$

for all $x \in J$. Thus, by Definition 3, the sequence of functions defined by $\{f[x, \phi_n(x)]\}$ converges uniformly on J to the function defined by $f[x, \phi(x)]$ on J. It follows from the continuity of f and ϕ_n that each $f[x, \phi_n(x)]$ is continuous on J. It follows from Theorem 1 that $f[x, \phi(x)]$ is also continuous on J. Using (51) and Theorem 2, we obtain

$$\phi(x) = \lim_{n \to \infty} \phi_n(x)$$

$$= \lim_{n \to \infty} \left[y_0 + \int_{x_0}^{x} f[t, \phi_{n-1}(t)]dt \right] = y_0 + \lim_{n \to \infty} \int_{x_0}^{x} f[t, \phi_{n-1}(t)]dt$$

$$= y_0 + \int_{x_0}^{x} \lim_{n \to \infty} f[t, \phi_{n-1}(t)]dt = y_0 + \int_{x_0}^{x} f[t, \phi(t)]dt, \tag{69}$$

for all $x \in J$. From (69), we readily see that $\phi(x_0) = y_0$. Thus, the limit function ϕ is a solution of the integral equation (69) on J and is such that $\phi(x_0) = y_0$. In view of Theorem 5, the limit function ϕ is a solution of the differential equation (45) on J satisfying the initial condition (46). Part (iv) is now established.

Note that since

$$\frac{d\phi(x)}{dx} = f[x, \phi(x)], \tag{70}$$

it follows that the function ϕ is differentiable for all $x \in J$.

(v) *The function ϕ is the unique solution of the differential equation (45) on J which satisfies the initial condition (46).*

PROOF OF (v). Suppose there exists *another* differentiable function ψ on J such that for $x \in J$, the point $[x, \psi(x)] \in D$; and

$$\begin{cases} \dfrac{d\psi(x)}{dx} = f[x, \psi(x)], & x \in J, \tag{71} \\[2mm] \psi(x_0) = y_0. \tag{72} \end{cases}$$

We shall show that $\psi(x) = \phi(x)$ for all $x \in J$ which will establish part (v). Integrating (70) and (71) and using (46) and (72) to evaluate the constants of integrations, we obtain, respectively,

$$\phi(x) = y_0 + \int_{x_0}^{x} f[t, \phi(t)]dt \tag{73}$$

and

$$\psi(x) = y_0 + \int_{x_0}^{x} f[t, \psi(t)]dt, \tag{74}$$

for all $x \in J$. Subtracting (73) from (74) and then taking absolute values, we obtain

$$|\psi(x) - \phi(x)| \leq \int_{x_0}^{x} |f[t, \psi(t)] - f[t, \phi(t)]| \, dt. \tag{75}$$

Since the functions ψ and ϕ are continuous on J, so is the function g, where $g(x) = \psi(x) - \phi(x)$. Thus [*see* Theorem 4 with $n = 1$], we have for some constant $\beta > 0$

$$|\psi(x) - \phi(x)| \leqq \beta, \tag{76}$$

for all $x \in J$. Utilizing inequality (43) with $y_1 = \psi(x)$ and $y_2 = \phi(x)$, and inequality (76), inequality (75) becomes

$$|\psi(x) - \phi(x)| \leqq K \int_{x_0}^{x} |\psi(t) - \phi(t)| \, dt \leqq K\beta \int_{x_0}^{x} dt = K\beta(x - x_0), \tag{77}$$

for all $x \in J$.

We shall now prove by mathematical induction that for $n = 1, 2, \ldots,$

$$|\psi(x) - \phi(x)| \leqq \beta K^n \frac{(x - x_0)^n}{n!}, \tag{78}$$

for all $x \in J$. In view of (77), inequality (78) is true for $n = 1$. Assume that inequality (78) is also true for $n = k$, that is,

$$|\psi(x) - \phi(x)| \leqq \beta K^k \frac{(x - x_0)^k}{k!}, \tag{79}$$

for all $x \in J$. Then utilizing (75), (43), and (79), we obtain

$$|\psi(x) - \phi(x)| \leqq \int_{x_0}^{x} |f[t, \psi(t)] - f[t, \phi(t)]| \, dt \leqq K \int_{x_0}^{x} |\psi(t) - \phi(t)| \, dt$$

$$\leqq K\beta \frac{K^k}{k!} \int_{x_0}^{x} (t - x_0)^k dt = \beta \frac{K^{k+1}}{k!} \frac{1}{k+1} (t - x_0)^{k+1} \Big|_{x_0}^{x}$$

$$= \beta K^{k+1} \frac{(x - x_0)^{k+1}}{(k+1)!},$$

for all $x \in J$, and the inductive proof of inequality (78) is complete.

Since for all $x \in J$, $0 \leqq x - x_0 \leqq h$, inequality (78) becomes

$$|\psi(x) - \phi(x)| \leqq \beta \frac{(Kh)^n}{n!}. \tag{80}$$

The term $(Kh)^n/n!$ is the general term of the series expansion e^{Kh} which converges for all finite values of Kh. A necessary condition for a series to converge is that its general term (or nth term) approaches zero as $n \to \infty$. Hence, $\beta(Kh)^n/n! \to 0$ as $n \to \infty$. Consequently, in view of (80), we have for all $x \in J$,

$$0 \leqq |\psi(x) - \phi(x)| \leqq \lim_{n \to \infty} \beta \frac{(Kh)^n}{n!} = 0.$$

Therefore, $|\psi(x) - \phi(x)| = 0$ for all $x \in J$. Hence, $\psi(x) = \phi(x)$ for all $x \in J$. Thus, the function ϕ is the unique solution of the differential equation (45) on J satisfying the initial condition (46). Part (v) is now established.

Since similar arguments can be applied to the interval $x_0 - h \leqq x \leqq x_0$, the differentiable function ϕ is the unique solution of the differential equation (45) on I: $|x - x_0| \leqq h$, satisfying the initial condition (46). Thus, Theorem 6 is now established.

PROOF OF THEOREM 1.4.1 (OF CHAPTER 1). Let the function f, the domain D, and the point (x_0, y_0) be as in the hypothesis of Theorem 1.4.1. Let $R \subset D$ be a closed rectangle with center at (x_0, y_0) and sides parallel to the x, y-axes. Let D_0 be the interior of R. (Thus, D_0 is an open rectangle consisting of R minus its boundary. Clearly, D_0 is a convex domain.)

Since the function f_y is given continuous in D, it is bounded on the compact set $R \subset D$, and so is also bounded in D_0 ($\subset R$). Thus [*see* Remark 6], the function f satisfies a Lipschitz condition (with respect to y) in D_0. We may now apply Theorem 6 to the function f, domain D_0, and the point (x_0, y_0) to obtain an interval

$$I: \quad |x - x_0| \leqq h, \qquad h > 0,$$

and a function ϕ defined on I such that $[x, \phi(x)] \in D_0$ for all $x \in I$, and such that ϕ is a solution on I of the initial-value problem in Theorem 1.4.1.

Concerning uniqueness. Suppose that the function ψ is also a solution on I of the initial-value problem in Theorem 1.4.1, with $[x, \psi(x)] \in D$ for all $x \in I$. Let

$$I_0: \quad |x - x_0| \leqq h_0 \tag{81}$$

be the largest interval with center x_0 such that $\psi(x) = \phi(x)$ for all $x \in I$. Since the functions ψ and ϕ are continuous on I, and $\psi(x_0) = \phi(x_0)(= y_0)$, we have that h_0 exists (with $h_0 \geqq 0$). *Suppose that $h_0 < h$.* Let $x_1 = x_0 + h_0$, $y_1 = \phi(x_0 + h_0) = \psi(x_0 + h_0)$. Now, $(x_1, y_1) \in D_0$, $y_1 = \psi(x_1)$, and ψ is continuous at $x = x_1$. Thus, there is an interval

$$I_1: \quad |x - x_1| < h_1, \qquad h_1 > 0,$$

such that $[x, \psi(x)] \in D_0$ for all $x \in I_1$. Let the set D_1 be defined by $(x, y) \in D_1$ if and only if $x \in I_1$ and $(x, y) \in D_0$. Thus, D_1 is an open rectangle containing (x_1, y_1); and $D_1 \subset D_0$. Observe that $[x, \psi(x)] \in D_1$ and $[x, \phi(x)] \in D_1$ for all $x \in I_1$.

Applying again Theorem 6 to the function f, domain D_1, and the point (x_1, y_1), we obtain an interval

$$I_2: \quad |x - x_1| \leqq h_2, \qquad h_2 > 0, \tag{82}$$

such that $\psi(x) = \phi(x)$ for all $x \in I_2$.

Similarly, as above, if we let $x_2 = x_0 - h_0$, $y_2 = \phi(x_0 - h_0) = \psi(x_0 - h_0)$, we obtain an interval

$$I_3: \quad |x - x_2| \leqq h_3, \qquad h_3 > 0, \tag{83}$$

such that $\psi(x) = \phi(x)$ for all $x \in I_3$. Thus, since $x_1 = x_0 + h_0$ and $x_2 = x_0 - h_0$, and $\psi(x) = \phi(x)$ for x in I_0, I_2, and I_3, we have [*see* (81), (82), and (83)], $\psi(x) = \phi(x)$ for all x in the interval

$$I_4: \quad |x - x_0| \leqq h_0 + \min(h_2, h_3). \tag{84}$$

However, this contradicts the maximality of I_0. Thus, $h_0 \geqq h$, and consequently $\psi(x) = \phi(x)$ for all $x \in I$. Theorem 1.4.1 is now established.

Remark 9. Arguments similar to those used above to prove Theorems 1.4.1 (of Chapter 1) and 6 may be utilized to prove the following theorem.

Theorem. *Let the function f be defined and continuous in a domain* $D \subset R^2$, *and let f satisfy in D a local Lipschitz condition (with respect to y) [that is, for each* $(x, y) \in D$, *there is a neighborhood* $N = N(x, y) \subset D$ *of* (x, y), *and a constant* $K = K(x, y) > 0$ *such that (with respect to y) the function f satisfies in N a Lipschitz condition with constant K.] Let* (x_0, y_0) *be a point in D. Then*:

(i) *There exists an interval on the x-axis containing the point* x_0 *in its interior on which there is a solution for the initial-value problem*

$$\begin{cases} \dfrac{dy}{dx} = f(x, y), \\[2mm] y(x_0) = y_0. \end{cases} \tag{85}$$

(ii) *On any interval on the x-axis containing the point* x_0, *there is at most one solution of the initial-value problem* (85).

Since a global Lipschitz condition implies a local one, the theorem above is a generalization of Theorem 6. The continuity of f_y in D would imply that the function f satisfies a local Lipschitz condition (with respect to y) in D. Thus, the theorem above is also a generalization of Theorem 1.4.1 (of Chapter 1).

Remark 10. Let us replace the domain D in the hypothesis of Theorems 1.4.1 and 6 by a closed rectangle R with center (x_0, y_0), and let the theorems so modified be numbered 1.4.1A and 6A, respectively. The proof of Theorem 6A is similar to the proof of Theorem 6. The continuity of f_y on R implies [*see* Theorem 4 and Remark 6] that the function f satisfies a Lipschitz condition (with respect to y) on R. Thus, Theorem 6A is a generalization of Theorem 1.4.1A.

Remark 11. As mentioned in Remark 1.4.3 of Chapter 1, Theorems 1.4.1 and 6 provide us with a sufficient (but not a necessary) condition for the existence of a unique solution of the initial-value problem (45) and (46) on the interval I: $|x - x_0| \leq h$. Also, Theorems 1.4.1 and 6 are known as *local* existence theorems, since they assure us a unique solution of the initial-value problem (45) and (46) only for x near the initial point x_0. Existence theorems which guarantee us a unique solution of the initial-value problem (45) and (46) on the interval $|x - x_0| \leq a$ (where a may be a relatively large number) are known as *nonlocal* or *global* existence theorems.

Example 10. Let us consider the initial-value problem given in Example 1.4.7 of Chapter 1, namely,

$$\begin{cases} y' = 4 + y^2, & (86) \\ y(0) = 0. & (87) \end{cases}$$

Let us determine an interval I: $|x - 0| = |x| \leq h$ on which there exists a unique solution.

SOLUTION.   Since $f(x, y) = 4 + y^2$ and $\partial f / \partial y = 2y$, we see that the functions f and f_y are continuous in any domain $D \subset R^2$. Let $R \subset D$ be a rectangle defined by

$$R: \quad |x| \leq 4, \qquad |y| \leq 6. \tag{88}$$

[*Note:* The center of the rectangle R is at $(0, 0)$.] In view of Remark 6, the function f satisfies a Lipschitz condition (with respect to y) in any bounded region in D and hence in R. On the compact set R we have

$$|f(x, y)| = |4 + y^2| \leq 4 + y^2 \leq 4 + 36 = 40.$$

With $M = 40$, $a = 4$, and $b = 6$, we observe that $a > b/M$. Thus, $h = \min \left(a, \dfrac{b}{M} \right) = \dfrac{b}{M} = \dfrac{6}{40} = 0.15$.   Hence, by Theorem 6 we are guaranteed a unique solution of the given initial-value problem for all $x \in I$, where

$$I: \quad |x| \leq 0.15. \tag{89}$$

Let us now compute some of the Picard iterants leading to the solution ϕ of the initial-value problem (86) and (87).

$$\begin{cases} \phi_0(x) = y_0 = 0, \\[6pt] \phi_1(x) = y_0 + \displaystyle\int_0^x f[t, y_0]dt = \int_0^x f[t, 0]dt = \int_0^x 4 \, dt = 4x, \\[6pt] \phi_2(x) = \displaystyle\int_0^x f[t, \phi_1(t)]dt = \int_0^x (4 + 16t^2)dt = 4x + \tfrac{16}{3}x^3, \\[6pt] \phi_3(x) = \displaystyle\int_0^x f[t, \phi_2(t)]dt = \int_0^x [4 + (4t + \tfrac{16}{3}t^3)^2]dt \\[6pt] \qquad\quad = 4x + \tfrac{16}{3}x^3 + \tfrac{128}{15}x^5 + \tfrac{256}{63}x^7. \end{cases} \tag{90}$$

From 1.4.30 of Chapter 1, the solution ϕ of the given initial-value problem is given by

$$\phi(x) = 2 \tan 2x. \tag{91}$$

The Maclaurin series expansion of $2 \tan 2x$ is given by

$$2 \tan 2x = 4x + \tfrac{16}{3} x^3 + \tfrac{128}{15} x^5 + \tfrac{2176}{315} x^7 + \cdots, \tag{92}$$

which converges for $|x| < \dfrac{\pi}{4}$, or $|x| < 0.785$. Observe from (90) and (92) that $\phi_3(x)$ agrees with the expansion of $2 \tan 2x$ up to the third term. If we compute more Picard iterants, namely, $\phi_4(x)$, $\phi_5(x), \ldots, \phi_n(x)$, we would see that with increasing n, the approximating function ϕ_n would contain more terms agreeing with the expansion of $2 \tan 2x$. In fact, as $n \to \infty$, $\phi_n(x) \to \phi(x) = 2 \tan 2x$.

Observe that while Theorem 6 assures us a unique solution of our initial-value problem on the interval I: $\ |x| \leq 0.15$, the above analysis reveals that the solution actually exists in a larger interval I^*: $\ |x| < 0.785$. For the present example, we were fortunate in obtaining an explicit solution, namely, $\phi(x) = 2 \tan 2x$. However, if this solution was not known, all that we could say when applying Theorem 6 is that, if $x = 0.1$, for example, then $\phi_n(0.1)$, for n sufficiently large, is a very close approximation to $2 \tan 0.2$. But, nothing could be said about $\phi_n(0.3)$, for all n sufficiently large, being a close approximation to the actual solution $2 \tan 0.6$ unless we find a larger h. As mentioned in Remark 11, the existence Theorem 6 assures a unique solution of a given initial-value problem only for x near the initial point x_0.

5. Existence and Uniqueness Theorems for a System of First-Order Differential Equations, for Differential Equations of Order Greater than One, for a System of Linear First-Order Differential Equations, and for Linear Differential Equations of Order Greater than One

The reader is urged to review the material of Section 8.2, Chapter 8. Consider the system of n first-order equations

$$\left\{ \begin{aligned} \frac{dx_1}{dt} &= F_1(t, x_1, x_2, \ldots, x_n), \\[2mm] \frac{dx_2}{dt} &= F_2(t, x_1, x_2, \ldots, x_n), \\ &\ \vdots \\ \frac{dx_n}{dt} &= F_n(t, x_1, x_2, \ldots, x_n). \end{aligned} \right. \tag{93}$$

We shall now prove Theorem 8.2.1 as a corollary of the following theorem.

Theorem 7. *Let each function F_i $(i = 1, 2, \ldots, n)$ of the system (93) be continuous in some domain D of $(n + 1)$-dimensional $(t, x_1, x_2, \ldots, x_n)$ space. In D, let each function F_i $(i = 1, 2, \ldots, n)$ satisfy a Lipschitz condition*

$$|F_i(t, x_1, x_2, \ldots, x_n) - F_i(t, \bar{x}_1, \bar{x}_2, \ldots, \bar{x}_n)|$$
$$\leq K[|x_1 - \bar{x}_1| + |x_2 - \bar{x}_2| + \cdots + |x_n - \bar{x}_n|], \qquad (94)$$

where $(t, x_1, x_2, \ldots, x_n)$ and $(t, \bar{x}_1, \bar{x}_2, \ldots, \bar{x}_n)$ are any two points in D and $K > 0$. Let $(t_0, x_1{}^0, x_2{}^0, \ldots, x_n{}^0)$ be a point in D. Then an interval

$$I: \quad |t - t_0| \leq h, \qquad h > 0, \qquad (95)$$

exists on which there is a unique solution

$$x_1 = \phi_1(t), \quad x_2 = \phi_2(t), \ldots, x_n = \phi_n(t) \qquad (96)$$

satisfying the system (93) and the initial conditions

$$\phi_1(t_0) = x_1{}^0, \quad \phi_2(t_0) = x_2{}^0, \ldots, \phi_n(t_0) = x_n{}^0. \qquad (97)$$

PROOF OF THEOREM 7.   Since the proof of this theorem follows the same line of reasoning as that of Theorem 6, we shall merely sketch the proof and leave the details as an excellent exercise for the reader.

(i) *Determination of the interval I.*   Let $R \subset D$ be an $(n + 1)$-dimensional rectangle defined by

$$R: \quad |t - t_0| \leq a, \quad |x_1 - x_1{}^0| \leq b_1,$$
$$|x_2 - x_2{}^0| \leq b_2, \ldots, |x_n - x_n{}^0| \leq b_n, \qquad (98)$$

where $a, b_1, \ldots, b_n$ are positive numbers.   Note that R exists, since D is an *open* connected set.   Since R is a compact set, there exists a constant $M > 0$ such that

$$|F_i(t, x_1, x_2, \ldots, x_n)| \leq M \qquad (99)$$

for all $i = 1, 2, \ldots, n$ and for all $(t, x_1, x_2, \ldots, x_n) \in R$.   Let us define h by

$$h = \min\left(a, \frac{b_1}{M}, \frac{b_2}{M}, \ldots, \frac{b_n}{M}\right). \qquad (100)$$

Clearly, $h > 0$.   For the interval I under consideration, we may take

$$I: \quad |t - t_0| \leq h. \qquad (101)$$

Let us consider the Picard iterants

$$\phi_{i,0}(t) = x_i{}^0,$$

$$\phi_{i,j}(t) = x_i{}^0 + \int_{t_0}^{t} F_i[u, \phi_{1,j-1}(u), \ldots, \phi_{n,j-1}(u)]du, \qquad (102)$$

$$i = 1, 2, \ldots, n; \quad j = 1, 2, \ldots.$$

(ii) *Each function $\phi_{i,j}$ defined by (102) exists, is continuous, and satisfies the inequality*

$$|\phi_{i,j}(t) - x_i^0| \leq b_i, \qquad i = 1, 2, \ldots, n; \quad j = 1, 2, \ldots, \tag{103}$$

for all $t \in I$: $|t - t_0| \leq h$.

The proof of (ii) is similar to that of part (ii) in the proof of Theorem 6.

(iii) *For each $i = 1, 2, \ldots, n$, the sequence $\{\phi_{i,j}\}$ defined by (102) converges uniformly on I to a continuous limit function ϕ_i on I.*

One proves this result by first establishing by induction the inequality

$$|\phi_{i,j}(t) - \phi_{i,j-1}(t)| \leq \frac{M(Kn)^{j-1}}{j!} |t - t_0|^j \leq \frac{M}{Kn} \frac{(Knh)^j}{j!}, \tag{104}$$

$i = 1, 2, \ldots, n; j = 1, 2, \ldots,$ and all $t \in I$. Then, observing that $\phi_{i,j}(t)$ can be written as

$$\phi_{i,j}(t) = \phi_{i,0}(t) + \sum_{r=1}^{j} [\phi_{i,r}(t) - \phi_{i,r-1}(t)], \tag{105}$$

one completes the proof by following an argument similar to that given in part (iii) of the proof of Theorem 6.

(iv) *For each $i = 1, 2, \ldots, n$, the limit function ϕ_i is a solution of the integral equation*

$$\phi_i(t) = x_i^0 + \int_{t_0}^{t} F_i[u, \phi_1(u), \ldots, \phi_n(u)]du \tag{106}$$

on the interval I and satisfies the initial conditions (97).

The proof of this result is similar to the one used to establish part (iv) of the proof of Theorem 6.

(v) *For each $i = 1, 2, \ldots, n$, the function ϕ_i is the unique solution of the system (93) for all $t \in I$ satisfying the initial conditions (97).*

To establish this result, follow the method utilized in part (v) of the proof of Theorem 6.

PROOF OF THEOREM 8.2.1.   In Theorem 8.2.1 each of the functions $\partial F_i/\partial x_j$, $i, j = 1, 2, \ldots, n$ are continuous in the domain D. Thus, each F_i, $i = 1, 2, \ldots, n$, satisfies in D a (common) local Lipschitz condition similar to (94). Theorem 8.2.1 now follows from Theorem 7 by an argument similar to the one used to obtain Theorem 1.4.1 (of Chapter 1) from Theorem 6.

[*Note:* Analogs of Remarks 9 and 10 are valid here.]

Example 11.   Consider the first-order system

$$\begin{cases} \dfrac{dx_1}{dt} = x_1 + x_2, \\[2mm] \dfrac{dx_2}{dt} = tx_1, \end{cases} \tag{107}$$

with initial conditions

$$x_1(0) = x_1{}^0 = 1, \qquad x_2(0) = x_2{}^0 = 1. \tag{108}$$

For this problem, let us find for $i = 1, 2$, the first four Picard iterants $\phi_{i,j}(t)$. Also, let us determine an interval I: $|t - 0| = |t| \leq h$, on which there exists a unique solution.

SOLUTION.   Comparing (107) with (93) we see that

$$F_1(t, x_1, x_2) = x_1 + x_2, \qquad F_2(t, x_1, x_2) = tx_1. \tag{109}$$

Also, from (102) and (108), we have

$$\phi_{1,0}(t) = x_1{}^0 = 1, \qquad \phi_{2,0}(t) = x_2{}^0 = 1, \tag{110}$$

and

$$\phi_{i,j}(t) = x_i{}^0 + \int_0^t F_i[u, \phi_{1,j-1}(u), \phi_{2,j-1}(u)]du, \qquad i = 1, 2; \; j = 1, 2, \ldots. \tag{111}$$

Thus, with $j = 1$, (111) reads

$$\phi_{1,1}(t) = x_1{}^0 + \int_0^t F_1[u, \phi_{1,0}(u), \phi_{2,0}(u)]du,$$

$$\phi_{2,1}(t) = x_2{}^0 + \int_0^t F_2[u, \phi_{1,0}(u), \phi_{2,0}(u)]du. \tag{112}$$

In view of (109) and (110), we have

$$\begin{aligned}
F_1[u, \phi_{1,0}(u), \phi_{2,0}(u)] &= \phi_{1,0}(u) + \phi_{2,0}(u) = 1 + 1 = 2, \\
F_2[u, \phi_{1,0}(u), \phi_{2,0}(u)] &= u\phi_{1,0}(u) = u \cdot 1 = u.
\end{aligned} \tag{113}$$

Substituting (113) into (112), we obtain

$$\phi_{1,1}(t) = 1 + \int_0^t 2 \, du = 1 + 2t,$$

$$\phi_{2,1}(t) = 1 + \int_0^t u \, du = 1 + \frac{t^2}{2}. \tag{114}$$

With $j = 2$ in (111), we obtain using (109) and (114)

$$\phi_{1,2}(t) = 1 + \int_0^t F_1[u, \phi_{1,1}(u), \phi_{2,1}(u)]du = 1 + \int_0^t \left(2 + 2u + \frac{u^2}{2}\right)du$$

$$= 1 + 2t + t^2 + \frac{t^3}{6},$$

$$\phi_{2,2}(t) = 1 + \int_0^t F_2[u, \phi_{1,1}(u), \phi_{2,1}(u)]du = 1 + \int_0^t (u + 2u^2)du$$

$$= 1 + \frac{t^2}{2} + \frac{2t^3}{3}. \tag{115}$$

With $j = 3$ in (111), we obtain using (109) and (115)

$$\phi_{1,3}(t) = 1 + \int_0^t F_1[u, \phi_{1,2}(u), \phi_{2,2}(u)]du = 1 + \int_0^t \left(2 + 2u + \frac{3u^2}{2} + \frac{5u^3}{6}\right)du$$

$$= 1 + 2t + t^2 + \frac{t^3}{2} + \frac{5t^4}{24},$$

$$\phi_{2,3}(t) = 1 + \int_0^t F_2[u, \phi_{1,2}(u), \phi_{2,2}(u)]du = 1 + \int_0^t \left(u + 2u^2 + u^3 + \frac{u^4}{6}\right)du$$

$$= 1 + \frac{t^2}{2} + \frac{2t^3}{3} + \frac{t^4}{4} + \frac{t^5}{30}.$$

From (109), we see that the functions F_1 and F_2 are continuous for all (finite) values of t, x_1, x_2. For definiteness, let us take for the 3-dimensional rectangle [*see* (98)]

$$R: \quad |t - 0| = |t| \leq 2, \quad |x_1 - 1| \leq 3, \quad |x_2 - 1| \leq 5.$$

Thus, for all points $(t, x_1, x_2) \in R$, we have $|t| \leq 2$, $|x_1| \leq 4$, $|x_2| \leq 6$, and

$$|F_1| = |x_1 + x_2| \leq |x_1| + |x_2| \leq 4 + 6 = 10,$$
$$|F_2| = |tx_1| = |t|\,|x_1| \leq 8.$$

With $M = 10$, $a = 2$, $b_1 = 3$, $b_2 = 5$, we obtain from (100)

$$h = \min\left(2, \tfrac{3}{10}, \tfrac{5}{10}\right) = 0.3.$$

Thus, for all $t \in I$: $|t| \leq 0.3$, we have $\lim_{j \to \infty} \phi_{i,j}(t) = \phi_i(t)$ $(i = 1, 2)$. Consequently, the set of functions $x_1 = \phi_1(t)$, $x_2 = \phi_2(t)$ is the unique solution of the system (107) for all $t \in I$ satisfying the initial conditions (108).

Consider the differential equation of order n

$$y^{(n)} = F[t, y, y', \ldots, y^{(n-1)}], \tag{116}$$

where the function F is defined and continuous in some domain D of $(n + 1)$-dimensional $(t, y, y', \ldots, y^{(n-1)})$ space.

Definition 8. A function ϕ defined on an interval I: $\alpha \leq t \leq \beta$ is said to be a *solution* of Equation (116), if for each $t \in I$:

$$\text{(i)} \quad \phi(t), \phi'(t), \ldots, \phi^{(n-1)}(t) \text{ exist,}$$
$$\text{(ii)} \quad \text{the point } [t, \phi(t), \phi'(t), \ldots, \phi^{n-1}(t)] \in D, \tag{117}$$
$$\text{(iii)} \quad \phi^{(n)}(t) = F[t, \phi(t), \phi'(t), \ldots, \phi^{n-1}(t)].$$

In Section 8.2, we saw that under the change of variables

$$x_1 = y, \quad x_2 = y', \quad x_3 = y'', \quad \ldots, \quad x_{n-1} = y^{(n-2)}, \quad x_n = y^{(n-1)}, \tag{118}$$

Equation (116) was equivalent to the system

$$\begin{cases} x_1' = x_2, \\ x_2' = x_3, \\ \quad \vdots \\ x_{n-1}' = x_n, \\ x_n' = F(t, x_1, x_2, \ldots, x_n). \end{cases} \tag{119}$$

The system (119) is a special case of the system (93) with $F_1 = x_2$, $F_2 = x_3$, $\ldots, F_{n-1} = x_n$, and $F_n = F(t, x_1, x_2, \ldots, x_n)$.

In the hint to Exercise 8.2.1, we established the equivalence of (116) and (119). We observed that if $\phi_1, \phi_2, \ldots, \phi_n$ is a solution of the system (119) on I, then setting $\phi(t) = \phi_1(t)$, we obtained

$$\phi^{(n)}(t) = F[t, \phi(t), \phi'(t), \ldots, \phi^{(n-1)}(t)] \tag{120}$$

for all $t \in I$. Thus, $\phi(t)$ is a solution of (116) on I.

Let a set of initial conditions associated with Equation (116) on I be given by

$$y(t_0) = a_0, \quad y'(t_0) = a_1, \quad \ldots \quad , y^{(n-1)}(t_0) = a_{n-1} \tag{121}$$

where $t_0 \in I$. Then in view of (118), a set of initial conditions associated with the equivalent system (119) is given by

$$x_1(t_0) = a_0, \quad x_2(t_0) = a_1, \quad \ldots \quad , x_n(t_0) = a_{n-1}. \tag{122}$$

In view of Theorem 7 and the above observations, we have essentially established the following existence and uniqueness theorem for nonlinear differential equations of order n. In particular, Theorem 1.4.2 of Chapter 1 may be obtained as a corollary of Theorem 8.

Theorem 8.   *Consider the differential equation*

$$y^{(n)} = F[t, y, y', \ldots, y^{(n-1)}], \tag{116}$$

where the function F is defined and continuous in a domain D of $(n+1)$-dimensional $(t, y, y', \ldots, y^{(n-1)})$ space. In D, let F satisfy a Lipschitz condition

$$|F[t, y, y', \ldots, y^{(n-1)}] - F[t, \bar{y}, \bar{y}', \ldots, \bar{y}^{(n-1)}]|$$
$$\leq K[|y - \bar{y}| + |y' - \bar{y}'| + \cdots + |y^{(n-1)} - \bar{y}^{(n-1)}|], \tag{123}$$

where $(t, y, y', \ldots, y^{(n-1)})$ and $(t, \bar{y}, \bar{y}', \ldots, \bar{y}^{(n-1)})$ are any two points in D and $K > 0$. Let $(t_0, a_0, a_1, \ldots, a_{n-1})$ be a point in D. Then an interval

$$I: \quad |t - t_0| \leq h, \quad h > 0, \tag{124}$$

exists on which there is a unique solution $\phi(t)$ satisfying Equation (116) and the initial conditions (121).

PROOF.   Let $R \subset D$ be an $(n + 1)$-dimensional rectangle about the point $(t_0, a_0, a_1, \ldots, a_{n-1})$.   The reader may easily verify that in R the equivalent system (119) satisfies the hypotheses of Theorem 7.   Consequently, an interval I: $|t - t_0| \leq h$ exists on which there is a unique solution $\phi_1, \phi_2, \ldots, \phi_n$ of the system (119) which satisfies the initial conditions (122).   By setting $\phi(t) = \phi_1(t)$ for all $t \in I$, $\phi(t)$ becomes the unique solution of Equation (116) on I satisfying the initial conditions (121).   Thus, the theorem is established.

Theorem 1.4.2 of Chapter 1 follows from Theorem 8 in essentially the same way that Theorem 8.2.1 follows from Theorem 7.

[*Note:*   Analogs of Remarks 9 and 10 are valid here.]

We shall now establish an existence and uniqueness theorem concerning a system of n *linear* first-order equations, namely,

$$\begin{cases} x_1' = a_{11}(t)x_1 + a_{12}(t)x_2 + \cdots + a_{1n}(t)x_n + g_1(t), \\ x_2' = a_{21}(t)x_1 + a_{22}(t)x_2 + \cdots + a_{2n}(t)x_n + g_2(t), \\ \qquad\qquad\qquad\qquad \vdots \\ x_n' = a_{n1}(t)x_1 + a_{n2}(t)x_2 + \cdots + a_{nn}(t)x_n + g_n(t), \end{cases} \tag{125}$$

where the functions a_{ij} and g_i $(i, j = 1, 2, \ldots, n)$ are defined and continuous on some closed interval I: $\alpha \leq t \leq \beta$.   The system (125) can be written compactly as

$$x_i' = F_i(t, x_1, x_2, \ldots, x_n), \tag{126}$$

where

$$F_i(t, x_1, x_2, \ldots, x_n) = \sum_{j=1}^{n} a_{ij}(t)x_j + g_i(t), \qquad i = 1, 2, \ldots, n. \tag{127}$$

Thus the linear system (125) is a special case of the system (93).

Remark 12.   We shall now see that for a linear system (125) in which the functions a_{ij} and g_i $(i, j = 1, 2, \ldots, n)$ are continuous on the interval I: $\alpha \leq t \leq \beta$, each function F_i given in (127) satisfies a Lipschitz condition on a set S of the form

$$S: \quad \alpha \leq t \leq \beta, \quad -\infty < x_i < \infty, \qquad i = 1, 2, \ldots, n. \tag{128}$$

Since the functions a_{ij} are continuous on (the compact set) I, they are bounded. Thus, for all $t \in I$, there exist constants $K_{ij} > 0$ such that

$$|a_{ij}(t)| \leq K_{ij}, \qquad i, j = 1, 2, \ldots, n. \tag{129}$$

Let

$$K = \max_{i,j} K_{ij}, \tag{130}$$

for all $t \in I$, and let $\bar{x}_1, \bar{x}_2, \ldots, \bar{x}_n$ and $x_1^*, x_2^*, \ldots, x_n^*$ be any two sets of n real numbers. Then, from (127) through (130), we obtain

$$|F_i(t, \bar{x}_1, \bar{x}_2, \ldots, \bar{x}_n) - F_i(t, x_1^*, x_2^*, \ldots, x_n^*)|$$

$$= \left| \sum_{j=1}^{n} a_{ij}(t)\bar{x}_j + g_i(t) - \sum_{j=1}^{n} a_{ij}(t)x_j^* - g_i(t) \right|$$

$$= \left| \sum_{j=1}^{n} a_{ij}(t)(\bar{x}_j - x_j^*) \right| \leqq \sum_{j=1}^{n} |a_{ij}(t)| \, |\bar{x}_j - x_j^*| \leqq \sum_{j=1}^{n} K |\bar{x}_j - x_j^*|$$

$$= K \sum_{j=1}^{n} |\bar{x}_j - x_j^*|, \tag{131}$$

which is a Lipschitz condition for each function F_i $(i = 1, 2, \ldots, n)$ on the set S.

In view of this remark, a Lipschitz condition need not be postulated in the hypothesis of the following theorem, which is Theorem 8.2.2 of Chapter 8. Also, since the proof of this theorem is essentially the same as that of Theorem 7, we shall merely sketch it and leave the details as an exercise for the reader.

Theorem 9. *Consider a system of n linear first-order equations*

$$x_i' = \sum_{j=1}^{n} a_{ij}(t)x_j + g_i(t), \qquad i = 1, 2, \ldots, n, \tag{132}$$

where the functions a_{ij} and g_i are defined and continuous on an interval I: $\alpha \leqq t \leqq \beta$, containing the point $t = t_0$. Let $x_1^0, x_2^0, \ldots, x_n^0$ be any set of n real numbers. Then there exists a unique solution

$$x_1 = \phi_1(t), \quad x_2 = \phi_2(t), \ldots, x_n = \phi_n(t) \tag{133}$$

satisfying the system (132) for each t in I and the initial conditions

$$\phi_1(t_0) = x_1^0, \quad \phi_2(t_0) = x_2^0, \ldots, \phi_n(t_0) = x_n^0. \tag{134}$$

PROOF. On the interval I: $\alpha \leqq t \leqq \beta$, consider the Picard iterants

$$\phi_{i,0}(t) = x_i^0,$$

$$\phi_{i,j}(t) = x_i^0 + \int_{t_0}^{t} F_i[u, \phi_{1,j-1}(u), \phi_{2,j-1}(u), \ldots, \phi_{n,j-1}(u)]du$$

$$= x_i^0 + \int_{t_0}^{t} [a_{i1}(u)\phi_{1,j-1}(u) + a_{i2}(u)\phi_{2,j-1}(u) + \cdots$$

$$+ a_{in}(u)\phi_{n,j-1}(u) + g_i(u)]du, \tag{135}$$

$i = 1, 2, \ldots, n; j = 1, 2, \ldots$. With $j = 1$ in (135), we have

$$\phi_{i,1}(t) - x_1{}^0 = \int_{t_0}^{t} [a_{i1}(u)\phi_{1,0}(u)$$

$$+ a_{i2}(u)\phi_{2,0}(u) + \cdots + a_{in}(u)\phi_{n,0}(u) + g_i(u)du$$

$$= \int_{t_0}^{t} [a_{i1}(u)x_1{}^0 + a_{i2}(u)x_2{}^0 + \cdots + a_{in}(u)x_n{}^0 + g_i(u)]du.$$

$$(136)$$

Since the functions a_{ij} and g_i are continuous on I, there exists a constant $M > 0$ such that

$$|a_{i1}(t)x_1{}^0 + a_{i2}(t)x_2{}^0 + \cdots + a_{in}(t)x_n{}^0 + g_i(t)| \leqq M, \qquad (137)$$

$i = 1, 2, \ldots, n$ for all $t \in I$. Taking absolute values in (136) and then substituting (137), we obtain

$$|\phi_{i,1}(t) - x_i{}^0| \leqq \int_{t_0}^{t} M \, du = M|t - t_0|,$$

$i = 1, 2, \ldots, n$ for all $t \in I$. One may now show by mathematical induction that each function $\phi_{i,j}$ defined by (135) exists, is continuous, and satisfies the inequality

$$|\phi_{i,j}(t) - x_i{}^0| \leqq M|t - t_0|, \qquad (138)$$

$i = 1, 2, \ldots, n; \quad j = 1, 2, \ldots$ for all $t \in I$.

In view of Remark 12, each function F_i satisfies a Lipschitz condition on $S: \alpha \leqq t \leqq \beta, \quad -\infty < x_i < \infty, \quad i = 1, 2, \ldots, n$. Consequently, we may establish by mathematical induction the inequality

$$|\phi_{i,j}(t) - \phi_{i,j-1}(t)| \leqq \frac{M(Kn)^{j-1}}{j!} |t - t_0|^j \leqq \frac{M}{Kn} \frac{(KnL)^j}{j!}, \qquad (139)$$

$i = 1, 2, \ldots, n; \quad j = 1, 2, \ldots$ for all $t \in I$, and where $L = \max[|\alpha - t_0|, |\beta - t_0|]$.

Similar results, such as (iii), (iv), and (v) of Theorem 7, may now be established, thus completing the proof of the theorem.

We shall now establish an existence and uniqueness theorem concerning linear differential equations of order n, namely, Theorem 6.2.1 of Chapter 6. Clearly, Theorem 4.2.1 of Chapter 4 is a special case of Theorem 6.2.1. In the following theorem, $x^{(n)}$ denotes $\dfrac{d^n x}{dt^n}$.

Theorem 10. *Consider the linear differential equation of order n*

$$a_0(t)x^{(n)} + a_1(t)x^{(n-1)} + \cdots + a_{n-1}(t)x' + a_n(t)x = F(t), \qquad (140)$$

where the functions $a_0, \ldots, a_n$, and F are defined and continuous on a common interval I: $\alpha \leq t \leq \beta$, and $a_0(t) \neq 0$ for every t in I.   Let t_0 be a point in I, and let $x_0, x_0', \ldots, x_0^{(n-1)}$ be n arbitrary real constants.   Then there exists a unique solution $\phi(t)$ satisfying Equation (140) for every t in I and the initial conditions

$$\phi(t_0) = x_0, \quad \phi'(t_0) = x_0', \quad \ldots, \quad \phi^{(n-1)}(t_0) = x_0^{(n-1)}. \tag{141}$$

PROOF.   Consider the initial-value problem on I:

$$a_0(t)x^{(n)} + a_1(t)x^{(n-1)} + \cdots + a_{n-1}(t)x' + a_n(t)x = F(t), \tag{140}$$

$$x(t_0) = x_0, \quad x'(t_0) = x_0', \quad \ldots, \quad x^{(n-1)}(t_0) = x_0^{(n-1)}. \tag{142}$$

By an extension of the result of Exercise 8.2.3, under the change of variables,

$$x_1 = x, \quad x_2 = x', \quad \ldots, \quad x_{n-1} = x^{(n-2)}, \quad x_n = x^{(n-1)}, \tag{143}$$

the above initial-value problem is equivalent on I to the initial-value problem:

$$\begin{cases} x_1' = x_2, \\ x_2' = x_3, \\ \quad \vdots \\ x_{n-1}' = x_n, \\ x_n' = -\dfrac{a_n(t)}{a_0(t)} x_1 - \dfrac{a_{n-1}(t)}{a_0(t)} x_2 - \cdots - \dfrac{a_1(t)}{a_0(t)} x_n + \dfrac{F(t)}{a_0(t)}, \end{cases} \tag{144}$$

$$x_1(t_0) = x_0, \quad x_2(t_0) = x_0', \ldots, x_n(t_0) = x_0^{(n-1)}. \tag{145}$$

Clearly the system (144) is a special case of the system (125).   Consequently, Theorem 9 is applicable to the initial-value problem (144), (145).   Thus, there exists a unique solution

$$x_1 = \phi_1(t), \quad x_2 = \phi_2(t), \quad \ldots, \quad x_n = \phi_n(t) \tag{146}$$

satisfying the system (144) for each $t \in I$ and the initial conditions

$$\phi_1(t_0) = x_0, \quad \phi_2(t_0) = x_0', \quad \ldots, \quad \phi_n(t_0) = x_0^{(n-1)}. \tag{147}$$

In view of the equivalence on I of the initial-value problems (140), (142) and (144), (145), by setting $\phi(t) = \phi_1(t)$ for every $t \in I$, we have that $\phi(t)$ is the unique solution of Equation (140) for every $t \in I$ satisfying the initial conditions (141).   The proof of the theorem is now completed.

Suggested Readings

Brauer and Nohel [10]   Lefschetz [27]
Coddington [12]   Murray and Miller [29]
Cronin-Scanlon [13]   Ross [40]
Hurewicz [22]   Tricomi [47]
Ince [23]

BIBLIOGRAPHY

1. Abramowitz, M., and Stegun, I. A., Editors, *Handbook of Mathematical Functions with Formulas, Graphs, and Mathematical Tables*, National Bureau of Standards, Applied Mathematics Series· 55, 1964.
2. Agnew, R. P., *Differential Equations*, 2nd ed., McGraw-Hill Book Company, New York, 1960.
3. Apostol, T. M., *Mathematical Analysis*, Addison-Wesley Publishing Company, Inc., Reading, Mass., 1957.
4. Ayres, F. Jr., *Theory and Problems of Matrices*, Schaum Publishing Company, New York, 1962.
5. Berg, P. W., and McGregor, J. L., *Elementary Partial Differential Equations*, Holden-Day, Inc., San Francisco, 1966.
6. Betz, H., Burcham, P. B., and Ewing, G. M., *Differential Equations with Applications*, Harper and Brothers Publishers, New York, 1954.
7. Bliss, G. A., *Calculus of Variations*, The Mathematical Association of America, 1944.
8. Bowman, F., *Introduction to Bessel Functions*, Dover Publications, Inc., New York, 1958. ·
9. Boyce, W. E., and DiPrima, R. C., *Elementary Differential Equations and Boundary Value Problems*, 2nd ed., John Wiley & Sons, Inc., New York, 1969.
10. Brauer, F., and Nohel, J. A., *Ordinary Differential Equations*, W. A. Benjamin, Inc., New York, 1967.
11. Brenner, J. L., *Problems in Differential Equations*, W. H. Freeman and Company, San Francisco, 1963.
12. Coddington, E. A., *An Introduction to Ordinary Differential Equations*, Prentice-Hall, Inc., Englewood Cliffs, New Jersey, 1961.
13. Cronin-Scanlon, J., *Advanced Calculus*, D. C. Heath and Company, Lexington, Mass., 1969.
14. DeRusso, P. M., Roy, R. J., and Close, C. M., *State Variables for Engineers*, John Wiley & Sons, Inc., New York, 1967.
15. Friedman, B., *Principles and Techniques of Applied Mathematics*, John Wiley & Sons, Inc., New York, 1956.
16. Gladstone, S., and Lewis, D., *Elements of Physical Chemistry*, 2nd ed., D. Van Nostrand Company, Inc., New York, 1960.

17. Goursat, E., *A Course in Mathematical Analysis, Vol. II, Part Two, Differential Equations*, Dover Publications, Inc., New York, 1959.

18. Greenspan, D., *Introduction to Partial Differential Equations*, McGraw-Hill Book Company, New York, 1961.

19. Hayt, W. H. Jr., and Kemmerly, J. E., *Engineering Circuit Analysis*, McGraw-Hill Book Company, New York, 1962.

20. Hitchcock, F. L., and Robinson, C. S., *Differential Equations in Applied Chemistry*, 2nd ed., John Wiley & Sons, Inc., New York, 1936.

21. Hohn, F. E., *Elementary Matrix Algebra*, 2nd ed., The Macmillan Company, New York, 1964.

22. Hurewicz, W., *Lectures on Ordinary Differential Equations*, The M.I.T. Press, Cambridge, Mass., 1958.

23. Ince, E. L., *Ordinary Differential Equations*, Dover Publications, Inc., New York, 1956.

24. Kaplan, W., *Advanced Calculus*, Addison-Wesley Publishing Company, Inc., Cambridge, Mass., 1952.

25. Kiselev, A. I., Krasnov, M. L., and Makarenko, G. I., *Ordinary Differential Equations*, Frederick Ungar Publishing Company, New York, 1967.

26. Kittsley, S. L., *Physical Chemistry*, 2nd ed., Barnes and Noble, Inc., New York, 1963.

27. Lefschetz, S., *Differential Equations, Geometric Theory*, Interscience Publishers, Inc., New York, 1957.

28. Leighton, W., *Ordinary Differential Equations*, Wadsworth Publishing Company, Inc., Belmont, California, 1963.

29. Murray, F. J., and Miller, K. S., *Existence Theorems for Ordinary Differential Equations*, New York University Press, Washington Square, New York, 1954.

30. Nielsen, K. L., *Differential Equations*, 2nd ed., Barnes and Noble, Inc., New York, 1966.

31. Paige, L. J., and Swift, J. D., *Elements of Linear Algebra*, Ginn and Company, Boston, Mass., 1961.

32. Peirce, B. O., *A Short Table of Integrals*, 4th ed., Ginn and Company, Boston, Mass., 1957.

33. Pennisi, L. L., *Elements of Complex Variables* (with the collaboration of L. I. Gordon and S. Lasher), Holt, Rinehart and Winston, Inc., New York, 1963.

34. Pipes, L. A., *Applied Mathematics for Engineers and Physicists*, 2nd ed., McGraw-Hill Book Company, New York, 1958.

35. Rainville, E. D., *Intermediate Differential Equations*, 2nd ed., The Macmillan Company, New York, 1964.

36. Reddick, H. W., and Kibbey, D. E., *Differential Equations*, 3rd ed., John Wiley & Sons, Inc., New York, 1956.

37. Reddick, H. W., and Miller, F. H., *Advanced Mathematics for Engineers*, 3rd ed., John Wiley & Sons, Inc., New York, 1955.

38. Richards, J. A., Sears, F. W., Wehr, M. R., and Zemansky, M. W., *Modern University Physics*, Addison-Wesley Publishing Company, Inc., Reading, Mass., 1960.

39. Ritger, P. D., and Rose, N. J., *Differential Equations with Applications*, McGraw-Hill Book Company, New York, 1968.

40. Ross, S. L., *Differential Equations*, Blaisdell Publishing Company, New York, 1964.

41. Schaum, D., and Van Der Merwe, C. W., Editor, *Theory and Problems of College Physics*, 6th ed., Schaum Publishing Company, New York, 1961.

42. Skilling, H. H., *Electrical Engineering Circuits*, 2nd ed., John Wiley & Sons, Inc., New York, 1965.

43. Spiegel, M. R., *Applied Differential Equations*, 2nd ed., Prentice-Hall, Inc., Englewood Cliffs, New Jersey, 1969.

44. Steen, F. H., *Differential Equations*, Ginn and Company, Boston, Mass., 1955.

45. Taylor, A. E., *Advanced Calculus*, Ginn and Company, Boston, Mass., 1955.

46. Tenenbaum, M., and Pollard, H., *Ordinary Differential Equations*, Harper and Row, Publishers, New York, 1963.

47. Tricomi, F. G., *Differential Equations*, Hafner Publishing Company, Inc., New York, 1961.

48. Tricomi, F. G., *Integral Equations*, Interscience Publishers, Inc., New York, 1957.

49. Wayland, H., *Differential Equations Applied in Science and Engineering*, D. Van Nostrand Company, Inc., Princeton, New Jersey, 1957.

50. Webster, A. G., *The Dynamics of Particles and of Rigid, Elastic, and Fluid Bodies*, 3rd ed., Hafner Publishing Company, Inc., New York, 1949.

51. Weinstock, R., *Calculus of Variations with Applications to Physics and Engineering*, McGraw-Hill Book Company, New York, 1952.

52. Weisner, L., *Introduction to the Theory of Equations*, The Macmillan Company, New York, 1938.

53. Weiss, M. J., *Higher Algebra for the Undergraduate*, John Wiley & Sons, Inc., New York, 1949.

54. Whittaker, E. T., *A Treatise on the Analytical Dynamics of Particles and Rigid Bodies*, 4th ed., Cambridge University Press, New York, 1959.

55. Yosida, K., *Lectures on Differential and Integral Equations*, John Wiley & Sons, Inc., New York, 1960.

56. Zadeh, L. A., and Desoer, C. A., *Linear System Theory: The State Space Approach*, McGraw-Hill Book Company, New York, 1963.

ANSWERS AND HINTS

Exercises 1.1, page 6

1. (a) ordinary;   (b) nonlinear;   (c) order one and degree one.
2. (a) ordinary;   (b) nonlinear;   (c) order one and degree one.
3. (a) ordinary;   (b) nonlinear;   (c) order four and degree two.
4. (a) ordinary;   (b) nonlinear;   (c) order three and degree one.
5. (a) partial;   (b) nonlinear;   (c) order one and degree one.
6. (a) ordinary;   (b) linear;   (c) order two and degree one.
7. (a) ordinary;   (b) linear;   (c) order two and degree one.
8. (a) ordinary;   (b) nonlinear;   (c) order three and degree one.
9. (a) partial;   (b) linear;   (c) order two and degree one.
10. (a) partial;   (b) nonlinear;   (c) order two and degree one.
11. (a) ordinary;   (b) nonlinear;   (c) order two and degree nine.
12. (a) ordinary;   (b) nonlinear;   (c) order two and degree one.
13. (a) ordinary;   (b) nonlinear;   (c) order one and degree one.
14. (a) ordinary;   (b) linear;   (c) order four and degree one.
15. (a) partial;   (b) nonlinear;   (c) order two and degree one.
16. (a) ordinary;   (b) linear;   (c) order n and degree one.
17. (a) partial;   (b) nonlinear;   (c) order two and degree two.
18. (a) partial;   (b) nonlinear;   (c) order two and degree two.

Exercises 1.2, page 16

1. $y'' - y = 0.$
2. $y''' + 2y'' - y' - 2y = 0.$
3. $y'' + 4y' + 4y = 0.$
4. $y''' - 6y'' + 12y' - 8y = 0.$
5. $y^{iv} - 2y''' - y'' + 2y' = 0.$
6. $y'' + \beta^2 y = 0.$
7. $y'' - 2\alpha y' + (\alpha^2 + \beta^2)y = 0.$
8. $y^{iv} - 4\alpha y''' + 2(3\alpha^2 + \beta^2)y'' - 4\alpha(\alpha^2 + \beta^2)y' + (\alpha^2 + \beta^2)^2 y = 0.$
9. $\theta\, dr + r\, d\theta = 0.$
10. $2\theta\, dr + r\, d\theta = 0.$
11. $\theta\, dr - r \ln r\, d\theta = 0.$
12. $\sin 2\theta\, dr - 2r \cos 2\theta\, d\theta = 0.$

13. $\cos 2\theta \, dr + r \sin 2\theta \, d\theta = 0.$

14. $\theta \left(\dfrac{dr}{d\theta} \right)^2 + 2\theta \, \dfrac{dr}{d\theta} - r = 0.$

15. $(x^2 - y^2)dx + 2xy \, dy = 0.$

16. $y' + y \, e^x \tanh e^x = 0.$

17. $y(y^2 + 3x^2)dx - 2x^3 \, dy = 0.$

18. $x(xy' + 2y)^3 + 4y^4 y' = 0.$

19. $\cos \theta \, dr - (r \sin \theta + \tan \theta)d\theta = 0.$

20. $(x^2 - 2xy)(y')^2 - 2xyy' + y^2 - 2xy = 0.$

21. $x^2(y'')^2 - 2xy'y''[1 + (y')^2] - [1 + (y')^2]^2 = 0.$

22. $\sin \theta \, \dfrac{d^2 r}{d\theta^2} - \cos \theta \, \dfrac{dr}{d\theta} = 0.$

23. $(y^4 + 4x^2 y^2 - x^4)dx - 4x^3 y \, dy = 0.$

24. $y(2 - 2x - y)dx + 2(x + y)dy = 0.$

25. $(xy^3 - y \, e^{2x})dx + e^{2x} \, dy = 0.$

26. $(yy' + x)(xy' - y) = (\alpha^2 - \beta^2)y'.$

27. $xy' - y = 0.$

28. $y' - xy(1 + x^2 y^2) = 0.$

29. $x(2y^3 - x^3)y' + y(2x^3 - y^3) = 0.$

30. $x \, e^x \, dx - 2y(x + 1)^2 \, dy = 0.$

31. $xy' - y\left[1 + \ln \left(\dfrac{y}{x} \right) \right] = 0.$

32. $y' - (1 + y^2) = 0.$

33. $y' + (1 - y)y \sin x = 0.$

34. $yy''' + 3y'y'' = 0.$

35. $(x - y)(y')^2(y''')^2 - [3(x - y)y'(y'')^2 + 3(y')^3 y'' - 3(y')^2 y'']y''' + 9(y')^2(y'')^3 = 0.$

36. $\cos \theta \, dr - r(k + \sin \theta)d\theta = 0 \quad$ or $\quad \dfrac{dr}{d\theta} = r(k \sec \theta + \tan \theta).$

37. $xy' = y - k\sqrt{x^2 + y^2} \quad$ or $\quad x^2(y')^2 - 2xyy' + y^2 - k^2(x^2 + y^2) = 0.$

Exercises 1.3, page 21

9. (e) (i) $f(x) = c_1 x^3 + c_2 x^{-1}.$ (ii) $f(x) = c_1 x^{1/2} + c_2 x^{-1}.$
   (iii) $f(x) = c_1 x^4 + c_2 x^{-5/3}.$

Exercises 1.4, page 33

7. $y = \sin x.$

8. $y = \dfrac{2 e^{2x}}{e^{2x} + 1}.$

9. $y = \dfrac{x^3}{3} + x - 1.$

10. $y = \dfrac{1}{2} e^{2x} + x + 1.$

11. $y = \dfrac{x}{2} \sqrt{x^2 + 1} + \dfrac{1}{2} \ln (x + \sqrt{x^2 + 1}).$

13. (d) $u = u(x, y)$, $x = r \cos \theta$, $y = r \sin \theta$.

(1) $\dfrac{\partial u}{\partial r} = \dfrac{\partial u}{\partial x}\dfrac{\partial x}{\partial r} + \dfrac{\partial u}{\partial y}\dfrac{\partial y}{\partial r} = \cos\theta\,\dfrac{\partial u}{\partial x} + \sin\theta\,\dfrac{\partial u}{\partial y}.$

(2) $\dfrac{\partial u}{\partial \theta} = \dfrac{\partial u}{\partial x}\dfrac{\partial x}{\partial \theta} + \dfrac{\partial u}{\partial y}\dfrac{\partial y}{\partial \theta} = -r \sin\theta\,\dfrac{\partial u}{\partial x} + r \cos\theta\,\dfrac{\partial u}{\partial y}.$

(3) $\dfrac{\partial^2 u}{\partial r^2} = \cos^2\theta\,\dfrac{\partial^2 u}{\partial x^2} + 2 \sin\theta\cos\theta\,\dfrac{\partial^2 u}{\partial x\,\partial y} + \sin^2\theta\,\dfrac{\partial^2 u}{\partial y^2}.$

(4) $\dfrac{\partial^2 u}{\partial \theta^2} = -r \cos\theta\,\dfrac{\partial u}{\partial x} + r^2 \sin^2\theta\,\dfrac{\partial^2 u}{\partial x^2} - 2r^2 \sin\theta\cos\theta\,\dfrac{\partial^2 u}{\partial x\,\partial y}$

$$- r \sin\theta\,\dfrac{\partial u}{\partial y} + r^2 \cos^2\theta\,\dfrac{\partial^2 u}{\partial y^2}.$$

Multiplying (4) by $1/r^2$ and then adding it to (3), we obtain upon using the identity $\sin^2\theta + \cos^2\theta = 1$

(5)
$$\dfrac{\partial^2 u}{\partial r^2} + \dfrac{1}{r^2}\dfrac{\partial^2 u}{\partial \theta^2} = \dfrac{\partial^2 u}{\partial x^2} + \dfrac{\partial^2 u}{\partial y^2} - \dfrac{\cos\theta}{r}\dfrac{\partial u}{\partial x} - \dfrac{\sin\theta}{r}\dfrac{\partial u}{\partial y}.$$

Multiplying (1) by $1/r$ and then adding it to (5), we get

(6)
$$\dfrac{\partial^2 u}{\partial r^2} + \dfrac{1}{r}\dfrac{\partial u}{\partial r} + \dfrac{1}{r^2}\dfrac{\partial^2 u}{\partial \theta^2} = \dfrac{\partial^2 u}{\partial x^2} + \dfrac{\partial^2 u}{\partial y^2},$$

from which the result may now be deduced.

15. (c) $c_2 = 1$, $c_3 = -3$, $c_4 = 6$, $c_1 = c_5 = c_6 = \cdots = c_m = 0$.

17. (a) (1) $\quad\;\; u_{xy}(x, y) = 0,$
    (2) $\quad\;\; u(0, y) = y^2,$
    (3) $\quad\;\; u(x, 0) = x^2.$

Integrating (1) with respect to x and keeping y fixed, we obtain

(4) $\qquad u_y(x, y) = g(y).$

Integrating (4) with respect to y and holding x fixed, we get

(5) $\qquad u(x, y) = \displaystyle\int_0^y g(\xi)\,d\xi + f(x).$

From (3) and (5), we obtain

$$x^2 = u(x, 0) = \int_0^0 g(\xi)\,d\xi + f(x) = f(x).$$

Thus, we may write (5) as

(6) $\qquad u(x, y) = \displaystyle\int_0^y g(\xi)\,d\xi + x^2.$

From (2) and (6), we get

$$y^2 = u(0, y) = \int_0^y g(\xi)d\xi + 0^2 = \int_0^y g(\xi)d\xi.$$

Thus, the function u that satisfies the given boundary-value problem is given by

$$\text{(7)} \qquad\qquad u(x, y) = y^2 + x^2.$$

(b) $u(x, y) = x^3y + y + \sin x.$ (c) $u(x, y) = x \sin x + y \sin x + \cos y.$

(d) $u(x, y) = x^3y - \cos x + y + 1.$ (e) $u(x, y) = x^2y + xy^2$

$$- \cos 2y + \cos \frac{x}{2} + 1.$$

(f) $u(x, y) = y(x - x^2) + x^2.$

(g) $u(x, y) = x^2yz + x(yz - y^2 - z^2) + y^2 + z^2.$

(h) $u(x, y) = x^2y^2z^2 + y^3z^3 + x \sin z + y \sin x.$

18. (b) Yes.

19. (a) $y = y(\xi, \zeta), \quad \xi = x + at, \quad \zeta = x - at.$

$$\frac{\partial y}{\partial x} = \frac{\partial y}{\partial \xi}\frac{\partial \xi}{\partial x} + \frac{\partial y}{\partial \zeta}\frac{\partial \zeta}{\partial x} = \frac{\partial y}{\partial \xi} \cdot 1 + \frac{\partial y}{\partial \zeta} \cdot 1 = \frac{\partial y}{\partial \xi} + \frac{\partial y}{\partial \zeta}.$$

$$\frac{\partial y}{\partial t} = \frac{\partial y}{\partial \xi}\frac{\partial \xi}{\partial t} + \frac{\partial y}{\partial \zeta}\frac{\partial \zeta}{\partial t} = \frac{\partial y}{\partial \xi} \cdot a + \frac{\partial y}{\partial \zeta} \cdot -a = a\left(\frac{\partial y}{\partial \xi} - \frac{\partial y}{\partial \zeta}\right).$$

$$\text{(A)} \quad \frac{\partial^2 y}{\partial x^2} = \frac{\partial^2 y}{\partial \xi^2} + 2\frac{\partial^2 y}{\partial \xi \partial \zeta} + \frac{\partial^2 y}{\partial \zeta^2}.$$

$$\text{(B)} \quad \frac{\partial^2 y}{\partial t^2} = a^2\left(\frac{\partial^2 y}{\partial \xi^2} - 2\frac{\partial^2 y}{\partial \xi \partial \zeta} + \frac{\partial^2 y}{\partial \zeta^2}\right).$$

Substituting (A) and (B) into $y_{tt} = a^2 y_{xx}$, we obtain the result.

(c) (A) $\qquad\qquad y(x, t) = F(x + at) + G(x - at).$

Since $y(x, 0) = f(x)$, we have in view of (A)

(B) $\qquad\qquad F(x) + G(x) = f(x).$

From (A), we obtain upon differentiation with respect to t

(C) $\qquad\qquad y_t(x, t) = aF'(x + at) - aG'(x - at).$

Since $y_t(x, 0) = 0$, we have in view of (C)

(D) $\qquad\qquad F'(x) - G'(x) = 0.$

Integrating Equation (D), we get

(E) $\qquad\qquad F(x) - G(x) = k,$

where k is an arbitrary constant of integration. Solving simultaneously Equations (B) and (E), we find

(F) $\qquad\qquad F(x) = \tfrac{1}{2}f(x) + \tfrac{1}{2}k, \quad G(x) = \tfrac{1}{2}f(x) - \tfrac{1}{2}k.$

Thus, from Equation (F), we obtain

$$(G) \quad \begin{cases} F(x+at) = \tfrac{1}{2}f(x+at) + \tfrac{1}{2}k, \\ G(x-at) = \tfrac{1}{2}f(x-at) - \tfrac{1}{2}k. \end{cases}$$

In view of (G), Equation (A) becomes

$$y(x,t) = \tfrac{1}{2}[f(x+at) + f(x-at)].$$

(d)   (i) $y(x,t) = \tfrac{1}{2}[f(x+at) + f(x-at)] = \tfrac{1}{2}[(x+at)^2 + (x-at)^2]$
$= x^2 + a^2 t^2.$

(ii) $y(x,t) = \sin x \cos at.$

(iii) $y(x,t) = x \sin x \cos at + at \cos x \sin at.$

(iv) $y(x,t) = (x^2 + a^2 t^2) \cosh x \cosh at + 2\,axt \sinh x \sinh at.$

(v) $y(x,t) = \tfrac{1}{4}(3 \sin x \cos at - \sin 3x \cos 3at).$

(vi) $y(x,t) = \dfrac{x}{4}(\cos 3x \cos 3at + 3 \cos x \cos at)$

$$-\frac{at}{4}(\sin 3x \sin 3at + 3 \sin x \sin at).$$

Exercises 2.2, page 56

1. $x^2 - 4xy = C.$

2. Not exact.

3. Not exact.

4. $x^2 - 2y \sin x = C.$

5. $\tan x \tan y = C.$

6. $e^{-x} + (1+y)e^{-y} = C.$

7. $3yx - (x^2 + y^2)^{3/2} = C.$

8. $yx^2 + 2x + y^3 = Cy.$

9. $x^3 + x^2 + \sin(xy) - \cos y = C.$

10. $\sinh 2x \cosh 2y = C.$

11. Not exact.

12. Not exact.

13. $e^{x^2 y} + e^{xy^2} + x^2 + y^2 = C.$

14. $x e^{y^2} + \csc y \cot x = C.$

15. $x^2 + xy^2 + y^2 + \ln\left(\dfrac{x+y}{x}\right) = C.$

16. $x^2 - 3xy + 5x + 2y^2 - 2y - 3 = 0.$

17. $e^{xy} + 4xy^3 - y^2 + 3 = 0.$

18. $x^3 \ln x - xy + 5 = 0.$

20. 154.

21. 9.718.

22. 0.614.

23. 3.

24. Since $x = r \cos\theta$ and $y = r \sin\theta$, the result is $dx = \cos\theta\, dr - r \sin\theta\, d\theta$, $dy = \sin\theta\, dr + r\cos\theta\, d\theta$, $(x^2 + y^2)^2 = (r^2\cos^2\theta + r^2\sin^2\theta)^2 = r^4$, $x^2 y\, dx = r^3 \cos^3\theta \sin\theta\, dr - r^4 \cos^2\theta \sin^2\theta\, d\theta$, and $x^3\, dy = r^3 \cos^3\theta \sin\theta\, dr + r^4 \cos^4\theta\, d\theta$. Thus,

$$M\,dx + N\,dy = \frac{x^2 y}{(x^2 + y^2)^2}\,dx - \frac{x^3}{(x^2 + y^2)^2}\,dy = -\frac{r^4 \cos^2\theta}{r^4}\,d\theta = -\cos^2\theta\, d\theta.$$

Hence,

$$\int_\Gamma M\,dx + N\,dy = -\int_0^{2\pi} \cos^2\theta\, d\theta = -\int_0^{2\pi}\left(\frac{1}{2} + \frac{1}{2}\cos 2\theta\right)d\theta = -\pi.$$

25. See hint to Exercise 27.

26. (1)
$$M\,dx + N\,dy = \frac{x^2 y}{(x^2 + y^2)^2}\,dx - \frac{x^3}{(x^2 + y^2)^2}\,dy.$$

Thus,

(2)
$$M = \frac{x^2 y}{(x^2 + y^2)^2}, \quad N = -\frac{x^3}{(x^2 + y^2)^2}, \quad \frac{\partial M}{\partial y} = \frac{x^4 - 3x^2 y^2}{(x^2 + y^2)^3},$$

$$\frac{\partial N}{\partial x} = \frac{x^4 - 3x^2 y^2}{(x^2 + y^2)^3}.$$

Let R be any rectangular region having the point $(0, 0)$ in its exterior and having the points $(0, 1)$ and $(3, 3)$ in its interior. In view of (2), the differential form given in (1) is exact in R. [*See* Theorem 2.2.2.] Utilizing Equation (2.2.48) with $(x_0, y_0) = (0, 1)$, we get

(3)
$$u(x, y) = \int_0^x \frac{ys^2}{(s^2 + y^2)^2}\,ds = \frac{1}{2}\left(\arctan\frac{x}{y} - \frac{xy}{x^2 + y^2}\right).$$

Let Γ be any simple curve with initial and terminal points at $(0, 1)$ and $(3, 3)$, respectively, and contained in R. In view of Formula (2.2.50) and (3), we find

$$\int_\Gamma M\,dx + N\,dy = u(x_2, y_2) - u(x_1, y_1) = u(3, 3) - u(0, 1) = \tfrac{1}{2}(\arctan 1 - \tfrac{1}{2})$$

$$= \frac{1}{2}\left(\frac{\pi}{4} + 2n\pi - \frac{1}{2}\right) = \frac{\pi}{8} + n\pi - \frac{1}{4}, \quad n = 0, \pm 1, \pm 2, \ldots.$$

27. Let the parametric equations of the curve Γ be given by [*see* Definition 27 of Appendix 1]

$$x = x(t), \quad y = y(t), \quad t_1 \leq t \leq t_2.$$

From Definition 28 of Appendix 1, the points $(x_1, y_1) = (x(t_1), y(t_1))$ and $(x_2, y_2) = (x(t_2), y(t_2))$ are, respectively, the initial and terminal points of Γ. Since Γ is closed, we have $(x(t_1), y(t_1)) = (x(t_2), y(t_2))$ or $x_1 = x_2$ and $y_1 = y_2$. [*See* Definition 29 of Appendix 1.] Utilizing Formula (2.2.50), we see that

$$\int_\Gamma M\,dx + N\,dy = u(x_2, y_2) - u(x_1, y_1) = u(x_2, y_2) - u(x_2, y_2) = 0.$$

28. 0.   29. 0.   30. 0.

Exercises 2.3, page 62

1. General solution: $\sqrt{x} - \sqrt{y} = C.$ Singular solutions: $x = 0, \quad y = 0.$
2. General solution: $x - 1 = Ce^{1/y}.$ Singular solution: $y = 0.$
   Particular solution: $x = 1.$

3. General solution: $(x^2 - 1)e^{2y} = Cx^2$.   Singular solution: $x = 0$.
Particular solutions: $x = \pm 1$.

4. General solution: $(1 + x)(1 + y) = C(1 - x)(1 - y)$ or $(1 - x)(1 - y) = B(1 + x)(1 + y)$.
Particular solutions: $x = \pm 1$, $y = \pm 1$.

5. General solution: $(1 + y) = C(1 - y)e^{1/x^2}$ or $(1 - y)e^{1/x^2} = B(1 + y)$.
Singular solution: $x = 0$.   Particular solutions: $y = \pm 1$.

6. General solution: $\sin^{-1} x + \sin^{-1} y = C$ or $x\sqrt{1 - y^2} + y\sqrt{1 - x^2} = B$.
Singular solutions: $x = \pm 1$, $y = \pm 1$.

7. General solution: $\tan^{-1} x + \tan^{-1} y = C$ or $x + y = B(1 - xy)$.

8. General solution: $(6 - x)^3 = Cxy^2$ or $xy^2 = B(6 - x)^3$.
Particular solutions: $x = 0$, $x = 6$, $y = 0$.

9. General solution: $e^x = (C + y^2)(x + 1)$.   Singular solution: $x = -1$.

10. General solution: $y(x + 1) = Cx^{x/(x+1)}$.   Particular solutions: $x = -1$, $y = 0$.

11. General solution: $e^x - 2 \ln (e^x + 1) - \dfrac{1}{e^x + 1} - y = C$.

12. General solution: $y = C \ln \tan x$ or $\tan x = e^{By}$.   Particular solutions:
$y = 0$, $x = \dfrac{\pi}{4} + n\pi$, $n = 0, \pm 1, \pm 2, \ldots$.

Exercises 2.4, page 67

1.
$$F\left(\frac{x}{y}\right) \equiv \frac{1 - \dfrac{x}{y} + \left(\dfrac{x}{y}\right)^2}{1 + \dfrac{x}{y} + \left(\dfrac{x}{y}\right)^2} + \sin\left(\frac{x}{y}\right),$$

$$F\left(\frac{x}{y}\right) \equiv e^{(x/y) - 2} + \cos\left[\frac{\left(\dfrac{x}{y}\right)^3 + 1}{\dfrac{x}{y}}\right] + \ln\left(\frac{\dfrac{x}{y} + 1}{\dfrac{x}{y} - 1}\right) + \frac{\dfrac{x}{y} + 1}{\sqrt{\left(\dfrac{x}{y}\right)^2 - 1}}.$$

3. $y - x = Ce^{x/(y-x)}$.

4. $\ln x + e^{-y/x} = C$.

5. $\ln x - \arcsin y/x = C$.

6. $x = C \ln y/x$.

7. $x^2(2y^2 - x^2) = C$.

8. $x = e^{(y-x)/y}$.

9. $\ln x + \tan y/x = 1$.

10. $(y + x)(4y - x)^4 = 4\left(\dfrac{xy}{3}\right)^5$.

12. $\ln (4x - 8y + 5) = 4(x + 2y) + C$.

13. $9 \ln (14x + 21y + 22) = 42y - 21x + C$.

14. $(y - x - \tfrac{3}{4})^3(y + 6x - \tfrac{4}{3})^4 = C$.

15. Let $u = x - y$. $x + \cot (x - y) = C$.

16. Let $u = x - y$. $x - y + 1 = \tan (C + x)$.

17. Since $f(x, y)$ is homogeneous of degree zero, we may write f as a function of y/x only. Thus, $dy/dx = F(y/x)$. Since $x = r \cos \theta$, $y = r \sin \theta$, we have $dx = \cos \theta \, dr - r \sin \theta \, d\theta$, $dy = \sin \theta \, dr + r \cos \theta \, d\theta$. Thus $dy/dx = F(y/x)$ may be written as

$$\frac{\sin \theta \, dr + r \cos \theta \, d\theta}{\cos \theta \, dr - r \sin \theta \, d\theta} = F\left(\frac{r \sin \theta}{r \cos \theta}\right) = F(\tan \theta).$$

The above equation can be put into the form

$$\frac{dr}{r} = \frac{\tan \theta \, F(\tan \theta) + 1}{F(\tan \theta) - \tan \theta} \, d\theta,$$

and thus the variables are separated.

18. (a) $x^2 = Ce^{(y/x)^2}$.   (b) $x = C \ln \dfrac{y}{x}$.

19. (a) Suppose that the function $f = f(x, y, z, \ldots)$ is homogeneous of degree k. Thus,

(1) $$f(tx, ty, tz, \ldots) = t^k f(x, y, z, \ldots).$$

Let us write (1) as

(2) $$f(u, v, w, \ldots) = t^k f(x, y, z, \ldots),$$

where

(3) $$u = tx, \quad v = ty, \quad w = tz, \ldots .$$

Differentiating both sides of (2) with respect to t, and utilizing the chain rule, we get

(4) $$\frac{\partial f}{\partial u} \frac{\partial u}{\partial t} + \frac{\partial f}{\partial v} \frac{\partial v}{\partial t} + \frac{\partial f}{\partial w} \frac{\partial w}{\partial t} + \cdots = k t^{k-1} f(x, y, z, \ldots).$$

Since

$$\frac{\partial u}{\partial t} = x, \quad \frac{\partial v}{\partial t} = y, \quad \frac{\partial w}{\partial t} = z, \ldots,$$

Equation (4) becomes

(5) $$x \frac{\partial f}{\partial u} + y \frac{\partial f}{\partial v} + z \frac{\partial f}{\partial w} + \cdots = k t^{k-1} f(x, y, z, \ldots).$$

Setting $t = 1$ in (5) and (3), we obtain

(6) $$x \frac{\partial f}{\partial x} + y \frac{\partial f}{\partial y} + z \frac{\partial f}{\partial z} + \cdots = k f(x, y, z, \ldots).$$

Thus, Euler's relation is established.

(b) Conversely, suppose that Euler's relation holds. We wish to show that the function $f = f(x, y, z, \ldots)$ is homogeneous of degree k. If the function f is homogeneous of degree k, then

(7)
$$\frac{1}{x^k} f(x, y, z, \ldots)$$

depends only on the ratios y/x, z/x, $\ldots$. (Why?) It is sufficient to show that it follows from (6), if new variables

(8) $\qquad u = x, \quad v = y/x, \quad w = z/x, \quad \ldots, \quad$ or $\quad u = x, \quad y = vu, \quad z = wu, \quad \ldots$

are introduced in (7), the expression

(9) $\qquad \dfrac{1}{x^k} f(x, y, z, \ldots) = \dfrac{1}{u^k} f(u, vu, wu, \ldots) = F(u, v, w, \ldots)$

no longer depends on the variable u. That is, the equation

(10)
$$\frac{\partial F}{\partial u} = 0$$

is an identity. Utilize the chain rule to obtain

(11)
$$\frac{\partial F}{\partial u} = \frac{1}{u^k}\left(\frac{\partial f}{\partial u} + \frac{\partial f}{\partial y}\frac{\partial y}{\partial u} + \frac{\partial f}{\partial z}\frac{\partial z}{\partial u} + \cdots\right) - \frac{k}{u^{k+1}} f(u, vu, wu, \ldots)$$

$$= \frac{1}{u^k}\left(\frac{\partial f}{\partial u} + v\frac{\partial f}{\partial y} + w\frac{\partial f}{\partial z} + \cdots\right) - \frac{k}{u^{k+1}} f(u, vu, wu, \ldots)$$

$$= \frac{1}{x^k}\left(\frac{\partial f}{\partial x} + \frac{y}{x}\frac{\partial f}{\partial y} + \frac{z}{x}\frac{\partial f}{\partial z} + \cdots\right) - \frac{k}{x^{k+1}} f(x, y, z, \ldots)$$

$$= \frac{1}{x^{k+1}}\left(x\frac{\partial f}{\partial x} + y\frac{\partial f}{\partial y} + z\frac{\partial f}{\partial z} + \cdots\right) - \frac{k}{x^{k+1}} f(x, y, z, \ldots)$$

$$= \frac{1}{x^{k+1}}\left[x\frac{\partial f}{\partial x} + y\frac{\partial f}{\partial y} + z\frac{\partial f}{\partial z} + \cdots - k f(x, y, z, \ldots)\right].$$

Since the relation given in (6) is assumed to hold, we see that $\partial F/\partial u = 0$.

20. Follow a method similar to that given in the hint to Exercise 19.

Exercises 2.5, page 76

2. $x^4(3 \ln x - 1) - 9y = Cx.$
         3. $x^2 - 2y^2 = Cy^3.$

4. $2x^3 y^3 + 3x[\sin(\ln x) + \cos(\ln x)] = C.$

5. $y^4 + y^3 x + 2x = Cy^2.$
         6. $2 \arctan \dfrac{x}{y} + e^x(\sin x - \cos x) = C.$

7. $x^2 y^3 + e^{x^3} = C.$
         8. $4 \ln \dfrac{x}{y} + e^{2x}(2x^2 - 2x + 1) - 4y = C.$

9. $16xy + 2x^4 + 8\tan 2y - 2x(2x^2 - 3)\sin 2x - 3(2x^2 - 1)\cos 2x = C.$

10. $xe^{x+y} - \ln(\cos e^x) = C.$

11. (a) We must show that

$$\frac{M}{Mx + Ny}\,dx + \frac{N}{Mx + Ny}\,dy = 0, \quad Mx + Ny \neq 0,$$

is exact, that is, we must show that

$$\frac{\partial}{\partial y}\left(\frac{M}{Mx + Ny}\right) = \frac{\partial}{\partial x}\left(\frac{N}{Mx + Ny}\right).$$

Now,

$$\frac{\partial}{\partial y}\left(\frac{M}{Mx + Ny}\right) = \frac{Ny\,\dfrac{\partial M}{\partial y} - MN - My\,\dfrac{\partial N}{\partial y}}{(Mx + Ny)^2},$$

and

$$\frac{\partial}{\partial x}\left(\frac{N}{Mx + Ny}\right) = \frac{Mx\,\dfrac{\partial N}{\partial x} - MN - Nx\,\dfrac{\partial M}{\partial x}}{(Mx + Ny)^2}.$$

Thus,

$$(1)\quad \frac{\partial}{\partial y}\left(\frac{M}{Mx + Ny}\right) - \frac{\partial}{\partial x}\left(\frac{N}{Mx + Ny}\right)$$

$$= \frac{N\left(x\,\dfrac{\partial M}{\partial x} + y\,\dfrac{\partial M}{\partial y}\right) - M\left(x\,\dfrac{\partial N}{\partial x} + y\,\dfrac{\partial N}{\partial y}\right)}{(Mx + Ny)^2}.$$

From Euler's relation [*see* Exercise 2.4.19(a)], we know that

$$(2)\quad x\,\frac{\partial M}{\partial x} + y\,\frac{\partial M}{\partial y} = kM \quad \text{and} \quad x\,\frac{\partial N}{\partial x} + y\,\frac{\partial N}{\partial y} = kN.$$

The result now follows from Equations (1) and (2).

(b) If $Mx + Ny \equiv 0$, then $M/N = -y/x$. Thus, $M\,dx + N\,dy = 0$ becomes

$$\frac{dx}{x} = \frac{dy}{y},$$

and an integration gives

$$\ln y = \ln x + \ln C \quad \text{or} \quad y = Cx.$$

12. (a) $y^3 = 3x^3(C + \ln x),$

(b) $e^{-y/x} + \ln x = C,$ (c) $x = C\cosh^2 \dfrac{y}{x}.$

13. (a) (1)
$$M\,dx + N\,dy \equiv yf(xy)dx + xg(xy)dy = 0.$$

Thus,
$$M = yf(xy), \quad N = xg(xy).$$

Also,
$$Mx - Ny = xyf(xy) - xyg(xy) = xy[f(xy) - g(xy)].$$

In order to show that $1/(Mx - Ny)$ is an integrating factor of Equation (1), we must show
$$\frac{\partial}{\partial y}\left\{\frac{yf(xy)}{xy[f(xy) - g(xy)]}\right\} = \frac{\partial}{\partial x}\left\{\frac{xg(xy)}{xy[f(xy) - g(xy)]}\right\},$$

or
$$\frac{\partial}{\partial y}\left[\frac{f}{x(f-g)}\right] = \frac{\partial}{\partial x}\left[\frac{g}{y(f-g)}\right].$$

Now,
$$\frac{\partial}{\partial y}\left[\frac{f}{x(f-g)}\right] = \frac{-g\dfrac{\partial f}{\partial y} + f\dfrac{\partial g}{\partial y}}{x(f-g)^2}$$

and
$$\frac{\partial}{\partial x}\left[\frac{g}{y(f-g)}\right] = \frac{f\dfrac{\partial g}{\partial x} - g\dfrac{\partial f}{\partial x}}{y(f-g)^2}.$$

Thus,

(2)
$$\frac{\partial}{\partial y}\left[\frac{f}{x(f-g)}\right] - \frac{\partial}{\partial x}\left[\frac{g}{y(f-g)}\right]$$
$$= \frac{f\left(y\dfrac{\partial g}{\partial y} - x\dfrac{\partial g}{\partial x}\right) + g\left(-y\dfrac{\partial f}{\partial y} + x\dfrac{\partial f}{\partial x}\right)}{xy(f-g)^2}.$$

But

(3)
$$y\frac{\partial f(xy)}{\partial y} = x\frac{\partial f(xy)}{\partial x}, \quad y\frac{\partial g(xy)}{\partial y} = x\frac{\partial g(xy)}{\partial x}. \quad \text{(Why?)}$$

From Equations (2) and (3), the result now follows.

(b) If $Mx - Ny \equiv 0$, then $M/N = y/x$. Thus, Equation (1) may be written as
$$\frac{dx}{x} + \frac{dy}{y} = 0,$$

and an integration gives
$$\ln x + \ln y = C \quad \text{or} \quad xy = C.$$

14. (a) $x = Cy^3 e^{1/x^3 y^3}$,   (b) $xe^{xy} = Cy(xy + 1)$.

15. In order that $x^r y^s$ be an integrating factor of

(1) $$x^\alpha y^\beta (Ay\,dx + Bx\,dy) + x^\gamma y^\delta (ay\,dx + bx\,dy) = 0,$$

we must have

(2) $$\frac{\partial}{\partial y}\left(Ax^{r+\alpha}y^{s+\beta+1} + ax^{r+\gamma}y^{s+\delta+1}\right) = \frac{\partial}{\partial x}\left(Bx^{r+\alpha+1}y^{s+\beta} + bx^{r+\gamma+1}y^{s+\delta}\right).$$

Performing the indicated partial differentiations in (2), we obtain

(3) $$A(s+\beta+1)x^{r+\alpha}y^{s+\beta} + a(s+\delta+1)x^{r+\gamma}y^{s+\delta}$$
$$= B(r+\alpha+1)x^{r+\alpha}y^{s+\beta} + b(r+\gamma+1)x^{r+\gamma}y^{s+\delta}.$$

The relation given in (3) will be satisfied provided that

(4) $A(s+\beta+1) = B(r+\alpha+1)$ and $a(s+\delta+1) = b(r+\gamma+1)$.

Thus,

(5) $$\begin{cases} As - Br = B(\alpha+1) - A(\beta+1), \\ as - br = b(\gamma+1) - a(\delta+1). \end{cases}$$

Solving the system (5) for r and s, we find

(6) $$r = \frac{A[a(\delta+1) - b(\gamma+1)] - a[A(\beta+1) - B(\alpha+1)]}{Ab - Ba},$$

(7) $$s = \frac{B[a(\delta+1) - b(\gamma+1)] - b[A(\beta+1) - B(\alpha+1)]}{Ab - Ba}.$$

Letting

(8) $$k = a(\delta+1) - b(\gamma+1), \qquad K = A(\beta+1) - B(\alpha+1),$$

we have

(9) $$r = \frac{kA - Ka}{Ab - Ba}, \qquad s = \frac{kB - Kb}{Ab - Ba}.$$

In case $Ab - Ba = 0$, then $A/B = a/b = p$, where p is a constant. Thus, $A = pB$, $a = pb$, and Equation (1) may be written as

(10) $$(Bx^\alpha y^\beta + bx^\gamma y^\delta)(py\,dx + x\,dy) = 0,$$

from which the general solution may be obtained.

16. (a) $7x^2 + 17y^3 = Cx^{21/2}y^{17}$,   (b) $x^{2/9}y^{12/9}(x^2 y^2 - 5) = C$.

18. Suppose that $f(x, y) = C$ is the general solution of the equation

(1) $$M(x, y)dx + N(x, y)dy = 0.$$

Since $f(x, y) = C$, its total differential df is equal to zero, that is,

(2) $$\frac{\partial f}{\partial x}\,dx + \frac{\partial f}{\partial y}\,dy = 0.$$

In order that (2) be identically equal to (1), we must have that

$$\frac{\partial f}{\partial x} = k\,M, \quad \frac{\partial f}{\partial y} = kN,$$

where k is a constant.   Thus, the common value of the two ratios given by

$$\frac{\dfrac{\partial f}{\partial x}}{M} = \frac{\dfrac{\partial f}{\partial y}}{N}$$

is an integrating factor of Equation (1).

19. Concerning (a).   Suppose μ_1 and μ_2 are integrating factors of the equation

(1) $$M(x, y)dx + N(x, y)dy = 0.$$

Thus, $\mu_1 M\,dx + \mu_1 N\,dy = 0$ is an exact differential equation.   In view of Definition 2.2.1, there exists a function u such that

(2) $$du = \mu_1 M\,dx + \mu_1 N\,dy.$$

Multiply both sides of (2) by μ_2/μ_1, and we obtain

(3) $$\frac{\mu_2}{\mu_1}\,du = \mu_2 M\,dx + \mu_2 N\,dy.$$

Since μ_2 is also an integrating factor of Equation (1), the right-hand side of (3) is an exact differential, and thus the left-hand side of (3) is also an exact differential.   Hence, μ_2/μ_1 is a function of u only.

Concerning (b).   Suppose μ_2/μ_1 is not a constant.   From the result of part (a), μ_2/μ_1 is a function of u only, say $f(u)$.   Clearly $f'(u) \neq 0$.   Multiplying (2) by $f'(u)$, we obtain

(4) $$\mu_1 f'(u)M\,dx + \mu_1 f'(u)N\,dy = f'(u)du = df.$$

Thus, the solution of (1) is given by $df = 0$ (Why?), or

$$f = \frac{\mu_1}{\mu_2} = C,$$

where C is an arbitrary constant.

Exercises 2.6, page 81

1. $y = x - 1 + Ce^{-x}$.

2. $2y = \sin x - \cos x + Ce^{-x}$.

3. $8ye^{4x} = \ln(3 + 2e^{4x}) + C$.

4. $2x = y^2 - 1 + Ce^{-y^2}$.

5. $xe^y = y^y(e^y + C)$.

6. $y(x^2 + 1) = e^x(x^2 - 2x + 2) + C$.

7. $xe^{2y} = \tan y + C$.

8. $x(y^2 + 1)^3 = y - \arctan y + C$.

9. $y = -x \arctan 2x + x^2 \ln\left(\dfrac{x^2}{4x^2 + 1}\right) + Cx^2$.

10. $y \ln x = (\ln x)^2 + C.$
$\qquad\qquad\qquad$ 11. $x = 2(\sin y - 1) + Ce^{-\sin y}.$
12. $y = x + C \sin x.$

13. $yx^n = \dfrac{1}{na} \sec^n ax + C, \quad a \neq 0, \quad n \neq 0. \qquad yx^n = C, \quad a = 0, \quad n \neq 0.$

14. $r \sin^3 \theta + 2 \sin^5 \theta = C.$
$\qquad\qquad\qquad$ 15. $yx = \sin x + C \cos x.$

16. $2y\sqrt{x^2 - 1} = x \ln \left(\dfrac{x-1}{x+1}\right) + Cx.$
$\qquad$ 17. $y = u(x) - 1 + Ce^{-u(x)}.$

18. $y = \tan x - 1 + Ce^{-\tan x}.$
$\qquad\qquad$ 19. $y = \ln (\ln x) - 1 + C(\ln x)^{-1}.$
20. $r = \sin 2\theta - 1 + Ce^{-\sin 2\theta}.$
22. (a) $y^2(1 - x^2 + Ce^{-x^2}) = 1.$
$\qquad$ (b) $y^{1/2} = -\frac{1}{3}(1 - x^2) + C(1 - x^2)^{1/4}.$
$\quad$ (c) $\cot y + x \ln \sin y = Cx.$
$\qquad$ (d) $(x^2 + 1)y^4[2 \ln (x^2 + 1) - 2x^2 + C] = 1.$

23. (a) (1)
$\qquad\qquad\qquad\qquad\qquad$ $\begin{cases} y = Cf(x) + g(x), \\ y' = Cf'(x) + g'(x). \end{cases}$
$\qquad$ (2)

$\qquad$ Solving for C in (1) and then substituting its value into (2), we get

$\qquad$ (3) $\quad y' = \left[\dfrac{y - g(x)}{f(x)}\right] f'(x) + g'(x) = \dfrac{f'(x)}{f(x)} y + \dfrac{f(x)g'(x) - f'(x)g(x)}{f(x)}$

$\qquad$ Letting

$$P(x) = -\frac{f'(x)}{f(x)}, \quad Q(x) = \frac{f(x)g'(x) - f'(x)g(x)}{f(x)},$$

$\qquad$ Equation (3) becomes

$$y' + P(x)y = Q(x),$$

$\qquad$ which is a linear equation of the first order.
$\quad$ (b) From Equation (2.6.1), we have

$\qquad$ (1) $\qquad\qquad\qquad y' + P(x)y = Q(x).$

$\qquad$ Since y_1 is a particular solution of (1), we have

$\qquad$ (2) $\qquad\qquad\qquad y'_1 + P(x)y_1 = Q(x).$

$\qquad$ Subtracting (2) from (1), we get

$\qquad$ (3) $\qquad\qquad\qquad \dfrac{y' - y'_1}{y - y_1} = -P(x).$

$\qquad$ An integration of (3) gives

$$\ln (y - y_1) = \int -P(x)dx + \ln C_1,$$

$\qquad$ or

$\qquad$ (4) $\qquad\qquad\qquad y - y_1 = C_1 \exp\left[-\int P(x)dx\right].$

Similarly, since y_2 is a particular solution of (1), we find

$$(5) \qquad y - y_2 = C_2 \exp\left[-\int P(x)dx\right].$$

Thus, in view of (4) and (5), we obtain

$$(6) \qquad \frac{y - y_1}{y - y_2} = \frac{C_1}{C_2} = K,$$

where K is an arbitrary constant.

(c) From Equation (2.6.1), we have

$$(1) \qquad y' + P(x)y = Q(x).$$

Since y_1 is a particular solution of (1), we have

$$(2) \qquad y_1' + P(x)y_1 = Q(x).$$

Let

$$(3) \qquad y = y_1(x) + Y(x).$$

Then,

$$(4) \qquad y' = y_1' + Y'.$$

Substituting y and y' of (3) and (4) into (1), we get

$$(5) \qquad y_1' + P(x)y_1 + Y' + P(x)Y = Q(x).$$

In view of (2), Equation (5) becomes

$$(6) \qquad Y' + P(x)Y = 0.$$

An integration of (6) gives

$$(7) \qquad Y = C \exp\left[-\int P(x)dx\right].$$

In view of (7) and (3), the general solution of Equation (2.6.1) is

$$y = y_1 + C \exp\left[-\int P(x)dx\right].$$

24. $y = x + C \sin x.$

Exercises 2.7, page 89

1. $y = -\dfrac{1}{x} + \dfrac{\exp\left[\int xP(x)dx\right]}{x^2\left[C - \displaystyle\int \dfrac{\exp\left[\int xP(x)dx\right]}{x^2}\,dx\right]}.$

2. $yx^2 + x + 1 = Cxe^{1/x}(yx + 1).$

3. $y = -\cos x + \dfrac{e^{-\sin x}}{C - \int e^{-\sin x}dx}$.

4. $yx = Cx(yx - 2) + 1$.

5. $y + x = Ce^{x^2}(y - x)$.

6. $(Cx - x \ln x - 1)(yx - 1) = 1$.

7. $8x = (C + 4x - x^4)(yx^2 - 1)$.

8. $(Ce^{\cos x} - 1)(y - \sin x) = 1$.

9. $2e^{x^2} = (C - e^{x^2})(y - \ln x)$.

10. $y = \dfrac{1 - C \cot x}{C + \cot x}$ or $y = \dfrac{B - \cot x}{1 + B \cot x}$.

11. $y = x + \dfrac{2x}{C + x^2}$ or $y = x + \dfrac{2Bx}{Bx^2 + 1}$.

12. (a) (1)
$$\frac{d^2y}{dx^2} + Q(x)\frac{dy}{dx} + P(x)y = 0.$$

Let

(2)
$$y = \exp\left[\int u(x)dx\right].$$

Then,

$$y' = \exp\left[\int u(x)dx \; u(x) = yu\right]$$

and

$$y'' = y'u + yu' = yu^2 + yu' = y(u^2 + u').$$

Substituting these derivatives into (1), we get

$$y(u^2 + u') + Q(x)yu + P(x)y = 0.$$

In view of (2), $y \neq 0$.   Thus, the above equation may be written as

(3)
$$\frac{du}{dx} = -P(x) - Q(x)u - u^2.$$

(b) Conversely, to show that every Ricatti equation can be transformed into a homogeneous linear equation of the second order, we proceed as follows. Let

$$y = \exp\left[\int u(x)dx\right].$$

Then,

$$y' = yu \quad \text{or} \quad u = \frac{y'}{y},$$

and

$$u' = \frac{yy'' - (y')^2}{y^2}.$$

Substituting u and u' into Equation (3) of part (a), we get

$$\frac{yy'' - (y')^2}{y^2} = -P(x) - Q(x)\frac{y'}{y} - \left(\frac{y'}{y}\right)^2.$$

which simplifies to the desired result, namely,

$$y'' + Q(x)y' + P(x)y = 0.$$

(c) (1) $\qquad y'' + Q(x)y' + P(x)y = 0.$

Since y_1 and y_2 are solutions of (1), we have

(A) $\qquad y_1'' + Q(x)y_1' + P(x)y_1 = 0, \quad y_2'' + Q(x)y_2' + P(x)y_2 = 0.$

To show that

(4) $\qquad y = c_1 y_1(x) + c_2 y_2(x)$

is also a solution of (1), we proceed as follows. From (4), we obtain by differentiations

(B) $\qquad y' = c_1 y_1' + c_2 y_2', \quad y'' = c_1 y_1'' + c_2 y_2''.$

If we substitute y, y', y'' of (4) and (B) in the left-hand member of (1), we get

$$c_1[y_1'' + Q(x)y_1' + P(x)y_1] + c_2[y_2'' + Q(x)y_2' + P(x)y_2].$$

In view of (A), this expression is equal to zero. Thus, $y = c_1 y_1(x) + c_2 y_2(x)$ is a solution of (1).

13. $y = c_1 x + c_2/x^2.$

Exercises 2.8, page 91

1. $4yx^3 + 12y^3 x + 6x^2 \ln x - 3x^2 - 12e^y(y - 1) = C.$

2. $3x^2 y^3 + x^3 y^4 = C.$ $\qquad\qquad$ 3. $(1 + x^2)(1 + y^2) = Cx^2.$

4. $y = -\dfrac{1}{x} - 1 + Ce^{1/x}.$ $\qquad\qquad$ 5. $\ln x + e^{y/x} = C.$

6. $2 \sin x - \ln(\sec y + \tan y) = C.$ $\qquad$ 7. $x^2 \sin y + y^2 \sin x = C.$

8. $2 \ln(y + \sqrt{y^2 + 1}) = x\sqrt{x^2 + 1} + \ln(x + \sqrt{x^2 + 1}) + C.$

9. $2x^3 + 3xy^2 + 3y^3 = C.$

10. $r(\sec\theta + \tan\theta) = 2\ln\sec\theta + 2\ln(\sec\theta + \tan\theta) - 2\sin\theta + C.$

11. $(x^2 + 1)(y + 1) = C(y - 1)e^{1/y}.$

12. $\left\{C - 3\left[\tan x \sec x + \ln\tan\left(\dfrac{x}{2} + \dfrac{\pi}{4}\right)\right]\right\}y^3\cos^3 x = 1.$

13. $y = e^{x+1}[C + \ln(e^x + 1)] - e^x.$ $\qquad$ 14. $y \ln x = x(\ln x)^2 - 2x\ln x + 2x + C.$

15. $y^2(x^4 - 1)[3\ln(x - 1) - \ln(x + 1) - \ln(x^2 + 1) + C] = 1.$

16. $yx^2(2x + y + 2) = 12.$ $\qquad\qquad$ 17. $y = -e^{2x}[e^{-x}\sin(e^{-x})$
$\qquad\qquad\qquad\qquad\qquad\qquad\qquad + \cos(e^{-x}) + 1 - \sin 1 - \cos 1].$

18. $y(x^3 + 1 + 3 \ln x) = 9x$.

19. $y = \arcsin x - \sqrt{1 - x^2}$.

20. $(y + x)^2(y + 2x)^3 + 32 = 0$.

21. $2y = 2xe^{2x} - (e^{2x} + 1) \ln (e^{2x} + 1)$.

22. $y = \dfrac{a}{2}\left[\left(\dfrac{x}{a}\right)^{1-k} - \left(\dfrac{x}{a}\right)^{1+k}\right]$.

23. $r \cos \theta = a(\sec \theta + \tan \theta)^k$.

24. $3x = y(x^3 \cos x^3 - \sin x^3 + 3 + \pi)$.

25. $(yx + 1)(1 - 3 \ln x) = 3$.

26. $\ln [(y - 1)^2 - 2(y - 1)(x - 3) + 2(x - 3)^2] + 4 \arctan \left(\dfrac{y - x + 2}{x - 3}\right) + \pi$

$\quad - \ln 2 = 0$.

27. $(y - x^2)(3x^4 + 1) = 12x$.

28. $y = \dfrac{x^2}{4(\alpha^2 + 1)} \{\alpha^2 + 1 - [\cos (2\alpha \ln x) + \alpha \sin (2\alpha \ln x)]\} - \dfrac{\alpha^2 b^2}{4(\alpha^2 + 1)}$.

29. Verification of (5):

$$\begin{cases} P(u) = K(u^2 - 1) \\ P(v) = K(v^2 - 1), \end{cases} \quad \text{where } K \text{ is a constant.}$$

From (3), $P(x) = cx^2 + dx + e = c\left(x^2 + \dfrac{d}{c}x + \dfrac{d^2}{4c^2}\right) - \dfrac{d^2}{4c} + e = c\left(x + \dfrac{d}{2c}\right)^2$

$-\dfrac{d^2}{4c} + e$.   From (4), $x = \xi + h$.   Let $h = -d/2c$.   Thus $x + d/2c = \xi$, and we may write $P(x)$ as

$$P(\xi) = c\xi^2 - h^2c + e.$$

From (4), $\xi = ku$.   Let $k^2 = \dfrac{h^2 c - e}{c}$.   Thus, the above expression may be written as

$$P(u) = (h^2c - e)u^2 - (h^2c - e) = K(u^2 - 1),$$

where

$$K = h^2c - e = \dfrac{d^2 - 4ce}{4c}.$$

Similarly, we show that $P(y) = cy^2 + dy + e$ can be written as

$$P(v) = K(v^2 - 1).$$

30. (a) $u(x, y) = 2 + [\cos (x + y) - 2]e^y$.

(b) $u(x, y) = \tfrac{1}{4}(x + 2y) + \tfrac{1}{4}e^{4y}[4e^{x + 2y} - (x + 2y)]$.

(c) $u(x, y) = e^{-(c/b)y}u_0\left(x - \dfrac{a}{b}y\right)$.

(d) $u(x, y) = e^{y/2}(x + \tfrac{3}{4}y)^2 \cos (x + \tfrac{3}{4}y)$.

(e) $u(x, y) = e^{\sin (x + (3/4)y)}$.   (f) $u(x, y) = x^2 e^{4y}$.

(g) $u(x, y) = e^y[e^x - e^{x+y} + \sinh (x + y)]$.

[*Note:* In the answers from (h) through (k), C is an arbitrary constant and g is any function that satisfies $g(0) = C.$]
(h) $u_0(x) = Ce^{-(c/a)x}.$ $u(x, y) = g(y)e^{-(c/a)x}.$
(i) $u_0(x) = Ce^{-x} + 2.$ $u(x, y) = g(y)e^{-x} + 2.$
(j) $u_0(x) = Ce^{-x} + x - 1.$ $u(x, y) = g(y)e^{-x} + x + y - 1.$
(k) $u_0(x) = Ce^{2x}.$ $u(x, y) = g(y)e^{2x} + \frac{1}{2}ye^{3x}.$

Exercises 3.1, page 106

1. 12,346 bacteria.   2. 24,691 bacteria.   3. 32.29 years.

4. Show that the solution of the initial-value problem: $\dfrac{dx}{dt} = kx, \;\; x(0) = n_0$,

   is given by

   (1) $$x = n_0 e^{kt}.$$

   Since $x(t_1) = n_1$ and $x(t_2) = n_2$, utilize (1) to show that

   $$\left(\frac{n_1}{n_0}\right)^{1/t_1} = e^k = \left(\frac{n_2}{n_0}\right)^{1/t_2},$$

   from which the given result may now be deduced.

5. 6.70 percent.   6. 14.75 grams.   7. 84.93 percent.

8. Show that the solution of $dx/dt = -kx$, subject to the conditions that $x(0) = x_0$, $x(t_1) = x_1$, and $x(t_2) = x_2$, is given by

   (1) $$x = x_0\left(\frac{x_2}{x_1}\right)^{t/(t_2 - t_1)}.$$

   For half-life, $x = \frac{1}{2}x_0$. Substitute this value of x into (1), and then solve for t to obtain the given result.

9. 2.16 hours.   10. 5.49 percent.

11. (a) Let $x =$ amount of money at time t. Then,

   (1) $$\frac{dx}{dt} = \frac{k}{100} x - B.$$

   Show that the solution of Equation (1), subject to the initial condition $x(0) = A$, is given by

   (2) $$x = \frac{B}{\dfrac{k}{100}} - \left(\frac{B}{\dfrac{k}{100}} - A\right)e^{(k/100)t}.$$

   To obtain the given result, set $x = 0$ into (2) and then solve for t.
   (b) 35.83 years.

12. Let $x =$ amount of money at time t.   Then,

(1)
$$\frac{dx}{dt} = kx + B,$$

where $k = \dfrac{5.5}{100} = \dfrac{55}{1000}$, and B is the amount of money the man saves in the bank in a year, namely, 3650 dollars.   Show that the solution of Equation (1) subject to the initial condition $x(0) = 0$, is given by

(2)
$$t = \frac{1}{k} \ln \left(1 + \frac{kx}{B} \right).$$

Thus, when $x = 100,000$ dollars, $t = 16.71$ years.

13. Solve the initial-value problem

$$\begin{cases} \dfrac{dx}{dt} = (\text{rate of increase} - \text{rate of decrease}) \cdot x = \left[\dfrac{(a + bt)}{100} - \dfrac{(\alpha + \beta t)}{100} \right] x, \\[2mm] x(0) = x_0. \end{cases}$$

Exercises 3.2, page 114

1. 40 min.

2. 83. 33°C.

3. (a) 63.47°F,
   (b) 34.77 min.

4. 5 min.

5. 1:19 p.m.

6. 63.63 min.

7. (a) 22.56 lb.   (b) 50 lb.

8. 20.08 gal.

9. 296.30 lb;   0.494 lb/gal.

10. Let:

$x =$ concentration of CO_2 (carbon dioxide) by volume at time t,
$A =$ cubic feet/min of fresh air entering the room,
$V =$ volume of room,
$x_0 =$ the initial concentration of CO_2,
$B =$ the concentration of CO_2 in fresh air.

In dt time: the amount of inflow $= AB\,dt$,

the amount of outflow $= Ax\,dt$.

The total CO_2 at time t is Vx.   Hence, the change in time dt is $d(Vx) = V\,dx$, since V is a constant.   Therefore, $V\,dx =$ input minus the output.   Thus,

(1)
$$V\,dx = AB\,dt - Ax\,dt = -A(x - B)dt.$$

Integrate (1) between the limits $t = 0$, $x = x_0$ and $t = t$, $x = x$, and show that

(2)
$$A = \frac{V}{t} \ln \left(\frac{x_0 - B}{x - B} \right).$$

Now, substitute the given data into (2) to obtain $A = 7486$ ft³/min.

11. $x(t) = \dfrac{p}{2} [1 + e^{-(2r/G)t}], \qquad y(t) = \dfrac{p}{2} [1 - e^{-(2r/G)t}].$

Exercises 3.3, page 122

3. Utilize Exercise 2 with $k = 0.03092/\text{min}$, $t = t_2 = 38$ min, $t_1 = 0$, $x_1 = x(0) = 5.11$ g, and $x_2 = x(t) = x$ to obtain $x = 1.5784$ g.

5. Utilize Exercise 4(a) with $x = (0.20)(2) = 0.4$ mole/l, $\alpha = 2$ moles/l, and $t = 20$ min to find $k = 0.00625$ $(\text{min})^{-1}\text{l}\,(\text{mole})^{-1}$. Now, use Exercise 4(b) to find that $t_{1/2} = 80$ min.

6. Utilize Formula (3.3.24) with $k = 0.00140$ $(\text{min})^{-1}\text{l}\,(\text{mole})^{-1}, \alpha = 0.5638$ mole/l, $\beta = 0.3114$ mole/l, and $t = 855$ min to find that $x = 0.1372$ mole/l. Thus, the number of grams (per liter) of ethyl alcohol when $t = 855$ min is $(0.1372$ mole/l$) \cdot (46.068$ g/mole$) = 6.3205$ g/l.

8. (b) 1.73 cc.

10. (a) Let $x =$ amount of C formed at time t. The formation of x grams of C requires $\frac{4}{7}x$ grams of A and $\frac{3}{7}x$ grams of B. Hence, when x grams of C have been formed, there are present

$$(48 - \tfrac{4}{7}x) \text{ grams of } A \quad \text{and} \quad (24 - \tfrac{3}{7}x) \text{ grams of } B.$$

By the Law of Mass Action, we then have

$$(1) \qquad \frac{dx}{dt} = k\left(48 - \frac{4}{7}x\right)\left(24 - \frac{3}{7}x\right).$$

Show that the solution of Equation (1), subject to the conditions that $x(0) = 0$ and $x(3) = 36$, is given by

$$x = 168\,\frac{(\tfrac{8}{5})^{t/3} - 1}{3(\tfrac{8}{5})^{t/3} - 2}.$$

(b) 4.42 min.

Exercises 3.4, page 132

1. $I(t) = 0.7947 \sin(120\,\pi t - \phi) + 0.7935 e^{-20t}$, where

$$\phi = \tan^{-1} 6\pi = 86.96°.$$

2. (c) $I(t) = 14.87 \cos(120\pi t - \phi) - 1.955 e^{-50t}$, where

$$\phi = \tan^{-1} 2.4\pi = 82.44°.$$

3. (c) $Q(t) = 15te^{-(20/3)t}$, $I(t) = 15(1 - \tfrac{20}{3}t)e^{-(20/3)t}$. $Q_{\max} = 0.8277$ coulomb.

4. (b) 3.93 amps. 5. (b) 0.0364 coulomb, -0.182 amp.
    (c) 0.0554 coulomb.

6. (b) 5.704 coulombs, 8.694 amps. 7. $t = RC \ln\left(\dfrac{RC + 1}{RC}\right)$ sec.

8. (b) 0.035 sec and 0.230 sec.

9. (b) $I(t) = 0.7949e^{-2t}\sin(120\pi t - \phi) + 0.7940e^{-20t}$, where

$$\phi = \tan^{-1}\left(\frac{60\pi}{9}\right) = 87.27°.$$

10. 7.388 amps.

Exercises 3.5, page 147

1. (a) Show that the solution of the initial-value problem

$$\begin{cases} m\dfrac{dv}{dt} = mg - Kv^2, \\[2mm] v(0) = 0, \end{cases}$$

is given by

$$(1) \qquad \frac{\sqrt{\dfrac{mg}{K}} - v}{\sqrt{\dfrac{mg}{K}} + v} = e^{-2\sqrt{(Kg/m)}\,t}.$$

Since $\lim\limits_{t\to\infty} v(t) = U$, in view of (1), show that $U = \sqrt{\dfrac{mg}{K}}$. Thus, (1) becomes

$$(2) \qquad \frac{U - v}{U + v} = e^{-(2g/U)t}.$$

To obtain the given result, substitute $v = kU$ into (2) and then solve for t.
(b) 3.41 sec.

2. (a) From (2) of hint to Exercise 1(a), we have

$$(1) \qquad \frac{U - v}{U + v} = e^{-(2g/U)t}.$$

From (1), show that

$$(2) \qquad v = \frac{U(1 - e^{-(2g/U)t})}{1 + e^{-(2g/U)t}}.$$

Now,

$$(3) \qquad \tanh\theta = \frac{\sinh\theta}{\cosh\theta} = \frac{\dfrac{e^{\theta} - e^{-\theta}}{2}}{\dfrac{e^{\theta} + e^{-\theta}}{2}} = \frac{1 - e^{-2\theta}}{1 + e^{-2\theta}}.$$

In view of (3), show that (2) may be written as

$$(4) \qquad v = U \tanh \left(\frac{g}{U} t \right).$$

Since $v = ds/dt$ and $s = 0$ when $t = 0$, show that the solution of (4) is given by

$$(5) \qquad s = \frac{U^2}{g} \ln \left[\cosh \left(\frac{g}{U} t \right) \right].$$

To obtain the desired result, substitute $s = H - h$ into (5).

(b) 20.9 sec.

3. (a) Solve the initial-value problem

$$\begin{cases} m \dfrac{dv}{dt} = A \sin \omega t - kv, \\[2mm] v(0) = 0. \end{cases}$$

(b) 10.75 ft/sec.

Concerning the units; k: slug rad/sec, ω: rad/sec, A: slug ft rad/sec^2.

4. Follow the method given in Example 3.5.5.

5. See the method given in Example 3.5.6.

6. (b) 197.03 lb.   7. (b) 3/2.

8. (b) $v = 6.87$ m/sec, $s = 8.98$ m.   (c) No.

12. (a) Let r (measured outward) be the variable distance from the center of the planet to the particle. Then,

$$\frac{GMm}{r^2} = - m \frac{dv}{dt} = - mv \frac{dv}{dr},$$

and $v = 0$ when $r = d$. Show that

$$v = \frac{dr}{dt} = - \sqrt{2GM} \sqrt{\frac{1}{r} - \frac{1}{d}}.$$

Thus,

$$T = \int_0^T dt = - \frac{1}{\sqrt{2GM}} \lim_{\varepsilon \to 0+} \int_{d-\varepsilon}^R \frac{dr}{\sqrt{\dfrac{1}{r} - \dfrac{1}{d}}}$$

$$= - \sqrt{\frac{d}{2GM}} \lim_{\varepsilon \to 0+} \int_{d-\varepsilon}^R \frac{r \, dr}{\sqrt{\left(\dfrac{d}{2} \right)^2 - \left(r - \dfrac{d}{2} \right)^2}},$$

from which the result may be deduced.

13. We shall first derive the equation of motion. At a given instant of time, let the mass of the rocket be m and its speed v, relative to the station. In a small interval Δt, the rocket has (i) expelled a mass of $\Delta m < 0$ with a speed

$u > 0$, relative to itself; and (ii) gained a velocity increment of Δv. There are no external forces, and we can assume that momentum is conserved. We equate the initial momentum of the rocket mv, with the momentum of the rocket (plus that of the expelled matter) at the end of the time interval Δt, and we have

$$(1) \qquad mv = (m + \Delta m)(v + \Delta v) - \Delta m(v - u).$$

Simplifying, dividing by Δt, and then letting $\Delta t \to 0$, show that (1) becomes

$$(2) \qquad m\frac{dv}{dt} = -u\frac{dm}{dt}.$$

Concerning (a). In view of (2), we have

$$\int_0^v dv = -u \int_{m_0}^m \frac{dm}{m},$$

or

$$(3) \qquad v = u \ln \left(\frac{m_0}{m}\right).$$

If v_1 is actually attained, we have

$$(4) \qquad v_1 = u \ln \left(\frac{m_0}{m_1}\right),$$

where the initial mass m_0 has (by virtue of expelled fuel) been reduced to m_1. We must therefore regard m_1 as the initial mass for the return fire.

Since we have actually attained our critical speed v_1, we must necessarily conclude that our remaining fuel is precisely that which is required to cancel the separation. Thus, we can attain a speed of v_1 relative to a frame, itself receding from the station at v_1. Our speed with respect to the *station* is therefore $v_1 - v_1 = 0$, and we will in fact be "standing still."

We now apply (3) to our *receding frame*. We take m_1 as our initial mass; $m_0 - P$ (being the mass of the rocket less its fuel) is, of course, taken as our final mass. Thus, (3) becomes

$$(5) \qquad v_1 = u \ln \left(\frac{m_1}{m_0 - P}\right).$$

Utilizing (4) and (5), show that

$$(6) \qquad m_1 = \sqrt{m_0} \sqrt{m_0 - P}.$$

Substituting m_1 of (6) into (4), show that

$$v_1 = \ln \left[1 - \frac{P}{m_0}\right]^{-u/2}.$$

Concerning (b). The fuel remaining for retro-fire is $P - (m_0 - m_1)$, where $m_0 - m_1$ has already been spent. The fraction r' remaining for return fire is

$$(7) \qquad r' = 1 - \frac{m_0}{P} + \frac{m_1}{P}.$$

In view of (6), show that r' may be expressed as

$$(8) \qquad r' = \sqrt{\frac{m_0}{P} - 1} \left[\sqrt{\frac{m_0}{P}} - \sqrt{\frac{m_0}{P} - 1} \right].$$

We must have $r > r'$. Hence,

$$r > \sqrt{\frac{m_0}{P} - 1} \left[\sqrt{\frac{m_0}{P}} - \sqrt{\frac{m_0}{P} - 1} \right].$$

Exercises 3.6, page 161

1. $y = kx$.
2. $2x^2 + 3y^2 = k^2$.
3. $x^2 + y^2 = ky$.
4. $2x + my^2 = k$.
5. $x^{2/3} - y^{2/3} = k^{2/3}$.
6. $2y^2 = kx^4 + x^2$.
7. $y^2 = 2kx + k^2$.
8. $r = ke^{-\sin \theta}$.
9. $r = ke^{-(\theta^2/2 + \theta^3/3)}$.
10. $r[k + \ln (\sec n\theta + \tan n\theta)] = n^2$.
11. $r^n = k^n \sin n\theta$.
12. $r^2 = k \sin \theta \tan \theta$.

13. $\ln [k^2(x^2 + y^2)] + 2\sqrt{3} \arctan \frac{y}{x} = 0; \quad \ln [c^2(x^2 + y^2)] - 2\sqrt{3} \arctan \frac{y}{x} = 0$.

14. $\ln [k^2(2x^2 - xy + y^2)] - \frac{6}{\sqrt{7}} \arctan \left(\frac{2y - x}{\sqrt{7}x} \right) = 0;$

$\ln [c^2(2x^2 + xy + y^2)] + \frac{6}{\sqrt{7}} \arctan \left(\frac{2y + x}{\sqrt{7}x} \right) = 0$.

15. $r = k \cos \left(\theta + \frac{\pi}{4} \right); \quad r = c \cos \left(\theta - \frac{\pi}{4} \right)$.

16. $r^2 = k \left(2\theta - \frac{\sqrt{3}}{3} \right)^{-4/3} e^{-(2\sqrt{3}/3)\theta}; \quad r^2 = c \left(2\theta + \frac{\sqrt{3}}{3} \right)^{-4/3} e^{(2\sqrt{3}/3)\theta}$.

17. 3.
18. 1/2.

19. From Exercise 1.2.26, the differential equation of the given family is

$$(1) \qquad (xy' - y)(yy' + x) = (a^2 - b^2)y'.$$

The differential equation of the orthogonal trajectories of the given family is obtained by replacing y' by $-1/y'$ in Equation (1). Show that this substitution leads back to Equation (1), and thus, the given family is self-orthogonal.

20. (1)
$$x^2 + y^2 + c^2 = 1 + 2cxy.$$

Differentiate (1) with respect to x and then solve for c. Show that $c = (x + yy')/(y + xy')$. Substitute c into (1) and show that

(2)
$$(y')^2(x^2 - 1) = (y^2 - 1).$$

To find the families of orthogonal trajectories, replace y' in (2) by $-1/y'$. Thus, we find

(3) $$\sqrt{y^2 - 1}\, dy = \sqrt{x^2 - 1}\, dx \quad \text{and} \quad \sqrt{y^2 - 1}\, dy = -\sqrt{x^2 - 1}\, dx.$$

Integrate each equation in (3), and show that

$$\ln\left[(y + \sqrt{y^2 - 1})(x - \sqrt{x^2 - 1})\right] = y\sqrt{y^2 - 1} - x\sqrt{x^2 - 1} + k$$

and

$$\ln\left[(y + \sqrt{y^2 - 1})(x + \sqrt{x^2 - 1})\right] = y\sqrt{y^2 - 1} + x\sqrt{x^2 - 1} + K,$$

where k and K are arbitrary constants. [*See* also Exercise 2.8.29, Equations (6) and (10).]

21. (a) Let the level curve $u(x, y) = c_1$ intersect the level curve $v(x, y) = k_1$ at a point (x_0, y_0) in D. Suppose that $u_x(x_0, y_0)$ and $v_y(x_0, y_0)$ do not vanish simultaneously, say, $u_y(x_0, y_0) \neq 0$. Since $u(x, y) = c_1$ and $v(x, y) = k_1$, we have

$$u_x\, dx + u_y\, dy = 0 \quad \text{or} \quad \frac{dy}{dx} = -\frac{u_x}{u_y},$$

and

$$v_x\, dx + v_y\, dy = 0 \quad \text{or} \quad -\frac{dx}{dy} = \frac{v_y}{v_x}.$$

[*Note:* $v_x(x_0, y_0) \neq 0$, since by the Cauchy-Riemann differential equations: $v_x(x_0, y_0) = -u_y(x_0, y_0) \neq 0$.] In order that the curve $v(x, y) = k_1$ be orthogonal to the curve $u(x, y) = c_1$ at (x_0, y_0), the slope of $v(x, y) = k_1$ must be the negative reciprocal of the slope of $u(x, y) = c_1$ at (x_0, y_0). Thus, in view of the above equations, this condition of orthogonality is given by

$$-\frac{u_x(x_0, y_0)}{u_y(x_0, y_0)} = \frac{v_y(x_0, y_0)}{v_x(x_0, y_0)},$$

which can be written as

(A)
$$u_y(x_0, y_0)v_y(x_0, y_0) + u_x(x_0, y_0)v_x(x_0, y_0) = 0.$$

Since f is analytic at z_0: (x_0, y_0), the Cauchy-Riemann differential equations,

$$u_x(x_0, y_0) = v_y(x_0, y_0), \quad u_y(x_0, y_0) = -v_x(x_0, y_0),$$

imply Equation (A). Thus, the level curve $u(x, y) = c_1$ is orthogonal to the level curve $v(x, y) = k_1$ at the point (x_0, y_0) in D.

(b) Since $u_x = v_y$ and $u_y = -v_x$, we have that $u_{xx} = v_{yx}$ and $u_{yy} = -v_{xy}$. From calculus, $v_{yx} = v_{xy}$. Thus, $u_{xx} + u_{yy} = 0$. Similarly, we show that $v_{xx} + v_{yy} = 0$.

22. (a) The level curves $u = c$: $x = c$; $v = k$: $y = k$.

(b) $u = c$ is a family of vertical lines; $v = k$ is a family of horizontal lines.

23. (a) The level curves $u = c$: $x^2 - y^2 = c$; $v = k$: $2xy = k$.

(b) $u = c$ is a family of hyperbolas; $v = k$ is a family of equilateral hyperbolas.

24. (a) The level curves $u = c$: $\dfrac{x}{x^2 + y^2} = c$; $v = k$: $-\dfrac{y}{x^2 + y^2} = k$.

(b) Since $x/(x^2 + y^2) = c$, we have $x^2 - \dfrac{1}{c}x + y^2 = 0$. Completing the square of the latter expression, show that

(A)
$$\left(x - \frac{1}{2c}\right)^2 + y^2 = \left(\frac{1}{2c}\right)^2.$$

Equation (A) represents a family of circles.   Centers: $\left(\dfrac{1}{2c}, 0\right)$ and radii $\left|\dfrac{1}{2c}\right|$.

Also, show that $-y/(x^2 + y^2) = k$ can be written as

(B)
$$x^2 + \left(y + \frac{1}{2k}\right)^2 = \left(\frac{1}{2k}\right)^2.$$

Equation (B) also represents a family of circles.   Centers: $\left(0, -\dfrac{1}{2k}\right)$ and radii: $\left|\dfrac{1}{2k}\right|$.

25. (a) The level curves $u = c$: $\dfrac{x^2 + x + y^2}{x^2 + y^2} = c$; $v = k$: $-\dfrac{y}{x^2 + y^2} = k$.

(b) The level curves $u = c$: $\left(x + \dfrac{1}{2(1 - c)}\right)^2 + y^2 = \left(\dfrac{1}{2(1 - c)}\right)^2$ represent a family of circles.   Centers: $\left(-\dfrac{1}{2(1 - c)}, 0\right)$, and radii: $\left|\dfrac{1}{2(1 - c)}\right|$.

The level curves $v = k$: $x^2 + \left(y + \dfrac{1}{2k}\right)^2 = \left(\dfrac{1}{2k}\right)^2$ represent a family of circles.   Centers: $\left(0, -\dfrac{1}{2k}\right)$ and radii: $\left|\dfrac{1}{2k}\right|$.

26. (a) The level curves $u = c$: $\dfrac{x^2 + y^2 - 1}{(x + 1)^2 + y^2} = c$; and $v = k$: $\dfrac{2y}{(x + 1)^2 + y^2} = k$.

(b) The level curves $u = c$: $\left(x - \dfrac{c}{1-c}\right)^2 + y^2 = \left(\dfrac{1}{1-c}\right)^2$ represent a family

of circles. Centers: $\left(\dfrac{c}{1-c}, 0\right)$ and radii: $\left|\dfrac{1}{1-c}\right|$. The level curves

$v = k$: $(x+1)^2 + \left(y - \dfrac{1}{k}\right)^2 = \left(\dfrac{1}{k}\right)^2$ represent a family of circles. Centers:

$\left(-1, \dfrac{1}{k}\right)$ and radii: $\left|\dfrac{1}{k}\right|$.

27. $v(x, y) = 3x^2 y - y^3 + C.$

Exercises 3.7, page 170

1. In view of Remark 3.7.3, we may omit the constant 2π from the integrand. Thus, $f = y\sqrt{1 + y'^2}$. The function f is explicitly independent of the independent variable x. Utilize Equation (3.7.15) with constant $= k$, $f = y\sqrt{1 + y'^2}$ and $\dfrac{\partial f}{\partial y'} = \dfrac{yy'}{\sqrt{1 + y'^2}}$. Show that

(1)
$$\frac{dy}{\sqrt{\left(\dfrac{y}{k}\right)^2 - 1}} = dx.$$

Show that the solution of Equation (1) is given by

(2)
$$y = k \cosh\left(\frac{x + c}{k}\right),$$

where the constants k and c are determined by the conditions that the curve passes through the points (x_1, y_1) and (x_2, y_2). Equation (2) is an arc of a catenary curve.

3. Utilize Equation (3.7.17) with constant $= k$, $f = x\sqrt{1 + y'^2}$, and $\dfrac{\partial f}{\partial y'} = \dfrac{xy'}{\sqrt{1 + y'^2}}$. Show that

(1)
$$y = k \cosh^{-1}\left(\frac{x}{k}\right) + c,$$

where the constants k and c are determined by the conditions that the curve passes through the points $(0, 0)$ and (x_2, y_2). Equation (1) is an arc of a catenary curve.

4. (a) Utilize Equation (3.7.15) with constant $= k$, $f = g(y)\sqrt{1+y'^2}$, and

$$\frac{\partial f}{\partial y'} = \frac{g(y)y'}{\sqrt{1+y'^2}}.$$

5. (a) Utilize Equation (3.7.17) with constant $= k$, $f = g(x)\sqrt{1+y'^2}$ and

$$\frac{\partial f}{\partial y'} = \frac{g(x)y'}{\sqrt{1+y'^2}}.$$

6. Utilize Equation (3.7.15) with $f = \sqrt{y}\,\sqrt{1-y'^2}$.

7. (a) Utilize Equation (3.7.2) with $f = py'^2 + 2qyy' + ry^2$.

8. Utilize Equation (3.7.2) with $f = p^2y'^2 + r^2y^2$. Show that $r^2y = \dfrac{d}{dx}(p^2y')$.

Thus, $r^2y^2 = y\dfrac{d}{dx}(p^2y')$. Substituting r^2y^2 in the given integral I, we obtain

$$I = \int_{x_1}^{x_2}\left[p^2y'^2 + y\frac{d}{dx}(p^2y')\right]dx = \int_{x_1}^{x_2}\frac{d}{dx}(p^2yy')dx = (p^2yy')\Big|_{x=x_1}^{x=x_2}.$$

Exercises 4.3, page 184

1. If the functions f and g are linearly dependent on I, then there exist constants c_1 and c_2, not both zero, such that $c_1f(x) + c_2g(x) = 0$ for every x in I. For definiteness, suppose that $c_2 \neq 0$. Thus, $g(x) = kf(x)$ for every x in I, where $k = -c_1/c_2$. Conversely, if $g(x) = kf(x)$ for every x in I, where k is a constant, then there exist constants c_1 and c_2, not both zero, namely, $c_1 = -k$ and $c_2 = 1$ such that $c_1f(x) + c_2g(x) = 0$ for every x in I. Thus, the functions f and g are linearly dependent on I.

2. If the functions v, u_1, u_2, $\ldots$, u_n are linearly dependent on I, then there exist constants c, c_1, c_2, $\ldots$, c_n, not all zero, such that $cv(x) + c_1u_1(x) + c_2u_2(x) + \cdots + c_nu_n(x) = 0$ for every x in I. If $c = 0$, then the functions u_1, u_2, $\ldots$, u_n would be linearly dependent on I, contrary to the hypothesis. Hence, $c \neq 0$, and therefore

$$v(x) = -\frac{c_1}{c}u_1(x) - \frac{c_2}{c}u_2(x) - \cdots - \frac{c_n}{c}u_n(x).$$

Thus, v can be written as a linear combination of u_1, u_2, $\ldots$, u_n. [The converse of this result is left for the reader.]

3. The functions given in (a), (b), (d), (e), (f), (h), (i), (l), (m), (p), and (r) are linearly independent. The functions given in (c), (g), (j), (k), (n), (o), and (q) are linearly dependent.

4. (a) Let u_{n_1}, u_{n_2}, $\ldots$, u_{n_k} be any subcollection of the set of functions $u_1, u_2, \ldots, u_n$. Since the functions $u_1, u_2, \ldots, u_k, u_{k+1}, \ldots, u_n$ are linearly independent on I, the only constants $c_1, c_2, \ldots, c_k, c_{k+1}, \ldots, c_n$, such that

$$c_1u_1(x) + c_2u_2(x) + \cdots + c_ku_k(x) + c_{k+1}u_{k+1}(x) + \cdots + c_nu_n(x) = 0$$

for every x in I, are the constants $c_1 = c_2 = \cdots = c_k = c_{k+1} = \cdots = c_n = 0$. Without any loss of generality, we may take $u_{n_1} = u_1,\ u_{n_2} = u_2,\ \ldots,\ u_{n_k} = u_k$. If the subcollection $u_1,\ u_2,\ \ldots,\ u_k$ were linearly dependent on I, so would the n functions $u_1,\ u_2,\ \ldots,\ u_n$ be linearly dependent on I, which is contrary to the hypothesis.

5. To prove this result, note that there exist constants $c_1,\ c_2,\ \ldots,\ c_n$, not all zero, such that $c_1 u_1(x) + c_2 u_2(x) + \cdots + c_n u_n(x) = 0$ for every x in I. If c_1, for example, is not zero, we have

$$u_1(x) = -\frac{c_2}{c_1}\, u_2(x) - \cdots - \frac{c_n}{c_1}\, u_n(x),$$

which expresses u_1 as a linear combination of the remaining functions.

6. The functions f and g defined by $f(x) = x|x|$ and $g(x) = x^2$ are linearly independent on the interval $I:\ -\infty < x < \infty$. However, they are linearly dependent on the interval $J:\ 0 \leq x < \infty,\ J \subset I$.

7. To prove this result, note that there exist constants c_1 and c_2, not both zero, such that $c_1 f(x) + c_2 g(x) = 0$ for every x in I. In particular, $c_1 f(x) + c_2 g(x) = 0$ for every x in $J \subset I$. Thus, the functions f and g are linearly dependent on J.

8. $A(c_1 f + c_2 g) = a_0(x)(c_1 f + c_2 g)^{(n)} + a_1(x)(c_1 f + c_2 g)^{(n-1)} + \cdots$

$\qquad\qquad + a_n(x)(c_1 f + c_2 g)$

$\qquad = a_0(x)(c_1 f^{(n)} + c_2 g^{(n)}) + a_1(x)(c_1 f^{(n-1)} + c_2 g^{(n-1)}) + \cdots$

$\qquad\qquad + a_n(x)(c_1 f + c_2 g)$

$\qquad = c_1[a_0(x)f^{(n)} + a_1(x)f^{(n-1)} + \cdots + a_n(x)f]$

$\qquad\qquad + c_2[a_0(x)g^{(n)} + a_1(x)g^{(n-1)} + \cdots + a_n(x)g]$

$\qquad = c_1 Af + c_2 Ag.$

[*Note:* This result may be obtained directly by observing that since each term in $L[y]$ is a linear operator, the sum is also a linear operator by an extension of Theorem 4.3.2.]

10. $A[c_1 f + c_2 g] = [c_1 f(t, y, z) + c_2 g(t, y, z)] + [c_1 f(x, t, z) + c_2 g(x, t, z)]$

$\qquad\qquad + [c_1 f(x, y, t) + c_2 g(x, y, t)]$

$\qquad = c_1[f(t, y, z) + f(x, t, z) + f(x, y, t)] + c_2[g(t, y, z)$

$\qquad\qquad + g(x, t, z) + g(x, y, t)]$

$\qquad = c_1 Af + c_2 Ag.$

12. $A(c_1 f + c_2 g) = \displaystyle\int_a^b G(x, t)[c_1 f(x, t) + c_2 g(x, t)]\,dt$

$\qquad = c_1 \displaystyle\int_a^b G(x, t)f(x, t)\,dt + c_2 \displaystyle\int_a^b G(x, t)g(x, t)\,dt$

$\qquad = c_1 Af + c_2 Ag.$

Exercises 4.4, page 197

1. (c) No.   (d) Yes.
2. (a) If the functions y_1 and y_2 are linearly dependent on I, then there exist constants c_1 and c_2, not both zero, such that

(1) $$c_1 y_1(x) + c_2 y_2(x) = 0$$

for every x in I. Also

(2) $$c_1 y_1'(x) + c_2 y_2'(x) = 0$$

for every x in I. Let x_0 be any fixed point in I. Thus, Equations (1) and (2) become

(3) $$\begin{cases} y_1(x_0)c_1 + y_2(x_0)c_2 = 0, \\ y_1'(x_0)c_1 + y_2'(x_0)c_2 = 0. \end{cases}$$

Since the homogeneous linear system (3) has a nontrivial solution $(c_1, c_2) \neq (0, 0)$ it follows that [*see* Theorem II]

$$W(y_1, y_2; x_0) = \begin{vmatrix} y_1(x_0) & y_2(x_0) \\ y_1'(x_0) & y_2'(x_0) \end{vmatrix} = 0.$$

Since x_0 was any fixed point in I, it follows that $W(y_1, y_2; x) = 0$ for every x in I.
   (b) Observe that

$$\frac{d}{dx}\left[\frac{y_2(x)}{y_1(x)}\right] = \frac{y_1(x)y_2'(x) - y_2(x)y_1'(x)}{y_1{}^2(x)} = \frac{0}{y_1{}^2(x)} = 0.$$

Thus,

$$\frac{y_2(x)}{y_1(x)} = c \quad \text{or} \quad y_2(x) = cy_1(x),$$

where c is a constant. The result now follows from Exercise 4.3.1.
   (c) The proof of this result is similar to that given for the sufficiency proof of Theorem 4.4.3.
3. (a) $y = c_1 + c_2 x.$ (b) $y = c_1 e^{\alpha x} + c_2 e^{-\alpha x}.$ (c) $y = c_1 \cos \beta x + c_2 \sin \beta x.$
   (d) $y = (c_1 + c_2 x)e^{\alpha x}.$ (e) $y = c_1 e^{\alpha x} + c_2 e^{\beta x}.$
   (f) $y = (c_1 \cos \beta x + c_2 \sin \beta x)e^{\alpha x}.$
4. (a) $y = c_1 x^3 + c_2 x^5.$ (b) $y = c_1 x + c_2 x e^x.$ (c) $y = c_1 x e^x + c_2 x e^{-x}.$

5. (a) $$W(f, g; x) = \begin{vmatrix} a_1 y_1(x) + a_2 y_2(x) & b_1 y_1(x) + b_2 y_2(x) \\ a_1 y_1'(x) + a_2 y_2'(x) & b_1 y_1'(x) + b_2 y_2'(x) \end{vmatrix}$$
$$= (a_1 b_2 - a_2 b_1)[y_1(x)y_2'(x) - y_2(x)y_1'(x)]$$
$$= \Delta W(y_1, y_2; x).$$

(b) Suppose to the contrary that the functions f and g are linearly dependent on I.   By Exercise 4.3.1, we then have that $f(x) = kg(x)$ for all x in I, where k is a constant.   Thus, $(a_1 - kb_1)y_1(x) + (a_2 - kb_2)y_2(x) = 0$ for all x in I. Since the functions y_1 and y_2 are linearly independent on I, we then must have that $a_1 - kb_1 = 0$ and $a_2 - kb_2 = 0$. Thus, $\dfrac{a_1}{b_1} = k = \dfrac{a_2}{b_2}$.   Hence, $a_1 b_2 - a_2 b_1 = 0$, which contradicts the hypothesis that $\Delta \neq 0$.   Consequently, the functions f and g are linearly independent on I.

6. (a) Linearly independent.   (b) Linearly independent.
   (c) Linearly dependent.   (d) Linearly independent.
7. (a) $y = 11e^{3x} - 5e^{5x}$.   (b) $y = -6e^{3x} + 4e^{5x}$.
8. $y = 5 \cos (2 \ln x) + 2 \sin (2 \ln x)$.
9. From Definition 4.4.2, we have

$$\begin{cases} y_1(x_0) = 1, & y_1'(x_0) = 0, \\ y_2(x_0) = 0, & y_2'(x_0) = 1. \end{cases}$$

Thus, the Wronskian of the functions y_1 and y_2 at x_0 is different from zero, that is,

$$W(y_1, y_2; x_0) = \begin{vmatrix} y_1(x_0) & y_2(x_0) \\ y_1'(x_0) & y_2'(x_0) \end{vmatrix} = \begin{vmatrix} 1 & 0 \\ 0 & 1 \end{vmatrix} = 1 \neq 0.$$

Concerning (i).   If $y_1(x_0) = 0$ and $y_2(x_0) = 0$, then

$$W(y_1, y_2; x_0) = \begin{vmatrix} y_1(x_0) & y_2(x_0) \\ y_1'(x_0) & y_2'(x_0) \end{vmatrix} = \begin{vmatrix} 0 & 0 \\ y_1'(x_0) & y_2'(x_0) \end{vmatrix} = 0.$$

Concerning (ii).   Suppose that $y_1(x)$ and $y_2(x)$ have a maximum or a minimum at x_0.   Then $y_1'(x_0) = 0$ and $y_2'(x_0) = 0$.   Thus,

$$W(y_1, y_2; x_0) = \begin{vmatrix} y_1(x_0) & y_2(x_0) \\ y_1'(x_0) & y_2'(x_0) \end{vmatrix} = \begin{vmatrix} y_1(x_0) & y_2(x_0) \\ 0 & 0 \end{vmatrix} = 0.$$

10. We shall give the sufficiency proof, that is, if $G = 0$ on I, then the functions y_1 and y_2 are linearly dependent on I.   Since G, a determinant with constant terms, is equal to zero, we can choose constants c_1 and c_2, not both zero, such that [*see* Theorem II]

(1)
$$\left\{ c_1 \int_a^b y_1{}^2(x)dx + c_2 \int_a^b y_1(x)y_2(x)dx = 0, \right.$$

(2)
$$\left. c_1 \int_a^b y_2(x)y_1(x)dx + c_2 \int_a^b y_2{}^2(x)dx = 0. \right.$$

Multiplying (1) and (2) by c_1 and c_2, respectively, and then adding, we obtain

(3)
$$\int_a^b [c_1 y_1(x) + c_2 y_2(x)]^2 \, dx = 0.$$

From (3), we may now infer that

(4)
$$c_1 y_1(x) + c_2 y_2(x) = 0$$

for every x in I. (Why?) Since c_1 and c_2 are not both zero, Equation (4) asserts that the functions y_1 and y_2 are linearly dependent on I.

11. (a) Separate the variables in (4.4.56), and then integrate from x_0 to x.
    (b) $3/x^5$.

12. (c) The graphs of the functions y_1 and y_2 are the same for $x \geq 0$; however, for $x < 0$, the graph of the function y_2 is obtained by reflecting in the x-axis the graph of the function y_1.

Exercises 4.5, page 204

2. $y_2 = xe^{-x}$.   3. $y_2 = x^5$.   4. $y_2 = \dfrac{1}{x-1}$.

5. $y_2 = \dfrac{1}{x}$.   6. $y_2 = 1 - \dfrac{\sqrt{1-x^2}}{x}\,\text{arc sin } x$.

7. (b) $y_2 = \dfrac{1}{2}\ln\left(\dfrac{1+x}{1-x}\right)$, $n=0$.

$y_2 = \dfrac{x}{2}\ln\left(\dfrac{1+x}{1-x}\right) - 1$, $n=1$.

$y_2 = \dfrac{3x^2-1}{4}\ln\left(\dfrac{1+x}{1-x}\right) - \dfrac{3}{2}x$, $n=2$.

$y_2 = \dfrac{5x^3-3x}{4}\ln\left(\dfrac{1+x}{1-x}\right) - \dfrac{5}{2}x^2 + \dfrac{2}{3}$, $n=3$.

(c) Yes. [*Note:* $P_2(x) = \frac{1}{2}[3xP_1(x) - P_0(x)]$, $P_3(x) = \frac{1}{3}[5xP_2(x) - 2P_1(x)]$ and $P_4(x) = \frac{1}{4}[7xP_3(x) - 3P_2(x)]$.]

(h) $C_0 = 8$, $C_1 = 32$, $C_2 = 16$, $C_3 = 16$, and $C_4 = 8$.

8. (c) Yes. [*Note:* $J_{3/2}(x) = \dfrac{1}{x}J_{1/2}(x) - J_{-1/2}(x)$.]

Exercises 4.6, page 214

2. Verification of $e^{i\theta_1} \cdot e^{i\theta_2} = e^{i(\theta_1+\theta_2)}$.

$$e^{i\theta_1} \cdot e^{i\theta_2} = (\cos\theta_1 + i\sin\theta_1)(\cos\theta_2 + i\sin\theta_2)$$
$$= \cos\theta_1\cos\theta_2 - \sin\theta_1\sin\theta_2 + i(\sin\theta_2\cos\theta_1 + \sin\theta_1\cos\theta_2)$$
$$= \cos(\theta_1 + \theta_2) + i\sin(\theta_1 + \theta_2) = e^{i(\theta_1+\theta_2)}.$$

3. (a) Note that $e^{\beta x} = \cosh\beta x + \sinh\beta x$ and $e^{-\beta x} = \cosh\beta x - \sinh\beta x$. Thus,

$$y(x) = c_1 e^{\beta x} + c_2 e^{-\beta x} = (c_1 + c_2)\cosh\beta x + (c_1 - c_2)\sinh\beta x.$$

Letting $A = c_1 + c_2$ and $B = c_1 - c_2$, we obtain the desired result.

4. Find the general solution of Equation (1) when $\lambda = 0$. Does this solution agree with that given by Equation (3) as $\lambda \to 0$ or by Equation (4) when $\lambda = 0$?

5. $y = e^{-x} + e^{x-2}$.

6. $y = \frac{12}{17}\left(e^{(2/3)x} - e^{-(3/4)x}\right)$.

7. $y = e^{x/2}(4 - x)$.

8. $y = -\frac{1}{2}\cos x + \frac{3}{2}\sin x$.

9. $y = e^{3x}(\cos 4x - \sin 4x)$.

10. $y = e^{\sqrt{3}x}(2\sqrt{3}\cos 2x - 4\sin 2x)$.

11. $y(x) = 0$ (trivial solution).

12. $x = 4.582e^{-1.127t} - 0.582e^{-8.873t}$.

13. $y_1 = \dfrac{\sqrt{2}}{2}(\cos 2x + \sin 2x), \quad y_2 = \dfrac{\sqrt{2}}{4}(-\cos 2x + \sin 2x)$.

15. (c) $Q = \dfrac{\sqrt{2}}{20}e^{-50t}\cos(50t - \phi)$, where $\phi = \tan^{-1}1 = \dfrac{\pi}{4}$.

18. (b) $\dfrac{\sqrt{3}}{2} + \dfrac{i}{2}$,   (c) $\dfrac{1}{2} - i\dfrac{\sqrt{3}}{2}$,   (d) i,   (e) $-\dfrac{\sqrt{3}}{2} + \dfrac{i}{2}$,   (f) -1,

(g) $-i$,   (h) $0.9234 + 0.3827i$,   (i) $0.3827 - 0.9234i$.

19. (a) $y = c_1 e^{m_1 x} + c_2 e^{m_2 x}$, where $m_1 = \dfrac{1}{2}(2 + \sqrt{6}) + i\dfrac{\sqrt{2}}{2}$, $m_2 = \dfrac{1}{2}(2 - \sqrt{6}) - i\dfrac{\sqrt{2}}{2}$.

(b) $y = c_1 e^{m_1 x} + c_2 e^{m_2 x}$, where $m_1 = \dfrac{\sqrt{2}}{2} - \dfrac{i}{2}(2 - \sqrt{2})$, $m_2 = -\dfrac{\sqrt{2}}{2} - \dfrac{i}{2}(2 + \sqrt{2})$.

(c) $y = (c_1 + c_2 x)e^{ix}$.

(d) $y = c_1 e^{m_1 x} + c_2 e^{m_2 x}$, where $m_1 = i + \sqrt[4]{2}\,e^{i(3/8)\pi}$, $m_2 = i - \sqrt[4]{2}\,e^{i(3/8)\pi}$.

(e) $y = c_1 e^{m_1 x} + c_2 e^{m_2 x}$, where $m_1 = i + \sqrt[4]{2}\,e^{i\pi/8}$, $m_2 = i - \sqrt[4]{2}\,e^{i\pi/8}$.

(f) $y = (c_1 + c_2 x)e^{mx}$, where $m = 2 + i\sqrt{3}$.

(g) $y = c_1 e^{m_1 x} + c_2 e^{m_2 x}$, where $m_1 = \sqrt{3} + i$, $m_2 = 1 + i$.

(h) $y = c_1 e^{m_1 x} + c_2 e^{m_2 x}$, where $m_1 = e^{i(3/4)\pi}$, $m_2 = -e^{i(3/4)\pi}$.

(i) The functions u and v individually do not satisfy the equation $L[y] = 0$.

Exercises 4.7, page 221

1. Utilize Theorem 4.7.3.

4. $y = c_1 e^x + c_2 e^{-x} - 4x^2 - 8 + 2\sin 2x + (-4 + 3x)e^{2x}$.

5. Suppose that $y(x) = c_1 y_1(x) + c_2 y_2(x)$ is a solution of $L[y] = F(x)$. Then

$$\begin{aligned} F(x) &= L[c_1 y_1 + c_2 y_2] = c_1 L[y_1] + c_2 L[y_2] \\ &= c_1 F(x) + c_2 F(x) = (c_1 + c_2)F(x). \end{aligned}$$

Thus, $c_1 + c_2 = 1$. This relation is satisfied if we take $c_1 = \frac{1}{2} + c$ and $c_2 = \frac{1}{2} - c$, where c is an arbitrary constant. Hence

$$y(x) = (\tfrac{1}{2} + c)y_1(x) + (\tfrac{1}{2} - c)y_2(x).$$

6. (a) Since $y_p(x) = u(x) + iv(x)$ is a solution of $L[y] = U(x) + iV(x)$, we have

$$L[y_p] = P(x)u'' + Q(x)u' + R(x)u + i[P(x)v'' + Q(x)v' + R(x)v]$$
$$= U(x) + iV(x).$$

Also, $L[y_p] = L[u + iv] = L[u] + iL[v]$. Thus,

$$L[u] = P(x)u'' + Q(x)u' + R(x)u = U(x)$$

and $\qquad L[v] = P(x)v'' + Q(x)v' + R(x)v = V(x).$

Hence, $\mathscr{R}[y_p(x)]$ and $\mathscr{I}[y_p(x)]$ are solutions of $L[y] = U(x)$ and $L[y] = V(x)$, respectively.

(b) Yes.

Exercises 4.8, page 237

1. We shall establish Formula (4.8.67). From Case 2(i), a particular solution $y_p(x)$ of Equation (4.8.65) has the form $y_p(x) = B_0 e^{\alpha x}$. Thus, $y_p'(x) = B_0 \alpha e^{\alpha x}$ and $y_p''(x) = B_0 \alpha^2 e^{\alpha x}$. Substituting y_p, y_p', and y_p'' in (4.8.65), we obtain $B_0(a\alpha^2 + b\alpha + c) = k$. Utilizing (4.8.66), $B_0 = k/p(\alpha)$ if $p(\alpha) \neq 0$. Hence,

$$y_p(x) = \frac{k}{p(\alpha)}\, e^{\alpha x}, \quad \text{if } p(\alpha) \neq 0.$$

3. We shall establish Formula (4.8.86). Consider the differential equation

$$(1) \qquad\qquad L[w] = aw'' + bw' + cw = ke^{(\alpha + i\beta)x},$$

where $w = y + iv$. Note that $\mathscr{R}\{L[w]\} = ke^{\alpha x}\cos\beta x$ and $\mathscr{R}(w) = y$. Let $w_p = y_p + iv_p$ be a particular solution of Equation (1). Note that $\mathscr{R}(w_p) = y_p$. Utilizing Formula (4.8.67), we then have that

$$w_p = \frac{ke^{(\alpha + i\beta)x}}{p(\alpha + i\beta)}, \quad \text{if } p(\alpha + i\beta) \neq 0.$$

Thus, a particular solution $y_p(x)$ of Equation (4.8.85) is given by [*see* Exercise 4.7.6]

$$y_p(x) = \mathscr{R}(w_p) = \mathscr{R}\left[\frac{ke^{(\alpha + i\beta)x}}{p(\alpha + i\beta)}\right], \quad \text{if } p(\alpha + i\beta) \neq 0.$$

4. We shall establish Formula (4.8.90). Consider the differential equation

$$(1) \qquad\qquad L[w] = aw'' + bw' + cw = ke^{(\alpha + i\beta)x},$$

where $w = u + iy$. Note that $\mathcal{I}\{L[w]\} = ke^{\alpha x} \sin \beta x$ and $\mathcal{I}(w) = y$. Let $w_p = u_p + iy_p$ be a particular solution of Equation (1). Note that $\mathcal{I}(w_p) = y_p$. Utilizing Formula (4.8.68), we then have that

$$w_p = \frac{kxe^{(\alpha + i\beta)x}}{p'(\alpha + i\beta)}, \quad \text{if } p(\alpha + i\beta) = 0.$$

Thus, a particular solution $y_p(x)$ of Equation (4.8.88) is given by [*see* Exercise 4.7.6]

$$y_p(x) = \mathcal{I}(w_p) = \mathcal{I}\left[\frac{kxe^{(\alpha + i\beta)x}}{p'(\alpha + i\beta)}\right], \quad \text{if } p(\alpha + i\beta) = 0.$$

5. We shall establish Formula (4.8.92). From Case 3(i), a particular solution $y_p(x)$ of Equation (4.8.91) has the form

(1) $$y_p(x) = B_0 \cos \beta x + C_0 \sin \beta x$$

if $p(\alpha + i\beta) = p(i\beta) = -\beta^2 + \lambda^2 \neq 0$, that is, if $\beta \neq \pm\lambda$. Thus,

$$y_p'(x) = -B_0 \beta \sin \beta x + C_0 \beta \cos \beta x$$

and

$$y_p''(x) = -B_0 \beta^2 \cos \beta x - C_0 \beta^2 \sin \beta x.$$

Substituting y_p and y_p'' in (4.8.91), we obtain

(2) $$B_0(\lambda^2 - \beta^2) \cos \beta x + C_0(\lambda^2 - \beta^2) \sin \beta x = k \cos \beta x.$$

Equation (2) will be an identity provided that

(3) $$B_0(\lambda^2 - \beta^2) = k \quad \text{and} \quad C_0(\lambda^2 - \beta^2) = 0.$$

Since $\beta \neq \pm\lambda$, (3) implies that

$$B_0 = \frac{k}{\lambda^2 - \beta^2} \quad \text{and} \quad C_0 = 0.$$

Substituting these values in (1), we obtain Formula (4.8.92).

6. $y = c_1 e^{3x} + c_2 e^{2x} + 2x + 3.$

7. $y = c_1 e^{2x} + c_2 e^{-2x} - 2e^x - 2x.$

8. $y = c_1 e^{-3x} + c_2 e^{-x} - \frac{9}{5} \cos x + \frac{12}{5} \sin x + 3x^2 - 8x - 2.$

9. $y = c_1 + c_2 e^{3x} + (3x^2 - 2x)e^{3x} + x^3 + x^2 + \frac{2}{3}x.$

10. $y = c_1 \cos x + c_2 \sin x + x \cos x + x^2 \sin x.$

11. $y = c_1 \cos 2x + c_2 \sin 2x - 2x \cos 2x.$

12. $y = c_1 e^{3x} + c_2 e^{-3x} - (6 \cos x + \sin x)e^{3x}.$

13. $y = c_1 \cos 2x + c_2 \sin 2x + x \sin 2x - 2x^2 \cos 2x.$

14. $y = c_1 \cos 4x + c_2 \sin 4x - 2x \cos 4x + x \sin 4x.$

15. $y = (c_1 \cos x + c_2 \sin x + \frac{1}{3} \cos 2x + 1)e^{2x}.$

16. $y = (c_1 \cos x + c_2 \sin x)e^{2x} + 5x \sin xe^{2x} + \frac{1}{4}(2 \cos x - \sin x)e^{-2x}.$

17. $y = (c_1 \cos x + c_2 \sin x + x \sin x - x^2 \cos x)e^x.$

18. $y = c_1 \cos 2x + c_2 \sin 2x + (\sin 2x - 2 \cos 2x)e^{2x} + 2x \sin 2x.$

19. $y = c_1 + c_2 e^{-x} + xe^{-x} + x^4 + x^3 + 2x^2 - 4x.$

20. $y = (c_1 + c_2 x + x^4 - x^3 + x^2)e^{-x} + x^4 + x^2 + x - 1.$

21. $y = (-3 \cos x + 4 \sin x)e^{-x} + 3 \cos 2x - 6 \sin 2x$.

22. $y = 3e^{5x} - 10e^{2x} + 10x + 7$.

23. $y = 6e^{-x} + (2 \cos 3x + 3 \sin 3x - 4)e^x$.

24. $y = 2(1 + x + x^4)e^x$.

25. $y = (x - \pi) \cos 2x + (2x^2 - \frac{1}{2}) \sin 2x$.

26. $y = (\cos x + x \cos x + 2 \sin x + x^2 \sin x)e^{-x}$.

28. (a) $y = 2 \cos 3x - \cos 6x$.

Exercises 4.9, page 244

1. $y = c_1 \cos x + c_2 \sin x + x \sin x + \cos x \ln \cos x$.

2. $y = c_1 e^x + c_2 e^{-x} - \frac{1}{10} \cos 2x - \frac{1}{2}$.

3. $y = (c_1 + c_2 x - \ln \cos x)e^{2x}$.

4. $y = \left[c_1 + c_2 x + \dfrac{x^{n+2}}{(n+1)(n+2)} \right] e^{2x}, \qquad n \neq -1, \quad n \neq -2$.

5. $y = [c_1 + c_2 x + \frac{1}{2}x^2(\ln x - \frac{3}{2})]e^{-x}$.

6. $y = c_1 e^{-x} + c_2 e^{-3x} + e^{-3x} \sin e^x - e^{-2x} \cos e^x$.

7. $y = c_1 e^x + c_2 e^{-x} + xe^x - \cosh x \ln (e^{2x} + 1)$.

8. $y = (c_1 + c_2 x - \ln x)e^{-x}$.

9. $y = c_1 e^x + c_2 e^{-x} + \frac{1}{3} \cosh 2x - \sin e^{-x} - e^x \cos e^{-x}$.

10. $y = [c_1 \cos x + c_2 \sin x + \sin x \ln (\sec x + \tan x) - 2]e^x$.

11. $y = \dfrac{e^x}{2} \ln \dfrac{2}{e^{2x} + 1} + e^{2x} \arctan e^x$.

12. $y = c_1(x^2 - 1) + c_2 x + \frac{1}{6}x^4 + \frac{1}{2}x^2$.

13. $y = [c_1 + c_2 x^2 + \frac{2}{9}x^3(3 \ln x - 4)]e^x$.

14. (a) This exercise is a special case of Exercise 15(a). [*See* hint to Exercise 15(a).]

    (b) $y = e^x - x - 1$.

15. (a) Utilizing Equations (4.9.16) and (4.9.15), the general solution of

$$(1) \qquad L[y] = P(x)y'' + Q(x)y' + R(x)y = F(x)$$

may be expressed as

$$(2) \qquad y(x) = c_1 y_1(x) + c_2 y_2(x) - y_1(x) \int_a^x G_1(t)\,dt + y_2(x) \int_a^x G_2(t)\,dt,$$

where

$$(3) \qquad G_1(t) = \frac{y_2(t)F(t)}{P(t)W(y_1, y_2; t)}, \qquad G_2(t) = \frac{y_1(t)F(t)}{P(t)W(y_1, y_2; t)}.$$

We may also write (2) as

$$(4) \qquad y(x) = c_1 y_1(x) + c_2 y_2(x) - \int_a^x y_1(x)G_1(t)\,dt + \int_a^x y_2(x)G_2(t)\,dt.$$

Thus,

$$(5) \qquad y'(x) = c_1 y_1'(x) + c_2 y_2'(x) - \int_a^x y_1'(x)G_1(t)dt + \int_a^x y_2'(x)G_2(t)dt.$$

(Why?)

Since $y(x_0) = 0$ and $y'(x_0) = 0$, Equations (4) and (5) give

$$(6) \qquad y_1(x_0)c_1 + y_2(x_0)c_2 = \int_a^{x_0} y_1(x_0)G_1(t)dt - \int_a^{x_0} y_2(x_0)G_2(t)dt,$$

$$(7) \qquad y_1'(x_0)c_1 + y_2'(x_0)c_2 = \int_a^{x_0} y_1'(x_0)G_1(t)dt - \int_a^{x_0} y_2'(x_0)G_2(t)dt.$$

Solving Equations (6) and (7) simultaneously for c_1 and c_2, we find

$$(8) \qquad\qquad c_1 = \int_a^{x_0} G_1(t)dt \quad \text{and} \quad c_2 = -\int_a^{x_0} G_2(t)dt.$$

Substituting these values of c_1 and c_2 in (2), we obtain

$$(9) \quad y(x) = y_1(x)\int_a^{x_0} G_1(t)dt - y_2(x)\int_a^{x_0} G_2(t)dt - \int_a^x y_1(x)G_1(t)dt$$

$$+ \int_a^x y_2(x)G_2(t)dt = -\int_{x_0}^x y_1(x)G_1(t)dt + \int_{x_0}^x y_2(x)G_2(t)dt$$

$$= \int_{x_0}^x [y_2(x)G_2(t) - y_1(x)G_1(t)]dt.$$

In view of (3), we may write (9) as

$$(10) \qquad\qquad y(x) = \int_{x_0}^x \frac{y_2(x)y_1(t) - y_1(x)y_2(t)}{P(t)W(y_1, y_2; t)} F(t)dt.$$

(b) $y = \dfrac{e^x}{9}(6x^3 \ln x - 8x^3 + 9x^2 - 1)$.

16. (a) Follow a similar method as given in the hint to Exercise 15(a).

(b) $y = \dfrac{e^x}{9}\left[(6 \ln x - 8)x^3 + \dfrac{9}{e}(1 + e)x^2 + \dfrac{9 - e}{e}\right]$.

17. $y = \dfrac{3}{4}x^3 + \dfrac{1}{2}x^2 + \dfrac{1}{2}x + \dfrac{1}{2} - \dfrac{1}{4x}$.

18. $y = 3(x^2 - 2x + 2)e^{2x}$.

Exercises 4.10, page 250

1. $y = x^{-2}[c_1 \cos(3 \ln x) + c_2 \sin(3 \ln x)]$.
2. $y = c_1 \cos(2 \ln x) + c_2 \sin(2 \ln x)$.

3. $y = x^{3/2}(c_1 + c_2 \ln x).$
4. $y = c_1 + c_2 x^4 - \frac{8}{3}x.$
5. $y = c_1 x^2 + c_2 x^3 + x^2 \ln x + 2.$
6. $y = (c_1 + c_2 \ln x + 2 \ln^2 x)x^{-3}.$
7. $y = c_1 x + 1 + (c_2 x + 1 + \ln x) \ln x.$
8. $y = c_1 x + c_2 x^2 + \frac{1}{6}x^2 \ln x (\ln^2 x - 3 \ln x + 6).$

9. $y = c_1 x + \dfrac{c_2}{x} - \cosh (\ln x) \, \text{arc tan} \, x - \dfrac{1}{2}.$

10. $y = x[c_1 + c_2 \ln x + \ln x \, \text{arc tan} \, (\ln x) - \frac{1}{2} \ln (1 + \ln^2 x)].$

11. (1) $$a(\alpha + \beta x)^2 y'' + b(\alpha + \beta x)y' + cy = F(x).$$

Since

$$\alpha + \beta x = e^u \quad \text{or} \quad u = \ln (\alpha + \beta x),$$

we see that

$$\frac{dy}{dx} = \frac{dy}{du} \cdot \frac{du}{dx} = \beta e^{-u} \frac{dy}{du}$$

and

$$\frac{d^2 y}{dx^2} = \frac{d}{du}\left(\frac{dy}{dx}\right)\frac{du}{dx} = \beta^2 e^{-2u}\left(\frac{d^2 y}{du^2} - \frac{dy}{du}\right).$$

Substituting dy/dx and $d^2 y/dx^2$ in Equation (1), we obtain the desired result.
12. $y = (2 + 3x)[c_1 + c_2 \ln (2 + 3x) + \frac{3}{2} \ln^2 (2 + 3x)] - 6.$

13. $y = 3(x + 1) - \dfrac{2}{x + 1} - 2 \ln (x + 1).$

14. $y = \cos (\ln x) + \frac{1}{2}(1 + \ln x) \sin (\ln x).$

15. (a) (1) $$P(x)y'' + Q(x)y' + R(x)y = F(x).$$

If Equation (1) is exact, then

$$\frac{d}{dx}[M(x)y' + N(x)y] = F(x),$$

or

(2) $$M(x)y'' + [M'(x) + N(x)]y' + N'(x)y = F(x).$$

In view of (1) and (2), we then have that

$$M(x) = P(x), \quad M'(x) + N(x) = Q(x), \quad \text{and} \quad N'(x) = R(x).$$

Thus,

(3) $$M(x) = P(x) \quad \text{and} \quad N(x) = Q(x) - P'(x).$$

(b) For Equation (1) to be exact, we must have that

$$M(x) = P(x), \quad N(x) = Q(x) - P'(x), \quad \text{and} \quad N'(x) = R(x).$$

Thus,

$$(4) \qquad 0 = P''(x) - Q'(x) + N'(x) = P''(x) - Q'(x) + R(x).$$

(c) Exact. $(x^2 + 2x)y = c_1 + c_2 x - \sin x + \dfrac{x^2}{2}\left(\ln x - \dfrac{3}{2}\right).$

(d) Exact. $y = \dfrac{c_1}{x^2} + \dfrac{c_2}{x} + x + \dfrac{1}{x}\ln x.$

(e) Not exact.

Exercises 5.3, page 261

1. (a) $x = \dfrac{1}{2}\cos 8t.$ (b) $\dfrac{1}{2}$ ft, $\dfrac{\pi}{4}$ sec, $\dfrac{4}{\pi}$ cycles/sec.

(c) -0.073 ft, -3.957 ft/sec, 4.656 ft/sec².

2. (a) $x = -\dfrac{1}{2}\cos 2t.$ (b) $\dfrac{1}{2}$ ft, π sec, $\dfrac{1}{\pi}$ cycle/sec.   (c) $\dfrac{1}{2}n\pi$ sec,

$n = 1, 2, \ldots,$ $\frac{1}{4}(2n + 1)\pi$ sec, $n = 0, 1, \ldots$.

3. (a) 28.28 cm/sec, -10 cm/sec².   (b) $(2n + \frac{1}{6})\pi$ sec, $n = 0, 1, \ldots$.

4. 25 lb.

6. (a) $x = \cos\left(8t - \dfrac{5}{6}\pi\right).$ (b) 1 ft, $\dfrac{\pi}{4}$ sec, $\dfrac{4}{\pi}$ cycles/sec.

(c) $\frac{1}{16}(4n + 3)\pi$ sec, $n = 0, 1, \ldots$.

7. $\dfrac{7\pi}{48}$ sec.

8. 0.233 sec, 0.555 sec.

9. (a) 0.314 ft,   (b) -0.893 ft/sec,   (c) -20.10 ft/sec².

10. (b) $\theta = \dfrac{\sqrt{2}}{16}\cos\left(4t + \dfrac{\pi}{4}\right),$ $\dfrac{\sqrt{2}}{16}$ rad, $\dfrac{\pi}{2}$ sec, $\dfrac{2}{\pi}$ cycle/sec, $-\dfrac{\pi}{4}$ rad.

(c) $\dfrac{\sqrt{2}}{2}\pi$ sec, $\pm\dfrac{\sqrt{2}}{8}$ rad/sec.   (d) ± 0.445 ft/sec.

(e) $\frac{1}{32}$ rad, $\frac{1}{8}(2n + 1)\pi$ sec, $n = 0, 1, \ldots$.   (f) 89 times. (g) $\dfrac{T}{12}.$

12. (a) 2.815 ft.   (b) 3.242 ft.

Exercises 5.4, page 277

1. -13.10 ft/sec^2.   2. 76.10 percent.

3. (a) $x = \dfrac{\sqrt{2}}{4}\, e^{-8\sqrt{2}\,t} \cos\left(8\sqrt{2}\,t - \dfrac{\pi}{4}\right)$.   (b) $\dfrac{\sqrt{2}}{4}\, e^{-8\sqrt{2}\,t}$ ft, $\dfrac{\sqrt{2}}{8}\,\pi$ sec,

$\dfrac{4\sqrt{2}}{\pi}$ cycles/sec.

   (c) 0.052 ft, -1.176 ft/sec, 13.312 ft/sec^2.   (d) 0.0613 sec.
4. (a) Yes.   (b) $x = \frac{1}{4}(1 + 16t)e^{-8t}$.   (c) 0.303 ft.
   (d) 0.273 sec.   (e) $\frac{3}{16}$ sec.
5. (a) Yes.   (b) $x = \frac{7}{4}e^{-t} - \frac{5}{4}e^{-3t}$.   (c) 0.797 ft.   (d) 0.93 sec.
   (e) 0.057 sec, 0.96 sec.
6. (a) -4.54 ft/sec.   (b) -3 ft/sec.
7. 0.285 sec.   8. 1.196 ft. $K = 0.59$ lb sec/ft.

Exercises 5.5, page 285

1. (b) $x = 2e^{-t} - \dfrac{1}{2}\, e^{-3t} + \dfrac{\sqrt{5}}{2}\, \sin\,(t - 1.107)$.

2. $x = \dfrac{13}{52}\, e^{-t} + \dfrac{17}{52}\, e^{-3t} + \dfrac{\sqrt{65}}{13}\, \cos\,(2t - 1.695)$.

3. $x = \frac{3}{2}e^{-2t} - e^{-3t} - \frac{1}{2}$.
4. (a) $x = (1 + 2\sqrt{g}\,t)e^{-2\sqrt{g}\,t} - \cos 2\sqrt{g}\,t$.

   (b) $\dfrac{1}{2\sqrt{g}}\,(2n + 1)\pi$ sec, $n = 0, 1, \ldots$.   (c) 0.736 ft.

5. $x = \dfrac{\sqrt{10}}{5}\, e^{-2t} \cos\,(2t - 1.249) + \dfrac{\sqrt{5}}{5}\, \cos\,(4t - 2.035)$.

6. -0.4 ft.

Exercises 5.6, page 291

1. (a) $x = 2.286 \sin 0.165\,t \sin 2.665\,t$.
   (b) $x = 2.424 \sin 0.165\,t \sin 2.5\,t$.
2. (a) $x = e^{-2t} \cos 2\sqrt{2}\,t + \sin 2t - \cos 2t$.

   (b) $2\sqrt{3}$ rad/sec. $x = \dfrac{2}{3} \sin 2\sqrt{3}\,t - \dfrac{4\sqrt{3}}{3}\,t \cos 2\sqrt{3}\,t$.

3. (a) $0 \leq K < 4.$ (c) $2\sqrt{2}$ rad/sec.   (d) $\sqrt{6}$ rad/sec.

(e) $\dfrac{\sqrt{30}}{4\pi}$ cycle/sec.   (f) $x = \dfrac{\sqrt{6}}{7} e^{-t} \cos \sqrt{7}\, t + \dfrac{1}{7} (\sin \sqrt{6}\, t - \sqrt{6} \cos \sqrt{6}\, t).$

Exercises 5.7, page 299

4. (a) $\theta = \dfrac{\pi}{2} e^{-t/2} \left(\cos \dfrac{t}{2} + \sin \dfrac{t}{2} \right).$ (b) 1.386 rad.

5. (a) $\frac{1}{5}$ sec.   (b) 0.368 coulomb.

6. (a) $Q(t) = Q_0 \cos \sqrt{\dfrac{1}{LC}}\, t$

$$+ \sqrt{LC} \left(I_0 - \dfrac{E_0 C\omega}{1 - LC\omega^2} \right) \sin \sqrt{\dfrac{1}{LC}}\, t + \dfrac{E_0 C}{1 - LC\omega^2} \sin \omega t,$$

$$I(t) = -\dfrac{Q_0}{\sqrt{LC}} \sin \sqrt{\dfrac{1}{LC}}\, t$$

$$+ \left(I_0 - \dfrac{E_0 C\omega}{1 - LC\omega^2} \right) \cos \sqrt{\dfrac{1}{LC}}\, t + \dfrac{E_0 C\omega}{1 - LC\omega^2} \cos \omega t.$$

(b) $Q(t) = \left(Q_0 - \dfrac{E_0}{2L\omega} t \right) \cos \omega t + \dfrac{1}{\omega} \left(I_0 + \dfrac{E_0}{2L\omega} \right) \sin \omega t,$

$$I(t) = I_0 \cos \omega t - \omega \left(Q_0 - \dfrac{E_0}{2L\omega} t \right) \sin \omega t.$$

(c) The network will break down.
(d) 7.224 coulombs,   4.366 amperes.

7. (a) $Q(t) = 6 - 6e^{-5t} (\cos 5t + \sin 5t),$
    $I(t) = 60e^{-5t} \sin 5t.$
(c) 6.259 coulombs,   19.341 amperes.

9. (a) $I_s(t) = 4.851 \cos (50t - 1.816) + 4.472 \cos (100t - 2.68).$
(c) 0.736 ampere.

10. (a) $I(t) = 1.815 \cos (120\pi t - 0.721) - (1.364 + 1815t)e^{-1000t}.$
(c) 1000 rad/sec.   (d) 2.75 amperes.   (e) 3.518×10^{-4} farad.

11. $I(t) = I_c(t) + I_{s_1}(t) + I_{s_2}(t),$
where the transient current $I_c(t)$ and the steady-state currents $I_{s_1}(t)$ and $I_{s_2}(t)$
are given by

$$I_c(t) = 0.5618 e^{-100t} \cos (200t - 1.249),$$
$$I_{s_1}(t) = 0.3483 \cos (120\pi t - 2.456),$$
$$I_{s_2}(t) = 0.2569 \cos (60\pi t - 1.204).$$

Exercises 6.3, page 304

3. Suppose to the contrary that the given functions are linearly dependent on the interval I: $-\infty < x < \infty$. Then, in view of Definition 4.3.5, there exist constants $a_0, a_1, \ldots, a_{n_1}, b_0, b_1, \ldots, b_{n_2}, \ldots, c_0, c_1, \ldots, c_{n_i}$, not all zero, such that

$$(1) \quad a_0 e^{m_1 x} + a_1 x e^{m_1 x} + \cdots + a_{n_1} x^{n_1} e^{m_1 x} + b_0 e^{m_2 x} + b_1 x e^{m_2 x} + \cdots$$
$$+ b_{n_2} x^{n_2} e^{m_2 x} + \cdots + c_0 e^{m_i x} + c_1 x e^{m_i x} + \cdots + c_{n_i} x^{n_i} e^{m_i x} = 0$$

for every x in I. Equation (1) may be written as

$$(2) \qquad P_1(x) e^{m_1 x} + P_2(x) e^{m_2 x} + \cdots + P_i(x) e^{m_i x} = 0,$$

where $P_1(x)$, $P_2(x)$, $\ldots$, $P_i(x)$ are polynomials (or constants), and at least one of these polynomials, say $P_i(x)$, is not identically zero in I. Since $e^{m_1 x} \neq 0$ for every x in I, Equation (2) may be expressed as

$$(3) \qquad P_1(x) + P_2(x) e^{(m_2 - m_1)x} + \cdots + P_i(x) e^{(m_i - m_1)x} = 0.$$

Since $P_1(x)$ is a polynomial (or possibly a constant), if we differentiate Equation (3) a sufficient number of times [say, beyond the degree of $P_1(x)$], we see that $P_1(x)$ reduces to zero, while the other terms in (3) are not reduced to zero. (Why?) Thus, the following relation results

$$(4) \qquad Q_2(x) e^{(m_2 - m_1)x} + \cdots + Q_i(x) e^{(m_i - m_1)x} = 0,$$

where $Q_2(x)$, $\ldots$, $Q_i(x)$ are certain polynomials (or constants) and $Q_i(x)$ is not identically zero in I. Since $e^{(m_2 - m_1)x} \neq 0$ for every x in I, Equation (4) can be written as

$$(5) \qquad Q_2(x) + \cdots + Q_i(x) e^{(m_i - m_2)x} = 0.$$

Differentiating Equation (5) a sufficient number of times, we see that $Q_2(x)$ reduces to zero. This procedure is continued until a relation of the form

$$(6) \qquad R_i(x) e^{(m_i - m_{i-1})x} = 0$$

is derived, where $R_i(x)$ is a polynomial (or a constant different from zero). However, since $e^{(m_i - m_{i-1})x} \neq 0$ for every x in I, it follows from (6) that $R_i(x) = 0$ for every x in I, which is a contradiction. Thus, the given functions are linearly independent on I.

Exercises 6.4, page 314

4. (a) If the functions y_1, y_2, $\ldots$, y_n are linearly dependent on I, then there exist constants $c_1, c_2, \ldots, c_n$, not all equal to zero, such that

$$(1) \qquad c_1 y_1(x) + c_2 y_2(x) + \cdots + c_n y_n(x) = 0$$

for every x in I. Differentiating Equation (1) $n-1$ times in succession, we obtain

$$\begin{cases} c_1 y_1'(x) + c_2 y_2'(x) + \cdots + c_n y_n'(x) = 0, \\ c_1 y_1''(x) + c_2 y_2''(x) + \cdots + c_n y_n''(x) = 0, \\ \qquad\qquad \cdots \\ c_1 y_1^{(n-1)}(x) + c_2 y_2^{(n-1)}(x) + \cdots + c_n y_n^{(n-1)}(x) = 0. \end{cases}$$

Let x_0 be any fixed point in I.   By hypothesis, the system

$$\begin{cases} c_1 y_1(x_0) + c_2 y_2(x_0) + \cdots + c_n y_n(x_0) = 0, \\ c_1 y_1'(x_0) + c_2 y_2'(x_0) + \cdots + c_n y_n'(x_0) = 0, \\ \quad \cdots \\ c_1 y_1^{(n-1)}(x_0) + c_2 y_2^{(n-1)}(x_0) + \cdots + c_n y_n^{(n-1)}(x_0) = 0, \end{cases}$$

has a nontrivial solution $(c_1, c_2, \ldots, c_n) \neq (0, 0, \ldots, 0)$.   Thus,

$$\begin{vmatrix} y_1(x_0) & y_2(x_0) & \cdots & y_n(x_0) \\ y_1'(x_0) & y_2'(x_0) & \cdots & y_n'(x_0) \\ \cdots & \cdots & \cdots & \cdots \\ y_1^{(n-1)}(x_0) & y_2^{(n-1)}(x_0) & \cdots & y_n^{(n-1)}(x_0) \end{vmatrix} = 0.$$

Since x_0 is any fixed point in I, it follows that $W(y_1, y_2, \ldots, y_n; x) = 0$ for every x in I.

5. Concerning the Vandermonde determinant.   Let us denote by D_n the determinant given in (1).   Clearly, the result is true for $n = 2$, for

$$D_2 = \begin{vmatrix} 1 & 1 \\ x_2 & x_1 \end{vmatrix} = x_1 - x_2 = \prod_{1 \leq i < j \leq 2} (x_i - x_j).$$

Assume the result is true for $n = k$, that is,

$$D_k = \prod_{1 \leq i < j \leq k} (x_i - x_j).$$

If one can show that the result is also true for $n = k + 1$, then the result will be true for all positive integral values of n.   Show that

$$D_{k+1} = (x_1 - x_{k+1})(x_2 - x_{k+1}) \cdots (x_k - x_{k+1}) D_k$$
$$= (x_1 - x_{k+1})(x_2 - x_{k+1}) \cdots (x_k - x_{k+1}) \prod_{1 \leq i < j \leq k} (x_i - x_j)$$
$$= \prod_{1 \leq i < j \leq k+1} (x_i - x_j).$$

10. (a) We shall give the sufficiency proof, that is, if $G = 0$ on I, then the functions $f_1, f_2, \ldots, f_n$ are linearly dependent on I.   Since G, a determinant with constant terms, is equal to zero, we can choose constants (complex numbers) $c_1, c_2, \ldots, c_n$, not all equal to zero, such that [*see* Theorem II of Section 4.4]

$$(1) \quad c_1 \int_a^b f_1(x)\overline{f_1(x)}\,dx + c_2 \int_a^b f_1(x)\overline{f_2(x)}\,dx + \cdots + c_n \int_a^b f_1(x)\overline{f_n(x)}\,dx = 0,$$

$$(2) \quad c_1 \int_a^b f_2(x)\overline{f_1(x)}\,dx + c_2 \int_a^b f_2(x)\overline{f_2(x)}\,dx + \cdots + c_n \int_a^b f_2(x)\overline{f_n(x)}\,dx = 0,$$

$$\cdots$$

$$(3) \quad c_1 \int_a^b f_n(x)\overline{f_1(x)}\,dx + c_2 \int_a^b f_n(x)\overline{f_2(x)}\,dx + \cdots + c_n \int_a^b f_n(x)\overline{f_n(x)}\,dx = 0.$$

Multiplying Equations (1), (2), $\ldots$, (3) by $\overline{c_1}$, $\overline{c_2}$, $\ldots$, $\overline{c_n}$, respectively, and then adding, we obtain

$$(4) \quad \int_a^b \{[c_1\overline{c_1}f_1(x)\overline{f_1(x)} + c_2\overline{c_1}f_1(x)\overline{f_2(x)} + \cdots + c_n\overline{c_1}f_1(x)\overline{f_n(x)}]$$

$$+ [c_1\overline{c_2}f_2(x)\overline{f_1(x)} + c_2\overline{c_2}f_2(x)\overline{f_2(x)} + \cdots + c_n\overline{c_2}f_2(x)\overline{f_n(x)}] + \cdots$$

$$+ [c_1\overline{c_n}f_n(x)\overline{f_1(x)} + c_2\overline{c_n}f_n(x)\overline{f_2(x)} + \cdots + c_n\overline{c_n}f_n(x)\overline{f_n(x)}]\}dx = 0.$$

Equation (4) can be written as

$$(5) \quad 0 = \int_a^b [(\overline{c_1}f_1(x) + \overline{c_2}f_2(x) + \cdots + \overline{c_n}f_n(x))(c_1\overline{f_1}(x) + c_2\overline{f_2}(x) + \cdots$$

$$+ c_n\overline{f_n}(x))]dx$$

$$= \int_a^b |\overline{c_1}f_1(x) + \overline{c_2}f_2(x) + \cdots + \overline{c_n}f_n(x)|^2 dx.$$

From Equation (5), we may infer that

$$(6) \qquad \overline{c_1}f_1(x) + \overline{c_2}f_2(x) + \cdots + \overline{c_n}f_n(x) = 0$$

for every x in I. (Why?) Since $c_1, c_2, \ldots, c_n$ are not all equal to zero, so $\overline{c_1}, \overline{c_2}, \ldots, \overline{c_n}$ are not all equal to zero. Thus, the relation in (6) asserts that the functions $f_1, f_2, \ldots, f_n$ are linearly dependent on I.

Exercises 6.5, page 319

1. xe^x, e^{3x}.   2. x^2, x^3.

Exercises 6.6, page 328

1. $y = c_1 e^{-2x} + c_2 e^{2x} + c_3 e^{-3x} + c_4 e^{3x}$.
2. $y = c_1 e^{-x} + (c_2 + c_3 x)e^{3x}$.
3. $y = (c_1 + c_2 x + c_3 x^2)e^x$.
4. $y = c_1 e^{4x} + c_2 e^{-(3/2)x} + c_3 e^{(3/2)x}$.
5. $y = c_1 e^{-2x} + e^{3x}(c_1 \cos 2x + c_2 \sin 2x)$.
6. $y = e^x[(c_1 + c_2 x)\cos x + (c_3 + c_4 x)\sin x]$.
7. $y = (c_1 + c_2 x)e^{-x/3} + (c_3 + c_4 x)e^{x/2}$.
8. $y = c_1 e^x + (c_2 + c_3 x)e^{-2x} + e^{2x}(c_4 \cos 3x + c_5 \sin 3x)$.
9. $y = c_1 e^{3x} + (c_2 + c_3 x)e^{2ix}$.
10. $y = c_1 + c_2 x + c_3 e^{(1+2i)x} + c_4 e^{(2-i)x}$.
11. $y = c_1 e^{3x} + c_2 e^{-4x} + c_3 e^{(3-4i)x}$.
12. $y = -e^{-2x} + 3e^{2x} - e^{3x}$.
13. $y = 2e^{2x} + e^{-2x}(\cos 2x + \sin 2x)$.
14. $y = -e^{-x} + (2 - 4x + 2x^2)e^x$.

15. $y = 2 - 2x + 3x^2 - x^3 + e^{x-1}$.
16. $y = (2 - x)e^{-2x}$.
17. $y = (2x^2 - 1)e^{-x}$.

Exercises 6.8, page 333

1. We shall carry out the proof for (iii).   The other cases are treated in a similar fashion.   From Case 2(iii), a particular solution $y_p(x)$ of

(1) $L[y] = a_0 y^{(n)} + a_1 y^{(n-1)} + \cdots + a_{n-2} y'' + a_{n-1} y' + a_n y = k e^{\alpha x}$

has the form

(2) $$y_p(x) = B_0 x^2 e^{\alpha x}.$$

Let $u = e^{\alpha x}$ and $v = x^2$.   Then, $u' = \alpha e^{\alpha x}$, $u'' = \alpha^2 e^{\alpha x}$, $\ldots$, $u^{(n)} = \alpha^n e^{\alpha x}$, and $v' = 2x$, $v'' = 2$, $v''' = 0, \ldots, v^{(n)} = 0$.   Utilizing Formula (6.5.1), we obtain

$$(3)\ \begin{cases} y_p^{(n)}(x) = B_0 \sum_{k=0}^{n} C_{n\,k}\, u^{(n-k)} v^{(k)} \\ \qquad = B_0(C_{n,\,0}\, u^{(n)} v + C_{n,\,1} u^{(n-1)} v' + C_{n,\,2} u^{(n-2)} v'' + \cdots + C_{n,\,n} u v^{(n)}) \\ \qquad = B_0(\alpha^n x^2 + 2n\alpha^{n-1} x + n(n-1)\alpha^{n-2}) e^{\alpha x}, \\ y_p^{(n-1)}(x) = B_0(\alpha^{n-1} x^2 + 2(n-1)\alpha^{n-2} x + (n-1)(n-2)\alpha^{n-3}) e^{\alpha x}, \ldots, \\ y_p''(x) = B_0(\alpha^2 x^2 + 4\alpha x + 2) e^{\alpha x}, \quad y_p'(x) = B_0(\alpha x^2 + 2x) e^{\alpha x}. \end{cases}$$

Substituting the derivatives found in (3) and y_p given in (2) into Equation (1), we obtain

$$(4)\quad B_0\,[(a_0 \alpha^n + a_1 \alpha^{n-1} + \cdots + a_{n-2} \alpha^2 + a_{n-1} \alpha + a_n) x^2 + 2(n a_0 \alpha^{n-1}$$
$$+ (n-1)a_1 \alpha^{n-2} + \cdots + 2a_{n-2}\alpha + a_{n-1})x + n(n-1)a_0 \alpha^{n-2}$$
$$+ (n-1)(n-2)a_1 \alpha^{n-3} + \cdots + 2a_{n-2}] e^{\alpha x} = k e^{\alpha x}.$$

Since $p(m) = a_0 m^n + a_1 m^{n-1} + \cdots + a_{n-2} m^2 + a_{n-1} m + a_n$, we see that

$$(5)\ \begin{cases} p(\alpha) = a_0 \alpha^n + a_1 \alpha^{n-1} + \cdots + a_{n-2} \alpha^2 + a_{n-1} \alpha + a_n, \\ p'(\alpha) = n a_0 \alpha^{n-1} + (n-1)a_1 \alpha^{n-2} + \cdots + 2a_{n-2}\alpha + a_{n-1}, \\ p''(\alpha) = n(n-1)a_0 \alpha^{n-2} + (n-1)(n-2)a_1 \alpha^{n-3} + \cdots + 2a_{n-2}. \end{cases}$$

We are given that $p(\alpha) = 0$, $p'(\alpha) = 0$, and $p''(\alpha) \neq 0$.   Thus, in view of (5), the expression in (4) reduces to

$$B_0\, p''(\alpha) e^{\alpha x} = k e^{\alpha x}.$$

Hence,

$$B_0 = \frac{k}{p''(\alpha)}.$$

Substituting the above value of B_0 into (2), we obtain the desired particular solution.

2. Consider the differential equation

$$(1) \qquad L[w] = a_0 w^{(n)} + a_1 w^{(n-1)} + \cdots + a_{n-1} w' + a_n w = k e^{i\beta x},$$

where

$$w = y + iv, \qquad i^2 = -1.$$

Note that

$$\mathscr{R}\{L[w]\} = k \cos \beta x \qquad \text{and} \qquad \mathscr{R}(w) = y.$$

Let

$$w_p = y_p + i v_p$$

be a particular solution of Equation (1). Note that $\mathscr{R}(w_p) = y_p$. The characteristic polynomial of $L[w] = 0$ is

$$p(m) = a_0 m^n + a_1 m^{n-1} + \cdots + a_{n-1} m + a_n.$$

Utilizing the results of Exercise 1, we then have

$$\begin{cases} w_p = \dfrac{k e^{i\beta x}}{p(i\beta)}, & \text{if } p(i\beta) \neq 0; \\[2ex] w_p = \dfrac{k x e^{i\beta x}}{p'(i\beta)}, & \text{if } p(i\beta) = 0, \quad p'(i\beta) \neq 0; \\[1ex] \quad\vdots \\[1ex] w_p = \dfrac{k x^l e^{i\beta x}}{p^{(l)}(i\beta)}, & \text{if } p(i\beta) = 0, \quad p'(i\beta) = 0, \ldots, p^{(l-1)}(i\beta) = 0, \quad p^{(l)}(i\beta) \neq 0. \end{cases}$$

Thus, a particular solution $y_p(x)$ of $L[y] = k \cos \beta x$ is given by

$$\begin{cases} y_p(x) = \mathscr{R}(w_p) = \mathscr{R}\left[\dfrac{k e^{i\beta x}}{p(i\beta)}\right], & \text{if } p(i\beta) \neq 0; \\[2ex] y_p(x) = \mathscr{R}(w_p) = \mathscr{R}\left[\dfrac{k x e^{i\beta x}}{p'(i\beta)}\right], & \text{if } p(i\beta) = 0, \quad p'(i\beta) \neq 0; \\[1ex] \quad\vdots \\[1ex] y_p(x) = \mathscr{R}(w_p) = \mathscr{R}\left[\dfrac{k x^l e^{i\beta x}}{p^{(l)}(i\beta)}\right], & \text{if } p(i\beta) = 0, \quad p'(i\beta) = 0, \ldots, \\[1ex] & \qquad p^{(l-1)}(i\beta) = 0, \quad p^{(l)}(i\beta) \neq 0. \end{cases}$$

4. Consider the differential equation

$$(1) \qquad L[w] = a_0 w^{(n)} + a_1 w^{(n-1)} + \cdots + a_{n-1} w' + a_n w = k e^{(\alpha + i\beta)x},$$

where

$$w = y + iv, \qquad i^2 = -1.$$

Note that

$$\mathscr{R}\{L[w]\} = k e^{\alpha x} \cos \beta x \quad \text{and} \quad \mathscr{R}(w) = y.$$

Let

$$w_p = y_p + i v_p$$

be a particular solution of Equation (1). Note that $\mathscr{R}(w_p) = y_p$. Now proceed as in the hint to Exercise 2.

6. $y = c_1 e^{-x} + c_2 e^x + c_3 + c_4 x + c_5 x^2 - 320x^3 + 40x^4 - 32x^5 + 2x^6 - x^7$.

7. $y = (c_1 + c_2 x + x^2)e^x + (c_3 + c_4 x - 2x^2)e^{2x}$.

8. $y = (c_1 + c_2 \cos 3x + c_3 \sin 3x - 2x \sin 3x)e^{2x}$.

9. $y = (c_1 + c_2 x)e^{-2x} + (c_3 + c_4 x)e^{-x} + 2 \cos 2x - 11 \sin 2x$.

10. $y = (c_1 + c_2 x)e^{-3x} + c_3 e^{-4x} + c_4 e^{4x} - (x^2 - x + 1)e^x$.

11. $y = c_1 e^{-4x} + c_2 e^{4x} + (c_3 + c_4 x + c_5 x^2 - 2x^3)e^{-3x}$.

12. $y = e^{2x}[(c_1 + c_2 x + c_3 x^2 + 2x^3) \cos x + (c_4 + c_5 x + c_6 x^2 - x^3) \sin x]$.

13. $y = (c_1 + c_2 x)e^{-x} + c_3 e^x + (x^2 - 6x - 2) \cos x - (2x^2 + 4x - 1) \sin x$.

14. $y = c_1 + c_2 x + \frac{1}{4}x^2 - \frac{1}{4}x^3 + \frac{1}{2}x^4 + c_3 \cos 2x + (c_4 - 2x) \sin 2x$.

15. $y = c_1 + 2x - 3x^2 + (c_2 + c_3 x + c_4 x^2 - 40x^3 + 10x^4 - 2x^5)e^x$.

16. $y = (x - 2)e^x + 3 \cos x + 4 \sin x$.

17. $y = e^{-x}[(\frac{3}{2} - x) \sin 2x - (1 + 2x) \cos 2x] + 2$.

18. $y = 4e^{-x} + e^x - 2e^{2x} + x^2 + 3x - 1$.

19. $y = 2e^{-2x} - e^{2x} - \frac{19}{2} e^{-x} + \frac{17}{2} e^x + \cos x - 2 \sin x$.

20. $y = -9 \cos x + 3 \sin x + \cos 2x + 5 \sin 2x + \cos 3x$.

Exercises 6.9, page 340

1. $y = c_1 + c_2 \cos x + c_3 \sin x - \ln \cos x - \sin x \ln (\sec x + \tan x)$.

2. $y = c_1 + c_2 \cos x + c_3 \sin x - x \cos x + \sin x \ln \cos x + \ln (\sec x + \tan x)$
$+ x^3 + 2x^2 - 6x$.

3. $y = c_1 + c_2 \cos x + c_3 \sin x - \ln (\csc x - \cot x) + \cos x \ln \sin x$.

4. $y = \left[c_1 + c_2 x + c_3 x^2 + \frac{x^3}{36} (6 \ln x - 11) \right] e^{-x}$.

5. $y = \left[c_1 + c_2 x + c_3 x^2 + \frac{x^2}{8} (2 \ln^2 x - 6 \ln x) \right] e^x$.

6. $y = c_1 e^x + c_2 e^{2x} + c_3 e^{3x} + \frac{e^x}{4} \ln (e^{2x} + 1) - e^{2x} \arctan e^x - \frac{e^{3x}}{4} [\ln (e^{2x} + 1) - 2x]$.

7. $y = (c_1 + c_2 x + c_3 x^2 + c_4 x^3)e^x + \frac{1}{6} \left(-\frac{1}{4 - n} + \frac{3}{3 - n} - \frac{3}{2 - n} + \frac{1}{1 - n} \right) \frac{e^x}{x^{n-4}}$,

provided that $n \neq 1, 2, 3, 4$.

8. $y = (c_1 + c_2 x) \cos x + (c_3 + c_4 x) \sin x + \frac{1}{2} \cos x \ln \cos x$.

9. $y = \ln 2 + (1 + 2 \ln 2)e^x + (x + \ln 2)e^{2x} - (1 + e^x)^2 \ln (e^x + 1) + x^2 - x$.

10. $y = 2 + 18x - 36x^2 + 48x^3 - 25x^4 + 12x^4 \ln x$.

12. Utilize Exercise 11 to find a particular solution of $y^{(n)} = f(x)$.

13. $y = x - 3x^2 + 2x^3 + \frac{x^4}{48} (12 \ln x - 25)$.

15. (a) $u(x) = \cos^2 x + 1 - 2x + \frac{x^2}{2} + \int_0^x \left[3 - 3(x - t) + \frac{1}{2} (x - t)^2 \right] u(t) dt$.

(b) $u(x) = g(x) + 1 - x + x^2 - x^3 + \frac{1}{6} \int_0^x (x - t)^3 u(t) dt$.

(c) $u(x) = g(x) - 2 - 2x^2 + \int_0^x [xt + (x - t)^2] u(t) dt$.

17. $u(x) = \frac{1}{2}(\cos x + \sin x + e^x)$.
18. $u(x) = \frac{1}{4}(3 + 2x)e^{-x} + \frac{1}{4}e^x - \cos x$.
19. $u(x) = (6 - 5x + 2x^2)e^x - 6$.
20. $u(x) = \frac{1}{4}(\cosh x + 3\cos x - x\sin x)$.

Exercises 6.10, page 348

2. $y = c_1 x^2 + c_2 x^3 + c_3 x^4 + (2x^4 - x^2)\ln x - 2$.
3. $y = (c_1 + c_2 \ln x + \ln^2 x)x^{-1} + (c_3 + c_4 \ln x + 2\ln^2 x)x^{-2}$.
4. $y = (c_1 \cos \ln x + c_2 \sin \ln x)x + c_3 x + c_4 x^{-1} - 1 + 14\ln x + \ln^2 x - 4\ln^3 x$
$\quad - \ln^4 x$.
5. $y = c_1 + c_2 x + c_3 x^2 + x^2 \ln x - (1 + x)^2 \ln(x + 1)$.
6. $y = (c_1 + c_2 \ln x - \ln^2 x)\cos \ln x + (c_3 + c_4 \ln x)\sin \ln x$.
7. $y = c_1 + c_2 \ln x + c_3 \ln^2 x + c_4 \ln^3 x + [c_5 + 12\ln(\ln x)]\ln^4 x$.
8. $y = c_1 \cos \ln x + c_2 \sin \ln x + (c_3 + c_4 \ln x + 2\ln^2 x)x^4 + 3\cos(2\ln x)$
$\quad - 4\sin(2\ln x) + \frac{75}{4}$.
9. $y = c_1 \cos \ln x + c_2 \sin \ln x + c_3 x^{-1} + c_4 x + (\ln^5 x + \ln^4 x + 2\ln^3 x$
$\quad - 3\ln^2 x - 9\ln x)x$.
10. Follow a similar method as given in the proof of Theorem 6.10.1.   Under the
transformation $\alpha + \beta x = e^u$ or $u = \ln(\alpha + \beta x)$, $x > -\dfrac{\alpha}{\beta}$, we have the following
useful differentiation formula:

$$\frac{d^k}{dx^k} = \beta^k e^{-ku}\frac{d}{du}\left(\frac{d}{du} - 1\right)\left(\frac{d}{du} - 2\right)\cdots\left(\frac{d}{du} - k + 1\right).$$

12. $y = [c_1 + c_2 \ln(4 + x)](4 + x)^{-1} + [c_3 + c_4 \ln(4 + x)](4 + x) + (4 + x)^2$
$\quad - 9(4 + x)\ln^2(4 + x) + 2\ln(4 + x) + 144$.
13. $y = \frac{1}{4}(x^2 - 1) + \frac{1}{2}\ln x$.
14. $y = -\frac{1}{18}(14 - 27x + 13x^3) + \frac{1}{3}(3x^3 + 2)\ln x$.

15. $y = -\dfrac{1}{33}\left\{-11 + \dfrac{9}{x} + x^2\left[2\cos(\sqrt{2}\ln x) + \dfrac{5\sqrt{2}}{2}\sin(\sqrt{2}\ln x)\right]\right\}$.

Exercises 7.2, page 364

1. (a) 1,   (b) $\frac{1}{2}$,   (c) infinity,   (d) 1,   (e) $\frac{1}{2}$,   (f) infinity,
   (g) infinity,   (h) $\frac{1}{4}$,   (i) e,   (j) infinity,   (k) 1,   (l) 1.

2. (a) $\dfrac{1}{1-x} = \displaystyle\sum_{n=0}^{\infty} x^n$ when $|x| < 1$,

   (b) $\dfrac{1}{2x+3} = \dfrac{1}{3}\cdot\dfrac{1}{1+\frac{2}{3}x} = \dfrac{1}{3}\displaystyle\sum_{n=0}^{\infty}(-1)^n\left(\dfrac{2}{3}x\right)^n = \displaystyle\sum_{n=0}^{\infty}\dfrac{(-1)^n 2^n x^n}{3^{n+1}}$ when $|x| < \dfrac{3}{2}$,

   (c) $\dfrac{1}{1+x^2} = \displaystyle\sum_{n=0}^{\infty}(-1)^n x^{2n}$ when $|x| < 1$,

(d) $e^{x-1} = \sum_{n=0}^{\infty} \dfrac{e(x-2)^n}{n!}$ when $|x| < \infty$,

(e) $\cosh x = \sum_{n=0}^{\infty} \dfrac{x^{2n}}{(2n)!}$ when $|x| < \infty$,

(f) $\dfrac{1}{x+3} = \dfrac{1}{1+(x+2)} = \sum_{n=0}^{\infty} (-1)^n (x+2)^n$ when $|x+2| < 1$,

(g) $\dfrac{x}{1-x} = x \cdot \dfrac{1}{1-x} = x \sum_{n=0}^{\infty} x^n = \sum_{n=0}^{\infty} x^{n+1}$ when $|x| < 1$,

(h) $\dfrac{1}{x^2} = \sum_{n=0}^{\infty} (-1)^n (n+1)(x-1)^n$ when $|x-1| < 1$,

(i) $\dfrac{1}{(x-3)(x-4)} = \dfrac{1}{x-4} - \dfrac{1}{x-3} = -\dfrac{1}{4} \cdot \dfrac{1}{1-\dfrac{x}{4}} + \dfrac{1}{3} \cdot \dfrac{1}{1-\dfrac{x}{3}}$

$$= -\dfrac{1}{4} \sum_{n=0}^{\infty} \left(\dfrac{x}{4}\right)^n + \dfrac{1}{3} \sum_{n=0}^{\infty} \left(\dfrac{x}{3}\right)^n = \sum_{n=0}^{\infty} \left(\dfrac{1}{3^{n+1}} - \dfrac{1}{4^{n+1}}\right) x^n \quad \text{when } |x| < 3,$$

(j) $\sin x = \sum_{n=0}^{\infty} \dfrac{(-1)^n \left(x - \dfrac{\pi}{2}\right)^{2n}}{(2n)!}$ when $|x| < \infty$,

(k) $\ln(x+1) = \sum_{n=0}^{\infty} \dfrac{(-1)^n x^{n+1}}{n+1}$ when $-1 < x \leqq 1$,

(l) $\dfrac{1}{(1-x)^k} = 1 + \sum_{n=1}^{\infty} \dfrac{[k(k+1)\cdots(k+n-1)]x^n}{n!}, \quad k = 1, 2, 3, \ldots,$

   when $|x| < 1$.

4. Consider: $\left(\sum_{n=0}^{\infty} a_n x^{2n}\right)\left(\sum_{n=0}^{\infty} b_n x^n\right)$.
   We wish to collect the nth power of x from all possible products $a_k x^{2k} b_j x^j$.
   Thus, $j + 2k = n$ or $j = n - 2k$. Hence, $a_k x^{2k} b_j x^j = a_k x^{2k} b_{n-2k} x^{n-2k}$, where
   $k \geqq 0$ and $n - 2k \geqq 0$. Since n runs over all nonnegative integers, it follows
   that $k = \left[\dfrac{n}{2}\right]$. (Why?) Consequently, we have that

$$\left(\sum_{n=0}^{\infty} a_n x^{2n}\right)\left(\sum_{n=0}^{\infty} b_n x^n\right) = \sum_{n=0}^{\infty} \sum_{k=0}^{[n/2]} a_k x^{2k} b_{n-2k} x^{n-2k} = \sum_{n=0}^{\infty} \sum_{k=0}^{[n/2]} a_k b_{n-2k} x^n.$$

5. (a) $e^x \cosh x = \sum_{n=0}^{\infty} \sum_{k=0}^{[n/2]} \dfrac{x^n}{(2k)!(n-2k)!}$ when $|x| < \infty$,

   (b) $\dfrac{\sinh x}{1-x} = \sum_{n=0}^{\infty} \sum_{k=0}^{[n/2]} \dfrac{x^{n+1}}{(2k+1)!}$ when $|x| < 1$,

(c) $\dfrac{e^{-x}}{1-x^2} = \sum_{n=0}^{\infty} \sum_{k=0}^{[n/2]} \dfrac{(-1)^n x^n}{(n-2k)!}$ when $|x| < 1$,

(d) $e^x \sin 2x = \sum_{n=0}^{\infty} \sum_{k=0}^{[n/2]} \dfrac{(-1)^k 2^{2k+1} x^{n+1}}{(2k+1)!(n-2k)!}$ when $|x| < \infty$.

6. $\tan x = x + \dfrac{x^3}{3} + \dfrac{2x^5}{15} + \dfrac{17x^7}{315} + \cdots$,   when $|x| < \dfrac{\pi}{2}$,

$\sec x = 1 + \dfrac{x^2}{2} + \dfrac{5x^4}{24} + \dfrac{61x^6}{720} + \cdots$,   when $|x| < \dfrac{\pi}{2}$.

Exercises 7.3, page 381

3. $y = a_0 \left[1 + \sum_{k=1}^{\infty} \dfrac{x^{2k}}{1 \cdot 3 \cdots (2k-1)} \right] + a_1 \left[x + \sum_{k=1}^{\infty} \dfrac{x^{2k+1}}{2^k k!} \right]$,

or

$y = a_0 \left[1 + \sum_{k=1}^{\infty} \dfrac{x^{2k}}{1 \cdot 3 \cdots (2k-1)} \right] + a_1 x e^{x^2/2}$; converges for all finite x.

4. $y = a_0 \left[1 + \sum_{k=1}^{\infty} \dfrac{(-1)^k [1 \cdot 4 \cdots (3k-2)] x^{3k}}{(3k)!} \right]$

$\quad + a_1 \left[x + \sum_{k=1}^{\infty} \dfrac{(-1)^k [2 \cdot 5 \cdots (3k-1)] x^{3k+1}}{(3k+1)!} \right]$;

converges for all finite x.

5. $y = a_0 \left[1 + \sum_{k=1}^{\infty} \dfrac{[1 \cdot 4 \cdots (3k-2)] x^{3k}}{3^k k! [2 \cdot 5 \cdots (3k-1)]} \right] + a_1 \left[x + \sum_{k=1}^{\infty} \dfrac{[2 \cdot 5 \cdots (3k-1)] x^{3k+1}}{3^k k! [4 \cdot 7 \cdots (3k+1)]} \right]$;

or

$y = a_0 \left[1 + \sum_{k=1}^{\infty} \dfrac{[1 \cdot 4 \cdots (3k-2)]^2 x^{3k}}{(3k)!} \right] + a_1 \left[x + \sum_{k=1}^{\infty} \dfrac{[2 \cdot 5 \cdots (3k-1)]^2 x^{3k+1}}{(3k+1)!} \right]$;

converges for all finite x.

6. $y = a_1 x + a_0 \left[1 + \sum_{k=1}^{\infty} \dfrac{(-1)^{k-1} x^{2k}}{2k-1} \right]$;   or $y = a_1 x + a_0 (1 + x \arctan x)$;

converges for $|x| \le 1$.

7. $y = a_0 (1 - 3x^2) + a_1 \left[x - \sum_{k=1}^{\infty} \dfrac{(k+1) x^{2k+1}}{4k^2 - 1} \right]$;   converges for $|x| < 1$.

8. $y = a_0 (1 + x^3) + a_1 \left[x + \sum_{k=1}^{\infty} \dfrac{(-1)^{k+1} 2 x^{3k+1}}{(3k-2)(3k+1)} \right]$;   converges for $|x| < 1$.

9. $y = a_0\left[1 + \sum_{k=1}^{\infty} \dfrac{x^{3k}}{3^k k!}\right] + a_1\left[x + \sum_{k=1}^{\infty} \dfrac{x^{3k+1}}{4 \cdot 7 \cdots (3k+1)}\right]$

$\quad + a_2\left[x^2 + \sum_{k=1}^{\infty} \dfrac{x^{3k+2}}{5 \cdot 8 \cdots (3k+2)}\right];$

converges for all finite x.

10. $y = a_0 \sum_{k=0}^{\infty} \dfrac{x^{4k}}{(4k)!} + a_1 \sum_{k=0}^{\infty} \dfrac{x^{4k+1}}{(4k+1)!} + a_2 \sum_{k=0}^{\infty} \dfrac{2x^{4k+2}}{(4k+2)!} + a_3 \sum_{k=0}^{\infty} \dfrac{6x^{4k+3}}{(4k+3)!};$

converges for all finite x.

11. $y = a_0\left[1 + \sum_{k=1}^{\infty} \dfrac{[2 \cdot 5 \cdots (3k-1)]x^{3k}}{2^k 3^k k!}\right] + a_1\left[x + \sum_{k=1}^{\infty} \dfrac{3^k k! x^{3k+1}}{2^k[4 \cdot 7 \cdots (3k+1)]}\right];$

converges for $-\sqrt[3]{2} \leqq x < \sqrt[3]{2}$.

12. $y = a_0\left[1 - \dfrac{1}{3}x^3 - \dfrac{1}{3}\sum_{k=2}^{\infty} \dfrac{(-1)^{k-1}x^{3k}}{6 \cdot 9 \cdots 3k}\right] + a_1\left[x - \dfrac{1}{4}x^4 - \dfrac{1}{4}\sum_{k=2}^{\infty} \dfrac{(-1)^{k-1}x^{3k+1}}{7 \cdot 10 \cdots (3k+1)}\right]$

$\quad + x^2 + \sum_{k=2}^{\infty} \dfrac{(-1)^{k-1}x^{3k-1}}{5 \cdot 8 \cdots (3k-1)} + x^3 + \sum_{k=2}^{\infty} \dfrac{(-1)^{k-1}x^{3k}}{6 \cdot 9 \cdots 3k}$

$\quad + x^4 + \sum_{k=2}^{\infty} \dfrac{(-1)^{k-1}x^{3k+1}}{7 \cdot 10 \cdots (3k+1)};$

converges for all finite x.

13. $y = 1 - x + \sum_{k=1}^{\infty} \dfrac{[1 \cdot 4 \cdots (3k-2)]x^{3k}}{(3k)!} - \sum_{k=1}^{\infty} \dfrac{[2 \cdot 5 \cdots (3k-1)]x^{3k+1}}{(3k+1)!};$

converges for all finite x.

14. $y = 2x + \sum_{k=1}^{\infty} \dfrac{(-1)^k[2 \cdot 5 \cdots (3k-1)]x^{3k+1}}{(3k+1)!};$ converges for all finite x.

15. Recurrence formula (R.F.) $a_{n+2} = \dfrac{a_{n+1} + a_n}{n+2}, \qquad n \geqq 1.$

$y = 1 + x + x^2 + \frac{2}{3}x^3 + \frac{5}{12}x^4 + \frac{13}{60}x^5 + \frac{19}{180}x^6 + \cdots;$ converges for all finite x.

16. R.F. $a_{n+2} = \dfrac{(n+1)a_n + a_{n-1}}{(n+1)(n+2)}, \qquad n \geqq 1.$

$y = 2 - x + x^2 + \frac{1}{6}x^4 + \frac{1}{20}x^5 + \frac{1}{36}x^6 + \cdots;$ converges for all finite x.

17. R.F. $a_{n+2} = -\dfrac{(n-2)(n+2)a_n - a_{n-1}}{3(n+1)(n+2)}, \qquad n \geqq 2.$

$y = 3 + 5x + 2x^2 + x^3 + \frac{5}{36}x^4 - \frac{1}{20}x^5 - \frac{1}{135}x^6 - \cdots;$ converges for at least $|x| < \sqrt{3}$.

18. R.F. $a_{n+2} = -\dfrac{(n+1)^2 a_{n+1} + (n-1)a_n}{2(n+1)(n+2)}$, $n \geq 1$.

$y = 2\left(1 - \dfrac{x}{1!} + \dfrac{x^2}{2!} - \dfrac{x^3}{3!} + \dfrac{x^4}{4!} - \dfrac{x^5}{5!} + \dfrac{x^6}{6!} - \cdots\right) = 2e^{-x};$ converges for all finite x.

19. R.F. $a_{2n+2} = \dfrac{a_{2n}}{2(n+1)}$, $n \geq 1$, $2n(2n+1)a_{2n+1} - 2na_{2n-1} = \dfrac{(-1)^{n-1}}{(2n-1)!}$, $n \geq 1$

$y = 2 + x + x^2 + \frac{1}{2}x^3 + \frac{1}{4}x^4 + \frac{11}{120}x^5 + \frac{1}{24}x^6 + \cdots;$ converges for all finite x.

20. R.F. $a_{n+3} = \dfrac{na_{n+1} + a_n}{(n+2)(n+3)}$, $n \geq 4$.

$y = 6 + 4x + 2x^2 + 2x^3 + \frac{1}{2}x^4 + \frac{3}{10}x^5 + \frac{1}{5}x^6 + \cdots;$ converges for all finite x.

21. Let $v = x - 1$. Then $y'' + (1 - x)y' - y = 0$ becomes $\dfrac{d^2y}{dv^2} - v\dfrac{dy}{dv} - y = 0$.

R.F. $a_{n+2} = \dfrac{a_n}{n+2}$, $n \geq 1$.

$y = a_0\left[1 + \sum_{k=1}^{\infty} \dfrac{(x-1)^{2k}}{2^k k!}\right] + a_1\left[(x-1) + \sum_{k=1}^{\infty} \dfrac{(x-1)^{2k+1}}{3 \cdot 5 \cdots (2k+1)}\right];$

converges for all finite x.

22. Let $v = x - 2$. Then $y'' + (x - 1)y' + y = 0$ becomes $\dfrac{d^2y}{dv^2} + (v+1)\dfrac{dy}{dv} + y = 0$.

R.F. $a_{n+2} = -\dfrac{a_{n+1} + a_n}{n+2}$, $n \geq 1$.

$y = a_0\left[1 - \frac{1}{2}(x-2)^2 + \frac{1}{6}(x-2)^3 + \frac{1}{12}(x-2)^4 - \frac{1}{20}(x-2)^5 - \frac{1}{180}(x-2)^6 \right.$
$\left. + \cdots\right] + a_1[(x-2) - \frac{1}{2}(x-2)^2 - \frac{1}{6}(x-2)^3 + \frac{1}{6}(x-2)^4 - \frac{1}{36}(x-2)^6$
$+ \cdots];$ converges for all finite x.

23. Let $v = x + 1$. Then $x^2y'' + (x+1)y' - 4y = 0$ becomes $(v-1)^2\dfrac{d^2y}{dv^2} + v\dfrac{dy}{dv}$
$\qquad - 4y = 0.$

R.F. $a_{n+2} = \dfrac{2n(n+1)a_{n+1} - (n-2)(n+2)a_n}{(n+1)(n+2)}$, $n \geq 2$.

$y = a_0[1 + 2(x+1)^2 + \frac{4}{3}(x+1)^3 + \frac{4}{3}(x+1)^4 + \frac{19}{15}(x+1)^5 + \frac{52}{45}(x+1)^6$
$\quad + \cdots] + a_1[(x+1) + \frac{1}{2}(x+1)^3 + \frac{1}{2}(x+1)^4 + \frac{19}{40}(x+1)^5 + \frac{13}{30}(x+1)^6$
$\quad + \cdots];$ converges for $-2 < x < 0$.

24. Let $v = x - 3$. Then $y'' + (x-4)y' + (x-2)y = 0$ becomes $\dfrac{d^2y}{dv^2} + (v-1)\dfrac{dy}{dv}$
$\qquad + (v+1)y = 0.$

R.F. $a_{n+2} = \dfrac{(n+1)a_{n+1} - (n+1)a_n - a_{n-1}}{(n+1)(n+2)}$, $n \geq 1$.

$$y = a_0 \left[1 - \tfrac{1}{2}(x-3)^2 - \tfrac{1}{3}(x-3)^3 + \tfrac{1}{24}(x-3)^4 + \tfrac{1}{10}(x-3)^5 + \tfrac{1}{48}(x-3)^6 \right.$$
$$\left. + \cdots\right] + a_1\left[(x-3) + \tfrac{1}{2}(x-3)^2 - \tfrac{1}{6}(x-3)^3 - \tfrac{1}{4}(x-3)^4 - \tfrac{1}{24}(x-3)^5 \right.$$
$$\left. + \tfrac{29}{720}(x-3)^6 + \cdots\right]; \quad \text{converges for all finite } x.$$

25. R.F. $(n+1)(n+2)a_{n+2} - \displaystyle\sum_{k=0}^{n} \dfrac{a_{n-k}}{k!} = 0, \qquad n = 0, 1, 2, \ldots.$

$$y = a_0\left(1 + \tfrac{1}{2}x^2 + \tfrac{1}{6}x^3 + \tfrac{1}{12}x^4 + \tfrac{1}{24}x^5 + \tfrac{13}{720}x^6 + \cdots\right)$$
$$+ a_1\left(x + \tfrac{1}{6}x^3 + \tfrac{1}{12}x^4 + \tfrac{1}{30}x^5 + \tfrac{1}{72}x^6 + \cdots\right); \quad \text{converges for all finite } x.$$

26. R.F. $(n+1)(n+2)a_{n+2} = \displaystyle\sum_{k=0}^{[n/2]} \dfrac{(-1)^k a_{n-2k}}{(2k)!} + \dfrac{1}{n!}, \qquad n = 0, 1, 2, \ldots.$

$$y = a_0\left(1 + \tfrac{1}{2}x^2 - \tfrac{1}{144}x^6 + \cdots\right) + a_1\left(x + \tfrac{1}{6}x^3 - \tfrac{1}{60}x^5 - \cdots\right)$$
$$+ \tfrac{1}{2}x^2 + \tfrac{1}{6}x^3 + \tfrac{1}{12}x^4 + \tfrac{1}{60}x^5 - \tfrac{1}{240}x^6 - \cdots; \quad \text{converges for all finite } x.$$

Exercises 7.4, page 408

1. $x = -1$ is a regular singular point (R.S.P.); $x = 0$ and $x = 1$ are irregular singular points (I.S.P.)
2. $x = -1$ is a R.S.P.
3. $x = -2$ is a R.S.P.; $x = 0$ and $x = -3$ are I.S.P.
4. $x = 3$ is a R.S.P.; $x = 0$ is an I.S.P.
5. $x = -2$ is a R.S.P.; $x = 0$ and $x = 2$ are I.S.P.
6. $x = -1$ is a R.S.P.; $x = 1$ is an I.S.P.
7. $x = 0$ is a R.S.P.   8. $x = 0$ is a R.S.P.
9. $x = 0$ is an I.S.P.

10. R.F. $a_n(r) = -\dfrac{a_{n-1}(r)}{2(r+n)}, \qquad n \geq 1.$

$$y_1 = |x|^{1/2} \sum_{n=0}^{\infty} \frac{(-1)^n x^n}{1 \cdot 3 \cdots (2n+1)}; \quad y_2 = \sum_{n=0}^{\infty} \frac{(-1)^n x^n}{2^n n!}. \quad \text{Intervals: } -\infty < x < 0,$$
$$0 < x < \infty.$$

11. R.F. $a_n(r) = -\dfrac{2a_{n-1}(r)}{(2r+2n-1)}, \qquad n \geq 1. \quad y_1 = |x|^{1/2} \displaystyle\sum_{n=0}^{\infty} \dfrac{(-1)^n x^n}{n!};$

$$y_2 = 1 + \sum_{n=1}^{\infty} \frac{(-1)^n 2^n x^n}{1 \cdot 3 \cdots (2n-1)} = \sum_{n=0}^{\infty} \frac{(-1)^n 2^{2n} n! x^n}{(2n)!}. \quad \text{Intervals: } -\infty < x < 0,$$
$$0 < x < \infty.$$

12. R.F. $a_n(r) = -\dfrac{a_{n-2}(r)}{(3r+3n-1)(2r+2n-1)}, \qquad n \geq 2.$

$$y_1 = |x|^{1/2}\left[1 + \sum_{n=1}^{\infty} \frac{(-1)^n x^{2n}}{2^n n! [13 \cdot 25 \cdots (12n+1)]}\right];$$

$$y_2 = |x|^{1/3}\left[1 + \sum_{n=1}^{\infty} \frac{(-1)^n x^{2n}}{2^n n! [11 \cdot 23 \cdots (12n-1)]}\right].$$

Intervals: $-\infty < x < 0, \quad 0 < x < \infty.$

13. R.F. $a_n(r) = -\dfrac{2a_{n-1}(r)}{(2r+2n-1)(r+n+1)}$, $n \geq 1.$

$$y_1 = |x|^{1/2}\left[1 + \sum_{n=1}^{\infty} \frac{(-1)^n 2^n x^n}{n![5 \cdot 7 \cdots (2n+3)]}\right];$$

$$y_2 = |x|^{-1}\left[1 + \sum_{n=1}^{\infty} \frac{(-1)^n 2^n x^n}{n![-1 \cdot 1 \cdot 3 \cdots (2n-3)}\right].$$

Intervals: $-\infty < x < 0,$ $0 < x < \infty.$

14. R.F. $a_n(r) = -a_{n-1}(r),$ $n \geq 1.$ $y_1 = |x|\sum_{n=0}^{\infty}(-1)^n x^n;$
$y_2 = |x|^{1/2}\sum_{n=0}^{\infty}(-1)^n x^n.$ Intervals: $-1 < x < 0,$ $0 < x < 1.$

15. R.F. $a_n(r) = \dfrac{(3r+3n-5)(3r+3n-4)a_{n-1}(r)}{2(3r+3n-1)(3r+3n+1)}$, $n \geq 1.$ $y_1 = |x|^{1/3};$

$y_2 = |x|^{-1/3}(1+x).$ Intervals: $-\infty < x < 0,$ $0 < x < \infty.$

16. $a_n(r) = \dfrac{6(r+n-1)(r+n-2)a_{n-1}(r)}{(3r+3n-1)(2r+2n-1)}$, $n \geq 1.$

$$y_1 = |x|^{1/2}\left[1 + \sum_{n=1}^{\infty} \frac{3^n[-1 \cdot 1 \cdots (2n-3)][1 \cdot 3 \cdots (2n-1)]x^n}{2^n n![7 \cdot 13 \cdots (6n+1)]}\right];$$

$$y_2 = |x|^{1/3}\left[1 + \sum_{n=1}^{\infty} \frac{2^n[-2 \cdot 1 \cdots (3n-5)][1 \cdot 4 \cdots (3n-2)]x^n}{3^n n! [5 \cdot 11 \cdots (6n-1)]}\right].$$

Intervals: $-1 \leq x < 0,$ $0 < x < 1.$

17. R.F. $a_n(r) = -\dfrac{(r+n-5)a_{n-2}(r)}{2r+2n+1}$, $n \geq 2.$

$$y_1 = 1 + \sum_{n=1}^{\infty} \frac{(-1)^n[(-3)(-1) \cdot 1 \cdot 3 \cdots (2n-5)]x^{2n}}{5 \cdot 9 \cdots (4n+1)};$$

$$y_2 = |x|^{-1/2}\left[1 + \sum_{n=1}^{\infty} \frac{(-1)^n[(-7)(-3) \cdot 1 \cdot 5 \cdots (4n-11)]x^{2n}}{2^{3n}n!}\right].$$

Intervals: $-\sqrt{2} \leq x < 0,$ $0 < x \leq \sqrt{2}.$

18. R.F. $a_n(r) = -\dfrac{a_{n-2}(r)}{(r+n)^2}$, $n \geq 2.$ $y_1 = \sum_{n=0}^{\infty} \dfrac{(-1)^n x^{2n}}{2^{2n}(n!)^2};$

$$y_2 = y_1 \ln|x| - \sum_{n=1}^{\infty} \frac{(-1)^n H_n x^{2n}}{2^{2n}(n!)^2}.$$ Intervals: $-\infty < x < 0,$ $0 < x < \infty.$

19. R.F. $a_n(r) = \dfrac{4a_{n-2}(r)}{(r+n)^2}$, $n \geq 2.$ $y_1 = \sum_{n=0}^{\infty} \dfrac{x^{2n}}{(n!)^2};$

$$y_2 = y_1 \ln|x| - \sum_{n=1}^{\infty} \frac{H_n x^{2n}}{(n!)^2}.$$ Intervals: $-\infty < x < 0,$ $0 < x < \infty.$

20. R.F. $a_n(r) = \dfrac{4a_{n-2}(r)}{(r+n-2)^2}$, $n \geq 2$. $y_1 = x^2 \displaystyle\sum_{n=0}^{\infty} \dfrac{x^{2n}}{(n!)^2}$;

$y_2 = y_1 \ln |x| - x^2 \displaystyle\sum_{n=1}^{\infty} \dfrac{H_n x^{2n}}{(n!)^2}$.   Intervals: $-\infty < x < 0$, $0 < x < \infty$.

21. R.F. $a_n(r) = \dfrac{(r+n-1)a_{n-2}(r)}{(r+n-2)^2}$, $n \geq 2$.

$y_1 = x^2 + x^2 \displaystyle\sum_{n=1}^{\infty} \dfrac{[3 \cdot 5 \cdots (2n+1)]x^{2n}}{2^{2n}(n!)^2} = x^2 \displaystyle\sum_{n=0}^{\infty} \dfrac{(2n+1)!x^{2n}}{2^{3n}(n!)^3}$;

$y_2 = y_1 \ln |x| + x^2 \displaystyle\sum_{n=1}^{\infty} \dfrac{(2n+1)!\left[\left(\dfrac{1}{3} + \dfrac{1}{5} + \cdots + \dfrac{1}{2n+1}\right) - H_n\right]x^{2n}}{2^{3n}(n!)^3}$

$= y_1 \ln |x| + x^2 \displaystyle\sum_{n=1}^{\infty\prime} \dfrac{(2n+1)!(H_{2n+1} - \frac{3}{2}H_n - 1)x^{2n}}{2^{3n}(n!)^3}$.

Intervals: $-\infty < x < 0$, $0 < x < \infty$.
[*Note:*

$$H_{2n+1} = 1 + \dfrac{1}{2} + \dfrac{1}{3} + \dfrac{1}{4} + \cdots + \dfrac{1}{2n} + \dfrac{1}{2n+1}.$$

Thus,

$$H_{2n+1} - \dfrac{1}{2}H_n = 1 + \dfrac{1}{2} + \dfrac{1}{3} + \dfrac{1}{4} + \cdots + \dfrac{1}{2n}$$

$$+ \dfrac{1}{2n+1} - \dfrac{1}{2}\left(1 + \dfrac{1}{2} + \dfrac{1}{3} + \dfrac{1}{4} + \cdots + \dfrac{1}{n}\right)$$

$$= 1 + \dfrac{1}{3} + \dfrac{1}{5} + \cdots + \dfrac{1}{2n+1}.$$

Hence,

$$\dfrac{1}{3} + \dfrac{1}{5} + \cdots + \dfrac{1}{2n+1} = H_{2n+1} - \dfrac{1}{2}H_n - 1.$$

22. R.F. $a_n(r) = -\dfrac{(r+n)a_{n-1}(r)}{(2r+2n+1)^2}$, $n \geq 1$. $y_1 = |x|^{-1/2} \displaystyle\sum_{n=0}^{\infty} \dfrac{(-1)^n(2n)!x^n}{2^{4n}(n!)^3}$;

$y_2 = y_1 \ln |x| + |x|^{-1/2} \displaystyle\sum_{n=1}^{\infty} \dfrac{(-1)^n(2n)![2H_{2n} - 3H_n]x^n}{2^{4n}(n!)^3}$.

Intervals: $-\infty < x < 0$, $0 < x < \infty$.

23. R.F. $a_n(r) = -\dfrac{4(r+n)a_{n-1}(r)}{(4r+4n+1)^2}, \qquad n \geqq 1.$

$$y_1 = |x|^{-1/4}\left[1 + \sum_{n=1}^{\infty} \frac{(-1)^n[3 \cdot 7 \cdots (4n-1)]x^n}{4^{2n}(n!)^2}\right];$$

$$y_2 = y_1 \ln|x|$$

$$+|x|^{-1/4}\sum_{n=1}^{\infty} \frac{(-1)^n[3 \cdot 7 \cdots (4n-1)]\left[4\left(\dfrac{1}{3}+\dfrac{1}{7}+\cdots+\dfrac{1}{4n-1}\right)-2H_n\right]x^n}{4^{2n}(n!)^2}.$$

Intervals: $-\infty < x < 0, \quad 0 < x < \infty.$

24. R.F. $a_n(r) = -\dfrac{2(r+n)a_{n-2}(r)}{(2r+2n+3)^2}, \qquad n \geqq 2.$

$$y_1 = |x|^{-3/2}\left[1 + \sum_{n=1}^{\infty} \frac{(-1)^n[1 \cdot 5 \cdots (4n-3)]x^{2n}}{4^{2n}(n!)^2}\right];$$

$$y_2 = y_1 \ln|x|$$

$$+|x|^{-3/2}\sum_{n=1}^{\infty} \frac{(-1)^n[1 \cdot 5 \cdots (4n-3)]\left[2\left(1+\dfrac{1}{5}+\cdots+\dfrac{1}{4n-3}\right)-H_n\right]x^{2n}}{4^{2n}(n!)^2}.$$

Intervals: $-\infty < x < 0, \quad 0 < x < \infty.$

25. R.F. $a_n(r) = -\dfrac{(r+n)^2 a_{n-2}(r)}{(r+n+4)^2}, \qquad n \geqq 2.$

$y_1 = x^{-4}(1-x^2); \quad y_2 = y_1 \ln|x| + 2x^{-2}.$ Intervals: $-\infty < x < 0,$
$0 < x < \infty.$

26. R.F. $(r+n-3)(r+n-1)a_n(r) + (r+n-3)a_{n-1}(r) = 0, \quad n \geqq 1.$

$$y_1 = |x|(1-x); \quad y_2 = |x|^3\left[1 + \sum_{n=1}^{\infty} \frac{(-1)^n x^n}{3 \cdot 4 \cdots (n+2)}\right].$$

Intervals: $-\infty < x < 0, \quad 0 < x < \infty.$

27. R.F. $(r+n-4)(r+n+1)a_n(r) + (r+n-4)(r+n-1)a_{n-2}(r) = 0, \quad n \geqq 2.$

$$y_1 = |x|^{-1}; \quad y_2 = |x|^{-1}x^5 \sum_{n=0}^{\infty} \frac{(-1)^n 5 x^{2n}}{2n+5}.$$

Intervals: $-1 \leqq x < 0, \quad 0 < x \leqq 1.$

28. R.F. $(2r+2n-5)(2r+2n+5)a_n(r) - (2r+2n-5)a_{n-1}(r) = 0, \qquad n \geqq 1.$

$$y_1 = |x|^{-5/2}\left(1 + \frac{1}{2}x + \frac{1}{8}x^2 + \frac{1}{48}x^3 + \frac{1}{384}x^4\right);$$

$$y_2 = |x|^{5/2}\left[1 + \sum_{n=1}^{\infty} \frac{x^n}{2^n[6 \cdot 7 \cdots (n+5)]}\right].$$

Intervals: $-\infty < x < 0, \quad 0 < x < \infty.$

29. R.F. $(r+n+1)(r+n+4)a_n(r) - (r+n+1)a_{n-2}(r) = 0$, $n \geq 2$.

$$y_1 = x^{-4} \sum_{n=0}^{\infty} \frac{x^{2n}}{2^n n!} \; ; \quad y_2 = |x|^{-1} \left[1 + \sum_{n=1}^{\infty} \frac{x^{2n}}{5 \cdot 7 \cdots (2n+3)} \right].$$

Intervals: $-\infty < x < 0$, $0 < x < \infty$.

30. R.F. $a_n(r) = -\frac{a_{n-1}(r)}{(r+n-1)(r+n)}$, $n \geq 1$.

$$y_1 = \sum_{n=1}^{\infty} \frac{(-1)^n x^n}{(n-1)!n!} \; ; \quad y_2 = y_1 \ln |x| + 1 - \sum_{n=1}^{\infty} \frac{(-1)^n (H_{n-1} + H_n) x^n}{(n-1)!n!}.$$

Intervals: $-\infty < x < 0$, $0 < x < \infty$.

31. R.F. $a_n(r) = \frac{4 a_{n-2}(r)}{(r+n-4)(r+n-2)}$, $n \geq 2$.

$$y_1 = 2x^2 \sum_{n=1}^{\infty} \frac{x^{2n}}{(n-1)!n!} \; ; \quad y_2 = y_1 \ln |x| + x^2 \left[1 - \sum_{n=1}^{\infty} \frac{(H_{n-1} + H_n) x^{2n}}{(n-1)!n!} \right].$$

Intervals: $-\infty < x < 0$, $0 < x < \infty$.

32. R.F. $a_n(r) = \frac{(r+n)a_{n-1}(r)}{(r+n+1)(r+n+2)}$, $n \geq 1$.

$$y_1 = -|x|^{-1}; \quad y_2 = y_1 \ln |x| + 2|x|^{-1} + x^{-2} \left[1 - \sum_{n=2}^{\infty} \frac{x^n}{(n-1)!n!} \right].$$

Intervals: $-\infty < x < 0$, $0 < x < \infty$.

33. R.F. $(r+n)(r+n+2)a_n(r) - 4(r+n+3)a_{n-1}(r) = 0$, $n \geq 1$.

$$y_1 = -x^{-2} \sum_{n=2}^{\infty} \frac{4^n(n+1)x^n}{(n-2)!} \; ;$$

$$y_2 = y_1 \ln |x| - 8|x|^{-1} + x^{-2} \left[1 - \sum_{n=2}^{\infty} \frac{4^n(n+1)\left(\dfrac{1}{n+1} - H_{n-2} \right) x^n}{(n-2)!} \right].$$

Intervals: $-\infty < x < 0$, $0 < x < \infty$.

34. R.F. $a_n(r) = \frac{16 a_{n-4}(r)}{(r+n-3)(r+n+1)}$, $n \geq 4$.

$$y_1 = 4|x|^{-1} \sum_{n=1}^{\infty} \frac{x^{4n}}{(n-1)!n!} \; ;$$

$$y_2 = y_1 \ln |x| + |x|^{-1} \left[1 - \sum_{n=1}^{\infty} \frac{(H_{n-1} + H_n) x^{4n}}{(n-1)!n!} \right].$$

Intervals: $-\infty < x < 0$, $0 < x < \infty$.

35. R.F. $(3r + 3n - 2)(3r + 3n + 1)a_n(r) + (3r + 3n + 1)a_{n-1}(r) = 0,$ $n \geq 1.$

$$y_1 = |x|^{-1/3} \sum_{n=1}^{\infty} \frac{(-1)^n x^n}{3^{n-1}(n-1)!} \, ;$$

$$y_2 = y_1 \ln |x| + |x|^{-1/3} \left[3 - \sum_{n=2}^{\infty} \frac{(-1)^n H_{n-1} x^n}{3^{n-1}(n-1)!} \right].$$

Intervals: $-\infty < x < 0, \quad 0 < x < \infty.$

36. R.F. $(r + n + 1)(r + n + 2)a_n(r) - (r + n - 1)(r + n + 2)a_{n-1}(r) = 0,$ $n \geq 1.$
$y_1 = -2|x|^{-1} + 2;$

$$y_2 = y_1 \ln |x| + |x|^{-1} - 5 + x^{-2} \left[1 + \sum_{n=3}^{\infty} \frac{2x^n}{(n-2)(n-1)} \right].$$

Intervals: $-1 \leq x < 0, \quad 0 < x \leq 1.$

37. R.F. $a_n(r) = -\dfrac{a_{n-1}(r)}{(r + n - 1)(2r + 2n - 1)},$ $n \geq 1.$

$$y_1 = |x - 1| \left[1 + \sum_{n=1}^{\infty} \frac{(-1)^n (x - 1)^n}{n![3 \cdot 5 \cdots (2n + 1)]} \right];$$

$$y_2 = |x - 1|^{1/2} \left[1 + \sum_{n=1}^{\infty} \frac{(-1)^n (x - 1)^n}{n![1 \cdot 3 \cdots (2n - 1)]} \right].$$

38. R.F. $a_n(r) = \dfrac{2a_{n-1}(r) - a_{n-2}(r)}{(r + n)(3r + 3n + 1)},$ $n \geq 2.$

$$y_1 = 1 + \frac{1}{2}(x + 2) - \frac{1}{60}(x + 2)^2 - \frac{1}{1560}(x + 2)^4 + \frac{1}{5200}(x + 2)^5$$

$$+ \frac{1}{111150}(x + 2)^6 - \cdots;$$

$$y_2 = |x + 2|^{-1/3} \left[1 + (x + 2) + \frac{1}{10}(x + 2)^2 - \frac{1}{30}(x + 2)^3 - \frac{1}{264}(x + 2)^4 \right.$$

$$\left. + \frac{17}{46200}(x + 2)^5 + \frac{19}{428400}(x + 2)^6 + \cdots \right].$$

39. R.F. $a_n(r) = -\dfrac{(r + n - 2)^2 a_{n-1}(r) + a_{n-2}(r)}{(r + n - 2)(2r + 2n - 1)},$ $n \geq 2.$

$$y_1 = (x - 2)^2 \left[1 - \frac{1}{5}(x - 2) - \frac{1}{70}(x - 2)^2 + \frac{23}{1890}(x - 2)^3 - \frac{31}{7560}(x - 2)^4 \right.$$

$$\left. + \frac{683}{491400}(x - 2)^5 - \cdots \right];$$

$$y_2 = |x - 2|^{1/2} \left[1 + \frac{1}{4}(x - 2) - \frac{17}{32}(x - 2)^2 + \frac{121}{1152}(x - 2)^3 - \frac{577}{92160}(x - 2)^4 \right.$$

$$\left. - \frac{10447}{12902400}(x - 2)^5 - \cdots \right].$$

40. R.F. $a_n(r) = -\dfrac{(r+n-2)a_{n-3}(r) + a_{n-4}(r)}{(r+n)(r+n-2)}$, $n \geq 4$.

$$y_1 = 1 - \frac{1}{3}(x-1)^3 - \frac{1}{8}(x-1)^4 + \frac{1}{18}(x-1)^6 + \frac{23}{840}(x-1)^7 + \cdots;$$

$$y_2 = (x-1)^2\left[1 - \frac{1}{5}(x-1)^3 - \frac{1}{24}(x-1)^4 + \frac{1}{40}(x-1)^6 + \frac{59}{7560}(x-1)^7 + \cdots\right].$$

Exercises 7.6, page 425

1. (a)   6.9111,   (b) -1.3580.
8. Use Exercise 6(f).
9. Utilize Exercise 6(e), (f) with n replaced by $1 - m$.
12. (a) Use Exercise 11(c) with n replaced by $2n$, Example 7.6.3 and Exercise 8(b).

Exercises 7.7, page 440

1. Assume that Formula (7.7.45) is true for $k = s$, that is,

(1) $$b_{2m+2s} = -\frac{cc_{2s}}{2}(H_s + H_{m+s}).$$

We would like to show that Formula (7.7.45) is also true for $k = s + 1$. Substituting $k = s + 1$ in (7.7.35) and then using (1), we obtain

(2) $$b_{2m+2s+2} = -\frac{cc_{2s+2}}{2}\left(\frac{1}{s+1} + \frac{1}{m+s+1}\right) - \frac{b_{2m+2s}}{2^2(s+1)(m+s+1)}$$

$$= -\frac{cc_{2s+2}}{2}\left(\frac{1}{s+1} + \frac{1}{m+s+1}\right) + \frac{cc_{2s}}{2}\frac{(H_s + H_{m+s})}{2^2(s+1)(m+s+1)}.$$

Substituting $k = s + 1$ in (7.7.23), we obtain

(3) $$c_{2s+2} = -\frac{c_{2s}}{2^2(s+1)(s+m+1)}.$$

Solving for c_{2s} in (3) and then substituting it in (2), we obtain

$$b_{2m+2s+2} = -\frac{cc_{2s+2}}{2}\left(H_s + \frac{1}{s+1} + H_{m+s} + \frac{1}{m+s+1}\right)$$

$$= -\frac{cc_{2s+2}}{2}(H_{s+1} + H_{m+s+1}).$$

2. We shall restrict our attention to $x > 0$. With $\alpha = m$, (7.7.17) becomes

(1) $\displaystyle J_{-m}(x) = \sum_{n=0}^{\infty} \frac{(-1)^n}{n!\,\Gamma(n-m+1)} \left(\frac{x}{2}\right)^{-m+2n} = \sum_{n=m}^{\infty} \frac{(-1)^n}{n!\,\Gamma(n-m+1)} \left(\frac{x}{2}\right)^{-m+2n}$

since by assumption, $\dfrac{1}{\Gamma(n-m+1)} = 0$ for $n = 0,\ 1,\ \ldots,\ m-1$. Let $n = k + m$. Then, we may write (1) as

$$J_{-m}(x) = \sum_{k=0}^{\infty} \frac{(-1)^{k+m}}{(k+m)!\,\Gamma(k+1)} \left(\frac{x}{2}\right)^{m+2k}$$

But $(k+m)!\,\Gamma(k+1) = k!(k+m)(k+m-1)\cdots(k+2)(k+1)\Gamma(k+1)$
$= k!\,\Gamma(k+m+1)$. Thus,

$$J_{-m}(x) = (-1)^m \sum_{k=0}^{\infty} \frac{(-1)^k}{k!\,\Gamma(k+m+1)} \left(\frac{x}{2}\right)^{m+2k} = (-1)^m J_m(x).$$

3. We shall restrict our attention to $x > 0$. We have

$$J_0(x) = \sum_{k=0}^{\infty} \frac{(-1)^k x^{2k}}{2^{2k}(k!)^2} \quad \text{and} \quad J_1(x) = \sum_{k=0}^{\infty} \frac{(-1)^k x^{2k+1}}{2^{2k+1} k!(k+1)!}.$$

Thus,

$$J_0'(x) = \frac{d}{dx}\left[\sum_{k=0}^{\infty} \frac{(-1)^k x^{2k}}{2^{2k}(k!)^2}\right] = \sum_{k=1}^{\infty} \frac{(-1)^k 2k x^{2k-1}}{2^{2k}(k!)^2}$$

$$= \sum_{k=0}^{\infty} \frac{(-1)^{k+1} 2(k+1) x^{2k+1}}{2^{2k+2}(k+1)!(k+1)!} = -J_1(x).$$

5. $\displaystyle [xJ_1(x)]' = \frac{d}{dx}\left[\sum_{k=0}^{\infty} \frac{(-1)^k x^{2k+2}}{2^{2k+1} k!(k+1)!}\right] = x \sum_{k=0}^{\infty} \frac{(-1)^k x^{2k}}{2^{2k}(k!)^2} = xJ_0(x).$

6. $\displaystyle [x^{\alpha+1} J_{\alpha+1}(x)]' = \frac{d}{dx}\left[\sum_{n=0}^{\infty} \frac{(-1)^n x^{2n+2\alpha+2}}{2^{2n+\alpha+1} n!\,\Gamma(n+\alpha+2)}\right]$

$$= \sum_{n=0}^{\infty} \frac{(-1)^n x^{2n+2\alpha+1}}{2^{2n+\alpha} n!\,\Gamma(n+\alpha+1)} = x^{\alpha+1} J_\alpha(x).$$

8. $\displaystyle J_{\alpha-1}(x) - \frac{\alpha}{x} J_\alpha(x) = \sum_{n=0}^{\infty} \frac{(-1)^n x^{2n+\alpha-1}}{2^{2n+\alpha-1} n!\,\Gamma(n+\alpha)} - \alpha \sum_{n=0}^{\infty} \frac{(-1)^n x^{2n+\alpha-1}}{2^{2n+\alpha} n!\,\Gamma(n+\alpha+1)}$

$$= \sum_{n=0}^{\infty} \frac{(-1)^n (2n+\alpha) x^{2n+\alpha-1}}{2^{2n+\alpha} n!\,\Gamma(n+\alpha+1)} = J_\alpha'(x).$$

9. $\displaystyle \frac{\alpha}{x} J_\alpha(x) - J_{\alpha+1}(x) = \alpha \sum_{n=0}^{\infty} \frac{(-1)^n x^{2n+\alpha-1}}{2^{2n+\alpha} n!\,\Gamma(n+\alpha+1)} - \sum_{n=0}^{\infty} \frac{(-1)^n x^{2n+\alpha+1}}{2^{2n+\alpha+1} n!\,\Gamma(n+\alpha+2)}$

$$= \frac{\alpha x^{\alpha-1}}{2^\alpha \Gamma(\alpha+1)} + \sum_{n=1}^{\infty} \frac{(-1)^n \alpha x^{2n+\alpha-1}}{2^{2n+\alpha} n!\,\Gamma(n+\alpha+1)} + \sum_{n=1}^{\infty} \frac{(-1)^n x^{2n+\alpha-1}}{2^{2n+\alpha-1}(n-1)!\,\Gamma(n+\alpha+1)}$$

$$= \frac{\alpha x^{\alpha-1}}{2^\alpha \Gamma(\alpha+1)} + \sum_{n=1}^{\infty} \frac{(-1)^n (2n+\alpha) x^{2n+\alpha-1}}{2^{2n+\alpha} n!\,\Gamma(n+\alpha+1)} = J_\alpha'(x).$$

10. Utilize Exercises 8 and 9.

11. Utilize Exercises 10, 9, and 8.

12. $[xJ_0(x)J_1(x)]' = J_0(x)J_1(x) + xJ_0'(x)J_1(x) + xJ_0(x)J_1'(x).$ Now utilize Exercise 3 and Exercise 8 with $\alpha = 1$.

13. (b) Utilize Exercise 10 with $\alpha = \frac{1}{2}$.

14. $J_0(\frac{1}{2}) = 0.9385,$ $J_0(1) = 0.7652,$ $J_0(2) = 0.2239,$
$J_1(\frac{1}{2}) = 0.2423,$ $J_1(1) = 0.4400,$ $J_1(2) = 0.5767,$
$J_2(\frac{1}{2}) = 0.0306,$ $J_2(1) = 0.1149,$ $J_2(2) = 0.3528.$

15. (a) 0.3367,   (b) 0.1149,   (c) $-\dfrac{4}{\pi^2}$,   (d) 0.3222.

16. (a) Utilize the chain rule and Exercise 3 to show that $J_0'(\beta x) = -\beta J_1(\beta x)$. Also note that $\beta = 0$ is not a root of the equation $J_0(x) = 0$, since $J_0(0) = 1$.

18. (a) Utilize Exercises 3 and 5.

   (b) Let $u = \beta x$. Then, $xy'' + xy' + \beta^2 xy = 0$ becomes $u\dfrac{d^2y}{du^2} + \dfrac{dy}{du} + uy = 0$.
   The result may now be inferred from part (a).

19. Utilize Exercises 6 and 9 with α replaced by $\alpha - 1$.

23. (a) Utilize (8) of Exercise 22.
    (c) Utilize (5) and (6) of Exercise 22.

24. Let β be a root of the equation $f(x) = 0$. Then, $\beta \neq 0$. (Why?) Suppose that β is a repeated root. Then, $f(\beta) = 0$ and $f'(\beta) = 0$. Since $f(x)$ satisfies the equation

(1) $$xf''(x) + f'(x) + xf(x) = 0,$$

it follows that $\beta f''(\beta) = 0$. Hence, $f''(\beta) = 0$. Differentiating (1), we obtain

$$xf'''(x) + 2f''(x) + xf'(x) + f(x) = 0.$$

Thus, $f'''(\beta) = 0$. Continuing in this fashion, we obtain

$$f^{(n)}(\beta) = 0, \quad n = 0, 1, 2, \ldots.$$

Consequently, we have that

$$f(x) = \sum_{n=0}^{\infty} \frac{f^{(n)}(\beta)}{n!} x^n \equiv 0,$$

which is a contradiction.

25. Use Exercises 23(c) and 24.

26. Put $x = i\beta$ ($\beta \neq 0$) in (7.7.15). We obtain

$$J_0(i\beta) = \sum_{n=0}^{\infty} \frac{\beta^{2n}}{2^{2n}(n!)^2} = 1 + \frac{\beta^2}{2^2} + \frac{\beta^4}{2^4(2!)^2} + \frac{\beta^6}{2^6(3!)^2} + \cdots.$$

Thus, $J_0(i\beta) \neq 0$.

27. Suppose that $a + ib$ ($a \neq 0$, $b \neq 0$) was a root of the equation $J_0(x) = 0$. Since the coefficients of $J_0(x) = 0$ are all real numbers, it follows that $a - ib$ is also a root of $J_0(x) = 0$. Clearly, $(a + ib)^2 \neq (a - ib)^2$. In view of Exercise 21 (11), we then have that

$$\int_0^1 xJ_0[(a + ib)x]J_0[(a - ib)x]dx = 0,$$

which gives us a contradiction, since the value of this integral is greater than zero.   (Why?)

28. *Rolle's Theorem* asserts: *Let the function* g *be continuous for* $a \leq x \leq b$ *and differentiable for* $a < x < b$. *If* $g(a) = g(b)$, *then* $g'(x_1) = 0$ *for at least one* x_1 *such that* $a < x_1 < b$.   Since the functions J_0 and J_0' are continuous, it follows from Rolle's Theorem that $J_0'(x) = 0$ has at least one root between every pair of roots of the equation $J_0(x) = 0$.   The result now follows from Exercise 25.

29. Since $J_1(x) = -J_0'(x)$, the result now follows from Exercise 28.

30. In view of Exercise 25 and (6.6.12), the equations $J_0(x) = 0$ and $J_0'(x) = 0$ have no common root.   The result now follows, since $J_1(x) = -J_0'(x)$.

31. (a) Use Rolle's Theorem and Exercise 7.
    (b) Use Rolle's Theorem and Exercise 6.

32. Concerning (6). $x \operatorname{ber}' x = \displaystyle\sum_{k=1}^{\infty} \frac{(-1)^k 4k x^{4k}}{2^{4k}(2k!)^2}$.   Thus,

$$\frac{d}{dx}(x \operatorname{ber}' x) = \sum_{k=1}^{\infty} \frac{(-1)^k (4k)^2 x^{4k-1}}{2^{4k}(2k!)^2} = \sum_{k=1}^{\infty} \frac{(-1)^k (4k)^2 x^{4k-1}}{2^{4k}(2k)^2[(2k-1)!]^2}$$

$$= \sum_{k=1}^{\infty} \frac{(-1)^k x^{4k-1}}{2^{4k-2}[(2k-1)!]^2} = -x \sum_{k=1}^{\infty} \frac{(-1)^{k+1} x^{4k-2}}{2^{4k-2}[(2k-1)!]^2} = -x \operatorname{bei} x.$$

Exercises 7.8, page 454

2. Let $D^n \equiv \dfrac{d^n}{dx^n}$ so that (7.8.33) becomes

(1) $$2^n n! \int_{-1}^{1} x^n P_n(x)\,dx = \int_{-1}^{1} x^n D^n(x^2 - 1)^n \, dx.$$

Let $u = x^n$ and $dv = D^n(x^2 - 1)^n \, dx$; then $du = nx^{n-1}\,dx$ and $v = D^{n-1}(x^2 - 1)^n$.
Thus,

(2) $$\int_{-1}^{1} x^n D^n(x^2 - 1)^n \, dx = \left[x^n D^{n-1}(x^2 - 1)^n \right]_{-1}^{1} - n \int_{-1}^{1} x^{n-1} D^{n-1}(x^2 - 1)^n \, dx.$$

Since the $(n-1)$th derivative of $(x^2 - 1)^n$ has $(x^2 - 1)$ as a factor, the integrated term in (2) vanishes for $x = 1$ and $x = -1$.   Hence,

$$\int_{-1}^{1} x^n D^n(x^2 - 1)^n \, dx = -n \int_{-1}^{1} x^{n-1} D^{n-1}(x^2 - 1)^n \, dx.$$

Continuing the process of integrations by parts and noticing that each $D^{n-2}(x^2 - 1)^n$, $D^{n-3}(x^2 - 1)^n$, ..., $D'(x^2 - 1)^n$ has $(x^2 - 1)$ as a factor, we obtain

$$2^n n! \int_{-1}^{1} x^n P_n(x)\,dx = (-1)^{n-1} n(n-1) \cdots 2 \left\{ \left[x(x^2 - 1)^n \right]_{-1}^{1} - \int_{-1}^{1} (x^2 - 1)^n \, dx \right\}$$

$$= (-1)^n n! \int_{-1}^{1} (x^2 - 1)^n \, dx = (-1)^n n! \int_{-1}^{1} (-1)^n (1 - x^2)^n \, dx$$

$$= n! \int_{-1}^{1} (1 - x^2)^n \, dx.$$

3. As in hint to Exercise 2, we have for $k < n$ that

$$(1) \qquad 2^n n! \int_{-1}^{1} x^k P_n(x)dx = (-1)^k k! \left[D^{n-k-1}(x^2-1)^n \right]_{-1}^{1} = 0.$$

Since $P_k(x)$ is a polynomial of degree k, it follows from (1) that

$$(2) \qquad \int_{-1}^{1} P_k(x)P_n(x)dx = 0, \qquad k < n.$$

Interchanging the role of k and n in (2), the result may now be inferred.

4. *Concerning* (4). Let $v + j$ have a fixed value n. As j assumes the values of $0, 1, \ldots, k, \ldots, n$, we see that v assumes the values of $n, n-1, \ldots, n-k, \ldots, 0$, so that the total coefficient of $r^{v+j} = r^n$ is

$$\sum_{k=0}^{N} \frac{(-1)^k(2n-2k)! \, x^{n-2k}}{2^n k!(n-k)!(n-2k)!} = P_n(x),$$

where $N = n/2$ when n is even and $N = (n-1)/2$ when n is odd. (Why?)

5. Differentiating (4) of Exercise 4 with respect to r, show that

$$(1) \qquad (1 - 2xr + r^2)^{-3/2}(x-r) = \sum_{n=1}^{\infty} nP_n(x)r^{n-1}.$$

Show that (1) may be written as

$$(2) \qquad (x-r)\sum_{n=0}^{\infty} P_n(x)r^n = (1 - 2xr + r^2)\sum_{n=1}^{\infty} nP_n(x)r^{n-1}.$$

Show that (2) can be expressed in the form

$$\sum_{n=2}^{\infty} [(n+1)P_{n+1}(x) - (2n+1)xP_n(x) + nP_{n-1}(x)]r^n = 0,$$

from which the result now follows in view of Theorem 7.2.7.

6. Differentiating (4) of Exercise 4 with respect to r and with respect to x, we have

$$(1) \qquad (1 - 2xr + r^2)^{-3/2}(x-r) = \sum_{n=1}^{\infty} nP_n(x)r^{n-1}$$

and

$$(2) \qquad (1 - 2xr + r^2)^{-3/2} = \sum_{n=1}^{\infty} P_n'(x)r^{n-1}.$$

In view of (1) and (2), we see that

$$(3) \qquad \sum_{n=1}^{\infty} nP_n(x)r^{n-1} = (x-r)\sum_{n=1}^{\infty} P_n'(x)r^{n-1}.$$

Show that (3) can be expressed as

$$\sum_{n=2}^{\infty} [nP_n(x) - xP_n'(x) + P_{n-1}'(x)]r^{n-1} = 0,$$

from which the result now follows from Theorem 7.2.7.

7. Differentiate the result of Exercise 5, and then substitute into it the value of $xP'_n(x)$ of Exercise 6.

8. Differentiate the result of Exercise 5, and then substitute into it the value of $P'_{n-1}(x)$ of Exercise 6.

9. Use the fact that $P_{n+1}(x)$ and $P_{n-1}(x)$ satisfy the Legendre Equation (7.8.1) with $\alpha = n+1$ and $\alpha = n-1$, and show that

$$(1)\qquad (1-x^2)[P''_{n+1}(x) - P''_{n-1}(x)] - 2x[P'_{n+1}(x) - P'_{n-1}(x)]$$
$$+ (n+1)(n+2)P_{n+1}(x) - n(n-1)P_{n-1}(x) = 0.$$

Substitute $P'_{n+1}(x) - P'_{n-1}(x)$ of Exercise 7 and its derivative $P''_{n+1}(x) - P''_{n-1}(x) = (2n+1)P'_n(x)$ into (1); in the result substitute $(2n+1)\,xP_n(x)$ of Exercise 5.

10. Utilize Exercise 7 with n replaced successively by $(n-1)$, $(n-2)$, $(n-3)$, ...

11. Multiplying the result of Exercise 10 by x, we obtain

$$(1)\qquad xP'_n(x) = (2n-1)xP_{n-1}(x) + (2n-5)xP_{n-3}(x) + (2n-9)xP_{n-5}(x) + \cdots.$$

From Exercise 5, we have

$$(2)\qquad (2n+1)xP_n(x) = (n+1)P_{n+1}(x) + nP_{n-1}(x).$$

In (2), replace n successively by $(n-1)$, $(n-3)$, $(n-5)$, ...   With these values substituted into (1), we obtain the result.

12. Utilize (4) of Exercise 4.   Also use Theorem 7.2.7.

13. (a) Utilize Exercise 5 and Formula (7.8.26).
    (b) Utilize Exercises 9, 10, and Formula (7.8.26).
    (c) Utilize Formula (7.8.26).
    (d) Utilize Exercises 7 and 12 (a).

16. (a) Follow a method similar to the one used in establishing the first result in (7.8.26).
    (b) With m replaced by $m+1$ in Formula (4) of Exercise 15, we have

$$P_n^{m+1}(x) = (1-x^2)^{(m+1)/2}D^{m+1}[P_n(x)] = (1-x^2)^{(m+1)/2}D\{D^m[P_n(x)]\}$$
$$= (1-x^2)^{(m+1)/2}D[(1-x^2)^{-m/2}P_n^m(x)],$$

from which the result may now be obtained by carrying out the differentiation.

    (c) We have

$$(1)\qquad D\{(1-x^2)D[P_n^m(x)]\} = (1-x^2)D^2[P_n^m(x)] - 2xD[P_n^m(x)].$$

Since $P_n^m(x)$ satisfies Equation (3) of Exercise 15, we have

$$(2)\quad (1-x^2)D^2[P_n^m(x)] - 2xD[P_n^m(x)] + \left[n(n+1) - \frac{m^2}{1-x^2}\right]P_n^m(x) = 0.$$

The result now follows from (1) and (2).

(d) In view of part (b), we see that

(1)
$$\int_{-1}^{1} [P_n^{m+1}(x)]^2\, dx = \int_{-1}^{1} (1-x^2)\{D[P_n^m(x)]\}^2\, dx$$

$$+ 2m \int_{-1}^{1} xP_n^m(x)D[P_n^m(x)]dx + m^2 \int_{-1}^{1} \frac{x^2[P_n^m(x)]^2}{1-x^2}\, dx.$$

Integrating by parts the first two integrals in the right-hand side of (1), show that

(2)
$$\int_{-1}^{1} (1-x^2)\{D[P_n^m(x)]\}^2\, dx = -\int_{-1}^{1} P_n^m(x)D\{(1-x^2)D[P_n^m(x)]\}dx,$$

(3)
$$\int_{-1}^{1} xP_n^m(x)D[P_n^m(x)]dx = -\frac{1}{2}\int_{-1}^{1} [P_n^m(x)]^2\, dx.$$

Substituting (2) and (3) into (1), and then using part (c), the result may now be inferred.

(e) Replacing m by $m-1$ and m by $m-2$ in the result of part (d), we obtain

(1)
$$\int_{-1}^{1} [P_n^m(x)]^2\, dx = (n-m+1)(n+m)\int_{-1}^{1} [P_n^{m-1}(x)]^2\, dx$$

and

(2)
$$\int_{-1}^{1} [P_n^{m-1}(x)]^2\, dx = (n-m+2)(n+m-1)\int_{-1}^{1} [P_n^{m-2}(x)]^2\, dx.$$

Substituting (2) into (1), we get

(3)
$$\int_{-1}^{1} [P_n^m(x)]^2\, dx = [(n-m+1)(n-m+2)]$$

$$\times [(n+m)(n+m-1)]\int_{-1}^{1} [P_n^{m-2}(x)]^2\, dx.$$

Replacing m by $m-3$ in part (d), and then substituting this result into (3), show that

$$\int_{-1}^{1} [P_n^m(x)]^2\, dx = [(n-m+1)(n-m+2)(n-m+3)]$$

$$\times [(n+m)(n+m-1)(n+m-2)]\int_{-1}^{1} [P_n^{m-3}(x)]^2\, dx.$$

Continuing in this fashion, show that

$$\int_{-1}^{1} [P_n^m(x)]^2\, dx = [(n-m+1)(n-m+2)\cdots n]$$

$$\times [(n+m)(n+m-1)\cdots(n+1)]\int_{-1}^{1} [P_n(x)]^2\, dx$$

$$= \frac{(n+m)!}{(n-m)!}\int_{-1}^{1} [P_n(x)]^2 dx.$$

Exercises 7.9, page 459

1. (a) $y_1(x) = |x|^k \left[1 + \sum_{n=1}^{\infty} \dfrac{(-1)^n k(k-1) \cdots (k-2n+1)}{2^n n! (2k-1)(2k-3) \cdots (2k-2n+1)} x^{-2n} \right],$

$y_2(x) = |x|^{-(k+1)} \left[1 + \sum_{n=1}^{\infty} \dfrac{(k+1)(k+2) \ldots (k+2n)}{2^n n! (2k+3)(2k+5) \ldots (2k+2n+1)} x^{-2n} \right],$

$$|x| > 1.$$

(b) $\qquad y_1(x) = \sum_{n=0}^{[k/2]} \dfrac{(-1)^n k(k-1) \ldots (k-2n+1)}{2^n n! (2k-1)(2k-3) \ldots (2k-2n+1)} x^{k-2n},$

where $[p]$ means the greatest integer less than or equal to the real number p. [*See* Exercise 7.2.4.]   This is a polynomial of degree k, and when multiplied by the constant factor $(2k)!/2^k(k!)^2$ gives the Legendre polynomial of degree k:

$$P_k(x) = \frac{(2k)!}{2^k(k!)^2} y_1(x).$$

3. (b) $\qquad xF(1, 1, 2; -x) = x \left[1 + \sum_{k=1}^{\infty} \dfrac{(1 \cdot 2 \cdots k)(1 \cdot 2 \cdots k)}{k! 2 \cdot 3 \cdots (k+1)} (-x)^k \right]$

$$= x + \sum_{k=1}^{\infty} \frac{(-1)^k}{k+1} x^{k+1} = \ln(1+x).$$

(f) $F'(\alpha, \beta, \gamma; x) = \displaystyle\sum_{k=1}^{\infty} \dfrac{\alpha(\alpha+1) \cdots (\alpha+k-1)\beta(\beta-1) \cdots (\beta+k-1)}{k!\gamma(\gamma+1) \ldots (\gamma+k-1)} kx^{k-1}$

$= \dfrac{\alpha\beta}{\gamma} + \displaystyle\sum_{k=2}^{\infty} \dfrac{\alpha(\alpha+1) \cdots (\alpha+k-1)\beta(\beta-1) \cdots (\beta+k-1)}{(k-1)!\gamma(\gamma+1) \cdots (\gamma+k-1)} x^{k-1}$

$= \dfrac{\alpha\beta}{\gamma} + \dfrac{\alpha\beta}{\gamma} \displaystyle\sum_{k=1}^{\infty} \dfrac{(\alpha+1) \cdots (\alpha+k)(\beta+1) \cdots (\beta+k)}{k!(\gamma+1) \cdots (\gamma+k)} x^k$

$= \dfrac{\alpha\beta}{\gamma} F(\alpha+1, \beta+1, \gamma+1; x).$

7. (c) We have from (9):

(I) $\qquad x[L_n^{(\mu)}(x)]'' + (\mu+1-x)[L_n^{(\mu)}(x)]' + nL_n^{(\mu)}(x) = 0,$

(II) $\qquad x[L_m^{(\mu)}(x)]'' + (\mu+1-x)[L_m^{(\mu)}(x)]' + mL_m^{(\mu)}(x) = 0.$

Multiplying (I) and (II) by $x^\mu e^{-x}$, show that

(I)′ $\qquad \dfrac{d}{dx}(e^{-x}x^{\mu+1}[L_n^{(\mu)}(x)]') + ne^{-x}x^\mu L_n^{(\mu)}(x) = 0,$

(II)′ $\qquad \dfrac{d}{dx}(e^{-x}x^{\mu+1}[L_m^{(\mu)}(x)]') + me^{-x}x^\mu L_m^{(\mu)}(x) = 0.$

Multiplying (I)′ and (II)′ by $L_m^{(\mu)}(x)$ and $L_n^{(\mu)}(x)$, respectively, and then subtracting, we have

$$(m - n)e^{-x}x^\mu L_m^{(\mu)}(x)L_n^{(\mu)}(x) = L_m^{(\mu)}(x) \frac{d}{dx}(e^{-x}x^{\mu+1}[L_n^{(\mu)}(x)]')$$

$$-L_n^{(\mu)}(x) \frac{d}{dx}(e^{-x}x^{\mu+1}[L_m^{(\mu)}(x)]').$$

Thus,

$$\text{(III)} \quad (m - n)\int_a^b e^{-x}x^\mu L_m^{(\mu)}(x)L_n^{(\mu)}(x)dx$$

$$= \int_a^b L_m^{(\mu)}(x) \frac{d}{dx}(e^{-x}x^{\mu+1}[L_n^{(\mu)}(x)]')dx$$

$$- \int_a^b L_n^{(\mu)}(x) \frac{d}{dx}(e^{-x}x^{\mu+1}[L_m^{(\mu)}(x)]')dx.$$

Integrating by parts the integrals in the right-hand side of (III), show that

$$\text{(IV)} \quad (m - n)\int_a^b e^{-x}x^\mu L_m^{(\mu)}(x)L_n^{(\mu)}(x)dx$$

$$= \left[e^{-x}x^{\mu+1}\{L_m^{(\mu)}(x)[L_n^{(\mu)}(x)]' - L_n^{(\mu)}(x)[L_m^{(\mu)}(x)]'\}\right]_{x=a}^{x=b}.$$

Infer the result from (IV).

(d) In view of Leibnitz's rule [*see* (6.5.1)], we have

$$\frac{d^n}{dx^n}(x^n e^{-x}) = C_{n,\,0}(x^n)^{(n)}(e^{-x})^{(0)} + C_{n,\,1}(x^n)^{(n-1)}(e^{-x})' + C_{n,\,2}(x^n)^{(n-2)}(e^{-x})''$$

$$+ \cdots + C_{n,\,n-1}(x^n)'(e^{-x})^{(n-1)} + C_{n,\,n}(x^n)^{(0)}(e^{-x})^{(n)}$$

$$= C_{n,\,0}n!e^{-x} - C_{n,\,1}n(n-1)\cdots 2xe^{-x} + C_{n,\,2}n(n-1)\cdots 3x^2 e^{-x}$$

$$- \cdots + (-1)^{n-1}C_{n,\,n-1}nx^{n-1}e^{-x} + (-1)^n C_{n,\,n}x^n e^{-x}$$

$$= C_{n,0}n!e^{-x} - \frac{C_{n,\,1}}{1!}n!xe^{-x} + \frac{C_{n,\,2}}{2!}n!x^2 e^{-x} - \cdots$$

$$+ (-1)^{n-1}\frac{C_{n,\,n-1}}{(n-1)!}n!x^{n-1}e^{-x} + (-1)^n\frac{C_{n,\,n}}{n!}n!x^n e^{-x}$$

$$= n!e^{-x}\sum_{k=0}^n \frac{(-1)^k C_{n,\,k}}{k!}x^k$$

$$= n!e^{-x}\sum_{k=0}^n \frac{(-1)^k n!}{(k!)^2(n-k)!}x^k = n!e^{-x}L_n(x).$$

(e) $\displaystyle\int_0^\infty e^{-x}L_n{}^2(x)dx = \int_0^\infty e^{-x}\frac{e^x}{n!}D^n(x^n e^{-x})L_n(x)dx = \frac{1}{n!}\int_0^\infty D^n(x^n e^{-x})L_n(x)dx,$

where $D^n = d^n/dx^n$.   By n successive integrations by parts of the last integral above, show that

$$\int_0^\infty e^{-x}L_n{}^2(x)dx = \frac{(-1)^n}{n!}\int_0^\infty x^n e^{-x}D^n[L_n(x)]dx$$

$$= \frac{(-1)^n}{n!}\int_0^\infty x^n e^{-x}D^n\left[1 - nx + \cdots + \frac{(-1)^n}{n!}x^n\right]dx$$

$$= \frac{1}{n!}\int_0^\infty x^n e^{-x}\,dx = \frac{\Gamma(n+1)}{n!} = 1.$$

(f) *Concerning* (13).

$$xL_n'(x) = x\frac{d}{dx}\sum_{k=0}^n \frac{(-1)^k n!}{(k!)^2(n-k)!}x^k = \sum_{k=1}^n \frac{(-1)^k n! k}{(k!)^2(n-k)!}x^k.$$

$$nL_n(x) - nL_{n-1}(x) = \sum_{k=0}^n \frac{(-1)^k n! n}{(k!)^2(n-k)!}x^k - \sum_{k=0}^{n-1}\frac{(-1)^k(n-1)! n}{(k!)^2(n-1-k)!}x^k$$

$$= n + \sum_{k=1}^{n-1}\frac{(-1)^k n! n}{(k!)^2(n-k)!}x^k + \frac{(-1)^n n! n}{(n!)^2}x^n - n$$

$$\qquad - \sum_{k=1}^{n-1}\frac{(-1)^k(n-1)! n}{(k!)^2(n-1-k)!}x^k$$

$$= \sum_{k=1}^{n-1}\frac{(-1)^k}{(k!)^2}n\left[\frac{n!}{(n-k)!} - \frac{(n-1)!}{(n-1-k)!}\right]x^k + \frac{(-1)^n n! n}{(n!)^2}x^n$$

$$= \sum_{k=1}^n \frac{(-1)^k n! k}{(k!)^2(n-k)!}x^k = xL_n'(x).$$

Concerning (15).   Use (13) and (14).
Concerning (16).   Utilize (14).

8. (d) $\displaystyle e^{2xt - t^2} = e^{2xt}\cdot e^{-t^2} = \left(\sum_{n=0}^\infty \frac{(2x)^n}{n!}t^n\right)\left(\sum_{n=0}^\infty \frac{(-1)^n}{n!}t^{2n}\right).$

Utilizing Exercise 7.2.4, we obtain

$$e^{2xt - t^2} = \sum_{n=0}^\infty \sum_{k=0}^{[n/2]}\frac{(-1)^k(2x)^{n-2k}}{k!(n-2k)!}t^n = \sum_{n=0}^\infty \frac{1}{n!}\left(\sum_{k=0}^{[n/2]}\frac{(-1)^k n!(2x)^{n-2k}}{k!(n-2k)!}\right)t^n$$

$$= \sum_{n=0}^\infty \frac{H_n(x)}{n!}t^n.$$

(e) Using part (d), if x is held constant, then in view of the Maclaurin expansion of the generating function $e^{2xt - t^2}$, we have

$$H_n(x) = \left[\frac{d^n}{dt^n}(e^{2xt - t^2})\right]_{t=0}.$$

Also,

$$e^{-x^2} H_n(x) = \left\{ \frac{d^n}{dt^n} \left[e^{-(x-t)^2} \right] \right\}_{t=0}.$$

Letting $\xi = x - t$, the above expression becomes

$$e^{-x^2} H_n(x) = (-1)^n \left[\frac{d^n}{d\xi^n} (e^{-\xi^2}) \right]_{\xi=x} = (-1)^n \frac{d^n}{dx^n} (e^{-x^2}).$$

(f) Follow similar methods as given for the hints to Exercise 7(c) and (e).
(g) *Concerning* (8).   Consider the generating function

$$(4) \qquad\qquad e^{2xt - t^2} = \sum_{n=0}^{\infty} \frac{H_n(x)}{n!} t^n.$$

Differentiating (4) with respect to t, we obtain

$$2(x - t)e^{2xt - t^2} = \sum_{n=1}^{\infty} \frac{H_n(x)}{(n-1)!} t^{n-1}.$$

Show that the above expression can be written as

$$\sum_{n=1}^{\infty} \frac{2xH_{n-1}(x)}{(n-1)!} t^{n-1} - \sum_{n=2}^{\infty} \frac{2H_{n-2}(x)}{(n-2)!} t^{n-1} - \sum_{n=1}^{\infty} \frac{H_n(x)}{(n-1)!} t^{n-1} = 0.$$

Thus,

$$2xH_0(x) - H_1(x) + \sum_{n=2}^{\infty} \left[\frac{2xH_{n-1}(x)}{(n-1)!} - \frac{2H_{n-2}(x)}{(n-2)!} - \frac{H_n(x)}{(n-1)!} \right] t^{n-1} = 0,$$

from which the result may be deduced, in view of part (c) and Theorem 7.2.7.
Concerning (9).   Differentiating (4) with respect to x, we obtain

$$2te^{2xt - t^2} = \sum_{n=1}^{\infty} \frac{H_n'(x)}{n!} t^n.$$

Thus,

$$\sum_{n=1}^{\infty} \frac{2H_{n-1}(x)}{(n-1)!} t^{n-1} = \sum_{n=1}^{\infty} \frac{H_n'(x)}{n!} t^{n-1},$$

from which the result may be inferred.
Concerning (10).   Use (8) and (9).
(h) *Concerning* (12) and (13).   Utilize the generating function given in (4) and Theorem 7.2.7.
(i) *Concerning* (14).   We have

$$e^{2xt} = e^{t^2} \sum_{n=0}^{\infty} \frac{H_n(x)}{n!} t^n.$$

Thus,

$$\sum_{n=0}^{\infty} \frac{(2x)^n}{n!} t^n = \left(\sum_{n=0}^{\infty} \frac{t^{2n}}{n!} \right) \left(\sum_{n=0}^{\infty} \frac{H_n(x)}{n!} t^n \right) = \sum_{n=0}^{\infty} \sum_{k=0}^{[n/2]} \frac{H_{n-2k}(x)}{k!(n-2k)!} t^n.$$

Hence,

$$\frac{(2x)^n}{n!} = \sum_{k=0}^{[n/2]} \frac{H_{n-2k}(x)}{k!(n-2k)!},$$

and the result now follows.

Exercises 8.2, page 473

1. Let $y = \phi(t)$ be a solution of Equation (8.2.14) on an interval I, $\alpha \leqq t \leqq \beta$. Thus, we have

(1) $$\phi^{(n)} = F(t, \phi, \phi', \ldots, \phi^{(n-1)}).$$

Setting

(2) $$\psi_1(t) = \phi(t), \quad \psi_2(t) = \phi'(t), \quad \psi_3(t) = \phi''(t), \ldots,$$
$$\psi_{n-1}(t) = \phi^{(n-2)}(t), \quad \psi_n(t) = \phi^{(n-1)}(t),$$

and noticing that each $\psi_i(t)$ $(i = 1, \ldots, n)$ is defined on I, we have

(3) $$\begin{cases} \psi_1' = \phi' = \psi_2, \\ \psi_2' = \phi'' = \psi_3, \\ \quad \vdots \\ \psi_{n-1}' = \phi^{(n-1)} = \psi_n. \end{cases}$$

In view of (1) and (2), we see that

(4) $$\psi_n' = \phi^{(n)} = F(t, \psi_1, \psi_2, \ldots, \psi_n)$$

for each t in I. From (3) and (4), we have

(5) $$\begin{cases} \psi_1' = \psi_2, \\ \psi_2' = \psi_3, \\ \quad \vdots \\ \psi_n' = F(t, \psi_1, \psi_2, \ldots, \psi_n), \end{cases}$$

for each t in I. Thus, the set of functions $x_i = \psi_i(t)$, $i = 1, \ldots, n$ is a solution of the system (8.2.16) for each t in I.

Conversely, suppose that the set of functions $x_i = \psi_i(t)$, $i = 1, \ldots, n$ is a solution of the system (8.2.16) for each t in I. Setting $\phi(t) = \psi_1(t)$ and utilizing (5), we see that

(6) $$\begin{cases} \phi' = \psi_1' = \psi_2, \\ \phi'' = \psi_2' = \psi_3, \\ \quad \vdots \\ \phi^{(n)} = \psi_n' = F(t, \phi, \phi', \ldots, \phi^{(n-1)}), \end{cases}$$

for each t in I. In view of the last equation in (6), we claim that $y = \phi(t)$ is a solution of Equation (8.2.14) for each t in I. (Why?)

Exercises 8.3, page 479

1. $x_1 = \frac{11}{24}$, $x_2 = \frac{5}{12}$.
2. $x_1 = 3$, $x_2 = 1$, $x_3 = 2$.
3. $x_1 = 1$, $x_2 = 2$, $x_3 = 3$.
4. $x_1 = -2$, $x_2 = -1$, $x_3 = 2$, $x_4 = 3$.
5. No solution.
6. Infinitely many solutions: $x_1 = -\frac{3}{5}x_3$, $x_2 = \frac{8}{5}x_3 + 1$, $x_3 = x_3$.
7. Infinitely many solutions: $x_1 = -\frac{7}{2}x_4$, $x_2 = 2x_4$, $x_3 = \frac{3}{2}x_4 + 2$, $x_4 = x_4$.
13. Utilize the principle of mathematical induction.
14. Utilize Exercise 13.
15. Utilize the principle of mathematical induction.

Exercises 8.4, page 493

1. $x_1 = -c_1 t + c_2 + \cos t + 2 \sin t$, $x_2 = c_1 - \cos t$.
2. $x_1 = c_3 + c_2 t + c_1(\frac{1}{2}t^2 - t) + t^4 + t^5$, $x_2 = c_2 + c_1 t + t^4 + e^t$.
3. $x_1 = c_1 e^{-t} + \frac{1}{3}e^{2t}$, $x_2 = (c_2 + 2c_1 t)e^{-t} - \frac{1}{9}e^{2t} + 1$.
4. $x_1 = c_3 + c_2 t + \frac{1}{2}c_1 t^2$, $x_2 = \frac{1}{2}c_2 + \frac{1}{4}c_1(1 + 2t) + c_4 e^{2t} + \frac{1}{2}e^{4t}$.
5. $x_1 = c_1 + (3t - 2)e^t$, $x_2 = c_1 + (t - 2)e^t$.
6. $x = 3c_1 e^t - 2c_2 e^{-4t}$, $y = c_1 e^t + c_2 e^{-4t}$.
7. $x = (c_1 + t)e^{2t} + c_2 e^{-3t}$, $y = (3c_1 + 1 + 3t)e^{2t} - 2c_2 e^{-3t} - \cos t$.
8. $x = 2c_1 e^{-2t} - 2t + \frac{1}{2}$, $y = c_1 e^{-2t} + \frac{1}{2}t + \frac{1}{4}$.
9. $x = c_1 + c_2 e^{-t} + c_3 e^{-2t}$, $y = 2c_1 + 3c_2 e^{-t} + 6c_3 e^{-2t}$.
10. $x = c_1 \cos t + c_2 \sin t + (c_3 + t)e^t + c_4 e^{-t} - 2$,
    $y = -c_1 \cos t - c_2 \sin t + (c_3 + 2 + t)e^t + c_4 e^{-t} - t^2$.
11. $x_1 = c_1 + c_2 e^{-2t} + \frac{2}{3}t^3 - \frac{1}{2}t^2 + \frac{1}{2}t$,
    $x_2 = c_3 + 3c_1 t - \frac{5}{2}c_2 e^{-2t} + \frac{1}{2}t^4 - \frac{5}{6}t^3 + \frac{5}{4}t^2 + \frac{1}{2}t$.
12. $x_1 = 2c_1 + (c_3 - \frac{3}{2}c_2 t)e^t + (c_4 - \frac{9}{4}t)e^{-t} - 4t - 2$,
    $x_2 = c_1 + c_2 e^t - \frac{1}{2}e^{-t} - 2t$.
13. $x = c_1 e^t + (c_2 + \cos t - 3 \sin t)e^{-t} + c_3 e^{-2t}$,
    $y = 3c_1 e^t + (3c_2 + 8 \cos t - 4 \sin t)e^{-t} + 6c_3 e^{-2t}$.
14. $x = 5c_1 e^t + 5c_2 e^{-t} + (3c_3 + 2) \cos t + (3c_4 + 9t)\sin t$,
    $y = c_1 e^t + c_2 e^{-t} + c_3 \cos t + (c_4 + 3t) \sin t$.
15. $x = \frac{2}{3}c_1 e^t + (3c_2 - \frac{1}{4})e^{2t} + \frac{36}{5}c_3 e^{3t} + (c_5 - 4c_4 t)e^{-2t} + 3 \cos t - \sin t$,
    $y = c_1 e^t + c_2 e^{2t} + c_3 e^{3t} + c_4 e^{-2t} - 5 \cos t$.
16. $x = c_1 e^{2t} + c_2 e^{-t}$, $y = c_1 e^{2t} + c_3 e^{-t}$, $z = c_1 e^{2t} - c_2 e^{-t} - c_3 e^{-t}$.
17. $x = -c_1 e^{2t} + c_3 e^t + \frac{1}{2}c_2 e^{-t}$, $y = c_1 e^{2t} + c_2 e^{-t}$, $z = c_1 e^{2t} - c_3 e^t - \frac{7}{2}c_2 e^{-t}$.
18. $x = c_1 \exp \sqrt{2}t + c_2 \exp(-\sqrt{2}t) + c_3 \cos t + c_4 \sin t$,
    $y = c_1 \exp \sqrt{2}t + c_2 \exp(-\sqrt{2}t) + c_5 \cos t + c_6 \sin t$,
    $z = c_1 \exp \sqrt{2}t + c_2 \exp(-\sqrt{2}t) - (c_3 + c_5) \cos t - (c_4 + c_6) \sin t$.
19. $x = c_1 e^{-t} + e^t + t - 1$, $y = c_1 e^{-t} + c_2 e^{-2t} + e^t - \frac{1}{2}$,
    $z = -c_1 e^{-t} - c_2 e^{-2t} + e^t + t + \frac{1}{2}$.
20. $x = c_1 e^{-t/2}$, $y = -c_1 e^{-t/2} + c_2 e^{-t} + t^2$, $z = -\frac{3}{2}c_1 e^{-t/2} + c_2 e^{-t} + t^2 - 2t$.
21. Infinitely many. $x_2(t) = e^t - (D + 1)[x_1(t)]$,   where $x_1(t)$ is arbitrary.
22. Infinitely many. $x_2(t) = \sin t + \frac{1}{3}t^3 - \int (D + 2)[x_1(t)]dt$,   where $x_1(t)$ is arbitrary.

23. No solution.

24. Infinitely many. $x(t) = (D+1)[y(t)] - D[z(t)] + t,\quad y(t) = \int z(t)dt - e^t - t,$
    where $z(t)$ is arbitrary.

25. No solution.

Exercises 8.5, page 501

1. $x = \dfrac{Em}{H^2 e}\left[1 - \cos\left(\dfrac{He}{m}t\right)\right],\quad y = \dfrac{Em}{H^2 e}\left[\dfrac{He}{m}t - \sin\left(\dfrac{He}{m}t\right)\right].$

2. $x = \dfrac{V_0 \cos\alpha}{k}(1 - e^{-kt}),\quad y = \dfrac{1}{k}\left[\left(\dfrac{g}{k} + V_0\sin\alpha\right)(1 - e^{-kt}) - gt\right].$

3. $x_1 = -10(\sqrt{10} + 5)e^{((-5+\sqrt{10})/100)t} + 10(\sqrt{10} - 5)e^{-((5+\sqrt{10})/100)t} + 100,$

   $x_2 = -25(\sqrt{10} + 2)e^{((-5+\sqrt{10})/100)t} + 25(\sqrt{10}-2)\,e^{((-5+\sqrt{10})/100)t} + 100.$

4. (a) $x_1 = Pe^{-(r/G)t}\cosh\left(\dfrac{r}{G}t\right),\quad x_2 = Pe^{-(r/G)t}\sinh\left(\dfrac{r}{G}t\right).$

   (b) $x_1 \to \dfrac{P}{2}$ lb of salt, $\quad x_2 \to \dfrac{P}{2}$ lb of salt.

5. $x_1 = c_1\cos\alpha t + c_3\cos\beta t,$

   $$x_2 = \frac{m_1}{k_2}\left[\left(\frac{k_1+k_2}{m_1} - \alpha^2\right)c_1\cos\alpha t + \left(\frac{k_1+k_2}{m_1} - \beta^2\right)c_3\cos\beta t\right],$$

   where

   $$\alpha^2 = \left|\frac{-\left(\dfrac{k_1+k_2}{m_1} + \dfrac{k_2}{m_2}\right) + \sqrt{\left(\dfrac{k_1+k_2}{m_1} + \dfrac{k_2}{m_2}\right)^2 - 4\dfrac{k_1 k_2}{m_1 m_2}}}{2}\right|,\quad \alpha > 0,$$

   $$\beta^2 = \left|\frac{-\left(\dfrac{k_1+k_2}{m_1} + \dfrac{k_2}{m_2}\right) - \sqrt{\left(\dfrac{k_1+k_2}{m_1} + \dfrac{k_2}{m_2}\right)^2 - 4\dfrac{k_1 k_2}{m_1 m_2}}}{2}\right|,\quad \beta > 0,$$

   and c_1 and c_3 are determined by

   $$c_1 = \frac{\begin{vmatrix} s & 1 \\[2mm] \dfrac{k_2}{m_1}s & \dfrac{k_1+k_2}{m_1} - \beta^2 \end{vmatrix}}{\begin{vmatrix} 1 & 1 \\[2mm] \dfrac{k_1+k_2}{m_1} - \alpha^2 & \dfrac{k_1+k_2}{m_1} - \beta^2 \end{vmatrix}},\quad c_3 = \frac{\begin{vmatrix} 1 & s \\[2mm] \dfrac{k_1+k_2}{m_1} - \alpha^2 & \dfrac{k_2}{m_1}s \end{vmatrix}}{\begin{vmatrix} 1 & 1 \\[2mm] \dfrac{k_1+k_2}{m_1} - \alpha^2 & \dfrac{k_1+k_2}{m_1} - \beta^2 \end{vmatrix}}.$$

6. (a) $x_1 = \frac{1}{15}(8\cos 10t + 7\cos 40t)$, $x_2 = \frac{1}{15}(16\cos 10t - \cos 40t)$;

   (b) $x_1 = -\frac{1}{30}(\cos 10t + 14\cos 40t)$, $x_2 = -\frac{1}{15}(\cos 10t - \cos 40t)$;

   (c) $x_1 = \frac{7}{300}(4\sin 10t - \sin 40t)$, $x_2 = \frac{1}{300}(56\sin 10t + \sin 40t)$;

   (d) $x_1 = \frac{1}{60}(11\cos 10t - 56\cos 40t)$, $x_2 = \frac{1}{30}(11\cos 10t + 4\cos 40t)$;

   (e) $x_1 = -\frac{1}{60}(17\cos 10t + 28\cos 40t) + \frac{1}{150}(13\sin 10t - 7\sin 40t)$,

       $x_2 = \frac{1}{60}(-34\cos 10t + 4\cos 40t) + \frac{1}{150}(26\sin 10t + \sin 40t)$.

7. $i_1 = -\frac{1}{3}(2e^{-1000t} + e^{-4000t}) + 1$, $i_2 = \frac{1}{3}(e^{-1000t} - e^{-4000t})$.

8. $i_1 = \dfrac{5\sqrt{3}}{12}[(\sqrt{3}-1)e^{-500(2-\sqrt{3})t} + (\sqrt{3}+1)e^{-500(2+\sqrt{3})t}]$,

   $i_2 = \dfrac{5\sqrt{3}}{6}[e^{-500(2-\sqrt{3})t} - e^{-500(2+\sqrt{3})t}]$.

9. $i_1 = \frac{5}{7}e^{-50t}(7\cos 350t + 24\sin 350t)$, $i_2 = 1 - \frac{125}{7}e^{-50t}\sin 350t$.

10. (a) $i_1 = \dfrac{E_0}{R_1}\sin\omega t$, $i_2 = \dfrac{E_0}{Z_c}\cos(\omega t - \phi) - \dfrac{E_0}{\omega C Z_c^2}e^{-t/R_2C}$,

     where

$$Z_c = \sqrt{R_2^2 + \frac{1}{(\omega C)^2}}, \quad \phi = \tan^{-1}(\omega R_2 C), \quad 0 \le \phi < \frac{\pi}{2}.$$

   (b) $i_1 = 3\sin 120\pi t$, $i_2 = 14.87\cos(120\pi t - \phi) - 1.955e^{-50t}$,

      where $\phi = 82.44°$.

11. (a) $Q = EC + (Q_0 - EC)e^{-t/R_1C}$, $i_1 = \left(\dfrac{E}{R_1} - \dfrac{Q_0}{R_1 C}\right)e^{-t/R_1C}$,

     $i_2 = \dfrac{E}{R_2}(1 - e^{-(R_2/L)t})$, $i = i_1 + i_2$.

   (b) $Q = 2 - e^{-t}$, $i_1 = e^{-t}$, $i_2 = 10(1 - e^{-(25/4)t})$, $i = e^{-t} + 10(1 - e^{-(25/4)t})$

12. $Q = \dfrac{E_0}{\omega Z_c}\sin(\omega t - \phi_1) + \left(Q_0 + \dfrac{E_0 R_1}{\omega Z_c^2}\right)e^{-t/R_1C}$,

   $i_1 = \dfrac{E_0}{Z_c}\cos(\omega t - \phi_1) - \dfrac{1}{R_1 C}\left(Q_0 + \dfrac{E_0 R_1}{\omega Z_c^2}\right)e^{-t/R_1C}$,

   $i_2 = \dfrac{E_0}{Z_L}\sin(\omega t - \phi_2) + \dfrac{E_0 \omega L}{Z_L^2}e^{-(R_2/L)t}$,

   where

$$Z_c = \sqrt{R_1^2 + \frac{1}{(\omega C)^2}}, \quad Z_L = \sqrt{R_2^2 + \omega^2 L^2},$$

$$\phi_1 = \tan^{-1}(\omega R_1 C), \quad \phi_2 = \tan^{-1}\left(\frac{\omega L}{R_2}\right), \quad 0 \le \phi_1 < \frac{\pi}{2}, \quad 0 \le \phi_2 < \frac{\pi}{2}.$$

14. $i_1 = \frac{36}{13}(1 - e^{-500t})$, $\quad i_2 = \frac{6}{13}(1 - e^{-500t})$, $\quad i_3 = \frac{30}{13}(1 - e^{-500t})$.

15. (b) $i = 5e^{-(25/3)t}$.

16. $i_1 = -36e^{-2t} + 33e^{-t/2} + 3\cos t + 2\sin t$, $\quad i_2 = -18e^{-2t} - 33e^{-t/2}$
$\qquad + \cos t + 50.$

17. (a) $i_1 = -\dfrac{E}{R_1 + R_3}\left[1 - \dfrac{\alpha R_3}{R} e^{-t/RC}\right]$, $\quad i_2 = \dfrac{E}{R_2 + R_4}\left[1 + \dfrac{\alpha R_4}{R} e^{-t/RC}\right]$,

$\qquad i_3 = \dfrac{\alpha E}{R} e^{-t/RC}$,

where

$$R = \frac{(R_1 + R_3)(R_2 + R_4)(R_3 + R_4) - R_3{}^2(R_2 + R_4) - R_4{}^2(R_1 + R_3)}{(R_1 + R_3)(R_2 + R_4)}$$

and

$$\alpha = \frac{R_1 R_4 - R_2 R_3}{(R_1 + R_3)(R_2 + R_4)}.$$

(b) Current through R_4: $\quad |i_3 - i_2| = \left|-\frac{30}{19} e^{-(1400/19)t} - 5 + \frac{5}{19} e^{-(1400/19)t}\right|$
$\qquad\qquad\qquad\qquad\qquad = 5 + \frac{25}{19} e^{-(1400/19)t}.$

18. Current through L: $\quad |i_2 - i_3| = |3(1 + e^{-5t} - 2e^{-(20/3)t}) - 3(1 - e^{-5t})|$
$\qquad\qquad\qquad\qquad = 6|e^{-5t} - e^{-(20/3)t}| = 6(e^{-5t} - e^{-(20/3)t}).$

19. Current through C_1: $\quad |\frac{2}{5}(-3e^{-(3/4)t} + 3\cos t + 4\sin t)|$,
Current through C_2: $\quad |\frac{2}{5}(-3e^{-(3/4)t} + 3\cos t + 4\sin t)|$.

20. (b) $V_1 = -\frac{1}{4}e^{-t} + \frac{1}{4}\cos t + \frac{1}{2}\sin t$, $\quad V_2 = -\frac{1}{4}e^{-t} + \frac{1}{4}\cos t + \frac{1}{4}\sin t$,
$\qquad V_3 = \frac{1}{4}\sin t$.

21. (b) $x_1 = c_1 + c_2 t - \dfrac{M}{2m} c_3 \cos \beta t - \dfrac{M}{2m} c_4 \sin \beta t - c_5 \cos \alpha t - c_6 \sin \alpha t$,

$\qquad x_2 = c_1 + c_2 t - \dfrac{M}{2m} c_3 \cos \beta t - \dfrac{M}{2m} c_4 \sin \beta t + c_5 \cos \alpha t + c_6 \sin \alpha t$,

$\qquad x_3 = c_1 + c_2 t + c_3 \cos \beta t + c_4 \sin \beta t$,

where

$$\beta = \sqrt{\frac{k(M + 2m)}{Mm}} \quad \text{and} \quad \alpha = \sqrt{\frac{3k}{m}}.$$

(c) $x_1 = \dfrac{M}{M + 2m}\left(s + v_0 t - s \cos \beta t - \dfrac{v_0}{\beta} \sin \beta t\right)$,

$\qquad x_2 = \dfrac{M}{M + 2m}\left(s + v_0 t - s \cos \beta t - \dfrac{v_0}{\beta} \sin \beta t\right)$,

$\qquad x_3 = \dfrac{M}{M + 2m}(s + v_0 t) + \dfrac{2m}{M + 2m}\left(s \cos \beta t + \dfrac{v_0}{\beta} \sin \beta t\right).$

Exercises 8.6, page 518

2. *Concerning the associative law:* $\mathbf{A(BC)} = \mathbf{(AB)C}$. Let $\mathbf{A} = [a_{ij}]$ be an $m \times n$ matrix, $\mathbf{B} = [b_{jk}]$ be an $n \times p$ matrix, and $\mathbf{C} = [c_{kr}]$ be an $p \times q$ matrix. Then, $\mathbf{AB} = [d_{ik}]$, where $d_{ik} = \sum_{j=1}^{n} a_{ij} b_{jk}$, and $\mathbf{(AB)C} = [e_{ir}]$, where $e_{ir} = \sum_{k=1}^{p} d_{ik} c_{kr} = \sum_{k=1}^{p} \sum_{j=1}^{n} a_{ij} b_{jk} c_{kr}$. Now, $\mathbf{BC} = [f_{jr}]$, where $f_{jr} = \sum_{k=1}^{p} b_{jk} c_{kr}$, and $\mathbf{A(BC)} = [g_{ir}]$, where $g_{ir} = \sum_{j=1}^{n} a_{ij} f_{jr} = \sum_{j=1}^{n} a_{ij} \sum_{k=1}^{p} b_{jk} c_{kr} = \sum_{j=1}^{n} \sum_{k=1}^{p} a_{ij} b_{jk} c_{kr} = \sum_{k=1}^{p} \sum_{j=1}^{n} a_{ij} b_{jk} c_{kr} = e_{ir}$. Thus, $\mathbf{A(BC)} = \mathbf{(AB)C}$.

   Concerning the distributive law: $\mathbf{A(B+C)} = \mathbf{AB} + \mathbf{AC}$. Let $\mathbf{A} = [a_{ij}]$ be an $m \times n$ matrix, $\mathbf{B} = [B_{jk}]$ and $\mathbf{C} = [c_{jk}]$ be $n \times p$ matrices. Now, $\mathbf{B+C} = [b_{jk} + c_{jk}] = [g_{jk}]$, and $\mathbf{A(B+C)} = [h_{ik}]$, where $h_{ik} = \sum_{j=1}^{n} a_{ij} g_{jk} = \sum_{j=1}^{n} a_{ij} (b_{jk} + c_{jk}) = \sum_{j=1}^{n} a_{ij} b_{jk} + \sum_{j=1}^{n} a_{ij} c_{jk} = d_{ik} + \alpha_{ik}$. But, $[d_{ik}] = \mathbf{AB}$ and $[\alpha_{ik}] = \mathbf{AC}$. Thus, $\mathbf{A(B+C)} = \mathbf{AB} + \mathbf{AC}$.

3. *Concerning* $\mathbf{(A+B)}^T = \mathbf{A}^T + \mathbf{B}^T$: Let $\mathbf{A} = [a_{ij}]$ and $\mathbf{B} = [b_{ij}]$ be $m \times n$ matrices. Then $\mathbf{A+B} = [c_{ij}]$, were $c_{ij} = a_{ij} + b_{ij}$. Thus, $\mathbf{(A+B)}^T = [d_{ji}]$, where $d_{ji} = c_{ij} = a_{ij} + b_{ij}$. Now $\mathbf{A}^T = [e_{ji}]$, where $e_{ji} = a_{ij}$ and $\mathbf{B}^T = [f_{ji}]$, where $f_{ji} = b_{ij}$. Thus, $\mathbf{A}^T + \mathbf{B}^T = [h_{ji}]$, where $h_{ji} = e_{ji} + f_{ji} = a_{ij} + b_{ij} = c_{ij} = d_{ji}$. Hence, $\mathbf{(A+B)}^T = \mathbf{A}^T + \mathbf{B}^T$.

   Concerning $\mathbf{(AB)}^T = \mathbf{B}^T \mathbf{A}^T$: Let $\mathbf{A} = [a_{ij}]$ be an $m \times n$ matrix and $\mathbf{B} = [b_{jk}]$ be an $n \times p$ matrix. Then $\mathbf{A}^T = [c_{ji}]$, where $c_{ji} = a_{ij}$ and $\mathbf{B}^T = [d_{kj}]$, where $d_{kj} = b_{jk}$. Thus, $\mathbf{B}^T \mathbf{A}^T = [e_{ki}]$, where $e_{ki} = \sum_{j=1}^{n} d_{kj} c_{ji} = \sum_{j=1}^{n} a_{ij} b_{jk} = e_{ik}$. However, $\mathbf{AB} = [e_{ik}]$. Thus, $\mathbf{(AB)}^T = [e_{ki}] = \mathbf{B}^T \mathbf{A}^T$.

4. *Concerning* $\dfrac{d}{dt}(\mathbf{C}\mathbf{A}(t)) = \mathbf{C}\dfrac{d\mathbf{A}(t)}{dt}$: Let $\mathbf{C} = [c_{ij}]$ be an $m \times n$ constant matrix, and $\mathbf{A}(t) = [a_{jk}(t)]$ be an $n \times p$ matrix. Then $\mathbf{C}\mathbf{A}(t) = [e_{ik}(t)]$, where

$$e_{ik}(t) = \sum_{j=1}^{n} c_{ij} a_{jk}(t).$$

Thus,

$$\frac{d}{dt} e_{ik}(t) = \frac{d}{dt} \sum_{j=1}^{n} c_{ij} a_{jk}(t) = \sum_{j=1}^{n} \frac{d}{dt}(c_{ij} a_{jk}(t)) = \sum_{j=1}^{n} c_{ij} \frac{d}{dt} a_{jk}(t).$$

Hence, $\dfrac{d}{dt}(\mathbf{C}\mathbf{A}(t)) = \mathbf{C}\dfrac{d}{dt}\mathbf{A}(t)$.

5. (a) $\begin{bmatrix} 8 & 1 & 3 \\ -1 & 6 & 8 \\ 9 & 8 & 1 \end{bmatrix}$,   (b) $\begin{bmatrix} -1 & 7 & 0 \\ -7 & 7 & -1 \\ -4 & -2 & 6 \end{bmatrix}$,   (c) $\begin{bmatrix} 3 & 1 & 7 \\ -1 & -1 & 1 \\ 5 & 3 & -3 \end{bmatrix}$,

  (d) $\begin{bmatrix} -9 & 7 & 2 \\ -10 & 13 & 2 \\ -4 & 15 & -14 \end{bmatrix}$,   (f) $\begin{bmatrix} 2 & 4 & 3 \\ -3 & 9 & 9 \\ 2 & 8 & 4 \end{bmatrix}$,   (h) $\begin{bmatrix} 4 & 2 & 7 \\ -1 & 4 & 4 \\ 5 & 3 & 0 \end{bmatrix}$.

6. (a) $\begin{bmatrix} 1 & 2 \\ -1 & -5 \\ 1 & 4 \end{bmatrix}$,   (b) $[-3]$,   (c) $\begin{bmatrix} a_1 & a_2 \\ b_1 & b_2 \\ c_1 & c_2 \\ d_1 & d_2 \end{bmatrix}$,

(d) $\begin{bmatrix} \alpha a_1 & \alpha a_2 & \alpha a_3 \\ \alpha b_1 & \alpha b_2 & \alpha b_3 \\ \alpha c_1 & \alpha c_2 & \alpha c_3 \end{bmatrix}$,   (e) $\begin{bmatrix} -2 & 1 & 2 & 6 \\ 4 & -2 & -4 & -12 \\ 6 & -3 & -6 & -18 \\ -2 & 1 & 2 & 6 \end{bmatrix}$,

(f) $\begin{bmatrix} i & 1+i \\ 1+i & 2-i \end{bmatrix}$.

9. $0, -3$.   10. (a) Yes.   (b) $\begin{bmatrix} \alpha_1{}^p & 0 & 0 \\ 0 & \alpha_2{}^p & 0 \\ 0 & 0 & \alpha_3{}^p \end{bmatrix}$.

11. In general, $\mathbf{AB} \neq \mathbf{BA}$.

12. (a) $\begin{bmatrix} e^t - e^{-t} & e^t - e^{-t} \\ 1 - \sin t & 2t + \cos t \end{bmatrix}$,

(b) $\begin{bmatrix} (1+t)e^t & (2t+t^2)e^t \\ (t - e^{-t})\cos t + (1 - e^{-t})\sin t & (t^2 - e^{-t})\cos t + (2t - e^{-t})\sin t \end{bmatrix}$,

(c) $\begin{bmatrix} e - e^{-1} & e - e^{-1} \\ \frac{1}{2} + \sin 1 & \frac{4}{3} - \cos 1 \end{bmatrix}$.

13. We demonstrate the result for the case when $r = 2$ and $s = 3$. The general proof may be given by using mathematical induction. In view of the Associative Law (8.6.11), we see that $\mathbf{A}^2 = \mathbf{AA}$ and $\mathbf{B}^3 = \mathbf{BBB}$. Thus $\mathbf{A}^2\mathbf{B}^3 = \mathbf{AABBB} = \mathbf{ABABB} = \mathbf{ABBAB} = \mathbf{ABBBA} = \mathbf{BABBA} = \mathbf{BBABA} = \mathbf{BBBAA} = \mathbf{B}^3\mathbf{A}^2$.

14. Use the principle of mathematical induction and follow a similar proof as given for the binomial expansion in the algebra of real numbers.

15. *Concerning* (a): $\operatorname{tr}(\mathbf{A} + \mathbf{B}) = \sum_{i=1}^{n}(a_{ii} + b_{ii}) = \sum_{i=1}^{n} a_{ii} + \sum_{i=1}^{n} b_{ii} = \operatorname{tr}\mathbf{A} + \operatorname{tr}\mathbf{B}$. *Concerning* (d): The product $\mathbf{AB}$ is the $m \times m$ matrix $\mathbf{C} = [c_{ij}]$, where $c_{ij} = \sum_{k=1}^{n} a_{ik}b_{kj}$, $i, j = 1, 2, \ldots, m$. The product $\mathbf{BA}$ is the $n \times n$ matrix $\mathbf{D} = [d_{ij}]$, where $d_{ij} = \sum_{k=1}^{m} b_{ik}a_{kj}$, $i, j = 1, 2, \ldots, n$. Thus, $\operatorname{tr}(\mathbf{AB}) = \sum_{i=1}^{m} c_{ii} = \sum_{i=1}^{m} \sum_{k=1}^{n} a_{ik}b_{ki}$. Also, $\operatorname{tr}(\mathbf{BA}) = \sum_{i=1}^{n} d_{ii} = \sum_{i=1}^{n} \sum_{k=1}^{m} b_{ik}a_{ki}$. The result now follows, since $\sum_{i=1}^{m} \sum_{k=1}^{n} a_{ik}b_{ki} = \sum_{i=1}^{n} \sum_{k=1}^{m} b_{ik}a_{ki}$. (Why?)

Exercises 8.7, page 529

1. Suppose that there exist two inverses $\mathbf{B}$ and $\mathbf{C}$ for the matrix $\mathbf{A}$. Then, $\mathbf{AB} = \mathbf{BA} = \mathbf{I}$ and $\mathbf{CA} = \mathbf{AC} = \mathbf{I}$. Thus, $\mathbf{AB} = \mathbf{CA}$. But $\mathbf{AB} = \mathbf{BA}$. Consequently, $\mathbf{BA} = \mathbf{CA}$, $(\mathbf{BA})\mathbf{B} = (\mathbf{CA})\mathbf{B}$, and in view of (8.6.11), $\mathbf{B}(\mathbf{AB}) = \mathbf{C}(\mathbf{AB})$. But $\mathbf{AB} = \mathbf{I}$. Hence, $\mathbf{BI} = \mathbf{CI}$ or $\mathbf{B} = \mathbf{C}$.

2. (a), (b). Since $\mathbf{A}$ and $\mathbf{B}$ are nonsingular matrices, $\mathbf{A}^{-1}$ and $\mathbf{B}^{-1}$ exist. Hence,

$$(\mathbf{B}^{-1}\mathbf{A}^{-1})(\mathbf{AB}) = \mathbf{B}^{-1}(\mathbf{A}^{-1}(\mathbf{AB})) = \mathbf{B}^{-1}((\mathbf{A}^{-1}\mathbf{A})\mathbf{B}) = \mathbf{B}^{-1}(\mathbf{IB}) = \mathbf{B}^{-1}\mathbf{B} = \mathbf{I}.$$

Similarly, $(\mathbf{AB})(\mathbf{B}^{-1}\mathbf{A}^{-1}) = \mathbf{I}$. Thus, $\mathbf{AB}$ is a nonsingular matrix, and $\mathbf{B}^{-1}\mathbf{A}^{-1}$ is its inverse.

3. Since $\mathbf{A}$ is nonsingular, $\mathbf{A}^{-1} \equiv \mathbf{B}$ exists.   Then, $\mathbf{AB} = \mathbf{I}$, and $\mathbf{IB}^{-1} = (\mathbf{AB})\mathbf{B}^{-1}$ $= \mathbf{A}(\mathbf{BB}^{-1}) = \mathbf{AI}$.   Thus, $\mathbf{A} = \mathbf{B}^{-1} = (\mathbf{A}^{-1})^{-1}$.

4. $\begin{bmatrix} 3 & -2 \\ -1 & 1 \end{bmatrix}$.

5. $\begin{bmatrix} -\dfrac{i}{2} & \dfrac{1}{2} \\ \dfrac{1}{2} & -\dfrac{i}{2} \end{bmatrix}$.

6. $\begin{bmatrix} 5 & -12 & 3 \\ 2 & -4 & 1 \\ -2 & 5 & -1 \end{bmatrix}$.

7. $\begin{bmatrix} -\dfrac{29}{6} & \dfrac{19}{6} & -\dfrac{1}{6} \\ \dfrac{1}{6} & \dfrac{1}{6} & -\dfrac{1}{6} \\ \dfrac{11}{6} & -\dfrac{7}{6} & \dfrac{1}{6} \end{bmatrix}$.

8. $\begin{bmatrix} -\dfrac{1}{10} & \dfrac{3}{10} & \dfrac{3}{5} \\ -\dfrac{1}{20} & \dfrac{3}{20} & -\dfrac{1}{5} \\ -\dfrac{1}{2} & \dfrac{1}{2} & 1 \end{bmatrix}$.

9. $\begin{bmatrix} 5 & -\dfrac{5}{2} & 2 \\ -1 & \dfrac{3}{2} & -1 \\ 1 & 0 & 0 \end{bmatrix}$.

10. $\begin{bmatrix} 15 & 6 & -15 & 4 \\ 11 & 5 & -12 & 3 \\ 4 & 2 & -4 & 1 \\ -4 & -2 & 5 & -1 \end{bmatrix}$.

11. $\begin{bmatrix} \dfrac{8}{15} & \dfrac{3}{5} & -\dfrac{2}{5} & \dfrac{1}{3} \\ -\dfrac{2}{5} & \dfrac{4}{5} & -\dfrac{1}{5} & 0 \\ -\dfrac{2}{5} & -\dfrac{1}{5} & -\dfrac{1}{5} & 0 \\ -\dfrac{1}{3} & 1 & 0 & -\dfrac{1}{3} \end{bmatrix}$.

12. $\begin{bmatrix} \dfrac{1}{\alpha_1} & 0 & 0 & 0 & 0 \\ 0 & \dfrac{1}{\alpha_2} & 0 & 0 & 0 \\ 0 & 0 & \dfrac{1}{\alpha_3} & 0 & 0 \\ 0 & 0 & 0 & \dfrac{1}{\alpha_4} & 0 \\ 0 & 0 & 0 & 0 & \dfrac{1}{\alpha_5} \end{bmatrix}$.

13. $\begin{bmatrix} 1 & -1 & 0 & 0 & 0 \\ 0 & 1 & -1 & 0 & 0 \\ 0 & 0 & 1 & -1 & 0 \\ 0 & 0 & 0 & 1 & -1 \\ 0 & 0 & 0 & 0 & 1 \end{bmatrix}$.

14. 2.

15. 1.

16. 2.

17. 1.

18. 3.

19. 2.

21. *Concerning* (a): Since $\mathbf{A}$ is nonsingular, $\mathbf{A}^{-1}$ exists. Since $\mathbf{AB} = \mathbf{AC}$, we have $\mathbf{A}^{-1}(\mathbf{AB}) = \mathbf{A}^{-1}(\mathbf{AC})$. In view of (8.6.11), we then have $(\mathbf{A}^{-1}\mathbf{A})\mathbf{B} = (\mathbf{A}^{-1}\mathbf{A})\mathbf{C}$. Thus, $\mathbf{IB} = \mathbf{IC}$ or $\mathbf{B} = \mathbf{C}$.
    Concerning (b): Since $\mathbf{AB} = \mathbf{BA}$, it follows that $\mathbf{A}^{-1}(\mathbf{AB})\mathbf{A}^{-1} = \mathbf{A}^{-1}(\mathbf{BA})\mathbf{A}^{-1}$. But $\mathbf{A}^{-1}(\mathbf{AB})\mathbf{A}^{-1} = (\mathbf{A}^{-1}\mathbf{A})(\mathbf{BA}^{-1}) = \mathbf{I}(\mathbf{BA}^{-1}) = \mathbf{BA}^{-1}$; and $\mathbf{A}^{-1}(\mathbf{BA})\mathbf{A}^{-1} = (\mathbf{A}^{-1}\mathbf{B})(\mathbf{AA}^{-1}) = (\mathbf{A}^{-1}\mathbf{B})\mathbf{I} = \mathbf{A}^{-1}\mathbf{B}$. Thus, $\mathbf{A}^{-1}\mathbf{B} = \mathbf{BA}^{-1}$.
    Concerning (d): Since $\mathbf{AB} = \mathbf{BA}$, it follows that $\mathbf{A}^{-1}\mathbf{B}^{-1}(\mathbf{AB})\mathbf{A}^{-1}\mathbf{B}^{-1} = \mathbf{A}^{-1}\mathbf{B}^{-1}(\mathbf{BA})\mathbf{A}^{-1}\mathbf{B}^{-1}$. But $\mathbf{A}^{-1}\mathbf{B}^{-1}(\mathbf{AB})\mathbf{A}^{-1}\mathbf{B}^{-1} = \mathbf{A}^{-1}(\mathbf{B}^{-1}\mathbf{A})(\mathbf{BA}^{-1})\mathbf{B}^{-1} = \mathbf{A}^{-1}(\mathbf{AB}^{-1})(\mathbf{A}^{-1}\mathbf{B})\mathbf{B}^{-1} = (\mathbf{A}^{-1}\mathbf{A})(\mathbf{B}^{-1}\mathbf{A}^{-1})(\mathbf{BB}^{-1}) = \mathbf{I}(\mathbf{B}^{-1}\mathbf{A}^{-1})\mathbf{I} = \mathbf{B}^{-1}\mathbf{A}^{-1}$. Also, $\mathbf{A}^{-1}\mathbf{B}^{-1}(\mathbf{BA})\mathbf{A}^{-1}\mathbf{B}^{-1} = \mathbf{A}^{-1}\mathbf{B}^{-1}$. Thus, $\mathbf{A}^{-1}\mathbf{B}^{-1} = \mathbf{B}^{-1}\mathbf{A}^{-1}$.

22. *Concerning* (a): Now, $\mathbf{A}^{-1}\mathbf{A} = \mathbf{I} = \mathbf{I}^T = (\mathbf{AA}^{-1})^T = (\mathbf{A}^{-1})^T\mathbf{A}^T = (\mathbf{A}^{-1})^T\mathbf{A}$. Thus, $(\mathbf{A}^{-1}\mathbf{A})\mathbf{A}^{-1} = ((\mathbf{A}^{-1})^T\mathbf{A})\mathbf{A}^{-1}$, from which we deduce that $(\mathbf{A}^{-1})^T = \mathbf{A}^{-1}$.
    Concerning (d): Utilizing Exercise 21(d), Formula (8.6.18), and part (a), we obtain $(\mathbf{A}^{-1}\mathbf{B}^{-1})^T = (\mathbf{B}^{-1}\mathbf{A}^{-1})^T = (\mathbf{A}^{-1})^T(\mathbf{B}^{-1})^T = \mathbf{A}^{-1}\mathbf{B}^{-1}$.

23. (a) Utilize (8.7.1) and (8.6.24).

(b) $\begin{bmatrix} -\sin t & \cos t \\ -\cos t & -\sin t \end{bmatrix}.$

Exercises 8.8, page 537

1. *Concerning* (8.8.7): $\mathbf{x}(\mathbf{y} + \mathbf{z})^T = \sum_{i=1}^n x_i(y_i + z_i) = \sum_{i=1}^n x_i y_i + \sum_{i=1}^n x_i z_i = \mathbf{xy}^T + \mathbf{xz}^T$.

2. *Concerning* (8.8.13): $(\mathbf{x}, \mathbf{y} + \mathbf{z}) = \sum_{k=1}^n \bar{x}_k(y_k + z_k) = \sum_{k=1}^n \bar{x}_k y_k + \sum_{k=1}^n \bar{x}_k z_k = (\mathbf{x}, \mathbf{y}) + (\mathbf{x}, \mathbf{z})$.

3. (a) $[0, 2, 4, 2 + 2i]$,   (b) $[-2i, 2i, 2i, 2 - 2i]$,   (c) $[1 - i, 2i, 1 + 3i, 2 + 2i]$,
   (d) $[2, -2 - 2i, -2 - 4i, 4]$,   (e) $8 + 4i$,   (f) $8 + 4i$,   (g) $2 - 2i$,
   (h) $2 + 2i$,   (i) 12,   (j) 12,   (k) $-4 - 4i$,   (l) $4 + 4i$.

4. (a) Linearly independent.   (b) Linearly dependent. $\mathbf{x}^{(3)} = 2\mathbf{x}^{(1)} + \mathbf{x}^{(2)}$.
   (c) Linearly dependent. $\mathbf{x}^{(2)} = \frac{1}{2}\mathbf{x}^{(1)} + 2\mathbf{x}^{(3)}$.
   (d) Linearly dependent. $\mathbf{x}^{(1)} = 2\mathbf{x}^{(2)} + \frac{2}{3}\mathbf{x}^{(3)}$.   (e) Linearly dependent.
       $\mathbf{x}^{(2)} = \frac{1}{2}\mathbf{x}^{(1)} - 3\mathbf{x}^{(3)}$, $\mathbf{x}^{(4)} = \frac{3}{2}\mathbf{x}^{(1)} + 3\mathbf{x}^{(3)}$.   (f) Linearly independent.
   (g) Linearly dependent. $\mathbf{x}^{(1)} = \frac{1}{2}\mathbf{x}^{(2)} + \frac{1}{3}\mathbf{x}^{(4)}$, $\mathbf{x}^{(3)} = 2\mathbf{x}^{(2)} + \frac{5}{3}\mathbf{x}^{(4)}$.
   (h) Linearly independent.   (i) Linearly dependent. $\mathbf{x}^{(3)} = 2\mathbf{x}^{(1)} + i\mathbf{x}^{(2)}$.
   (j) Linearly dependent. $\mathbf{x}^{(1)} = 2\mathbf{x}^{(3)} - \mathbf{x}^{(4)}$, $\mathbf{x}^{(2)} = -2i\mathbf{x}^{(3)} + 2\mathbf{x}^{(4)}$.

5. (a) Linearly independent.   (b) Linearly dependent. $\mathbf{x}^{(3)}(t) = 2\mathbf{x}^{(1)}(t) + 3\mathbf{x}^{(2)}(t)$.
   (c) Linearly dependent. $\mathbf{x}^{(1)}(t) = \mathbf{x}^{(2)}(t) - \mathbf{x}^{(3)}(t)$.
   (d) Linearly independent.   (e) Linearly dependent. $\mathbf{x}^{(2)}(t) = \mathbf{x}^{(1)}(t) + 2\mathbf{x}^{(3)}(t)$.   (f) Linearly independent.   (g) Linearly dependent.
       $\mathbf{x}^{(3)}(t) = -2\mathbf{x}^{(1)}(t) - \mathbf{x}^{(2)}(t)$.   (h) Linearly independent.

6. Without loss of generality, we may take the q n-vectors $\mathbf{x}^{(1)}$, $\mathbf{x}^{(2)}$, $\ldots$, $\mathbf{x}^{(q)}$, $q < k$, as an arbitrary subcollection of the k n-vectors $\mathbf{x}^{(1)}$, $\mathbf{x}^{(2)}$, $\ldots$, $\mathbf{x}^{(k)}$. Since this subcollection of vectors is linearly dependent, there exist constants

$c_1, c_2, \ldots, c_q$, not all zero, such that $c_1\mathbf{x}^{(1)} + c_2\mathbf{x}^{(2)} + \cdots + c_q\mathbf{x}^{(q)} = \mathbf{0}$.   Then, also

$$(1) \qquad c_1\mathbf{x}^{(1)} + c_2\mathbf{x}^{(2)} + \cdots + c_q\mathbf{x}^{(q)} + 0 \cdot \mathbf{x}^{(q+1)} + \cdots + 0 \cdot \mathbf{x}^{(k)} = \mathbf{0}.$$

Since not all of the c's are zero in (1), the k n-vectors $\mathbf{x}^{(1)}$, $\mathbf{x}^{(2)}$, $\ldots$, $\mathbf{x}^{(k)}$ are linearly dependent.

7. Let the q n-vectors $\mathbf{x}^{(1)}$, $\mathbf{x}^{(2)}$, $\ldots$, $\mathbf{x}^{(q)}$, $q < k$, be an arbitrary subcollection of the k n-vectors $\mathbf{x}^{(1)}$, $\mathbf{x}^{(2)}$, $\ldots$, $\mathbf{x}^{(k)}$.   If the subcollection of vectors were linearly dependent, so would the k n-vectors $\mathbf{x}^{(1)}$, $\mathbf{x}^{(2)}$, $\ldots$, $\mathbf{x}^{(k)}$ be linearly dependent [*see* Exercise 6], which is contrary to the hypothesis.   Thus, the q n-vectors $\mathbf{x}^{(1)}$, $\mathbf{x}^{(2)}$, $\ldots$, $\mathbf{x}^{(q)}$ are linearly independent.

8. We shall establish the "only if" part of the result.   That is, suppose that the k n-vectors $\mathbf{x}^{(1)}$, $\mathbf{x}^{(2)}$, $\ldots$, $\mathbf{x}^{(k)}$ are linearly dependent.   Then there exist constants $c_1, c_2, \ldots, c_k$, not all zero, such that $c_1\mathbf{x}^{(1)} + c_2\mathbf{x}^{(2)} + \cdots + c_k\mathbf{x}^{(k)} = \mathbf{0}$. If c_1, for example, is not zero, then

$$\mathbf{x}^{(1)} = -\frac{c_2}{c_1}\,\mathbf{x}^{(2)} - \cdots - \frac{c_k}{c_1}\,\mathbf{x}^{(k)}.$$

Thus, $\mathbf{x}^{(1)}$ is a linear combination of the remaining vectors.

9. We shall establish the "if" part of the result by proving its negative converse. Suppose that the $(k+1)$ n-vectors $\mathbf{x}^{(1)}$, $\mathbf{x}^{(2)}$, $\ldots$, $\mathbf{x}^{(k)}$, $\mathbf{x}^{(k+1)}$ were linearly dependent.   Then there would exist constants $c_1, c_2, \ldots, c_k, c_{k+1}$, not all zero, such that $c_1\mathbf{x}^{(1)} + c_2\mathbf{x}^{(2)} + \cdots + c_k\mathbf{x}^{(k)} + c_{k+1}\mathbf{x}^{(k+1)} = \mathbf{0}$. If $c_{k+1} = 0$, then the k n-vectors $\mathbf{x}^{(1)}$, $\mathbf{x}^{(2)}$, $\ldots$, $\mathbf{x}^{(k)}$ would be linearly dependent, contrary to the hypothesis.   Thus, $c_{k+1} \neq 0$, and

$$\mathbf{x}^{(k+1)} = -\frac{c_1}{c_{k+1}}\,\mathbf{x}^{(1)} - \frac{c_2}{c_{k+1}}\,\mathbf{x}^{(2)} - \cdots - \frac{c_k}{c_{k+1}}\,\mathbf{x}^{(k)}.$$

Hence, $\mathbf{x}^{(k+1)}$ is expressible as a linear combination of the k n-vectors $\mathbf{x}^{(1)}$, $\mathbf{x}^{(2)}$, $\ldots$, $\mathbf{x}^{(k)}$.   Consequently, if $\mathbf{x}^{(k+1)}$ cannot be expressed as a linear combination of the remaining vectors, then the $(k+1)$ vectors are linearly independent.

10. *Concerning* (a):   Let $\mathbf{A} = [a_{ij}]$ be an $m \times n$ matrix,

$$\mathbf{x} = \begin{bmatrix} x_{11} \\ x_{21} \\ \vdots \\ x_{n1} \end{bmatrix}, \quad \mathbf{y} = \begin{bmatrix} y_{11} \\ y_{21} \\ \vdots \\ y_{n1} \end{bmatrix},$$

$\mathbf{Z} = [z_{ij}] = [\alpha x_{ij} + \beta y_{ij}]$ be an $n \times 1$ matrix, and $\mathbf{C} = [c_{ij}]$ be an $m \times 1$ matrix. In view of Definition 8.6.4, $\mathbf{AZ} = \mathbf{C}$ asserts that $c_{ij} = \sum_{k=1}^{n} a_{ik} z_{kj}$, $j = 1$.   Thus,

$$c_{ij} = \sum_{k=1}^{n} a_{ik}(\alpha x_{kj} + \beta y_{kj}) = \alpha \sum_{k=1}^{n} a_{ik} x_{kj} + \beta \sum_{k=1}^{n} a_{ik} y_{kj}, \qquad j = 1.$$

Hence,

$$\mathbf{A}(\alpha\mathbf{x} + \beta\mathbf{y}) = \alpha\mathbf{A}\mathbf{x} + \beta\mathbf{A}\mathbf{y}.$$

Exercises 8.9, page 543

1. We shall prove the "if" part of the result.   That is, suppose that the columns of the matrix $\mathbf{A}$ are linearly dependent.   Then, by Exercise 8.8.8, at least one of the columns of $\mathbf{A}$, say the first one, may be expressed as a linear combination of the rest.   Thus, by adding appropriate multiples of the other columns respectively, to the first column, we may reduce all the elements in the first column to 0 without changing the value of the det $\mathbf{A}$.   Thus, det $\mathbf{A} = 0$ and the matrix $\mathbf{A}$ is singular.

2. $x_1 = \frac{9}{7}$, $x_2 = -\frac{13}{7}$.

3. $x_1 = 5$, $x_2 = -2$, $x_3 = -5$.

4. $x_1 = 4$, $x_2 = -1$, $x_3 = -7$.

5. $x_1 = 2$, $x_2 = 2$, $x_3 = 3$, $x_4 = 2$.

6. Inconsistent.

7. Consistent. $x_1 = 3t + 2$, $x_2 = t + 1$, $x_3 = -t$, where t is arbitrary.

8. Consistent. $x_1 = t_1 + t_2$, $x_2 = t_1 - t_2 - 3$, $x_3 = -t_1$, $x_4 = -t_2$,   where t_1, t_2 are arbitrary.

9. Consistent. $x_1 = 2$, $x_2 = t$, $x_3 = -t$, $x_4 = 4t + 2$, where t is arbitrary.

10. Inconsistent.

11. Since $\mathbf{Ax} = \mathbf{0}$, we have the following system of equations

$$(1) \quad \begin{cases} a_{11}x_1 + a_{12}x_2 + \cdots + a_{1n}x_n = 0, \\ a_{21}x_1 + a_{22}x_2 + \cdots + a_{2n}x_n = 0, \\ \quad\vdots \qquad\quad \vdots \qquad\qquad\quad \vdots \\ a_{m1}x_1 + a_{m2}x_2 + \cdots + a_{mn}x_n = 0. \end{cases}$$

By hypothesis, the system (1) must be satisfied no matter what n-vector $\mathbf{x}$ we take.   Take in succession for the vector $\mathbf{x}$, the following vectors.

$$\begin{bmatrix} 1 \\ 0 \\ 0 \\ \vdots \\ 0 \end{bmatrix}, \quad \begin{bmatrix} 0 \\ 1 \\ 0 \\ \vdots \\ 0 \end{bmatrix}, \quad \begin{bmatrix} 0 \\ 0 \\ 1 \\ \vdots \\ 0 \end{bmatrix}, \quad \ldots, \quad \begin{bmatrix} 0 \\ 0 \\ 0 \\ \vdots \\ 1 \end{bmatrix},$$

and then substitute individually into (1).   Show that $a_{ij} = 0$, $i = 1, 2, \ldots, m$, $j = 1, 2, \ldots, n$.   Thus, $\mathbf{A}$ is a zero matrix.

Exercises 8.10, page 552

1. Utilize Theorem 8.10.1.

2. Follow the proof given for Theorem 6.4.2.   Replace Equation (6.4.3) by

$$\mathbf{x}^{(1)}(t_0) = \begin{bmatrix} 1 \\ 0 \\ \vdots \\ 0 \end{bmatrix}, \quad \mathbf{x}^{(2)}(t_0) = \begin{bmatrix} 0 \\ 1 \\ \vdots \\ 0 \end{bmatrix}, \quad \ldots, \quad \mathbf{x}^{(n)}(t_0) = \begin{bmatrix} 0 \\ 0 \\ \vdots \\ 1 \end{bmatrix}.$$

In the proof, utilize Definition 8.8.2, Exercise 8.8.9, and Theorems 8.10.1, 8.9.6, 8.10.3, and 8.10.2.

3. Follow the proof given in Theorem 4.4.4 with some modifications.

6. *Concerning* (b): If the vectors are linearly dependent on J, then there exist constants c_1, $c_2, \ldots, c_n$, not all zero, such that $c_1 \mathbf{x}^{(1)}(t) + c_2 \mathbf{x}^{(2)}(t) + \cdots + c_n \mathbf{x}^{(n)}(t) = \mathbf{0}$ for every t in J. This linear combination can be written equivalently as $\mathbf{X}(t)\mathbf{c} = \mathbf{0}$, where $\mathbf{c}^T = [c_1, c_2, \ldots, c_n]$. Since $\mathbf{c} \neq \mathbf{0}$, it follows from Theorem 8.9.5 that $\det \mathbf{X}(t) = W(\mathbf{x}^{(1)}, \mathbf{x}^{(2)}, \ldots, \mathbf{x}^{(n)}; t) = 0$ for every t in J. (c) Yes.   (d) Yes.

Exercises 8.11, page 562

1. We shall establish the result by contradiction. Suppose that the eigenvectors $\boldsymbol{\xi}^{(1)}, \boldsymbol{\xi}^{(2)}, \ldots, \boldsymbol{\xi}^{(n)}$ are linearly dependent. Then there exist constants c_1, $c_2, \ldots, c_n$, not all zero, such that

$$(1) \qquad c_1 \boldsymbol{\xi}^{(1)} + c_2 \boldsymbol{\xi}^{(2)} + \cdots + c_n \boldsymbol{\xi}^{(n)} = \mathbf{0}.$$

Multiply (1) by the matrix $\mathbf{A}$ associated with the system (8.11.2). We have

$$c_1 \mathbf{A}\boldsymbol{\xi}^{(1)} + c_2 \mathbf{A}\boldsymbol{\xi}^{(2)} + \cdots + c_n \mathbf{A}\boldsymbol{\xi}^{(n)} = \mathbf{0}.$$

In view of (8.11.9), $\mathbf{A}\boldsymbol{\xi}^{(1)} = \lambda_1 \boldsymbol{\xi}^{(1)}$, $\mathbf{A}\boldsymbol{\xi}^{(2)} = \lambda_2 \boldsymbol{\xi}^{(2)}, \ldots, \mathbf{A}\boldsymbol{\xi}^{(n)} = \lambda_n \boldsymbol{\xi}^{(n)}$ (up to a scalar multiple). Thus, we may write the above equation as

$$(2) \qquad c_1 \lambda_1 \boldsymbol{\xi}^{(1)} + c_2 \lambda_2 \boldsymbol{\xi}^{(2)} + \cdots + c_n \lambda_n \boldsymbol{\xi}^{(n)} = \mathbf{0}.$$

Multiply (2) by the matrix $\mathbf{A}$, and we have

$$c_1 \lambda_1 \mathbf{A}\boldsymbol{\xi}^{(1)} + c_2 \lambda_2 \mathbf{A}\boldsymbol{\xi}^{(2)} + \cdots + c_n \lambda_n \mathbf{A}\boldsymbol{\xi}^{(n)} = \mathbf{0}.$$

Again, in view of (8.11.9), we may write the above equation as

$$(3) \qquad c_1 \lambda_1^2 \boldsymbol{\xi}^{(1)} + c_2 \lambda_2^2 \boldsymbol{\xi}^{(2)} + \cdots + c_n \lambda_n^2 \boldsymbol{\xi}^{(n)} = \mathbf{0}.$$

Continuing in this fashion, we obtain

$$(4) \qquad c_1 \lambda_1^{n-1} \boldsymbol{\xi}^{(1)} + c_2 \lambda_2^{n-1} \boldsymbol{\xi}^{(2)} + \cdots + c_n \lambda_n^{n-1} \boldsymbol{\xi}^{(n)} = \mathbf{0}.$$

We may write Equations (1), $\ldots$, (4) as

$$(5) \qquad \begin{bmatrix} 1 & 1 & \cdots & 1 \\ \lambda_1 & \lambda_2 & \cdots & \lambda_n \\ \lambda_1^2 & \lambda_2^2 & \cdots & \lambda_n^2 \\ \vdots & \vdots & & \vdots \\ \lambda_1^{n-1} & \lambda_2^{n-1} & \cdots & \lambda_n^{n-1} \end{bmatrix} \begin{bmatrix} c_1 \boldsymbol{\xi}^{(1)} \\ c_2 \boldsymbol{\xi}^{(2)} \\ \vdots \\ c_n \boldsymbol{\xi}^{(n)} \end{bmatrix} = \begin{bmatrix} 0 \\ 0 \\ \vdots \\ 0 \end{bmatrix}.$$

Let
$$\mathbf{B} = \begin{bmatrix} 1 & 1 & \cdots & 1 \\ \lambda_1 & \lambda_2 & \cdots & \lambda_n \\ \lambda_1^2 & \lambda_2^2 & \cdots & \lambda_n^2 \\ \vdots & \vdots & & \vdots \\ \lambda_1^{n-1} & \lambda_2^{n-1} & \cdots & \lambda_n^{n-1} \end{bmatrix}.$$

Since the eigenvalues $\lambda_1, \lambda_2, \ldots, \lambda_n$ are simple, it follows from Exercise 6.4.5 that $\det \mathbf{B} \neq 0$. Hence, $\mathbf{B}$ is nonsingular and $\mathbf{B}^{-1}$ exists. Multiply (5) by $\mathbf{B}^{-1}$, and we obtain

$$\begin{bmatrix} c_1 \boldsymbol{\xi}^{(1)} \\ c_2 \boldsymbol{\xi}^{(2)} \\ \vdots \\ c_n \boldsymbol{\xi}^{(n)} \end{bmatrix} = \begin{bmatrix} 0 \\ 0 \\ \vdots \\ 0 \end{bmatrix}.$$

Thus, $c_1 \boldsymbol{\xi}^{(1)} = \mathbf{0}$, $c_2 \boldsymbol{\xi}^{(2)} = \mathbf{0}, \ldots, c_n \boldsymbol{\xi}^{(n)} = \mathbf{0}$. Hence, $c_1 = c_2 = \cdots = c_n = 0$, which is a contradiction. Consequently, the eigenvectors $\boldsymbol{\xi}^{(1)}, \boldsymbol{\xi}^{(2)}, \ldots, \boldsymbol{\xi}^{(n)}$ are linearly independent.

2. *Concerning* (a): $\overline{\mathbf{AB}} = \sum_{k=1}^{n} \overline{a_{ik} b_{kj}} = \sum_{k=1}^{n} \bar{a}_{ik} \bar{b}_{kj} = \overline{\mathbf{A}}\,\overline{\mathbf{B}}$.

Concerning (b): Since $\boldsymbol{\xi}^{(1)}$ and $\boldsymbol{\xi}^{(2)}$ are the eigenvectors corresponding to λ_1, λ_2, we have

(1) $$\mathbf{A}\boldsymbol{\xi}^{(1)} = \lambda_1 \boldsymbol{\xi}^{(1)},$$

and

(2) $$\mathbf{A}\boldsymbol{\xi}^{(2)} = \lambda_2 \boldsymbol{\xi}^{(2)}.$$

Taking the complex conjugate of (2), and then utilizing the result in part (a), we obtain

(3) $$\overline{\mathbf{A}}\,\overline{\boldsymbol{\xi}}^{(2)} = \overline{\lambda}_2 \overline{\boldsymbol{\xi}}^{(2)}.$$

Since $\mathbf{A}$ is real, $\overline{\mathbf{A}} = \mathbf{A}$. By hypothesis, $\overline{\lambda}_2 = \lambda_1$. Thus, (3) becomes

(4) $$\mathbf{A}\overline{\boldsymbol{\xi}}^{(2)} = \lambda_1 \overline{\boldsymbol{\xi}}^{(2)}.$$

Comparing (4) and (1), we see that $\boldsymbol{\xi}^{(1)}$ and $\overline{\boldsymbol{\xi}}^{(2)}$ satisfy the same eigenvalue equation. Therefore, one is a scalar multiple of the other.

3. $\mathbf{x} = c_1 \begin{bmatrix} 4 \\ -1 \\ 0 \end{bmatrix} + c_2 \begin{bmatrix} 4 \\ 0 \\ -1 \end{bmatrix} e^t + c_3 \begin{bmatrix} 2 \\ -1 \\ 0 \end{bmatrix} e^{2t}.$

4. $\mathbf{x} = c_1 \begin{bmatrix} 0 \\ -2 \\ 1 \end{bmatrix} e^t + c_2 \begin{bmatrix} 1 \\ 1 \\ 0 \end{bmatrix} e^{2t} + c_3 \begin{bmatrix} 2 \\ 2 \\ 1 \end{bmatrix} e^{3t}.$

5. $\mathbf{x} = c_1 \begin{bmatrix} 1 \\ 1 \\ 2 \end{bmatrix} e^{-t} + c_2 \begin{bmatrix} 1 \\ 1 \\ 1 \end{bmatrix} e^t + c_3 \begin{bmatrix} 1 \\ 2 \\ 1 \end{bmatrix} e^{2t}.$

6. $\mathbf{x} = c_1 \begin{bmatrix} 1 \\ 0 \\ -1 \end{bmatrix} e^{2t} + c_2 \begin{bmatrix} 1 \\ 1 \\ 1 \end{bmatrix} e^{3t} + c_3 \begin{bmatrix} 1 \\ -2 \\ 1 \end{bmatrix} e^{6t}.$

7. $\mathbf{x} = c_1 \begin{bmatrix} 1 \\ -2 \\ 1 \end{bmatrix} e^{-2t} + c_2 \begin{bmatrix} 1 \\ 1 \\ 1 \end{bmatrix} e^{t} + c_3 \begin{bmatrix} 1 \\ 0 \\ -1 \end{bmatrix} e^{2t}.$

8. $\mathbf{x} = c_1 \begin{bmatrix} 2 \\ 0 \\ -1 \end{bmatrix} e^{-3t} + c_2 \begin{bmatrix} 5 \\ -6 \\ 2 \end{bmatrix} e^{3t} + c_3 \begin{bmatrix} 2 \\ 0 \\ 1 \end{bmatrix} e^{5t}.$

9. $\mathbf{x} = c_1 \begin{bmatrix} 0 \\ 1 \\ -1 \end{bmatrix} e^{-t} + c_2 \begin{bmatrix} \cos t - \sin t \\ -\cos t \\ -\cos t \end{bmatrix} + c_3 \begin{bmatrix} \cos t + \sin t \\ -\sin t \\ -\sin t \end{bmatrix}.$

10. $\mathbf{x} = c_1 \begin{bmatrix} 2 \\ 2 \\ 5 \end{bmatrix} e^{5t} + c_2 \begin{bmatrix} 5\cos t \\ 3\cos t - \sin t \\ 10\cos t \end{bmatrix} e^{2t} + c_3 \begin{bmatrix} 5\sin t \\ 3\sin t + \cos t \\ 10\sin t \end{bmatrix} e^{2t}.$

11. $\mathbf{x} = c_1 \begin{bmatrix} 1 \\ 0 \\ 0 \\ 0 \end{bmatrix} e^{t} + c_2 \begin{bmatrix} 1 \\ 2 \\ 0 \\ 0 \end{bmatrix} e^{-t} + c_3 \begin{bmatrix} 1 \\ 2 \\ 3 \\ 0 \end{bmatrix} e^{2t} + c_4 \begin{bmatrix} 1 \\ 2 \\ 3 \\ 4 \end{bmatrix} e^{-2t}.$

12. $\mathbf{x} = \dot{c}_1 \begin{bmatrix} 1 \\ 0 \\ 0 \\ \vdots \\ 0 \end{bmatrix} e^{\alpha_1 t} + c_2 \begin{bmatrix} 0 \\ 1 \\ 0 \\ \vdots \\ 0 \end{bmatrix} e^{\alpha_2 t} + c_3 \begin{bmatrix} 0 \\ 0 \\ 1 \\ \vdots \\ 0 \end{bmatrix} e^{\alpha_3 t} + \cdots + c_n \begin{bmatrix} 0 \\ 0 \\ 0 \\ \vdots \\ 1 \end{bmatrix} e^{\alpha_n t}.$

13. $\mathbf{x} = - \begin{bmatrix} 1 \\ -1 \end{bmatrix} e^{-t} + 2 \begin{bmatrix} 1 \\ 2 \end{bmatrix} e^{5t}.$ 14. $\mathbf{x} = 2 \begin{bmatrix} \cos t \\ -\sin t \end{bmatrix} e^{2t} - 3 \begin{bmatrix} \sin t \\ \cos t \end{bmatrix} e^{2t}.$

15. $\mathbf{x} = -2 \begin{bmatrix} 1 \\ -1 \\ 0 \end{bmatrix} e^{-2t} + \begin{bmatrix} 1 \\ 1 \\ -1 \end{bmatrix} e^{-t} + 2 \begin{bmatrix} 1 \\ 1 \\ 2 \end{bmatrix} e^{2t}.$

16. $\mathbf{x} = -2 \begin{bmatrix} \cos t - \sin t \\ \cos t - \sin t \\ -2\cos t \end{bmatrix} + 4 \begin{bmatrix} \cos t + \sin t \\ \cos t + \sin t \\ -2\sin t \end{bmatrix} - 3 \begin{bmatrix} 1 \\ 0 \\ -1 \end{bmatrix} e^{t}.$

18. *Concerning* (b): By hypothesis, $\det [\mathbf{A} - \lambda\mathbf{I}] = (\lambda - \lambda_1)(\lambda - \lambda_2) \cdots (\lambda - \lambda_n)$. Thus, $\det [(\mathbf{A} - k\mathbf{I}) - \lambda\mathbf{I}] = \det [\mathbf{A} - (\lambda + k)\mathbf{I}] = (\lambda + k - \lambda_1)(\lambda + k - \lambda_2) \cdots (\lambda + k - \lambda_n)$, from which the result may now be deduced.

19. *Concerning* (b): $\det [\mathbf{A}^{-1}\mathbf{B}\mathbf{A} - \lambda\mathbf{I}] = \det [\mathbf{A}^{-1}(\mathbf{B}\mathbf{A} - \lambda\mathbf{A})] = \det \mathbf{A}^{-1} \cdot \det [\mathbf{B}\mathbf{A} - \lambda\mathbf{A}] = \det \mathbf{A}^{-1} \det [(\mathbf{B} - \lambda\mathbf{I})\mathbf{A}] = \det \mathbf{A}^{-1} \det [\mathbf{B} - \lambda\mathbf{I}] \det \mathbf{A} = \det [\mathbf{B} - \lambda\mathbf{I}]$, from which the result may now be deduced.

20. Utilize (8.11.38).

22. *Concerning* (a): Since $\mathbf{A} = \mathbf{A}^* = \overline{\mathbf{A}}^T$, we have $\det \mathbf{A} = \det \overline{\mathbf{A}}^T = \det \overline{\mathbf{A}} = \overline{\det \mathbf{A}}$, from which the result may now be deduced.
Concerning (b): From (8.8.10), we have $(\mathbf{x}, \mathbf{y}) = \mathbf{x}^*\mathbf{y}$. Thus, $(\mathbf{A}\mathbf{x}, \mathbf{y}) = (\mathbf{A}\mathbf{x})^*\mathbf{y} = \mathbf{x}^*\mathbf{A}^*\mathbf{y} = \mathbf{x}^*\mathbf{A}\mathbf{y} = (\mathbf{x}, \mathbf{A}\mathbf{y})$.
Concerning (c): Utilizing part (b) with $\mathbf{x} = \xi$ and $\mathbf{y} = \xi$, we have $(\mathbf{A}\xi, \xi) = (\xi, \mathbf{A}\xi)$. By hypothesis, $\mathbf{A}\xi = \lambda\xi$. Thus, $(\lambda\xi, \xi) = (\xi, \lambda\xi)$, and in view of (8.8.14), we obtain $\overline{\lambda}(\xi, \xi) = \lambda(\xi, \xi)$. Hence, $\overline{\lambda} = \lambda$.

Concerning (d): Utilizing part (b) with $\mathbf{x} = \boldsymbol{\xi}^{(1)}$ and $\mathbf{y} = \boldsymbol{\xi}^{(2)}$, we have $(\mathbf{A}\boldsymbol{\xi}^{(1)}, \boldsymbol{\xi}^{(2)}) = (\boldsymbol{\xi}^{(1)}, \mathbf{A}\boldsymbol{\xi}^{(2)})$. Since $\mathbf{A}\boldsymbol{\xi}^{(1)} = \lambda_1\boldsymbol{\xi}^{(1)}$ and $\mathbf{A}\boldsymbol{\xi}^{(2)} = \lambda_2\boldsymbol{\xi}^{(2)}$, we have $(\lambda_1\boldsymbol{\xi}^{(1)}, \boldsymbol{\xi}^{(2)}) = (\boldsymbol{\xi}^{(1)}, \lambda_2\boldsymbol{\xi}^{(2)})$. The result may now be obtained by utilizing part (c) and the fact that $\lambda_1 \neq \lambda_2$.

Concerning (e): Let $\mathbf{B} = \mathbf{xy}^* + \mathbf{yx}^*$. Clearly, $\mathbf{B}$ is an $n \times n$ matrix. $\mathbf{B}^* = (\mathbf{xy}^* + \mathbf{yx}^*)^* = (\mathbf{y}^*)^*\mathbf{x}^* + (\mathbf{x}^*)^*\mathbf{y}^* = \mathbf{yx}^* + \mathbf{xy}^* = \mathbf{xy}^* + \mathbf{yx}^* = \mathbf{B}$. Thus, $\mathbf{B}$ is Hermitian.

(f) Yes.

23. $\mathbf{x} = c_1 \begin{bmatrix} 0 \\ 3-4i \\ -5 \end{bmatrix} e^{-5t} + c_2 \begin{bmatrix} 0 \\ 3-4i \\ 5 \end{bmatrix} e^{5t} + c_3 \begin{bmatrix} 1 \\ 0 \\ 0 \end{bmatrix} e^t.$

24. $\mathbf{x} = c_1 \begin{bmatrix} 1-i \\ -2+i \\ -1 \end{bmatrix} e^{-t} + c_2 \begin{bmatrix} 1+i \\ i \\ 1 \end{bmatrix} e^t + c_3 \begin{bmatrix} 1+3i \\ 2+i \\ -5 \end{bmatrix} e^{3t}.$

26. (a) Observe that $\mathbf{I}^T = \mathbf{I}$ and $(-\lambda\mathbf{I})^T = -\lambda\mathbf{I}$. Thus, in view of (8.6.17) we have $(\mathbf{A} - \lambda\mathbf{I})^T = \mathbf{A}^T - \lambda\mathbf{I}$. Let us denote $\det \mathbf{C}$ by $|\mathbf{C}|$. For an $n \times n$ matrix $\mathbf{C}$, we know that $|\mathbf{C}^T| = |\mathbf{C}|$. Consequently, $|\mathbf{A}^T - \lambda\mathbf{I}| = |(\mathbf{A} - \lambda\mathbf{I})^T| = |\mathbf{A} - \lambda\mathbf{I}|.$

(b) We shall establish the result for the case when the matrix $\mathbf{B}$ is nonsingular. The remaining cases will be left to the reader. Consider the characteristic equation of $\mathbf{AB}$, namely, $|\mathbf{AB} - \lambda\mathbf{I}| = 0$. Thus, $0 = |\mathbf{B}| \cdot |\mathbf{AB} - \lambda\mathbf{I}| = |\mathbf{BAB} - \lambda\mathbf{B}| = |\mathbf{BA} - \lambda\mathbf{I}| \, |\mathbf{B}|$. Hence, $|\mathbf{BA} - \lambda\mathbf{I}| = 0.$

27. $|\mathbf{A} - \lambda\mathbf{I}| = |\mathbf{P}^{-1}\mathbf{BP} - \lambda\mathbf{I}| = |\mathbf{P}^{-1}| \, |\mathbf{B} - \lambda\mathbf{I}| \, |\mathbf{P}| = |\mathbf{B} - \lambda\mathbf{I}|.$

Exercises 8.12, page 571

1. $\mathbf{x} = c_1 \begin{bmatrix} 1 \\ 1 \end{bmatrix} e^{3t} + c_2 \left(\begin{bmatrix} 1 \\ 1 \end{bmatrix} te^{3t} + \begin{bmatrix} 0 \\ 1 \end{bmatrix} e^{3t} \right).$

2. $\mathbf{x} = c_1 \begin{bmatrix} 1 \\ -1 \end{bmatrix} e^{2t} + c_2 \left(\begin{bmatrix} 1 \\ -1 \end{bmatrix} te^{2t} + \begin{bmatrix} 0 \\ -1 \end{bmatrix} e^{2t} \right).$

3. $\mathbf{x} = c_1 \begin{bmatrix} 1 \\ 0 \\ -1 \end{bmatrix} e^t + c_2 \begin{bmatrix} 0 \\ 1 \\ 0 \end{bmatrix} e^t + c_3 \left(\begin{bmatrix} 1 \\ 0 \\ -1 \end{bmatrix} te^t + \begin{bmatrix} 1 \\ 0 \\ -2 \end{bmatrix} e^t \right).$

4. $\mathbf{x} = c_1 \begin{bmatrix} 1 \\ -2 \\ 0 \end{bmatrix} e^t + c_2 \begin{bmatrix} 1 \\ -2 \\ -1 \end{bmatrix} e^{2t} + c_3 \left(\begin{bmatrix} 1 \\ -2 \\ -1 \end{bmatrix} te^{2t} + \begin{bmatrix} -1 \\ 1 \\ 0 \end{bmatrix} e^{2t} \right).$

5. $\mathbf{x} = c_1 \begin{bmatrix} 1 \\ -1 \\ 0 \end{bmatrix} e^t + c_2 \begin{bmatrix} 1 \\ 0 \\ -1 \end{bmatrix} e^t + c_3 \begin{bmatrix} 1 \\ 2 \\ 1 \end{bmatrix} e^{5t}.$

6. $\mathbf{x} = c_1 \begin{bmatrix} -1 \\ 1 \\ 0 \end{bmatrix} e^t + c_2 \begin{bmatrix} 1 \\ -1 \\ 1 \end{bmatrix} e^{2t} + c_3 \left(\begin{bmatrix} 1 \\ -1 \\ 1 \end{bmatrix} te^{2t} + \begin{bmatrix} 1 \\ 0 \\ 0 \end{bmatrix} e^{2t} \right).$

7. $\mathbf{x} = c_1 \begin{bmatrix} 1 \\ 2 \\ 0 \\ -3 \end{bmatrix} e^t + c_2 \begin{bmatrix} 1 \\ 2 \\ -6 \\ 0 \end{bmatrix} e^t + c_3 \begin{bmatrix} 1 \\ 1 \\ 1 \\ -2 \end{bmatrix} e^{2t} + c_4 \begin{bmatrix} 2 \\ 2 \\ 2 \\ -3 \end{bmatrix} e^{3t}.$

8. $\mathbf{x} = \begin{bmatrix} 2+t \\ -1-t \end{bmatrix} e^{3t}.$
                                 9. $\mathbf{x} = \begin{bmatrix} -e^{2t}+4e^t \\ e^t \\ e^{2t}-2e^t \end{bmatrix}.$

10. $\mathbf{x} = \begin{bmatrix} -t+t^2 \\ -1 \\ 1+t+t^2 \end{bmatrix} e^{2t}.$

11. *Concerning* (d): Since $\lambda = r$ is a root of multiplicity $s = 2$ of $p(\lambda) = 0$, we have $p(r) = 0$, $p'(r) = 0$, and $p''(r) \neq 0$. But $p''(r) = 2!$ times the sum of the principal minors of order $3 - 2 = 1$ of $\mathbf{A} - r\mathbf{I}$. Thus, not every principal minor is equal to zero, and $\mathbf{A} - r\mathbf{I}$ is of rank at least one.

Exercises 8.13, page 577

1. $\mathbf{x} = c_1 \begin{bmatrix} \cos 2t \\ -\cos 2t - 2\sin 2t \end{bmatrix} e^{4t} + c_2 \begin{bmatrix} \sin 2t \\ 2\cos 2t - \sin 2t \end{bmatrix} e^{4t}.$

2. $\mathbf{x} = c_1 \begin{bmatrix} 4\cos t \\ \cos t + \sin t \\ -2\cos t \end{bmatrix} e^{3t} + c_2 \begin{bmatrix} 4\sin t \\ \sin t - \cos t \\ -2\sin t \end{bmatrix} e^{3t} + c_3 \begin{bmatrix} 1 \\ 0 \\ -1 \end{bmatrix} e^{4t}.$

3. $\mathbf{x} = c_1 \begin{bmatrix} \cos t \\ \sin t \\ 0 \\ 0 \end{bmatrix} + c_2 \begin{bmatrix} \sin t \\ -\cos t \\ 0 \\ 0 \end{bmatrix} + c_3 \begin{bmatrix} 0 \\ 0 \\ \cos t \\ \sin t \end{bmatrix} + c_4 \begin{bmatrix} 0 \\ 0 \\ \sin t \\ -\cos t \end{bmatrix}.$

4. $\mathbf{x} = \begin{bmatrix} 1 \\ -2 \end{bmatrix} e^{3t}.$
                                 5. $\mathbf{x} = \begin{bmatrix} 4\cos 2t + 2\sin 2t \\ 2\cos 2t + \sin 2t \\ -2\cos 2t + 4\sin 2t \end{bmatrix} e^{2t}.$

6. $\mathbf{x} = \begin{bmatrix} 7t^2 + 23t + 19 \\ 7t^2 + 51t + 51 \\ 7t^2 + 65t + 46 \\ 7t^2 + 23t + 33 \end{bmatrix} e^t - 6 \begin{bmatrix} 3 \\ 8 \\ 8 \\ 5 \end{bmatrix} e^{2t}.$
   7. $\mathbf{x} = \begin{bmatrix} 2\cos t \\ \cos t + \sin t \\ -2\cos 2t \\ -\cos 2t - \sin 2t \end{bmatrix}.$

8. $\mathbf{x} = \begin{bmatrix} 2\cos 2t - 4\sin t \\ -\cos t - \sin 2t \\ \cos t + 4\sin 2t \\ -2\cos 2t + \sin t \end{bmatrix}.$

9. $\mathbf{x} = \begin{bmatrix} -1 + 3\cos t - \sin t \\ -1 + 4\cos t + 2\sin t \\ -1 \\ 1 - 3\cos t + \sin t \\ 1 \end{bmatrix} e^t + \begin{bmatrix} -\sin 2t \\ -\sin 2t - \cos 2t \\ 0 \\ \sin 2t + \cos 2t \\ \sin 2t \end{bmatrix} e^{2t}.$

10. $\mathbf{x} = \dfrac{1}{30}\begin{bmatrix} 60 \\ 988\cos t + 3166\sin t \\ 282\cos t + 1134\sin t \\ 378\cos t - 94\sin t \\ 236\cos t + 142\sin t \end{bmatrix} e^t - \dfrac{1}{30}\begin{bmatrix} 0 \\ 928\cos 2t + 94\sin 2t \\ 312\cos 2t + 306\sin 2t \\ 408\cos 2t - 416\sin 2t \\ 206\cos 2t - 2\sin 2t \end{bmatrix} e^{2t}.$

11. (g) *Concerning* $\boldsymbol{\xi}^{(i)} = k\mathbf{P}^{-1}\boldsymbol{\eta}^{(i)}$ $(i = 1, \ldots, n)$. Since $\mathbf{A}$ and $\mathbf{B}$ are similar matrices, they have the same eigenvalues $\lambda_i (i = 1, \ldots, n)$. In view of (8.11.9), we have $(\mathbf{A} - \lambda_i\mathbf{I})\boldsymbol{\xi}^{(i)} = \mathbf{0}$ and $(\mathbf{B} - \lambda_i\mathbf{I})\boldsymbol{\eta}^{(i)} = \mathbf{0}$ $(i = 1, \ldots, n)$. Since $\mathbf{A} = \mathbf{P}^{-1}\mathbf{B}\mathbf{P}$, it follows that $\mathbf{B} = \mathbf{P}\mathbf{A}\mathbf{P}^{-1}$. Thus, $\mathbf{0} = (\mathbf{B} - \lambda_i\mathbf{I})\boldsymbol{\eta}^{(i)}$ $= (\mathbf{P}\mathbf{A}\mathbf{P}^{-1} - \lambda_i\mathbf{I})\boldsymbol{\eta}^{(i)} = \mathbf{P}(\mathbf{A} - \lambda_i\mathbf{I})\mathbf{P}^{-1}\boldsymbol{\eta}^{(i)}$. Since the matrix $\mathbf{P}$ is nonsingular, it follows that $(\mathbf{A} - \lambda_i\mathbf{I})\mathbf{P}^{-1}\boldsymbol{\eta}^{(i)} = \mathbf{0}$. But, we also have that $(\mathbf{A} - \lambda_i\mathbf{I})\boldsymbol{\xi}^{(i)} = \mathbf{0}$. Consequently, $\boldsymbol{\xi}^{(i)} = k\mathbf{P}^{-1}\boldsymbol{\eta}^{(i)}$. (Why?)

Concerning $\boldsymbol{\Phi}_{\mathbf{A}}(t) = k\mathbf{P}^{-1}\boldsymbol{\Phi}_{\mathbf{B}}(t)$. Consider the systems

$$(1)\quad \mathbf{x}' = \mathbf{A}\mathbf{x} \quad\text{and}\quad (2)\quad \mathbf{y}' = \mathbf{B}\mathbf{y}.$$

Let $\mathbf{x}^{(1)}, \mathbf{x}^{(2)}, \ldots, \mathbf{x}^{(n)}$ and $\mathbf{y}^{(1)}, \mathbf{y}^{(2)}, \ldots, \mathbf{y}^{(n)}$ be n linearly independent solutions of the systems (1) and (2) on an interval J, respectively. In view of (8.13.2) and (8.13.3), we have

$$\boldsymbol{\Phi}_{\mathbf{A}}(t) = [\mathbf{x}^{(1)}\,\mathbf{x}^{(2)}\,\cdots\,\mathbf{x}^{(n)}] \quad\text{and}\quad \boldsymbol{\Phi}_{\mathbf{B}}(t) = [\mathbf{y}^{(1)}\,\mathbf{y}^{(2)}\,\cdots\,\mathbf{y}^{(n)}].$$

Thus,
$$\boldsymbol{\Phi}_{\mathbf{A}}(t) = [\mathbf{x}^{(1)}\,\mathbf{x}^{(2)}\,\cdots\,\mathbf{x}^{(n)}]$$
$$= [k\mathbf{P}^{-1}\mathbf{y}^{(1)}\,k\mathbf{P}^{-1}\mathbf{y}^{(2)}\,\cdots\,k\mathbf{P}^{-1}\mathbf{y}^{(n)}] \quad\text{(Why?)}$$
$$= k\mathbf{P}^{-1}[\mathbf{y}^{(1)}\,\mathbf{y}^{(2)}\,\cdots\,\mathbf{y}^{(n)}] = k\mathbf{P}^{-1}\boldsymbol{\Phi}_{\mathbf{B}}(t),$$

for every t in J.

Exercises 8.14, page 585

1. We have $\mathbf{u}' = \mathbf{A}(t)\mathbf{u} + \mathbf{g}(t)$ and $\mathbf{v}' = \mathbf{A}(t)\mathbf{v} + \mathbf{g}(t)$. Thus, $\mathbf{u}' - \mathbf{v}' = \mathbf{A}(t)\mathbf{u} - \mathbf{A}(t)\mathbf{v}$, from which the result may now be deduced by utilizing Exercise 8.8.10.

2. Follow the proof given in Theorem 4.7.2 with proper changes of notation and reference.

3. Utilize Exercise 8.8.10. [*See* also the proof of Theorem 4.7.3.]

4. (1) $$\mathbf{x}_p = \boldsymbol{\Phi}(t)\int \boldsymbol{\Phi}^{-1}(t)\mathbf{g}(t)dt.$$

Differentiating (1), we obtain

(2) $$\mathbf{x}'_p = \boldsymbol{\Phi}'(t)\int \boldsymbol{\Phi}^{-1}(t)\mathbf{g}(t)dt + \boldsymbol{\Phi}(t)\boldsymbol{\Phi}^{-1}(t)\mathbf{g}(t) = \boldsymbol{\Phi}'(t)\int \boldsymbol{\Phi}^{-1}(t)\mathbf{g}(t)dt + \mathbf{g}(t).$$

Since $\boldsymbol{\Phi}$ is a fundamental matrix of (8.14.2), in view of (8.14.6), we may write (2) as

$$(3) \qquad \mathbf{x}_p' = \mathbf{A}(t)\boldsymbol{\Phi}(t)\int \boldsymbol{\Phi}^{-1}(t)\mathbf{g}(t)dt + \mathbf{g}(t).$$

From (1) and (3), the result now follows.

5. $\mathbf{x} = c_1 \begin{bmatrix} 1 \\ -1 \end{bmatrix} e^{-3t} + c_2 \begin{bmatrix} 1 \\ -2 \end{bmatrix} e^{-t} + \begin{bmatrix} 0 \\ 1 \end{bmatrix} e^{t} + \begin{bmatrix} -1 \\ 4 \end{bmatrix} e^{3t}.$

6. $\mathbf{x} = c_1 \begin{bmatrix} \cos t \\ 2\cos t + \sin t \end{bmatrix} e^{-t} + c_2 \begin{bmatrix} \sin t \\ 2\sin t - \cos t \end{bmatrix} e^{-t}$
$$+ \frac{1}{2} \begin{bmatrix} (1+2t)\sin t - 2t\cos t \\ 2(1+t)\sin t + (1-6t)\cos t \end{bmatrix} e^{-t}.$$

7. $\mathbf{x} = c_1 \begin{bmatrix} 1 \\ 2 \end{bmatrix} e^{t} + c_2 \begin{bmatrix} t \\ 2t+1 \end{bmatrix} e^{t} + \begin{bmatrix} 2t + \ln t - 2t\ln t \\ 4t - 4t\ln t \end{bmatrix} e^{t}.$

8. $\mathbf{x} = c_1 \begin{bmatrix} -\cos t \\ 2\cos t + \sin t \end{bmatrix} + c_2 \begin{bmatrix} -\sin t \\ 2\sin t - \cos t \end{bmatrix}$
$$+ \begin{bmatrix} (\sin t - 2\cos t)\ln(\sec t + \tan t) \\ 5\cos t\ln(\sec t + \tan t) - 2 \end{bmatrix}.$$

9. $\mathbf{x} = c_1 \begin{bmatrix} 1 \\ -1 \\ 0 \end{bmatrix} e^{t} + c_2 \begin{bmatrix} 4 \\ -2 \\ -1 \end{bmatrix} e^{2t} + c_3 \begin{bmatrix} 2 \\ -1 \\ -1 \end{bmatrix} e^{3t}$
$$+ \begin{bmatrix} -(4t+1)e^{2t} + (2t-3)e^{3t} - 4(2t-3)e^{4t} \\ 2te^{2t} - (t-1)e^{3t} + 2(2t-3)e^{4t} \\ (t+1)e^{2t} - (t-1)e^{3t} + (6t-7)e^{4t} \end{bmatrix}.$$

10. $\mathbf{x} = c_1 \begin{bmatrix} 1 \\ -1 \end{bmatrix} e^{-2t} + c_2 \begin{bmatrix} 1 \\ 4 \end{bmatrix} e^{3t} + \begin{bmatrix} e^{t} + e^{-t} \\ -4e^{t} \end{bmatrix}.$

11. $\mathbf{x} = c_1 \begin{bmatrix} 2\cos 2t + 2\sin 2t \\ -\cos 2t \end{bmatrix} e^{t} + c_2 \begin{bmatrix} 2\sin 2t - 2\cos 2t \\ -\sin 2t \end{bmatrix} e^{t}$
$$+ \frac{1}{3} \begin{bmatrix} -7\cos t - 2\sin t \\ 2\cos t \end{bmatrix} e^{t}.$$

12. $\mathbf{x} = c_1 \begin{bmatrix} 1 \\ -1 \end{bmatrix} e^{2t} + c_2 \left(\begin{bmatrix} 1 \\ -1 \end{bmatrix} te^{2t} + \frac{1}{3} \begin{bmatrix} 1 \\ 0 \end{bmatrix} e^{2t} \right)$
$$+ \begin{bmatrix} 1 \\ -3 \end{bmatrix} t^2 + \frac{1}{2} \begin{bmatrix} -1 \\ 3 \end{bmatrix} t + \frac{3}{8} \begin{bmatrix} -3 \\ 5 \end{bmatrix}.$$

13. $\mathbf{x} = \left\{ c_1 \begin{bmatrix} 1 \\ -1 \end{bmatrix} + c_2 \left(\begin{bmatrix} 1 \\ -1 \end{bmatrix} t + \frac{1}{2} \begin{bmatrix} 1 \\ 0 \end{bmatrix} \right) + \begin{bmatrix} 1 \\ -1 \end{bmatrix} t^2 + \begin{bmatrix} 1 \\ 0 \end{bmatrix} t \right\} e^{t}$
$$+ \begin{bmatrix} 1 \\ -2 \end{bmatrix} \sin t + \begin{bmatrix} 1 \\ -1 \end{bmatrix} \cos t.$$

14. $\mathbf{x} = c_1 \begin{bmatrix} 1 \\ -2 \\ 0 \end{bmatrix} e^{-3t} + c_2 \begin{bmatrix} 1 \\ 4 \\ -2 \end{bmatrix} e^{-t} + c_3 \begin{bmatrix} 1 \\ 2 \\ 0 \end{bmatrix} e^t + \frac{1}{2} \begin{bmatrix} 1 \\ 0 \\ 0 \end{bmatrix} e^t + \begin{bmatrix} 1 \\ 2 \\ 0 \end{bmatrix} te^t + \begin{bmatrix} 2 \\ 5 \\ 0 \end{bmatrix}$

$$+ 3 \begin{bmatrix} 1 \\ 1 \\ 0 \end{bmatrix} t + \frac{1}{5} \begin{bmatrix} \sin t - 7 \cos t \\ -2 \sin t - 16 \cos t \\ 5 \sin t + 5 \cos t \end{bmatrix}.$$

15. $\mathbf{x} = \begin{bmatrix} -\frac{4}{5} e^t + \frac{1}{2}t^2 + 6t + \frac{9}{5} \\ -e^t + \frac{1}{2}t^2 + \frac{31}{5}t + 3 \end{bmatrix} e^{3t}.$

16. $\mathbf{x} = \frac{1}{2} \begin{bmatrix} 2t \cos t + 2 \\ (t-2) \cos t + (t-1) \sin t + 4e^{-t} + 2 \\ t \cos t + (t+1) \sin t - 4e^{-t} + 2 \end{bmatrix}.$

18. $\mathbf{x} = \begin{bmatrix} 3t^4 - t^3 - 10t^2 + 9t \\ 5t^4 - 2t^3 - 10t^2 + 6t \end{bmatrix}.$

20. $\mathbf{x} = \begin{bmatrix} 2t \ln t + t^3 - t^2 + t \\ 2t \ln t + 2t^3 - 2t^2 + 2t \end{bmatrix}.$

21. $\mathbf{x} = \begin{bmatrix} t \ln^2 t + 3t \ln t + 2t \\ t \ln t + 3t \end{bmatrix}.$

Exercises 8.15, page 601

2. Utilizing Theorem 7.2.8 and Exercise 8.6.14, we obtain

$$e^\mathbf{A} e^\mathbf{B} = \sum_{n=0}^\infty \frac{\mathbf{A}^n}{n!} \sum_{n=0}^\infty \frac{\mathbf{B}^n}{n!} = \sum_{n=0}^\infty \sum_{k=0}^n \frac{\mathbf{A}^k \mathbf{B}^{n-k}}{k!(n-k)!} = \sum_{n=0}^\infty \frac{1}{n!} \sum_{k=0}^n \frac{n!}{k!(n-k)!} \mathbf{A}^k \mathbf{B}^{n-k}$$

$$= \sum_{n=0}^\infty \frac{(\mathbf{A}+\mathbf{B})^n}{n!} = e^{\mathbf{A}+\mathbf{B}}.$$

3. The theorems for power series extend to matrix power series. Show that the series

$$e^{t\mathbf{A}} = \sum_{k=0}^\infty \frac{t^k \mathbf{A}^k}{k!}$$

converges for all finite values of t. Utilizing Theorem 7.2.6, we obtain for $-\infty < t < \infty$,

$$\frac{d}{dt}(e^{t\mathbf{A}}) = \sum_{k=0}^\infty \frac{d}{dt}\left(\frac{t^k \mathbf{A}^k}{k!}\right) = \sum_{k=1}^\infty \frac{t^{k-1} \mathbf{A}^k}{(k-1)!} = \mathbf{A} \sum_{k=1}^\infty \frac{t^{k-1} \mathbf{A}^{k-1}}{(k-1)!} = \mathbf{A} e^{t\mathbf{A}}.$$

6. Let $f(\lambda) = a_0 + a_1\lambda + a_2\lambda^2 + \cdots + a_s\lambda^s$ be a polynomial of degree s in λ. Thus, $f(\mathbf{A}) = a_0\mathbf{I} + a_1\mathbf{A} + a_2\mathbf{A}^2 + \cdots + a_s\mathbf{A}^s$. Let λ_k $(1 \le k \le n)$ be *any*

eigenvalue of the matrix $\mathbf{A}$, that is, $|\mathbf{A} - \lambda_k\mathbf{I}| = 0$. (Here " $|\ |$ " denotes determinant.)   We would like to show that $|f(\mathbf{A}) - f(\lambda_k)\mathbf{I}| = 0$.

$$
\begin{aligned}
|f(\mathbf{A}) - f(\lambda_k)\mathbf{I}| &= |a_0\mathbf{I} + a_1\mathbf{A} + a_2\mathbf{A}^2 + \cdots + a_s\mathbf{A}^s - a_0\mathbf{I} \\
&\quad - a_1\lambda_k\mathbf{I} - a_2\lambda_k^2\mathbf{I} - \cdots - a_s\lambda_k^s\mathbf{I}| \\
&= |a_1(\mathbf{A} - \lambda_k\mathbf{I}) + a_2(\mathbf{A}^2 - \lambda_k^2\mathbf{I}) + \cdots + a_s(\mathbf{A}^s - \lambda_k^s\mathbf{I})| \\
&= |\mathbf{A} - \lambda_k\mathbf{I}|\,|a_1 + a_2(\mathbf{A} + \lambda_k\mathbf{I}) + \cdots + a_s(\mathbf{A}^{s-1} \\
&\quad + \cdots + \lambda_k^{s-1}\mathbf{I})|.
\end{aligned}
$$

Since $|\mathbf{A} - \lambda_k\mathbf{I}| = 0$, it follows that $|f(\mathbf{A}) - f(\lambda_k)\mathbf{I}| = 0$.   Thus, we have shown that to *each* eigenvalue λ_k of the matrix $\mathbf{A}$ there corresponds an eigenvalue $f(\lambda_k)$ of the matrix $f(\mathbf{A})$.

7. $\mathbf{x} = \begin{bmatrix} 2e^{2t} - e^{3t} & e^{2t} - e^{3t} \\ -2e^{2t} + 2e^{3t} & -e^{2t} + 2e^{3t} \end{bmatrix}\mathbf{c}.$

8. $\mathbf{x} = \begin{bmatrix} -e^t + 2e^{2t} - 2te^{2t} & -e^t + e^{2t} - te^{2t} & e^t - e^{2t} \\ 2e^t - 2e^{2t} + 4te^{2t} & 2e^t - e^{2t} + 2te^{2t} & -2e^t + 2e^{2t} \\ 2te^{2t} & te^{2t} & e^{2t} \end{bmatrix}\mathbf{c}.$

9. $\mathbf{x} = \begin{bmatrix} e^t & 0 & 0 & 0 \\ 0 & e^t & 0 & 0 \\ 0 & 0 & \cos t + \sin t & -2\sin t \\ 0 & 0 & \sin t & \cos t - \sin t \end{bmatrix}\mathbf{c}.$

10. $\mathbf{x} = \begin{bmatrix} 4e^{2t} + 8e^{-t} - 9 \\ -9e^{3t} + 16e^{2t} + 8e^{-t} - 18 \end{bmatrix}.$ 11. $\mathbf{x} = \begin{bmatrix} (t+1)\cos t - \sin t \\ -(t+1)\sin t - \cos t \end{bmatrix}e^{2t}.$

12. $\mathbf{x} = \begin{bmatrix} t^2 + 1 \\ t^3 + 2 \\ t^4 + 3 \end{bmatrix}e^{-t}.$ 13. $\mathbf{x} = \dfrac{1}{4}\begin{bmatrix} 2te^t + 2te^{-t} - e^{-t} \\ 2t^2e^t \\ 2te^t - 2te^{-t} + 2e^t - e^{-t} \end{bmatrix}.$

14. $\mathbf{x} = \begin{bmatrix} 2t^2 - t - 2t^2\ln t \\ 4\ln t \\ -2t^2 - t + 2 + 2(t^2 + 2t)\ln t \end{bmatrix}e^{-2t}.$

15. $\mathbf{x} = \begin{bmatrix} (4t + 3)e^{5t} - (4t + 1)e^t \\ (8t - 2)e^{5t} + (8t - 2)e^t \\ (4t - 1)e^{5t} - (4t + 1)e^t \end{bmatrix}.$

16. $\mathbf{x} = \dfrac{1}{5}\begin{bmatrix} 6\cos t + 2\sin t \\ 2\cos t + 4\sin t - 5 \\ 12\cos t - 6\sin t - 5 \end{bmatrix}e^t.$

18. (a) Show that the characteristic roots of the matrix $\mathbf{A} = \begin{bmatrix} \dfrac{\pi}{4} & 1 \\ 0 & \dfrac{5}{4}\pi \end{bmatrix}$ are $\lambda_1 = \dfrac{\pi}{4}$

and $\lambda_2 = \frac{5}{4}\pi$. Show that the eigenvectors $\xi^{(1)}$ and $\xi^{(2)}$ associated with the characteristic roots λ_1 and λ_2 are, respectively, $\xi^{(1)} = \begin{bmatrix} 1 \\ 0 \end{bmatrix}$ and $\xi^{(2)} = \begin{bmatrix} 1 \\ \pi \end{bmatrix}$.

Consider the matrix $\mathbf{P} = [\xi^{(1)} \ \xi^{(2)}] = \begin{bmatrix} 1 & 1 \\ 0 & \pi \end{bmatrix}$. Show that $\mathbf{P}^{-1} = \begin{bmatrix} 1 & -\dfrac{1}{\pi} \\ 0 & \dfrac{1}{\pi} \end{bmatrix}$.

Show that $\mathbf{P}^{-1}\mathbf{AP} = \begin{bmatrix} \dfrac{\pi}{4} & 0 \\ 0 & \dfrac{5}{4}\pi \end{bmatrix} \equiv \mathbf{D}$. Thus, $\mathbf{A} = \mathbf{PDP}^{-1}$. Consequently,

$$\sin \mathbf{A} = \sin(\mathbf{PDP}^{-1}) = \lim_{m \to \infty} \sum_{k=0}^{m} \frac{(-1)^k}{(2k+1)!} (\mathbf{PDP}^{-1})^{2k+1}$$

$$= \lim_{m \to \infty} \sum_{k=0}^{m} \frac{(-1)^k}{(2k+1)!} \mathbf{PD}^{2k+1}\mathbf{P}^{-1} \qquad \text{(Why?)}$$

$$= \mathbf{P}\left(\lim_{m \to \infty} \sum_{k=0}^{m} \frac{(-1)^k}{(2k+1)!} \mathbf{D}^{2k+1} \right)\mathbf{P}^{-1}$$

$$= \mathbf{P}\left(\lim_{m \to \infty} \sum_{k=0}^{m} \frac{(-1)^k}{(2k+1)!} \begin{bmatrix} \left(\dfrac{\pi}{4}\right)^{2k+1} & 0 \\ 0 & \left(\dfrac{5}{4}\pi\right)^{2k+1} \end{bmatrix} \right)\mathbf{P}^{-1} \qquad \text{(Why?)}$$

$$= \mathbf{P}\begin{bmatrix} \displaystyle\sum_{k=0}^{\infty} \frac{(-1)^k}{(2k+1)!}\left(\dfrac{\pi}{4}\right)^{2k+1} & 0 \\ 0 & \displaystyle\sum_{k=0}^{\infty} \frac{(-1)^k}{(2k+1)!}\left(\dfrac{5}{4}\pi\right)^{2k+1} \end{bmatrix}\mathbf{P}^{-1} \qquad \text{(Why?)}$$

$$= \mathbf{P}\begin{bmatrix} \sin\dfrac{\pi}{4} & 0 \\ 0 & \sin\dfrac{5}{4}\pi \end{bmatrix}\mathbf{P}^{-1} = \begin{bmatrix} 1 & 1 \\ 0 & \pi \end{bmatrix}\begin{bmatrix} \dfrac{1}{\sqrt{2}} & 0 \\ 0 & -\dfrac{1}{\sqrt{2}} \end{bmatrix}\begin{bmatrix} 1 & -\dfrac{1}{\pi} \\ 0 & \dfrac{1}{\pi} \end{bmatrix}$$

$$= \frac{1}{\sqrt{2}}\begin{bmatrix} 1 & -\dfrac{2}{\pi} \\ 0 & -1 \end{bmatrix}.$$

Exercises of Appendix 1, page 620

1. (a) It is a function.   Domain: $\{1, 2, 3, 4, 5\}$, range: $\{1, 2, 4, 7, 11\}$.
   (b) It is a function.   Domain: $\{-4, -3, -2, -1, 0\}$, range:
       $\{-7, -5, -3, -2, 0\}$.
   (c) It is a function.   Domain: $\{-1, 0, 2, 3\}$, range: $\{-2, 1, 2\}$.
   (d) It is a function.   Domain: $\{-3, -2, -1, 4\}$, range: $\{-1, 1\}$.
   (e) It is a function.   Domain: $\{-4, -2, -1, 0\}$, range: $\{0, \frac{1}{2}, 1, \frac{3}{2}\}$.
   (f) It is not a function due to the ordered pairs $(0, -2)$, $(0, 0)$.
2. $D_f = \{0, 1, 2, 3, 4, 5, 6\}$, $f(x) = x^3$, $x \in D_f$.
   $D_g = \{0, 1, 4, 9, 16, 25\}$, $g(x) = \sqrt{x}$, $x \in D_g$.
   $D_h = \{-3, -2, -1, 0, 1, 2, 3\}$, $h(x) = |x|$, $x \in D_h$.
   $D_F = \{-2, -1, 0, 1, 2\}$, $F(x) = x^2 - 4$, $x \in D_F$.
3. Range of $f \equiv R_f = \{0, 1, 8, 27, 64, 125, 216\}$. $R_g = \{0, 1, 2, 3, 4, 5\}$.
   $R_h = \{0, 1, 2, 3\}$. $R_F = \{-4, -3, 0\}$.
4. (a) 0,   (b) 2,   (c) 13,   (d) 13,   (e) 1,   (f) 5.
5. (a) Yes.   (b) No.   (c) Yes.   (d) No.   (e) Yes.   (f) No.
   (g) No.

7. (a)

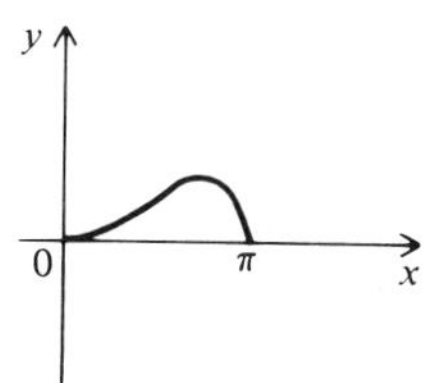

(b)

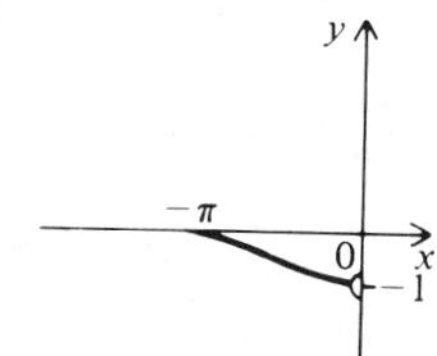

(c)

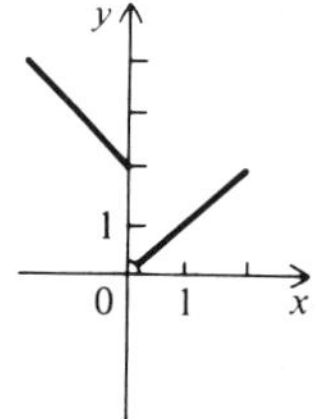

(d)

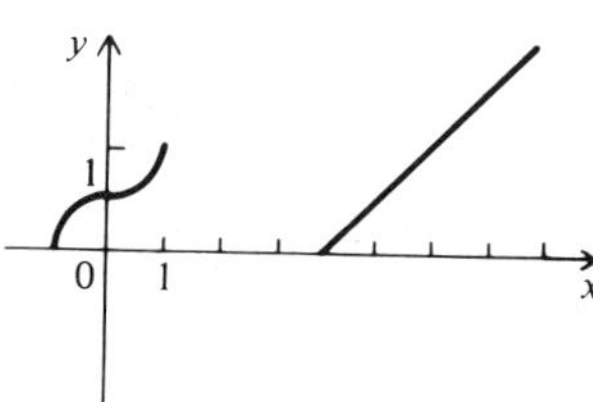

(e)

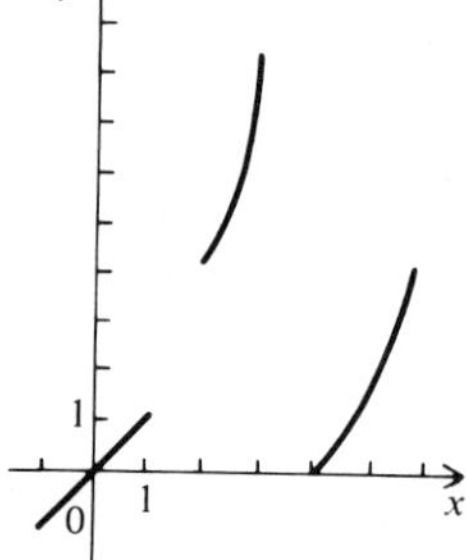

(f)

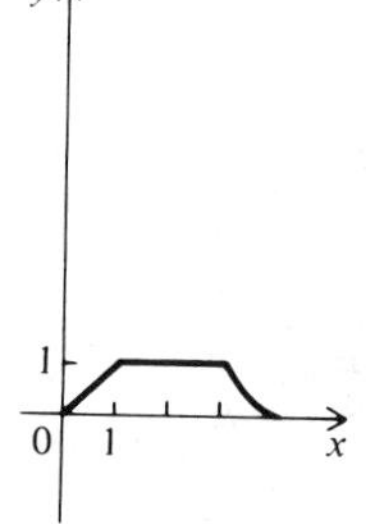

(g)

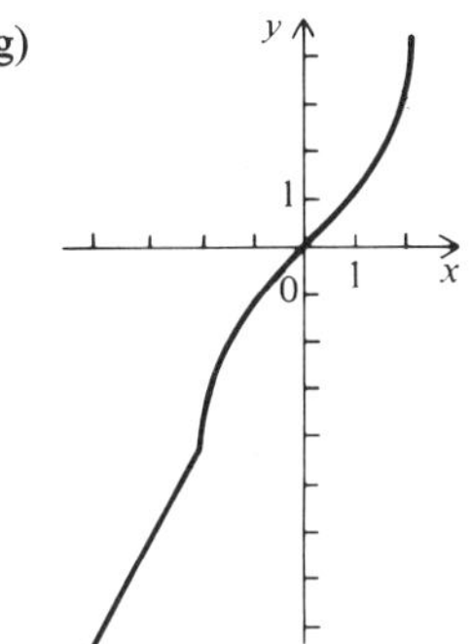

8. No.

9. (b) $B(S) = \{(x, y) \mid x^2 + y^2 = 4\} \cup \{(x, y) \mid x = 0, 0 \leq y \leq 2\}$.

   (c) S is open.

   (d) S is an (open) region or domain.

10. (b) $B(S) = \{(x, y) \mid x = y^2, -1 \leq y \leq 1\} \cup \{(x, y) \mid x = 1, |y| \geq 1\}$.

    (c) S is closed.

    (d) S is a closed region.

11. (b) $B(S) = \{(x, y) \mid xy = 4, x > 0\} \cup \{(x, y) \mid x = 0, y \geq 0\} \cup \{(x, y) \mid y = 0, x \geq 0\}$.

    (c) S is neither open nor closed.

    (d) S is a region.

12. (b) $B(S) = \{(x, y) \mid x = 0, \quad -2 \leq y \leq 2\}$.

    (c) S is neither open nor closed.

    (d) S is a·region.

13. (b) $B(S) = \{(x, y) \mid x^2 + y^2 = 9, \quad y < 0\} \cup \{(x, y) \mid x = 0,$
    $-1 \leq y < 0\} \cup \{(x, y) \mid y = 0, \quad -3 \leq x \leq 3\} \cup \{(x, y) \mid x = 1/n,$
    $n = 1, 2, \ldots, -1 \leq y < 0\}$.

    (c) S is neither open nor closed.

    (d) S is a region.

14. (a) S is closed. S does not have any interior points.

    (b) No.

16. (a) If $P \in R \cap S$, then $P \in R$ and $P \in S$.   Therefore, $R \cap S \subseteq R$ and $R \cap S \subseteq S$.

    (c) If $P \in R$, then $P \in S$.   If $P \in S$, then $P \in T$.   Thus, if $P \in R$, then $P \in T$. Therefore, $R \subset T$.

17. *Necessity.*   Given that S is closed in R^2, we must show that S contains all of its limit points.   We shall establish this result by contradiction.   Assume there exists a limit point P of S but $P \in C(S)$.   Since S is closed, $C(S)$ is open.   Thus, P is an interior point of $C(S)$.   Hence, for some $r > 0$, there exists a circular neighborhood $N(P, r) \subset C(S)$.   Since $N(P, r)$ contains no points of S, it follows that P cannot be a limit point of S.   Hence, our assumption that $P \in C(S)$ is false, and S must contain all of its limit points.

Sufficiency.   Given that S contains all of its limit points, we must show that S is closed in R^2; or equivalently, we must show that $C(S)$ is open.   However, to show that $C(S)$ is open, it is sufficient to show that $C(S)$ contains only interior points.   Suppose then that there exists a point $P \in C(S)$ but P is not an interior point of $C(S)$. Then every circular neighborhood, $N(P, r)$, $r > 0$, contains a point $P_1 \in S$. Clearly, $P \neq P_1$, since $P \in C(S)$, $P_1 \in S$, and $C(S) \cap S$ is the null set. Thus, P is a limit point of S, which is a contradiction. Hence, $C(S)$ contains only interior points, and thus S is closed in R^2.

18. (b) If S or T is the null set, the result follows. (Why?) Suppose then that S and T are nonempty closed sets in R^2. Let P be a limit point of $U = S \cup T$. We must show that $P \in U$. Since P is a limit point of U, for every $r = 1/n$, $n = 1, 2, \ldots$, there exists a point P_n, $P_n \neq P$, such that $P_n \in [U \cap N(P, r)]$. The sequence $\{P_n\} \subset S \cup T$. Thus, $S \cup T$ contains an infinite number of distinct points of the sequence. Case 1. Suppose that S contains the infinite number of distinct points of the sequence. Then $P \in S \subset U$, and thus U is closed. (Why?) Case 2. Replace S by T in Case 1. We find that $P \in T \subset U$, and again U is closed. Case 3. Suppose that the infinite number of distinct points in the sequence is contained in both S and T. Then U is closed. (Why?)

19. (b) If S or T is the null set, the result follows. (Why?) Suppose then that S and T are nonempty open sets in R^2. To show that $U = S \cup T$ is open, we must show that every point $P \in U$ has a circular neighborhood $N(P, r) \subset U, r > 0$. If $P \in U$, then $P \in S$ or $P \in T$ (or $P \in$ both S and T). If $P \in S$, then since S is open, there exists $N(P, r) \subset S \subset U$. If $P \in T$, then since T is open, there exists $N(P, r) \subset T \subset U$.

20. (b) Let P_1, P_2 be points and r_1, r_2 be positive numbers such that $S \subset N(P_1, r_1)$ and $T \subset N(P_2, r_2)$. Since the sets S and T are bounded, P_1, P_2, r_1, and r_2 exist. Let P_0 be the midpoint of the line segment P_1P_2, d_0 denote the distance between P_0 and P_1, and $r_0 = \max\ (r_1, r_2) + d_0$. Then $S \cup T \subset N(P_0, r_0)$. Thus, in view of Definition 23, the set $U = S \cup T$ is bounded [*Note:* If we choose P_1 and P_2 to be the same point, as we may, then $P_0 = P_1$ and $d_0 = 0$.]

22. To show that $S = S_1 \cup S_2$ is not necessarily a connected set, it is sufficient to give a counterexample. Let S_1 and S_2 consist of all points (x, y) whose coordinates satisfy $1 \leq x^2 + y^2 \leq 2$ and $3 \leq x^2 + y^2 \leq 4$, respectively. The sets S_1 and S_2 are each connected. Now, show that $S = S_1 \cup S_2$ is a disconnected set.

23. (a) Let G denote a collection of open sets in R^2, and let $T = \bigcup_{S \in G} S$. Assume that an arbitrary point $P \in T$. Then P must belong to at least one of the sets of G, say $P \in S$. Since by hypothesis S is an open set, there exists a circular neighborhood $N(P, r) \subset S$. But $S \subset T$, so that $N(P, r) \subset T$, and hence P is an interior point of T. Thus, T is an open set.

(c) Let $G = \{S_\alpha\}$ denote the given collection of connected sets in R^2, with $P_0 \in S_\alpha$ for each $S_\alpha \in G$. Suppose that $S = \bigcup_{S_\alpha \in G} S_\alpha$ can be separated into two sets U and V with $P_0 \in U$. We have $S_\alpha = (S_\alpha \cap U) \cup (S_\alpha \cap V)$. If $S_\alpha \cap V$ were nonempty, then S_α would not be a connected set. (Why?) Hence, $S_\alpha \subset U$ for each $S_\alpha \in G$. Thus, V is the empty set, which is a contradiction.

(e) Let x be any point $\in [0, 1]$. Then, $x \in S_k$ for each k. Thus, $x \in T$. Hence, $[0, 1] \subseteq T$. If $x \notin [0, 1]$ so that $x < 0$ or $x > 1$, then $x \notin S_k$ for some k. Hence, $x \notin T$. Thus, $T \subseteq [0, 1]$. [*Note:* The following three conditions are equivalent: (1) $x \notin [0, 1]$ implies $x \notin T$.   (2) $x \in T$ implies $x \in [0, 1]$.   (3) $T \subseteq [0, 1]$.]   Since $[0, 1] \subseteq T$ also, we have that $T = [0, 1]$. [*Note:* Each S_k is an open set, since it consists only of interior points. T is not an open set, since $0, 1 \in T$, but $0, 1$ are not interior points of T.]

25. (a) Let $U = \bigcup_{S \in G} S$ and $V = \bigcap_{S \in G} C(S)$.   If $P \in C(U)$, then $P \notin U$.   Hence, it is not true that P belongs to at least one S in G.   Thus, P belongs to no S in G.   Hence, for every S in G, $P \in C(S)$, and consequently $P \in V$.   Thus, $C(U) \subseteq V$.   Reversing these steps, we see that $V \subseteq C(U)$.   In view of Definition 10 of Appendix 1, the result now follows.

INDEX

* Numbers in parentheses refer to exercises.